Finite Rotation Shells

Basic Equations and Finite Elements for Reissner Kinematics

Lecture Notes on Numerical Methods in Engineering and Sciences

Aims and Scope of the Series

This series publishes text books on topics of general interest in the field of computational engineering sciences.

The books will focus on subjects in which numerical methods play a fundamental role for solving problems in engineering and applied sciences. Advances in finite element, finite volume, finite differences, discrete and particle methods and their applications to classical single discipline fields and new multidisciplinary domains are examples of the topics covered by the series.

The main intended audience is the first year graduate student. Some books define the current state of a field to a highly specialised readership; others are accessible to final year undergraduates, but essentially the emphasis is on accessibility and clarity.

The books will be also useful for practising engineers and scientists interested in state of the art information on the theory and application of numerical methods.

Titles:

1. E. Oñate, Structural Analysis with the Finite Element Method. Linear Statics. Volume 1. Basis and Solids, 2009
2. K. Wiśniewski, Finite Rotation Shells. Basic Equations and Finite Elements for Reissner Kinematics, 2010

Finite Rotation Shells

Basic Equations and Finite Elements for Reissner Kinematics

K. Wiśniewski

Institute of Fundamental Technological Research
Polish Academy of Sciences
Warsaw, Poland

ISBN: 978-94-007-3198-1
ISBN: 978-90-481-8761-4 (eBook)

Depósito legal: B-10834-2010

Lecture Notes Series Manager: **Mª Jesús Samper,** CIMNE, Barcelona, Spain

Cover page: **Pallí Disseny i Comunicació,** www.pallidisseny.com

Printed by: **Artes Gráficas Torres S.L.**
Morales 17, 08029 Barcelona, España
www.agraficastorres.es

Printed on elemental chlorine-free paper

Finite Rotation Shells. Basic Equations and Finite Elements for Reissner Kinematics
K. Wiśniewski

First edition, March 2010

To my wife Ewa,
my parents, and my children

Preface

The objective of this book is to provide a comprehensive introduction to finite rotation shells and to non-linear shell finite elements.

It is divided into 5 parts: I. Preliminaries (20 pages), II. Shell equations (104 pages), III. Finite rotations for shells (103 pages), IV. Four-node shell elements (189 pages), and V. Numerical examples (41 pages). Additional numerical examples are presented in Parts III and IV. The bibliography includes 270 entries.

The book is intended for both teaching and self-study, and emphasizes fundamental aspects and techniques of the subject. Some familiarity with non-linear mechanics and the finite element method is assumed.

Shell elements are a subject of active research which results in many publications every year and several conferences and sessions are held regularly, among them, two large international conferences: "Computation of Shell and Spatial Structures" and "Shell Structures. Theory and Applications" (SSTA). The literature is voluminous, not easy to follow and evaluate, and the subject is difficult to comprehend. I hope that this will be facilitated by the book.

I would like to express my gratitude to several persons who helped me in my professional life, in this way contributing to the book.

I thank Prof. R.L. Taylor from the University of California at Berkeley, Prof. B. Schrefler from the University of Padua, and Prof. J.T. Santos from the Instituto Superior Tecnico at Lisbon, for hosting and supporting me when I was a post-doctoral researcher.

I am very grateful to Prof. M. Kleiber from the Institute of Fundamental Technological Research of the Polish Academy of Sciences (IPPT PAN) in Warsaw, where I have been working since my Ph.D., for continuous support and stimulating discussions.

Special thanks go to Prof. W. Pietraszkiewicz from the Institute of Fluid Flow Machinery of the Polish Academy of Sciences (IMP PAN) in

Gdańsk for his excellent papers on shells which generated my interest in this field of mechanics when I was a student.

Finally, I would like to express my appreciation to many colleagues from the Institute of Fundamental Technological Research, for providing a friendly research environment and for collaboration.

K. Wiśniewski
March 2010

Contents

V NUMERICAL EXAMPLES

Part I
PRELIMINARIES

1 Introduction

1.1 Subject of the book

An imprecise but intuitively clear definition of the *shell structure* is that it is a 3D body, with one dimension much smaller that the other two. In other words, it is a surface in a 3D space equipped with a thickness which is much smaller than the size of the surface. There are numerous examples of shells among engineering structures, such as bodies of cars, hulls of ships, and fuselages of airplanes. Also some civil engineering structures, such as tanks and roofs can often be considered and analyzed as shells.

However, we must see the difference between shell structures and the shells defined in mechanics.

In mechanics, the *shell* designates the 2D governing equations obtained from equations for a 3D body when using some kinematical assumptions. The most often used assumption is either the Kirchhoff hypothesis or the Reissner hypothesis, and each of them implies a different structure of governing equations. The first hypothesis is preferred in theoretical works, while the second one is used in computational mechanics, and also in this book. When the shell equations are treated by the finite element method, which is the most popular and convenient method for engineering analysis, we then obtain *shell elements*. Mostly due to the presence of the rotational degrees of freedom, the shell elements involve specific problems not encountered in displacement-type elements.

This book is concerned with finite elements for Reissner shells and its subject is defined by the following *keywords*:

1. Computational mechanics of non-linear shells,
2. Shell equations: Reissner kinematics, finite rotations, finite strains,

3. Shell finite elements: four-node, enhanced or mixed or mixed/enhanced,
4. Drilling rotation: drill rotation constraint or Allman's shape functions,
5. Normal strain: recovered or parameterized,
6. Constitutive equations: incremental, plane stress or 3D.

The basic information on linear shell elements may be found in some textbooks on finite elements, but this book contains several advanced topics related to non-linear shells such as, e.g., the parametrization of finite rotations, the methods of inclusion of the drilling rotation, various methods of treating the normal strain, and the mixed/enhanced finite elements. Some of these topics have been the subject of our research for years and all the described methods have been implemented in our own elements and tested.

A wide range of applications of shell elements implies that they should be versatile, i.e. account for finite rotations and strains, admit the incorporation of various constitutive laws, and enable convenient linking with other elements. This is a serious challenge which requires various aspects of the element's formulation to be well advanced.

We must be aware that the use of the Reissner hypothesis has not only positive but also negative consequences. The positive ones are that the shell elements admit large thickness/size ratios, adequately represent bending modes, and can be based on C^o approximations. The negative consequences are related to the normal strain and the drilling rotation.

1. *Normal strain.* The standard Reissner (or Kirchhoff) hypothesis yields the normal strain component equal to zero which provides an unrealistic constraint on bending deformation so that the 3D constitutive laws cannot be directly used. For this reason, the normal strain must be either recovered from a suitable auxiliary condition or the shell kinematics must be enhanced by additional stretch parameters.
2. *Drilling rotation.* The shell strains obtained from the Green strain by using the Reissner hypothesis do not depend on the drilling rotation which is the rotation about the vector normal to the shell surface. The drilling rotation is needed to use three parameters for increments of rotations and conveniently link the shell elements with 3D beam and shell elements. For this reason, we derive shell equations from the 3D mixed equations which incorporate the drilling rotation as an independent variable.

The current state of the shell equations and shell elements has been achieved gradually through the efforts of many researchers which yielded thousands of works on the subject. This book benefitted from many of them, although not all of them have been cited here.

1.2 Notation

General rules of notation

1. Small bold letters - vectors, e.g. $\mathbf{v}$.
2. Capital bold letters - second-rank tensors, e.g. $\mathbf{A}$.
3. Open-face letters - fourth-rank tensors, e.g. $\mathbb{C}$.
4. Arrays of components of vectors and tensors are denoted by the same letters as vectors and tensors. Sometimes, sans serif fonts are used, e.g. v, A.
5. Superscript asterisk (*) - forward-rotated objects, e.g. $\mathbf{A}^*$,
6. Subscript asterisk ($_*$) - backward-rotated objects, e.g. $\mathbf{A}_*$,
7. "$\cdot$", "$\times$", "$\otimes$" - scalar product, cross product, tensorial product,
8. Symmetric part - $\operatorname{sym}\mathbf{T} \doteq \frac{1}{2}(\mathbf{T}+\mathbf{T}^T)$. Besides, $\mathbf{T}_s \doteq \operatorname{sym}\mathbf{T}$,
9. Skew-symmetric part - $\operatorname{skew}\mathbf{T} \doteq \frac{1}{2}(\mathbf{T}-\mathbf{T}^T)$. Besides, $\mathbf{T}_a \doteq \operatorname{skew}\mathbf{T}$,
10. Components of tensors - $A_{ij} = \mathbf{A}\cdot(\mathbf{t}_i \otimes \mathbf{t}_j) = (\mathbf{A}\mathbf{t}_j)\cdot\mathbf{t}_i$,
11. Gradient of a scalar A and gradient of a vector $\{v_i\}$ w.r.t. coordinates S^k,

$$\left[\frac{\partial A}{\partial S^k}\right] = \begin{bmatrix} \frac{\partial A}{\partial S^1} \\ \frac{\partial A}{\partial S^2} \\ \frac{\partial A}{\partial S^3} \end{bmatrix}, \qquad \left[\frac{\partial v_i}{\partial S^k}\right] = \begin{bmatrix} \frac{\partial v_1}{\partial S^1} & \frac{\partial v_2}{\partial S^1} & \frac{\partial v_3}{\partial S^1} \\ \frac{\partial v_1}{\partial S^2} & \frac{\partial v_2}{\partial S^2} & \frac{\partial v_3}{\partial S^2} \\ \frac{\partial v_1}{\partial S^3} & \frac{\partial v_2}{\partial S^3} & \frac{\partial v_3}{\partial S^3} \end{bmatrix}, \qquad i,k=1,2,3. \tag{1.1}$$

List of symbols

1. h - initial shell thickness,
2. $\zeta \in [-h/2, +h/2]$ - thickness coordinate,
3. $\mathbf{y}$ - position vector in the non-deformed (initial) configuration,
4. $\mathbf{x}$ - position vector in the deformed (current) configuration,
5. χ: $\mathbf{x} = \chi(\mathbf{y})$ - deformation function,
6. $\mathbf{F} \doteq \partial\mathbf{x}/\partial\mathbf{y}$ - deformation gradient,
7. $\mathbf{Q}, \mathbf{R} \in \mathrm{SO}(3)$ - orthogonal (rotation) tensors,
8. $\mathbf{Q}_0 \in \mathrm{SO}(3)$ - rotation constant over the shell thickness,
9. $\mathbf{C} \doteq \mathbf{F}^T\mathbf{F}$ - Cauchy–Green tensor,
10. $\mathfrak{C} \doteq \operatorname{skew}(\mathbf{Q}^T\mathbf{F})$ - Rotation Constraint (RC),

11. $\mathbf{E} \doteq \frac{1}{2}(\mathbf{C} - \mathbf{I})$ - Green strain,
12. $\mathbf{n}_{\alpha}^{B}, \mathbf{m}_{\alpha}^{B}$ - shell stress and couple resultant vectors,
13. $\boldsymbol{\varepsilon}_{\alpha}, \boldsymbol{\kappa}_{\alpha}$ - strain vectors of zeroth and first order,
14. E, ν, G - Young's modulus, Poisson's ratio, shear modulus,
15. λ, μ - Lamé coefficients.

Reference bases

1. $\{\mathbf{i}_k\}$ - global reference ortho-normal basis, $k = 1, 2, 3$,
2. $\{\mathbf{g}_{\alpha}\}$ - local natural basis at the middle surface for the initial configuration, $\alpha = 1, 2$,
3. $\{\mathbf{t}_k\}$ - local ortho-normal basis at the middle surface for the initial configuration,
4. $\{\mathbf{a}_k\}$ - local ortho-normal forward-rotated basis at the middle surface for the deformed configuration.

Abbreviations

1. AD - Automatic Differentiation
2. AMB - Angular Momentum Balance
3. BC - Boundary Conditions
4. BVP - Boundary Value Problem
5. CL - Constitutive Law
6. DK - Discrete Kirchhoff (elements)
7. dof - degree of freedom
8. FE - Finite Element
9. FD - Finite Difference
10. HR - Hellinger–Reissner (functional)
11. HW - Hu–Washizu (functional)
12. LMB - linear momentum balance
13. PE - Potential Energy
14. PS - Pian–Sumihara (element)
15. RBF - Residual Bending Flexibility (correction)
16. RC - Rotation Constraint
17. RI - Reduced Integration
18. VW - Virtual Work
19. ZNS - Zero Normal Stress (condition)
20. 1D, 2D, 3D - one-dimensional, two-dimensional, three-dimensional
21. 1-F, 2-F, 3-F, 4-F - one-field, two-field, three-field, four-field
22. 2nd PK - second Piola–Kirchhoff (stress)

2

Operations on tensors and their representations

In this chapter, the transformations of tensor representations between various bases are described; they are needed in the derivation of the shell FEs.

Short introductions to the subject of operations on tensors and their representations are typically provided in books on continuum mechanics and on the FE method. A comprehensive introduction to vector and tensor algebra and analysis can be found, e.g., in [33, 34], where other references are also listed.

2.1 Cartesian bases

Product of vector and tensor in a Cartesian basis. Let $\{\mathbf{i}_k\}$, $k = 1,2,3$, be an ortho-normal (Cartesian) basis. A vector $\mathbf{v}$ and a 2nd rank tensor $\mathbf{A}$ can be represented in the basis as follows:

$$\mathbf{v} = v_k\, \mathbf{i}_k, \qquad \mathbf{A} = A_{jk}\, \mathbf{i}_j \otimes \mathbf{i}_k, \tag{2.1}$$

where the components are $v_k \doteq \mathbf{v} \cdot \mathbf{i}_k$ and $A_{jk} \doteq \mathbf{i}_j \cdot (\mathbf{A}\mathbf{i}_k)$. Operations on vectors and tensors have direct counterparts in operations on their components

$$\mathbf{A}\mathbf{v} = (A_{jk}\, \mathbf{i}_j \otimes \mathbf{i}_k)(v_l\, \mathbf{i}_l) = (A_{jl} v_l)\, \mathbf{i}_j,$$

when using $(\mathbf{i}_j \otimes \mathbf{i}_k)\, \mathbf{i}_l = (\mathbf{i}_k \cdot \mathbf{i}_l)\mathbf{i}_j = \delta_{kl}\mathbf{i}_j$. Denote the vector resulting from the above multiplication by $\bar{\mathbf{v}}$ and decompose it as follows:

$$\bar{\mathbf{v}} = \bar{v}_k\, \mathbf{i}_k. \tag{2.2}$$

From the equality of terms for $\mathbf{i}_1, \mathbf{i}_2, \mathbf{i}_3$ in the above two formulas, we obtain three equations which we can write in a matrix form as follows:

$$\begin{bmatrix} \bar{v}_1 \\ \bar{v}_2 \\ \bar{v}_3 \end{bmatrix} = \begin{bmatrix} A_{11} & A_{12} & A_{13} \\ A_{21} & A_{22} & A_{23} \\ A_{31} & A_{32} & A_{33} \end{bmatrix} \begin{bmatrix} v_1 \\ v_2 \\ v_3 \end{bmatrix}. \tag{2.3}$$

This equation contains only components and can be written concisely as

$$\bar{\mathsf{v}} = \mathsf{A}\,\mathsf{v}, \tag{2.4}$$

where $\mathsf{v} \doteq [v_k]$, $\bar{\mathsf{v}} \doteq [\bar{v}_k]$, and $\mathsf{A} \doteq [A_{mk}]$.

Remark. In general, we should always distinguish an object (vector or tensor) from an array of its components. However, it is common that they are both denoted by the same letter, typically a bold one. Although, undeniably, it is an abuse of notation, it is acceptable when the meaning of the symbol (or relation) is clear. Hence, often, instead of two forms, $\bar{\mathbf{v}} = \mathbf{A}\mathbf{v}$ and $\bar{\mathsf{v}} = \mathsf{A}\,\mathsf{v}$, only the first one is used.

Transformation of vector and tensor components between Cartesian bases. Consider two ortho-normal (Cartesian) bases: the global $\{\mathbf{i}_k\}$ and the local $\{\mathbf{t}_k\}$, $k = 1, 2, 3$. Assume that $\mathbf{t}_k = \mathbf{R}\,\mathbf{i}_k$, where $\mathbf{R} \in \mathrm{SO}(3)$ is a rotation tensor which describes the angular position of $\{\mathbf{t}_k\}$ relative to $\{\mathbf{i}_k\}$. Transformations of components of vectors and second-rank tensors between these bases are performed as described below.

Vectors. An arbitrary vector $\mathbf{v}$ can be represented in these two bases as follows

$$\mathbf{v} = v_k\,\mathbf{i}_k = \bar{v}_k\,\mathbf{t}_k, \tag{2.5}$$

where the components are $v_k \doteq \mathbf{v} \cdot \mathbf{i}_k$ and $\bar{v}_k \doteq \mathbf{v} \cdot \mathbf{t}_k$. Because

$$\mathbf{t}_k = \mathbf{R}\,\mathbf{i}_k = (R_{mn}\mathbf{i}_m \otimes \mathbf{i}_n)\,\mathbf{i}_k = R_{mk}\mathbf{i}_m, \tag{2.6}$$

thus, eq. (2.5) yields

$$v_k\,\mathbf{i}_k = \bar{v}_k\,R_{mk}\,\mathbf{i}_m, \tag{2.7}$$

which implies the relation between components

$$v_k = R_{km}\bar{v}_m \qquad \text{or} \qquad \mathsf{v} = \mathsf{R}\,\bar{\mathsf{v}}, \tag{2.8}$$

where $\mathsf{v} \doteq [v_k]$, $\bar{\mathsf{v}} \doteq [\bar{v}_k]$, and the rotation matrix $\mathsf{R} \doteq [R_{mk}]$. To verify this equation, we write each side of eq. (2.7) separately,

$$
\begin{aligned}
v_k \mathbf{i}_k &= v_1 \mathbf{i}_1 + v_2 \mathbf{i}_2, \\
\bar{v}_k R_{mk}\, \mathbf{i}_m &= \bar{v}_1 (R_{11}\, \mathbf{i}_1 + R_{21}\, \mathbf{i}_2) + \bar{v}_2 (R_{12}\, \mathbf{i}_1 + R_{22}\, \mathbf{i}_2) \\
&= (\bar{v}_1 R_{11} + \bar{v}_2 R_{12})\, \mathbf{i}_1 + (\bar{v}_1 R_{21} + \bar{v}_2 R_{22})\, \mathbf{i}_2.
\end{aligned}
$$

Then, from equality of terms for $\mathbf{i}_1$ and $\mathbf{i}_2$, we obtain

$$
\begin{bmatrix} v_1 \\ v_2 \end{bmatrix} = \begin{bmatrix} R_{11} & R_{12} \\ R_{21} & R_{22} \end{bmatrix} \begin{bmatrix} \bar{v}_1 \\ \bar{v}_2 \end{bmatrix}, \tag{2.9}
$$

which is the second form of eq. (2.8). The components of $\mathbf{v}$ in the local $\{\mathbf{t}_k\}$ are computed from eq. (2.8), i.e.

$$
\bar{\mathsf{v}} = \mathsf{R}^{-1}\, \mathsf{v} = \mathsf{R}^T\, \mathsf{v}. \tag{2.10}
$$

Second-rank tensors. An arbitrary second-rank tensor $\mathbf{T}$ can be represented in the two considered bases as follows:

$$
\mathbf{A} = A_{jk}\, \mathbf{i}_j \otimes \mathbf{i}_k = \bar{A}_{jk}\, \mathbf{t}_j \otimes \mathbf{t}_k, \tag{2.11}
$$

where the components are $A_{jk} \doteq \mathbf{i}_j \cdot (\mathbf{A}\mathbf{i}_k)$ and $\bar{A}_{jk} \doteq \mathbf{t}_j \cdot (\mathbf{A}\mathbf{t}_k)$. By eq. (2.6), $\mathbf{t}_j = R_{mj}\mathbf{i}_m$ and $\mathbf{t}_k = R_{nk}\mathbf{i}_n$, and hence

$$
\bar{A}_{jk}\mathbf{t}_j \otimes \mathbf{t}_k = \bar{A}_{jk} R_{mj} R_{nk} \mathbf{i}_m \otimes \mathbf{i}_n.
$$

Using this formula, and by the change of summation indices, eq. (2.11) provides $A_{jk} = \bar{A}_{mn} R_{jm} R_{kn}$, or, in terms of the earlier introduced rotation matrix $\mathsf{R} \doteq [R_{mk}]$,

$$
\mathsf{A} = \mathsf{R}\, \bar{\mathsf{A}}\, \mathsf{R}^T, \tag{2.12}
$$

where $\mathsf{A} \doteq [A_{jk}]$ and $\bar{\mathsf{A}} \doteq [\bar{A}_{jk}]$ are matrices of components. If the components of $\mathbf{A}$ and $\mathbf{R}$ in $\{\mathbf{i}_k\}$ are given, then we can compute the components of $\mathbf{A}$ in the local basis $\{\mathbf{t}_k\}$ as follows:

$$
\bar{\mathsf{A}} = \mathsf{R}^T\, \mathsf{A}\, \mathsf{R}. \tag{2.13}
$$

This formula can be checked in a similar manner as the one for vectors but, due to a large number of terms, a symbolic manipulator is helpful.

Forward- and backward-rotated objects in Cartesian bases. Instead of transforming components from one Cartesian basis to another, as described earlier, we can operate on forward- or backward-rotated objects and use only one basis, i.e. the reference basis $\{\mathbf{i}_k\}$. Then, the use of the same symbol for an object and an array of its components typically does not cause errors. Below, we identify two situations in which the use of rotated objects is particularly useful.

1. Define the forward-rotated basis $\{\mathbf{t}_k\}$, such that $\mathbf{t}_k \doteq \mathbf{R}\,\mathbf{i}_k$, and $\mathbf{R} \in \mathrm{SO}(3)$. Consider the vector $\mathbf{v}$ and the second-rank tensor $\mathbf{A}$, and define their backward-rotated counterparts,

$$\mathbf{v}_* \doteq \mathbf{R}^T \mathbf{v}, \qquad \mathbf{A}_* \doteq \mathbf{R}^T \mathbf{A}\,\mathbf{R}. \tag{2.14}$$

Let us calculate components of the original vector and tensor in the forward-rotated basis:

$$\mathbf{v} \cdot \mathbf{t}_k = \mathbf{v} \cdot (\mathbf{R}\,\mathbf{i}_k) = (\mathbf{R}^T \mathbf{v}) \cdot \mathbf{i}_k = \mathbf{v}_* \cdot \mathbf{i}_k, \tag{2.15}$$

$$\mathbf{t}_j \cdot (\mathbf{A}\mathbf{t}_k) = (\mathbf{R}\mathbf{i}_j) \cdot (\mathbf{A}\mathbf{R}\mathbf{i}_k) = \mathbf{i}_j \cdot (\mathbf{R}^T \mathbf{A}\mathbf{R}\,\mathbf{i}_k) = \mathbf{i}_j \cdot (\mathbf{A}_*\,\mathbf{i}_k). \tag{2.16}$$

We see that that they are equal to components of the backward-rotated objects in the reference basis.

This property can be applied, e.g., to strain $\mathbf{E}$, which is computed in the global $\{\mathbf{i}_k\}$ but we need its components in the local $\{\mathbf{t}_k\}$. By eq. (2.16), we can use the components of the backward-rotated $\mathbf{E}_* \doteq \mathbf{R}^T \mathbf{E}\,\mathbf{R}$ in the global $\{\mathbf{i}_k\}$.

2. Define the forward-rotated basis $\{\mathbf{a}_k\}$, such that $\mathbf{a}_k \doteq \mathbf{Q}\,\mathbf{t}_k$ and $\mathbf{Q} \in \mathrm{SO}(3)$. Consider the vector $\mathbf{v}$ and the 2nd rank tensor $\mathbf{A}$ and define their forward-rotated counterparts,

$$\mathbf{v}^* \doteq \mathbf{Q}\,\mathbf{v}, \qquad \mathbf{A}^* \doteq \mathbf{Q}\,\mathbf{A}\,\mathbf{Q}^T. \tag{2.17}$$

Note that

$$\mathbf{v} \cdot \mathbf{t}_k = \mathbf{v} \cdot (\mathbf{Q}^T \mathbf{a}_k) = (\mathbf{Q}\,\mathbf{v}) \cdot \mathbf{a}_k = \mathbf{v}^* \cdot \mathbf{a}_k, \tag{2.18}$$

$$\mathbf{t}_k \cdot (\mathbf{T}\,\mathbf{t}_l) = (\mathbf{Q}^T \mathbf{a}_k) \cdot (\mathbf{A}\,\mathbf{Q}^T \mathbf{a}_l) = \mathbf{a}_k \cdot (\mathbf{Q}\,\mathbf{A}\,\mathbf{Q}^T \mathbf{a}_l) = \mathbf{a}_k \cdot (\mathbf{A}^*\,\mathbf{a}_l), \tag{2.19}$$

i.e. they are equal to the components of the forward-rotated objects in the forward-rotated basis. This property enables the use of co-rotational local bases in the finite rotation problems.

Remark. Note that components of tensors can be arranged as vectors of tensorial components which is particularly useful when transforming a constitutive matrix from one basis to another, e.g. for orthotropic materials. For such vectors, the rotate-forward operation, $\mathbf{R}\,(\cdot)\,\mathbf{R}^T$, can be replaced by the operator $\mathbf{T}$. For symmetric tensors, we have to use two operators; the respective formulas can be derived as a special case of a transformation between a non-orthogonal and a Cartesian basis, see eqs. (2.38) and (2.39).

2.2 Normal bases

In shells, the normal basis is defined as the one in which the normal vector $\mathbf{t}_3$ is perpendicular to the tangent vectors. Various tangent vectors are used.

Transformation of contravariant in-plane components of vector and tensor. Components of vectors and tensors in the ortho-normal basis $\{\mathbf{t}_k\}$ are obtained from contravariant components in the natural (non-orthogonal) basis $\{\mathbf{g}_k\}$, assuming that the normal vectors are identical, i.e. $\mathbf{t}_3 = \mathbf{g}_3$. In other words, the vectors $\mathbf{t}_\alpha$ and $\mathbf{g}_\alpha$ $(\alpha = 1, 2)$ belong to the same local tangent plane, see Fig. 2.1.

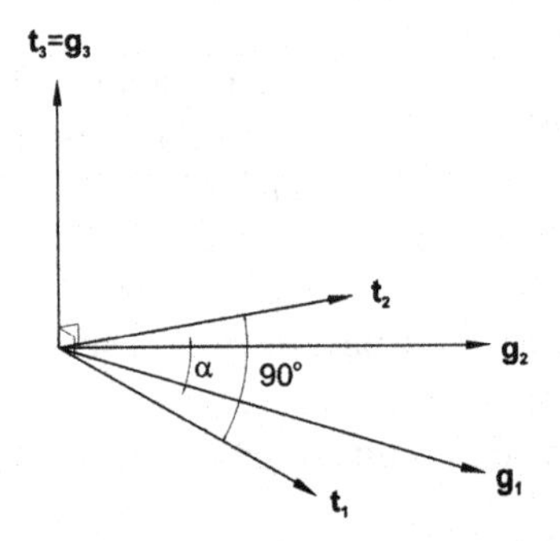

Fig. 2.1 Cartesian basis $\{\mathbf{t}_k\}$ and natural basis $\{\mathbf{g}_k\}$ share the normal vector.

The transformation procedure is slightly complicated because the components of a vector $\mathbf{v}$ in the non-orthogonal $\{\mathbf{g}_\alpha\}$ cannot be obtained by orthogonal projections because $v_1 \neq \mathbf{v} \cdot \mathbf{g}_1$ and $v_2 \neq \mathbf{v} \cdot \mathbf{g}_2$, see Fig. 2.2. However, by the orthogonal projections, we can obtain the components in $\{\mathbf{t}_\alpha\}$, which we subsequently use to obtain the components in the non-orthogonal $\{\mathbf{g}_\alpha\}$.

Vectors. A transformation formula for the components of a vector $\mathbf{v}$ is obtained in the following way. First, we calculate the components in $\{\mathbf{t}_\alpha\}$, i.e. $v_\alpha^S \doteq \mathbf{v} \cdot \mathbf{t}_\alpha$. Next, we transform them to the skew co-basis $\{\mathbf{g}^\alpha\}$, using eq. $(10.52)_2$ rewritten as follows:

$$\mathbf{t}_1 = J_{11}\,\mathbf{g}^1 + J_{12}\,\mathbf{g}^2, \qquad \mathbf{t}_2 = J_{21}\,\mathbf{g}^1 + J_{22}\,\mathbf{g}^2,$$

where $J_{\alpha\beta}$ are components of the 2×2 Jacobian matrix of the local mapping $\xi^\beta \mapsto S^\alpha$, see eq. (11.13). Finally, we identify the contravariant

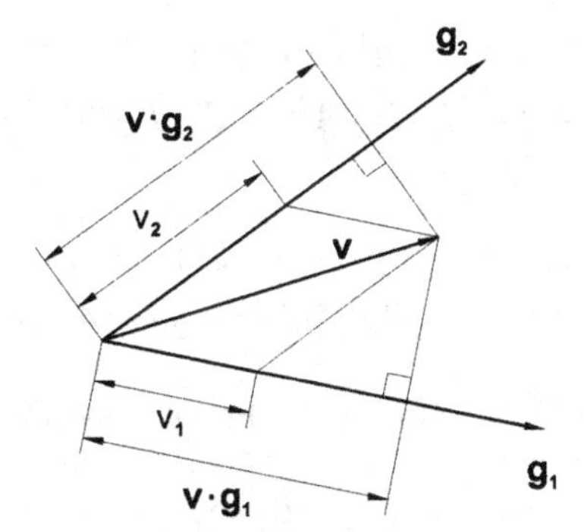

Fig. 2.2 Components of $\mathbf{v}$ in a natural basis $\{\mathbf{g}_\alpha\}$ cannot be obtained by orthogonal projections.

components $v^\alpha \doteq \mathbf{v} \cdot \mathbf{g}^\alpha$ of the vector $\mathbf{v} = v^\alpha \, \mathbf{g}_\alpha$. The transformations are performed as follows,

$$v_1^S \doteq \mathbf{v} \cdot \mathbf{t}_1 = \mathbf{v} \cdot (J_{11}\mathbf{g}^1 + J_{12}\mathbf{g}^2) = J_{11}(\mathbf{v} \cdot \mathbf{g}^1) + J_{12}(\mathbf{v} \cdot \mathbf{g}^2) = J_{11}\, v^1 + J_{12}\, v^2,$$

$$v_2^S \doteq \mathbf{v} \cdot \mathbf{t}_2 = \mathbf{v} \cdot (J_{21}\mathbf{g}^1 + J_{22}\mathbf{g}^2) = J_{21}(\mathbf{v} \cdot \mathbf{g}^1) + J_{22}(\mathbf{v} \cdot \mathbf{g}^2) = J_{21}\, v^1 + J_{22}\, v^2.$$

The matrix form is

$$\begin{bmatrix} v_1^S \\ v_2^S \end{bmatrix} = \mathbf{J} \begin{bmatrix} v^1 \\ v^2 \end{bmatrix}. \tag{2.20}$$

Second rank tensors. The transformation formulas for the 2nd rank tensor $\mathbf{A}$ can be obtained similarly as for the vector. First, we calculate the components $A_{\alpha\beta}^S = \mathbf{t}_\alpha \cdot (\mathbf{A}\, \mathbf{t}_\beta)$ in $\{\mathbf{t}_i\}$. Next, we transform them to the skew co-basis $\{\mathbf{g}^\alpha\}$, using eq. $(10.52)_2$ rewritten as

$$\mathbf{t}_1 = J_{11}\, \mathbf{g}^1 + J_{12}\, \mathbf{g}^2, \qquad \mathbf{t}_2 = J_{21}\, \mathbf{g}^1 + J_{22}\, \mathbf{g}^2,$$

where $J_{\alpha\beta}$ are components of the Jacobian matrix, $\mathbf{J}$. Finally, we identify contravariant components $A^{\alpha\beta} \doteq \mathbf{g}^\alpha \cdot (\mathbf{A}\, \mathbf{g}^\beta)$ of the tensor $\mathbf{A} = A^{\alpha\beta}\, \mathbf{g}_\alpha \otimes \mathbf{g}_\beta$. The transformations are performed as follows:

$$\begin{aligned} A_{11}^S &\doteq \mathbf{t}_1 \cdot (\mathbf{A}\, \mathbf{t}_1) = (J_{11}\mathbf{g}^1 + J_{12}\mathbf{g}^2) \cdot [\mathbf{A}\, (J_{11}\mathbf{g}^1 + J_{12}\mathbf{g}^2)] \\ &= J_{11}^2\, A^{11} + J_{12}^2\, A^{22} + J_{11}J_{12}\, A^{12} + J_{11}J_{12}\, A^{21}, \end{aligned}$$

$$\begin{aligned} A_{22}^S &\doteq \mathbf{t}_2 \cdot (\mathbf{A}\, \mathbf{t}_2) = (J_{21}\, \mathbf{g}^1 + J_{22}\, \mathbf{g}^2) \cdot [\mathbf{A}\, (J_{21}\, \mathbf{g}^1 + J_{22}\, \mathbf{g}^2)] \\ &= J_{21}^2\, A^{11} + J_{22}^2\, A^{22} + J_{22}J_{21}\, A^{12} + J_{22}J_{21}\, A^{21}, \end{aligned}$$

$$A^S_{12} \doteq \mathbf{t}_1 \cdot (\mathbf{A}\,\mathbf{t}_2) = (J_{11}\mathbf{g}^1 + J_{12}\mathbf{g}^2) \cdot [\mathbf{A}\ (J_{21}\,\mathbf{g}^1 + J_{22}\,\mathbf{g}^2)]$$
$$= J_{11}J_{21}\,A^{11} + J_{12}J_{22}\,A^{22} + J_{11}J_{22}\,A^{12} + J_{12}J_{21}\,A^{21},$$

$$A^S_{21} \doteq \mathbf{t}_2 \cdot (\mathbf{A}\,\mathbf{t}_1) = (J_{21}\,\mathbf{g}^1 + J_{22}\,\mathbf{g}^2) \cdot [\mathbf{A}\ (J_{11}\mathbf{g}^1 + J_{12}\mathbf{g}^2)]$$
$$= J_{11}J_{21}\,A^{11} + J_{12}J_{22}\,A^{22} + J_{12}J_{21}\,A^{12} + J_{11}J_{22}\,A^{21}.$$

We can rewrite the above formulas in the vector-matrix form as

$$\mathbf{A}^S_v = \mathbf{T}\ \mathbf{A}^\xi_v, \tag{2.21}$$

where

$$\mathbf{T} \doteq \begin{bmatrix} J^2_{11} & J^2_{12} & J_{11}J_{12} & J_{11}J_{12} \\ J^2_{21} & J^2_{22} & J_{21}J_{22} & J_{21}J_{22} \\ J_{11}J_{21} & J_{12}J_{22} & J_{11}J_{22} & J_{12}J_{21} \\ J_{11}J_{21} & J_{12}J_{22} & J_{12}J_{21} & J_{11}J_{22} \end{bmatrix},$$

$$\mathbf{A}^S_v \doteq \begin{bmatrix} A^S_{11} \\ A^S_{22} \\ A^S_{12} \\ A^S_{21} \end{bmatrix}, \qquad \mathbf{A}^\xi_v \doteq \begin{bmatrix} A^{11} \\ A^{22} \\ A^{12} \\ A^{21} \end{bmatrix}.$$

If $\mathbf{A}^S_v$ and $\mathbf{A}^\xi_v$ are rewritten as 2×2 matrices then, instead of using the transformation operator $\mathbf{T}$, we can write

$$\mathbf{A}^S = \mathbf{J}\,\mathbf{A}^\xi\,\mathbf{J}^T. \tag{2.22}$$

The equivalence of $\mathbf{A}^S$ and $\mathbf{A}^S_v$ can be verified by comparing their corresponding components, i.e. $(1,1) \leftrightarrow (1)$, $(2,2) \leftrightarrow (2)$, $(1,2) \leftrightarrow (3)$ and $(2,1) \leftrightarrow (4)$.

Transformation of covariant in-plane components of vector and tensor. Components of vectors and tensors in the ortho-normal basis $\{\mathbf{t}_i\}$ are obtained from the covariant components in the co-basis $\{\mathbf{g}^i\}$, assuming that the normal vectors are identical, i.e. $\mathbf{t}_3 = \mathbf{g}^3$. Again, the components in $\{\mathbf{g}^i\}$ cannot be obtained by orthogonal projections.

Vectors. The transformation formula for the vector $\mathbf{v}$ is obtained in the following steps. First, we calculate the components $v^S_\alpha \doteq \mathbf{v} \cdot \mathbf{t}_\alpha$ in $\{\mathbf{t}_i\}$. Next, we transform them to the skew co-basis $\{\mathbf{g}_\alpha\}$, using eq. $(10.44)_2$ rewritten as follows:

$$\mathbf{t}_1 = J^{-1}_{11}\,\mathbf{g}_1 + J^{-1}_{21}\,\mathbf{g}_2, \qquad \mathbf{t}_2 = J^{-1}_{12}\,\mathbf{g}_1 + J^{-1}_{22}\,\mathbf{g}_2,$$

where J^{-1}_{ij} are components of the 2×2 Jacobian matrix, $\mathbf{J}^{-1}$, defined in eq. (10.49). Finally, we identify the covariant components $v_\alpha \doteq \mathbf{v} \cdot \mathbf{g}_\alpha$ of the vector $\mathbf{v} = v_\alpha\, \mathbf{g}^\alpha$. The transformations are performed as follows:

$$v^S_1 \doteq \mathbf{v}\cdot\mathbf{t}_1 = \mathbf{v}\cdot(J^{-1}_{11}\mathbf{g}^1 + J^{-1}_{21}\mathbf{g}^2) = J^{-1}_{11}\,(\mathbf{v}\cdot\mathbf{g}^1) + J^{-1}_{21}\,(\mathbf{v}\cdot\mathbf{g}^2) = J^{-1}_{11}\,v_1 + J^{-1}_{21}\,v_2,$$

$$v^S_2 \doteq \mathbf{v}\cdot\mathbf{t}_2 = \mathbf{v}\cdot(J^{-1}_{12}\mathbf{g}^1 + J^{-1}_{22}\mathbf{g}^2) = J^{-1}_{12}\,(\mathbf{v}\cdot\mathbf{g}^1) + J^{-1}_{22}\,(\mathbf{v}\cdot\mathbf{g}^2) = J^{-1}_{12}\,v_1 + J^{-1}_{22}\,v_2.$$

In matrix form, we have

$$\begin{bmatrix} v^S_1 \\ v^S_2 \end{bmatrix} = (\mathbf{J}^{-1})^T \begin{bmatrix} v_1 \\ v_2 \end{bmatrix}. \tag{2.23}$$

Second-rank tensors. The transformation formulas for the second-rank tensor $\mathbf{A}$ can be obtained similarly as for a vector. Now we shall identify the covariant components $A_{\alpha\beta} \doteq \mathbf{g}_\alpha \cdot (\mathbf{A}\,\mathbf{g}_\beta)$ of the tensor $\mathbf{A} = A_{\alpha\beta}\,\mathbf{g}^\alpha \otimes \mathbf{g}^\beta$. By analogy to eq. (2.22), we obtain

$$\mathbf{A}^S = (\mathbf{J}^{-1})^T\, \mathbf{A}_\xi\, \mathbf{J}^{-1}. \tag{2.24}$$

By analogy to eq. (2.21), we can rewrite eq. (2.24) in the vector-matrix form

$$\mathbf{A}^S_v = \mathbf{T}^*\, \mathbf{A}_{v\xi}, \tag{2.25}$$

where

$$\mathbf{T}^* \doteq \begin{bmatrix} (J^{-1}_{11})^2 & (J^{-1}_{21})^2 & J^{-1}_{11} J^{-1}_{21} & J^{-1}_{11} J^{-1}_{21} \\ (J^{-1}_{12})^2 & (J^{-1}_{22})^2 & J^{-1}_{12} J^{-1}_{22} & J^{-1}_{12} J^{-1}_{22} \\ J^{-1}_{11} J^{-1}_{12} & J^{-1}_{21} J^{-1}_{22} & J^{-1}_{11} J^{-1}_{22} & J^{-1}_{21} J^{-1}_{12} \\ J^{-1}_{11} J^{-1}_{12} & J^{-1}_{21} J^{-1}_{22} & J^{-1}_{21} J^{-1}_{12} & J^{-1}_{11} J^{-1}_{22} \end{bmatrix},$$

$$\mathbf{A}^S_v \doteq \begin{bmatrix} A^S_{11} \\ A^S_{22} \\ A^S_{12} \\ A^S_{21} \end{bmatrix}, \qquad \mathbf{A}_{v\xi} \doteq \begin{bmatrix} A_{11} \\ A_{22} \\ A_{12} \\ A_{21} \end{bmatrix}.$$

Note that $\mathbf{T}^*$ can be obtained from $\mathbf{T}$ by replacing $J_{\alpha\beta}$ with $J^{-1}_{\beta\alpha}$, in which the order of indices $\alpha\beta$ is reversed. Hence, $\mathbf{T}^* = \mathbf{T}^{-T}$, as can be readily checked. The equivalence of $\mathbf{A}^S$ and $\mathbf{A}^S_v$ can be verified by comparing their corresponding components, i.e. $(1,1) \leftrightarrow (1)$, $(2,2) \leftrightarrow (2)$, $(1,2) \leftrightarrow (3)$, and $(2,1) \leftrightarrow (4)$.

Check: Invariance of scalar product. We can check the correctness of the tensorial formulas of eqs. (2.22) and (2.24) by calculating a scalar product of two tensors,

$$\mathbf{A}^S = \mathbf{J}\,\mathbf{A}^{\xi}\,\mathbf{J}^T, \qquad \mathbf{B}^S = (\mathbf{J}^{-1})^T\,\mathbf{B}_{\xi}\,\mathbf{J}^{-1}.$$

Such a product appears, e.g., in the strain energy, $\mathcal{W} \doteq \frac{1}{2}\boldsymbol{\sigma}\cdot\boldsymbol{\varepsilon}$. Using eq. (2.22) and (2.24), we obtain

$$\mathbf{A}^S\cdot\mathbf{B}^S = \mathrm{tr}[\mathbf{A}^S(\mathbf{B}^S)^T] = \mathrm{tr}(\mathbf{J}\,\mathbf{A}^{\xi}\,\mathbf{J}^T\mathbf{J}^{-T}\,\mathbf{B}_{\xi}^T\,\mathbf{J}^{-1}) = \mathrm{tr}(\mathbf{A}^{\xi}\,\mathbf{B}_{\xi}^T) = \mathbf{A}^{\xi}\cdot\mathbf{B}_{\xi}.$$

The same can be repeated for the vectors of tensorial components of eqs. (2.21) and (2.25), i.e.

$$\mathbf{A}_v^S = \mathbf{T}\,\mathbf{A}_v^{\xi}, \qquad \mathbf{B}_v^S = \mathbf{T}^*\,\mathbf{B}_{v\xi},$$

in which the 4×4 operators $\mathbf{T}$ and $\mathbf{T}^*$ appear. The scalar product of the vectors is

$$\mathbf{A}_v^S\cdot\mathbf{B}_v^S = \mathbf{A}_v^{\xi}\cdot(\mathbf{T}^T\mathbf{T}^*\,\mathbf{B}_{v\xi}) = \mathbf{A}_v^{\xi}\cdot\mathbf{B}_{v\xi},$$

using $\mathbf{T}^T\mathbf{T}^* = \mathbf{T}^T\mathbf{T}^{-T} = \mathbf{I}$. We see that the scalar product is invariant for both ways of calculation, as required.

Transformation of transverse ($\alpha 3$ and 3α) components. The transverse components in the ortho-normal basis $\{\mathbf{t}_i\}$ can be obtained from the components either in the basis $\{\mathbf{g}_i\}$ or in the co-basis $\{\mathbf{g}^i\}$. We assume that the normal vectors are identical, i.e. $\mathbf{t}_3 = \mathbf{g}_3 = \mathbf{g}^3$. The derivation is analogous to the one for the in-plane components.

Contravariant components. The transformations are as follows:

$$A_{13}^S \doteq \mathbf{t}_1\cdot(\mathbf{A}\,\mathbf{t}_3) = (J_{11}\mathbf{g}^1 + J_{12}\mathbf{g}^2)\cdot(\mathbf{A}\,\mathbf{g}^3) = J_{11}\,A^{13} + J_{12}\,A^{23},$$

$$A_{23}^S \doteq \mathbf{t}_2\cdot(\mathbf{A}\,\mathbf{t}_3) = (J_{21}\,\mathbf{g}^1 + J_{22}\,\mathbf{g}^2)\cdot(\mathbf{A}\,\mathbf{g}^3) = J_{21}\,A^{13} + J_{22}\,A^{23}$$

and

$$A_{31}^S \doteq \mathbf{t}_3\cdot(\mathbf{A}\,\mathbf{t}_1) = \mathbf{g}^3\cdot(\mathbf{A}\,[J_{11}\mathbf{g}^1 + J_{12}\mathbf{g}^2)] = J_{11}\,A^{31} + J_{12}\,A^{32},$$

$$A_{32}^S \doteq \mathbf{t}_3\cdot(\mathbf{A}\,\mathbf{t}_2) = \mathbf{g}^3\cdot[\mathbf{A}\,(J_{21}\,\mathbf{g}^1 + J_{22}\,\mathbf{g}^2)] = J_{21}\,A^{31} + J_{22}\,A^{32},$$

where the contravariant components are $A^{\alpha 3} \doteq \mathbf{g}^{\alpha}\cdot(\mathbf{A}\,\mathbf{g}^3)$ and $A^{3\alpha} \doteq \mathbf{g}^3\cdot(\mathbf{A}\,\mathbf{g}^{\alpha})$. We can rewrite the above formulas in the vector-matrix form as follows:

$$\begin{bmatrix} A_{13}^S \\ A_{23}^S \end{bmatrix} = \mathbf{J}\begin{bmatrix} A^{13} \\ A^{23} \end{bmatrix}, \qquad \begin{bmatrix} A_{31}^S \\ A_{32}^S \end{bmatrix} = \mathbf{J}\begin{bmatrix} A^{31} \\ A^{32} \end{bmatrix}. \tag{2.26}$$

Covariant components. The transformations are

$$A_{13}^S \doteq \mathbf{t}_1 \cdot (\mathbf{A}\,\mathbf{t}_3) = (J_{11}^{-1}\,\mathbf{g}_1 + J_{21}^{-1}\,\mathbf{g}_2) \cdot (\mathbf{A}\,\mathbf{g}_3) = J_{11}^{-1}\,A_{13} + J_{21}^{-1}\,A_{23},$$

$$A_{23}^S \doteq \mathbf{t}_2 \cdot (\mathbf{A}\,\mathbf{t}_3) = (J_{12}^{-1}\,\mathbf{g}_1 + J_{22}^{-1}\,\mathbf{g}_2) \cdot (\mathbf{A}\,\mathbf{g}_3) = J_{12}^{-1}\,A_{13} + J_{22}^{-1}\,A_{23}$$

and

$$A_{31}^S \doteq \mathbf{t}_3 \cdot (\mathbf{A}\,\mathbf{t}_1) = \mathbf{g}^3 \cdot (\mathbf{A}\,[J_{11}^{-1}\mathbf{g}^1 + J_{21}^{-1}\mathbf{g}^2)] = J_{11}^{-1}\,A_{31} + J_{21}^{-1}\,A_{32},$$

$$A_{32}^S \doteq \mathbf{t}_3 \cdot (\mathbf{A}\,\mathbf{t}_2) = \mathbf{g}^3 \cdot [\mathbf{A}\,(J_{12}^{-1}\,\mathbf{g}^1 + J_{22}^{-1}\,\mathbf{g}^2)] = J_{12}^{-1}\,A_{31} + J_{22}^{-1}\,A_{32},$$

where the covariant components are $A_{\alpha 3} \doteq \mathbf{g}_\alpha \cdot (\mathbf{A}\,\mathbf{g}_3)$ and $A_{3\alpha} \doteq \mathbf{g}_3 \cdot (\mathbf{A}\,\mathbf{g}_\alpha)$. We can rewrite the above formulas in the vector-matrix form as follows:

$$\begin{bmatrix} A_{13}^S \\ A_{23}^S \end{bmatrix} = \mathbf{J}^{-T} \begin{bmatrix} A_{13} \\ A_{23} \end{bmatrix}, \qquad \begin{bmatrix} A_{31}^S \\ A_{32}^S \end{bmatrix} = \mathbf{J}^{-T} \begin{bmatrix} A_{31} \\ A_{32} \end{bmatrix}. \tag{2.27}$$

Symmetric second-rank tensors. Now we consider the case when $\mathbf{A}^\xi$ and $\mathbf{A}_\xi$ for the in-plane ($\alpha\beta$) components are symmetric tensors. The tensorial eq. (2.22) and (2.24) remain valid, but we may simplify the formulas involving the 4×4 operators $\mathbf{T}$ and $\mathbf{T}^*$ as follows.

Contravariant components. For a symmetric $\mathbf{A}^\xi$, when $A^{21} = A^{12}$, we may simplify eq. (2.21) by adding the third and fourth column of $\mathbf{T}$, and by omitting the row for A_{21}^S which is identical as the row for A_{12}^S. Then,

$$\mathbf{T} \doteq \begin{bmatrix} J_{11}^2 & J_{12}^2 & 2J_{11}J_{12} \\ J_{21}^2 & J_{22}^2 & 2J_{21}J_{22} \\ J_{11}J_{21} & J_{12}J_{22} & J_{11}J_{22} + J_{12}J_{21} \end{bmatrix}, \tag{2.28}$$

$$\mathbf{A}_v^S \doteq \begin{bmatrix} A_{11}^S \\ A_{22}^S \\ A_{12}^S \end{bmatrix}, \qquad \mathbf{A}_v^\xi \doteq \begin{bmatrix} A^{11} \\ A^{22} \\ A^{12} \end{bmatrix}.$$

This form is useful in three applications in which vectors of the contravariant components appear:

1. in the strain energy $\mathcal{W} \doteq \frac{1}{2}\boldsymbol{\sigma}\cdot\boldsymbol{\varepsilon}$, for the stresses $\boldsymbol{\sigma} \doteq [\sigma^{11}, \sigma^{22}, \sigma^{12}]^T$,
2. in the Assumed Stress methods, for the stresses $\boldsymbol{\sigma} \doteq [\sigma^{11}, \sigma^{22}, \sigma^{12}]^T$,
3. in the Assumed Strain methods, for the strains $\boldsymbol{\varepsilon} \doteq [\varepsilon^{11}, \varepsilon^{22}, 2\varepsilon^{12}]^T$.

Covariant components. Re-defined operator. For a symmetric $\mathbf{A}_\xi$, when $A_{21} = A_{12}$, we can add the third and fourth columns and retain only A^S_{12}, because $A^S_{21} = A^S_{12}$. Then, in eq. (2.25) we have

$$\mathbf{T}^* \doteq \begin{bmatrix} (J^{-1}_{11})^2 & (J^{-1}_{21})^2 & 2J^{-1}_{11}J^{-1}_{21} \\ (J^{-1}_{12})^2 & (J^{-1}_{22})^2 & 2J^{-1}_{12}J^{-1}_{22} \\ J^{-1}_{11}J^{-1}_{12} & J^{-1}_{21}J^{-1}_{22} & J^{-1}_{11}J^{-1}_{22} + J^{-1}_{21}J^{-1}_{12} \end{bmatrix}, \tag{2.29}$$

$$\mathbf{A}^S_v \doteq \begin{bmatrix} A^S_{11} \\ A^S_{22} \\ A^S_{12} \end{bmatrix}, \qquad \mathbf{A}_{v\xi} \doteq \begin{bmatrix} A_{11} \\ A_{22} \\ A_{12} \end{bmatrix}.$$

However, the above 3×3 operator $\mathbf{T}^*$ is deficient because it is not equal to $\mathbf{T}^{-T}$ for $\mathbf{T}$ of eq. (2.28). Recall that this property holds for the 4×4 operators and is necessary to maintain invariance of the scalar product. Unfortunately, it was destroyed while accounting for the symmetry of the tensors. To remedy the problem, we do not use $\mathbf{T}^*$ of eq. (2.29), but we define

$$\mathbf{T}^{**} \doteq \mathbf{T}^{-T} = \begin{bmatrix} (J^{-1}_{11})^2 & (J^{-1}_{21})^2 & J^{-1}_{11}J^{-1}_{21} \\ (J^{-1}_{12})^2 & (J^{-1}_{22})^2 & J^{-1}_{12}J^{-1}_{22} \\ 2J^{-1}_{11}J^{-1}_{12} & 2J^{-1}_{21}J^{-1}_{22} & J^{-1}_{11}J^{-1}_{22} + J^{-1}_{21}J^{-1}_{12} \end{bmatrix}, \tag{2.30}$$

where $\mathbf{T}$ is specified in eq. (2.28). Using components of $\mathbf{J}$ instead of $\mathbf{J}^{-1}$, we can obtain another, equivalent, form

$$\mathbf{T}^{**} = \frac{1}{(\det \mathbf{J})^2} \begin{bmatrix} J^2_{22} & J^2_{21} & -J_{21}J_{22} \\ J^2_{12} & J^2_{11} & -J_{11}J_{12} \\ -2J_{12}J_{22} & -2J_{11}J_{21} & J_{11}J_{22} + J_{12}J_{21} \end{bmatrix}, \tag{2.31}$$

where $\det \mathbf{J} \doteq J_{11}J_{22} - J_{12}J_{21}$.

Note that eq. (2.29), but with $\mathbf{T}^{**}$ of eq. (2.30) or (2.31) used instead of $\mathbf{T}^*$, can be applied to the strain energy, $\mathcal{W} \doteq \frac{1}{2}\boldsymbol{\sigma} \cdot \boldsymbol{\varepsilon}$, for the vector of covariant strain components $\boldsymbol{\varepsilon} \doteq [\varepsilon_{11}, \varepsilon_{22}, 2\varepsilon_{12}]^T$.

Check: Invariance of strain energy. For the symmetric tensors, the tensorial formulas are identical as for non-symmetric ones for which we already checked the invariance of the scalar product. Using the formulas with the 3×3 operators $\mathbf{T}$ and $\mathbf{T}^{**}$, we have

$$\boldsymbol{\sigma}^S_v = \mathbf{T}\, \boldsymbol{\sigma}_v, \qquad \boldsymbol{\varepsilon}^S_v = \mathbf{T}^{**}\, \boldsymbol{\varepsilon}_v,$$

where the vector of contravariant stresses $\boldsymbol{\sigma}_v^\xi \doteq [\sigma^{11}, \sigma^{22}, \sigma^{12}]^T$ and the vector of covariant strains $\boldsymbol{\varepsilon}_{v\xi} \doteq [\varepsilon_{11}, \varepsilon_{22}, 2\varepsilon_{12}]^T$. The strain energy can be expressed as $\mathcal{W} \doteq \frac{1}{2}\boldsymbol{\sigma}_v^S \cdot \boldsymbol{\varepsilon}_v^S$, where the scalar product of the vectors is

$$\boldsymbol{\sigma}_v^S \cdot \boldsymbol{\varepsilon}_v^S = \boldsymbol{\sigma}_v^\xi \cdot (\mathbf{T}^T\mathbf{T}^{**}\ \boldsymbol{\varepsilon}_{v\xi}) = \boldsymbol{\sigma}_v^\xi \cdot \boldsymbol{\varepsilon}_{v\xi} = \sigma^{11}\varepsilon_{11} + \sigma^{22}\varepsilon_{22} + 2\sigma^{12}\varepsilon_{12}.$$

An identical invariant form is obtained by tensorial operations.

Transformation of second-rank tensors between two Cartesian bases. Now we assume that $\mathbf{g}_3 = \mathbf{g}^3 = \mathbf{t}_3$, and that $\{\mathbf{g}_k\}$ is an ortho-normal basis, so $\mathbf{g}^k = \mathbf{g}_k$. We obtain the components in the ortho-normal basis $\{\mathbf{t}_i\}$ by simplification of the earlier derived general formulas. Then, the Jacobian matrix is simply a proper orthogonal matrix, i.e.

$$\mathbf{J} = \begin{bmatrix} c & -s \\ s & c \end{bmatrix} = \mathbf{R}, \qquad c \doteq \cos\alpha, \qquad s \doteq \sin\alpha, \tag{2.32}$$

where α is the angle of rotation about $\mathbf{t}_3$. Note that $\mathbf{J}^{-1} = \mathbf{R}^T$ and $\det \mathbf{J} = 1$.

Non-symmetric tensors. For the in-plane $(\alpha\beta)$ components, the transformation rules of eqs. (2.22) and (2.24) become identical,

$$\mathbf{A}^S = \mathbf{R}\,\mathbf{A}^\xi\,\mathbf{R}^T, \qquad \mathbf{A}^S = \mathbf{R}\,\mathbf{A}_\xi\,\mathbf{R}^T. \tag{2.33}$$

Also, the 4×4 operators become identical, i.e.

$$\mathbf{T} \doteq \begin{bmatrix} J_{11}^2 & J_{12}^2 & J_{11}J_{12} & J_{11}J_{12} \\ J_{21}^2 & J_{22}^2 & J_{21}J_{22} & J_{21}J_{22} \\ J_{11}J_{21} & J_{12}J_{22} & J_{11}J_{22} & J_{12}J_{21} \\ J_{11}J_{21} & J_{12}J_{22} & J_{12}J_{21} & J_{11}J_{22} \end{bmatrix} = \begin{bmatrix} c^2 & s^2 & -sc & -sc \\ s^2 & c^2 & sc & sc \\ sc & -sc & c^2 & -s^2 \\ sc & -sc & -s^2 & c^2 \end{bmatrix},$$

$$\mathbf{T}^* \doteq \begin{bmatrix} (J_{11}^{-1})^2 & (J_{21}^{-1})^2 & J_{11}^{-1}J_{21}^{-1} & J_{11}^{-1}J_{21}^{-1} \\ (J_{12}^{-1})^2 & (J_{22}^{-1})^2 & J_{12}^{-1}J_{22}^{-1} & J_{12}^{-1}J_{22}^{-1} \\ J_{11}^{-1}J_{12}^{-1} & J_{21}^{-1}J_{22}^{-1} & J_{11}^{-1}J_{22}^{-1} & J_{21}^{-1}J_{12}^{-1} \\ J_{11}^{-1}J_{12}^{-1} & J_{21}^{-1}J_{22}^{-1} & J_{21}^{-1}J_{12}^{-1} & J_{11}^{-1}J_{22}^{-1} \end{bmatrix} = \begin{bmatrix} c^2 & s^2 & -sc & -sc \\ s^2 & c^2 & sc & sc \\ sc & -sc & c^2 & -s^2 \\ sc & -sc & -s^2 & c^2 \end{bmatrix},$$

so $\mathbf{T}^* = \mathbf{T}$, and the transformation rules of eqs. (2.21) and (2.25) become identical,

$$\mathbf{A}_v^S = \mathbf{T}\,\mathbf{A}_v^\xi, \qquad \mathbf{B}_v^S = \mathbf{T}\,\mathbf{B}_{v\xi}. \tag{2.34}$$

For the transverse ($\alpha 3$ and 3α) components, we have $\mathbf{J} = \mathbf{J}^{-T} = \mathbf{R}$, and the transformation rules of eqs. (2.26) and (2.27) become identical,

$$\begin{bmatrix} A^S_{13} \\ A^S_{23} \end{bmatrix} = \mathbf{R} \begin{bmatrix} A^{13} \\ A^{23} \end{bmatrix}, \qquad \begin{bmatrix} A^S_{31} \\ A^S_{32} \end{bmatrix} = \mathbf{R} \begin{bmatrix} A^{31} \\ A^{32} \end{bmatrix}, \tag{2.35}$$

$$\begin{bmatrix} A^S_{13} \\ A^S_{23} \end{bmatrix} = \mathbf{R} \begin{bmatrix} A_{13} \\ A_{23} \end{bmatrix}, \qquad \begin{bmatrix} A^S_{31} \\ A^S_{32} \end{bmatrix} = \mathbf{R} \begin{bmatrix} A_{31} \\ A_{32} \end{bmatrix}. \tag{2.36}$$

Symmetric tensors. For the in-plane ($\alpha\beta$) components, the operators $\mathbf{T}$ of eq. (2.28) and $\mathbf{T}^{**}$ of eq. (2.31) assume the forms

$$\mathbf{T} \doteq \begin{bmatrix} J_{11}^2 & J_{12}^2 & 2J_{11}J_{12} \\ J_{21}^2 & J_{22}^2 & 2J_{21}J_{22} \\ J_{11}J_{21} & J_{12}J_{22} & (J_{11}J_{22} + J_{12}J_{21}) \end{bmatrix} = \begin{bmatrix} c^2 & s^2 & -2sc \\ s^2 & c^2 & 2sc \\ sc & -sc & c^2 - s^2 \end{bmatrix},$$

$$\mathbf{T}^{**} \doteq \frac{1}{(\det \mathbf{J})^2} \begin{bmatrix} J_{22}^2 & J_{21}^2 & -J_{21}J_{22} \\ J_{12}^2 & J_{11}^2 & -J_{11}J_{12} \\ -2J_{12}J_{22} & -2J_{11}J_{21} & J_{11}J_{22} + J_{12}J_{21} \end{bmatrix}$$
$$= \begin{bmatrix} c^2 & s^2 & -sc \\ s^2 & c^2 & sc \\ 2sc & -2sc & c^2 - s^2 \end{bmatrix} = \mathbf{T}^{-T},$$

and eqs. (2.21) and (2.25) become

$$\mathbf{A}^S_v = \mathbf{T}\,\mathbf{A}^\xi_v, \qquad \mathbf{B}^S_v = \mathbf{T}^{-T}\,\mathbf{B}_{v\xi}.$$

The scalar product of them is invariant, i.e.

$$\mathbf{A}^S_v \cdot \mathbf{B}^S_v = \mathbf{A}^\xi_v \cdot [(\mathbf{T}^T\mathbf{T}^{-T})\,\mathbf{B}_{v\xi}] = \mathbf{A}^\xi \cdot \mathbf{B}_\xi.$$

We can check that $\det \mathbf{T} = 1$, but $\mathbf{T}^{-1} \neq \mathbf{T}^T$, i.e. $\mathbf{T}$ is not an orthogonal matrix. Besides, $\mathbf{T}^{-1}(\alpha) = \mathbf{T}(-\alpha)$.

For the symmetric transverse components ($\alpha 3 = 3\alpha$), eqs. (2.26) and (2.27) yield

$$\begin{bmatrix} A^S_{13} \\ A^S_{23} \end{bmatrix} = \mathbf{R} \begin{bmatrix} A^{13} \\ A^{23} \end{bmatrix}, \qquad \begin{bmatrix} A^S_{31} \\ A^S_{32} \end{bmatrix} = \mathbf{R} \begin{bmatrix} A^{31} \\ A^{32} \end{bmatrix}. \tag{2.37}$$

Combining the above formulas for in-plane ($\alpha\beta$) components and the transverse ($\alpha 3$) components, yields the following transformation operators for a symmetric tensor:

- for "contravariant" components,

$$\begin{bmatrix} A^S_{11} \\ A^S_{22} \\ A^S_{33} \\ \hline A^S_{12} \\ A^S_{13} \\ A^S_{23} \end{bmatrix} = \left[\begin{array}{ccc|ccc} c^2 & s^2 & 0 & -2sc & 0 & 0 \\ s^2 & c^2 & 0 & 2sc & 0 & 0 \\ 0 & 0 & 1 & 0 & 0 & 0 \\ \hline sc & -sc & 0 & c^2 - s^2 & 0 & 0 \\ 0 & 0 & 0 & 0 & c & -s \\ 0 & 0 & 0 & 0 & s & c \end{array}\right] \begin{bmatrix} A^{11} \\ A^{22} \\ A^{33} \\ \hline A^{12} \\ A^{13} \\ A^{23} \end{bmatrix}, \tag{2.38}$$

- for "covariant" components,

$$\begin{bmatrix} A^S_{11} \\ A^S_{22} \\ A^S_{33} \\ \hline A^S_{12} \\ A^S_{13} \\ A^S_{23} \end{bmatrix} = \left[\begin{array}{ccc|ccc} c^2 & s^2 & 0 & -sc & 0 & 0 \\ s^2 & c^2 & 0 & sc & 0 & 0 \\ 0 & 0 & 1 & 0 & 0 & 0 \\ \hline 2sc & -2sc & 0 & c^2 - s^2 & 0 & 0 \\ 0 & 0 & 0 & 0 & c & -s \\ 0 & 0 & 0 & 0 & s & c \end{array}\right] \begin{bmatrix} A_{11} \\ A_{22} \\ A_{33} \\ \hline A_{12} \\ A_{13} \\ A_{23} \end{bmatrix}. \tag{2.39}$$

Note that the terms "contravariant" and "covariant" components are used above only to maintain their relation to the non-orthogonal bases and to indicate that applications of these formulas are similar to those of the subsection *Symmetric second-rank tensors.*

2.3 Gradients and derivatives

Representation of a gradient of vector. The gradient of vector $\mathbf{a} \doteq a^l\, \mathbf{i}_l$ w.r.t. the vector $\mathbf{b} \doteq b^k\, \mathbf{i}_k$ is the second-rank tensor, which in the basis $\{\mathbf{i}_k\}$, assumes the form

$$\mathbf{G} \doteq \frac{\partial \mathbf{a}}{\partial \mathbf{b}} = \frac{\partial a^l}{\partial b^k}\, \mathbf{i}_l \otimes \mathbf{i}_k, \qquad l, k = 1, 2, 3. \tag{2.40}$$

The matrix of its components is

$$\mathsf{G} \doteq \left[\frac{\partial a^l}{\partial b^k}\right] = \begin{bmatrix} \frac{\partial a^1}{\partial b^1} & \frac{\partial a^1}{\partial b^2} & \frac{\partial a^1}{\partial b^3} \\ \frac{\partial a^2}{\partial b^1} & \frac{\partial a^2}{\partial b^2} & \frac{\partial a^2}{\partial b^3} \\ \frac{\partial a^3}{\partial b^1} & \frac{\partial a^3}{\partial b^2} & \frac{\partial a^3}{\partial b^3} \end{bmatrix}. \tag{2.41}$$

Note that, accepting the standard abuse of notation, we can also use the same letter $\mathbf{G}$ to denote the tensor and the array of its components.

Derivatives of shape functions w.r.t. Cartesian coordinates. The shape functions are functions of natural coordinates $\xi, \eta \in [-1, +1]$ but, using a chain rule, we can calculate their derivatives w.r.t. the Cartesian coordinates x, y. Then, we obtain

$$\begin{bmatrix} \frac{\partial N_I(\xi,\eta)}{\partial x} \\ \frac{\partial N_I(\xi,\eta)}{\partial y} \end{bmatrix} = \begin{bmatrix} \frac{\partial N_I}{\partial \xi}\frac{\partial \xi}{\partial x} + \frac{\partial N_I}{\partial \eta}\frac{\partial \eta}{\partial x} \\ \frac{\partial N_I}{\partial \xi}\frac{\partial \xi}{\partial y} + \frac{\partial N_I}{\partial \eta}\frac{\partial \eta}{\partial y} \end{bmatrix} = \begin{bmatrix} \frac{\partial \xi}{\partial x} & \frac{\partial \eta}{\partial x} \\ \frac{\partial \xi}{\partial y} & \frac{\partial \eta}{\partial y} \end{bmatrix} \begin{bmatrix} \frac{\partial N_I}{\partial \xi} \\ \frac{\partial N_I}{\partial \eta} \end{bmatrix}. \tag{2.42}$$

The derivatives $\partial N_I/\partial \xi$ and $\partial N_I/\partial \eta$ can be explicitly calculated, while for the matrix

$$\mathsf{A} \doteq \begin{bmatrix} \frac{\partial \xi}{\partial x} & \frac{\partial \eta}{\partial x} \\ \frac{\partial \xi}{\partial y} & \frac{\partial \eta}{\partial y} \end{bmatrix} \tag{2.43}$$

we shall find its relation to the Jacobian matrix and its inverse defined as

$$\mathsf{J} \doteq \begin{bmatrix} \frac{\partial x}{\partial \xi} & \frac{\partial x}{\partial \eta} \\ \frac{\partial y}{\partial \xi} & \frac{\partial y}{\partial \eta} \end{bmatrix}, \qquad \mathsf{J}^{-1} = \frac{1}{\det \mathsf{J}} \begin{bmatrix} \frac{\partial y}{\partial \eta} & -\frac{\partial x}{\partial \eta} \\ -\frac{\partial y}{\partial \xi} & \frac{\partial x}{\partial \xi} \end{bmatrix}. \tag{2.44}$$

We can verify that

$$\mathsf{J}^T \mathsf{A} = \begin{bmatrix} A & C \\ C & B \end{bmatrix} = \mathsf{I}, \tag{2.45}$$

where

$$A \doteq \frac{\partial \xi}{\partial x}\frac{\partial x}{\partial \xi} + \frac{\partial \xi}{\partial y}\frac{\partial y}{\partial \xi} = \frac{d\xi}{d\xi} = 1, \quad B \doteq \frac{\partial \eta}{\partial x}\frac{\partial x}{\partial \eta} + \frac{\partial \eta}{\partial y}\frac{\partial y}{\partial \eta} = \frac{d\eta}{d\eta} = 1,$$

$$C \doteq \frac{\partial \xi}{\partial x}\frac{\partial x}{\partial \eta} + \frac{\partial \xi}{\partial y}\frac{\partial y}{\partial \eta} = \frac{d\xi}{d\eta} = 0.$$

Hence, $\mathsf{A} = \mathsf{J}^{-T}$, and the derivatives of shape functions w.r.t. Cartesian coordinates can be calculated by the formula

$$\begin{bmatrix} \frac{\partial N_I(\xi,\eta)}{\partial x} \\ \frac{\partial N_I(\xi,\eta)}{\partial y} \end{bmatrix} = \mathsf{J}^{-T} \begin{bmatrix} \frac{\partial N_I}{\partial \xi} \\ \frac{\partial N_I}{\partial \eta} \end{bmatrix} = \frac{1}{\det \mathsf{J}} \begin{bmatrix} \frac{\partial y}{\partial \eta} & -\frac{\partial y}{\partial \xi} \\ -\frac{\partial x}{\partial \eta} & \frac{\partial x}{\partial \xi} \end{bmatrix} \begin{bmatrix} \frac{\partial N_I}{\partial \xi} \\ \frac{\partial N_I}{\partial \eta} \end{bmatrix}. \tag{2.46}$$

Finally, we note that we can use as $\{x, y\}$ and J, the Cartesian coordinates and the Jacobian matrices introduced in Sect. 10.3:

1. for the reference Cartesian coordinates $\{y^1, y^2\}$, J is the upper 2×2 matrix of J_G of eq. (10.38), or
2. for the local Cartesian coordinates $\{S^1, S^2\}$, J is the upper 2×2 matrix of J_L of eq. (10.39). An alternative form of this matrix is given in eq. (11.13).

Part II
SHELL EQUATIONS

3

Rotations for 3D Cauchy continuum

In this chapter we consider the *classical configuration space* of the non-polar Cauchy continuum, defined as

$$\mathcal{C} \doteq \{\chi\colon\ B \to R^3\}, \tag{3.1}$$

where χ is the deformation function defined over the reference configuration of the body B. The rotations are calculated from the deformation gradient, $\mathbf{F} \doteq \nabla\chi$, and are not independent variables.

3.1 Polar decomposition of deformation gradient

In the non-polar Cauchy media, the rotations associated with deformation can be obtained by the polar decomposition of the deformation gradient, which appears in two forms: right and left. The right polar decomposition is given by the formula

$$\mathbf{F} = \mathbf{R}\mathbf{U}, \tag{3.2}$$

where $\mathbf{U} \doteq (\mathbf{F}^T\mathbf{F})^{\frac{1}{2}}$ is the right stretching tensor (symmetric and positive definite) and $\mathbf{R} = \mathbf{F}\mathbf{U}^{-1} \in \mathrm{SO}(3)$ is a rotation tensor. The left polar decomposition is

$$\mathbf{F} = \mathbf{V}\mathbf{R}, \tag{3.3}$$

where $\mathbf{V} \doteq (\mathbf{F}\mathbf{F}^T)^{\frac{1}{2}}$ is the left stretching tensor, also symmetric and positive definite. Then the rotation tensor is calculated as $\mathbf{R} = \mathbf{V}^{-1}\mathbf{F} \in \mathrm{SO}(3)$.

Uniqueness of polar decomposition. The proof of uniqueness of the decomposition (3.2) follows the standard lines, and is given, e.g., in [115], p. 77 or [159], p. 93.

Properties of $\mathbf{U}$. The properties of $\mathbf{U}$ result from the properties of $\mathbf{C} = \mathbf{F}^T\mathbf{F}$. First, $\mathbf{C}$ is symmetric as $(\mathbf{F}^T\mathbf{F})^T = \mathbf{F}^T(\mathbf{F}^T)^T = \mathbf{F}^T\mathbf{F}$. Symmetry of $\mathbf{U}$ can be shown directly by using the Cayley–Hamilton theorem, see [236], eq. (2.7). Another way is to note that $\mathbf{C}^{\frac{1}{2}}$ is an isotropic function, i.e.

$$\mathbf{Q}\mathbf{C}^{\frac{1}{2}}\mathbf{Q}^T = (\mathbf{Q}\mathbf{C}\mathbf{Q}^T)^{\frac{1}{2}}, \tag{3.4}$$

for an arbitrary $\mathbf{Q} \in \mathrm{SO}(3)$, thus, by Serrin's theorem, it can be represented as $\mathbf{C}^{\frac{1}{2}} = a_0\mathbf{I} + a_1\mathbf{C} + a_2\mathbf{C}^2$. Hence, $\mathbf{U} = \mathbf{C}^{\frac{1}{2}}$ is symmetric as a polynomial of symmetric tensors.

Next, $\mathbf{C}$ is positive definite, which can be shown by considering a line element, $\mathrm{d}\mathbf{x} = \mathbf{F}\,\mathrm{d}\mathbf{X}$, where $\det \mathbf{F} \neq 0$ and $\mathrm{d}\mathbf{X} \neq \mathbf{0}$. A square of a length of the line element is positive, i.e.

$$\mathrm{d}\mathbf{x} \cdot \mathrm{d}\mathbf{x} = (\mathbf{F}\mathrm{d}\mathbf{X}) \cdot (\mathbf{F}\mathrm{d}\mathbf{X}) = \mathrm{d}\mathbf{X} \cdot (\mathbf{F}^T\mathbf{F}\,\mathrm{d}\mathbf{X}) > 0, \tag{3.5}$$

where the last form is a definition of positive definiteness of $\mathbf{F}^T\mathbf{F} = \mathbf{C}$. By the definition of a square root function in spectral representations, also $\mathbf{U} = \mathbf{C}^{\frac{1}{2}}$ is positive definite.

Algorithm for calculation of $\mathbf{U}$ for given $\mathbf{C}$. The eigenvalues of $\mathbf{C}$ are real and positive and its eigenvectors are pairwise orthogonal. Denote these eigenvalues as λ_i^2 and the eigenvectors as $\mathbf{v}_i$ $(i = 1, 2, 3)$, and arrange them as matrices as follows:

$$\boldsymbol{\Lambda} = \mathrm{diag}\left[\lambda_1^2, \lambda_2^2, \lambda_2^2\right], \qquad \mathbf{Q} = [\mathbf{v}_1|\mathbf{v}_2|\mathbf{v}_3],$$

where $\boldsymbol{\Lambda}$ is a diagonal matrix and $\mathbf{Q}$ is an orthogonal matrix. Then $\mathbf{C}$ is represented as

$$\mathbf{C} = \mathbf{Q}^T\boldsymbol{\Lambda}\mathbf{Q}, \tag{3.6}$$

where the position of $\mathbf{Q}$ and $\mathbf{Q}^T$ can be interchanged. The standard steps to calculate $\mathbf{U} = \mathbf{C}^{\frac{1}{2}}$ are shown in Table 3.1. Note that $\mathbf{U}$ is not diagonal but is symmetric and positive definite. The algorithm to find a square root of a symmetric positive definite 3×3 matrix is described, e.g., in [75]. Note that for $\mathbf{U} = \mathbf{Q}^T\boldsymbol{\Lambda}^{\frac{1}{2}}\mathbf{Q}$, we obtain $\mathbf{F} = \mathbf{R}\mathbf{U} = \mathbf{R}\,\mathbf{Q}^T\boldsymbol{\Lambda}^{\frac{1}{2}}\mathbf{Q}$, which is a relation for the deformation gradient in which the stretches and the orthogonal tensors are separated.

Table 3.1 Calculation of square root of $\mathbf{C}$.

1. calculate eigenvectors of $\mathbf{C}$:	$\mathbf{Q}$
2. rotate forward $\mathbf{C}$:	$\mathbf{Q}\mathbf{C}\mathbf{Q}^T = \boldsymbol{\Lambda}$
3. calculate square root of $\boldsymbol{\Lambda}$:	$\boldsymbol{\Lambda}^{\frac{1}{2}} = \mathrm{diag}\left[\sqrt{\lambda_1^2}, \sqrt{\lambda_2^2}, \sqrt{\lambda_2^2}\right]$
4. rotate backward $\boldsymbol{\Lambda}^{\frac{1}{2}}$:	$\mathbf{Q}^T\boldsymbol{\Lambda}^{\frac{1}{2}}\mathbf{Q} = \mathbf{C}^{\frac{1}{2}} = \mathbf{U}.$

Properties of $\mathbf{R}$. Orthogonality of $\mathbf{R}$ is shown as

$$\mathbf{R}^T\mathbf{R} = (\mathbf{U}^{-1})^T\mathbf{F}^T\mathbf{F}\mathbf{U}^{-1} = \mathbf{U}^{-1}\mathbf{U}^2\mathbf{U}^{-1} = \mathbf{I}. \tag{3.7}$$

From this condition we obtain $\det(\mathbf{R}^T\mathbf{R}) = (\det \mathbf{R})^2 = 1$, i.e. a relation for the square of the determinant. To establish the sign of $\det \mathbf{R}$, we note that $\det \mathbf{R} = \det(\mathbf{F}\mathbf{U}^{-1}) = (\det \mathbf{F})/(\det \mathbf{U})$. Positive definiteness of $\mathbf{U}$ implies that the principal stretches $\lambda_i > 0$ and, hence, $\det \mathbf{U} = \lambda_1\lambda_2\lambda_3 > 0$. Besides, we must take $\det \mathbf{F} > 0$ to exclude annihilation of line elements and negative volumes, see [159], pp. 85 and 87. Hence, $(\det \mathbf{F})/(\det \mathbf{U}) > 0$ and, therefore, $\det \mathbf{R} = +1$.

3.2 Rotation Constraint equation

Instead of calculating $\mathbf{R}$ as $\mathbf{F}\mathbf{U}^{-1}$, we can find a tensor $\mathbf{Q} \in \mathrm{SO}(3)$, by solving the RC equation

$$\mathfrak{C} \doteq \mathrm{skew}(\mathbf{Q}^T\mathbf{F}) = \mathbf{0}. \tag{3.8}$$

This it permitted because the equations $\mathbf{Q}^T\mathbf{F} = \mathbf{U}$ and $\mathrm{skew}(\mathbf{Q}^T\mathbf{F}) = \mathbf{0}$, are equivalent, which is shown below.

1. $\mathrm{skew}(\mathbf{Q}^T\mathbf{F}) = \mathbf{0} \quad \Rightarrow \quad \mathbf{Q}^T\mathbf{F} = \mathbf{U}$.

From $\mathrm{skew}(\mathbf{Q}^T\mathbf{F}) = \mathbf{0}$ we have $\mathbf{Q}^T\mathbf{F} = \mathbf{F}^T\mathbf{Q} = (\mathbf{Q}^T\mathbf{F})^T$, i.e. $\mathbf{Q}^T\mathbf{F}$ is symmetric. Using this symmetry, we have

$$(\mathbf{Q}^T\mathbf{F})^2 = (\mathbf{Q}^T\mathbf{F})(\mathbf{Q}^T\mathbf{F}) = (\mathbf{Q}^T\mathbf{F})^T(\mathbf{Q}^T\mathbf{F}) = \mathbf{F}^T\mathbf{Q}\mathbf{Q}^T\mathbf{F} = \mathbf{U}^2. \tag{3.9}$$

It remains to show that $\mathbf{Q}^T\mathbf{F} \neq -\mathbf{U}$. Because $\mathbf{U}^2 = \mathbf{C}$ is positive definite, so also is $(\mathbf{Q}^T\mathbf{F})^2$. By the definition of a square root function in a spectral representation, also $\mathbf{Q}^T\mathbf{F} = [(\mathbf{Q}^T\mathbf{F})^2]^{\frac{1}{2}}$ is positive definite, similarly as $\mathbf{U}$ obtained from $\mathbf{C}$. This implies that eigenvalues of $\mathbf{Q}^T\mathbf{F}$ are positive, similarly as principal stretches λ_i, and that $\mathbf{Q}^T\mathbf{F} = +\mathbf{U}$.

Remark. We can also compare signs of $\mathbf{Q}^T\mathbf{F}$ and $\mathbf{U}$ by examining signs of their scalar invariants, but the results are not conclusive because of the first invariant. Both the third invariants are positive, as $\det(\mathbf{Q}^T\mathbf{F}) = (\det\mathbf{Q}^T)(\det\mathbf{F}) = \det\mathbf{F} > 0$ and also $\det\mathbf{U} = \lambda_1\lambda_2\lambda_3 > 0$, as principal stretches $\lambda_i > 0$. Also, the second invariants are positive, as

$$\mathrm{tr}(\mathbf{Q}^\mathrm{T}\mathbf{F})^2 = \mathrm{tr}(\mathbf{F}^\mathrm{T}\mathbf{F}) = \mathbf{F}\cdot\mathbf{F} > 0, \tag{3.10}$$

using eq. (3.9), and $\mathrm{tr}\mathbf{U}^2 = \lambda_1\lambda_2 + \lambda_2\lambda_3 + \lambda_3\lambda_1 > 0$ as $\lambda_i > 0$. For the first invariants, we have $\mathrm{tr}\mathbf{U} = \lambda_1 + \lambda_2 + \lambda_3 > 0$ as $\lambda_i > 0$, but there is a problem with $\mathrm{tr}(\mathbf{Q}^\mathrm{T}\mathbf{F})$. For instance, for the 2D case, when

$$\mathbf{Q} = \begin{bmatrix} \cos\omega & -\sin\omega \\ \sin\omega & \cos\omega \end{bmatrix}, \qquad \mathbf{F} = \begin{bmatrix} F_{11} & F_{12} \\ F_{21} & F_{22} \end{bmatrix}, \tag{3.11}$$

we obtain $\mathrm{tr}(\mathbf{Q}^\mathrm{T}\mathbf{F}) = \mathbf{Q}\cdot\mathbf{F} = \cos\omega(\mathrm{F}_{11}+\mathrm{F}_{22})+\sin\omega(\mathrm{F}_{21}-\mathrm{F}_{12})$. Noting that $\det\mathbf{F} = F_{11}F_{22} - F_{12}F_{21} > 0$, and even assuming small rotations, $\omega \approx 0$, for which $\mathrm{tr}(\mathbf{Q}^\mathrm{T}\mathbf{F}) \approx (\mathrm{F}_{11} + \mathrm{F}_{22}) + \omega(\mathrm{F}_{21} - \mathrm{F}_{12})$, it is still difficult to determine the sign of this invariant.

2. $\mathbf{Q}^T\mathbf{F} = \mathbf{U} \quad \Rightarrow \quad \mathrm{skew}(\mathbf{Q}^T\mathbf{F}) = \mathbf{0}.$

From the symmetry condition, $\mathbf{U} = \mathbf{U}^T$, we obtain $(\mathbf{Q}^T\mathbf{F}) = (\mathbf{Q}^T\mathbf{F})^T$, which can be rewritten as $\mathrm{skew}(\mathbf{Q}^T\mathbf{F}) = \mathbf{0}$.

This ends the proof that the conditions $\mathrm{skew}(\mathbf{Q}^T\mathbf{F}) = \mathbf{0}$ and $\mathbf{Q}^T\mathbf{F} = \mathbf{U}$ are fully equivalent. □

Finally, we note that the RC equation provides a link between the deformation gradient and the rotations, and can be used to derive mixed formulations including rotations, see Sect. 4.

***Relaxed* stretching tensors.** Using the product $\mathbf{Q}^T\mathbf{F}$, which was a basis of the RC equation, we can define two *relaxed* right stretching tensors:

1. the *symmetric relaxed* right stretching tensor

$$\tilde{\mathbf{U}} \doteq \mathrm{sym}(\mathbf{Q}^T\mathbf{F}). \tag{3.12}$$

If $\mathrm{skew}(\mathbf{Q}^T\mathbf{F}) = \mathbf{0}$, then $\mathrm{sym}(\mathbf{Q}^T\mathbf{F}) = \mathbf{U}$, as in the proof above.

2. the *non-symmetric relaxed* right stretching tensor

$$\tilde{\mathbf{U}}_n \doteq \mathbf{Q}^T\mathbf{F}. \tag{3.13}$$

If $\mathrm{skew}(\mathbf{Q}^T\mathbf{F}) = \mathbf{0}$, then $\mathbf{Q}^T\mathbf{F} = \mathrm{sym}(\mathbf{Q}^T\mathbf{F}) = \mathbf{U}$. The tensor $\tilde{\mathbf{U}}_n$ is used in the two-field formulation with unconstrained equations, see Sect. 4.6.

3.3 Interpretation of rotation Q

The rotation $\mathbf{Q}$ provided by the RC equation can also be physically interpreted in the range of large rotations, if we parameterize $\mathbf{F}$ in terms of rotations and stretches of a pair of initially ortho-normal vectors.

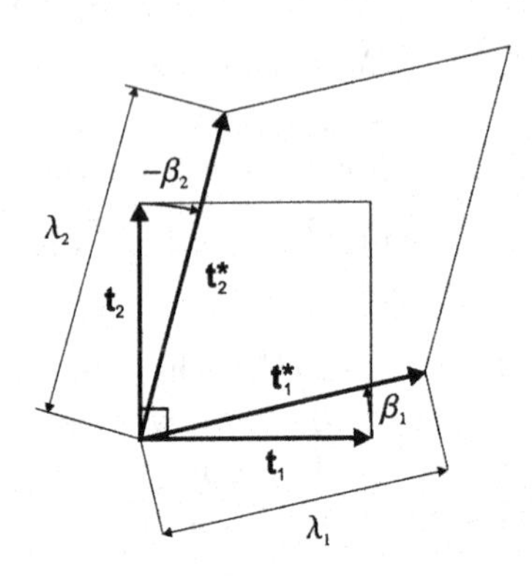

Fig. 3.1 Deformation of a pair of vectors $\mathbf{t}_1$ and $\mathbf{t}_2$.

Consider a 2D (planar) body, and denote by $\mathbf{t}_1$ and $\mathbf{t}_2$ two ortho-normal vectors associated with the non-deformed configuration, see Fig. 3.1. Under deformation, each of these vectors is rotated and stretched,

$$\mathbf{t}_1^* = \mathbf{F}\mathbf{t}_1 = \lambda_1 \mathbf{Q}_1 \mathbf{t}_1, \qquad \mathbf{t}_2^* = \mathbf{F}\mathbf{t}_2 = \lambda_2 \mathbf{Q}_2 \mathbf{t}_2, \tag{3.14}$$

where $\lambda_1, \lambda_2 > 0$ are scalar stretch parameters and $\mathbf{Q}_1$, $\mathbf{Q}_2$ are two rotation tensors, each depending on one rotation angle β_α, $\alpha = 1, 2$. The length of $\mathbf{t}_1$ and $\mathbf{t}_2$ is preserved if $\lambda_1 = \lambda_2 = 1$ and the angle between them remains unchanged for $\beta_1 = \beta_2$. Note that we can define the deformation gradient as $\mathbf{F} = \lambda_1 \mathbf{Q}_1 (\mathbf{t}_1 \otimes \mathbf{t}_1) + \lambda_2 \mathbf{Q}_2 (\mathbf{t}_2 \otimes \mathbf{t}_2)$ because products of it with $\mathbf{t}_1$ and $\mathbf{t}_2$ yield expressions of eq. (3.14).

For the representations

$$\begin{aligned} &\mathbf{F} = F_{\alpha\beta}\, \mathbf{t}_\alpha \otimes \mathbf{t}_\beta, \\ &\mathbf{Q}_\alpha(\beta_\alpha) = c_\alpha\, (\mathbf{t}_1 \otimes \mathbf{t}_1 + \mathbf{t}_2 \otimes \mathbf{t}_2) + s_\alpha\, (\mathbf{t}_2 \otimes \mathbf{t}_1 - \mathbf{t}_1 \otimes \mathbf{t}_2), \end{aligned} \tag{3.15}$$

where $s_\alpha = \sin\beta_\alpha$ and $c_\alpha = \cos\beta_\alpha$, from eq. (3.14) we obtain

$$F_{11} = \lambda_1\, c_1, \quad F_{12} = \lambda_2\, (-s_2), \quad F_{21} = \lambda_1\, s_1, \quad F_{22} = \lambda_2\, c_2, \tag{3.16}$$

which are algebraic and nonlinear formulas for $F_{\alpha\beta}$ in terms of four new parameters $\{\lambda_1, \lambda_2, \beta_1, \beta_2\}$.

Next, we shall find a relation between the rotation angles β_α and the angle ω of the rotation $\mathbf{Q}$ using the RC equation. For $\mathbf{Q}(\omega) = \cos\omega\,(\mathbf{t}_1\otimes\mathbf{t}_1+\mathbf{t}_2\otimes\mathbf{t}_2)+\sin\omega\,(\mathbf{t}_2\otimes\mathbf{t}_1-\mathbf{t}_1\otimes\mathbf{t}_2)$, the RC equation becomes

$$(F_{21} - F_{12})\cos\omega - (F_{11} + F_{22})\sin\omega = 0. \tag{3.17}$$

Using the components of the deformation gradients of eq. (3.16), we obtain

$$[\lambda_1 s_1 - \lambda_2(-s_2)]\cos\omega - (\lambda_1 c_1 + \lambda_2 c_2)\sin\omega = 0. \tag{3.18}$$

Assuming $\cos\omega \neq 0$, and $(\lambda_1 c_1 + \lambda_2 c_2) \neq 0$, we obtain

$$\tan\omega = \frac{\lambda_1 s_1 + \lambda_2 s_2}{\lambda_1 c_1 + \lambda_2 c_2} \approx \frac{s_1 + s_2}{c_1 + c_2}, \tag{3.19}$$

where the last form was obtained for small stretches, $\lambda_\alpha \approx 1$. Using trigonometric identities,

$$\tan\omega \approx \tan\frac{\beta_1 + \beta_2}{2}. \tag{3.20}$$

Hence,

$$\omega \approx \frac{1}{2}(\beta_1 + \beta_2) + k\pi, \qquad k = 0, \ldots, K, \tag{3.21}$$

i.e. the angle ω yielded by the RC is an average of rotations of vectors $\mathbf{t}_1$ and $\mathbf{t}_2$. This result, obtained for large rotations, provides a clear physical interpretation of $\mathbf{Q}$.

For rigid body rotation, the length of vectors $\mathbf{t}_1$ and $\mathbf{t}_2$ is constant, and their rotation angles are identical, i.e. $\lambda_1 = \lambda_2 = 1$, and $\beta_1 = \beta_2 = \beta$. Then, eq. (3.14) becomes

$$\mathbf{F}\mathbf{t}_1 = \mathbf{Q}(\beta)\,\mathbf{t}_1, \qquad \mathbf{F}\mathbf{t}_2 = \mathbf{Q}(\beta)\,\mathbf{t}_2 \tag{3.22}$$

and the deformation gradient $\mathbf{F} = \mathbf{Q}(\beta)\,(\mathbf{t}_1 \otimes \mathbf{t}_1 + \mathbf{t}_2 \otimes \mathbf{t}_2) = \mathbf{Q}(\beta)\mathbf{I} = \mathbf{Q}(\beta)$. Hence, eq. (3.21) becomes

$$\omega = \beta + k\pi, \tag{3.23}$$

i.e. the angle ω yielded by the RC is equal to the angle β of the rigid body rotation.

Summarizing the above two cases, we see that, in general, $\mathbf{Q}$ cannot be interpreted as a rigid rotation. This is also true for the rotation $\mathbf{R}$ yielded by the polar decomposition of $\mathbf{F}$, because $\mathbf{R} = \mathbf{Q}$.

3.4 Rate form of RC equation

The rate form of the RC equation has the advantage that it can be equivalently expressed in terms of the angular velocity and the spatial velocity gradient.

Differentiation w.r.t. time of the RC equation (3.8) yields

$$\operatorname{skew}(\dot{\mathbf{Q}}^T\mathbf{F} + \mathbf{Q}^T\dot{\mathbf{F}}) = \mathbf{0}. \tag{3.24}$$

From the definitions of the spatial (left) angular velocity, $\tilde{\omega}^* \doteq \dot{\mathbf{Q}}\mathbf{Q}^T$, and the spatial velocity gradient, $\nabla\mathbf{v} \doteq \dot{\mathbf{F}}\mathbf{F}^{-1}$, we obtain

$$\dot{\mathbf{Q}} = \tilde{\omega}^*\,\mathbf{Q}, \qquad \dot{\mathbf{F}} = \nabla\mathbf{v}\,\mathbf{F}, \tag{3.25}$$

for which, eq. (3.24) becomes

$$\operatorname{skew}\left(\mathbf{Q}^T\mathbf{A}\,\mathbf{F}\right) = \mathbf{0}, \qquad \text{where} \qquad \mathbf{A} \doteq -\tilde{\omega}^* + \nabla\mathbf{v}. \tag{3.26}$$

Next, we apply the rotation-forward operation, $\mathbf{Q}\,(\cdot)\,\mathbf{Q}^T$ and rewrite eq. (3.26) as

$$\mathbf{AV} - \mathbf{VA}^T = \mathbf{0}, \tag{3.27}$$

where $\mathbf{V} \doteq \mathbf{Q}\,\mathbf{U}\mathbf{Q}^T = \mathbf{F}\mathbf{Q}^T$ is the left stretching tensor, symmetric and positive definite. The split of $\mathbf{A}$ into a symmetric part and a skew-symmetric part yields $\mathbf{A} = \mathbf{A}_s + \mathbf{A}_a$, where $\mathbf{A}_s \doteq \operatorname{sym}\nabla\mathbf{v}$ and $\mathbf{A}_a \doteq -\tilde{\omega}^* + \operatorname{skew}\nabla\mathbf{v}$. Then eq. (3.26) becomes

$$(\mathbf{A}_s\mathbf{V} - \mathbf{V}\mathbf{A}_s) + (\mathbf{A}_a\mathbf{V} - \mathbf{V}\mathbf{A}_a^T) = \mathbf{0}. \tag{3.28}$$

Note that the first part,

$$\mathbf{A}_s\mathbf{V} - \mathbf{V}\mathbf{A}_s = 2\operatorname{skew}(\mathbf{A}_s\mathbf{V}) = \mathbf{0}, \tag{3.29}$$

because a skew part of a symmetric tensor is equal to zero. Hence, only the second part of eq. (3.28) remains,

$$\mathbf{A}_a\mathbf{V} + \mathbf{V}\mathbf{A}_a = \mathbf{0}, \tag{3.30}$$

in which $\mathbf{A}_a$ is skew-symmetric, and $\mathbf{V}$ is symmetric and positive definite. By Lemma 3.1 in [214], this equation is satisfied only if $\mathbf{A}_a = \mathbf{0}$, which yields

$$\tilde{\omega}^* - \operatorname{skew}\nabla\mathbf{v} = \mathbf{0}. \tag{3.31}$$

This relation is equivalent to the rate form of the RC equation of eq. (3.24).

3.5 Rotations calculated from the RC equation

Assume that $\mathbf{F}$ is given, and find the rotations from the RC equation.

The solution of the RC of eq. (3.8) is trivial for a rigid rotation, when $\mathbf{F} = \mathbf{Q}$, as then $\text{skew}(\mathbf{Q}^T\mathbf{F}) = \text{skew}(\mathbf{Q}^T\mathbf{Q}) = \text{skew}\,\mathbf{I} = \mathbf{0}$, i.e. the RC equation is an identity.

For large rotations, the RC of eq. (3.8) yields a system of non-linear equations for $\{\psi_1, \psi_2, \psi_3\}$, where $\psi_i \doteq (\boldsymbol{\psi})_i$ and $\boldsymbol{\psi}$ is the rotation vector, e.g. the canonical vector of eq. (8.79). Methods of solution of these nonlinear equations for a 2D problem are discussed in Sect. 12.1.

If we assume that rotations are small, then $\mathbf{Q} \approx \mathbf{I} + \tilde{\boldsymbol{\psi}}$, where $\tilde{\boldsymbol{\psi}} = \boldsymbol{\psi} \times \mathbf{I} \in \text{so}(3)$. Then the RC equation becomes

$$\mathfrak{C} \doteq \text{skew}(\mathbf{Q}^T\mathbf{F}) = \text{skew}\left[(\mathbf{I} + \tilde{\boldsymbol{\psi}})^T\mathbf{F}\right] = \text{skew}\mathbf{F} + \text{skew}(\tilde{\boldsymbol{\psi}}^T\mathbf{F}) = \mathbf{0}. \quad (3.32)$$

These are three equations which can be rewritten as

$$\begin{bmatrix} -F_{31} & -F_{32} & (F_{11}+F_{22}) \\ F_{21} & -(F_{11}+F_{33}) & F_{23} \\ (F_{22}+F_{33}) & -F_{12} & -F_{13} \end{bmatrix} \begin{bmatrix} \psi_1 \\ \psi_2 \\ \psi_3 \end{bmatrix} = - \begin{bmatrix} F_{12}-F_{21} \\ F_{13}-F_{31} \\ F_{23}-F_{32} \end{bmatrix}, \quad (3.33)$$

where $F_{ij} \doteq (\mathbf{F})_{ij}$. Note that for $\mathbf{F} = \mathbf{I}$, the determinant of the matrix is equal to 8 and the r.h.s. vector is equal to zero. Hence, a unique solution exists and is equal to zero.

A unique solution does not exist, e.g., when (i) the off-diagonal components are equal to zero, i.e. $F_{\alpha 3} = F_{3\alpha} = 0$ and $F_{12} = F_{21} = 0$, and (ii) at least one of the following conditions for the diagonal components is satisfied: $F_{11} = -F_{22}$ or $F_{11} = -F_{33}$ or $F_{22} = -F_{33}$.

4

3D formulations with rotations

In this chapter, the formulations including rotations as an independent (primary) variable are derived for a 3D continuum. The derived functionals amount to various forms of the potential energy modified by the Rotation Constraint; their extensions to the Hu–Washizu and Hellinger–Reissner functionals are provided in Sect. 12. Some of these 3D formulations are used in subsequent chapters as a basis for derivation of shell equations.

Extended configuration space. The classical configuration space of the non-polar Cauchy continuum is defined as

$$\mathcal{C} \doteq \{\chi : \ B \rightarrow R^3\}, \tag{4.1}$$

where χ is the deformation function defined over the reference configuration of the body B. In the present section, we consider the *extended configuration* space, defined in terms of the deformation function χ and rotations $\mathbf{R} \in \mathrm{SO}(3)$. We do not account for gradients of rotations, similarly as in the pseudo-Cosserat continuum, see [59, 128, 238]. The rotations are generated by the (left) skew-symmetric tensor $\delta\tilde{\boldsymbol{\theta}} \doteq \delta\mathbf{R}\mathbf{R}^T$ (in the sense explained for the weak form AMB equation), and are treated in two different ways:

- remain unconstrained, as in the Cosserat-type continuum. Then the extended configuration space is defined as

$$\mathcal{C}_{\mathrm{ext}} \doteq \{(\chi, \mathbf{R}) : \ B \rightarrow R^3 \times \mathrm{SO}(3)\}. \tag{4.2}$$

  Note that χ does not belong to the classical configuration space. This approach to rotations is quite popular in shells, which can be treated as pseudo-Cosserat surfaces, see e.g. [244, 52, 200] and the papers cited therein.

- are constrained, either by the polar decomposition of $\mathbf{F}$ equation (3.2) or the RC equation (3.8). Then the extended configuration space is defined as

$$\mathcal{C}_{\text{ext}} \doteq \{(\boldsymbol{\chi}, \mathbf{R}) : B \to R^3 \times \mathrm{SO}(3) \mid \boldsymbol{\chi} \in \mathcal{C}\}, \tag{4.3}$$

where $\mathcal{C}$ is the classical configuration space. Note that $\boldsymbol{\chi}$ is required to belong to $\mathcal{C}$, i.e. it is identical as for the classical non-polar Cauchy continuum, see [128, 238, 74, 175, 13]. This approach is used in [252, 253, 254], to define the second-order kinematics of shells.

The basic formulation of this chapter is given for the nominal stress from which the formulations for other types of stress are derived. The formulations based on the Biot stress, see [191, 42, 249], and the formulations based on the second Piola–Kirchhoff stress, see [99, 214], are presented. Several variational principles are also summarized in [9].

Four-field (4-F) formulation exploits the polar decomposition equation, while the three-field (3-F) formulations are based on the RC equation. Two-field (2-F) formulations are obtained by regularization of the 3-F functionals, except the one for unconstrained rotations.

Finally, we note that both approaches, i.e. with constrained and unconstrained rotations, can be applied to shells. The 3-F and 2-F formulations are used in subsequent chapters as the basis for derivation of shell equations.

4.1 Governing equations

Balance equations and boundary conditions. The local balance equations and the boundary conditions are

1. linear momentum balance (LMB):

$$\mathrm{Div}\mathbf{P} + \rho_{\mathrm{R}}\mathbf{b} = \mathbf{0} \quad \text{in} \quad \mathrm{B}, \tag{4.4}$$

where $\mathbf{P}$ is the nominal stress tensor (its transpose is the first Piola–Kirchhoff stress), ρ_R is the mass density for the reference (initial) configuration, and $\mathbf{b}$ is the body force.

2. angular momentum balance (AMB):

$$\mathbf{F} \times \mathbf{P} = \mathbf{0} \qquad \text{or} \qquad \mathrm{skew}(\mathbf{P}\mathbf{F}^T) = \mathbf{0} \quad \text{in} \quad \mathrm{B}, \tag{4.5}$$

where $\mathbf{F} = \mathrm{Grad}\boldsymbol{\chi}$ and $\det \mathbf{F} > 0$.

3. boundary conditions (BC):

$$\boldsymbol{\chi} = \hat{\boldsymbol{\chi}} \quad \text{on} \quad \partial_{\chi}\mathrm{B} \qquad \text{and} \qquad \mathbf{Pn} = \hat{\mathbf{p}} \quad \text{on} \quad \partial_{\sigma}\mathrm{B}, \tag{4.6}$$

where $\partial_{\chi}\mathrm{B}$ and $\partial_{\sigma}\mathrm{B}$ denote disjoint parts of the boundary $\partial\mathrm{B}$ on which the deformation and traction boundary conditions are specified. The outward normal vector is denoted by $\mathbf{n}$ and $\hat{\mathbf{p}}$ is the external load (surface traction) which we assume as not depending on deformation.

Weak form of basic equations. The weak form of eqs. (4.4)–(4.6) is obtained by calculating their scalar products with the respective admissible fields and integrating over the volume B or the surface traction BC area $\partial_{\sigma}\mathrm{B}$ of the initial configuration.

LMB. For eq. (4.4), we calculate the volume integral of its scalar product with the kinematically admissible variation of deformation $\delta\boldsymbol{\chi}$, i.e. such that $\delta\boldsymbol{\chi} = \mathbf{0}$ on $\partial_u\mathrm{B}$,

$$\int_B (\mathrm{Div}\mathbf{P} + \rho_{\mathrm{R}}\mathbf{b}) \cdot \delta\boldsymbol{\chi} \, \mathrm{d}V = 0. \tag{4.7}$$

Using the formula for the divergence of a product of two tensors, e.g. [33] eq. (5.5.19), we obtain

$$\mathrm{Div}\mathbf{P} \cdot \delta\boldsymbol{\chi} = \mathrm{Div}(\mathbf{P}^T \delta\boldsymbol{\chi}) - \mathbf{P} \cdot \nabla\delta\boldsymbol{\chi}. \tag{4.8}$$

For the first r.h.s. term, we use the divergence theorem, e.g. [33] eq. (5.8.11),

$$\int_B \mathrm{Div}(\mathbf{P}^T \delta\boldsymbol{\chi}) \, \mathrm{d}V = \int_{\partial B} (\mathbf{P}^T \delta\boldsymbol{\chi}) \cdot \mathbf{n} \, \mathrm{d}A = \int_{\partial B} (\mathbf{Pn}) \cdot \delta\boldsymbol{\chi} \, \mathrm{d}A. \tag{4.9}$$

For the second term, we note that $\nabla\delta\boldsymbol{\chi} = \delta\mathbf{F}$. Then the weak form of the LMB is

$$\int_B (\mathbf{P} \cdot \delta\mathbf{F} - \rho_R \mathbf{b} \cdot \delta\boldsymbol{\chi}) \, \mathrm{d}V - \int_{\partial B} (\mathbf{Pn}) \cdot \delta\boldsymbol{\chi} \, \mathrm{d}A. \tag{4.10}$$

AMB. For eq. (4.5), we calculate a volume integral of its scalar product with a skew-symmetric (left) tensor $\delta\tilde{\boldsymbol{\theta}}$,

$$\int_B \mathrm{skew}(\mathbf{P}\mathbf{F}^T) \cdot \delta\tilde{\boldsymbol{\theta}} \, \mathrm{d}V = 0. \tag{4.11}$$

If $\delta\tilde{\boldsymbol{\theta}}$ generates rotations, i.e. $\{\delta\tilde{\boldsymbol{\theta}} : \ \delta\mathbf{R} \doteq \delta\tilde{\boldsymbol{\theta}}\,\mathbf{R}\}$, which we can assume for the extended configuration space but not for the classical one, the weak form AMB becomes

$$\int_B \operatorname{skew}(\mathbf{P}\mathbf{F}^T) \cdot (\mathbf{R}^T \delta\mathbf{R})\, \mathrm{d}V = 0. \tag{4.12}$$

BCD. For the displacement BC, eq. (4.6), we calculate a surface integral of a scalar product with $\delta(\mathbf{Pn})$,

$$\int_{\partial_\chi \mathrm{B}} (\boldsymbol{\chi} - \hat{\boldsymbol{\chi}}) \cdot \delta(\mathbf{Pn})\, \mathrm{d}A = 0. \tag{4.13}$$

BCT. For the traction BC, eq. (4.6), we calculate a surface integral of a scalar product with a kinematically admissible variation $\delta\boldsymbol{\chi}$,

$$\int_{\partial_\sigma \mathrm{B}} (\mathbf{Pn} - \hat{\mathbf{p}}) \cdot \delta\boldsymbol{\chi}\, \mathrm{d}A = 0. \tag{4.14}$$

Remark. Note that, compared to the classical 1-F formulation in terms of $\boldsymbol{\chi}$ only, the weak form AMB equation is different while the other equations are the same. To explain the expression that $\delta\tilde{\boldsymbol{\theta}}$ *generates* rotations, we transform $\delta\mathbf{R} \doteq \delta\tilde{\boldsymbol{\theta}}\,\mathbf{R}$ by using $\delta(\ ) = \dot{(\ )}\,\delta t$, where the superimposed dot denotes the time-derivative. This yields a differential equation, $\dot{\mathbf{R}} - \dot{\tilde{\boldsymbol{\theta}}}\,\mathbf{R} = \mathbf{0}$, to which we append the initial condition $\mathbf{R}(t = 0) = \mathbf{R}_0$. From this equation we can calculate (*generate*) $\mathbf{R}$ for an assumed $\dot{\tilde{\boldsymbol{\theta}}}$; for details see Sect. 9.4.

Virtual work of stress, strain energy, constitutive law. Below, the VW of the nominal stress $\mathbf{P}$ is transformed to four equivalent forms. Next, the corresponding strain energy functionals are defined and the respective constitutive laws are derived.

a. Strain energy $\mathcal{W}(\mathbf{U})$. The VW of the nominal stress $\mathbf{P} \cdot \delta\mathbf{F}$ can be expressed as

$$\mathbf{P} \cdot \delta\mathbf{F} = \operatorname{sym}(\mathbf{R}^T\mathbf{P}) \cdot \delta\mathbf{U}, \tag{4.15}$$

where $\mathbf{R}$ and $\mathbf{U}$ are obtained from the polar decomposition equation $\mathbf{F} = \mathbf{RU}$. Taking a variation of this equation, we have $\delta\mathbf{F} = \delta\mathbf{R}(\mathbf{R}^T\mathbf{R})\mathbf{U} + \mathbf{R}\delta\mathbf{U}$, where $\delta\mathbf{R}\mathbf{R}^T \doteq \delta\tilde{\boldsymbol{\theta}}$, and $\delta\tilde{\boldsymbol{\theta}} = -\delta\tilde{\boldsymbol{\theta}}^T$, i.e. is skew-symmetric. Then

$$\mathbf{P} \cdot \delta\mathbf{F} = \mathbf{P} \cdot (\delta\tilde{\boldsymbol{\theta}}\mathbf{F}) + \mathbf{P} \cdot (\mathbf{R}\delta\mathbf{U}) = (\mathbf{P}\mathbf{F}^T) \cdot \delta\tilde{\boldsymbol{\theta}} + \text{sym}(\mathbf{R}^T\mathbf{P}) \cdot \delta\mathbf{U},$$

and the first term vanishes as a scalar product of a symmetric $\mathbf{P}\mathbf{F}^T$ (which is a consequence of the AMB: $\text{skew}(\mathbf{P}\mathbf{F}^T) = \mathbf{0}$) and a skew-symmetric $\delta\tilde{\boldsymbol{\theta}}$.

The strain deduced from the r.h.s. of eq. (4.15) as the work conjugate to $\text{sym}(\mathbf{R}^T\mathbf{P})$, is the right stretch strain,

$$\mathbf{H} \doteq (\mathbf{F}^T\mathbf{F})^{1/2} - (\mathbf{I}^T\mathbf{I})^{1/2}, \tag{4.16}$$

where $\mathbf{I}$ is consistent with $\mathbf{F}$, see eq. (5.17).

Assume that the strain energy density per unit non-deformed volume, $\mathcal{W}$, is a function of $\mathbf{U}$. $\mathcal{W}(\mathbf{U})$ satisfies the material objectivity (frame indifference) requirement, because $\mathbf{U}$ is a polynomial of $\mathbf{C}$, i.e. $\mathbf{U} = \mathbf{C}^{\frac{1}{2}} = a_0\mathbf{I} + a_1\mathbf{C} + a_2\mathbf{C}^2$, as discussed below eq. (3.2). A variation of the strain energy is

$$\delta\mathcal{W}(\mathbf{U}) = \partial_U\mathcal{W}(\mathbf{U}) \cdot \delta\mathbf{U}. \tag{4.17}$$

The term $\text{sym}(\mathbf{R}^T\mathbf{P}) \cdot \delta\mathbf{U}$ can be treated as $\delta\mathcal{W}$ and, hence, from eqs. (4.15) and (4.17), we obtain the constitutive law

$$\text{sym}(\mathbf{R}^T\mathbf{P}) = \partial_U\mathcal{W}(\mathbf{U}). \tag{4.18}$$

This CL is used in the 4-F formulation for the nominal stress.

b. Strain energy $\mathcal{W}(\mathbf{C})$. The VW of the nominal stress $\mathbf{P} \cdot \delta\mathbf{F}$ can be expressed as

$$\mathbf{P} \cdot \delta\mathbf{F} = \tfrac{1}{2}\mathbf{S} \cdot \delta\mathbf{C}, \tag{4.19}$$

where $\mathbf{S}$ is the second Piola-Kirchhoff stress tensor and $\mathbf{C} \doteq \mathbf{F}^T\mathbf{F}$ is the right Cauchy–Green deformation tensor. The above formula is obtained by using $\mathbf{P} = \mathbf{F}\mathbf{S}$ for which the AMB, $\text{skew}(\mathbf{P}\mathbf{F}^T) = \text{skew}(\mathbf{F}\mathbf{S}\mathbf{F}^T) = \mathbf{0}$, implies $\mathbf{S} = \mathbf{S}^T$. Equation (4.19) is obtained by the following transformations:

$$(\mathbf{F}\mathbf{S}) \cdot \delta\mathbf{F} = \mathbf{S} \cdot (\mathbf{F}^T\delta\mathbf{F}) = \mathbf{S} \cdot \text{sym}(\mathbf{F}^T\delta\mathbf{F}) = \mathbf{S} \cdot \delta(\tfrac{1}{2}\mathbf{F}^T\mathbf{F}) = \tfrac{1}{2}\mathbf{S} \cdot \delta\mathbf{C}.$$

The strain deduced from the r.h.s. of eq. (4.19) as the work conjugate to $\mathbf{S}$, is the Green strain,

$$\mathbf{E} \doteq \tfrac{1}{2}(\mathbf{F}^T\mathbf{F} - \mathbf{I}^T\mathbf{I}). \tag{4.20}$$

This strain can be obtained from the change of the square of the length of an infinitesimal line element, $\mathbf{dx} = (\partial \mathbf{x}/\partial \mathbf{y})\, \mathbf{dy} = \mathbf{F}\, \mathbf{dy}$,

$$\mathbf{dx} \cdot \mathbf{dx} - \mathbf{dy} \cdot \mathbf{dy} = (\mathbf{F}\mathbf{dy}) \cdot (\mathbf{F}\mathbf{dy}) - (\mathbf{I}\mathbf{dy}) \cdot (\mathbf{I}\mathbf{dy}) = 2\mathbf{dy} \cdot (\mathbf{E}\, \mathbf{dy}). \quad (4.21)$$

Note that $\mathbf{I}$ is consistent with $\mathbf{F}$, see eq. (5.17).

Assume that the strain energy density per unit non-deformed volume, $\mathcal{W}$, is a function of $\mathbf{C}$. To satisfy the frame indifference requirement, $\mathcal{W}$ must remain the same for the observer transformation $\mathbf{x}^+ = \mathbf{O}\mathbf{x} + \mathbf{c}$, where $\mathbf{O} \in SO(3)$, ([239] p. 44). The observer transformation yields $\mathbf{F}^+ = \mathbf{O}\mathbf{F}$ and the right Cauchy–Green tensor is invariant, i.e. $\mathbf{C}^+ = (\mathbf{F}^+)^T\mathbf{F}^+ = \mathbf{F}^T\mathbf{O}^T\mathbf{O}\mathbf{F} = \mathbf{F}^T\mathbf{F} = \mathbf{C}$. If $\mathcal{W}$ is a function of $\mathbf{C}$, then $\mathcal{W}(\mathbf{C}^+) = \mathcal{W}(\mathbf{C})$, and this requirement is satisfied. A variation of the strain energy is as follows

$$\delta\mathcal{W}(\mathbf{C}) = \partial_C\mathcal{W}(\mathbf{C}) \cdot \delta\mathbf{C}. \quad (4.22)$$

The term $\frac{1}{2}\mathbf{S} \cdot \delta\mathbf{C}$ can be treated as $\delta\mathcal{W}$ and, hence, from eqs. (4.19) and (4.22), we obtain the constitutive law

$$\mathbf{S} = 2\, \partial_C\mathcal{W}(\mathbf{C}). \quad (4.23)$$

This CL is used in the formulations for the second Piola–Kirchhoff stress.

c. Strain energy $\mathcal{W}(\mathbf{Q}^T\mathbf{F})$. The VW of the nominal stress $\mathbf{P} \cdot \delta\mathbf{F}$ can be expressed as

$$\mathbf{P} \cdot \delta\mathbf{F} = (\mathbf{Q}^T\mathbf{P}) \cdot \delta(\mathbf{Q}^T\mathbf{F}), \quad (4.24)$$

where $\mathbf{Q} \in SO(3)$, and $(\mathbf{Q}^T\mathbf{F})$ is non-symmetric.

Proof. First,

$$\mathbf{P} \cdot \delta\mathbf{F} = \mathrm{tr}(\mathbf{Q}\mathbf{Q}^T\mathbf{P}\delta\mathbf{F}^T) = \mathrm{tr}(\mathbf{Q}^T\mathbf{P}\delta\mathbf{F}^T\mathbf{Q}) = (\mathbf{Q}^T\mathbf{P}) \cdot (\mathbf{Q}^T\delta\mathbf{F}).$$

Next, we use $\mathbf{Q}^T\delta\mathbf{F} = \delta(\mathbf{Q}^T\mathbf{F}) - \delta\mathbf{Q}^T\mathbf{F}$. Then,

$$(\mathbf{Q}^T\mathbf{P}) \cdot (\mathbf{Q}^T\delta\mathbf{F}) = (\mathbf{Q}^T\mathbf{P}) \cdot \delta(\mathbf{Q}^T\mathbf{F}) - (\mathbf{Q}^T\mathbf{P}) \cdot (\delta\mathbf{Q}^T\mathbf{F}),$$

where the 2nd component,

$$(\mathbf{Q}^T\mathbf{P}) \cdot (\delta\mathbf{Q}^T\mathbf{F}) = \mathrm{tr}(\mathbf{Q}^T\mathbf{P}\mathbf{F}^T\delta\mathbf{Q}) = \mathrm{tr}(\delta\mathbf{Q}\mathbf{Q}^T\mathbf{P}\mathbf{F}^T) = \delta\tilde{\boldsymbol{\theta}} \cdot (\mathbf{P}\mathbf{F}^T) = \mathbf{0},$$

as $\delta \mathbf{Q}\mathbf{Q}^T \doteq \delta \tilde{\boldsymbol{\theta}}$ is skew-symmetric and $\mathbf{P}\mathbf{F}^T$ is symmetric as a consequence of the AMB: $\text{skew}(\mathbf{P}\mathbf{F}^T) = \mathbf{0}$. This ends the proof. □

The strain deduced from the r.h.s. of eq. (4.24), as the work conjugate to $(\mathbf{Q}^T\mathbf{P})$, is the *non-symmetric relaxed* right stretch strain

$$\tilde{\mathbf{H}}_n \doteq \mathbf{Q}^T\mathbf{F} - \mathbf{I}^T\mathbf{I}, \tag{4.25}$$

where $\mathbf{I}$ is consistent with $\mathbf{F}$, see eq. (5.17).

Assume that the strain energy density per unit non-deformed volume, $\mathcal{W}$, is a function of $\mathbf{Q}^T\mathbf{F}$. Note that if $\text{skew}(\mathbf{Q}^T\mathbf{F}) \to \mathbf{0}$, then $\mathbf{Q}^T\mathbf{F} \to \mathbf{U}$ and $\mathcal{W}(\mathbf{Q}^T\mathbf{F}) \to \mathcal{W}(\mathbf{U})$ which satisfies the material objectivity (frame indifference) requirement, as discussed earlier. A variation of the strain energy is

$$\delta\mathcal{W}(\mathbf{Q}^T\mathbf{F}) = \partial_{Q^TF}\mathcal{W}(\mathbf{Q}^T\mathbf{F}) \cdot \delta(\mathbf{Q}^T\mathbf{F}). \tag{4.26}$$

The term $(\mathbf{Q}^T\mathbf{P}) \cdot \delta(\mathbf{Q}^T\mathbf{F})$ can be treated as $\delta\mathcal{W}$ and, hence, from eqs. (4.24) and (4.26), we obtain the constitutive law

$$(\mathbf{Q}^T\mathbf{P}) = \partial_{Q^TF}\mathcal{W}(\mathbf{Q}^T\mathbf{F}). \tag{4.27}$$

We note that the above CL is applicable only to the unconstrained formulation, with rotations restricted neither by the polar decomposition equation nor by the RC equation.

d. Strain energy $\mathcal{W}(\text{sym}(\mathbf{Q}^T\mathbf{F}))$. The sum of the VW of the nominal stress and the weak form of the RC equation can be expressed as

$$\begin{aligned}\mathbf{P} \cdot \delta\mathbf{F} &+ \delta\text{skew}(\mathbf{Q}^T\mathbf{P}) \cdot \text{skew}(\mathbf{Q}^T\mathbf{F}) \\ &= \text{sym}(\mathbf{Q}^T\mathbf{P}) \cdot \delta\text{sym}(\mathbf{Q}^T\mathbf{F}) + \delta[\text{skew}(\mathbf{Q}^T\mathbf{P}) \cdot \text{skew}(\mathbf{Q}^T\mathbf{F})],\end{aligned} \tag{4.28}$$

where we applied eq. (4.24) to the first component and the split into symmetric and skew-symmetric parts,

$$(\mathbf{Q}^T\mathbf{P}) \cdot \delta(\mathbf{Q}^T\mathbf{F}) = \text{sym}(\mathbf{Q}^T\mathbf{P}) \cdot \text{sym}\delta(\mathbf{Q}^T\mathbf{F}) + \text{skew}(\mathbf{Q}^T\mathbf{P}) \cdot \text{skew}\delta(\mathbf{Q}^T\mathbf{F}).$$

Besides, commuting of the operations of taking a symmetric (or skew) part and taking a variation, i.e. $\text{sym}\delta(\cdot) = \delta\text{sym}(\cdot)$ and $\text{skew}\delta(\cdot) = \delta\text{skew}(\cdot)$, is accounted for.

The strain deduced from the first term on the r.h.s. of eq. (4.28), as the work conjugate to $\mathrm{sym}(\mathbf{Q}^T\mathbf{P})$, is the *symmetric relaxed* right stretch strain,

$$\tilde{\mathbf{H}} \doteq \mathrm{sym}(\mathbf{Q}^T\mathbf{F}) - \mathrm{sym}(\mathbf{I}^T\mathbf{I}), \tag{4.29}$$

where $\mathbf{I}$ is consistent with $\mathbf{F}$, see eq. (5.17).

Assume that the strain energy density per unit non-deformed volume, $\mathcal{W}$, is a function of $\mathrm{sym}(\mathbf{Q}^T\mathbf{F})$. Note that if $\mathrm{skew}(\mathbf{Q}^T\mathbf{F}) \to \mathbf{0}$, then $\mathrm{sym}(\mathbf{Q}^T\mathbf{F}) \to \mathbf{U}$ and $\mathcal{W}(\mathrm{sym}(\mathbf{Q}^T\mathbf{F})) \to \mathcal{W}(\mathbf{U})$, which satisfies the material objectivity (frame indifference) requirement, as discussed earlier. The variation of the strain energy is

$$\delta\mathcal{W}(\tilde{\mathbf{U}}) = \partial_{\tilde{U}}\mathcal{W}(\tilde{\mathbf{U}}) \cdot \delta\tilde{\mathbf{U}}, \tag{4.30}$$

where $\tilde{\mathbf{U}} \doteq \mathrm{sym}(\mathbf{Q}^T\mathbf{F})$ is the *relaxed* right stretch tensor. The first term of eq. (4.28), $\mathrm{sym}(\mathbf{Q}^T\mathbf{P}) \cdot \delta\mathrm{sym}(\mathbf{Q}^T\mathbf{F})$, can be treated as $\delta\mathcal{W}$ and, hence, from this first term and eq. (4.30), we obtain the constitutive law

$$\mathrm{sym}(\mathbf{Q}^T\mathbf{P}) = \partial_{\tilde{U}}\mathcal{W}(\tilde{\mathbf{U}}). \tag{4.31}$$

This CL is used in the 3-F formulation for the nominal stress.

Using the Biot stress $\mathbf{T}_s^B \doteq \mathrm{sym}(\mathbf{Q}^T\mathbf{P})$ of eq. (4.50), we can rewrite eq. (4.31) as follows:

$$\mathbf{T}_s^B = \partial_{\tilde{U}}\mathcal{W}(\tilde{\mathbf{U}}). \tag{4.32}$$

This CL is used in the formulations for the Biot stress.

4.2 4-F formulation for nominal stress

In this section, we describe a four-field formulation including rotations derived from the balance equations in terms of the nominal stress, see [74, 10], which has the following features:

- the rotations $\mathbf{Q}$ are constrained by the polar decomposition equation (3.2),
- the strain energy and the CL are defined for the right stretch strain of eq. (3.12).

To the set of governing equations (4.4)–(4.6), we append the following equations:

1. Polar decomposition equation:

$$\mathbf{F} - \mathbf{R}\mathbf{U} = \mathbf{0}, \tag{4.33}$$

2. Constitutive law of eq. (4.18):

$$\operatorname{sym}(\mathbf{R}^T\mathbf{P}) = \frac{\partial \mathcal{W}(\mathbf{U})}{\partial \mathbf{U}}, \tag{4.34}$$

which all furnish a mixed formulation in terms of four fields $\{\boldsymbol{\chi}, \mathbf{R}, \mathbf{U}, \mathbf{P}\}$.

Weak form of basic equations. A weak form of the governing equations, (4.4)–(4.6), is given by eqs. (4.10) and (4.12)–(4.14). For the polar decomposition equation, (4.33), we calculate a volume integral of a scalar product of this equation with $\delta\mathbf{P}$,

$$\int_B (\mathbf{F} - \mathbf{R}\mathbf{U}) \cdot \delta\mathbf{P}\, \mathrm{d}V = 0. \tag{4.35}$$

VW equation. Adding the above weak form (scalar) equations, we obtain the VW equation

$$\int_B \{\mathbf{P} \cdot \delta\mathbf{F} + \delta\left[\mathbf{P} \cdot (\mathbf{F} - \mathbf{R}\mathbf{U})\right]\}\, \mathrm{d}V - \delta F_{\text{ext}} = 0, \tag{4.36}$$

where

$$\delta F_{\text{ext}} \doteq \delta F_b + \delta F_\sigma + \delta F_\chi, \tag{4.37}$$

$$\delta F_b \doteq \int_B \rho_R \mathbf{b} \cdot \delta\boldsymbol{\chi}\, \mathrm{d}V, \qquad \delta F_\sigma \doteq \int_{\partial_\sigma \mathrm{B}} (\mathbf{P}\mathbf{n} - \hat{\mathbf{p}}) \cdot \delta\boldsymbol{\chi}\, \mathrm{d}A,$$

$$\delta F_\chi \doteq \int_{\partial_\chi \mathrm{B}} (\boldsymbol{\chi} - \hat{\boldsymbol{\chi}}) \cdot \delta(\mathbf{P}\mathbf{n})\, \mathrm{d}A.$$

Proof. The integrand of eq. (4.36) is obtained as follows. Adding the scalar equations (4.10) and (4.12)–(4.14), we obtain

$$\mathbf{P} \cdot \delta\mathbf{F} + \delta\mathbf{P} \cdot (\mathbf{F} - \mathbf{R}\mathbf{U}) + \operatorname{skew}(\mathbf{P}\mathbf{F}^T) \cdot (\delta\mathbf{R}^T\mathbf{R}), \tag{4.38}$$

which can be transformed to the following equivalent form

$$\mathbf{P} \cdot \delta\mathbf{F} + \delta\left[\mathbf{P} \cdot (\mathbf{F} - \mathbf{R}\mathbf{U})\right], \tag{4.39}$$

as follows. Note that the first terms of both equations are identical. The second term of eq. (4.39) can be rewritten as

$$\delta[\mathbf{P} \cdot (\mathbf{F} - \mathbf{R}\mathbf{U})] = \delta\mathbf{P} \cdot (\mathbf{F} - \mathbf{R}\mathbf{U}) + \mathbf{P} \cdot \delta\mathbf{F} - \mathbf{P} \cdot (\delta\mathbf{R}\mathbf{U}) - \mathbf{P} \cdot (\mathbf{R}\delta\mathbf{U}), \tag{4.40}$$

where the first term is equal to the second term of eq. (4.38). The second and fourth terms cancel out because

$$\begin{aligned}\mathbf{P} \cdot (\mathbf{R}\delta\mathbf{U}) &= \operatorname{tr}(\mathbf{P}\delta\mathbf{U}\mathbf{R}^{\mathrm{T}}) = \operatorname{tr}(\mathbf{R}^{\mathrm{T}}\mathbf{P}\delta\mathbf{U}) = (\mathbf{R}^{\mathrm{T}}\mathbf{P}) \cdot \delta\mathbf{U} \\ &= \operatorname{sym}(\mathbf{R}^{T}\mathbf{P}) \cdot \delta\mathbf{U} = \mathbf{P} \cdot \delta\mathbf{F},\end{aligned}$$

where the last form was obtained on use of eq. (4.15). The third term is equal to the third term of eq. (4.39) because

$$\begin{aligned}\mathbf{P} \cdot (\delta\mathbf{R}\mathbf{U}) &= \mathbf{P} \cdot (\delta\mathbf{R}\mathbf{R}^{T}\mathbf{F}) = \operatorname{tr}(\mathbf{P}\mathbf{F}^{\mathrm{T}}\mathbf{R}\delta\mathbf{R}^{\mathrm{T}}) \\ &= (\mathbf{P}\mathbf{F}^{T}) \cdot (\delta\mathbf{R}\mathbf{R}^{T}) = -\operatorname{skew}(\mathbf{P}\mathbf{F}^{T}) \cdot (\mathbf{R}^{T}\delta\mathbf{R}),\end{aligned}$$

which ends the proof. □

Four-field potential. On use of eqs. (4.15) and (4.18), we have $\mathbf{P} \cdot \delta\mathbf{F} = \operatorname{sym}(\mathbf{R}^{T}\mathbf{P}) \cdot \delta\mathbf{U} = \partial_{U}\mathcal{W}(\mathbf{U}) \cdot \delta\mathbf{U}$. Thus, from eq. (4.36) we can deduce the four-field functional

$$F_4^P(\chi, \mathbf{R}, \mathbf{U}, \mathbf{P}) \doteq \int_B [\mathcal{W}(\mathbf{U}) + \mathbf{P} \cdot (\mathbf{F} - \mathbf{R}\mathbf{U})]\, \mathrm{d}V - F_{\mathrm{ext}}, \tag{4.41}$$

where $\mathbf{P}$ is a Lagrange multiplier for the polar decomposition equation (4.33). Besides, the functional of external forces

$$F_{\mathrm{ext}} \doteq F_b + F_\sigma + F_\chi, \tag{4.42}$$

where the functionals for the body force, the (deformation independent) external loads and the displacement boundary conditions are defined as

$$F_b \doteq \int_B \rho_R \mathbf{b} \cdot \chi\, \mathrm{d}V, \quad F_\sigma \doteq \int_{\partial_\sigma \mathrm{B}} \hat{\mathbf{p}} \cdot \chi\, \mathrm{d}A, \quad F_\chi \doteq \int_{\partial_\chi \mathrm{B}} (\mathbf{P}\mathbf{n}) \cdot (\chi - \hat{\chi})\, \mathrm{d}A.$$

Remark. In this formulation, the right stretch $\mathbf{U}$ is not a function of χ but an independent tensorial variable. It must be parameterized in a way ensuring that it is symmetric and positive definite; the latter can be achieved by expressing $\mathbf{U}$ in terms of its principal values, taken as squares of some parameters, and a rotation tensor. We see that $\mathbf{U}$ introduces six additional variables in a complicated form, and that's why other simpler formulations were developed; they are presented in the following sections.

4.3 3-F formulation for nominal stress

In this section, we describe a three-field formulation including rotations, in terms of $\{\chi, \mathbf{Q}, \mathbf{T}_a\}$, which has the following features:

- the rotations $\mathbf{Q}$ are constrained by the Rotation Constraint (RC), eq. (3.8),
- the strain energy and the CL are defined for the *relaxed* right stretch, eq. (3.12).

To the set of governing equations (4.4)–(4.6), we append the following equations:

1. Rotation Constraint:

$$\mathfrak{C} \doteq \text{skew}(\mathbf{Q}^T \mathbf{F}) = \mathbf{0}, \tag{4.43}$$

2. Constitutive Law of eq. (4.31):

$$\text{sym}(\mathbf{Q}^T \mathbf{P}) = \frac{\partial \mathcal{W}(\tilde{\mathbf{U}})}{\partial \tilde{\mathbf{U}}}, \qquad \tilde{\mathbf{U}} = \text{sym}(\mathbf{Q}^T \mathbf{F}), \tag{4.44}$$

which furnish a formulation in terms of three fields $\{\chi, \mathbf{Q}, \mathbf{P}\}$. Comparing with the four-field formulation of the previous section, the right stretch tensor $\mathbf{U}$ is not present.

Strain energy and constitutive law. If $\mathbf{Q} = \mathbf{R}$, where $\mathbf{R} \in \text{SO}(3)$ satisfies the polar decomposition equation, then, by eq. (4.15), $\delta \mathcal{W} = \mathbf{P} \cdot \delta \mathbf{F} = \text{sym}(\mathbf{R}^T \mathbf{P}) \cdot \delta \mathbf{U}$, i.e. the tensor $\text{sym}(\mathbf{R}^T \mathbf{P})$ is work-conjugate to $\mathbf{U}$. Let us assume the existence of the strain energy $\mathcal{W}$ in terms of the *relaxed* stretch strain $\tilde{\mathbf{U}} = \text{sym}(\mathbf{Q}^T \mathbf{F})$. Using $\tilde{\mathbf{U}}$ in place of $\mathbf{U}$ in eq. (4.34), we obtain the constitutive law (4.18).

Weak form of basic equations. A weak form of the governing equations (4.4)–(4.6), yields eqs. (4.10) and (4.12)–(4.14). For the RC, eq. (4.43), we calculate a volume integral of a scalar product of this equation with a skew-symmetric tensor $\delta \text{skew}(\mathbf{Q}^T \mathbf{P})$,

$$\int_B \text{skew}(\mathbf{Q}^T \mathbf{F}) \cdot \delta \text{skew}(\mathbf{Q}^T \mathbf{P}) \, \mathrm{d}V = 0. \tag{4.45}$$

The reason for using here a variation of $\text{skew}(\mathbf{Q}^T \mathbf{P})$ will become obvious in the sequel.

VW equation. Adding the scalar eq. (4.10), (4.13)–(4.14) and (4.45), we obtain

$$\int_B \left[\mathbf{P}\cdot\delta\mathbf{F} + \delta\text{skew}(\mathbf{Q}^T\mathbf{P})\cdot\text{skew}(\mathbf{Q}^T\mathbf{F})\right]\,\mathrm{d}V - \delta F_{\text{ext}} = 0, \tag{4.46}$$

where δF_{ext} is defined in eq. (4.37). The integrand can be further transformed to the form given by eq. (4.28), i.e.

$$\int_B \{\text{sym}(\mathbf{Q}^T\mathbf{P})\cdot\text{sym}\delta(\mathbf{Q}^T\mathbf{F}) + \delta[\text{skew}(\mathbf{Q}^T\mathbf{P})\cdot\text{skew}(\mathbf{Q}^T\mathbf{F})]\}\,\mathrm{d}V \\ - \delta F_{\text{ext}} = 0. \tag{4.47}$$

Note that the AMB, eq. (4.5), was exploited in the derivation of eq. (4.28), and earlier of eq. (4.24), but its weak form of eq. (4.12) is not present in the integrand of eq. (4.46).

Three-field potential. On the basis of eq. (4.47), by using the CL of eq. (4.32), we can define the three-field potential

$$F_3^P(\chi,\mathbf{Q},\mathbf{P}) \doteq \int_B \left[\mathcal{W}(\text{sym}(\mathbf{Q}^T\mathbf{F})) + \text{skew}(\mathbf{Q}^T\mathbf{P})\cdot\text{skew}(\mathbf{Q}^T\mathbf{F})\right]\,\mathrm{d}V - F_{\text{ext}}, \tag{4.48}$$

where F_{ext} is defined in eq. (4.42). This also proves that the use of $\delta\text{skew}(\mathbf{Q}^T\mathbf{P})$ in eq. (4.45) was indeed correct.

Remark 1. The right stretch $\mathbf{U}$ can also be eliminated from the four-field formulation in another way. Note that we can rewrite the Lagrange term of the functional of eq. (4.41) as $\mathbf{P}\cdot(\mathbf{F}-\mathbf{R}\mathbf{U}) = (\mathbf{R}^T\mathbf{P})\cdot(\mathbf{R}^T\mathbf{F}-\mathbf{U})$, and further split it into a symmetric part and a skew part,

$$(\mathbf{R}^T\mathbf{P})\cdot(\mathbf{R}^T\mathbf{F}-\mathbf{U}) = \text{sym}(\mathbf{R}^T\mathbf{P})\cdot\left[\text{sym}(\mathbf{R}^T\mathbf{F}) - \mathbf{U}\right] \\ + \text{skew}(\mathbf{R}^T\mathbf{P})\cdot\text{skew}(\mathbf{R}^T\mathbf{F}). \tag{4.49}$$

If we assume that $\mathbf{U} \doteq \text{sym}(\mathbf{R}^T\mathbf{F})$, i.e. adopting the *relaxed* right stretch of eq. (3.12), then the first term of eq. (4.49) vanishes, and $F_4^P(\chi,\mathbf{R},\mathbf{U},\mathbf{P})$ of eq. (4.41) reduces to the three-field functional of eq. (4.48), with $\mathbf{R}$ in place of $\mathbf{Q}$.

Remark 2. If we use in eq. (4.47) the CL $\text{sym}(\mathbf{Q}^T\mathbf{P}) = \partial_{\tilde{U}}\mathcal{W}$ of eq. (4.31), then the nominal stress $\mathbf{P}$ remains only in the term $\text{skew}(\mathbf{Q}^T\mathbf{P})$. Hence, we can define a skew-symmetric tensor $\mathbf{T}_a \doteq \text{skew}(\mathbf{Q}^T\mathbf{P})$ with only three components and abandon using $\mathbf{P}$ with nine components. That is the basic motivation behind using the Biot stress in the next section.

4.4 3-F and 2-F formulations for Biot stress

In this section, we describe a three-field formulation in terms of $\{\chi, \mathbf{Q}, \mathbf{T}_a\}$, developed in [42, 191]. This formulation can be obtained from the three-field formulation for the nominal stress tensor, which is described in the previous section, just by introducing the definition of the Biot stress. A two-field formulation, which is valid only for an isotropic material, is also presented.

Biot stress. Define the tensor $\mathbf{T} \doteq \mathbf{Q}^T\mathbf{P}$, where $\mathbf{Q} \in \text{SO}(3)$, and split it into the symmetric and skew-symmetric parts, $\mathbf{T} = \mathbf{T}_s^B + \mathbf{T}_a$, where

$$\mathbf{T}_s^B \doteq \text{sym}\mathbf{T} = \text{sym}(\mathbf{Q}^T\mathbf{P}), \qquad \mathbf{T}_a \doteq \text{skew}\mathbf{T} = \text{skew}(\mathbf{Q}^T\mathbf{P}). \tag{4.50}$$

The symmetric part $\mathbf{T}_s^B$ is called the Biot stress, or the Biot–Lure stress, or the Jaumann stress. Having $\mathbf{T}_s^B$, $\mathbf{T}_a$, and $\mathbf{Q}$, we can uniquely calculate $\mathbf{P}$.

VW equation. Introducing the definitions of $\mathbf{T}_s^B$ and $\mathbf{T}_a$ of eq. (4.50) into eq. (4.47), we obtain the VW in the form

$$\int_B \{\mathbf{T}_s^B \cdot \text{sym}\delta(\mathbf{Q}^T\mathbf{F}) + \delta[\mathbf{T}_a \cdot \text{skew}(\mathbf{Q}^T\mathbf{F})]\} \, dV - \delta F_{\text{ext}} = 0, \tag{4.51}$$

where δF_{ext} is defined in eq. (4.37), but with $\mathbf{P}$ replaced by $\mathbf{QT}$.

Three-field potential. For the symmetric $\mathbf{T}_s^B$ we can use the CL, eq. (4.32), i.e. $\mathbf{T}_s^B = \partial_{\tilde{U}}\mathcal{W}$. On the basis of this CL and eq. (4.51), we can define the three-field potential

$$F_3^B(\chi, \mathbf{Q}, \mathbf{T}_a) \doteq \int_B [\mathcal{W}(\text{sym}(\mathbf{Q}^T\mathbf{F})) + \mathbf{T}_a \cdot \text{skew}(\mathbf{Q}^T\mathbf{F})] \, dV - F_{\text{ext}}, \tag{4.52}$$

in which $\mathbf{T}_a$ is the Lagrange multiplier for the RC equation. Besides, F_{ext} is defined in eq. (4.42), but with $\mathbf{P}$ replaced by $\mathbf{QT}$.

Euler–Lagrange equations. Because of this new variable, $\mathbf{T}_a$, we have to check whether the Euler–Lagrange equations for F_3^B yield the governing equations (4.4)–(4.6) and (4.43). Using eq. (4.30) and (4.32), we obtain the following variations of the strain energy (4.52):

$$\delta_\chi \mathcal{W}(\text{sym}(\mathbf{Q}^T\mathbf{F})) = \mathbf{T}_s^B \cdot \delta_\chi \tilde{\mathbf{U}} = \mathbf{T}_s^B \cdot (\mathbf{Q}^T \delta\mathbf{F}) = (\mathbf{Q}\mathbf{T}_s^B) \cdot \delta\mathbf{F},$$
$$\delta_Q \mathcal{W}(\text{sym}(\mathbf{Q}^T\mathbf{F})) = \mathbf{T}_s^B \cdot \delta_Q \tilde{\mathbf{U}} = \mathbf{T}_s^B \cdot (\delta\mathbf{Q}^T \mathbf{F}) = -\delta\tilde{\boldsymbol{\theta}} \cdot \text{skew}(\mathbf{Q}\mathbf{T}_s^B\mathbf{F}^T),$$

where $\delta\mathbf{Q} \doteq \delta\tilde{\boldsymbol{\theta}}\,\mathbf{Q}$. The variations of the RC term are

$$\begin{aligned} \delta_\chi[\mathbf{T}_a \cdot \text{skew}(\mathbf{Q}^T\mathbf{F})] &= \delta_\chi[(\mathbf{Q}\mathbf{T}_a)\cdot\mathbf{F}] = (\mathbf{Q}\mathbf{T}_a)\cdot\delta\mathbf{F}, \\ \delta_{T_a}[\mathbf{T}_a \cdot \text{skew}(\mathbf{Q}^T\mathbf{F})] &= \delta\mathbf{T}_a \cdot \text{skew}(\mathbf{Q}^T\mathbf{F}), \\ \delta_Q[\mathbf{T}_a \cdot \text{skew}(\mathbf{Q}^T\mathbf{F})] &= (\delta\mathbf{Q}\mathbf{T}_a)\cdot\mathbf{F} = \text{tr}(\delta\tilde{\boldsymbol{\theta}}\mathbf{Q}\mathbf{T}_a\mathbf{F}^{\mathrm{T}}) \\ &= -\delta\tilde{\boldsymbol{\theta}} \cdot \text{skew}(\mathbf{Q}\mathbf{T}_a\mathbf{F}^T). \end{aligned} \tag{4.53}$$

Hence, the first variation of F_3^P of eq. (4.48) is

$$\delta F_3^B(\chi, \mathbf{Q}, \mathbf{T}_a) = \int_B \left[\mathbf{A}\cdot\delta\mathbf{F} + \delta\mathbf{T}_a \cdot \text{skew}(\mathbf{Q}^T\mathbf{F}) - \delta\tilde{\boldsymbol{\theta}} \cdot \text{skew}(\mathbf{A}\mathbf{F}^T)\right] \mathrm{d}V - \delta F_{\text{ext}}, \tag{4.54}$$

where $\mathbf{A} \doteq \mathbf{Q}(\mathbf{T}_s^B + \mathbf{T}_a) = \mathbf{Q}\mathbf{T} = \mathbf{P}$. We see that δF_3^B is identical as a sum of eqs. (4.10), (4.12)–(4.14) and (4.45), which now is rewritten as $\int_B \text{skew}(\mathbf{Q}^T\mathbf{F}) \cdot \delta\mathbf{T}_a \, \mathrm{d}V = 0$. Hence, the Euler–Lagrange equation are identical to eqs. (4.4)–(4.6) and (4.43), and hence F_3^B is a correct potential for the formulation including rotations.

Remark. Another form of elimination of $\mathbf{U}$ from the four-field functional of eq. (4.41), is to use the Legendre transformation

$$\mathcal{W}(\mathbf{U}) - \mathbf{T}_s^B \cdot \mathbf{U} = -\mathcal{W}_c(\mathbf{T}_s^B), \tag{4.55}$$

where $\mathcal{W}_c$ is the complementary energy density, which requires the constitutive relation to be invertible, $\partial\mathcal{W}_c/\partial\mathbf{T}_s^B = \mathbf{U}$. This approach is described in [9], eq. (3.36).

AMB for isotropic material. For an isotropic material, we can show that the AMB equation is satisfied when the skew-symmetric stress vanishes, i.e. $\mathbf{T}_a = \mathbf{0}$. Rewrite the AMB eq. (4.5), as

$$\mathbf{Q}\mathbf{T}\mathbf{F}^T - \mathbf{F}\mathbf{T}^T\mathbf{Q}^T = \mathbf{0}. \tag{4.56}$$

If the RC is satisfied, then we have $\mathbf{F} = \mathbf{Q}\mathbf{F}^T\mathbf{Q}$ and $\mathbf{Q}^T\mathbf{F} = \mathbf{U}$. Using them in the AMB, we obtain $\mathbf{Q}(\mathbf{T}\mathbf{U} - \mathbf{U}\mathbf{T}^T)\mathbf{Q}^T = \mathbf{0}$ and the split $\mathbf{T} = \mathbf{T}_s^B + \mathbf{T}_a$ yields

$$\mathbf{T}_a\mathbf{U} + \mathbf{U}\mathbf{T}_a = \mathbf{T}_s^B\mathbf{U} - \mathbf{U}\mathbf{T}_s^B. \tag{4.57}$$

For an isotropic material, $\mathbf{T}_s^B$ and $\mathbf{U}$ are a work-conjugate pair, so they are co-axial and commute. Hence, the r.h.s. of eq. (4.57) vanishes and the AMB is reduced to

$$\mathbf{T}_a\mathbf{U} + \mathbf{U}\mathbf{T}_a = \mathbf{0}. \tag{4.58}$$

Note that $\mathbf{U}$ is symmetric and positive definite, while $\mathbf{T}_a$ is skew-symmetric, hence the assumptions of Lemma 3.1 in [214] are satisfied. Using this lemma, eq. (4.58) is satisfied only when $\mathbf{T}_a = \mathbf{0}$, which completes the proof. □

When the RC is not satisfied or the material is not isotropic, then $\mathbf{T}_a \neq \mathbf{0}$ and must remain in the functional.

Two-field functional for isotropic material. To obtain a two-field functional which does not depend on $\mathbf{T}_a^B$, we regularize F_3^B of eq. (4.52) in $\mathbf{T}_a$ as follows

$$\tilde{F}_3^B(\chi, \mathbf{Q}, \mathbf{T}_a) \doteq F_3^B(\chi, \mathbf{Q}, \mathbf{T}_a) - \frac{1}{2\gamma}\int_B \mathbf{T}_a \cdot \mathbf{T}_a \, \mathrm{d}V, \tag{4.59}$$

where $\gamma \in (0, \infty)$ is the regularization parameter. In the volume integral in $\tilde{F}_3^B$, which is affected by the regularization, the integrand is

$$\mathcal{W}(\mathrm{sym}(\mathbf{Q}^T\mathbf{F})) + \mathbf{T}_a \cdot \mathrm{skew}(\mathbf{Q}^T\mathbf{F}) - \frac{1}{2\gamma}\mathbf{T}_a \cdot \mathbf{T}_a. \tag{4.60}$$

A variation of $\tilde{F}_3^B$ w.r.t. $\mathbf{T}_a$ yields the Euler–Lagrange equation: $\gamma\,\mathrm{skew}(\mathbf{Q}^T\mathbf{F}) - \mathbf{T}_a = 0$, for $\delta\mathbf{T}_a$ in B. From this equation, we calculate $\mathbf{T}_a$ and use it in eq. (4.60), which becomes

$$\mathcal{W}(\mathrm{sym}(\mathbf{Q}^T\mathbf{F})) + \frac{\gamma}{2}\mathrm{skew}(\mathbf{Q}^T\mathbf{F}) \cdot \mathrm{skew}(\mathbf{Q}^T\mathbf{F}), \tag{4.61}$$

in which the second term is the RC equation $\mathrm{skew}(\mathbf{Q}^T\mathbf{F}) = \mathbf{0}$ imposed by the penalty method. Then, the two-field functional, not depending on $\mathbf{T}_a$, is defined as

$$\tilde{F}_2^B(\chi, \mathbf{Q}) \doteq \int_B \left[\mathcal{W}(\mathrm{sym}(\mathbf{Q}^T\mathbf{F})) + \frac{\gamma}{2}\,\mathrm{skew}(\mathbf{Q}^T\mathbf{F}) \cdot \mathrm{skew}(\mathbf{Q}^T\mathbf{F})\right] \mathrm{d}V - F_{\mathrm{ext}}. \tag{4.62}$$

The corresponding VW equation is

$$\int_B \left[\mathbf{T}_s^B \cdot \delta \mathrm{sym}(\mathbf{Q}^T\mathbf{F}) \;+\; \gamma\, \mathrm{skew}(\mathbf{Q}^T\mathbf{F}) \cdot \delta \mathrm{skew}(\mathbf{Q}^T\mathbf{F})\right] \mathrm{d}V - \delta F_{\mathrm{ext}} = 0. \tag{4.63}$$

This equation can be used only for isotropic material.

4.5 3-F and 2-F formulations for second Piola–Kirchhoff stress

In this section, we assume that the strain energy $\mathcal{W}$ is a function of the right Cauchy–Green tensor $\mathbf{C}$, and we obtain the formulations with rotations in terms of the second Piola–Kirchhoff stress $\mathbf{S}$. The governing equations for such a formulation are obtained by using $\mathbf{P} = \mathbf{FS}$ in eqs. (4.4)–(4.6), and by appending the RC of eq. (4.43). From the outset, we assume that $\mathbf{S} = \mathbf{S}^T$, i.e. that the LMB of eq. (4.5) is satisfied, as in [214]. The CL for $\mathbf{S}$ is given by eq. (4.23).

Weak form of basic equations. We modify the VW of eq. (4.47) as follows: (i) we use $\mathbf{P} = \mathbf{FS}$ in the term

$$\delta[\mathrm{skew}(\mathbf{Q}^T\mathbf{P}) \cdot \mathrm{skew}(\mathbf{Q}^T\mathbf{F})] = \delta[\mathrm{skew}(\mathbf{Q}^T\mathbf{FS}) \cdot \mathrm{skew}(\mathbf{Q}^T\mathbf{F})],$$

(ii) we use the strain energy $\mathcal{W}(\mathbf{C})$ in terms of the right Cauchy–Green tensor $\mathbf{C}$,

$$\partial_{\tilde{U}} \mathcal{W}(\tilde{U}) \cdot \mathrm{sym}\delta(\mathbf{Q}^T\mathbf{F}) = \partial_C \mathcal{W}(\mathbf{C}) \cdot \mathrm{sym}\delta(\mathbf{F}^T\mathbf{F}),$$

where eqs. (4.19) and (4.22) were used. This yields the VW equation,

$$\int_B \{\partial_C \mathcal{W}(\mathbf{C}) \cdot \mathrm{sym}\delta(\mathbf{F}^T\mathbf{F}) + \delta[\mathbf{T}_a \cdot \mathrm{skew}(\mathbf{Q}^T\mathbf{F})]\}\, \mathrm{d}V - \delta F_{\mathrm{ext}} = 0, \tag{4.64}$$

where $\mathbf{T}_a \doteq \mathrm{skew}(\mathbf{Q}^T\mathbf{FS})$ is used to change the variables, i.e. instead of $\mathbf{S}$ with six components, we use $\mathbf{T}_a$ with only three components. δF_{ext} is defined in eq. (4.37), but with $\mathbf{P}$ replaced by $\mathbf{FS}$.

Three-field potential. From eq. (4.64), we can deduce the three-field potential

$$F_3^{\mathrm{2PK}}(\chi, \mathbf{Q}, \mathbf{T}_a) \doteq \int_B \left[\mathcal{W}(\mathbf{F}^T\mathbf{F}) + \mathbf{T}_a \cdot \mathrm{skew}(\mathbf{Q}^T\mathbf{F})\right] \mathrm{d}V - F_{\mathrm{ext}}, \tag{4.65}$$

where $\mathbf{T}_a$ is the Lagrange multiplier for the RC equation. Besides, F_{ext} is defined in eq. (4.42) but with $\mathbf{P}$ replaced by $\mathbf{FS}$.

Euler–Lagrange equations. Because of this new variable, $\mathbf{T}_a$, we have to check whether the Euler-Lagrange equations for F_3^{2PK} of eq. (4.65) yield the governing equations (4.4)–(4.6) and (4.43).

By using eqs. (4.19) and (4.23), we obtain a variation of the strain energy $\delta_\chi \mathcal{W}(\mathbf{F}^T\mathbf{F}) = (\mathbf{FS}) \cdot \delta\mathbf{F}$. The variations of the RC term are given by eq. (4.53). Hence,

$$\delta F_3^{2PK}(\chi, \mathbf{Q}, \mathbf{T}_a) = \int_B \left[\mathbf{A} \cdot \delta\mathbf{F} + \delta\mathbf{T}_a \cdot \text{skew}(\mathbf{Q}^T\mathbf{F}) - \delta\tilde{\boldsymbol{\theta}} \cdot \text{skew}(\mathbf{Q}\mathbf{T}_a\mathbf{F}^T)\right] \mathrm{d}V - \delta F_{\text{ext}}, \tag{4.66}$$

where $\mathbf{A} \doteq \mathbf{FS} + \mathbf{Q}\mathbf{T}_a$ and $\delta\tilde{\boldsymbol{\theta}} \doteq \delta\mathbf{Q}\mathbf{Q}^T$. The term $\mathbf{A} \cdot \delta\mathbf{F}$ can be transformed further. Using the formula for the divergence of a product of two tensors, see [33] eq. (5.5.19), and $\delta\mathbf{F} = \nabla\delta\mathbf{u}$, we obtain

$$\text{Div}\mathbf{A} \cdot \nabla\delta\mathbf{u} = \text{Div}(\mathbf{A}^T\delta\mathbf{u}) - \mathbf{A} \cdot \nabla\delta\mathbf{u}.$$

The second term contributes to the equilibrium equation, while the first term is transformed on use of the divergence theorem, see [33] eq. (5.8.11), as follows

$$\int_B \text{Div}(\mathbf{A}^T\delta\mathbf{u})\, \mathrm{d}V = \int_{\partial B} (\mathbf{A}^T\delta\mathbf{u}) \cdot \mathbf{n}\, \mathrm{d}A = \int_{\partial B} (\mathbf{A}\mathbf{n}) \cdot \delta\mathbf{u}\, \mathrm{d}A.$$

We see that this term contributes to the traction BC. Finally, the following Euler–Lagrange equations are obtained

$$\begin{aligned} \text{Div}\mathbf{A} + \rho_R\mathbf{b} &= \mathbf{0} \quad \text{in} \quad \text{B}, \\ \text{skew}(\mathbf{Q}\mathbf{T}_a\mathbf{F}^T) &= \mathbf{0} \quad \text{in} \quad \text{B}, \\ \text{skew}(\mathbf{Q}^T\mathbf{F}) &= \mathbf{0} \quad \text{in} \quad \text{B}, \\ \mathbf{A}\mathbf{n} &= \hat{\mathbf{p}} \quad \text{on} \quad \partial_\sigma\text{B}. \end{aligned} \tag{4.67}$$

These equations will be equal to the governing equations (4.4), (4.6) and (4.43) when $\mathbf{T}_a = \mathbf{0}$, as then the second of the above equations is trivially satisfied and $\mathbf{A} = \mathbf{FS} = \mathbf{P}$. The proof that $\mathbf{T}_a = \mathbf{0}$ is given below.

Proof. Eq.$(4.67)_2$ is post-multiplied by $\mathbf{Q}$, and transformed as follows:

$$2\,\text{skew}(\mathbf{Q}\mathbf{T}_a\mathbf{F}^T)\,\mathbf{Q} = \mathbf{Q}\mathbf{T}_a\mathbf{F}^T\mathbf{Q} - \mathbf{F}\mathbf{T}_a^T\mathbf{Q}^T\mathbf{Q} = \mathbf{Q}\mathbf{T}_a\mathbf{F}^T\mathbf{Q} + \mathbf{F}\mathbf{T}_a. \tag{4.68}$$

From eq. $(4.67)_3$, we have $\mathbf{F}^T\mathbf{Q} = \mathbf{Q}^T\mathbf{F}$ and, on the left polar decomposition $\mathbf{F} = \mathbf{V}\mathbf{Q}$, where $\mathbf{V}$ is the left stretching tensor, we obtain

$$\mathbf{Q}\mathbf{T}_a\mathbf{Q}^T\mathbf{V}\mathbf{Q}+\mathbf{V}\mathbf{Q}\mathbf{T}_a = \mathbf{T}_a^*\mathbf{V}\mathbf{Q}+\mathbf{V}\mathbf{Q}\mathbf{Q}^T\mathbf{T}_a^*\mathbf{Q} = (\mathbf{T}_a^*\mathbf{V}+\mathbf{V}\mathbf{T}_a^*)\,\mathbf{Q}, \quad (4.69)$$

where $\mathbf{T}_a^* \doteq \mathbf{Q}\mathbf{T}_a\mathbf{Q}^T$ is the forward-rotated $\mathbf{T}_a$. Hence, eq. $(4.67)_2$ yields

$$\mathbf{T}_a^*\,\mathbf{V} + \mathbf{V}\,\mathbf{T}_a^* = \mathbf{0}. \quad (4.70)$$

Because $\mathbf{T}_a^*$ is skew-symmetric and $\mathbf{V}$ is symmetric and positive definite, the assumptions of Lemma 3.1 in [214] are satisfied. With this lemma, the above equation is satisfied only when $\mathbf{T}_a^* = \mathbf{0}$. Hence, $\mathbf{T}_a = \mathbf{Q}^T\mathbf{T}_a^*\mathbf{Q} = \mathbf{0}$, which ends the proof. □

Two-field functional. We can regularize the functional (4.65) in $\mathbf{T}_a$ as follows:

$$\tilde{F}_3^{2\mathrm{PK}}(\chi, \mathbf{Q}, \mathbf{T}_a) = F_3^{2\mathrm{PK}}(\chi, \mathbf{Q}, \mathbf{T}_a) - \frac{1}{2\gamma}\int_B \mathbf{T}_a \cdot \mathbf{T}_a \,\mathrm{d}V, \quad (4.71)$$

where $\gamma \in (0,\infty)$ is the regularization parameter. It can be shown that the correct Euler–Lagrange equations of $\tilde{F}_3^{2\mathrm{PK}}$ are obtained not only when $\gamma \to \infty$, but for any value of γ. A variation of $\tilde{F}_3^{2\mathrm{PK}}$ w.r.t. $\mathbf{T}_a$ yields the following Euler–Lagrange equation for $\delta\mathbf{T}_a$ in B,

$$\mathrm{skew}(\mathbf{Q}^T\mathbf{F}) - \frac{1}{\gamma}\mathbf{T}_a = 0. \quad (4.72)$$

From this equation we calculate $\mathbf{T}_a = \gamma\,\mathrm{skew}(\mathbf{Q}^T\mathbf{F})$, and use it in eq. (4.71). Then we can define a two-field functional

$$\tilde{F}_2^{2\mathrm{PK}}(\chi, \mathbf{Q}) \doteq \int_B \left[\mathcal{W}(\mathbf{F}^T\mathbf{F}) + \frac{\gamma}{2}\mathrm{skew}(\mathbf{Q}^T\mathbf{F}) \cdot \mathrm{skew}(\mathbf{Q}^T\mathbf{F})\right] \mathrm{d}V - F_{\mathrm{ext}}, \quad (4.73)$$

with the penalty term for the RC equation. We have to check that the Euler–Lagrange equations for $F_2^{2\mathrm{PK}}$ yield the governing equations (4.4), (4.6) and (4.43); the proof is given in [214], eqs. (22)–(25). This functional is typically used in numerical implementations.

Remark. In this formulation, the rotations $\mathbf{Q}$ appear only in the RC, but not in the other governing equations. Hence, we can first solve the problem for χ and determine $\mathbf{Q}$ afterwards, which can be done in two ways, using either the RC equation or the polar decomposition of $\mathbf{F}$, as discussed earlier. This method cannot be used in the Reissner-type shells, where $\mathbf{Q}$ appears also in the governing equations.

4.6 2-F formulation with unconstrained rotations

In this section, we describe a two-field formulation in terms of $\{\chi, \mathbf{Q}\}$, which has the following features:

- neither the polar decomposition equation (3.2), nor the RC equation (3.8) are used to constrain rotations $\mathbf{Q}$.
- the strain energy and the CL are defined for the *non-symmetric relaxed* right stretch eq. (3.12).

To the set of governing equations, eqs. (4.4)–(4.6), we only append the constitutive law, eq. (4.27),

$$(\mathbf{Q}^T\mathbf{P}) = \partial_{Q^TF}\mathcal{W}(\mathbf{Q}^T\mathbf{F}), \tag{4.74}$$

where $\mathbf{Q}^T\mathbf{F}$ is non-symmetric.

VW equation. Adding weak forms of the governing equations, i.e. eqs. (4.10), and (4.13)–(4.14), and applying eq. (4.24), we obtain

$$\int_B (\mathbf{Q}^T\mathbf{P}) \cdot \delta(\mathbf{Q}^T\mathbf{F})\, \mathrm{d}V - \delta F_{\text{ext}} = 0, \tag{4.75}$$

where δF_{ext} is defined in eq. (4.37). Note that the AMB, eq. (4.12), was exploited in derivation of eq. (4.24) and, hence, it does not appear explicitly in eq. (4.75).

Two-field potential. Using eq. (4.74), we obtain

$$(\mathbf{Q}^T\mathbf{P}) \cdot \delta(\mathbf{Q}^T\mathbf{F}) = \partial_{Q^TF}\mathcal{W}(\mathbf{Q}^T\mathbf{F}) \cdot \delta(\mathbf{Q}^T\mathbf{F}).$$

Hence, on the basis of this equation and eq. (4.75), we can define a two-field potential

$$F_2^*(\chi, \mathbf{Q}) \doteq \int_B \mathcal{W}(\mathbf{Q}^T\mathbf{F})\, \mathrm{d}V - F_{\text{ext}}, \tag{4.76}$$

where F_{ext} is defined in eq. (4.42).

Remark. Note that $F_2^*(\chi, \mathbf{Q})$ can be additionally constrained by the RC equation, $\text{skew}(\mathbf{Q}^T\mathbf{F}) = \mathbf{0}$, by using the penalty method. Then,

$$F_2^{**}(\chi, \mathbf{Q}) \doteq \int_B \mathcal{W}(\mathbf{Q}^T\mathbf{F}) + \frac{\gamma}{2}\text{skew}(\mathbf{Q}^T\mathbf{F}) \cdot \text{skew}(\mathbf{Q}^T\mathbf{F})\, \mathrm{d}V - F_{\text{ext}}, \tag{4.77}$$

where $\gamma \in (0, \infty)$ is the penalty parameter. Note that the argument of $\mathcal{W}$ is different to that in eq. (4.62).

5

Basic geometric definitions for shells

In this chapter, the basic geometric definitions needed to develop the shell FEs are provided. The shell can be intuitively but imprecisely defined as a 3D body, which has one dimension much smaller than the other two. More precisely, the shell is a surface in a 3D space equipped with a thickness, which is much smaller than the size of the surface. This implies a specific geometrical description of shells.

5.1 Coordinates and position vector

Normal coordinates for shells. For the initial configuration of a shell, we use the *normal* coordinates, see Fig. 5.1, the characteristic feature of which is that one coordinate is normal to the reference surface. The coordinates involved are defined as follows:

1. The reference shell surface is parameterized by the coordinates ϑ^α $(\alpha = 1, 2)$. This surface is selected arbitrarily, but most often the middle surface is used for this purpose; this is not suitable, e.g., for composites with non-symmetric stacking sequence of layers. The *middle* surface is equidistant from the top and bottom surfaces bounding the shell. Various types of coordinates can be used as ϑ^α.
2. The direction normal to the reference surface is parameterized by the coordinate $\zeta \in [-h/2, +h/2]$, where h is the initial shell thickness. We can also use the natural coordinate $\xi^3 \in [-1, +1]$, which is more convenient in numerical integration over the thickness. The relation between these coordinates is $\zeta = (h/2)\,\xi^3$.

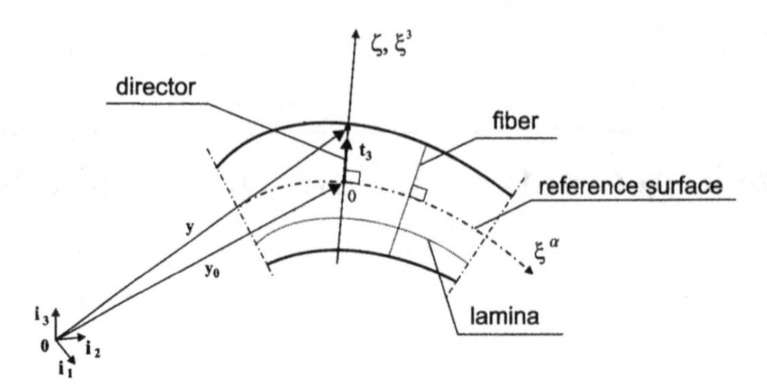

Fig. 5.1 Normal coordinates at a shell cross-section for initial configuration.

Selection of coordinates ϑ^α. Various types of coordinates can be used as ϑ^α.

1. In the FE method, the *natural* coordinates $\xi^\alpha \in [-1, +1]$ are used as ϑ^α. The corresponding tangent natural vectors $\mathbf{g}_1$ and $\mathbf{g}_2$ are skew, i.e. neither unit nor perpendicular. The natural coordinates are arguments of the shape functions for finite elements, see Chap. 10.
2. In analytical derivations, the *orthonormal* coordinates S^α can be used as ϑ^α, see Chap. 6. They are associated with the orthogonal and unit vectors $\mathbf{t}_1$ and $\mathbf{t}_2$, in the plane tangent to the reference surface. Using them, we do not have to distinguish between co-variant and contra-variant components of vectors and tensors, and derivations are simplified.

Position vectors for shells. The position vector in the initial configuration is split as follows:

$$\mathbf{y}(\vartheta^\alpha, \zeta) = \mathbf{y}_0(\vartheta^\alpha) + \zeta\ \mathbf{t}_3(\xi^\alpha), \qquad \alpha = 1, 2, \tag{5.1}$$

where $\mathbf{y}_0$ is the position of the reference surface and $\mathbf{t}_3$ is the vector normal to this surface, called the *director*, see Fig. 5.2. Besides, $\mathbf{y}(\vartheta^\alpha, \zeta = \text{const.})$ defines the *lamina* while $\mathbf{y}(\vartheta^\alpha = \text{const.}, \zeta)$ defines the *fiber* of a shell.

We also assume that the normal coordinates are *convected*, which means that a position of a selected point is identified by the same pair $(\vartheta^\alpha, \zeta)$ in the initial configuration and in each deformed configuration. The position vector in the deformed configuration is split as follows:

$$\mathbf{x}(\vartheta^\alpha, \zeta) = \mathbf{x}_0(\vartheta^\alpha) + \mathbf{d}(\vartheta^\alpha, \zeta), \tag{5.2}$$

where $\mathbf{x}_0$ is the current position of the reference surface and $\mathbf{d}$ is the out-of-plane vector defined by kinematical assumptions. Note that $\mathbf{d}$ is also called the deformed or current director. For the Reissner hypothesis, $\mathbf{d}$ is not normal to the current reference surface, see Sect. 6, but it is normal for the Kirchhoff hypothesis, see Sect. 6.3.4.

Various formalisms in shell description. Typically, the displacement and rotation vectors are represented in the reference ortho-normal basis $\{\mathbf{i}_k\}$, to enable linking of finite elements of various spatial orientation. Different formalisms are obtained as a result of the following two choices:

1. Various bases can be used to represent the position vectors $\mathbf{y}$ and $\mathbf{x}$.
   a) The local Cartesian basis $\{\mathbf{t}_k^c\}$ at the element center. Then, first, the displacement and rotation components must be transformed from the reference basis to this local basis and, later, the tangent stiffness matrix and the residual vector generated in this local basis must be transformed back to the reference basis $\{\mathbf{i}_k\}$.
   b) The reference Cartesian basis $\{\mathbf{i}_k\}$. Then, to apply various shell assumptions (and techniques related to the FE method), we must transform strain components to the local Cartesian basis $\{\mathbf{t}_k\}$.
2. Various coordinates can be used to parameterize the position vectors $\mathbf{y}$ and $\mathbf{x}$ and, as a consequence, as intermediate variables for differentiation in the deformation gradient:
   a) For natural coordinates $\{\xi^\alpha, \zeta\}$, the current position vector $\mathbf{x} = \mathbf{x}(\xi^\alpha(\mathbf{y}), \zeta(\mathbf{y}))$, and the deformation gradient is as follows:

$$\mathbf{F} \doteq \frac{\partial \mathbf{x}}{\partial \mathbf{y}} = \frac{\partial \mathbf{x}}{\partial \xi^\alpha} \otimes \frac{\partial \xi^\alpha}{\partial \mathbf{y}} + \frac{\partial \mathbf{x}}{\partial \zeta} \otimes \frac{\partial \zeta}{\partial \mathbf{y}}, \tag{5.3}$$

   this form is used, e.g., in Sect. 10.4.
   b) For orthonormal coordinates $\{S^\alpha, \zeta\}$, the current position vector $\mathbf{x} = \mathbf{x}(S^\alpha(\mathbf{y}), \zeta(\mathbf{y}))$, and the deformation gradient is as follows:

$$\mathbf{F} \doteq \frac{\partial \mathbf{x}}{\partial \mathbf{y}} = \frac{\partial \mathbf{x}}{\partial S^\alpha} \otimes \frac{\partial S^\alpha}{\partial \mathbf{y}} + \frac{\partial \mathbf{x}}{\partial \zeta} \otimes \frac{\partial \zeta}{\partial \mathbf{y}}, \tag{5.4}$$

   this form is used, e.g., in Chap. 6.

   Besides, the natural coordinate $\xi^3 \in [-1, +1]$ can be used instead of $\zeta \in [-h/2, +h/2]$.

5.2 Basic geometric definitions

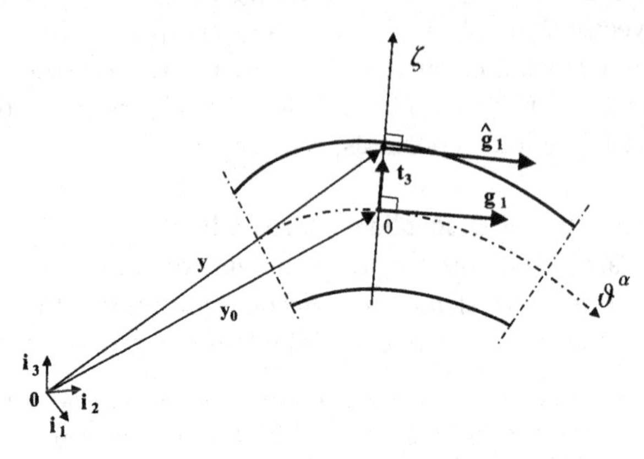

Fig. 5.2 Local bases $\{\hat{\mathbf{g}}_\alpha, \mathbf{t}_3\}$ and $\{\mathbf{g}_\alpha, \mathbf{t}_3\}$ for initial configuration.

Tangent basis varying over thickness. For the initial (non-deformed) configuration, the position vector $\mathbf{y} = \mathbf{y}(\vartheta^\alpha, \zeta)$ is given by eq. (5.1). The vectors tangent to the reference surface at arbitrary lamina ζ are obtained by differentiation of eq. (5.1),

$$\hat{\mathbf{g}}_\alpha(\zeta) \doteq \frac{\partial \mathbf{y}(\zeta)}{\partial \vartheta^\alpha} = \mathbf{g}_\alpha + \zeta\, \mathbf{t}_{3,\alpha}, \qquad \alpha = 1, 2, \tag{5.5}$$

where $\mathbf{g}_\alpha \doteq \partial \mathbf{y}_0 / \partial \vartheta^\alpha$. These vectors are neither unit nor mutually orthogonal, i.e.

$$\hat{\mathbf{g}}_\alpha(\zeta) \cdot \hat{\mathbf{g}}_\alpha(\zeta) = 1 + 2\zeta \mathbf{t}_{3,\alpha} \cdot \mathbf{g}_\alpha + \zeta^2 \mathbf{t}_{3,\alpha} \cdot \mathbf{t}_{3,\alpha} \neq 1 \quad \text{(no sum. over } \alpha\text{)},$$

$$\hat{\mathbf{g}}_1(\zeta) \cdot \hat{\mathbf{g}}_2(\zeta) = \zeta(\mathbf{t}_{3,1} \cdot \mathbf{g}_2 + \mathbf{t}_{3,2} \cdot \mathbf{g}_1) + \zeta^2 \mathbf{t}_{3,1} \cdot \mathbf{t}_{3,2} \neq 0,$$

but still $\hat{\mathbf{g}}_\alpha$ is normal to $\mathbf{t}_3$ because

$$\hat{\mathbf{g}}_\alpha \cdot \mathbf{t}_3 = \mathbf{g}_\alpha \cdot \mathbf{t}_3 + \zeta \mathbf{t}_{3,\alpha} \cdot \mathbf{t}_3 = 0,$$

where $\mathbf{g}_\alpha \cdot \mathbf{t}_3 = 0$ by definition, and $\mathbf{t}_{3,\alpha} \cdot \mathbf{t}_3 = 0$, as a result of differentiation of $\mathbf{t}_3 \cdot \mathbf{t}_3 = 1$ w.r.t. ϑ^α. Hence, $\hat{\mathbf{g}}_\alpha$ is parallel to $\mathbf{g}_\alpha$, and tangent to the reference surface.

Co-basis to tangent basis varying over thickness. The co-basis $\{\hat{\mathbf{g}}^\alpha, \mathbf{t}_3\}$ is also designated as the basis dual (or reciprocal) to $\{\hat{\mathbf{g}}_\alpha, \mathbf{t}_3\}$. The vectors $\hat{\mathbf{g}}^\alpha$ are defined as

$$\hat{\mathbf{g}}^\alpha \cdot \hat{\mathbf{g}}_\beta = \delta^\alpha_\beta, \qquad \hat{\mathbf{g}}^\alpha \cdot \mathbf{t}_3 = 0. \tag{5.6}$$

This definition provides three equations for $\hat{\mathbf{g}}^1$ and three for $\hat{\mathbf{g}}^2$, from which they can be directly determined. Alternatively, we can construct the co-basis as follows.

The conditions $\hat{\mathbf{g}}^1 \cdot \hat{\mathbf{g}}_2 = 0$ and $\hat{\mathbf{g}}^1 \cdot \mathbf{t}_3 = 0$ imply that $\hat{\mathbf{g}}^1$ is normal to $\hat{\mathbf{g}}_2$ and $\mathbf{t}_3$. Similarly, $\hat{\mathbf{g}}^2$ is normal to $\hat{\mathbf{g}}_1$ and $\mathbf{t}_3$. Hence, we can construct

$$\bar{\mathbf{g}}^1 = \bar{\mathbf{g}}_2 \times \mathbf{t}_3, \qquad \bar{\mathbf{g}}^2 = \mathbf{t}_3 \times \bar{\mathbf{g}}_1, \tag{5.7}$$

where $\bar{\mathbf{g}}_\alpha \doteq \hat{\mathbf{g}}_\alpha / \|\hat{\mathbf{g}}_\alpha\|$ are auxiliary unit vectors. The so-defined $\bar{\mathbf{g}}^\alpha$ have a proper direction, but their length is incorrect, i.e. $\bar{\mathbf{g}}^1 \cdot \hat{\mathbf{g}}_1 \neq 1$ and $\bar{\mathbf{g}}^2 \cdot \hat{\mathbf{g}}_2 \neq 1$. Hence, we define, $\hat{\mathbf{g}}^1 \doteq A\,\bar{\mathbf{g}}^1$ and $\hat{\mathbf{g}}^2 \doteq B\,\bar{\mathbf{g}}^2$, and from the conditions $\hat{\mathbf{g}}^1 \cdot \hat{\mathbf{g}}_1 = 1$ and $\hat{\mathbf{g}}^2 \cdot \hat{\mathbf{g}}_2 = 1$, we obtain $A = 1/(\bar{\mathbf{g}}^1 \cdot \hat{\mathbf{g}}_1)$ and $B = 1/(\bar{\mathbf{g}}^2 \cdot \hat{\mathbf{g}}_2)$. Finally, the vectors of the co-basis are as follows:

$$\hat{\mathbf{g}}^1 = \frac{\hat{\mathbf{g}}_2 \times \mathbf{t}_3}{(\hat{\mathbf{g}}_2 \times \mathbf{t}_3) \cdot \hat{\mathbf{g}}_1}, \qquad \hat{\mathbf{g}}^2 = \frac{\mathbf{t}_3 \times \hat{\mathbf{g}}_1}{(\mathbf{t}_3 \times \hat{\mathbf{g}}_1) \cdot \hat{\mathbf{g}}_2}, \tag{5.8}$$

and they belong to the plane spanned by $\hat{\mathbf{g}}_\alpha$.

From $\hat{\mathbf{g}}_\beta(\zeta) \doteq \partial \mathbf{y} / \partial \vartheta^\beta$ of eq. (5.5) and $\hat{\mathbf{g}}^\alpha \cdot \hat{\mathbf{g}}_\beta = \delta^\alpha_\beta$, we can deduce the following definition of a vector of the co-basis:

$$\hat{\mathbf{g}}^\alpha(\zeta) \doteq \frac{\partial \vartheta^\alpha}{\partial \mathbf{y}(\zeta)}. \tag{5.9}$$

Shifter (translation) tensor Z. The tangent vectors of eq. (5.5) can be alternatively expressed as

$$\hat{\mathbf{g}}_\alpha(\zeta) = \mathbf{g}_\alpha + \zeta \mathbf{t}_{3,\alpha} = (\mathbf{G}_0 - \zeta \mathbf{B})\, \mathbf{g}_\alpha = \mathbf{Z}(\zeta)\; \mathbf{g}_\alpha, \tag{5.10}$$

where $\mathbf{G}_0 \doteq \mathbf{g}_\alpha \otimes \mathbf{g}^\alpha$ is the metric tensor and $\mathbf{B} \doteq -\mathbf{t}_{3,\alpha} \otimes \mathbf{g}^\alpha$ is the curvature tensor, both for the reference surface and symmetric. Hence, the shifter tensor, $\mathbf{Z}(\zeta) \doteq \mathbf{G}_0 - \zeta \mathbf{B}$, maps the vectors $\mathbf{g}_\alpha$ at the reference surface onto the vectors $\hat{\mathbf{g}}_\alpha$ at an arbitrary lamina ζ, accounting for the curvature of the reference surface. For a flat geometry, i.e. when the curvature $\mathbf{B} = \mathbf{0}$, we have $\mathbf{Z}(\zeta) = \mathbf{G}_0$ i.e. the dependence on ζ vanishes.

The shifter tensor for the co-basis vectors $\hat{\mathbf{g}}^\alpha$ can be found by making use of the condition $\hat{\mathbf{g}}_\alpha(\zeta) \cdot \hat{\mathbf{g}}^\alpha(\zeta) = 1$ (no summation over α). Using the shifter tensor $\mathbf{Z}$ for $\hat{\mathbf{g}}_\alpha$ and an auxiliary (unknown) tensor $\mathbf{A}$ for $\hat{\mathbf{g}}^\alpha$, we have to satisfy the condition $(\mathbf{Z}\mathbf{g}_\alpha) \cdot (\mathbf{A}\mathbf{g}^\alpha) = 1$ or transforming further, $(\mathbf{A}^T\mathbf{Z}\,\mathbf{g}_\alpha) \cdot \mathbf{g}^\alpha = 1$. As $\mathbf{g}_\alpha \cdot \mathbf{g}^\alpha = 1$, hence $\mathbf{A}^T\mathbf{Z}\,\mathbf{g}_\alpha = \mathbf{g}_\alpha$ must hold. Therefore, a symmetric $\mathbf{A} \doteq \mathbf{Z}^{-1}$ is a shifter for the co-basis vectors, i.e.

$$\hat{\mathbf{g}}^\alpha(\zeta) = \mathbf{Z}^{-1}(\zeta)\, \mathbf{g}^\alpha. \tag{5.11}$$

The inverse $\mathbf{Z}^{-1}(\zeta)$ can be easily found in terms of components of $\mathbf{G}_0$ and $\mathbf{B}$,

$$(\mathbf{Z})_{ij} = \begin{bmatrix} G_{11} - \zeta B_{11} & G_{12} - \zeta B_{12} \\ \text{sym.} & G_{22} - \zeta B_{22} \end{bmatrix},$$

$$(\mathbf{Z})^{-1}_{ij} = \mu^{-1} \begin{bmatrix} G_{22} - \zeta B_{22} & -G_{12} + \zeta B_{12} \\ \text{sym.} & G_{11} - \zeta B_{11} \end{bmatrix}, \tag{5.12}$$

where $\mu \doteq \det \mathbf{Z} = \det \mathbf{G}_0 - \zeta(G_{11}B_{22} + G_{22}B_{11} - 2G_{12}B_{12}) + \zeta^2 \det \mathbf{B}$. The inverse of the shifter can be rewritten as

$$\begin{aligned}\mathbf{Z}^{-1}(\zeta) &= \mu^{-1}\left[(\det \mathbf{G}_0)\mathbf{G}_0^{-1} - \zeta(\det \mathbf{B})\,\mathbf{B}^{-1}\right] \\ &= \mu^{-1}\left[\text{tr}(\mathbf{G}_0 - \zeta\mathbf{B})\mathbf{I} - (\mathbf{G}_0 - \zeta\mathbf{B})\right],\end{aligned} \tag{5.13}$$

where the last form does not use the inverse of $\mathbf{G}_0$ and $\mathbf{B}$. It is obtained from the Cayley–Hamilton formula, which, e.g., for $\mathbf{B}$ is as follows:

$$\mathbf{B}^2 - I_1\,\mathbf{B} + I_2\,\mathbf{I} = \mathbf{0}, \tag{5.14}$$

where $I_1 = \text{tr}\mathbf{B} = 2H$ and $I_2 = \frac{1}{2}(\text{tr}\mathbf{B} - \text{tr}\mathbf{B}^2) = \det \mathbf{B} = \text{K}$. Besides, $H \doteq \frac{1}{2}\text{tr}\mathbf{B}$ is the mean curvature and $K \doteq \det \mathbf{B}$ is the Gaussian curvature. Multiplying eq. (5.14) by $\mathbf{B}^{-1}$, we obtain $I_2\,\mathbf{B}^{-1} = I_1\mathbf{I} - \mathbf{B}$, which provides the last form of eq. (5.13). In a similar way, we modify the term for $\mathbf{G}_0$.

For a flat geometry, i.e. when the curvature $\mathbf{B} = \mathbf{0}$, we obtain $\mu = \det \mathbf{G}_0$, and $\mathbf{Z}^{-1}(\zeta) = \mathbf{G}_0^{-1}$.

Deformation gradient and identity tensor. Assume the initial position vector of the shell as in eq. (5.1). For the current position vector $\mathbf{x} = \mathbf{x}(\vartheta^\alpha(\mathbf{y}), \zeta(\mathbf{y}))$, the deformation gradient can be written as

$$\mathbf{F} \doteq \frac{\partial \mathbf{x}}{\partial \mathbf{y}} = \frac{\partial \mathbf{x}}{\partial \vartheta^\alpha} \otimes \frac{\partial \vartheta^\alpha}{\partial \mathbf{y}} + \frac{\partial \mathbf{x}}{\partial \zeta} \otimes \frac{\partial \zeta}{\partial \mathbf{y}} = \mathbf{x}_{,\alpha} \otimes \hat{\mathbf{g}}^\alpha + \mathbf{x}_{,\zeta} \otimes \mathbf{t}^3, \tag{5.15}$$

where $\partial \vartheta^\alpha / \partial \mathbf{y}(\zeta) = \hat{\mathbf{g}}^\alpha$ by eq. (5.9) and $\partial \zeta / \partial \mathbf{y} = \mathbf{t}^3 = \mathbf{t}_3$ by eq. (5.1). Note that

$$\mathbf{F}\,\hat{\mathbf{g}}_\alpha = \mathbf{x}_{,\alpha}, \qquad \mathbf{F}\,\mathbf{t}_3 = \mathbf{x}_{,\zeta}. \tag{5.16}$$

The identity tensor is defined as the second-rank tensor obtained from the deformation gradient for the current position vector $\mathbf{x}$ assumed as equal to the initial position vector $\mathbf{y}$, i.e.

$$\mathbf{I} \doteq \mathbf{F}|_{\mathbf{x}=\mathbf{y}} = \hat{\mathbf{g}}_\alpha \otimes \hat{\mathbf{g}}^\alpha + \mathbf{t}_3 \otimes \mathbf{t}^3. \tag{5.17}$$

This definition guarantees that the approximations of $\mathbf{I}$ and $\mathbf{F}$ over ζ are consistent, and that the approximated $\mathbf{F} = \mathbf{I}$ for a rigid body motion. A similar reasoning can be applied to the rotation tensor $\mathbf{Q} \in SO(3)$, see the application in eqs. (6.13) and (6.15).

To express eq. (5.11) in the basis on the reference surface, we use $\hat{\mathbf{g}}^\alpha(\zeta) = \mathbf{Z}^{-1}(\zeta)\,\mathbf{g}^\alpha$ and then the simplicity of the above forms of $\mathbf{F}$ and $\mathbf{I}$ disappears.

Restriction on curvature of a shell. Let us estimate the contribution of the term related to the shell curvature to the norm of the tangent vector. Using eq. (5.5), we obtain

$$\|\,\hat{\mathbf{g}}_\alpha\| = \|\mathbf{g}_\alpha + \zeta \mathbf{t}_{3,\alpha}\| \leq \|\mathbf{g}_\alpha\| + \|\zeta \mathbf{t}_{3,\alpha}\|, \tag{5.18}$$

where $\|\mathbf{g}_\alpha\| = (\mathbf{g}_\alpha \cdot \mathbf{g}_\alpha)^{\frac{1}{2}}$, $\|\mathbf{t}_{3,\alpha}\| = (\mathbf{t}_{3,\alpha} \cdot \mathbf{t}_{3,\alpha})^{\frac{1}{2}}$. We may safely omit the second term, related to curvature, when

$$\frac{h}{2}\,\|\mathbf{t}_{3,\alpha}\| \ll \|\mathbf{g}_\alpha\|. \tag{5.19}$$

For a cylindrical surface, this restriction becomes

$$\frac{h}{2R} \ll 1, \tag{5.20}$$

see the example of Sect. 5.3 and eq. (5.36). If eq. (5.19) holds, then the ζ-dependent part of the shifter $\mathbf{Z}(\zeta)$ can be omitted, i.e. we use $\zeta \mathbf{B} \approx \mathbf{0}$, which implies

$$\mathbf{Z}(\zeta) \approx \mathbf{G}_0, \qquad \mathbf{Z}^{-1}(\zeta) \approx \mathbf{G}_0^{-1}, \qquad \mu \doteq \det \mathbf{Z} = \det \mathbf{G}_0. \tag{5.21}$$

Further simplifications are obtained for the orthonormal coordinates S^α, see the next paragraph.

Remark. The above restriction on the curvature of the reference surface is not used in the shell FEs derived in this work. It is used only in some analytical derivations, e.g. in Chap. 6.

Note, however, that there are FEs in use where this restriction is applied for efficiency. The curved shell structures can be analyzed by such elements provided that the discretization error is minimized by using a sufficiently large number of elements and by a suitable choice of their shapes and positions.

Simplifications for orthonormal coordinates S^α. The orthonormal coordinates S^α are often used as ϑ^α in analytical derivations, see e.g. Chap. 6. These coordinates are associated with the tangent orthonormal vectors $\mathbf{t}_\alpha$, which are used instead of $\mathbf{g}_\alpha$.

For the reference surface, $\zeta = 0$, we denote $\mathbf{g}_\alpha = \mathbf{t}_\alpha$, where $\mathbf{t}_\alpha$ are unit and orthogonal by the definition of coordinates S^α. Defining $\mathbf{t}^\alpha \doteq \hat{\mathbf{g}}^\alpha(\zeta)|_{\zeta=0}$, we obtain from eq. (5.8)

$$\mathbf{t}^1 = \frac{\mathbf{t}_2 \times \mathbf{t}_3}{(\mathbf{t}_2 \times \mathbf{t}_3) \cdot \mathbf{t}_1} = \mathbf{t}_1, \qquad \mathbf{t}^2 = \frac{\mathbf{t}_3 \times \mathbf{t}_1}{(\mathbf{t}_3 \times \mathbf{t}_1) \cdot \mathbf{t}_2} = \mathbf{t}_2, \tag{5.22}$$

i.e. the basis and the co-basis on the reference surface are identical. Hence, we do not distinguish between co-variant and contra-variant components of vectors and tensors, and derivations are simplified.

For the orthonormal coordinates, the metric tensor $\mathbf{G}_0 = \mathbf{I}$, and $\det \mathbf{G}_0 = 1$. The shifter tensor and its inverse of eq. (5.12) become simpler,

$$(\mathbf{Z})_{ij} = \begin{bmatrix} 1 - \zeta B_{11} & -\zeta B_{12} \\ \text{sym.} & 1 - \zeta B_{22} \end{bmatrix}, \quad (\mathbf{Z})^{-1}_{ij} = \mu^{-1} \begin{bmatrix} 1 - \zeta B_{22} & \zeta B_{12} \\ \text{sym.} & 1 - \zeta B_{11} \end{bmatrix} \tag{5.23}$$

or

$$\mathbf{Z}^{-1}(\zeta) = \mu^{-1} \left(\mathbf{I} - \zeta K \, \mathbf{B}^{-1} \right) = \mu^{-1} \left[\mathbf{I} - \zeta (2H\mathbf{I} - \mathbf{B}) \right], \tag{5.24}$$

where $\mu \doteq \det \mathbf{Z} = 1 - \zeta(2H) + \zeta^2 K$. For the restriction on curvature of eq. (5.19), we obtain $\mu \approx \det \mathbf{G}_0 = 1$, $\mu^{-1} = 1$, and $\mathbf{Z}^{-1}(\zeta) \approx \mathbf{G}_0^{-1} = \mathbf{G}_0 = \mathbf{I}$. As a consequence, some expressions are significantly simplified.

Remark. Geometry of the four-node finite element is approximated by the bilinear shape functions, so it is either flat (planar) or a hyperbolic paraboloid (h-p) surface. For a planar element, $H = 0$, and $K = 0$, i.e. it consists of only parabolic points. For the h-p element, H is a complicated function and $K < 0$, i.e. it consists of hyperbolic points only.

5.3 Example: Geometrical description of cylinder

Consider a cylindrical shell shown in Fig. 5.3. Its middle surface can be parameterized in a standard manner by cylindrical coordinates: the radius R, the angle θ (measured in the $\{\mathbf{i}_1, \mathbf{i}_3\}$-plane, and starting from $\mathbf{i}_1$) and the generator coordinate, t. The reference Cartesian basis is denoted by $\{\mathbf{i}_k\}$.

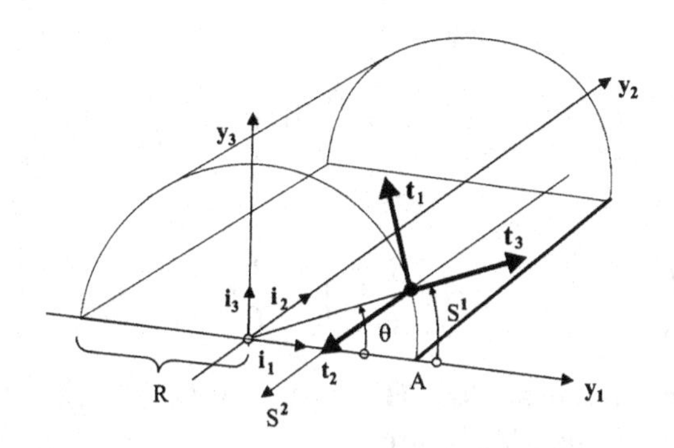

Fig. 5.3 Local basis $\{\mathbf{t}_k\}$ for a cylinder.

A position vector of an arbitrary point on the surface is given by $\mathbf{y} = y_k \mathbf{i}_k$, where $y_1 = R\cos\theta$, $y_2 = -t$, $y_3 = R\sin\theta$. The length of a circumferential arc on the cylinder is

$$S^1 = \int_0^\theta \sqrt{y_{1,\theta}^2 + y_{3,\theta}^2}\, \mathrm{d}\theta = \theta R. \tag{5.25}$$

Next, we introduce the arc-length surface coordinates: one along a circumference, $S^1 = \theta R$, and the other along a generator, $S^2 = t$. Then, the components of the position vector are

$$y_1 = R\cos\frac{S^1}{R}, \qquad y_2 = -S^2, \qquad y_3 = R\sin\frac{S^1}{R}, \tag{5.26}$$

and their non-zero derivatives are $\partial y_1/\partial S^1 = -\sin(S^1/R)$, $\partial y_2/\partial S^2 = -1$, $\partial y_3/\partial S^1 = \cos(S^1/R)$. Hence, the tangent vectors of the local basis associated with the arc-length coordinates are

$$\mathbf{t}_1 = \frac{\partial \mathbf{y}}{\partial S^1} = -\sin\frac{S^1}{R}\mathbf{i}_1 + \cos\frac{S^1}{R}\mathbf{i}_3, \qquad \mathbf{t}_2 = \frac{\partial \mathbf{y}}{\partial S^2} = -\mathbf{i}_2, \tag{5.27}$$

i.e. $\mathbf{t}_1$ and $\mathbf{t}_2$ are unit and orthogonal. Components of the metric tensor, $\mathbf{G}_0 \doteq \mathbf{t}_\alpha \otimes \mathbf{t}_\alpha$, are

$$G_{11} = \mathbf{t}_1 \cdot \mathbf{t}_1 = 1, \qquad G_{12} = \mathbf{t}_1 \cdot \mathbf{t}_2 = 0, \qquad G_{22} = \mathbf{t}_2 \cdot \mathbf{t}_2 = 1. \tag{5.28}$$

For the arc-length coordinates, a unit length of tangent vectors is a general property, see [230] p. 6, while their orthogonality is implied here by a specific choice of S^1 and S^2. The unit vector normal to the surface can be obtained as

$$\mathbf{t}_3 = \mathbf{t}_1 \times \mathbf{t}_2 = \cos\frac{S^1}{R}\mathbf{i}_1 + \sin\frac{S^1}{R}\mathbf{i}_3, \tag{5.29}$$

and its derivatives are

$$\mathbf{t}_{3,1} = \frac{1}{R}\left(-\sin\frac{S^1}{R}\mathbf{i}_1 + \cos\frac{S^1}{R}\mathbf{i}_3\right) = \frac{1}{R}\mathbf{t}_1, \qquad \mathbf{t}_{3,2} = \mathbf{0}. \tag{5.30}$$

Hence, the curvature tensor is $\mathbf{B} \doteq -\mathbf{t}_{3,\alpha} \otimes \mathbf{t}_\alpha = -\frac{1}{R}\mathbf{t}_1 \otimes \mathbf{t}_1$, at its components in the basis $\{\mathbf{t}_\alpha\}$ are

$$B_{11} = -\mathbf{t}_{3,1} \cdot \mathbf{t}_1 = -\frac{1}{R}, \qquad B_{12} = -\mathbf{t}_{3,1} \cdot \mathbf{t}_2 = 0, \qquad B_{22} = -\mathbf{t}_{3,2} \cdot \mathbf{t}_2 = 0. \tag{5.31}$$

Then, the mean curvature $H \doteq \frac{1}{2}\mathrm{tr}\mathbf{B} = -1/(2\mathrm{R})$ and the Gaussian curvature $K \doteq \det \mathbf{B} = 0$.

Let us construct a shell-like body by equipping the cylindrical surface with the thickness h. Then the position vector is $\mathbf{y}(\zeta) = \mathbf{y}_0 + \zeta\,\mathbf{t}_3$, where $\zeta \in [-h/2, +h/2]$. For an arbitrary ζ, the basis vectors defined by eq. (5.5) are

$$\hat{\mathbf{t}}_1(\zeta) = \left(1 + \frac{\zeta}{R}\right)\mathbf{t}_1, \qquad \hat{\mathbf{t}}_2(\zeta) = \mathbf{t}_2, \tag{5.32}$$

where the mid-surface tangent vectors of eq. (5.27) and the derivatives of eq. (5.30) are used. We see that the basis vector $\hat{\mathbf{t}}_1(\zeta)$ has a direction of $\mathbf{t}_1$, but its length varies with ζ, see Fig. 5.4. This has a obvious consequence that, e.g. for a displacement vector $\mathbf{u}$ constant over ζ, the component $u_1(\zeta) \doteq \mathbf{u} \cdot \hat{\mathbf{t}}_1(\zeta)$ varies with ζ. This also implies a nontrivial form of the shifter tensor of eq. (5.10), which becomes

$$\mathbf{Z}(\zeta) = \left(1 + \frac{\zeta}{R}\right)\mathbf{t}_1 \otimes \mathbf{t}_1 + \mathbf{t}_2 \otimes \mathbf{t}_2. \tag{5.33}$$

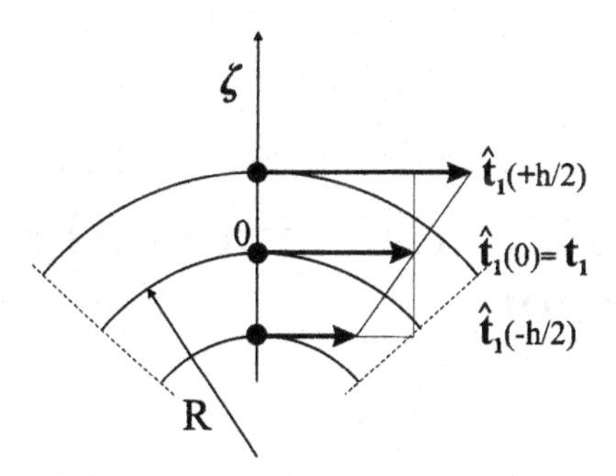

Fig. 5.4 Tangent vector $\hat{\mathbf{t}}_1$ at characteristic values of coordinate ζ.

We can easily check that $\hat{\mathbf{t}}_1(\zeta) = \mathbf{Z}(\zeta)\,\mathbf{t}_1$ and $\hat{\mathbf{t}}_2(\zeta) = \mathbf{Z}(\zeta)\,\mathbf{t}_2$, indeed. Besides, $\mu \doteq \det \mathbf{Z} = 1 + (\zeta/R)$.

Now, we can examine the basis vector $\hat{\mathbf{t}}_\alpha(\zeta) = \mathbf{t}_\alpha + \zeta \mathbf{t}_{3,\alpha}$, and estimate a contribution of the second term resulting from the shell curvature. Thus,

$$\|\,\hat{\mathbf{t}}_\alpha\| = \|\mathbf{t}_\alpha + \zeta \mathbf{t}_{3,\alpha}\| \leq \|\mathbf{t}_\alpha\| + \|\zeta \mathbf{t}_{3,\alpha}\|, \tag{5.34}$$

where $\|\mathbf{t}_1\| = (\mathbf{t}_1 \cdot \mathbf{t}_1)^{\frac{1}{2}} = 1$, $\|\mathbf{t}_2\| = (\mathbf{t}_2 \cdot \mathbf{t}_2)^{\frac{1}{2}} = 1$, and $\|\mathbf{t}_{3,1}\| = (\mathbf{t}_{3,1} \cdot \mathbf{t}_{3,1})^{\frac{1}{2}} = 1/R$, $\|\mathbf{t}_{3,2}\| = (\mathbf{t}_{3,2} \cdot \mathbf{t}_{3,2})^{\frac{1}{2}} = 0$. For $\zeta = \pm\frac{h}{2}$, we obtain

$$\|\hat{\mathbf{t}}_1\| \leq 1 + \frac{h}{2R}, \qquad \|\hat{\mathbf{t}}_2\| = 1. \tag{5.35}$$

The second term of $\|\hat{\mathbf{t}}_1\|$ is negligible when

$$\frac{h}{2R} \ll 1, \tag{5.36}$$

which illustrates the restriction of eq. (5.19).

The vector $\hat{\mathbf{t}}_2(\zeta)$ given by eq. (5.32) is a unit vector and, hence, we can easily obtain the co-basis, i.e.

$$\hat{\mathbf{t}}^1(\zeta) = \left(1 + \frac{\zeta}{R}\right)^{-1} \mathbf{t}_1, \qquad \hat{\mathbf{t}}^2(\zeta) = \mathbf{t}_2, \tag{5.37}$$

and check that $\hat{\mathbf{t}}^1 \cdot \hat{\mathbf{t}}_1 = 1$, $\hat{\mathbf{t}}^2 \cdot \hat{\mathbf{t}}_2 = 1$, $\hat{\mathbf{t}}^1 \cdot \hat{\mathbf{t}}_2 = 0$, and $\hat{\mathbf{t}}^2 \cdot \hat{\mathbf{t}}_1 = 0$, indeed. The inverse of the shifter is

$$\mathbf{Z}^{-1}(\zeta) = \left(1 + \frac{\zeta}{R}\right)^{-1} \mathbf{t}_1 \otimes \mathbf{t}_1 + \mathbf{t}_2 \otimes \mathbf{t}_2, \tag{5.38}$$

and, using eq. (5.33), we can check that indeed $\mathbf{Z}^{-1}(\zeta)\,\mathbf{Z}(\zeta) = \mathbf{I}$.

6

Shells with Reissner kinematics and drilling rotation

In this chapter, we derive shell equations with drilling rotation from the 3D mixed functionals with rotations of Sect. 4.4. The shell equations are based on the classical Reissner hypothesis and on the assumption regarding rotations. Various types of strain and stress tensors are used. To obtain equations of relative simplicity, we use the orthonormal coordinates, see Sect. 5.1 and the restriction on curvature of eq. (5.19).

6.1 Kinematics

Kinematical assumptions. The 3D mixed functionals with the rotations of Sect. 4.4 depend on the deformation vector χ and the rotation tensor $\mathbf{Q}$. For shells, we make assumptions regarding their dependence on the thickness coordinate, ζ, which are arbitrary, but must be properly balanced. The Reissner kinematics of first order is based on two assumptions:

A1. The rotations are constant over the shell thickness,

$$\mathbf{Q}(\zeta) \approx \mathbf{Q}_0, \tag{6.1}$$

where $\mathbf{Q}_0 \in \mathrm{SO}(3)$ is unrestricted in magnitude (finite).

A2. The current position vector is expressed by the classical Reissner hypothesis,

$$\mathbf{x}(\zeta) \approx \mathbf{x}_0 + \zeta\, \mathbf{a}_3, \tag{6.2}$$

where $\mathbf{a}_3 \doteq \mathbf{Q}_0 \mathbf{t}_3$ is the current director, defined as the forward-rotated initial director $\mathbf{t}_3$. The same $\mathbf{Q}_0 \in \mathrm{SO}(3)$ is used here as in **A1**.

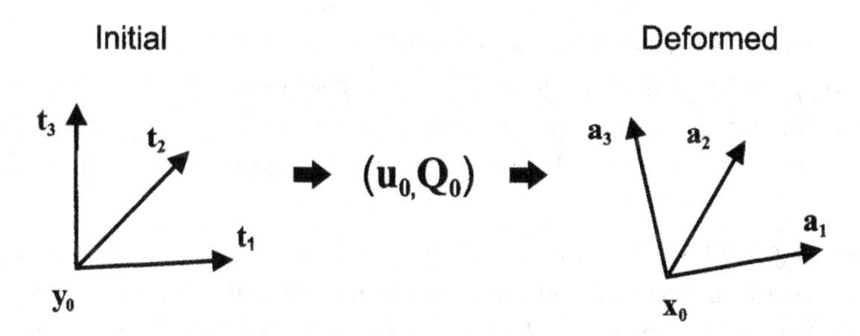

Fig. 6.1 The pairs (position, basis) for characteristic configurations.

Hence, the initial and current configurations of a shell are characterized by two pairs, $(\mathbf{y}_0, \{\mathbf{t}_k\})$ and $(\mathbf{x}_0, \{\mathbf{a}_k\})$, similarly as the Cosserat surface, see Fig. 6.1. The difference is, however, that here the rotations are linked with displacements by the RC, while in the Cosserat theory they are separate fields. Note that:

1. In the classical Reissner kinematics, only **A2** is applied to the standard 3D potential energy, which is a simplistic way of introducing rotations. Then, only two-parameter rotations can be used, excluding the drilling rotation.
2. When we apply **A1** and **A2** to the mixed 3D functional with rotations, then, the three-parameter rotations can be used and the drilling rotation is naturally retained in shell equations.

Forward-rotated basis and co-rotational basis. In the sequel, we use the forward-rotated vectors

$$\mathbf{a}_k \doteq \mathbf{Q}_0 \mathbf{t}_k, \qquad k = 1, 2, 3, \tag{6.3}$$

which form the forward-rotated basis $\{\mathbf{a}_k\}$. Because the vectors $\mathbf{t}_k$ are unit and orthogonal, hence the forward-rotated vectors $\mathbf{a}_k$ form the ortho-normal basis $\{\mathbf{a}_k\}$ associated with the deformed configuration. This basis is not tangent to the deformed mid-surface, unless the transverse shear strains are equal to zero. Note the difference between the forward-rotated vector and the convected vector which is obtained by using the deformation gradient

$$\underbrace{\mathbf{F}_0\, \mathbf{t}_k}_{\text{convected}} = \mathbf{V}_0 \underbrace{(\mathbf{Q}_0\, \mathbf{t}_k)}_{\text{forward-rotated}} \tag{6.4}$$

and the left polar decomposition formula of eq. (3.3).

The forward-rotated basis is also called the "co-rotational" basis. The basis $\{\mathbf{a}_k\}$ is local and associated with a point on the reference surface, so it is different from the co-rotational bases used, e.g., in [228, 60, 64], which are associated (bound) with the finite element and move with it.

Deformation gradient. The deformation function $\chi : \mathbf{x} = \chi(\mathbf{y})$ maps the reference configuration of the shell onto the current (deformed) one. For the orthonormal coordinates $\{S^\alpha, \zeta\}$, we have $\mathbf{x} = \chi(S^\alpha(\mathbf{y}), \zeta(\mathbf{y}))$, and the deformation gradient can be written as

$$\mathbf{F}(\zeta) \doteq \frac{\partial \mathbf{x}}{\partial \mathbf{y}} = \frac{\partial \mathbf{x}}{\partial S^\alpha} \otimes \frac{\partial S^\alpha}{\partial \mathbf{y}} + \frac{\partial \mathbf{x}}{\partial \zeta} \otimes \frac{\partial \zeta}{\partial \mathbf{y}}, \tag{6.5}$$

where $\partial S^\alpha / \partial \mathbf{y} = \hat{\mathbf{t}}^\alpha$ and $\partial \zeta / \partial \mathbf{y} = \mathbf{t}^3 = \mathbf{t}_3$, see Chap. 5 for basic geometrical definitions for a shell. For the local restriction on shell geometry, specified by eq. (5.19), we have $\mathbf{t}_{3,\alpha} \approx \mathbf{0}$ and $\hat{\mathbf{t}}^\alpha \approx \mathbf{t}_\alpha$.

For $\mathbf{x}(\zeta)$ of eq. (6.2) and the Taylor expansion w.r.t. ζ at $\zeta = 0$, we obtain

$$\mathbf{F}(\zeta) = \mathbf{F}_0 + \zeta (\mathbf{F}_{,\zeta})_0, \tag{6.6}$$

where the particular parts are

$$\mathbf{F}_0 \doteq \mathbf{x}_{0,\alpha} \otimes \mathbf{t}_\alpha + \mathbf{a}_3 \otimes \mathbf{t}_3, \qquad (\mathbf{F}_{,\zeta})_0 \doteq \mathbf{a}_{3,\alpha} \otimes \mathbf{t}_\alpha = (\bar{\boldsymbol{\omega}}_\alpha \times \mathbf{a}_3) \otimes \mathbf{t}_\alpha. \tag{6.7}$$

The derivative of the forward-rotated director $\mathbf{a}_3$ is obtained in the following way:

$$\mathbf{a}_{3,\alpha} = (\mathbf{Q}_0 \mathbf{t}_3)_{,\alpha} = \mathbf{Q}_{0,\alpha} \mathbf{t}_3 + \mathbf{Q}_0 \mathbf{t}_{3,\alpha} = \bar{\boldsymbol{\Omega}}_\alpha \mathbf{a}_3 = \bar{\boldsymbol{\omega}}_\alpha \times \mathbf{a}_3, \tag{6.8}$$

where the skew-symmetric tensor $\bar{\boldsymbol{\Omega}}_\alpha$ and its axial vector $\bar{\boldsymbol{\omega}}_\alpha$ are as follows:

$$\begin{aligned} \bar{\boldsymbol{\Omega}}_\alpha &= \boldsymbol{\Omega}_\alpha + \mathbf{Q}_0 \boldsymbol{\Omega}^0_\alpha \mathbf{Q}_0^T, & \bar{\boldsymbol{\omega}}_\alpha &= \boldsymbol{\omega}_\alpha + \mathbf{Q}_0 \boldsymbol{\omega}^0_\alpha, \\ \boldsymbol{\Omega}_\alpha &= \mathbf{Q}_{0,\alpha} \mathbf{Q}_0^T, & \boldsymbol{\Omega}_\alpha \mathbf{a}_3 &= \boldsymbol{\omega}_\alpha \times \mathbf{a}_3, \\ \boldsymbol{\Omega}^0_\alpha &= \mathbf{Q}^0_{0,\alpha} (\mathbf{Q}^0_0)^T, & \boldsymbol{\Omega}^0_\alpha \mathbf{a}_3 &= \boldsymbol{\omega}^0_\alpha \times \mathbf{a}_3. \end{aligned} \tag{6.9}$$

Here $\mathbf{Q}^0_0$ is a rotation tensor describing the position of $\mathbf{t}_3$, i.e. $\mathbf{t}_3 = \mathbf{Q}^0_0 \mathbf{i}_3$. Note that $\mathbf{t}_{3,\alpha} = \Gamma^k_{3\alpha} \mathbf{t}_k$, where $\Gamma^k_{3\alpha}$ is the second Christoffel symbol. For the locally restricted shell geometry, we have $\mathbf{Q}^0_{0,\alpha} \approx \mathbf{0}$ and $\bar{\boldsymbol{\Omega}}_\alpha \approx \boldsymbol{\Omega}_\alpha$, hence $\mathbf{a}_{3,\alpha} = (\mathbf{Q}_0 \mathbf{t}_3)_{,\alpha} \approx \mathbf{Q}_{0,\alpha} \mathbf{t}_3 = \boldsymbol{\Omega}_\alpha \mathbf{a}_3 = \boldsymbol{\omega}_\alpha \times \mathbf{a}_3$.

Product $\mathbf{Q}^T\mathbf{F}$. For the approximation of $\mathbf{F}(\zeta)$ by eq. (6.6) and $\mathbf{Q}(\zeta)$ by eq. (6.1), the product $\mathbf{Q}^T\mathbf{F}$ is

$$\mathbf{Q}^T(\zeta)\,\mathbf{F}(\zeta) = \mathbf{Q}_0^T\mathbf{F}_0 + \zeta\,\mathbf{Q}_0^T(\mathbf{F}_{,\zeta})_0, \tag{6.10}$$

where, using eq. (6.7),

$$\mathbf{Q}_0^T\mathbf{F}_0 = (\mathbf{Q}_0^T\mathbf{x}_{0,\alpha}) \otimes \mathbf{t}_\alpha + \mathbf{t}_3 \otimes \mathbf{t}_3, \qquad \mathbf{Q}_0^T(\mathbf{F}_{,\zeta})_0 = [(\mathbf{Q}_0^T\bar{\boldsymbol{\omega}}_\alpha) \times \mathbf{t}_3] \otimes \mathbf{t}_\alpha. \tag{6.11}$$

For the restricted shell geometry, we obtain a simpler form

$$\mathbf{Q}_0^T(\mathbf{F}_{,\zeta})_0 \approx [(\mathbf{Q}_0^T\boldsymbol{\omega}_\alpha) \times \mathbf{t}_3] \otimes \mathbf{t}_\alpha = \mathbf{Q}_0^T\mathbf{Q}_{0,\alpha}\mathbf{t}_3 \otimes \mathbf{t}_\alpha. \tag{6.12}$$

Identity tensor. By the definition of eq. (5.17), the identity tensor is consistent with the deformation gradient so we can use the above-derived forms of $\mathbf{F}(\zeta)$, and write $\mathbf{I}(\zeta) = \mathbf{I}_0 + \zeta(\mathbf{I}_{,\zeta})_0$, where

$$\begin{aligned} \mathbf{I}_0 &\doteq \mathbf{F}_0|_{\mathbf{u}=\mathbf{0}} = \mathbf{y}_{0,\alpha} \otimes \mathbf{t}_\alpha + \mathbf{t}_3 \otimes \mathbf{t}_3 = \mathbf{t}_k \otimes \mathbf{t}_k, \\ (\mathbf{I}_{,\zeta})_0 &\doteq (\mathbf{F}_{,\zeta})_0|_{\mathbf{a}_3=\mathbf{t}_3} = \mathbf{t}_{3,\alpha} \otimes \mathbf{t}_\alpha, \end{aligned} \tag{6.13}$$

where we used eq. (6.7) and $\mathbf{t}_\alpha \doteq \mathbf{y}_{0,\alpha}$. For the locally restricted shell geometry, see eq. (5.19), we have $\mathbf{t}_{3,\alpha} \approx \mathbf{0}$, which implies $(\mathbf{I}_{,\zeta})_0 \approx \mathbf{0}$. Hence, the product

$$\mathbf{I}^T(\zeta)\,\mathbf{I}(\zeta) = \mathbf{I}_0^T\mathbf{I}_0 + \zeta\left[\mathbf{I}_0^T(\mathbf{I}_{,\zeta})_0 + (\mathbf{I}_{,\zeta})_0^T\mathbf{I}_0\right] + \zeta^2(\mathbf{I}_{,\zeta})_0^T(\mathbf{I}_{,\zeta})_0 \approx \mathbf{I}_0^T\mathbf{I}_0 = \mathbf{I}_0. \tag{6.14}$$

A similar reasoning as that for $\mathbf{F}$ can be also applied to the rotation tensor $\mathbf{Q}_0 \in \mathrm{SO}(3)$. Note that the rotation tensor can be represented as $\mathbf{Q}_0 = \mathbf{a}_k \otimes \mathbf{t}_k$, for which we obtain $\mathbf{a}_i \doteq \mathbf{Q}_0\mathbf{t}_i$, as required. We can define the identity tensor as consistent with $\mathbf{Q}_0$ by assuming that the forward-rotated basis $\{\mathbf{a}_i\}$ is equal to the initial basis $\{\mathbf{t}_i\}$, which yields

$$\mathbf{I}(\zeta) \doteq \mathbf{Q}_0|_{\mathbf{a}_i=\mathbf{t}_i} = \mathbf{t}_\alpha \otimes \mathbf{t}_\alpha + \mathbf{t}_3 \otimes \mathbf{t}_3 = \mathbf{I}_0. \tag{6.15}$$

Hence, the product of eq. (6.10) is reduced to

$$\mathbf{Q}^T(\zeta)\,\mathbf{F}(\zeta)|_{\mathbf{x}=\mathbf{y},\,\mathbf{a}_i=\mathbf{t}_i} = \mathbf{I}_0^T\left[\mathbf{I}_0 + \zeta(\mathbf{I}_{,\zeta})_0\right] \approx \mathbf{I}_0^T\mathbf{I}_0 = \mathbf{I}_0, \tag{6.16}$$

obtained for the locally restricted shell geometry, eq. (5.19), for which $(\mathbf{I}_{,\zeta})_0 \approx \mathbf{0}$.

Variation of deformation gradient. The variation of the deformation gradient $\mathbf{F}$ can be expressed as $\delta\mathbf{F}(\zeta) = \delta\mathbf{F}_0 + \zeta\delta(\mathbf{F}_{,\zeta})_0$, where

$$\delta\mathbf{F}_0 \doteq \delta\mathbf{x}_{0,\alpha} \otimes \mathbf{t}_\alpha + \delta\mathbf{a}_3 \otimes \mathbf{t}_3, \quad \delta(\mathbf{F}_{,\zeta})_0 \doteq \delta\mathbf{a}_{3,\alpha} \otimes \mathbf{t}_\alpha = \delta[\bar{\boldsymbol{\omega}}_\alpha \times \zeta\mathbf{a}_3] \otimes \mathbf{t}_\alpha. \tag{6.17}$$

The particular terms are as follows:

$$\delta[\bar{\boldsymbol{\omega}}_\alpha \times \zeta\mathbf{a}_3] = \delta\bar{\boldsymbol{\omega}}_\alpha \times \zeta\mathbf{a}_3 + \bar{\boldsymbol{\omega}}_\alpha \times \zeta\delta\mathbf{a}_3, \tag{6.18}$$

$$\delta\mathbf{a}_3 = \delta\mathbf{Q}_0\mathbf{t}_3 = (\delta\mathbf{Q}_0\mathbf{Q}_0^T)\mathbf{a}_3 = \delta\tilde{\boldsymbol{\theta}}\,\mathbf{a}_3 = \delta\boldsymbol{\theta} \times \mathbf{a}_3, \quad \delta\tilde{\boldsymbol{\theta}} \doteq \delta\mathbf{Q}_0\mathbf{Q}_0^T, \tag{6.19}$$

where $\delta\tilde{\boldsymbol{\theta}} = \delta\boldsymbol{\theta} \times \mathbf{I}$ is a (left) skew-symmetric tensor and $\delta\boldsymbol{\theta}$ is its axial vector. Besides,

$$\delta\bar{\boldsymbol{\omega}}_\alpha = \delta\boldsymbol{\theta}_{,\alpha} + \delta\boldsymbol{\theta} \times \bar{\boldsymbol{\omega}}_\alpha, \tag{6.20}$$

which is proven below. Summarizing, $\delta\mathbf{F}(\zeta)$ is expressed by a variation of displacements, $\delta\mathbf{x}_0 = \delta\mathbf{u}_0$, and by a variation of the axial vector $\delta\boldsymbol{\theta}$.

Proof. Let us calculate the vector product of both sides of eq. (6.20) with $\mathbf{a}_i$,

$$\delta\bar{\boldsymbol{\omega}}_\alpha \times \mathbf{a}_i = \delta\boldsymbol{\theta}_{,\alpha} \times \mathbf{a}_i + (\delta\boldsymbol{\theta} \times \bar{\boldsymbol{\omega}}_\alpha) \times \mathbf{a}_i. \tag{6.21}$$

Applying the Lagrange identity for the triple cross-product of vectors, e.g. [33] p. 66, eq. (4.9.10), we can modify the last term

$$(\delta\boldsymbol{\theta} \times \bar{\boldsymbol{\omega}}_\alpha) \times \mathbf{a}_i = -(\bar{\boldsymbol{\omega}}_\alpha \times \mathbf{a}_i) \times \delta\boldsymbol{\theta} - (\mathbf{a}_i \times \delta\boldsymbol{\theta}) \times \bar{\boldsymbol{\omega}}_\alpha \tag{6.22}$$

and obtain

$$\delta\bar{\boldsymbol{\omega}}_\alpha \times \mathbf{a}_i = \delta\boldsymbol{\theta}_{,\alpha} \times \mathbf{a}_i - (\bar{\boldsymbol{\omega}}_\alpha \times \mathbf{a}_i) \times \delta\boldsymbol{\theta} - (\mathbf{a}_i \times \delta\boldsymbol{\theta}) \times \bar{\boldsymbol{\omega}}_\alpha. \tag{6.23}$$

By using $\mathbf{a}_{i,\alpha} = \bar{\boldsymbol{\omega}}_\alpha \times \mathbf{a}_i$ and $\delta\mathbf{a}_i = \delta\boldsymbol{\theta} \times \mathbf{a}_i$, we have

$$\delta\bar{\boldsymbol{\omega}}_\alpha \times \mathbf{a}_i - \delta\mathbf{a}_i \times \bar{\boldsymbol{\omega}}_\alpha = \delta\boldsymbol{\theta}_{,\alpha} \times \mathbf{a}_i - \mathbf{a}_{i,\alpha} \times \delta\boldsymbol{\theta}, \tag{6.24}$$

which can be simplified to

$$\delta(\bar{\boldsymbol{\omega}}_\alpha \times \mathbf{a}_i) = (\delta\boldsymbol{\theta} \times \mathbf{a}_i)_{,\alpha}. \tag{6.25}$$

Finally, we obtain

$$\delta(\mathbf{a}_{i,\alpha}) = (\delta\mathbf{a}_i)_{,\alpha}, \tag{6.26}$$

i.e. the condition that the operations of taking a variation and differentiation commute, which is generally true and ends the proof. □

6.2 Rotation Constraint for shells

In this section, we derive shell counterparts of the RC equation, $\mathfrak{C} \doteq \text{skew}(\mathbf{Q}^T\mathbf{F}) = \mathbf{0}$. We expand it into the Taylor series w.r.t. ζ at $\zeta = 0$ and, using eq. (6.11), we obtain

$$\mathfrak{C}(\zeta) = \mathfrak{C}_0 + \zeta(\mathfrak{C}_{,\zeta})_0 = \mathbf{0}. \tag{6.27}$$

This equation implies two RC equations:

$$\mathfrak{C}_0 \doteq \text{skew}(\mathbf{Q}_0^T\mathbf{F}_0) = \text{skew}\left[(\mathbf{Q}_0^T\mathbf{x}_{0,\alpha}) \otimes \mathbf{t}_\alpha\right] = \mathbf{0}, \tag{6.28}$$

$$(\mathfrak{C}_{,\zeta})_0 \doteq \text{skew}[\mathbf{Q}_0^T(\mathbf{F}_{,\zeta})_0] = \text{skew}\left[(\mathbf{Q}_0^T\mathbf{a}_{3,\alpha}) \otimes \mathbf{t}_\alpha\right] = \mathbf{0}, \tag{6.29}$$

which are analyzed separately below.

A. The zeroth order RC equation, $\mathfrak{C}_0 = \mathbf{0}$ of eq. (6.28), provides three scalar equations,

$$\mathbf{x}_{0,1} \cdot \mathbf{a}_2 - \mathbf{x}_{0,2} \cdot \mathbf{a}_1 = 0, \qquad \mathbf{x}_{0,\alpha} \cdot \mathbf{a}_3 = 0, \tag{6.30}$$

where the first equation is the RC for the drilling rotation.

To show this, we decompose the rotation tensor as follows: $\mathbf{Q}_0 \doteq \mathbf{Q}_d\,\mathbf{Q}_m$, where $\mathbf{Q}_m$ is a rotation around an axis tangent to the middle surface, while $\mathbf{Q}_d$ denotes a rotation around the normal vector $\mathbf{t}_3$. In the auxiliary basis $\{\mathbf{a}_i^m\}$, where $\mathbf{a}_i^m = \mathbf{Q}_m\mathbf{t}_i$, we have

$$\begin{aligned}\mathbf{Q}_d(\omega_d) = {} & \cos\omega_d(\mathbf{a}_1^m \otimes \mathbf{a}_1^m + \mathbf{a}_2^m \otimes \mathbf{a}_2^m) \\ & + \sin\omega_d(\mathbf{a}_2^m \otimes \mathbf{a}_1^m - \mathbf{a}_1^m \otimes \mathbf{a}_2^m) + \mathbf{a}_3^m \otimes \mathbf{a}_3^m.\end{aligned} \tag{6.31}$$

Using $\mathbf{a}_i = \mathbf{Q}_d\mathbf{Q}_m\mathbf{t}_i = \mathbf{Q}_d\,\mathbf{a}_i^m$, we have

$$\begin{aligned}\mathbf{a}_1 = \mathbf{Q}_d\mathbf{a}_1^m &= \cos\omega_d\mathbf{a}_1^m + \sin\omega_d\mathbf{a}_2^m, \\ \mathbf{a}_2 = \mathbf{Q}_d\mathbf{a}_2^m &= -\sin\omega_d\mathbf{a}_1^m + \cos\omega_d\mathbf{a}_2^m.\end{aligned} \tag{6.32}$$

Then, the first of eq. (6.30) becomes

$$\mathbf{x}_{0,1} \cdot (\cos\omega_d\,\mathbf{a}_2^m - \sin\omega_d\,\mathbf{a}_1^m) - \mathbf{x}_{0,2} \cdot (\cos\omega_d\,\mathbf{a}_1^m + \sin\omega_d\,\mathbf{a}_2^m) = 0 \tag{6.33}$$

and we can extract the drilling rotation

$$\tan\omega_d = \frac{\mathbf{x}_{0,1} \cdot \mathbf{a}_2^m - \mathbf{x}_{0,2} \cdot \mathbf{a}_1^m}{\mathbf{x}_{0,1} \cdot \mathbf{a}_1^m + \mathbf{x}_{0,2} \cdot \mathbf{a}_2^m}. \tag{6.34}$$

Note that the second equation of eq. (6.30) is related to transverse shear strains and, enforcing it, we obtain zero transverse shear strain, see eq. (6.74). Alternatively, we can neglect it and retain the transverse shear strains.

B. The first order RC equation, $(\mathfrak{C}_{,\zeta})_0 = \mathbf{0}$ of eq. (6.29), provides three scalar equations,

$$\mathbf{a}_{3,1} \cdot \mathbf{a}_2 - \mathbf{a}_{3,2} \cdot \mathbf{a}_1 = 0, \qquad \mathbf{a}_{3,\alpha} \cdot \mathbf{a}_3 = 0 \tag{6.35}$$

and, from the first one, we can extract the drilling rotation as follows:

$$\tan \omega_d = \frac{\mathbf{a}_{3,1} \cdot \mathbf{a}_2^m - \mathbf{a}_{3,2} \cdot \mathbf{a}_1^m}{\mathbf{a}_{3,1} \cdot \mathbf{a}_1^m + \mathbf{a}_{3,2} \cdot \mathbf{a}_2^m}. \tag{6.36}$$

For $\mathbf{a}_{3,\alpha} \approx \mathbf{Q}_{0,\alpha}\mathbf{t}_3$, the second of eq. (6.35) becomes

$$\mathbf{a}_{3,\alpha} \cdot \mathbf{a}_3 = \left(\mathbf{Q}_{0,\alpha}\mathbf{t}_3\right) \cdot \mathbf{a}_3 = \left(\mathbf{Q}_0^T \mathbf{Q}_{0,\alpha}\mathbf{t}_3\right) \cdot \mathbf{t}_3 = 0 \tag{6.37}$$

and both of the last two forms will be used in the sequel.

Remark. We note that eqs. (6.34) and (6.36) provide two formulas for the drilling rotation which, in general, can be contradictory. From the viewpoint of the kinematics described in the current section, it is difficult to judge which of these equations should be used. This issue was addressed in [252]. We shall use the first of eq. (6.30), while the first of eq. (6.35) is neglected.

6.3 Shell strains

In this section we derive shell counterparts of several types of the 3D strains. We assume that the local geometry of a shell is restricted by eq. (5.19).

6.3.1 Non-symmetric relaxed right stretch strain

The *non-symmetric relaxed* right stretch strain is defined by eq. (4.25), and for the defined shell kinematics, $\mathbf{H}_n(\zeta) \doteq \mathbf{Q}_0^T \mathbf{F}(\zeta) - \mathbf{I}_0^T \mathbf{I}(\zeta)$, where $\mathbf{Q}_0^T \mathbf{F}(\zeta)$ is defined by eq. (6.10) and $\mathbf{I}_0^T \mathbf{I}(\zeta)$ by eq. (6.16). Then we obtain $\mathbf{H}_n(\zeta) = \boldsymbol{\varepsilon} + \zeta\, \boldsymbol{\kappa}$, where

$$\boldsymbol{\varepsilon} = \boldsymbol{\varepsilon}_\alpha \otimes \mathbf{t}_\alpha, \qquad \boldsymbol{\kappa} = \boldsymbol{\kappa}_\alpha \otimes \mathbf{t}_\alpha. \tag{6.38}$$

The shell strain vectors are defined as

$$\boldsymbol{\varepsilon}_\alpha \doteq \mathbf{Q}_0^T \mathbf{x}_{0,\alpha} - \mathbf{t}_\alpha, \qquad \boldsymbol{\kappa}_\alpha \doteq (\mathbf{Q}_0^T \bar{\boldsymbol{\omega}}_\alpha) \times \mathbf{t}_3 \approx \mathbf{Q}_0^T \mathbf{a}_{3,\alpha}, \tag{6.39}$$

where the last form of $\boldsymbol{\kappa}_\alpha$ is for the restricted shell geometry, when $\mathbf{t}_{3,\alpha} \approx \mathbf{0}$.

Components of the shell strain tensors in the basis $\{\mathbf{t}_i\}$ are as follows:

- *0th order strains,*

$$\varepsilon_{11} = \mathbf{x}_{0,1} \cdot \mathbf{a}_1 - 1, \qquad \varepsilon_{22} = \mathbf{x}_{0,2} \cdot \mathbf{a}_2 - 1, \qquad \varepsilon_{33} = 0, \tag{6.40}$$

$$\varepsilon_{12} = \mathbf{x}_{0,1} \cdot \mathbf{a}_2, \qquad \varepsilon_{21} = \mathbf{x}_{0,2} \cdot \mathbf{a}_1, \qquad \varepsilon_{\alpha 3} = \mathbf{x}_{0,\alpha} \cdot \mathbf{a}_3, \qquad \varepsilon_{3\alpha} = 0,$$

- *1st order strains,*

$$\kappa_{11} = \mathbf{a}_{3,1} \cdot \mathbf{a}_1, \qquad \kappa_{22} = \mathbf{a}_{3,2} \cdot \mathbf{a}_2, \qquad \kappa_{33} = 0, \tag{6.41}$$

$$\kappa_{12} = \mathbf{a}_{3,1} \cdot \mathbf{a}_2, \qquad \kappa_{21} = \mathbf{a}_{3,2} \cdot \mathbf{a}_1, \qquad \kappa_{\alpha 3} = \tfrac{1}{2}\mathbf{a}_{3,\alpha} \cdot \mathbf{a}_3, \qquad \kappa_{3\alpha} = 0,$$

where $\varepsilon_{ij} = \boldsymbol{\varepsilon} \cdot (\mathbf{t}_i \otimes \mathbf{t}_j)$ and $\kappa_{ij} = \boldsymbol{\kappa} \cdot (\mathbf{t}_i \otimes \mathbf{t}_j)$. An alternative form of components of $\boldsymbol{\kappa}$ is obtained for $\mathbf{a}_{3,\alpha} \approx \mathbf{Q}_{0,\alpha}\mathbf{t}_3$, i.e. for the restriction regarding the local geometry.

The in-plane shear components, ε_{12} and ε_{21}, are non-symmetric but we can enforce the drill RC in order to symmetrize them. Besides, we can use the symmetrized transverse shear strains

$$\begin{aligned} \varepsilon^s_{\alpha 3} &= \varepsilon^s_{3\alpha} \doteq \tfrac{1}{2}(\varepsilon_{\alpha 3} + \varepsilon_{3\alpha}) = \tfrac{1}{2}\mathbf{x}_{0,\alpha} \cdot \mathbf{a}_3, \\ \kappa^s_{\alpha 3} &= \kappa^s_{3\alpha} \doteq \tfrac{1}{2}(\kappa_{\alpha 3} + \kappa_{3\alpha}) = \tfrac{1}{2}\mathbf{a}_{3,\alpha} \cdot \mathbf{a}_3. \end{aligned} \tag{6.42}$$

Then the transverse shear $(\alpha 3)$ components of the RC are abandoned; see the consequences of enforcing them specified by eq. (6.74) and the discussion therein.

Objectivity of strains. Superpose a rigid body motion on the current position vector $\mathbf{x}$ as follows: $\mathbf{x}^+ = \mathbf{c}_r + \mathbf{Q}_r\mathbf{x}$, where $\mathbf{c}_r$ denotes a rigid translation, and $\mathbf{Q}_r \in \mathrm{SO}(3)$ is a rigid rotation. For the Reissner hypothesis of eq. (6.2), we obtain

$$\mathbf{x}^+(\zeta) = \mathbf{c}_r + \mathbf{Q}_r(\mathbf{x}_0 + \zeta\mathbf{a}_3) = \mathbf{x}_0^+ + \zeta\mathbf{a}_3^+, \tag{6.43}$$

where $\mathbf{x}_0^+ \doteq \mathbf{c}_r + \mathbf{Q}_r\mathbf{x}_0$ and $\mathbf{a}_3^+ \doteq \mathbf{Q}_r\mathbf{a}_3$. Note that $\mathbf{a}_3^+ \doteq \mathbf{Q}_r\mathbf{a}_3 = \mathbf{Q}_r\mathbf{Q}_0\mathbf{t}_3 = \mathbf{Q}_0^+\mathbf{t}_3$, where $\mathbf{Q}_0^+ \doteq \mathbf{Q}_r\mathbf{Q}_0$. We see that the rigid rotation affects $\mathbf{x}_0$ and $\mathbf{Q}_0$, and that the $\mathbf{x}^+(\zeta)$ obeys the Reissner hypothesis.

Below, we check the strain vectors, eq. (6.39), for a superposed rigid motion, i.e. for the position vector and rotations indicated by "+". First, we evaluate the components

$$\mathbf{x}_{0,\alpha}^+ = \mathbf{Q}_r\mathbf{x}_{0,\alpha}, \qquad \mathbf{Q}_0^{+T}\mathbf{x}_{0,\alpha}^+ = \mathbf{Q}_0^{+T}\mathbf{Q}_r\mathbf{x}_{0,\alpha} = \mathbf{Q}_0^T\mathbf{x}_{0,\alpha}, \tag{6.44}$$

$$\mathbf{a}^+_{3,\alpha} = (\mathbf{Q}_r \mathbf{a}_3)_{,\alpha} = \mathbf{Q}_r \mathbf{a}_{3,\alpha}, \qquad \mathbf{Q}_0^{+T} \mathbf{a}^+_{3,\alpha} = \mathbf{Q}_0^{+T} \mathbf{Q}_r \mathbf{a}_{3,\alpha} = \mathbf{Q}_0^T \mathbf{a}_{3,\alpha}, \tag{6.45}$$

as $\mathbf{c}_{r,\alpha} = \mathbf{0}$ and $\mathbf{Q}_{r,\alpha} = \mathbf{0}$. Then, the strain vectors of eq. (6.39) are

$$\begin{aligned} \boldsymbol{\varepsilon}^+_\alpha &\doteq \mathbf{Q}_0^{+T} \mathbf{x}^+_{0,\alpha} - \mathbf{t}_\alpha = \mathbf{Q}_0^T \mathbf{x}_{0,\alpha} - \mathbf{t}_\alpha = \boldsymbol{\varepsilon}_\alpha \\ \boldsymbol{\kappa}^+_\alpha &\doteq \mathbf{Q}_0^{+T} \mathbf{a}^+_{3,\alpha} = \mathbf{Q}_0^T \mathbf{a}_{3,\alpha} = \boldsymbol{\kappa}_\alpha, \end{aligned} \tag{6.46}$$

i.e. they are invariant w.r.t. the superposed rigid motion. Thus, also the strain tensors of eq. (6.38) are invariant.

6.3.2 Symmetric relaxed right stretch strain

The *symmetric relaxed* right stretch strain is defined by eq. (4.29) and for the defined shell kinematics, $\mathbf{H}(\zeta) \doteq \mathrm{sym}[\mathbf{Q}_0^T \mathbf{F}(\zeta)] - \mathrm{sym}[\mathbf{I}_0^T \mathbf{I}(\zeta)]$, where $\mathbf{Q}_0^T \mathbf{F}(\zeta)$ is defined by eq. (6.10) and $\mathbf{I}_0^T \mathbf{I}(\zeta)$ by eq. (6.16). Note that $\mathbf{H}$ is the symmetric part of the earlier-defined non-symmetric $\mathbf{H}_n(\zeta)$ and, hence, $\mathbf{H}(\zeta) = \boldsymbol{\varepsilon} + \zeta \boldsymbol{\kappa}$, where

$$\boldsymbol{\varepsilon} = \mathrm{sym}(\boldsymbol{\varepsilon}_\alpha \otimes \mathbf{t}_\alpha), \qquad \boldsymbol{\kappa} = \mathrm{sym}(\boldsymbol{\kappa}_\alpha \otimes \mathbf{t}_\alpha). \tag{6.47}$$

The strain vectors $\boldsymbol{\varepsilon}_\alpha$ and $\boldsymbol{\kappa}_\alpha$ are defined in eq. (6.39). Components of the shell strain tensors in the basis $\{\mathbf{t}_i\}$ are as follows:

- *0th order strains,*

$$\varepsilon_{11} = \mathbf{x}_{0,1} \cdot \mathbf{a}_1 - 1, \qquad \varepsilon_{22} = \mathbf{x}_{0,2} \cdot \mathbf{a}_2 - 1, \qquad \varepsilon_{33} = 0, \tag{6.48}$$

$$\varepsilon_{12} = \varepsilon_{21} = \tfrac{1}{2} \left(\mathbf{x}_{0,1} \cdot \mathbf{a}_2 + \mathbf{x}_{0,2} \cdot \mathbf{a}_1 \right), \qquad \varepsilon_{\alpha 3} = \varepsilon_{3\alpha} = \tfrac{1}{2} \mathbf{x}_{0,\alpha} \cdot \mathbf{a}_3,$$

- *1st order strains,*

$$\kappa_{11} = \mathbf{a}_{3,1} \cdot \mathbf{a}_1, \qquad \kappa_{22} = \mathbf{a}_{3,2} \cdot \mathbf{a}_2, \qquad \kappa_{33} = 0, \tag{6.49}$$

$$\kappa_{12} = \kappa_{21} = \tfrac{1}{2} \left(\mathbf{a}_{3,1} \cdot \mathbf{a}_2 + \mathbf{a}_{3,2} \cdot \mathbf{a}_1 \right), \qquad \kappa_{\alpha 3} = \kappa_{3\alpha} = \tfrac{1}{2} \mathbf{a}_{3,\alpha} \cdot \mathbf{a}_3,$$

where $\varepsilon_{ij} = \boldsymbol{\varepsilon} \cdot (\mathbf{t}_i \otimes \mathbf{t}_j)$ and $\kappa_{ij} = \boldsymbol{\kappa} \cdot (\mathbf{t}_i \otimes \mathbf{t}_j)$. Comparing with the non-symmetric strain components of eqs. (6.40)–(6.41) with the symmetrization of eq. (6.42), the difference is that the in-plane shear components are equal, i.e. $\varepsilon_{12} = \varepsilon_{21}$.

6.3.3 Green strain

The Green strain tensor defined in eq. (4.20) is $\mathbf{E} \doteq \frac{1}{2}(\mathbf{F}^T\mathbf{F} - \mathbf{I}^T\mathbf{I})$. We shall use the approximations

$$\mathbf{F}(\zeta) = \mathbf{F}_0 + \zeta\mathbf{F}_1, \qquad \mathbf{I}(\zeta) = \mathbf{I}_0 + \zeta\mathbf{I}_1, \tag{6.50}$$

where $\mathbf{F}_0$ and $\mathbf{F}_1 \doteq (\mathbf{F}_{,\zeta})_0$ are defined in eq. (6.7), while $\mathbf{I}_0$ and $\mathbf{I}_1 \doteq (\mathbf{I}_{,\zeta})_0$ in eq. (6.13). Note that $\mathbf{I}$ is defined consistently with $\mathbf{F}$, see eq. (5.17). For these approximations, the products are

$$\begin{aligned} \mathbf{F}^T(\zeta)\,\mathbf{F}(\zeta) &= \mathbf{F}_0^T\mathbf{F}_0 + \zeta\left(\mathbf{F}_0^T\mathbf{F}_1 + \mathbf{F}_1^T\mathbf{F}_0\right) + \zeta^2\mathbf{F}_1^T\mathbf{F}_1, \\ \mathbf{I}^T(\zeta)\,\mathbf{I}(\zeta) &= \mathbf{I}_0^T\mathbf{I}_0 + \zeta\left(\mathbf{I}_0^T\mathbf{I}_1 + \mathbf{I}_1^T\mathbf{I}_0\right) + \zeta^2\mathbf{I}_1^T\mathbf{I}_1, \end{aligned} \tag{6.51}$$

where the approximations of both products over ζ are analogous. Then the Green strain can be expressed as the second order polynomial of ζ,

$$\mathbf{E}(\zeta) = \boldsymbol{\varepsilon} + \zeta\boldsymbol{\kappa} + \zeta^2\boldsymbol{\mu}, \tag{6.52}$$

where the shell strains are defined as

$$\begin{aligned} \boldsymbol{\varepsilon} &\doteq \tfrac{1}{2}(\mathbf{F}_0^T\mathbf{F}_0 - \mathbf{I}_0^T\mathbf{I}_0), \qquad \boldsymbol{\kappa} \doteq \tfrac{1}{2}\left(\mathbf{F}_0^T\mathbf{F}_1 + \mathbf{F}_1^T\mathbf{F}_0 - \mathbf{I}_0^T\mathbf{I}_1 - \mathbf{I}_1^T\mathbf{I}_0\right), \\ \boldsymbol{\mu} &\doteq \tfrac{1}{2}\left(\mathbf{F}_1^T\mathbf{F}_1 - \mathbf{I}_1^T\mathbf{I}_1\right). \end{aligned} \tag{6.53}$$

Components of the shell strain tensors in the basis $\{\mathbf{t}_i\}$ are:

- *0th order strains,*

$$\begin{aligned} &\varepsilon_{11} = \tfrac{1}{2}(\mathbf{x}_{0,1}\cdot\mathbf{x}_{0,1} - \mathbf{t}_1\cdot\mathbf{t}_1), \quad \varepsilon_{22} = \tfrac{1}{2}(\mathbf{x}_{0,2}\cdot\mathbf{x}_{0,2} - \mathbf{t}_2\cdot\mathbf{t}_2), \quad \varepsilon_{33} = 0, \\ &\varepsilon_{12} = \varepsilon_{21} = \tfrac{1}{2}\mathbf{x}_{0,1}\cdot\mathbf{x}_{0,2}, \qquad \varepsilon_{\alpha 3} = \varepsilon_{3\alpha} = \tfrac{1}{2}(\mathbf{x}_{0,\alpha}\cdot\mathbf{a}_3 - \mathbf{t}_\alpha\cdot\mathbf{t}_3), \end{aligned} \tag{6.54}$$

- *1st order strains,*

$$\begin{aligned} &\kappa_{11} = \tfrac{1}{2}\left(\mathbf{x}_{0,1}\cdot\mathbf{a}_{3,1} - \mathbf{t}_1\cdot\mathbf{t}_{3,1}\right), \quad \kappa_{22} = \tfrac{1}{2}\left(\mathbf{x}_{0,2}\cdot\mathbf{a}_{3,2} - \mathbf{t}_2\cdot\mathbf{t}_{3,2}\right), \quad \kappa_{33} = 0, \\ &\kappa_{12} = \kappa_{21} = \tfrac{1}{2}\left(\mathbf{x}_{0,1}\cdot\mathbf{a}_{3,2} + \mathbf{x}_{0,2}\cdot\mathbf{a}_{3,1} - \mathbf{t}_1\cdot\mathbf{t}_{3,2} - \mathbf{t}_2\cdot\mathbf{t}_{3,1}\right), \\ &\kappa_{\alpha 3} = \kappa_{3\alpha} = \tfrac{1}{2}(\mathbf{a}_{3,\alpha}\cdot\mathbf{a}_3 - \mathbf{t}_{3,\alpha}\cdot\mathbf{t}_3), \end{aligned} \tag{6.55}$$

- *2nd order strains,*

$$\begin{aligned} &\mu_{11} = \tfrac{1}{2}(\mathbf{a}_{3,1}\cdot\mathbf{a}_{3,1} - \mathbf{t}_{3,1}\cdot\mathbf{t}_{3,1}), \quad \mu_{22} = \tfrac{1}{2}(\mathbf{a}_{3,2}\cdot\mathbf{a}_{3,2} - \mathbf{t}_{3,2}\cdot\mathbf{t}_{3,2}), \quad \mu_{33} = 0, \\ &\mu_{12} = \mu_{21} = \tfrac{1}{2}\left(\mathbf{a}_{3,1}\cdot\mathbf{a}_{3,2} + \mathbf{a}_{3,2}\cdot\mathbf{a}_{3,1} - \mathbf{t}_{3,1}\cdot\mathbf{t}_{3,2} - \mathbf{t}_{3,2}\cdot\mathbf{t}_{3,1}\right), \\ &\mu_{\alpha 3} = \mu_{3\alpha} = 0, \end{aligned} \tag{6.56}$$

where $T_{ij} = \mathbf{T} \cdot (\mathbf{t}_i \otimes \mathbf{t}_j)$ for the second rank tensor $\mathbf{T}$. Obviously, $\mathbf{t}_1 \cdot \mathbf{t}_1 = 1$, $\mathbf{t}_2 \cdot \mathbf{t}_2 = 1$, and $\mathbf{t}_\alpha \cdot \mathbf{t}_3 = 0$, but the above forms are better suited to obtaining shell strain components in the natural co-basis. We note that for the locally restricted shell geometry, eq. (5.19), we should use $\mathbf{t}_{3,\alpha} \approx \mathbf{0}$ in the first and second order strains, which then become much simpler.

The quadratic term $\zeta^2 \boldsymbol{\mu}$ in eq. (6.52) is usually omitted in FEs and then the Green strain is linear in ζ, similarly as the *relaxed* right stretch strains, $\mathbf{H}_n$ and $\mathbf{H}$.

Covariant components of Green strain. Components of the Green strain in the ortho-normal basis $\{\mathbf{t}_i\}$ of eqs. (6.54)–(6.56) depend on the derivatives of the current position vector $\mathbf{x}_0$ w.r.t. the local ortho-normal coordinates S^α.

These derivatives are not directly available and must be computed from derivatives w.r.t. the natural coordinates $\{\xi, \eta\}$. Hence, to calculate eq. (6.54)–(6.56), we first determine the covariant components of the Green strain in the co-basis $\{\mathbf{g}^k\}$. By the chain rule

$$\mathbf{x}_{0,1} = \mathbf{x}_{0,\xi} \frac{\partial \xi}{\partial S^1} + \mathbf{x}_{0,\eta} \frac{\partial \eta}{\partial S^1}, \qquad \mathbf{x}_{0,2} = \mathbf{x}_{0,\xi} \frac{\partial \xi}{\partial S^2} + \mathbf{x}_{0,\eta} \frac{\partial \eta}{\partial S^2}, \tag{6.57}$$

which can be rewritten as

$$\mathbf{x}_{0,1} = J_{11}^{-1} \mathbf{x}_{0,\xi} + J_{21}^{-1} \mathbf{x}_{0,\eta}, \qquad \mathbf{x}_{0,2} = J_{12}^{-1} \mathbf{x}_{0,\xi} + J_{22}^{-1} \mathbf{x}_{0,\eta}, \tag{6.58}$$

where $J_{\alpha\beta}^{-1}$ are components of the inverse of the Jacobian $\mathbf{J}$ of eq.(10.41). We can transform $\mathbf{t}_\alpha$ similarly because $\mathbf{t}_\alpha \doteq \mathbf{y}_{0,\alpha}$, i.e. it also is a derivative w.r.t. S^α, see eq. (10.44). Hence, we have

$$\mathbf{t}_1 = J_{11}^{-1} \mathbf{g}_1 + J_{21}^{-1} \mathbf{g}_2, \qquad \mathbf{t}_2 = J_{12}^{-1} \mathbf{g}_1 + J_{22}^{-1} \mathbf{g}_2, \tag{6.59}$$

where the vectors of the natural basis, $\mathbf{g}_1 \doteq \mathbf{y}_{0,\xi}$ and $\mathbf{g}_2 \doteq \mathbf{y}_{0,\eta}$. Using the above relations, we can express the components of eqs. (6.54)–(6.56) in terms of the covariant strain components.

Membrane strain components. On use of the above expressions for derivatives, the components of the membrane shell strain of eq. (6.54) become

$$2\varepsilon_{11} = \mathbf{x}_{0,1} \cdot \mathbf{x}_{0,1} - \mathbf{t}_1 \cdot \mathbf{t}_1 = (J_{11}^{-1})^2 \, (2\varepsilon_{\xi\xi}) + (J_{21}^{-1})^2 \, (2\varepsilon_{\eta\eta}) + 2 J_{11}^{-1} J_{21}^{-1} \, (2\varepsilon_{\xi\eta}),$$

$$2\varepsilon_{22} = \mathbf{x}_{0,2} \cdot \mathbf{x}_{0,2} - \mathbf{t}_2 \cdot \mathbf{t}_2 = (J_{12}^{-1})^2 \, (2\varepsilon_{\xi\xi}) + (J_{22}^{-1})^2 \, (2\varepsilon_{\eta\eta}) + 2 J_{12}^{-1} J_{22}^{-1} \, (2\varepsilon_{\xi\eta}),$$

$$2\varepsilon_{12} = \mathbf{x}_{0,1} \cdot \mathbf{x}_{0,2} = J_{11}^{-1} J_{12}^{-1} (2\varepsilon_{\xi\xi}) + J_{21}^{-1} J_{22}^{-1} (2\varepsilon_{\eta\eta}) + (J_{11}^{-1} J_{22}^{-1} + J_{21}^{-1} J_{12}^{-1})(2\varepsilon_{\xi\eta}),$$

and $\varepsilon_{21} = \varepsilon_{12}$, or, in the matrix form,

$$\begin{bmatrix} \varepsilon_{11} \\ \varepsilon_{22} \\ \varepsilon_{12} \end{bmatrix} = \mathbf{T}^* \begin{bmatrix} \varepsilon_{\xi\xi} \\ \varepsilon_{\eta\eta} \\ \varepsilon_{\xi\eta} \end{bmatrix}, \tag{6.60}$$

where the operator $\mathbf{T}^*$ is defined in eq. (2.29). The covariant components of the membrane strain are defined as

$$\begin{gathered} 2\varepsilon_{\xi\xi} \doteq \mathbf{x}_{0,\xi} \cdot \mathbf{x}_{0,\xi} - \mathbf{g}_1 \cdot \mathbf{g}_1, \qquad 2\varepsilon_{\eta\eta} \doteq \mathbf{x}_{0,\eta} \cdot \mathbf{x}_{0,\eta} - \mathbf{g}_2 \cdot \mathbf{g}_2, \\ 2\varepsilon_{\xi\eta} \doteq \mathbf{x}_{0,\xi} \cdot \mathbf{x}_{0,\eta} - \mathbf{g}_1 \cdot \mathbf{g}_2. \end{gathered} \tag{6.61}$$

Note that $\mathbf{g}_1$ and $\mathbf{g}_2$ are derivatives of $\mathbf{y}_0$ w.r.t. ξ and η. In the FE method, $\mathbf{x}_0$ and $\mathbf{y}_0$ are approximated by shape functions in terms of ξ and η, so the above components can be directly computed.

Bending and twisting strain components. The transformation formulas for the derivatives of the directors $\mathbf{t}_3$ and $\mathbf{a}_3$ are similar to these derived earlier for the derivatives of $\mathbf{x}_0$, i.e.

$$\mathbf{t}_{3,1} = J_{11}^{-1}\,\mathbf{t}_{3,\xi} + J_{21}^{-1}\,\mathbf{t}_{3,\eta}, \qquad \mathbf{t}_{3,2} = J_{12}^{-1}\,\mathbf{t}_{3,\xi} + J_{22}^{-1}\,\mathbf{t}_{3,\eta}, \tag{6.62}$$

and

$$\mathbf{a}_{3,1} = J_{11}^{-1}\,\mathbf{a}_{3,\xi} + J_{21}^{-1}\,\mathbf{a}_{3,\eta}, \qquad \mathbf{a}_{3,2} = J_{12}^{-1}\,\mathbf{a}_{3,\xi} + J_{22}^{-1}\,\mathbf{a}_{3,\eta}. \tag{6.63}$$

Hence, for the bending and twisting strain components of eq. (6.55), we can write in matrix form

$$\begin{bmatrix} \kappa_{11} \\ \kappa_{22} \\ \kappa_{12} \end{bmatrix} = \mathbf{T}^* \begin{bmatrix} \kappa_{\xi\xi} \\ \kappa_{\eta\eta} \\ \kappa_{\xi\eta} \end{bmatrix}, \tag{6.64}$$

where the covariant components of the bending-twisting strain are defined as follows:

$$\begin{gathered} \kappa_{\xi\xi} = \mathbf{x}_{0,\xi} \cdot \mathbf{a}_{3,\xi} - \mathbf{y}_{0,\xi} \cdot \mathbf{t}_{3,\xi}, \qquad \kappa_{\eta\eta} = \mathbf{x}_{0,\eta} \cdot \mathbf{a}_{3,\eta} - \mathbf{y}_{0,\eta} \cdot \mathbf{t}_{3,\eta}, \\ \kappa_{\xi\eta} = \kappa_{\eta\xi} = \tfrac{1}{2}\left(\mathbf{x}_{0,\xi} \cdot \mathbf{a}_{3,\eta} + \mathbf{x}_{0,\eta} \cdot \mathbf{a}_{3,\xi} - \mathbf{y}_{0,\xi} \cdot \mathbf{t}_{3,\eta} - \mathbf{y}_{0,\eta} \cdot \mathbf{t}_{3,\xi}\right). \end{gathered} \tag{6.65}$$

Transverse shear zeroth and first order strains. For the transverse shear strain components of eqs. (6.54) and (6.55) we obtain, in a similar manner,

$$\begin{bmatrix} \varepsilon_{13} \\ \varepsilon_{23} \end{bmatrix} = \mathbf{J}^{-T} \begin{bmatrix} \varepsilon_{\xi 3} \\ \varepsilon_{\eta 3} \end{bmatrix}, \qquad \begin{bmatrix} \kappa_{13} \\ \kappa_{23} \end{bmatrix} = \mathbf{J}^{-T} \begin{bmatrix} \kappa_{\xi 3} \\ \kappa_{\eta 3} \end{bmatrix}, \tag{6.66}$$

see also eq. (2.27). The covariant transverse shear components are defined as follows:

$$\varepsilon_{\xi 3} = \varepsilon_{3\xi} \doteq \tfrac{1}{2}(\mathbf{x}_{0,\xi} \cdot \mathbf{a}_3 - \mathbf{g}_1 \cdot \mathbf{t}_3), \quad \varepsilon_{\eta 3} = \varepsilon_{3\eta} \doteq \tfrac{1}{2}(\mathbf{x}_{0,\eta} \cdot \mathbf{a}_3 - \mathbf{g}_2 \cdot \mathbf{t}_3), \tag{6.67}$$

$$\kappa_{\xi 3} = \kappa_{3\xi} = \tfrac{1}{2}(\mathbf{a}_{3,\xi} \cdot \mathbf{a}_3 - \mathbf{t}_{3,\xi} \cdot \mathbf{t}_3), \quad \kappa_{\eta 3} = \kappa_{3\eta} = \tfrac{1}{2}(\mathbf{a}_{3,\eta} \cdot \mathbf{a}_3 - \mathbf{t}_{3,\eta} \cdot \mathbf{t}_3), \tag{6.68}$$

where $\mathbf{g}_\alpha \cdot \mathbf{t}_3 = 0$, as $\mathbf{t}_3$ is perpendicular to the tangent plane spanned by $\mathbf{g}_\alpha$.

Green strain in the forward-rotated basis. We can derive in-plane components of the Green strain in the forward-rotated basis $\{\mathbf{a}_i\}$, where $\mathbf{a}_i \doteq \mathbf{Q}_0\, \mathbf{t}_i$. The derivative of the current position vector in the forward-rotated basis is

$$\mathbf{x}_{0,\alpha} = (\mathbf{x}_{0,\alpha} \cdot \mathbf{a}_1)\, \mathbf{a}_1 + (\mathbf{x}_{0,\alpha} \cdot \mathbf{a}_2)\, \mathbf{a}_2 + (\mathbf{x}_{0,\alpha} \cdot \mathbf{a}_3)\, \mathbf{a}_3 \tag{6.69}$$

and the scalar products of derivatives are

$$\begin{aligned} \mathbf{x}_{0,1} \cdot \mathbf{x}_{0,1} &= (\mathbf{x}_{0,1} \cdot \mathbf{a}_1)^2 + (\mathbf{x}_{0,1} \cdot \mathbf{a}_2)^2 + (\mathbf{x}_{0,1} \cdot \mathbf{a}_3)^2, \\ \mathbf{x}_{0,2} \cdot \mathbf{x}_{0,2} &= (\mathbf{x}_{0,2} \cdot \mathbf{a}_1)^2 + (\mathbf{x}_{0,2} \cdot \mathbf{a}_2)^2 + (\mathbf{x}_{0,2} \cdot \mathbf{a}_3)^2, \\ \mathbf{x}_{0,1} \cdot \mathbf{x}_{0,2} &= (\mathbf{x}_{0,1} \cdot \mathbf{a}_1)\, (\mathbf{x}_{0,2} \cdot \mathbf{a}_1) + (\mathbf{x}_{0,1} \cdot \mathbf{a}_2)\, (\mathbf{x}_{0,2} \cdot \mathbf{a}_2) \\ &\quad + (\mathbf{x}_{0,1} \cdot \mathbf{a}_3)\, (\mathbf{x}_{0,2} \cdot \mathbf{a}_3). \end{aligned} \tag{6.70}$$

Using them, the in-plane components of eq. (6.54) become

$$\begin{aligned} \varepsilon_{11} &= \tfrac{1}{2} \left[(\mathbf{x}_{0,1} \cdot \mathbf{a}_1)^2 + (\mathbf{x}_{0,1} \cdot \mathbf{a}_2)^2 + (\mathbf{x}_{0,1} \cdot \mathbf{a}_3)^2 - 1 \right], \\ \varepsilon_{22} &= \tfrac{1}{2} \left[(\mathbf{x}_{0,2} \cdot \mathbf{a}_1)^2 + (\mathbf{x}_{0,2} \cdot \mathbf{a}_2)^2 + (\mathbf{x}_{0,2} \cdot \mathbf{a}_3)^2 - 1 \right], \\ \varepsilon_{12} &= \tfrac{1}{2} \left[(\mathbf{x}_{0,1} \cdot \mathbf{a}_1)\, (\mathbf{x}_{0,2} \cdot \mathbf{a}_1) + (\mathbf{x}_{0,1} \cdot \mathbf{a}_2)\, (\mathbf{x}_{0,2} \cdot \mathbf{a}_2) \right. \\ &\quad \left. + (\mathbf{x}_{0,1} \cdot \mathbf{a}_3)\, (\mathbf{x}_{0,2} \cdot \mathbf{a}_3) \right] \end{aligned} \tag{6.71}$$

and $\varepsilon_{21} = \varepsilon_{12}$. Note that the terms $\mathbf{x}_{0,\alpha} \cdot \mathbf{a}_3$ are equal to the transverse shear components, i.e. $\mathbf{x}_{0,\alpha} \cdot \mathbf{a}_3 = 2\varepsilon_{\alpha 3} = 2\varepsilon_{3\alpha}$ by eq. (6.54) and, hence,

$$(\mathbf{x}_{0,1} \cdot \mathbf{a}_3)^2 = 4\varepsilon_{31}^2, \quad (\mathbf{x}_{0,2} \cdot \mathbf{a}_3)^2 = 4\varepsilon_{32}^2, \quad (\mathbf{x}_{0,1} \cdot \mathbf{a}_3)\, (\mathbf{x}_{0,2} \cdot \mathbf{a}_3) = 4\varepsilon_{31}\, \varepsilon_{32}. \tag{6.72}$$

If the transverse shear strains are small, then these terms can be neglected, and we obtain

$$\begin{aligned} \varepsilon_{11} &= \tfrac{1}{2}\left[(\mathbf{x}_{0,1}\cdot\mathbf{a}_1)^2 + (\mathbf{x}_{0,1}\cdot\mathbf{a}_2)^2 - 1\right], \\ \varepsilon_{22} &= \tfrac{1}{2}\left[(\mathbf{x}_{0,2}\cdot\mathbf{a}_1)^2 + (\mathbf{x}_{0,2}\cdot\mathbf{a}_2)^2 - 1\right], \\ \varepsilon_{12} = \varepsilon_{21} &= \tfrac{1}{2}\left[(\mathbf{x}_{0,1}\cdot\mathbf{a}_1)\,(\mathbf{x}_{0,2}\cdot\mathbf{a}_1) + (\mathbf{x}_{0,1}\cdot\mathbf{a}_2)\,(\mathbf{x}_{0,2}\cdot\mathbf{a}_2)\right]. \end{aligned} \tag{6.73}$$

Comparing with the in-plane components of the shell strains of eqs. (6.40) and (6.48), we see that they all contain the product $(\mathbf{x}_{0,\alpha}\cdot\mathbf{a}_\beta)$, and depend on the drilling rotation through $\mathbf{a}_1$ and $\mathbf{a}_2$.

Remark. Comparing transverse shear components for various shell strains of eqs. (6.42), (6.48)–(6.49) and (6.54)–(6.56), we see that $\varepsilon_{\alpha 3}$ and $\kappa_{\alpha 3}$ are identical. Regarding the normal strains of a shell, we note that, for all types of strains, they are equal to zero, i.e. $\varepsilon_{33}=0$ and $\kappa_{33}=0$, see eqs. (6.40), (6.48), (6.54), and (6.55). These zero values are unrealistic for most cases of deformation and more accurate values must be either recovered using some auxiliary conditions or obtained via the enhancement of kinematics.

6.3.4 Transverse shear strains satisfying RC. Kirchhoff kinematics

The shell forms of the RC are given by eqs. (6.30) and (6.35). Below, we specify the strains taking into account the $\alpha 3$-components of the RC equations.

Consider the second equations of eqs. (6.30) and (6.37), i.e. $\mathbf{x}_{0,\alpha}\cdot\mathbf{a}_3=0$ and $(\mathbf{Q}_{0,\alpha}\mathbf{t}_3)\cdot\mathbf{a}_3=0$. If these relations are used in the transverse shear components, which are identical for all the derived shell strains, then

$$\varepsilon_{\alpha 3} = \varepsilon_{3\alpha} = \tfrac{1}{2}\mathbf{x}_{0,\alpha}\cdot\mathbf{a}_3 = 0, \qquad \kappa_{\alpha 3} = \kappa_{3\alpha} = \tfrac{1}{2}(\mathbf{Q}_{0,\alpha}\mathbf{t}_3)\cdot\mathbf{a}_3 = 0, \tag{6.74}$$

i.e. the transverse shear strains of the zeroth and the first order are equal to zero. The zero shear strains are characteristic for the Kirchhoff-type kinematics which is described below. We note that

1. the conditions (6.74) also affect the strain vectors of eq. (6.39), because they imply that their normal components are equal to zero, i.e.

$$\boldsymbol{\varepsilon}_\alpha\cdot\mathbf{t}_3 = \mathbf{x}_{0,\alpha}\cdot\mathbf{a}_3 = 0, \qquad \boldsymbol{\kappa}_\alpha\cdot\mathbf{t}_3 = (\mathbf{Q}_{0,\alpha}\mathbf{t}_3)\cdot\mathbf{a}_3 = 0, \tag{6.75}$$

where we used $\boldsymbol{\kappa}_\alpha = \mathbf{Q}_0^T\mathbf{a}_{3,\alpha} \approx \mathbf{Q}_0^T\mathbf{Q}_{0,\alpha}\mathbf{t}_3$. Hence, the strain vectors belong to the tangent plane $\{\mathbf{t}_1,\mathbf{t}_2\}$ and, instead of strain vectors, we can use their projections on the tangent plane $\{\mathbf{t}_1, \mathbf{t}_2\}$, defined as follows:

$$\boldsymbol{\varepsilon}^\bullet_\alpha \doteq (\boldsymbol{\varepsilon}_\alpha \cdot \mathbf{t}_\beta)\, \mathbf{t}_\beta, \qquad \boldsymbol{\kappa}^\bullet_\alpha \doteq (\boldsymbol{\kappa}_\alpha \cdot \mathbf{t}_\beta)\, \mathbf{t}_\beta, \qquad \beta = 1, 2. \tag{6.76}$$

2. For $\varepsilon_{\alpha 3} = \varepsilon_{3\alpha} = 0$, the quadratic terms of eq. (6.72) are equal to zero and the components in the forward-rotated basis of eq. (6.74) are exact.

Kirchhoff kinematics. This kinematics was defined in [124], and is chronologically earlier than the Reissner kinematics. It dominated the shell literature when the computer era had not yet been in full bloom, and a simplicity of equations was essential to obtain any solution. Many theoretically vital results were obtained then; for an account we refer to [156, 78, 153, 127, 129, 171, 173, 174]. Still this kinematics is used in more mathematically-oriented papers, e.g. in the so-called asymptotic theories of shells, see [240, 134].

In the Kirchhoff hypothesis, the current position vector is defined by

$$\mathbf{x}(\zeta) \doteq \mathbf{x}_0 + \zeta\, \mathbf{a}_3, \qquad \mathbf{a}_3 \doteq \mathbf{n}, \tag{6.77}$$

where $\mathbf{n}$ is a unit vector normal to the deformed mid-surface. The current director $\mathbf{a}_3$ must remain normal to the deformed surface so this hypothesis is more restrictive than the Reissner hypothesis and implies zero shear strains.

The Kirchhoff kinematics is less popular in FE implementations than the Reissner kinematics for the following reasons:

1. the transverse shear strains are assumed equal to zero, which is not correct in some important practical applications such as layered composites.
2. the corresponding shell equations contain second derivatives, which require the C^1 approximation functions. This reduces the radius of convergence of the Newton method, and the stable time step in explicit dynamics, see [23].

However, in use still are very good elements based on the so-called Kirchhoff constraint, i.e. the condition that the transverse shear strain is zero. Such a condition is applied at some selected (discrete) points to modify the Hermitian shape functions which yields very good bending parts of two-node beam and four-node shell elements. The Discrete Kirchhoff

(DK) beam element is derived in Sect. 13.1.2, other DK elements are characterized in Sect. 13.2.4.

Finally, we note that having an element with the Reissner kinematics, we can obtain the Kirchhoff kinematics by enforcing the equations for the (α3)-components of the shell RC of eq. (6.30), i.e. the condition $\mathbf{x}_{0,\alpha} \cdot \mathbf{a}_3 = 0$.

6.3.5 Rotation as an intermediate variable symmetrizing strain

Symmetry of strains is important because it allows us to use standard constitutive equations. This question is undertaken in the early works of E. Reissner, cited in [205, 204] and later in [191, 42]. In [191], the rotation is chosen so that the $\mathbf{Q}^T\mathbf{F}$ product becomes symmetric. Besides, the symmetric part of the first Piola–Kirchhoff $\mathbf{P}^T$ is used, and the inverse constitutive equation $\mathbf{Q}^T\mathbf{F} = \partial\mathcal{W}_c(\text{sym}\mathbf{P})/\partial\text{sym}\mathbf{P}$, where $\mathcal{W}_c$ is the complementary energy.

The above idea was adapted for shells in [262]. The rotation is decomposed as $\mathbf{Q}_0 = \mathbf{Q}_2\mathbf{Q}_1\mathbf{Q}_d$, where $\mathbf{Q}_1$ and $\mathbf{Q}_2$ are rotations around the tangent vectors and $\mathbf{Q}_d$ is a rotation around the normal vector of the local basis. Then,

$$\mathbf{e}_\alpha \doteq \mathbf{Q}_0^T\mathbf{x}_{0,\alpha} = \mathbf{Q}_d^T\tilde{\mathbf{e}}_\alpha, \qquad \mathbf{n}_\alpha \doteq \mathbf{Q}_0^T\mathbf{n}_\alpha^* = \mathbf{Q}_d^T\tilde{\mathbf{n}}_\alpha, \tag{6.78}$$

where $\tilde{\mathbf{e}}_\alpha \doteq \mathbf{Q}_1^T\mathbf{Q}_2^T\mathbf{x}_{0,\alpha}$ and $\tilde{\mathbf{n}}_\alpha \doteq \mathbf{Q}_1^T\mathbf{Q}_2^T\mathbf{n}_\alpha^*$ are the vectors back-rotated by $\mathbf{Q}_2\mathbf{Q}_1$ and $\mathbf{n}_\alpha^*$ is the stress resultant obtained for the stress vector of $\mathbf{P}^T$.

The stretch and stress tensors are defined as follows: $\mathbf{U} \doteq \mathbf{Q}_d^T\tilde{\mathbf{e}}_\alpha \otimes \mathbf{t}_\alpha = \mathbf{Q}_d^T\tilde{\mathbf{U}}$ and $\mathbf{N} \doteq \mathbf{Q}_d^T\tilde{\mathbf{n}}_\alpha \otimes \mathbf{t}_\alpha = \mathbf{Q}_d^T\tilde{\mathbf{N}}$. The rotation $\mathbf{Q}_d$ is required to yield a symmetric $\mathbf{U}$, when applied to non-symmetric $\tilde{\mathbf{U}}$, i.e.

$$\mathbf{Q}_d : \qquad \mathbf{U} = \mathbf{Q}_d^T\tilde{\mathbf{U}}. \tag{6.79}$$

In other words, the angle of rotation ω_d of $\mathbf{Q}_d(\omega_d, \mathbf{t}_3)$ is determined from the condition $U_{12} = U_{21}$. The membrane strain is defined as $\mathbf{H} = (\mathbf{Q}_d^T\tilde{\mathbf{e}}_\alpha - \mathbf{t}_\alpha) \otimes \mathbf{t}_\alpha$.

A different procedure is used for the bending/twisting strain and the couple stress resultants. First, the vectors are back-rotated by $\mathbf{Q}_d$, i.e. $\bar{\boldsymbol{\kappa}}_\alpha = \mathbf{Q}_d^T\tilde{\boldsymbol{\kappa}}_\alpha$ and $\bar{\mathbf{m}}_\alpha = \mathbf{Q}_d^T\tilde{\mathbf{m}}_\alpha$. Then the tensors $\bar{\boldsymbol{\kappa}} = \bar{\boldsymbol{\kappa}}_\alpha \otimes \mathbf{t}_\alpha$ and $\bar{\mathbf{M}} = \bar{\mathbf{m}}_\alpha \otimes \mathbf{t}_\alpha$ are non-symmetric. Hence, only their symmetric parts are used in the virtual work, i.e.

$$\mathbf{N} \cdot \delta \mathbf{H} + \mathbf{M} \cdot \delta \boldsymbol{\kappa} \approx (\mathbf{Q}_d^T \tilde{\mathbf{N}}) \cdot \delta \mathbf{H} + \text{sym} \bar{\mathbf{M}} \cdot \text{sym} \delta \bar{\boldsymbol{\kappa}}, \tag{6.80}$$

and we see that the method is approximate. Note that ω_d is an intermediate variable depending on $\tilde{\mathbf{U}}$, i.e. $\omega_d(\tilde{\mathbf{U}})$, and cannot be used as a degree of freedom.

6.3.6 In-plane deformation with drilling rotation

Consider an in-plane deformation of a shell for which the drilling rotation ω_d about the normal vector $\mathbf{t}_3$ is non-zero, while the rotations about the tangent vector $\mathbf{Q}_m = \mathbf{I}$. Then $\mathbf{Q}_d$ of eq. (6.31) becomes

$$\mathbf{Q}_0 = \mathbf{Q}_d(\omega_d) = c\,(\mathbf{t}_1 \otimes \mathbf{t}_1 + \mathbf{t}_2 \otimes \mathbf{t}_2) + s\,(\mathbf{t}_2 \otimes \mathbf{t}_1 - \mathbf{t}_1 \otimes \mathbf{t}_2) + \mathbf{t}_3 \otimes \mathbf{t}_3, \tag{6.81}$$

where $s \doteq \sin \omega_d$ and $c \doteq \cos \omega_d$. Note that the normal vector $\mathbf{t}_3$ is the eigenvector of $\mathbf{Q}_d(\omega_d)$ associated with the eigenvalue $+1$. Hence, $\mathbf{a}_3 = \mathbf{Q}_0 \mathbf{t}_3 = \mathbf{Q}_d(\omega_d)\,\mathbf{t}_3 = \mathbf{t}_3$ and the normal vector $\mathbf{t}_3$ is unaltered by the drill rotation. But the tangent vectors $\mathbf{t}_\alpha$ are rotated about $\mathbf{t}_3$ by the angle ω_d, and $\mathbf{a}_\alpha = \mathbf{Q}_0 \mathbf{t}_\alpha = \mathbf{Q}_d(\omega_d)\,\mathbf{t}_\alpha$ yields

$$\mathbf{a}_1 = c\,\mathbf{t}_1 + s\,\mathbf{t}_2, \qquad \mathbf{a}_2 = -s\,\mathbf{t}_1 + c\,\mathbf{t}_2. \tag{6.82}$$

We see that the strains which depend on $\mathbf{a}_\alpha$ are affected by the drilling rotation! This includes the right stretch strains of eqs. (6.40) and (6.48) and the Green strain in the forward-rotated basis of eq. (6.74). On the other hand, the components of the Green strain in the initial basis $\{\mathbf{t}_i\}$, eqs. (6.54)–(6.56), are not affected by the drill rotation.

Let us consider the strain components in the initial basis $\{\mathbf{t}_i\}$ and compare the strains based on the products $(\mathbf{F}^T\mathbf{F})$ and $(\mathbf{Q}^T\mathbf{F})$. For eq. (6.81), the parts of $\mathbf{F}$ of eq. (6.7) become

$$\mathbf{F}_0 = \mathbf{x}_{0,\alpha} \otimes \mathbf{t}_\alpha + \mathbf{t}_3 \otimes \mathbf{t}_3, \qquad (\mathbf{F}_{,\zeta})_0 = \mathbf{t}_{3,\alpha} \otimes \mathbf{t}_\alpha \tag{6.83}$$

and we see that they do not depend on the drilling rotation, ω_d. The same is true for the Green strain, $\mathbf{E} \doteq \frac{1}{2}(\mathbf{F}^T\mathbf{F} - \mathbf{I})$, depending on the product $\mathbf{F}^T\mathbf{F}$.

On the other hand, the strains using the product $(\mathbf{Q}^T\mathbf{F})$ naturally contain the drilling rotation. For eq. (6.81), the components of $(\mathbf{Q}^T\mathbf{F})$ become

$$\mathbf{Q}_0^T \mathbf{F}_0 = (\mathbf{Q}_d^T \mathbf{x}_{0,\alpha}) \otimes \mathbf{t}_\alpha + \mathbf{t}_3 \otimes \mathbf{t}_3, \qquad \mathbf{Q}_0^T (\mathbf{F}_{,\zeta})_0 = (\mathbf{Q}_d^T \mathbf{Q}_{d,\alpha} + \boldsymbol{\Omega}_\alpha^0)\,\mathbf{t}_3 \otimes \mathbf{t}_\alpha \tag{6.84}$$

and they depend on ω_d. If we apply the drill RC of eq. (6.30) to such formulations, then, simply, the drilling rotation already existing within the formulation is constrained.

Below, we determine the drill RC and the strains for the displacement vector $\mathbf{u} = u_1 \mathbf{t}_1 + u_2 \mathbf{t}_2$.

Drill RC. In terms of the component displacements and the drilling rotation, eq. (6.30) becomes

$$-(u_{2,2} + u_{1,1} + 2)\, s + (u_{2,1} - u_{1,2})\, c = 0. \tag{6.85}$$

Linearization w.r.t. ω_d at $\omega_d = 0$ gives $\cos\omega_d \approx 1$ and $\sin\omega_d \approx \omega_d$, and for small strains, $u_{1,1} \approx 0$ and $u_{2,2} \approx 0$, we obtain the well-known expression for the infinitesimal drill rotation $\omega_d = \frac{1}{2}(u_{2,1} - u_{1,2})$.

Symmetric relaxed right stretch strain. The components of this strain of eq. (6.48) are

$$\varepsilon_{11} = (1 + u_{1,1})\, c + u_{2,1}\, s - 1, \qquad \varepsilon_{22} = -u_{1,2}\, s + (1 + u_{2,2})\, c - 1,$$

$$\varepsilon_{12} = \varepsilon_{21} = \tfrac{1}{2}\left[(u_{2,2} - u_{1,1})\, s + (u_{2,1} + u_{1,2})\, c\right] \tag{6.86}$$

and they depend on the drill rotation ω_d. Linearization w.r.t. ω_d at $\omega_d = 0$ gives $\cos\omega_d \approx 1$ and $\sin\omega_d \approx \omega_d$ and then

$$\varepsilon_{11} = u_{1,1} + u_{2,1}\, \omega_d, \qquad \varepsilon_{22} = -u_{1,2}\, \omega_d + u_{2,2},$$

$$\varepsilon_{12} = \varepsilon_{21} = \tfrac{1}{2}\left[(u_{2,2} - u_{1,1})\, \omega_d + (u_{2,1} + u_{1,2})\right]. \tag{6.87}$$

For an infinitesimal rotation, $\omega_d \approx 0$, we obtain $\varepsilon_{11} = u_{1,1}$, $\varepsilon_{22} = u_{2,2}$, and $\varepsilon_{12} = \varepsilon_{21} = \frac{1}{2}(u_{2,1} + u_{1,2})$.

Green strain. The components of the Green strain of eq. (6.54) are

$$\varepsilon_{11} = \tfrac{1}{2}\left[(1 + u_{1,1})^2 + u_{2,1}^2 - 1\right], \qquad \varepsilon_{22} = \tfrac{1}{2}\left[u_{1,2}^2 + (1 + u_{2,2})^2 - 1\right],$$

$$\varepsilon_{12} = \varepsilon_{21} = \tfrac{1}{2}\left[(1 + u_{1,1})u_{1,2} + u_{2,1}(1 + u_{2,2})\right] \tag{6.88}$$

and they do not depend on ω_d. By neglecting of the quadratic terms, we obtain the linearized formulas $\varepsilon_{11} = u_{1,1}$, $\varepsilon_{22} = u_{2,2}$, $\varepsilon_{12} = \varepsilon_{21} = \frac{1}{2}(u_{1,2} + u_{2,1})$, which are identical to the linearized components of the *symmetric relaxed* right stretch strain of eq. (6.87) for $\omega_d \approx 0$.

6.3.7 Forward-rotated shell strains variations

The forward-rotated *symmetric relaxed* right stretch strain is defined as $\mathbf{H}^* \doteq \mathbf{Q}_0 \, \mathbf{H} \, \mathbf{Q}_0^T$, where $\mathbf{H} \doteq \mathrm{sym}(\mathbf{Q}_0^T \mathbf{F}) - \mathbf{I}$. Hence,

$$\mathbf{H}^* \doteq \mathbf{Q}_0 \, \mathbf{H} \, \mathbf{Q}_0^T = \mathrm{sym}(\mathbf{F}\mathbf{Q}_0^T) - \mathbf{I}. \tag{6.89}$$

The forward-rotated shell strains, which are counterparts of $\mathbf{H}^*$, are defined as

$$\boldsymbol{\varepsilon}^* \doteq \mathbf{Q}_0 \, \boldsymbol{\varepsilon} \, \mathbf{Q}_0^T = \mathrm{sym}\left[(\mathbf{Q}_0 \boldsymbol{\varepsilon}_\alpha) \otimes (\mathbf{Q}_0 \mathbf{t}_\alpha)\right] = \mathrm{sym}\left(\boldsymbol{\varepsilon}_\alpha^* \otimes \mathbf{a}_\alpha\right), \tag{6.90}$$

$$\boldsymbol{\kappa}^* \doteq \mathbf{Q}_0 \, \boldsymbol{\kappa} \, \mathbf{Q}_0^T = \mathrm{sym}\left[(\mathbf{Q}_0 \boldsymbol{\kappa}_\alpha) \otimes (\mathbf{Q}_0 \mathbf{t}_\alpha)\right] = \mathrm{sym}\left(\boldsymbol{\kappa}_\alpha^* \otimes \mathbf{a}_\alpha\right), \tag{6.91}$$

where

$$\boldsymbol{\varepsilon}_\alpha^* = \mathbf{Q}_0 \, \boldsymbol{\varepsilon}_\alpha = \mathbf{Q}_0 (\mathbf{Q}_0^T \mathbf{x}_{0,\alpha} - \mathbf{t}_\alpha) = \mathbf{x}_{0,\alpha} - \mathbf{a}_\alpha, \tag{6.92}$$

$$\boldsymbol{\kappa}_\alpha^* = \mathbf{Q}_0 \, \boldsymbol{\kappa}_\alpha = \mathbf{Q}_0 (\mathbf{Q}_0^T \bar{\boldsymbol{\omega}}_\alpha - \mathbf{t}_{3,\alpha}) = \bar{\boldsymbol{\omega}}_\alpha - \mathbf{Q}_0 \mathbf{t}_{3,\alpha}. \tag{6.93}$$

For a shallow shell $\boldsymbol{\kappa}_\alpha^* = \boldsymbol{\omega}_\alpha$, because $\bar{\boldsymbol{\omega}}_\alpha \approx \boldsymbol{\omega}_\alpha$ and $\mathbf{t}_{3,\alpha} \approx \mathbf{0}$.

The forward-rotated variation of the zeroth order strain $\delta \boldsymbol{\varepsilon} = \mathrm{sym}(\delta \boldsymbol{\varepsilon}_i \otimes \mathbf{t}_i)$ is defined as

$$(\delta \boldsymbol{\varepsilon})^* \doteq \mathbf{Q}_0 \, \delta \boldsymbol{\varepsilon} \, \mathbf{Q}_0^T = \mathbf{Q}_0 \, \delta (\mathbf{Q}_0^T \boldsymbol{\varepsilon}^* \mathbf{Q}_0) \, \mathbf{Q}_0^T \doteq \overset{\circ}{\delta} \boldsymbol{\varepsilon}^*, \tag{6.94}$$

where the last form defines the co-rotational variation of the rotated strain, which has a form of the Green–McInnis–Naghdi objective rate, see [97]. The above definition yields

$$\overset{\circ}{\delta} \boldsymbol{\varepsilon}^* = \mathrm{sym}\left[(\mathbf{Q}_0 \delta \boldsymbol{\varepsilon}_\alpha) \otimes (\mathbf{Q}_0 \mathbf{t}_\alpha)\right] = \mathrm{sym}(\overset{\circ}{\delta} \boldsymbol{\varepsilon}_\alpha^* \otimes \mathbf{a}_\alpha), \tag{6.95}$$

where the co-rotational variation of the strain vector can be expressed as

$$\overset{\circ}{\delta} \boldsymbol{\varepsilon}_\alpha^* \doteq \mathbf{Q}_0 \delta \boldsymbol{\varepsilon}_\alpha = \mathbf{Q}_0 \left[\delta (\mathbf{Q}_0^T \boldsymbol{\varepsilon}_\alpha^*)\right]. \tag{6.96}$$

Relations of the results of particular operations to the vector $\boldsymbol{\varepsilon}_\alpha$ are as follows:

1. rotate-back $\boldsymbol{\varepsilon}_\alpha^*$,

$$(\mathbf{Q}_0^T \boldsymbol{\varepsilon}_\alpha^*) = \mathbf{Q}_0^T \mathbf{x}_{0,\alpha} - \mathbf{t}_\alpha \qquad (= \boldsymbol{\varepsilon}_\alpha),$$

2. calculate a variation,

$$\delta(\mathbf{Q}_0^T \boldsymbol{\varepsilon}_\alpha^*) = \delta\mathbf{Q}_0^T \mathbf{x}_{0,\alpha} + \mathbf{Q}_0^T \delta\mathbf{x}_{0,\alpha} \qquad (= \delta\boldsymbol{\varepsilon}_\alpha), \tag{6.97}$$

3. rotate-forward,

$$\mathbf{Q}_0 \left[\delta(\mathbf{Q}_0^T \boldsymbol{\varepsilon}_\alpha^*)\right] = (\mathbf{x}_{0,\alpha} \times \delta\boldsymbol{\theta}) + \delta\mathbf{x}_{0,\alpha} \qquad (= \mathbf{Q}_0 \delta\boldsymbol{\varepsilon}_\alpha = \overset{\circ}{\delta}\boldsymbol{\varepsilon}_\alpha^*).$$

In the last form, we used $\mathbf{Q}_0 \delta\mathbf{Q}_0^T = \delta\tilde{\boldsymbol{\theta}}^T$, in accord with the definition of $\delta\tilde{\boldsymbol{\theta}}$ in eq. (6.19). We proceed similarly for the first-order strain, and, finally, we have

$$\overset{\circ}{\delta}\boldsymbol{\varepsilon}_\alpha^* = \delta\mathbf{x}_{0,\alpha} + (\mathbf{x}_{0,\alpha} \times \delta\boldsymbol{\theta}), \qquad \overset{\circ}{\delta}\boldsymbol{\kappa}_\alpha^* = \delta\bar{\boldsymbol{\omega}}_\alpha + (\bar{\boldsymbol{\omega}}_\alpha \times \delta\boldsymbol{\theta}), \tag{6.98}$$

where both expressions depend on the axial vector $\delta\boldsymbol{\theta}$ but not on the rotation tensor.

Example. We compare the strain vectors $\boldsymbol{\varepsilon}_1$ and $\boldsymbol{\varepsilon}_1^*$ for a planar deformation.

Let us first specify the strain $\boldsymbol{\varepsilon}_1$ in the basis $\{\mathbf{t}_1, \mathbf{t}_3\}$, see Fig. 6.2. The displacement vector is decomposed as $\mathbf{u} = u\mathbf{t}_1 + w\mathbf{t}_3$. For the arc-length coordinate S_1, the derivative of the initial position vector is $\mathbf{y}_{0,1} = \mathbf{t}_1$. For $\mathbf{x}_0 = \mathbf{y}_0 + \mathbf{u}$, the derivative of the current position vector is

$$\mathbf{x}_{0,1} = (1 + u_{,1})\,\mathbf{t}_1 + w_{,1}\,\mathbf{t}_3. \tag{6.99}$$

The rotation tensor can be expressed as

$$\mathbf{Q}_0 = \cos\beta\,[\mathbf{t}_1 \otimes \mathbf{t}_1 + \mathbf{t}_3 \otimes \mathbf{t}_3] - \sin\beta\,[\mathbf{t}_1 \otimes \mathbf{t}_3 - \mathbf{t}_3 \otimes \mathbf{t}_1] + \mathbf{t}_2 \otimes \mathbf{t}_2, \tag{6.100}$$

where β is the angle of rotation. Then the strain vector $\boldsymbol{\varepsilon}_1 = \mathbf{Q}_0^T \mathbf{x}_{0,1} - \mathbf{t}_1$ is

$$\boldsymbol{\varepsilon}_1 = [(1 + u_{,1})\cos\beta + w_{,1}\sin\beta - 1]\,\mathbf{t}_1 + [-(1 + u_{,1})\sin\beta + w_{,1}\cos\beta]\,\mathbf{t}_3. \tag{6.101}$$

Next, we specify the strain $\boldsymbol{\varepsilon}_1^*$ in the basis $\{\mathbf{a}_1, \mathbf{a}_3\}$ for the same deformation. The displacement vector is decomposed as $\mathbf{u} = \bar{u}\mathbf{a}_1 + \bar{w}\mathbf{a}_3$. The derivative of the initial position vector is $\mathbf{y}_{0,1} = \mathbf{t}_1 = \mathbf{Q}_0\mathbf{a}_1 = \cos\bar{\beta}\mathbf{a}_1 - \sin\bar{\beta}\mathbf{a}_3$, where the rotation tensor

$$\mathbf{Q}_0 = \cos\bar{\beta}\,[\mathbf{a}_1 \otimes \mathbf{a}_1 + \mathbf{a}_3 \otimes \mathbf{a}_3] - \sin\bar{\beta}\,[\mathbf{a}_1 \otimes \mathbf{a}_3 - \mathbf{a}_3 \otimes \mathbf{a}_1] + \mathbf{a}_2 \otimes \mathbf{a}_2 \tag{6.102}$$

and $\bar{\beta}$ is an angle of rotation. For $\mathbf{x}_0 = \mathbf{y}_0 + \mathbf{u}$, we get a derivative of the current position vector,

$$\mathbf{x}_{0,1} = (\cos\bar{\beta} + \bar{u}_{,1})\,\mathbf{a}_1 + (\sin\bar{\beta} + \bar{w}_{,1})\,\mathbf{a}_3. \tag{6.103}$$

The forward-rotated strain vector, $\boldsymbol{\varepsilon}_1^* = \mathbf{x}_{0,1} - \mathbf{a}_1$, is

$$\boldsymbol{\varepsilon}_1^* = (\cos\bar{\beta} - 1 + \bar{u}_{,1})\,\mathbf{a}_1 + (-\sin\bar{\beta} + \bar{w}_{,1})\,\mathbf{a}_3. \tag{6.104}$$

Comparing eqs. (6.101) and (6.104), we note that expressions for $\boldsymbol{\varepsilon}_1^*$ are simpler than for $\boldsymbol{\varepsilon}_1$.

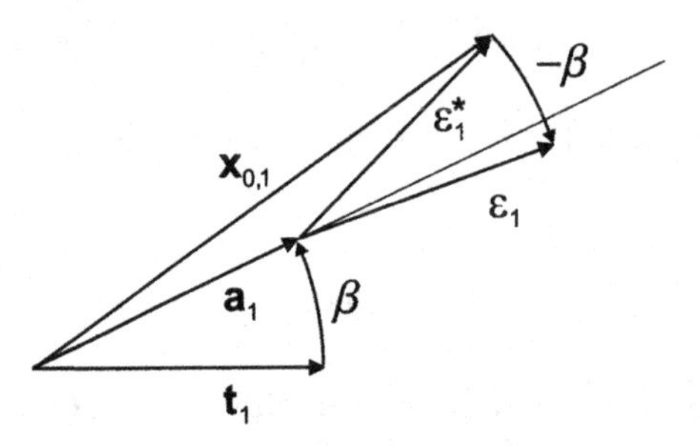

Fig. 6.2 Geometrical interpretation of strain vectors $\boldsymbol{\varepsilon}_1$ and $\boldsymbol{\varepsilon}_1^*$.

6.4 Virtual work equation for shell

In this section, the shell VW equation is derived from the 3D VW equations of Chap. 4, for the kinematics defined in Sect. 6.1. The shell stress and couple resultants are defined as integrals over the thickness.

6.4.1 Virtual work of Biot stress

The 3D VW of the Biot stress is given by the first term in eq. (4.51). The corresponding VW for a shell is defined as the integral of this term over the shell thickness,

$$\delta\Sigma \doteq \int_{-\frac{h}{2}}^{+\frac{h}{2}} \mathbf{T}_s^B \cdot \operatorname{sym}\delta(\mathbf{Q}^T\mathbf{F})\,\mu\,\mathrm{d}\zeta. \tag{6.105}$$

For $\operatorname{sym}\delta(\mathbf{Q}^T\mathbf{F}) = \delta\operatorname{sym}(\mathbf{Q}^T\mathbf{F}) = \delta\mathbf{H} = \delta\boldsymbol{\varepsilon} + \zeta\delta\boldsymbol{\kappa}$, using of eq. (6.47), where the strain vectors are given by eq. (6.39), the integrand is

$$\mathbf{T}_s^B \cdot \text{sym}\delta(\mathbf{Q}^T\mathbf{F}) = \mathbf{T}_s^B \cdot \text{sym}(\delta\boldsymbol{\varepsilon}_\alpha \otimes \mathbf{t}_\alpha) + \zeta\mathbf{T}_s^B \cdot \text{sym}(\delta\boldsymbol{\kappa}_\alpha \otimes \mathbf{t}_\alpha). \quad (6.106)$$

The stress tensor $\mathbf{T} \doteq \mathbf{Q}^T\mathbf{P}$ of eq. (4.50), can be expressed as $\mathbf{T} = \mathbf{f}_i \otimes \mathbf{t}_i$, where $\mathbf{f}_i$ is the stress vector corresponding to the vector $\mathbf{t}_i$ of the local Cartesian basis. The shell stress and couple resultant vectors are defined as

$$\mathbf{n}_i \doteq \int_{-\frac{h}{2}}^{+\frac{h}{2}} \mathbf{f}_i(\zeta)\,\mu\,\mathrm{d}\zeta, \qquad \mathbf{m}_i \doteq \int_{-\frac{h}{2}}^{+\frac{h}{2}} \zeta\,\mathbf{f}_i(\zeta)\,\mu\,\mathrm{d}\zeta, \quad (6.107)$$

and we can introduce the following shell stress and couple resultant tensors,

$$\mathbf{N} \doteq \mathbf{n}_i \otimes \mathbf{t}_i, \qquad \mathbf{M} \doteq \mathbf{m}_i \otimes \mathbf{t}_i. \quad (6.108)$$

Integrating eq. (6.106) over the thickness, we obtain the shell form of the VW for the Biot stress of eq. (6.105) in the following form:

$$\delta\Sigma = \mathbf{N}_s^B \cdot \delta\boldsymbol{\varepsilon} + \mathbf{M}_s^B \cdot \delta\boldsymbol{\kappa}, \quad (6.109)$$

where the strain tensors $\boldsymbol{\varepsilon}$ and $\boldsymbol{\kappa}$ are given by eq. (6.47) and are symmetric.

The 3D strain energy $\mathcal{W}(\tilde{\mathbf{U}})$ depends on the *symmetric relaxed* right stretch tensor $\tilde{\mathbf{U}} \doteq \text{sym}(\mathbf{Q}^T\mathbf{F})$ and the CL for the Biot stress is obtained by $\mathbf{T}_s^B = \partial_{\tilde{U}}\mathcal{W}$, see eq. (4.32). For the Reissner kinematics of first order, $\tilde{\mathbf{U}}$ is a linear polynomial of the thickness coordinate ζ, i.e. $\tilde{\mathbf{U}}(\zeta) = (\tilde{\mathbf{U}})_0 + \zeta(\tilde{\mathbf{U}}_{,\zeta})_0$, see eq. (6.10). The shell strain energy is defined as an integral of the 3D strain energy over the shell thickness, i.e. $\Sigma \doteq \int_{-\frac{h}{2}}^{+\frac{h}{2}} \mathcal{W}[\tilde{\mathbf{U}}(\zeta)]\,\mu\,\mathrm{d}\zeta$. Hence, $\Sigma = \Sigma\Big((\tilde{\mathbf{U}})_0,\, (\tilde{\mathbf{U}}_{,\zeta})_0\Big)$, and the CLs for the shell stress and couple resultants are defined as

$$\mathbf{N}_s^B = \frac{\partial\Sigma}{\partial(\tilde{\mathbf{U}})_0}, \qquad \mathbf{M}_s^B = \frac{\partial\Sigma}{\partial(\tilde{\mathbf{U}}_{,\zeta})_0}. \quad (6.110)$$

VW of stress for forward-rotated stress and couple resultants and shell strains. Define the forward-rotated stress and couple resultants as

$$\mathbf{N}_s^{B*} \doteq \mathbf{Q}_0\mathbf{N}_s^B\mathbf{Q}_0^T = \text{sym}[(\mathbf{Q}_0\mathbf{n}_i) \otimes (\mathbf{Q}_0\mathbf{t}_i)] = \text{sym}\,(\mathbf{n}_i^* \otimes \mathbf{a}_i)\,, \quad (6.111)$$

$$\mathbf{M}_s^{B*} \doteq \mathbf{Q}_0\mathbf{M}_s^B\mathbf{Q}_0^T = \text{sym}[(\mathbf{Q}_0\mathbf{m}_i) \otimes (\mathbf{Q}_0\mathbf{t}_i)] = \text{sym}\,(\mathbf{m}_i^* \otimes \mathbf{a}_i)\,, \quad (6.112)$$

where

$$\mathbf{n}_i^* \doteq \mathbf{Q}_0\mathbf{n}_i, \qquad \mathbf{m}_i^* \doteq \mathbf{Q}_0\mathbf{m}_i. \quad (6.113)$$

Note that these vectors become the shell counterparts of the nominal stress $\mathbf{P}$ when $\mathbf{Q} = \mathbf{Q}_0$ is used in $\mathbf{T} \doteq \mathbf{Q}^T\mathbf{P}$.

The forward-rotated shell membrane strain is defined in eq. (6.90), while the forward-rotated variation of this strain in eq. (6.94) and it is equal to the co-rotational variation of the rotated strain $\overset{\circ}{\delta} \boldsymbol{\varepsilon}^*$. Using the above forward-rotated tensors, the first term of the shell VW of eq. (6.109) becomes

$$\mathbf{N}_s^B \cdot \delta\boldsymbol{\varepsilon} = \mathrm{tr}(\mathbf{N}_s^B\, \delta\boldsymbol{\varepsilon}^T) = \mathrm{tr}(\mathbf{Q}_0 \mathbf{N}_s^B \mathbf{Q}_0^T\, \mathbf{Q}_0 \delta\boldsymbol{\varepsilon}^T \mathbf{Q}_0^T) = \mathbf{N}_s^* \cdot \overset{\circ}{\delta} \boldsymbol{\varepsilon}^*. \tag{6.114}$$

Similar relations can be obtained for the bending/twisting strain and, hence, the shell VW can be expressed in terms of the forward-rotated tensors as

$$\delta\Sigma = \mathbf{N}_s^{B*} \cdot \overset{\circ}{\delta} \boldsymbol{\varepsilon}^* + \mathbf{M}_s^{B*} \cdot \overset{\circ}{\delta} \boldsymbol{\kappa}^*. \tag{6.115}$$

For the 3D elasticity, it can be shown that the structure of a constitutive equation and a constitutive operator for the forward-rotated Biot stress are analogous to that for the Biot stress, but for $\mathbf{U}$ replaced by $\mathbf{U}^* = \mathbf{Q}\mathbf{U}\mathbf{Q}^T = \mathbf{V}$, see [251]. This result is also valid for shells and, hence, for the expansion $\tilde{\mathbf{U}}^*(\zeta) = (\tilde{\mathbf{U}}^*)_0 + \zeta(\tilde{\mathbf{U}}^*_{,\zeta})_0$ and the shell strain energy $\Sigma = \Sigma((\tilde{\mathbf{U}}^*)_0, (\tilde{\mathbf{U}}^*_{,\zeta})_0)$, we obtain

$$\mathbf{N}_s^{B*} = \frac{\partial \Sigma}{\partial (\tilde{\mathbf{U}}^*)_0}, \qquad \mathbf{M}_s^{B*} = \frac{\partial \Sigma}{\partial (\tilde{\mathbf{U}}^*_{,\zeta})_0}, \tag{6.116}$$

which is the counterpart of eq. (6.110).

6.4.2 Virtual work of second Piola–Kirchhoff stress

The 3D VW of the second Piola–Kirchhoff stress is given by the first term in eq. (4.64), which can be expressed as $\mathbf{S} \cdot \frac{1}{2}\delta \mathrm{sym}(\mathbf{Q}^T\mathbf{F})$. The corresponding VW for a shell is defined as the integral of this term over the shell thickness

$$\delta\Sigma \doteq \int_{-\frac{h}{2}}^{+\frac{h}{2}} \mathbf{S} \cdot \tfrac{1}{2}\mathrm{sym}\delta(\mathbf{F}^T\mathbf{F})\, \mu\, \mathrm{d}\zeta, \tag{6.117}$$

where $\frac{1}{2}\mathrm{sym}\delta(\mathbf{F}^T\mathbf{F}) = \delta\frac{1}{2}\mathrm{sym}(\mathbf{F}^T\mathbf{F}) = \delta\mathbf{E} = \delta\boldsymbol{\varepsilon} + \zeta\delta\boldsymbol{\kappa} + \zeta^2\delta\boldsymbol{\mu}$, by eq. (6.52). Then, the integrand of eq. (6.117) is

$$\mathbf{S} \cdot \tfrac{1}{2}\mathrm{sym}\delta(\mathbf{F}^T\mathbf{F}) = \mathbf{S} \cdot \delta\boldsymbol{\varepsilon} + \zeta\mathbf{S} \cdot \delta\boldsymbol{\kappa} + \zeta^2\mathbf{S} \cdot \delta\boldsymbol{\mu}. \tag{6.118}$$

The second Piola–Kirchhoff stress tensor can be expressed as $\mathbf{S} = \mathbf{s}_i \otimes \mathbf{t}_i$, where $\mathbf{s}_i$ is the stress vector corresponding to the vector $\mathbf{t}_i$ of the local Cartesian basis. The shell stress and couple resultant vectors are defined as

$$\mathbf{n}_i \doteq \int_{-\frac{h}{2}}^{+\frac{h}{2}} \mathbf{s}_i(\zeta)\, \mu\, d\zeta, \quad \mathbf{m}_i \doteq \int_{-\frac{h}{2}}^{+\frac{h}{2}} \zeta\, \mathbf{s}_i(\zeta)\, \mu\, d\zeta, \quad \mathbf{k}_i \doteq \int_{-\frac{h}{2}}^{+\frac{h}{2}} \zeta^2\, \mathbf{s}_i(\zeta)\, \mu\, d\zeta, \tag{6.119}$$

and we can introduce the shell stress and couple resultant tensors, $\mathbf{N} \doteq \mathbf{n}_i \otimes \mathbf{t}_i$, $\mathbf{M} \doteq \mathbf{m}_i \otimes \mathbf{t}_i$, and $\mathbf{K} \doteq \mathbf{k}_i \otimes \mathbf{t}_i$. Integrating eq. (6.118) over the thickness, we obtain the following shell form of the VW equation:

$$\delta\Sigma = \mathbf{N} \cdot \delta\boldsymbol{\varepsilon} + \mathbf{M} \cdot \delta\boldsymbol{\kappa} + \mathbf{K} \cdot \delta\boldsymbol{\mu}, \tag{6.120}$$

where the strain tensors $\boldsymbol{\varepsilon}$, $\boldsymbol{\kappa}$ and $\boldsymbol{\mu}$ are given by eq. (6.52).

Below, we derive the shell strain energy and define the CL for the shell stress and couple resultant tensors $\mathbf{N}$, $\mathbf{M}$, and $\mathbf{K}$. The 3D strain energy $\mathcal{W}(\mathbf{C})$ depends on the right Cauchy–Green tensor $\mathbf{C} \doteq \mathbf{F}^T\mathbf{F}$, and the CL for the second Piola–Kirchhoff stress is obtained as $\mathbf{S} = 2\,\partial_{\mathbf{C}}\mathcal{W}(\mathbf{C})$, see eq. (4.23). For the Reissner kinematics, $\mathbf{C}$ is a quadratic polynomial of the thickness coordinate ζ, i.e. $\mathbf{C}(\zeta) = (\mathbf{C})_0 + \zeta(\mathbf{C}_{,\zeta})_0 + \frac{1}{2}\zeta^2(\mathbf{C}_{,\zeta\zeta})_0$, see eq. (6.51). The shell strain energy is defined as an integral of the 3D strain energy over the shell thickness, i.e. $\Sigma \doteq \int_{-\frac{h}{2}}^{+\frac{h}{2}} \mathcal{W}(\mathbf{C}(\zeta))\, \mu\, d\zeta$. Hence, $\Sigma = \Sigma((\mathbf{C})_0, (\mathbf{C}_{,\zeta})_0, (\mathbf{C}_{,\zeta\zeta})_0)$, and the CLs for the shell stress and couple resultants are defined as follows:

$$\mathbf{N} = \frac{\partial\Sigma}{\partial(\mathbf{C})_0}, \qquad \mathbf{M} = \frac{\partial\Sigma}{\partial(\mathbf{C}_{,\zeta})_0}, \qquad \mathbf{K} = \frac{\partial\Sigma}{\partial(\mathbf{C}_{,\zeta\zeta})_0}. \tag{6.121}$$

Finally, we note that the VW of the forward-rotated stress and couple resultants and shell strains can be obtained in a similar manner as for the formulation based on the Biot stress.

6.4.3 Variation of RC term

The RC terms are identical for the formulations based on the Biot stress and the second Piola–Kirchhoff stress. We consider two forms of the RC term.

1. The term $\delta[\mathbf{T}_a \cdot \text{skew}(\mathbf{Q}^T\mathbf{F})]$ of eq. (4.51) or (4.64). Using $\text{skew}(\mathbf{Q}^T\mathbf{F}) = \mathfrak{C}_0 + \zeta(\mathfrak{C}_{,\zeta})_0$ of eq. (6.27) and by the integration over the thickness, we obtain

$$\delta F_{\mathrm{RC}} = \int_{-\frac{h}{2}}^{+\frac{h}{2}} \delta \left[\mathbf{T}_a \cdot \mathrm{skew}(\mathbf{Q}^T\mathbf{F})\right] \mu \, \mathrm{d}\zeta \; = \delta \left[\mathbf{N}_a \cdot \mathfrak{C}_0 + \mathbf{M}_a \cdot (\mathfrak{C}_{,\zeta})_0\right]. \tag{6.122}$$

For $\mathfrak{C}_0$ and $(\mathfrak{C}_{,\zeta})_0$ reduced to the first equations of eqs. (6.30) and (6.35), we obtain

$$\delta F_{\mathrm{RC}} = \; \delta \left[N_{12}\, C_{12} + M_{12}\, D_{12}\right] \approx \; \delta \left[N_{12}\, C_{12}\right], \tag{6.123}$$

where $N_{12} = (\mathbf{N}_a)_{12}$, $M_{12} = (\mathbf{M}_a)_{12}$, and

$$C_{12} \doteq \tfrac{1}{2} \left(\mathbf{x}_{0,1} \cdot \mathbf{a}_2 - \mathbf{x}_{0,2} \cdot \mathbf{a}_1\right), \qquad D_{12} \doteq \mathbf{a}_{3,1} \cdot \mathbf{a}_2 - \mathbf{a}_{3,2} \cdot \mathbf{a}_1. \tag{6.124}$$

2. The term $\gamma\, \mathrm{skew}(\mathbf{Q}^T\mathbf{F}) \cdot \delta \mathrm{skew}(\mathbf{Q}^T\mathbf{F})$ of eq. (4.63) or (4.73), where $\gamma \in (0, \infty)$. On use of $\mathrm{skew}(\mathbf{Q}^T\mathbf{F}) = \mathfrak{C}_0 + \zeta(\mathfrak{C}_{,\zeta})_0$ of eq. (6.27) and by the integration over the thickness, we obtain

$$\begin{aligned} \delta F_{\mathrm{RC}} &= \gamma \int_{-\frac{h}{2}}^{+\frac{h}{2}} \mathrm{skew}(\mathbf{Q}^T\mathbf{F}) \cdot \delta \mathrm{skew}(\mathbf{Q}^T\mathbf{F})\, \mu \, \mathrm{d}\zeta \\ &= \gamma \left[h\, \mathfrak{C}_0 \cdot \delta \mathfrak{C}_0 + \frac{h^3}{12} (\mathfrak{C}_{,\zeta})_0 \cdot \delta (\mathfrak{C}_{,\zeta})_0 \right], \end{aligned} \tag{6.125}$$

which, for $\mathfrak{C}_0$ and $(\mathfrak{C}_{,\zeta})_0$ reduced to the first equations of eqs. (6.30) and (6.35), can be rewritten as

$$\delta F_{\mathrm{RC}} \approx \gamma \left[h\, C_{12}\, \delta C_{12} + \frac{h^3}{12} D_{12}\, \delta D_{12} \right] \approx \gamma h\, C_{12}\, \delta C_{12}. \tag{6.126}$$

In both forms, the term with D_{12} is neglected for the reasons given in the *Remark* in Sect. 6.2.

6.4.4 Virtual work of body forces and external forces

For the Reissner kinematics of Sect. 6.1, the position vector in the deformed configuration is defined as $\mathbf{x}(\zeta) = \mathbf{x}_0 + \zeta \mathbf{a}_3$. Its variation is as follows:

$$\delta \mathbf{x} = \delta \mathbf{x}_0 + \zeta\, \delta \mathbf{a}_3, \tag{6.127}$$

where $\delta \mathbf{a}_3 = \delta \boldsymbol{\theta} \times \mathbf{a}_3$. In the above, $\delta \boldsymbol{\theta}$ is the axial vector of a left skew-symmetric tensor $\delta \tilde{\boldsymbol{\theta}} = \delta \boldsymbol{\theta} \times \mathbf{I}$ such that $\delta \mathbf{Q}_0 = \delta \tilde{\boldsymbol{\theta}}\, \mathbf{Q}_0$.

i. The VW of the body force is defined as $\delta F_b \doteq \int_B \rho_R \mathbf{b} \cdot \delta \mathbf{x}\, \mathrm{d}V$. For the shell, we assume that $\mathbf{b}$ is constant over the thickness and integrate over the thickness with $\mu \approx 1$, which yields

$$\delta F_b \doteq \int_{-\frac{h}{2}}^{+\frac{h}{2}} \rho_R\, \mathbf{b} \cdot \delta \mathbf{x}\, \mu \, \mathrm{d}\zeta \; = \; \rho_R h\, \mathbf{b} \cdot \delta \mathbf{x}_0. \tag{6.128}$$

ii. The VW of external forces acting on the top and bottom surface bounding the shell is defined as $\delta\mathcal{A} \doteq \int_{S^+} \hat{\mathbf{p}}^+ \cdot \delta\mathbf{x}^+ \, \mathrm{d}S^+ + \int_{S^-} \hat{\mathbf{p}}^- \cdot \delta\mathbf{x}^- \, \mathrm{d}S^-$. The subscript "+" indicates the top surface, at $\zeta = +h/2$, and "−" indicates the bottom surface, at $\zeta = -h/2$. The orientation of these surfaces is approximately defined by $\pm\mathbf{a}_3$, where $\mathbf{a}_3 = \mathbf{Q}_0\mathbf{t}_3$. Besides, $\hat{\mathbf{p}}$ denotes the external force corresponding to $\mathbf{a}_3$. Using $\delta\mathbf{x}$ of eq. (6.127) and $\zeta = \pm h/2$, the definition yields

$$\delta\mathcal{A} \doteq \hat{\mathbf{q}} \cdot \delta\mathbf{x}_0 + \hat{\mathbf{m}}_3 \cdot \delta\boldsymbol{\theta}, \tag{6.129}$$

where the external forces and moments for the shell are defined as follows

$$\hat{\mathbf{q}} \doteq \hat{\mathbf{p}}^+ + \hat{\mathbf{p}}^-, \qquad \hat{\mathbf{m}}_3 \doteq \frac{h}{2}\mathbf{a}_3 \times \left(\hat{\mathbf{p}}^+ - \hat{\mathbf{p}}^-\right). \tag{6.130}$$

Note that the projection of $\hat{\mathbf{m}}_3$ on $\mathbf{a}_3$ is equal to zero.

iii. The VW of external forces $\hat{\mathbf{p}}_\nu$ acting upon the lateral boundaries of the shell can be defined as $\delta\mathcal{A}_\nu \doteq \int_{-\frac{h}{2}}^{+\frac{h}{2}} \hat{\mathbf{p}}_\nu \cdot \delta\mathbf{x} \, \mu \, \mathrm{d}\zeta$. The lateral boundary of a non-deformed shell is a surface generated by $\mathbf{t}_3$ along a boundary curve ∂S. If $\mathbf{a}$ denotes a unit vector, tangent to the deformed ∂S, then the vector normal to the lateral boundary is $\boldsymbol{\nu} = \mathbf{a} \times \mathbf{a}_3$, where $\mathbf{a}_3$ generates the lateral surface for the deformed shell. Using eq. (6.127), the VW can be written as

$$\delta\mathcal{A}_\nu \doteq \hat{\mathbf{q}}_\nu \cdot \delta\mathbf{x}_0 + \hat{\mathbf{m}}_{\nu 3} \cdot \delta\boldsymbol{\theta}, \tag{6.131}$$

where the external load and external moment for the shell are

$$\hat{\mathbf{q}}_\nu \doteq \int_{-\frac{h}{2}}^{+\frac{h}{2}} \hat{\mathbf{p}}_\nu \, \mu \, \mathrm{d}\zeta, \qquad \hat{\mathbf{m}}_{\nu 3} \doteq \int_{-\frac{h}{2}}^{+\frac{h}{2}} \zeta\mathbf{a}_3 \times \hat{\mathbf{p}}_\nu \, \mu \, \mathrm{d}\zeta. \tag{6.132}$$

6.4.5 Virtual work equation for shell

Finally, the VW equation for a shell, comprising the contribution of the stress and couple resultants, the RC and the external loads, becomes

$$\delta\Pi_{sh} \doteq \int_S (\,\delta\Sigma + \delta F_{\mathrm{RC}} - \delta F_b - \delta\mathcal{A}) \, \mathrm{d}S - \int_{\partial_\sigma S} \delta\mathcal{A}_\nu \, \mathrm{d}\partial_\sigma S, \tag{6.133}$$

where S denotes the middle surface of the shell and $\partial_\sigma S$ is a traction part of the middle surface boundary. The particular components of eq. (6.133) are defined as follows:

1. The VW of the stress and couple resultants, $\delta\Sigma$, is given by eq. (6.109) for the Biot stress, and by eq. (6.120) for the second Piola–Kirchhoff stress,
2. The variation of the RC term, δF_{RC}, is given for the three-field shell formulation by eq. (6.123), while for the two-field formulation by eq. (6.126),
3. The VW of the body force, δF_b, is given by eq. (6.128),
4. The VW of the external forces acting on the lower and upper bounding surfaces, $\delta\mathcal{A}$, is given by eq. (6.129),
5. The VW of the external forces acting upon the lateral boundaries, $\delta\mathcal{A}_\nu$, is given by eq. (6.131).

6.5 Local shell equations

In this section, the local shell equations are obtained from the shell VW equation for the Biot stress. In several aspects, the derivation is similar to the derivation of the Euler–Lagrange equations for the 3-F potential in Sect. 4.4.

First, we consider the first two terms of the VW equation of eq. (6.133),

$$\delta\Sigma + \delta F_{\text{RC}} = \mathbf{N}^B_s \cdot \delta\boldsymbol{\varepsilon} + \mathbf{M}^B_s \cdot \delta\boldsymbol{\kappa} \; + \; \delta\left[\mathbf{N}_a \cdot \mathfrak{C}_0 + \mathbf{M}_a \cdot (\mathfrak{C}_{,\zeta})_0\right], \quad (6.134)$$

where eqs. (6.109) and (6.122) were used. They can be transformed to

$$\delta\Sigma + \delta F_{\text{RC}} = \mathbf{N} \cdot (\delta\boldsymbol{\varepsilon}_\alpha \otimes \mathbf{t}_\alpha) + \mathbf{M} \cdot (\delta\boldsymbol{\kappa}_\alpha \otimes \mathbf{t}_\alpha) + \delta\mathbf{N}_a \cdot \mathfrak{C}_0 + \delta\mathbf{M}_a \cdot (\mathfrak{C}_{,\zeta})_0, \quad (6.135)$$

where $\mathbf{N}$ and $\mathbf{M}$ are defined in eq. (6.108). The transformation which was used to obtain the latter form is similar to that of eq. (4.28), plus $\mathbf{P} \cdot \delta\mathbf{F} = (\mathbf{Q}^T\mathbf{P}) \cdot (\mathbf{Q}^T\delta\mathbf{F})$, to have a full analogy. This form is more convenient as it does not contain symmetric tensors, so we can use vectors instead of tensors. Below, we separately transform the first two terms and the last two terms on the r.h.s. of eq. (6.135).

First two terms. First, we consider the case when the RC equations are not satisfied and the strain vectors $\boldsymbol{\varepsilon}_\alpha$ and $\boldsymbol{\kappa}_\alpha$ are defined by eq. (6.39). The first two terms of eq. (6.135) can be transformed as follows:

$$\mathbf{N} \cdot (\delta\boldsymbol{\varepsilon}_\alpha \otimes \mathbf{t}_\alpha) = \mathbf{n}_\alpha \cdot \delta\boldsymbol{\varepsilon}_\alpha = \mathbf{n}_\alpha \cdot \delta(\mathbf{Q}_0^T \mathbf{x}_{0,\alpha}), \quad (6.136)$$

$$\begin{aligned}\mathbf{M} \cdot (\delta\boldsymbol{\kappa}_\alpha \otimes \mathbf{t}_\alpha) &= \mathbf{m}_\alpha \cdot \delta\boldsymbol{\kappa}_\alpha = \mathbf{m}_\alpha \cdot \left[\delta(\mathbf{Q}_0^T \bar{\boldsymbol{\omega}}_\alpha) \times \mathbf{t}_3\right] \\ &= (\mathbf{t}_3 \times \mathbf{m}_\alpha) \cdot \delta(\mathbf{Q}_0^T \bar{\boldsymbol{\omega}}_\alpha), \end{aligned} \quad (6.137)$$

where the stress and couple resultant vectors are defined in eq. (6.107). The last form of eq. (6.137) was obtained by using the identity $\mathbf{a}\cdot(\mathbf{b}\times\mathbf{c}) = \mathbf{b}\cdot(\mathbf{c}\times\mathbf{a})$, which is valid for arbitrary vectors $\mathbf{a}, \mathbf{b}, \mathbf{c}$. Note that

$$\mathbf{t}_3 \times \mathbf{m}_\alpha = \int_{-\frac{h}{2}}^{+\frac{h}{2}} \mathbf{t}_3 \times \zeta \mathbf{f}_\alpha(\zeta)\, \mu \, \mathrm{d}\zeta, \tag{6.138}$$

can also serve as a definition of the shell couple resultant vector, alternative to eq. (6.107).

Furthermore, we transform eqs. (6.136) and (6.137) using the forward-rotated stress and couple resultant vectors of eq. (6.113). First, we transform the respective terms according to the scheme $\mathbf{a}\cdot\delta\mathbf{b} = (\mathbf{Q}_0^T\mathbf{Q}_0\mathbf{a})\cdot\delta\mathbf{b} = (\mathbf{Q}_0\mathbf{a})\cdot(\mathbf{Q}_0\delta\mathbf{b})$, where $\mathbf{a}$ and $\mathbf{b}$ are vectors. Then we introduce the forward-rotated vectors, which yields

$$\mathbf{n}_\alpha \cdot \delta(\mathbf{Q}_0^T \mathbf{x}_{0,\alpha}) = \mathbf{n}_\alpha^* \cdot [\mathbf{Q}_0 \delta(\mathbf{Q}_0^T \mathbf{x}_{0,\alpha})] = \mathbf{n}_\alpha^* \cdot (\delta \mathbf{x}_{0,\alpha} - \delta\boldsymbol{\theta} \times \mathbf{x}_{0,\alpha}), \tag{6.139}$$

$$\begin{aligned}(\mathbf{t}_3 \times \mathbf{m}_\alpha)\cdot\delta(\mathbf{Q}_0^T\bar{\boldsymbol{\omega}}_\alpha) &= (\mathbf{a}_3 \times \mathbf{m}_\alpha^*)\cdot[\mathbf{Q}_0\delta(\mathbf{Q}_0^T\bar{\boldsymbol{\omega}}_\alpha)] \\ &= (\mathbf{a}_3 \times \mathbf{m}_\alpha^*)\cdot(\delta\bar{\boldsymbol{\omega}}_\alpha - \delta\boldsymbol{\theta}\times\bar{\boldsymbol{\omega}}_\alpha) \\ &= (\mathbf{a}_3 \times \mathbf{m}_\alpha^*)\cdot\delta\boldsymbol{\theta}_{,\alpha},\end{aligned} \tag{6.140}$$

where $\mathbf{n}_\alpha^*$ and $\mathbf{m}_\alpha^*$ are defined in eq. (6.113), and eq. (6.97) was applied to $[\mathbf{Q}_0\delta(\mathbf{Q}_0^T\mathbf{x}_{0,\alpha})]$ and $[\mathbf{Q}_0\delta(\mathbf{Q}_0^T\bar{\boldsymbol{\omega}}_\alpha)]$. The last form of eq. (6.140) is obtained by using the identity of eq. (6.20). We note that $\mathbf{a}_3 \times \mathbf{m}_\alpha^* = (\mathbf{Q}_0\mathbf{t}_3)\times(\mathbf{Q}_0\mathbf{m}_\alpha) = \mathbf{Q}_0(\mathbf{t}_3\times\mathbf{m}_\alpha)$, i.e. it is the forward-rotated shell couple resultant vector. Hence,

$$\delta\Sigma + \delta F_{\mathrm{RC}} = \mathbf{n}_\alpha^* \cdot (\delta\mathbf{x}_{0,\alpha} - \delta\boldsymbol{\theta}\times\mathbf{x}_{0,\alpha}) + (\mathbf{a}_3\times\mathbf{m}_\alpha^*)\cdot\delta\boldsymbol{\theta}_{,\alpha}. \tag{6.141}$$

Special case: RC equations satisfied. If the RC equations are satisfied, then the strain vectors $\boldsymbol{\varepsilon}_\alpha$ and $\boldsymbol{\kappa}_\alpha$ can be replaced by their projections $\boldsymbol{\varepsilon}_\alpha^\bullet$ and $\boldsymbol{\kappa}_\alpha^\bullet$ of eq. (6.76). Then the first two terms of the r.h.s. of eq. (6.135) can be transformed as follows:

$$\mathbf{N}\cdot(\delta\boldsymbol{\varepsilon}_\alpha^\bullet \otimes \mathbf{t}_\alpha) = \mathbf{n}_\alpha \cdot \delta\boldsymbol{\varepsilon}_\alpha^\bullet = \mathbf{n}_\alpha^\bullet \cdot \delta\boldsymbol{\varepsilon}_\alpha, \tag{6.142}$$

$$\mathbf{M}\cdot(\delta\boldsymbol{\kappa}_\alpha^\bullet \otimes \mathbf{t}_\alpha) = \mathbf{m}_\alpha \cdot \delta\boldsymbol{\kappa}_\alpha^\bullet = \mathbf{m}_\alpha^\bullet \cdot \delta\boldsymbol{\kappa}_\alpha, \tag{6.143}$$

where the projections of $\mathbf{n}_\alpha$ and $\mathbf{m}_\alpha$ of eq. (6.107) on the tangent plane $\{\mathbf{t}_1, \mathbf{t}_2\}$ are defined as

$$\mathbf{n}^{\bullet}_{\alpha} \doteq (\mathbf{n}_{\alpha} \cdot \mathbf{t}_{\beta})\, \mathbf{t}_{\beta}, \qquad \mathbf{m}^{\bullet}_{\alpha} \doteq (\mathbf{m}_{\alpha} \cdot \mathbf{t}_{\beta})\, \mathbf{t}_{\beta}. \tag{6.144}$$

Hence, we can use the projections of stress and couple resultant vectors instead of the projections of strain vectors, and in the final eq. (6.141), the asterisk denotes the forward-rotated projections, i.e. $\mathbf{n}^{*}_{\alpha} = \mathbf{Q}_0 \mathbf{n}^{\bullet}_{\alpha}$, and $\mathbf{m}^{*}_{\alpha} = \mathbf{Q}_0 \mathbf{m}^{\bullet}_{\alpha}$.

Last two terms. The last two terms of eq. (6.135) can be written as

$$\delta \mathbf{N}_a \cdot \mathfrak{C}_0 = \operatorname{skew}(\delta \mathbf{n}_{\alpha} \otimes \mathbf{t}_{\alpha}) \cdot \operatorname{skew}[(\mathbf{Q}_0^T \mathbf{x}_{0,\beta}) \otimes \mathbf{t}_{\beta}] \approx \delta N_{12}\, C_{12}, \tag{6.145}$$

$$\delta \mathbf{M}_a \cdot (\mathfrak{C}_{,\varsigma})_0 = \operatorname{skew}(\delta \mathbf{m}_{\alpha} \otimes \mathbf{t}_{\alpha}) \cdot \operatorname{skew}[(\mathbf{Q}_0^T \bar{\boldsymbol{\omega}}_{\beta}) \otimes \mathbf{t}_{\beta}] \approx \delta M_{12}\, D_{12}, \tag{6.146}$$

where $N_{12} = (\mathbf{N}_a)_{12}$, $M_{12} = (\mathbf{M}_a)_{12}$, and C_{12} and D_{12} are given by eq. (6.124). D_{12} is neglected by the reasons given in the *Remark* in Sect. 6.2.

Local shell equations. To obtain the local shell equations, we have to perform several additional transformations.

A. First, we transform the first term of eq. (6.141), i.e. $A \doteq \int_S \mathbf{n}^{*}_{\alpha} \cdot (\delta \mathbf{x}_{0,\alpha} - \delta \boldsymbol{\theta} \times \mathbf{x}_{0,\alpha})\, \mathrm{d}S$. On use of the divergence theorem, $\operatorname{Div}(\mathbf{T}^{\mathrm{T}} \mathbf{v}) = \mathbf{T} \cdot \operatorname{Grad}\mathbf{v} + (\operatorname{Div}\mathbf{T}) \cdot \mathbf{v}$, ([33], eq. (5.5.19)), where $\mathbf{T}$ is the second-rank tensor and $\mathbf{v}$ is a vector, we obtain

$$\mathbf{n}^{*}_{\alpha} \cdot \delta \mathbf{x}_{0,\alpha} = \mathbf{N}^{*} \cdot \operatorname{Grad}\delta \mathbf{x}_0 = \operatorname{Div}(\mathbf{N}^{*\mathrm{T}} \delta \mathbf{x}_0) - (\operatorname{Div}\mathbf{N}^{*}) \cdot \delta \mathbf{x}_0, \tag{6.147}$$

where $\operatorname{Grad}(\cdot) = (\cdot)_{,\alpha} \otimes \mathbf{a}_{\alpha}$, $\mathbf{N}^{*} = \mathbf{n}^{*}_{\alpha} \otimes \mathbf{a}_{\alpha}$, and $\operatorname{Div}\mathbf{N}^{*} = \mathbf{n}^{*}_{\alpha,\alpha} \otimes \mathbf{a}_{\alpha}$. The integral of its first component is

$$\int_S \operatorname{Div}(\mathbf{N}^{*\mathrm{T}} \delta \mathbf{x}_0)\, \mathrm{d}S = \int_{\partial S} (\mathbf{N}^{*\mathrm{T}} \delta \mathbf{x}_0) \cdot \boldsymbol{\nu}\, \mathrm{d}S = \int_{\partial S} (\mathbf{N}^{*} \boldsymbol{\nu}) \cdot \delta \mathbf{x}_0\, \mathrm{d}\partial S, \tag{6.148}$$

by the Gauss' integral identity ([33], eq. (5.8.11), p. 164), where ∂S denotes the boundary curve on the mid-surface, and $\boldsymbol{\nu}$ is a vector tangent to the mid-surface and normal to this curve. Besides, for the second part of A, by the cyclic permutation of vectors, see [33], eq. (4.9.8), p. 66, we obtain

$$-\mathbf{n}^{*}_{\alpha} \cdot (\delta \boldsymbol{\theta} \times \mathbf{x}_{0,\alpha}) = -\delta \boldsymbol{\theta} \cdot (\mathbf{x}_{0,\alpha} \times \mathbf{n}^{*}_{\alpha}). \tag{6.149}$$

Collecting the above formulas, we have

$$A = -\int_S \mathbf{n}^{*}_{\alpha,\alpha} \cdot \delta \mathbf{x}_0\, \mathrm{d}S - \int_S (\mathbf{x}_{0,\alpha} \times \mathbf{n}^{*}_{\alpha}) \cdot \delta \boldsymbol{\theta}\, \mathrm{d}S + \int_{\partial S} [(\mathbf{n}^{*}_{\alpha} \otimes \mathbf{a}_{\alpha}) \boldsymbol{\nu}] \cdot \delta \mathbf{x}_0\, \mathrm{d}\partial S. \tag{6.150}$$

B. The second term of eq. (6.141), i.e. $B \doteq \int_S (\mathbf{a}_3 \times \mathbf{m}_\alpha^*) \cdot \delta\boldsymbol{\theta}_{,\alpha} \, \mathrm{d}S$, is treated similarly as $\int_S \mathbf{n}_\alpha^* \cdot \delta\mathbf{x}_{0,\alpha} \, \mathrm{d}S$, and provides

$$B = -\int_S (\mathbf{a}_3 \times \mathbf{m}_\alpha^*)_{,\alpha} \cdot \delta\boldsymbol{\theta} \, \mathrm{d}S + \int_{\partial S} \{[(\mathbf{a}_3 \times \mathbf{m}_\alpha^*) \otimes \mathbf{a}_\alpha]\boldsymbol{\nu}\} \cdot \delta\boldsymbol{\theta} \, \mathrm{d}\partial S. \tag{6.151}$$

Finally, collecting all the terms, we obtain

$$\begin{aligned} &\int_S -(\mathbf{n}_{\alpha,\alpha}^* + \hat{\mathbf{q}} + h\rho_R \mathbf{b}) \cdot \delta\mathbf{x}_0 - [(\mathbf{a}_3 \times \mathbf{m}_\alpha^*)_{,\alpha} + (\mathbf{x}_{0,\alpha} \times \mathbf{n}_\alpha^*) + \hat{\mathbf{m}}_3] \cdot \delta\boldsymbol{\theta} \, \mathrm{d}S \\ &\quad + \int_S \tfrac{1}{2} (\mathbf{x}_{0,1} \cdot \mathbf{a}_2 - \mathbf{x}_{0,2} \cdot \mathbf{a}_1) \; \delta N_{12} \, \mathrm{d}S \\ &\quad + \int_{\partial S} [(\mathbf{n}_\alpha^* \otimes \mathbf{a}_\alpha)\boldsymbol{\nu}] \cdot \delta\mathbf{x}_0 + \{[(\mathbf{a}_3 \times \mathbf{m}_\alpha^*) \otimes \mathbf{a}_\alpha]\boldsymbol{\nu}\} \cdot \delta\boldsymbol{\theta} \, \mathrm{d}\partial S = 0, \end{aligned} \tag{6.152}$$

from which we obtain the following local equations:

$$\begin{aligned} \mathbf{n}_{\alpha,\alpha}^* + h\rho_R \mathbf{b} + \hat{\mathbf{q}} &= \mathbf{0} && \text{in S}, \\ (\mathbf{a}_3 \times \mathbf{m}_\alpha^*)_{,\alpha} + \mathbf{x}_{0,\alpha} \times \mathbf{n}_\alpha^* + \hat{\mathbf{m}}_3 &= \mathbf{0} && \text{in S}, \\ \mathbf{x}_{0,1} \cdot \mathbf{a}_2 - \mathbf{x}_{0,2} \cdot \mathbf{a}_1 &= 0 && \text{in S}, \\ (\mathbf{n}_\alpha^* \otimes \mathbf{a}_\alpha)\,\boldsymbol{\nu} = \mathbf{0}, \qquad [(\mathbf{a}_3 \times \mathbf{m}_\alpha^*) \otimes \mathbf{a}_\alpha]\,\boldsymbol{\nu} &= \mathbf{0} && \text{on } \partial\text{S}. \end{aligned} \tag{6.153}$$

These are the LMB and AMB equations, the drill RC equation in the form of eq. (6.30), and the natural BC for a shell with the Reissner kinematics of first order, respectively. These equations involve $\mathbf{n}_\alpha^*$ and $\mathbf{m}_\alpha^*$, which are the forward-rotated vectors of eq. (6.113).

The AMB equation $(6.153)_2$ contains only two scalar equations, because its projection on the director $\mathbf{a}_3$ is equal to zero. The proof is given below.

Proof. The projections of selected terms of eq. $(6.153)_2$ are as follows:

$$(\mathbf{a}_3 \times \mathbf{m}_\alpha^*)_{,\alpha} \cdot \mathbf{a}_3 = (\mathbf{a}_{3,\alpha} \times \mathbf{m}_\alpha^*) \cdot \mathbf{a}_3 + (\mathbf{a}_3 \times \mathbf{m}_{\alpha,\alpha}^*) \cdot \mathbf{a}_3 = (\mathbf{a}_{3,\alpha} \times \mathbf{m}_\alpha^*) \cdot \mathbf{a}_3, \tag{6.154}$$

$$\hat{\mathbf{m}}_3 \cdot \mathbf{a}_3 = \frac{h}{2} \left[\mathbf{a}_3 \times (\hat{\mathbf{p}}^+ - \hat{\mathbf{p}}^-)\right] \cdot \mathbf{a}_3 = 0, \tag{6.155}$$

by which, the projection of the whole eq. $(6.153)_2$ is

$$(\mathbf{a}_{3,\alpha} \times \mathbf{m}_\alpha^*) \cdot \mathbf{a}_3 + (\mathbf{x}_{0,\alpha} \times \mathbf{n}_\alpha^*) \cdot \mathbf{a}_3 = 0. \tag{6.156}$$

Using the definitions of eqs. (6.107) and (6.113), we have

$$\mathbf{n}_i^* = \int_{-\frac{h}{2}}^{+\frac{h}{2}} \mathbf{Q}_0 \mathbf{f}_i(\zeta)\, \mu\, d\zeta, \qquad \mathbf{m}_i^* = \int_{-\frac{h}{2}}^{+\frac{h}{2}} \zeta\, \mathbf{Q}_0 \mathbf{f}_i(\zeta)\, \mu\, d\zeta. \tag{6.157}$$

Inserting the above formula into eq. (6.156) and considering the integrands only, we obtain

$$[(\mathbf{x}_0 + \zeta \mathbf{a}_3)_{,\alpha} \times (\mathbf{Q}_0 \mathbf{f}_\alpha)] \cdot \mathbf{a}_3 = 0. \tag{6.158}$$

This equation can be rewritten in a different form. Note that $(\mathbf{x}_0 + \zeta \mathbf{a}_3)_{,\alpha} = \mathbf{x}(\zeta)_{,\alpha}$, by eq. (6.2), and that $\mathbf{Q}_0 \mathbf{f}_\alpha = \mathbf{p}_\alpha$. The latter results from the definition $\mathbf{T} \doteq \mathbf{Q}^T \mathbf{P}$, see Sect. 4.4, as well as $\mathbf{T} = \mathbf{f}_i \otimes \mathbf{t}_i$ and $\mathbf{P} = \mathbf{p}_i \otimes \mathbf{t}_i$, from which we obtain $\mathbf{f}_i = \mathbf{Q}^T \mathbf{p}_i$, where the stress vectors $\mathbf{f}_i$ and $\mathbf{p}_i$ correspond to the vector $\mathbf{t}_i$ of the mid-surface ortho-normal basis. Then eq. (6.158) can be rewritten as

$$[\mathbf{x}(\zeta)_{,\alpha} \times \mathbf{p}_\alpha] \cdot \mathbf{a}_3 = 0. \tag{6.159}$$

Now we recall the AMB equation, (4.5), $\mathbf{F} \times \mathbf{P} = \mathbf{0}$, which can be rewritten as $\mathbf{x}_{,i} \times \mathbf{p}_i = \mathbf{0}$, $(i = 1, 2, 3)$. From this equation we obtain $\mathbf{x}_{,\alpha} \times \mathbf{p}_\alpha = -\mathbf{x}_{,3} \times \mathbf{p}_3 = -\mathbf{a}_3 \times \mathbf{p}_3$, because $\mathbf{x}_{,3} = \mathbf{x}_{,\zeta} = \mathbf{a}_3$, for $\mathbf{x}(\zeta)$ of eq. (6.2). Then, eq. (6.159) becomes $-(\mathbf{a}_3 \times \mathbf{p}_3) \cdot \mathbf{a}_3 = 0$, i.e. is identically equal to zero, which ends the proof that a projection of eq. $(6.153)_2$ on $\mathbf{a}_3$ is equal to zero. □

Remark. We can count the number of scalar equations in the system of equations (6.153). The first equation contains three scalar equations, while the second one only two, because its projection on the director $\mathbf{a}_3$ is equal to zero, as shown above. Equation $(6.153)_3$ is the RC for the drilling rotation, ω_d, i.e. eq. (6.30), and it provides the sixth equation.

Summarizing, we have six scalar equations which, using the CL and kinematical relations for strains, can be written in terms of $\mathbf{x}_0$ and $\mathbf{Q}_0$. If these equations are written in an incremental form, then we can assume that the increment of the angle of rotation is smaller than π, so the rotation tensor $\mathbf{Q}_0$ can be parameterized using three parameters, see Chap. 8. The current position of the mid-surface $\mathbf{x}_0$ also provides three parameters, and hence, the number of equations is equal to the number of unknowns.

Local shell equations for shell resultants for second Piola–Kirchhoff stress. For the 3D continuum, when we express the governing equations (4.4)–(4.6) in terms of the second Piola–Kirchhoff stress $\mathbf{S}$, they become more complicated, because they depend not only on $\mathbf{S}$ but also on the deformation gradient $\mathbf{F}$, through $\mathbf{P} = \mathbf{F}\mathbf{S}$. A similar situation is in the case of shells when we introduce the shell stress and couple resultants corresponding to $\mathbf{S}$. Then the local shell equations also depend on the particular parts of $\mathbf{F}$, and are complicated. For this reason, they are not provided here.

6.6 Enhanced shell kinematics

A majority of the shell FEs used nowadays are based on the Reissner kinematics of the first order. However, a lot of research is being carried out to extend the range of applicability of shell elements, and this can be done only by enhancing the shell kinematics. This requires additional kinematical parameters which, unfortunately, render the equations to become very complicated and the FE approximations more difficult to control. As a consequence, very few of the extensions have proven manageable and remain as a permanent part of shell FEs.

Having this in mind, we separately present the most important enhancements and their geometrical meaning, as it is not difficult to combine these assumptions together. Then, however, the equations for the combined assumptions are not a simple sum of the equations for particular cases, but contain terms coupling various parameters. The judicious handling of these very complicated equations is necessary.

6.6.1 Two normal stretches

The Reissner kinematics with two additional normal stretches is based on the following assumptions:

A1. The position vector in the deformed configuration is a quadratic polynomial of the thickness coordinate

$$\mathbf{x}(\zeta) = \mathbf{x}_0 + \lambda(\zeta)\,\mathbf{a}_3, \qquad \lambda(\zeta) \doteq \zeta\lambda_0 + \frac{1}{2}\zeta^2\lambda_1, \tag{6.160}$$

where the scalars λ_0 and λ_1 are the normal stretch parameters, i.e. the normal components of the right stretching tensor, see Fig. 6.3. Comparing with eq. (6.2) for the Reissner kinematics of first order, we see that $\lambda(\zeta)$ is used in place of ζ.

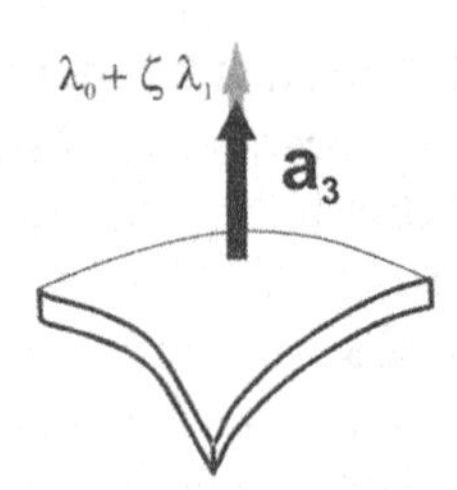

Fig. 6.3 Normal stretch.

A2. The rotations have the same form as for the Reissner kinematics of the first order, see eq. (6.1),

$$\mathbf{Q}(\zeta) \approx \mathbf{Q}_0, \tag{6.161}$$

The two-parameter scalar-valued function multiplying the director is used in [48, 49] for the Kirchhoff shell and, subsequently, a general form of this function is used for the Reissner kinematics in [147]. An alternative form of Reissner kinematics with two stretches is given in [254], eq. (50).

Note that for two normal stretch parameters, the order of approximation of the normal strain is the same as when recovered from the zero normal stress (ZNS) condition. The normal stretches λ_0 and λ_1 are needed for the following purposes:

1. to account for the thickness changes in stretching and for the shift of the initial middle surface in bending,
2. to avoid recovery of the normal strains from either the zero normal stress (ZNS) condition or the incompressibility condition, which are discussed in Sect. 7.2.

Finally, we note that the simplest way of handling the stretches in FEs is to treat them as the local elemental parameters. Then they are discontinues at the element's boundaries, but the boundary conditions for them are not required.

6.6.2 In-plane twist rotation

The in-plane twist rotation is performed about the shell director but is different from the drilling rotation, see Fig. 6.4.

The Reissner kinematics with the in-plane twist rotation is based on the following assumptions:

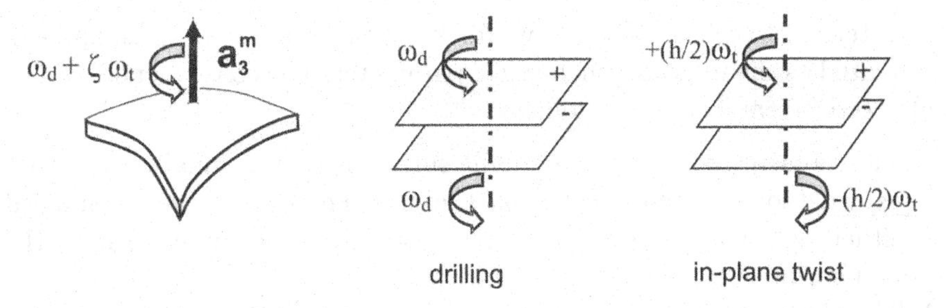

Fig. 6.4 Drilling and in-plane twist rotations.

A1. The position vector in the deformed configuration has the same form as for the classical Reissner hypothesis of eq. (6.2), i.e.

$$\mathbf{x}(\zeta) \doteq \mathbf{x}_0 + \zeta \mathbf{a}_3. \tag{6.162}$$

A2. The rotations are assumed as

$$\mathbf{Q}(\zeta) = \mathbf{Q}_t(\zeta)\,\mathbf{Q}_0, \tag{6.163}$$

where $\mathbf{Q}_t(\zeta)$ is the in-plane twist rotation tensor which is superimposed on $\mathbf{Q}_0$ used by the first-order kinematics, eq. (6.1).

We can split $\mathbf{Q}_0 = \mathbf{Q}_d\,\mathbf{Q}_m$, where $\mathbf{Q}_d$ is the drilling rotation tensor and $\mathbf{Q}_m$ is the rotation around an axis tangent to the middle surface and both are constant over the shell thickness. Then we can rewrite eq. (6.163) as

$$\mathbf{Q}(\zeta) = \mathbf{Q}_t(\zeta)\,\mathbf{Q}_0 = \mathbf{Q}_t(\zeta)\,\mathbf{Q}_d\,\mathbf{Q}_m, \tag{6.164}$$

where $\mathbf{Q}_t(\zeta)$ and $\mathbf{Q}_d$ are performed about the same forward-rotated director $\mathbf{a}_3^m \doteq \mathbf{Q}_m \mathbf{t}_3$. The normal rotation angle $\omega(\zeta)$ is approximated linearly,

$$\omega(\zeta) = \omega_d + \zeta\,\omega_t, \tag{6.165}$$

where ω_d is the drilling rotation and ω_t is the in-plane twist rotation, see Fig. 6.4. We assume that ω_t is small and use a linearized form of $\mathbf{Q}_t(\zeta)$.

The Reissner kinematics with the in-plane twist rotation was proposed in [252]. It was shown, using the first order RC equation, that the in-plane twist parameter can be expressed as

$$\omega_t = \frac{\mathbf{a}_{3,1} \cdot \mathbf{a}_2 - \mathbf{a}_{3,2} \cdot \mathbf{a}_1}{\mathbf{x}_{0,1} \cdot \mathbf{a}_1 + \mathbf{x}_{0,2} \cdot \mathbf{a}_2}, \tag{6.166}$$

i.e. in terms of vectors available within the first-order kinematics, and can be treated as the internal kinematical parameter. The consequences of this enhanced kinematics are as follows:

1. The moment normal to the middle surface is a variationally consistent external load, which means that the in-plane bending can be induced either by an in-plane force or by a twisting moment normal to the mid-surface.
2. The new first-order strain, $\boldsymbol{\rho}$, can be expressed in terms of the known bending and stretching strain vectors.
3. The presence of the twist correction should eliminate the in-plane twist rigid body mode in some elements, see [98], p. 333.

Finally, we note that treating the in-plane twist rotation as an internal parameter, we enrich the shell kinematics without increasing the number of dofs of a discrete model.

6.6.3 Warping parameters for cross-section

To account for the bubble-like warping of a cross-section, we have to enhance the position vector and the rotations simultaneously. The accordingly enhanced Reissner kinematics is based on the following assumptions:

A1. The current position vector is assumed as a quadratic polynomial of the thickness coordinate

$$\mathbf{x}(\zeta) = \mathbf{x}_0 + \zeta \mathbf{a}_3 + B(\zeta)(d_1 \mathbf{a}_1 + d_2 \mathbf{a}_2), \qquad B(\zeta) \doteq 1 - \frac{4}{h^2}\zeta^2, \quad (6.167)$$

where d_1 and d_2 are the scalar warping coefficients, and $B(\zeta)$ is a bubble function, such that $B(0) = 1$ and $B(\pm h/2) = 0$.

A2. The rotations are assumed as

$$\mathbf{Q}(\zeta) = \mathbf{Q}_1(\zeta)\, \mathbf{Q}_0, \qquad (6.168)$$

where $\mathbf{Q}_0$ was used in the first-order kinematics, eq. (6.1). The ζ-dependent rotation tensor $\mathbf{Q}_1$ is assumed in a linearized form

$$\mathbf{Q}_1(\zeta) \approx \mathbf{I} + \zeta \boldsymbol{\psi}^* \times \mathbf{I}, \qquad \boldsymbol{\psi}^* = \psi_1^*\, \mathbf{a}_1 + \psi_2^*\, \mathbf{a}_2, \qquad (6.169)$$

where ψ_α^* are rotation coefficients.

Note that $\boldsymbol{\psi}(\zeta) \doteq \zeta\, \boldsymbol{\psi}^*$ is equal to zero at $\zeta = 0$, and has opposite directions at $\zeta = \pm h/2$, see Fig. 6.5. We see that only the tangent

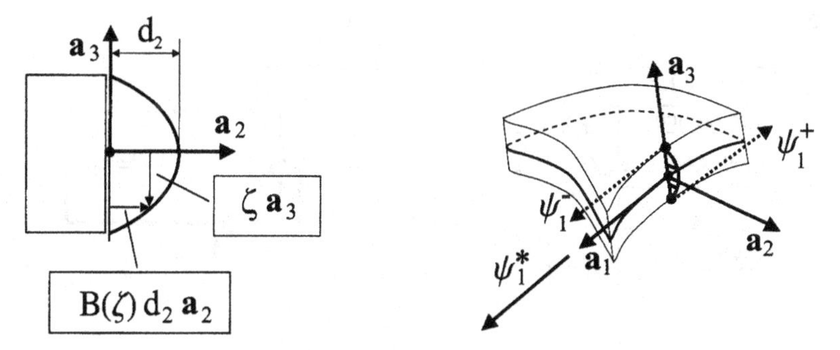

Fig. 6.5 Bubble-like warping of cross-section. Warping parameter d_2 and tangent rotation $\boldsymbol{\psi}_1^* = \psi_1^* \mathbf{a}_1$ are linked.

rotations correspond to the warping coefficients d_α. For this reason, the normal vector $\psi_3^* \mathbf{a}_3$ is omitted in eq. (6.169).

Note that the higher-order terms in eq. (6.167) are imposed *hierarchically* on the first order kinematics because $\mathbf{a}_\alpha \doteq \mathbf{Q}_0 \mathbf{t}_\alpha$ are the forward-rotated vectors. Similarly, $\mathbf{Q}_1(\zeta)$ is imposed hierarchically on $\mathbf{Q}_0$ in eq. (6.168). This kinematics in a similar but slightly different form was addressed in [253, 254].

6.6.4 Shift of the reference surface

The shift of the position of the reference surface is used in layered composites with non-symmetric stacking sequence of layers.

The total thickness of the laminate is designated by H, the normal coordinate $z \in [0, H]$, and the reference surface is located at $z = z_0$. The coordinates of the interfaces of layers are shown in Fig. 6.6a. The initial position vector is defined as

$$\mathbf{y}(\zeta) = \mathbf{y}_0 + (z - z_0)\, \mathbf{t}_3, \qquad z, z_0 \in [0, H], \tag{6.170}$$

where z_0 is the position of the reference surface. The value of z_0 is calculated as an average of positions of the neutral axes for unidirectional bending in two directions. For symmetric stacking sequence of layers, it coincides with the middle surface, but for non-symmetric ones, it is shifted. Note that z_0 is a geometric parameter but depends on particular properties of the laminate and affects its effective (integral) properties.

The Reissner kinematic with the shift parameter is based on the following assumptions:

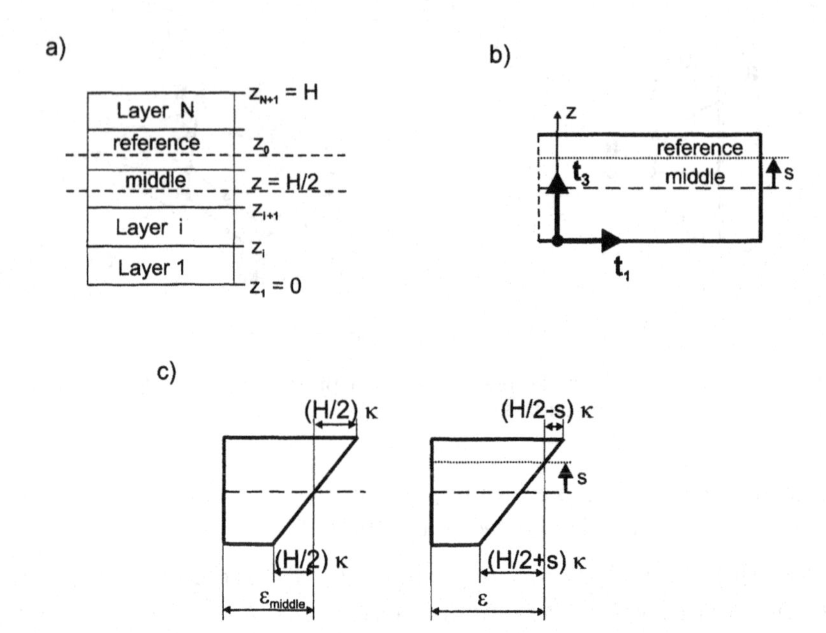

Fig. 6.6 Shift of the reference surface from the middle position. a) Coordinates of interfaces of layers. b) Shift s of the neutral axis. c) Parts of strain $E_{11}(z)$.

A1. The position vector in the deformed configuration is

$$\mathbf{x}(\zeta) = \mathbf{x}_0 + (z - z_0)\,\mathbf{a}_3, \qquad z, z_0 \in [0, H]. \tag{6.171}$$

Comparing with eq. (6.2) for the first-order Reissner kinematics, we see that $(z - z_0)$ is used in place of ζ.

A2. The rotations are identical as for the Reissner kinematics of first order, eq. (6.1),

$$\mathbf{Q}(\zeta) \approx \mathbf{Q}_0, \tag{6.172}$$

The shift of the reference surface from the middle position means that the split of the in-plane strains into the stretching and bending parts is different, see Fig. 6.6.

7

Shell-type constitutive equations

In this chapter we assume that the constitutive equation for a 3D body is given and we derive various forms of constitutive equation for shells. The elastic materials, compressible and incompressible, are considered.

The shell-type constitutive equations are discussed as follows: for the 3D shells in Sect. 7.1, while for the shells for which the normal strain is recovered in Sect. 7.2. The correction factors for the transverse shear are derived in Sect. 7.3.

7.1 Constitutive equations for 3D shells

Introduction. The 3D shells, by definition, have a non-zero normal strain and two formulations belong to this class:

1. the so-called "solid-shells", which have nodes on the top and bottom surfaces bounding the shell and use the translational degrees of freedom but not rotations. The "solid-shell" element, which is a counterpart of the four-node shell element, has eight nodes and three dofs per node, see [100, 199, 241]. The normal strain κ_{33} must be enhanced and properly approximated, see [31].
2. the shells based on Reissner kinematics with two additional normal stretch parameters, see Sect. 6.6.1. The normal stretch parameters enhance the shell kinematics but require additional equilibrium equations. They can be treated as elemental variables and eliminated at the element's level.

The 3D shells are not within the scope of this book but the constitutive equations for them have relatively simple forms and, hence, they are instructive. The constitutive equations for 3D shells can assume the following forms:

1. *The 3D constitutive equations*, written for 3D stresses and strains, can be used without any modification, see Sect. 7.1.1. They are particularly useful for non-linear complicated constitutive laws, e.g. for plasticity.
2. *The shell-type constitutive equations*, which are written for shell stress and couple resultants. They can be formulated either in the incremental form, see Sect. 7.1.2, or in the general form, see Sect. 7.1.3. They are particularly useful for linear constitutive laws, especially when they can be integrated over the thickness either analytically as for the linear SVK material or numerically as for layered composites.

7.1.1 Incremental 3D constitutive equations

Assume that the strain $\mathbf{E}$ is a polynomial of the normal coordinate $\zeta \in [+h/2, -h/2]$, i.e. $\mathbf{E}(\zeta)$, and all components of $\mathbf{E}$ are non-zero. Let $\mathbf{S}$ designate the stress which is work-conjugate to $\mathbf{E}$. The VW of the stress $\mathbf{S}$ is

$$\delta \mathcal{W} = \int_V \delta \mathbf{E} \cdot \mathbf{S} \, \mathrm{d}V. \tag{7.1}$$

To define the tangent matrix, we calculate the directional derivative of $\delta \mathcal{W}$ which yields

$$\Delta \delta \mathcal{W} = \int_V [\delta \mathbf{E} \cdot (\mathbb{C} \, \Delta \mathbf{E}) + \mathbf{S} \cdot \Delta \delta \mathbf{E}] \, \mathrm{d}V, \tag{7.2}$$

where $\mathbb{C} \doteq \partial \mathbf{S} / \partial \mathbf{E}$ denotes the 3D constitutive operator. Note that

1. To calculate the integral over the volume V we have to integrate over the thickness h and over the reference surface A, see Sect. 10.5.
2. The stress $\mathbf{S}$ is updated by using the incremental constitutive equation

$$\Delta \mathbf{S} = \mathbb{C} \, \Delta \mathbf{E} \tag{7.3}$$

at Gauss points. This form is general and applies to arbitrary non-linear materials.

In this approach, the modifications related to shells are minimal:

1. For shells based on the Reissner kinematics, the strain $\mathbf{E}(\zeta) = \boldsymbol{\varepsilon} + \zeta \boldsymbol{\kappa}$ is used,

2. The shell stress and couple resultants can be obtained in the post-processing phase but are not required to obtain the solution. The shell stress and couple resultants are defined as the following integrals:

$$\mathbf{N}_i \doteq \int_h \zeta^i \, \mathbf{S}(\zeta) \, \mu \, \mathrm{d}\zeta, \qquad i = 0, ..., L, \tag{7.4}$$

where ζ^i denotes the i-th power of ζ, $\mu \doteq \det \mathbf{Z}$, and $\mathbf{Z}$ is the shifter tensor, see eq. (5.10). The same Gauss points over the thickness are used as in the integration of $\Delta\delta\mathcal{W}$.

7.1.2 Incremental constitutive equations for shell resultants

Below, we derive the shell resultants and the constitutive (stiffness) matrices assuming that strain $\mathbf{E}$ is represented as the polynomial of the normal coordinate ζ,

$$\mathbf{E}(z) = \mathbf{E}_0 + \zeta \, \mathbf{E}_1 + ... + \zeta^L \, \mathbf{E}_L, \tag{7.5}$$

where the number of terms L is arbitrary. This form encompasses the first-order as well as the second-order kinematics of a shell as special cases. Separating the integration over the thickness from the integration over the reference surface and using eq. (7.5), the VW of stress $\mathbf{S}$ of eq. (7.1) becomes

$$\delta\mathcal{W} = \int_A [\delta\mathbf{E}_0, \delta\mathbf{E}_1, ..., \delta\mathbf{E}_L] \begin{bmatrix} \mathbf{N}_0 \\ \mathbf{N}_1 \\ \vdots \\ \mathbf{N}_L \end{bmatrix} \mathrm{d}A, \tag{7.6}$$

where A is the area of the reference surface. The shell stress resultants for the stress $\mathbf{S}$ are defined as

$$\mathbf{N}_i \doteq \int_h \zeta^i \, \mathbf{S}(\zeta) \, \mu \, \mathrm{d}\zeta, \qquad i = 0, ..., L. \tag{7.7}$$

The shell form of $\Delta\delta\mathcal{W}(\zeta)$ is defined as $\Delta\delta\Sigma \doteq \int_h \Delta\delta\mathcal{W}(\zeta) \, \mu \, \mathrm{d}\zeta$, and, upon integration of eq. (7.2) over the thickness and by using eq. (7.5), we obtain

$$\Delta\delta\Sigma = \int_A [\delta\mathbf{E}_0, \delta\mathbf{E}_1, ..., \delta\mathbf{E}_L] \begin{bmatrix} \mathbb{C}_0 & \mathbb{C}_1 & ... & \mathbb{C}_L \\ \mathbb{C}_1 & \mathbb{C}_2 & ... & \mathbb{C}_{L+1} \\ \vdots & \vdots & \ddots & \vdots \\ \mathbb{C}_L & \mathbb{C}_{L+1} & ... & \mathbb{C}_{L+L} \end{bmatrix} \begin{bmatrix} \Delta\mathbf{E}_0 \\ \Delta\mathbf{E}_1 \\ \vdots \\ \Delta\mathbf{E}_L \end{bmatrix} + [\mathbf{N}_0, \mathbf{N}_1, ..., \mathbf{N}_i] \begin{bmatrix} \Delta\delta\mathbf{E}_0 \\ \Delta\delta\mathbf{E}_1 \\ \vdots \\ \Delta\delta\mathbf{E}_i \end{bmatrix} \mathrm{d}A, \tag{7.8}$$

where the shell constitutive operators are defined as

$$\mathbb{C}_k \doteq \int_h \zeta^k \, \mathbb{C}(\zeta) \, \mu \, \mathrm{d}z, \qquad k = 0, ..., 2L, \tag{7.9}$$

where k indicates the power of the thickness coordinate ζ.

The stress and couple resultants are updated by the incremental constitutive equations

$$\begin{bmatrix} \Delta\mathbf{N}_0 \\ \Delta\mathbf{N}_1 \\ \vdots \\ \Delta\mathbf{N}_L \end{bmatrix} = \begin{bmatrix} \mathbb{C}_0 & \mathbb{C}_1 & ... & \mathbb{C}_L \\ \mathbb{C}_1 & \mathbb{C}_2 & ... & \mathbb{C}_{L+1} \\ \vdots & \vdots & \ddots & \vdots \\ \mathbb{C}_L & \mathbb{C}_{L+1} & ... & \mathbb{C}_{L+L} \end{bmatrix} \begin{bmatrix} \Delta\mathbf{E}_0 \\ \Delta\mathbf{E}_1 \\ \vdots \\ \Delta\mathbf{E}_L \end{bmatrix}. \tag{7.10}$$

This form is effective if we calculate the shell constitutive operators $\mathbb{C}_k$ only once, as for linear materials.

For strain linear over shell thickness. For shells based on the Reissner kinematics, we use a linear representation of strain

$$\mathbf{E}(\zeta) = \mathbf{E}_0 + \zeta\, \mathbf{E}_1 = \boldsymbol{\varepsilon} + \zeta \boldsymbol{\kappa} \tag{7.11}$$

and the shell stress and couple resultants are

$$\mathbf{N}_0 \doteq \int_h \mathbf{S}(\zeta) \, \mu \, \mathrm{d}\zeta = \mathbf{N}, \qquad \mathbf{N}_1 \doteq \int_h \zeta \, \mathbf{S}(\zeta) \, \mu \, \mathrm{d}\zeta = \mathbf{M}. \tag{7.12}$$

The shell form of the VW of stress, eq. (7.8), becomes

$$\Delta\delta\Sigma = \int_A \left\{ [\delta\boldsymbol{\varepsilon}, \delta\boldsymbol{\kappa}] \begin{bmatrix} \mathbb{C}_0 & \mathbb{C}_1 \\ \mathbb{C}_1 & \mathbb{C}_2 \end{bmatrix} \begin{bmatrix} \Delta\boldsymbol{\varepsilon} \\ \Delta\boldsymbol{\kappa} \end{bmatrix} + [\mathbf{N}, \mathbf{M}] \begin{bmatrix} \Delta\delta\boldsymbol{\varepsilon} \\ \Delta\delta\boldsymbol{\kappa} \end{bmatrix} \right\} \mathrm{d}A, \tag{7.13}$$

where the shell constitutive operators are

$$\mathbb{C}_0 \doteq \int_h \mathbb{C}(\zeta)\, \mu\, d\zeta, \quad \mathbb{C}_1 \doteq \int_h \zeta\, \mathbb{C}(\zeta)\, \mu\, d\zeta, \quad \mathbb{C}_2 \doteq \int_h \zeta^2\, \mathbb{C}(\zeta)\, \mu\, d\zeta. \tag{7.14}$$

In general, the integrals in the definitions of $\mathbf{N}$, $\mathbf{M}$, and $\mathbb{C}_k$ are evaluated numerically, although for simple materials analytical integration is also possible.

The stress and couple resultants can be updated using the incremental constitutive equations

$$\begin{bmatrix} \Delta\mathbf{N} \\ \Delta\mathbf{M} \end{bmatrix} = \begin{bmatrix} \mathbb{C}_0 & \mathbb{C}_1 \\ \mathbb{C}_1 & \mathbb{C}_2 \end{bmatrix} \begin{bmatrix} \Delta\boldsymbol{\varepsilon} \\ \Delta\boldsymbol{\kappa} \end{bmatrix}. \tag{7.15}$$

If cross-sectional properties are symmetric w.r.t. $\zeta = 0$, then $\mathbb{C}_1 = \mathbf{0}$ and the constitutive equations are uncoupled, which means that $\Delta\mathbf{N}$ depends only on $\Delta\boldsymbol{\varepsilon}$, and $\Delta\mathbf{M}$ only on $\Delta\boldsymbol{\kappa}$, i.e.

$$\begin{bmatrix} \Delta\mathbf{N} \\ \Delta\mathbf{M} \end{bmatrix} = \begin{bmatrix} \mathbb{C}_0 & \mathbf{0} \\ \mathbf{0} & \mathbb{C}_2 \end{bmatrix} \begin{bmatrix} \Delta\boldsymbol{\varepsilon} \\ \Delta\boldsymbol{\kappa} \end{bmatrix}. \tag{7.16}$$

Finally, we recall that this formulation requires non-zero normal components of $\boldsymbol{\varepsilon}$ and $\boldsymbol{\kappa}$, so it is suitable only for 3D shells.

7.1.3 General form of constitutive equations for shell resultants

In this section, the constitutive equations are derived in a general (non-incremental) form and the shell stress and couple resultants are used. The shell strain energy is obtained by the analytical integration over the thickness and two types of material are considered: (A) the linear SVK material and (B) the incompressible material.

A. Linear SVK material

The first-order isotropic elastic St.Venant–Kirchhoff (SVK) material is linear and is applicable only to small strain problems. The standard form of the strain energy function for the SVK material is

$$\mathcal{W}(\mathbf{E}) \doteq \tfrac{1}{2}\lambda\, (\mathrm{tr}\mathbf{E})^2 + G\, \mathrm{tr}\mathbf{E}^2, \tag{7.17}$$

where $\mathbf{E}$ is a symmetric strain, and λ, G are Lamé constants. The energy is defined per unit volume of the initial (non-deformed) configuration. The constitutive equations are $\mathbf{S} \doteq d\mathcal{W}(\mathbf{E})/d\mathbf{E}$ and, for the SVK material, we obtain the Hooke's law

$$\mathbf{S} = \lambda\,\mathrm{tr}(\mathbf{E})\mathbf{I} + 2G\,\mathbf{E}, \tag{7.18}$$

where the identities $\mathrm{d}(\mathrm{tr}\mathbf{E})/\mathrm{d}\mathbf{E} = \mathbf{I}$ and $\mathrm{d}(\mathrm{tr}\mathbf{E})^2/\mathrm{d}\mathbf{E} = 2\mathbf{E}$ were used.

The form of the strain energy of eq. (7.17) is only valid for a symmetric $\mathbf{E}$ because, for a non-symmetric $\mathbf{E}$, it yields $S_{12} = 2G\,E_{21}$. For non-symmetric $\mathbf{E}$, we should replace $\mathrm{tr}\mathbf{E}^2$ by $\mathrm{tr}(\mathbf{E}\mathbf{E}^{\mathrm{T}})$.

Strain energy for shell. Assume that the strain is expressed as a linear polynomial of the thickness coordinate ζ, i.e. $\mathbf{E} = \boldsymbol{\varepsilon} + \zeta\boldsymbol{\kappa}$, where $\boldsymbol{\varepsilon}$ and $\boldsymbol{\kappa}$ are symmetric shell strains. For this derivation, we assume that all components of these strain are non-zero. Then

$$\mathrm{tr}\mathbf{E} = \mathrm{tr}\boldsymbol{\varepsilon} + \zeta\mathrm{tr}\boldsymbol{\kappa}, \qquad (\mathrm{tr}\mathbf{E})^2 = (\mathrm{tr}\boldsymbol{\varepsilon})^2 + 2\zeta(\mathrm{tr}\boldsymbol{\varepsilon})(\mathrm{tr}\boldsymbol{\kappa}) + \zeta^2(\mathrm{tr}\boldsymbol{\kappa})^2, \tag{7.19}$$

$$\mathbf{E}^2 = \boldsymbol{\varepsilon}^2 + \zeta(\boldsymbol{\varepsilon}\boldsymbol{\kappa} + \boldsymbol{\kappa}\boldsymbol{\varepsilon}) + \zeta^2\boldsymbol{\kappa}^2, \qquad \mathrm{tr}\mathbf{E}^2 = \mathrm{tr}\boldsymbol{\varepsilon}^2 + 2\zeta\mathrm{tr}(\boldsymbol{\varepsilon}\boldsymbol{\kappa}) + \zeta^2\mathrm{tr}\boldsymbol{\kappa}^2. \tag{7.20}$$

Substituting these expressions into the strain energy (7.17) and integrating over the thickness, we obtain

$$\Sigma \doteq \int_{-\frac{h}{2}}^{+\frac{h}{2}} \mathcal{W}(\mathbf{E}(\zeta))\,\mu\,\mathrm{d}\zeta = h\,\mathcal{W}(\boldsymbol{\varepsilon}) + \frac{h^3}{12}\,\mathcal{W}(\boldsymbol{\kappa}), \tag{7.21}$$

which is the shell strain energy per unit area of the reference surface in the initial configuration. Note that the couplings $(\mathrm{tr}\boldsymbol{\varepsilon})(\mathrm{tr}\boldsymbol{\kappa})$ and $\mathrm{tr}(\boldsymbol{\varepsilon}\boldsymbol{\kappa})$ dropped out because the integral of terms depending linearly on ζ is zero.

Shell constitutive equations. A kinematically admissible variation of the shell strain energy is

$$\delta\Sigma = \frac{\partial\Sigma}{\partial\boldsymbol{\varepsilon}} \cdot \delta\boldsymbol{\varepsilon} + \frac{\partial\Sigma}{\partial\boldsymbol{\kappa}} \cdot \delta\boldsymbol{\kappa}. \tag{7.22}$$

The stress and couple resultants are defined as

$$\mathbf{N} \doteq \frac{\partial\Sigma}{\partial\boldsymbol{\varepsilon}} = h\,\frac{d\mathcal{W}(\boldsymbol{\varepsilon})}{d\boldsymbol{\varepsilon}}, \qquad \mathbf{M} \doteq \frac{\partial\Sigma}{\partial\boldsymbol{\kappa}} = \frac{h^3}{12}\frac{d\mathcal{W}(\boldsymbol{\kappa})}{d\boldsymbol{\kappa}} \tag{7.23}$$

and then the variation of the shell strain energy can be concisely written as

$$\delta\Sigma = \mathbf{N}\cdot\delta\boldsymbol{\varepsilon} + \mathbf{M}\cdot\delta\boldsymbol{\kappa}. \tag{7.24}$$

Because, the derivatives $\mathrm{d}\mathcal{W}(\boldsymbol{\varepsilon})/\mathrm{d}\boldsymbol{\varepsilon}$ and $\mathrm{d}\mathcal{W}(\boldsymbol{\kappa})/\mathrm{d}\boldsymbol{\kappa}$ have analogous forms as $\mathrm{d}\mathcal{W}(\mathbf{E})/\mathrm{d}\mathbf{E}$, hence, the constitutive equations for the stress resultants and the couple resultants are

$$\mathbf{N} = h\,[\lambda\,(\mathrm{tr}\boldsymbol{\varepsilon})\mathbf{I} + 2\mathrm{G}\,\boldsymbol{\varepsilon}]\,, \qquad \mathbf{M} = \frac{h^3}{12}\,[\lambda\,(\mathrm{tr}\boldsymbol{\kappa})\mathbf{I} + 2\mathrm{G}\,\boldsymbol{\kappa}]\,. \tag{7.25}$$

Note that the distribution of the shear stresses over the thickness is parabolic and the constitutive equations for the components $N_{\alpha 3}$ and $M_{\alpha 3}$ have to be corrected, see Sect. 7.3.

B. Incompressible Mooney–Rivlin material

Consider a class of second-order hyper-elastic materials which undergo an isochoric (or volume-preserving) deformation. Because the relation between the initial volume $\mathrm{d}V$ and the current volume $\mathrm{d}v$ is $\mathrm{d}v = \det\mathbf{F}\;\mathrm{d}V$, the incompressibility of the material is defined by the condition

$$\det\mathbf{F} = 1, \tag{7.26}$$

where $\mathbf{F}$ is the deformation gradient. This definition implies that the third invariant of the right Cauchy–Green tensor $\mathbf{C}$ is equal to one, $I_3(\mathbf{C}) \doteq \det\mathbf{C} = (\det\mathbf{F})^2 = 1$. Thus, the strain energy of incompressible materials depends only on the two first principal invariants of $\mathbf{C}$,

$$I_1(\mathbf{C}) \doteq \mathrm{tr}\mathbf{C}, \qquad \mathrm{I}_2(\mathbf{C}) \doteq \tfrac{1}{2}\left[(\mathrm{tr}\mathbf{C})^2 - \mathrm{tr}\mathbf{C}^2\right]. \tag{7.27}$$

Below, we define two classical incompressible materials:

1. The so-called neo-Hookean material is defined by the following strain energy function:
$$\tilde{\mathcal{W}}(I_1(\mathbf{C})) \doteq c_1\,[I_1(\mathbf{C}) - 3]\,, \tag{7.28}$$
where c_1 is a material constant. This energy function depends only on the first invariant of $\mathbf{C}$.
2. The Mooney–Rivlin is the material is defined by the following strain energy function,
$$\tilde{\mathcal{W}}(I_\alpha(\mathbf{C})) \doteq c_1\,[I_1(\mathbf{C}) - 3] + c_2\,[I_2(\mathbf{C}) - 3]\,, \qquad \alpha = 1, 2, \tag{7.29}$$
where c_1, c_2 are material constants. This energy function depends on the two first invariants of $\mathbf{C}$.

For more details on incompressible materials, see [81, 159, 93, 116].

For membranes, the strain energy for the incompressible material can also be expressed in terms of principal stretches. Then the incompressibility condition is applied to the Ogden's form of the strain energy of eq. (7.93).

Formulation for right stretching tensor U. The incompressibility condition can also be formulated in terms of the third invariant of the right stretching tensor $\mathbf{U}$,

$$I_3(\mathbf{U}) \doteq \det \mathbf{U} = 1. \tag{7.30}$$

This form is obtained from the polar decomposition $\mathbf{F} = \mathbf{QU}$, for which

$$\det \mathbf{F} = \det(\mathbf{RU}) = (\det \mathbf{R})(\det \mathbf{U}) = \det \mathbf{U} = 1, \tag{7.31}$$

as $\det \mathbf{R} = 1$ for $\mathbf{R} \in SO(3)$. Hence, for the incompressible material, the strain energy depends on the two first principal invariants of $\mathbf{U}$, i.e. $\tilde{\mathcal{W}} = \tilde{\mathcal{W}}(I_1(\mathbf{U}), I_2(\mathbf{U}))$, where the principal invariants of $\mathbf{U}$ are

$$I_1(\mathbf{U}) \doteq \mathrm{tr}\mathbf{U}, \qquad \mathrm{I}_2(\mathbf{U}) \doteq \tfrac{1}{2}\left[(\mathrm{tr}\mathbf{U})^2 - \mathrm{tr}\mathbf{U}^2\right]. \tag{7.32}$$

The constitutive equation for the symmetric Biot stress tensor is

$$\mathbf{T}_s^B \doteq \frac{\partial \mathcal{W}(\mathbf{U})}{\partial \mathbf{U}} + p\mathbf{I} = \frac{\partial \tilde{\mathcal{W}}(I_1(\mathbf{U}), I_2(\mathbf{U}))}{\partial \mathbf{U}} + p\mathbf{I}. \tag{7.33}$$

Using the chain rule, we obtain

$$\frac{\partial \tilde{\mathcal{W}}}{\partial \mathbf{U}} = \frac{\partial \tilde{\mathcal{W}}}{\partial I_1}\frac{\partial I_1}{\partial \mathbf{U}} + \frac{\partial \tilde{\mathcal{W}}}{\partial I_2}\frac{\partial I_2}{\partial \mathbf{U}}, \tag{7.34}$$

where $\partial I_1/\partial \mathbf{U} = \mathbf{I}$ and $\partial I_2/\partial \mathbf{U} = I_1\mathbf{I} - \mathbf{U}$. Thus, the constitutive equation can be rewritten as a linear polynomial of $\mathbf{U}$, i.e. $\mathbf{T}_s^B = \beta_0\mathbf{I} + \beta_1\mathbf{U}$, where β_0 and β_1 are scalar coefficients depending on the invariants.

The invariants of $\mathbf{C}$ in eq. (7.29) can be written as functions of the invariants of $\mathbf{U}$,

$$I_1(\mathbf{C}) = I_1^2(\mathbf{U}) - 2I_2(\mathbf{U}), \qquad I_2(\mathbf{C}) = I_2^2(\mathbf{U}) - 2I_1(\mathbf{U})\, I_3(\mathbf{U}), \tag{7.35}$$

where, for $I_3(\mathbf{U}) = 1$, the second one is reduced to $I_2(\mathbf{C}) = I_2^2(\mathbf{U}) - 2I_1(\mathbf{U})$. Thus, the Mooney–Rivlin strain energy of eq. (7.29) is, in terms of the invariants of $\mathbf{U}$, as follows:

$$\tilde{\mathcal{W}}(I_\alpha(\mathbf{U})) = c_1\left[I_1^2(\mathbf{U}) - 2I_2(\mathbf{U}) - 3\right] + c_2\left[I_2^2(\mathbf{U}) - 2I_1(\mathbf{U}) - 3\right]. \quad (7.36)$$

Assume that the right stretching tensor is a linear polynomial of the thickness coordinate ζ of the shell, i.e. $\mathbf{U} = \mathbf{e} + \zeta\mathbf{k}$, where $\mathbf{e}$ and $\mathbf{k}$ are the symmetric shell strains. Then the invariants of $\mathbf{U}$ can be expressed as

$$I_1(\mathbf{U}) = I_1(\mathbf{e}) + \zeta I_1(\mathbf{k}), \qquad I_2(\mathbf{U}) = I_2(\mathbf{e}) + \zeta A + \zeta^2 I_2(\mathbf{k}), \quad (7.37)$$

$$I_3(\mathbf{U}) = \frac{1}{6}\left[I_3(\mathbf{e}) + \zeta B(\mathbf{e},\mathbf{k}) + \zeta^2 B(\mathbf{k},\mathbf{e}) + \zeta^3 I_3(\mathbf{k})\right], \quad (7.38)$$

where the auxiliary scalars are

$$A \doteq I_1(\mathbf{e})\, I_1(\mathbf{k}) - \operatorname{tr}(\mathbf{ek}), \quad (7.39)$$

$$B(\mathbf{a},\mathbf{b}) \doteq 6\left[I_2(\mathbf{a})\, I_1(\mathbf{b}) + \operatorname{tr}(\mathbf{a}^2\mathbf{b}) - \mathrm{I}_1(\mathbf{a})\operatorname{tr}(\mathbf{ab})\right], \quad (7.40)$$

for the second rank tensors $\mathbf{a}$ and $\mathbf{b}$. Note the presence of the coupling terms in A and $B(\mathbf{a},\mathbf{b})$, which render that the second and third invariant of $\mathbf{U}$ are not expressible in terms of the invariants of $\mathbf{e}$ and $\mathbf{k}$. For the squares of invariants of $\mathbf{U}$, which are also present in eq. (122), we have

$$I_1^2(\mathbf{U}) = I_1^2(\mathbf{e}) + 2\zeta I_1(\mathbf{e})\, I_1(\mathbf{k}) + \zeta^2 I_1^2(\mathbf{k}), \quad (7.41)$$

$$I_2^2(\mathbf{U}) = I_2^2(\mathbf{e}) + \zeta^2 A^2 + \zeta^4 I_2^2(\mathbf{k}) + 2\zeta I_2(\mathbf{e})\, A + 2\zeta^2 I_2(\mathbf{e})\, I_2(\mathbf{k}) + 2\zeta^3 I_2(\mathbf{k})\, A, \quad (7.42)$$

where

$$A^2 = I_1^2(\mathbf{e})\, I_1^2(\mathbf{k}) - 2I_1(\mathbf{e})\, I_1(\mathbf{k}) \operatorname{tr}(\mathbf{ek}) + [\operatorname{tr}(\mathbf{ek})]^2. \quad (7.43)$$

Note that for the assumed approximations of $\mathbf{U}$, the strain energy is the second order polynomial of ζ for the neo-Hookean material and the fourth order polynomial for the Mooney–Rivlin material.

Strain energy for shell. Let us define the shell strain energy density per unit area of the middle surface in the initial configuration, as the integral of the strain energy over the thickness, i.e.

$$\tilde{\Sigma}(I_\alpha(\mathbf{U})) \doteq \int_{-\frac{h}{2}}^{+\frac{h}{2}} \tilde{\mathcal{W}}(I_\alpha(\mathbf{U}))\, \mu\, d\zeta. \quad (7.44)$$

The integration over the thickness renders that the terms of $\tilde{\mathcal{W}}(I_\alpha(\mathbf{U}))$ multiplied by even powers of ζ are equal to zero and the shell energy splits as follows:

$$\tilde{\Sigma} = c_1\,\tilde{\Sigma}_1 \;+\; c_2\,\tilde{\Sigma}_2, \tag{7.45}$$

where

$$\tilde{\Sigma}_1 = h\left[I_1^2(\mathbf{e}) - 2I_2(\mathbf{e}) - 3\right] + \frac{h^3}{12}\left[I_1^2(\mathbf{k}) - 2I_2(\mathbf{k})\right], \tag{7.46}$$

$$\tilde{\Sigma}_2 = h\left[I_2^2(\mathbf{e}) - 2I_1(\mathbf{e}) - 3\right] + \frac{h^3}{12}\left[A^2 + I_2(\mathbf{e})I_2(\mathbf{k})\right] + \frac{h^5}{80}I_2^2(\mathbf{k}). \tag{7.47}$$

In the neo-Hookean component $\tilde{\Sigma}_1$, the terms depending on $\mathbf{e}$ and $\mathbf{k}$ are separated, leading to uncoupled constitutive equations. On the other hand, the component $\tilde{\Sigma}_2$ contains coupling terms such as $I_1(\mathbf{e})I_1(\mathbf{k})$, $I_2(\mathbf{e})I_2(\mathbf{k})$, $\mathrm{tr}(\mathbf{ek})$, and some products and powers of them.

Shell constitutive equations. For symmetric $\mathbf{e}$, $\delta\mathbf{e}$, $\mathbf{k}$, and $\delta\mathbf{k}$, the variation of the shell strain energy may be written as

$$\delta\tilde{\Sigma}(\mathbf{e},\mathbf{k}) = \mathbf{N}_s^B \cdot \delta\mathbf{e} + \mathbf{M}_s^B \cdot \delta\mathbf{k}, \tag{7.48}$$

where the stress and couple resultants are defined as

$$\mathbf{N}_s^B \doteq \frac{\mathrm{d}\tilde{\Sigma}}{\mathrm{d}\mathbf{e}}, \qquad \mathbf{M}_s^B \doteq \frac{\mathrm{d}\tilde{\Sigma}}{\mathrm{d}\mathbf{k}}. \tag{7.49}$$

To facilitate further differentiation, we calculate the following derivatives:

$$\frac{\partial \mathrm{tr}(\mathbf{ek})}{\partial \mathbf{e}} = \mathbf{k}, \qquad \frac{\partial\left[\mathrm{tr}(\mathbf{ek})\right]^2}{\partial \mathbf{e}} = 2\mathrm{tr}(\mathbf{ek})\,\mathbf{k}, \qquad \frac{\partial \mathrm{A}^2}{\partial \mathbf{e}} = 2\mathrm{A}\,\mathbf{D}(\mathbf{k}), \tag{7.50}$$

$$\frac{\partial \mathrm{tr}(\mathbf{ek})}{\partial \mathbf{k}} = \mathbf{e}, \qquad \frac{\partial\left[\mathrm{tr}(\mathbf{ek})\right]^2}{\partial \mathbf{k}} = 2\mathrm{tr}(\mathbf{ek})\,\mathbf{e}, \qquad \frac{\partial \mathrm{A}^2}{\partial \mathbf{k}} = 2\mathrm{A}\,\mathbf{D}(\mathbf{e}), \tag{7.51}$$

where the auxiliary tensor is defined as

$$\mathbf{D}(\mathbf{A}) \doteq I_2(\mathbf{A})_{,\mathbf{A}} = I_1(\mathbf{A})\,\mathbf{I} - \mathbf{A}. \tag{7.52}$$

The derivatives of the shell strain energy are

$$\frac{\partial\tilde{\Sigma}_1}{\partial \mathbf{e}} = 2h\mathbf{e}, \qquad \frac{1}{2}\frac{\partial\tilde{\Sigma}_2}{\partial \mathbf{e}} = h\left[I_2(\mathbf{e})\,\mathbf{D}(\mathbf{e}) - \mathbf{I}\right] + \frac{h^3}{12}\pi(\mathbf{k},\mathbf{e}), \tag{7.53}$$

$$\frac{\partial \tilde{\Sigma}_1}{\partial \mathbf{k}} = 2\frac{h^3}{12}\mathbf{k}, \qquad \frac{1}{2}\frac{\partial \tilde{\Sigma}_2}{\partial \mathbf{k}} = \frac{h^3}{12}\pi(\mathbf{e},\mathbf{k}) + \frac{h^5}{80}I_2(\mathbf{k})\,\mathbf{D}(\mathbf{k}), \tag{7.54}$$

where the term which couples the contribution of $\mathbf{e}$ and $\mathbf{k}$ is defined as follows:

$$\pi(\mathbf{a},\mathbf{b}) \doteq A\,\mathbf{D}(\mathbf{b}) \;+\; \tfrac{1}{2}I_2(\mathbf{b})\,\mathbf{D}(\mathbf{a}). \tag{7.55}$$

Using the above equations, the following coupled constitutive equations for the shell are obtained:

$$\mathbf{N}_s^B = c_1\,[2h\mathbf{e}] + c_2 2\left[h\,(I_2(\mathbf{e})\,\mathbf{D}(\mathbf{e}) - \mathbf{I}) + \frac{h^3}{12}\pi(\mathbf{k},\mathbf{e})\right], \tag{7.56}$$

$$\mathbf{M}_s^B = c_1\left[2\frac{h^3}{12}\mathbf{k}\right] + c_2 2\left[\frac{h^3}{12}\pi(\mathbf{e},\mathbf{k}) + \frac{h^5}{80}I_2(\mathbf{k})\,\mathbf{D}(\mathbf{k})\right]. \tag{7.57}$$

Note that these constitutive equations for the hyper-elastic incompressible material have been obtained without any simplifications of the strain energy and have quite a complicated form. In numerical implementations, the incremental forms are much more convenient.

7.2 Reduced shell constitutive equations

The reduced shell constitutive equations are obtained by using the normal strain recovered from an auxiliary condition. This recovery is performed because the standard Reissner hypothesis yields the normal strains ε_{33} and κ_{33} equal to zero, as we can see in

1. Eqs. (6.40)–(6.41) for the *non-symmetric relaxed* right stretch strain,
2. Eqs. (6.48)–(6.49) for the *symmetric relaxed* right stretch strain, and
3. Eqs. (6.54)–(6.55) for the Green strain.

The zero values of the normal strains are non-physical and inaccurate, which can be easily shown for membranes, see eq. (7.73) and Fig. 7.1.

For the Kirchhoff shells, the components of strain are usually evaluated as follows:

$$\varepsilon_{\alpha\beta} \sim h\kappa_{(\alpha\beta)} = \mathrm{O}(\eta), \quad \varepsilon_{3\beta} \sim h\kappa_{3\beta} = \mathrm{O}(\eta\theta), \quad \varepsilon_{33} = \mathrm{O}(\nu\eta). \tag{7.58}$$

where η is the maximum eigenvalue of the Green in-plane strains and ν is the Poisson's ratio. The small parameter θ is defined in [171], p. 111, eq. (6.3.4). We note that a special methodology, proposed in [119, 120] and later successfully developed in [129, 173, 174], must be used to construct consistent approximations to the strain energy. Obviously, ε_{33} is not negligible compared to the other strain components.

To improve accuracy, we can calculate the normal strains from auxiliary conditions, such as (a) the zero normal stress (ZNS) condition, see Sect. 7.2.1, or (b) the incompressibility condition, see Sect. 7.2.2. This approach using the recovery is classical and most often used despite several difficulties involved such as:

1. the auxiliary conditions are not always fully physically justified, e.g. the ZNS condition in case of the multi-layer shells,
2. the reduced constitutive laws are often difficult to derive for some constitutive equations and are very complicated. For the ZNS condition, this problem can be alleviated by using the incremental form of the constitutive equations, see Sect. 7.2.1.
3. the reduced constitutive equations can be more difficult to solve than the original 3D equations, e.g. for the J_2 plasticity, where the 2D yield surface is not spherical and the radial return algorithm cannot be applied.

The recovery of the normal strain renders that the constitutive equations are more accurate but also more complicated.

7.2.1 Reduced constitutive equations for ZNS condition

For thin membranes, we can use the plane stress conditions

$$S_{31}(z) = 0, \qquad S_{32}(z) = 0, \qquad S_{33}(z) = 0, \tag{7.59}$$

where S_{31} and S_{32} are transverse shear stresses and S_{33} is the normal stress, all in the local Cartesian basis $\{\mathbf{t}_k\}$. However, for the Reissner shells, only the condition for the normal stress is acceptable,

$$S_{33}(z) = 0, \tag{7.60}$$

while the transverse shear strains must remain unconstrained. This ZNS condition was used for the Kirchhoff shells in the classical works [153, 171, 172].

A. Incremental formulation in stresses

In the 3D formulation of Sect. 7.1.1, the incremental constitutive equation (7.3) is written for stresses as $\Delta\mathbf{S} = \mathbb{C}\,\Delta\mathbf{E}$ and can be rewritten as

$$\begin{bmatrix} \Delta \mathbf{S}_v \\ \Delta S_{33} \end{bmatrix} = \begin{bmatrix} \mathbb{C}_{vv} & \mathbb{C}_{v3} \\ \mathbb{C}_{3v} & C_{33} \end{bmatrix} \begin{bmatrix} \Delta \mathbf{E}_v \\ \Delta E_{33} \end{bmatrix}, \tag{7.61}$$

where $(\cdot)_v$ denotes a vector of tangent components arranged in the order $\{11, 22, 12\}$. Besides, $\dim \mathbb{C}_{vv} = 3 \times 3$, $\dim \mathbb{C}_{v3} = 3 \times 1$, and $\dim \mathbb{C}_{3v} = 1 \times 3$. This equation involves only the tangent components and the normal components 33, while the transverse shear components were omitted for simplicity.

From the condition of a zero increment of the normal stress, i.e. $\Delta S_{33} = 0$, and the last (scalar) equation of eq. (7.61), we can calculate the normal strain increment

$$\Delta E_{33} = -\frac{1}{C_{33}} \mathbb{C}_{3v} \, \Delta \mathbf{E}_v, \tag{7.62}$$

for which the first (matrix) equation of eq. (7.61) becomes

$$\Delta \mathbf{S}_v = \mathbb{C}_{vv} \Delta \mathbf{E}_v + \mathbb{C}_{v3} \Delta E_{33} = \mathbb{C}^* \, \Delta \mathbf{E}_v, \tag{7.63}$$

where the constitutive matrix is defined as

$$\mathbb{C}^* \doteq \mathbb{C}_{vv} - \frac{1}{C_{33}} \mathbb{C}_{v3} \mathbb{C}_{3v}. \tag{7.64}$$

The above-reduced incremental constitutive equation (7.63) and the reduced constitutive matrix $\mathbb{C}^*$ are for tangent components and both account for the increment of normal strain, i.e. the change of thickness. The normal strain is updated as

$$E_{33}^{i} = E_{33}^{i-1} + \Delta E_{33}, \tag{7.65}$$

where ΔE_{33} is given by eq. (7.62), and E_{33}^{i-1} is the value for the previous iteration.

Remark. Note that this incremental procedure is quite general and can be applied to any non-linear hyper-elastic materials, e.g. to the compressible neo-Hookean materials, which are generalizations of the incompressible neo-Hookean material of eq. (7.28). For instance, in [212], the strain energy function has the form

$$\mathcal{W} \doteq \underbrace{\frac{\lambda}{2} (\ln J)^2 - G \ln J}_{\text{compressible part}} + \frac{G}{2} (\mathrm{tr}\mathbf{C} - 3), \tag{7.66}$$

where $J \doteq \det \mathbf{F}$. For $\mathbf{F} = \mathbf{I}$, we obtain $\mathcal{W} = 0$ and the constitutive operator is reduced to the one for the SVK material. A simpler form of the compressible part is obtained when the first term is replaced by the first term of its series expansion at $J = 1$, i.e. $(\ln J)^2 = (J-1)^2 + O(J-1)^3$; other forms of this part are listed in [116], p. 160.

Remark. A very simple scheme of treating the normal strain was used for shells in [101]. The nonlinear governing equations are solved iteratively and the normal strain is evaluated for the last available solution, so its value lags one iteration behind. This certainly somehow impairs the convergence rate, but it is acceptable as long as the iterations converge.

B. Incremental formulation in stress resultants and couple resultants

For the constitutive equations of Sect. 7.1.2, we use the recovery procedure for the shell stress and couple resultants, which is analogous to that for stresses. Note that the condition $S_{33}(z) = 0$ implies, by eq. (7.12), the zero values of shell resultants, i.e. $N_{33} = 0$ and $M_{33} = 0$. We use these conditions in the incremental form

$$\Delta N_{33} = 0, \qquad \Delta M_{33} = 0. \tag{7.67}$$

We assume that the shell constitutive equations are decoupled, as in eq. (7.16), and consider them separately.

For the first of eq. (7.16), $\Delta \mathbf{N} = \mathbb{C}_0\, \Delta \boldsymbol{\varepsilon}$, we use the condition $\Delta N_{33} = 0$, and the results are analogous to these for stresses, if we replace

$$\mathbf{E} \to \boldsymbol{\varepsilon}, \quad \Delta S_{33} \to \Delta N_{33}, \quad \Delta E_{33} \to \Delta \varepsilon_{33}, \qquad \mathbb{C} \to \mathbb{C}_0, \quad \mathbb{C}^* \to \mathbb{C}_0^*. \tag{7.68}$$

For the second of eq. (7.16), i.e. $\Delta \mathbf{M} = \mathbb{C}_2\, \Delta \boldsymbol{\kappa}$, we use the condition $\Delta M_{33} = 0$, and the results are analogous to these for stresses, if we replace

$$\mathbf{E} \to \boldsymbol{\kappa}, \quad \Delta S_{33} \to \Delta M_{33}, \quad \Delta E_{33} \to \Delta \kappa_{33}, \qquad \mathbb{C} \to \mathbb{C}_2, \quad \mathbb{C}^* \to \mathbb{C}_2^*. \tag{7.69}$$

For the recovered $\Delta \varepsilon_{33}$ and $\Delta \kappa_{33}$, the normal strain of a shell is linearly approximated in z,

$$\Delta E_{33}(z) = \Delta \varepsilon_{33} + z\, \Delta \kappa_{33}. \tag{7.70}$$

The example of the 2D beam indicates that the recovery of both normal strains is beneficial, although each one for a different deformation, see Table 7.1.

C. Constitutive equations for SVK material using ZNS condition

For the linear SVK material, the procedure of the previous section is reduced to the classical procedure using the general form of the ZNS condition. In terms of components, the SVK strain-energy function of eq. (7.17) has the following form:

$$\mathcal{W}(\mathbf{E}) \doteq \frac{\lambda}{2}\,(E_{11}+E_{22}+E_{33})^2 + G\left(E_{11}^2+2E_{12}^2+E_{22}^2+E_{33}^2+2E_{13}^2+2E_{23}^2\right) \tag{7.71}$$

and the constitutive equations are

$$S_{11} = \lambda\,(E_{11}+E_{22}+E_{33})+2G\,E_{11}, \qquad S_{22} = \lambda\,(E_{11}+E_{22}+E_{33})+2G\,E_{22},$$
$$S_{33} = \lambda\,(E_{11}+E_{22}+E_{33})+2G\,E_{33}, \tag{7.72}$$
$$S_{12} = 4G\,E_{12}, \qquad S_{13} = 4G\,E_{13}, \qquad S_{23} = 4G\,E_{23}.$$

For simplicity, in the sequel we neglect the transverse shear strain components. From the ZNS condition $S_{33}=0$ and for S_{33} of eq. (7.72), we can calculate the normal strain

$$E_{33} = -c_0\,(E_{11}+E_{22}), \qquad c_0 \doteq \frac{\lambda}{\lambda+2G} = \frac{\nu}{1-\nu}, \tag{7.73}$$

where c_0 is plotted for $\nu \in [0, \frac{1}{2}]$ in Fig. 7.1. We see that $0 \le c_0 \le 1$, and always is greater than ν, which is shown as a straight line in this figure. Note that ν is used in the estimation $\varepsilon_{33} = \mathrm{O}(\nu\eta)$ of eq. (7.58).

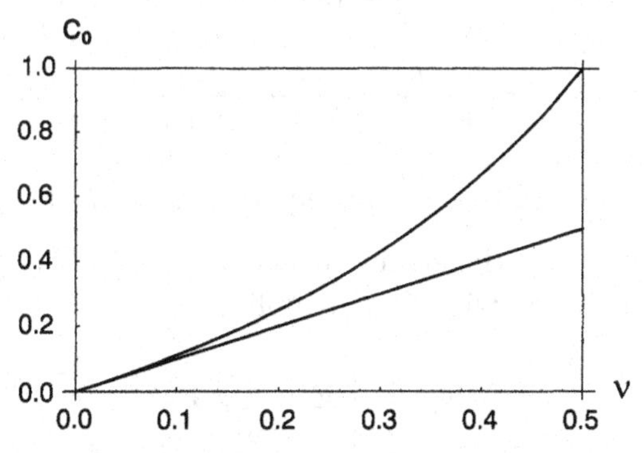

Fig. 7.1 Coefficient c_0 for $\nu \in [0, \frac{1}{2}]$.

Using eq. (7.73), the strain energy (7.71) becomes

$$\mathcal{W}^*(\mathbf{E}_v) \doteq \frac{1}{2}c_1(E_{11}^2 + E_{22}^2) + c_2\, E_{11}E_{22} + \frac{1}{2}c_3\, E_{12}^2, \tag{7.74}$$

where

$$c_1 \doteq \frac{4G(\lambda+G)}{\lambda+2G} = \frac{E}{1-\nu^2}, \quad c_2 \doteq \frac{2G\lambda}{\lambda+2G} = \frac{E\nu}{1-\nu^2}, \quad c_3 \doteq 4G = \frac{2E}{1+\nu}. \tag{7.75}$$

This form of strain energy depends only on the tangent components $\{11, 22, 12\}$.

The reduced constitutive equations $\mathbf{S}_v(\boldsymbol{\varepsilon}_v) \doteq \partial\mathcal{W}^*(\mathbf{E}_v)/\partial\mathbf{E}_v$ and the reduced constitutive matrix $\mathbb{C}_{vv} \doteq \partial\mathbf{S}_v(\mathbf{E}_v)/\partial\mathbf{E}_v$ are as follows:

$$\begin{bmatrix} S_{11} \\ S_{22} \\ S_{12} \end{bmatrix} = \begin{bmatrix} c_1\, E_{11} + c_2\, E_{22} \\ c_2\, E_{11} + c_1\, E_{22} \\ c_3\, E_{12} \end{bmatrix}, \qquad \mathbb{C}_{vv} = \begin{bmatrix} c_1 & c_2 & 0 \\ c_2 & c_1 & 0 \\ 0 & 0 & c_3 \end{bmatrix}. \tag{7.76}$$

The inverse of the constitutive matrix is

$$\mathbb{C}_{vv}^{-1} = \begin{bmatrix} d_1 & d_2 & 0 \\ d_2 & d_1 & 0 \\ 0 & 0 & d_3 \end{bmatrix}, \qquad d_1 = \frac{\lambda+G}{G(3\lambda+2G)},$$
$$d_2 = \frac{-\lambda}{2G(3\lambda+2G)}, \qquad d_3 = \frac{1}{4G}. \tag{7.77}$$

The eigenvalues of the constitutive matrix are

$$\begin{aligned} \operatorname{eigv}\mathbb{C}_{vv} &= \left\{ \frac{2G}{\lambda+2G}(3\lambda+2G),\ 4G,\ 2G \right\} \\ &= \left\{ \frac{E}{1-\nu}, \frac{2E}{1+\nu}, \frac{E}{1+\nu} \right\}. \end{aligned} \tag{7.78}$$

For $\nu \in [0, \frac{1}{2}]$, the smallest eigenvalue of $\mathbb{C}_{vv}$ is $E/(1+\nu) = 2G$.

Finally, we note that the strain energy of eq. (7.74) can be expressed using the constitutive matrix $\mathbb{C}_{vv}$ as follows:

$$\mathcal{W}^*(\mathbf{E}_v) \doteq \frac{1}{2}\mathbf{E}_v \cdot (\mathbb{C}_{vv}\mathbf{E}_v). \tag{7.79}$$

For in-plane strain linear over shell thickness. For shells, we use the in-plane strains which are linear in the normal coordinate, i.e. $\mathbf{E}_v(\zeta) = \boldsymbol{\varepsilon}_v + \zeta\boldsymbol{\kappa}_v$. Using the strains of this form in eq. (7.73), we obtain the normal strain of the shell as a linear polynomial of ζ,

$$E_{33}(\zeta) = \varepsilon_{33} + \zeta\kappa_{33}, \tag{7.80}$$

where $\varepsilon_{33} = c_0\,(\varepsilon_{11} + \varepsilon_{22})$ and $\kappa_{33} = c_0\,(\kappa_{11} + \kappa_{22})$. Hence, we obtain a linear approximation of the normal strain when the in-plane strains are linear in ζ.

Using $\mathbf{E}_v(\zeta) = \boldsymbol{\varepsilon}_v + \zeta\boldsymbol{\kappa}_v$, and integrating the strain energy of eq. (7.74) over the thickness, we obtain the shell strain energy as a sum of the membrane energy and the bending energy

$$\Sigma \doteq \int_{-\frac{h}{2}}^{+\frac{h}{2}} \mathcal{W}^*(\mathbf{E}_v(\zeta))\,\mu\,\mathrm{d}\zeta = h\,\mathcal{W}^*(\boldsymbol{\varepsilon}_v) + \frac{h^3}{12}\,\mathcal{W}^*(\boldsymbol{\kappa}_v). \tag{7.81}$$

We can rewrite eq. (7.76) as $\mathbf{S}_v = \mathbb{C}_{vv}\,\mathbf{E}_v$ and use it in the definition of eq. (7.12) to obtain the constitutive equations for the shell stress and couple resultants

$$\mathbf{N}_v \doteq \int_{-\frac{h}{2}}^{+\frac{h}{2}} \mathbf{S}_v(\zeta)\,\mu\,\mathrm{d}\zeta = h\,\mathbb{C}_{vv}\boldsymbol{\varepsilon}_v, \quad \mathbf{M}_v \doteq \int_{-\frac{h}{2}}^{+\frac{h}{2}} \zeta\,\mathbf{S}_v(\zeta)\,\mu\,\mathrm{d}\zeta = \frac{h^3}{12}\mathbb{C}_{vv}\boldsymbol{\kappa}_v. \tag{7.82}$$

D. Effects of normal strain recovery for 2D beam

Consider a straight 2D beam in the $\{\mathbf{t}_1, \mathbf{t}_3\}$-plane, where $\{\mathbf{t}_j\}$ $(j = 1, 3)$ is the local ortho-normal basis associated with the initial configuration. For the standard Reissner hypothesis, $\mathbf{x}(\zeta) = \mathbf{x}_0 + \zeta\mathbf{Q}_0\mathbf{t}_3$, we can split the *non-symmetric relaxed* right stretch strain, $\tilde{\mathbf{H}}_n \doteq \mathbf{Q}_0^T\mathbf{F} - \mathbf{I}$, as follows: $\tilde{\mathbf{H}}_n(\zeta) = \boldsymbol{\varepsilon} + \zeta\boldsymbol{\kappa}$, where the components in $\{\mathbf{t}_i\}$ are

$$\varepsilon_{11} = \mathbf{x}_{0,1}\cdot\mathbf{a}_1 - 1, \qquad 2\,\varepsilon_{13} = \mathbf{x}_{0,1}\cdot\mathbf{a}_3, \qquad \varepsilon_{33} = 0, \tag{7.83}$$

$$\kappa_{11} = \omega_{,1}, \qquad \kappa_{13} = 0, \qquad \kappa_{33} = 0,$$

where $\mathbf{a}_i \doteq \mathbf{Q}_0\,\mathbf{t}_i$. The 3D formulation is reduced to a 2D formulation by setting the 21 and 23 components of stress and strain to zero and recovering ε_{22} from the condition $\sigma_{22} = 0$. Then, for the SVK material, we obtain the following beam-type constitutive equations:

$$N_{11} = Ch\left(\varepsilon_{11} + \nu\varepsilon_{33}\right), \quad N_{33} = Ch\left(\nu\varepsilon_{11} + \varepsilon_{33}\right),$$
$$N_{13} = k\,Ch\frac{1-\nu}{2}(2\varepsilon_{13}), \tag{7.84}$$
$$M_{11} = C\frac{h^3}{12}\left(\kappa_{11} + \nu\kappa_{33}\right), \qquad M_{33} = C\frac{h^3}{12}\left(\nu\kappa_{11} + \kappa_{33}\right),$$

where $C = E/(1-\nu^2)$, E is the Young's modulus, ν is the Poisson's ratio, and k is the shear correction factor. Note that only the components $\{11, 33, 13\}$ are involved.

The normal strains ε_{33} and κ_{33} are equal to zero in eq. (7.83), but we can recover them as follows. The condition $N_{33} = 0$ yields $\varepsilon_{33} = -\nu\,\varepsilon_{11}$. Similarly, κ_{33} can be recovered using the condition $m_{33} = 0$, which yields $\kappa_{33} = -\nu\,\kappa_{11}$. Due to the recovery, the normal strain is linearly approximated over ζ,

$$\tilde{H}_{n33}(\zeta) = \varepsilon_{33} + \zeta\kappa_{33} = -\nu(\varepsilon_{11} + \zeta\kappa_{11}). \tag{7.85}$$

Using the recovered normal strains, the constitutive relations of eq. (7.84) become

$$N_{11} = Ch\left(1-\nu^2\right)\varepsilon_{11} = Eh\,\varepsilon_{11}, \quad N_{11} = C\frac{h^3}{12}\left(1-\nu^2\right)\kappa_{11} = E\frac{h^3}{12}\kappa_{11}. \tag{7.86}$$

Comparing these forms with N_{11} of eq. (7.84) for $\varepsilon_{33} = 0$ and M_{11} of eq. (7.84) for $\kappa_{33} = 0$, we see that, in both cases, the strain recovery renders that the stiffness is reduced by the factor $(1-\nu^2)$.

Numerical test. The slender cantilever test is described in Sect. 15.3.1, Fig. 15.13. The cantilever is modeled by 100 two-node beam elements and loaded by either the stretching force $P_x = 1$ or the bending moment $M_z = 1$, or by the transverse force $P_y = 1$.

The linear solutions are presented in Table 7.1, where the tip's displacement and rotation are reported. We see that for $P_x = 1$, the recovery of ε_{33} is beneficial, while the recovery of κ_{33} has no effect. On the other hand, for $M_z = 1$ and $P_y = 1$, the situation is opposite and only the recovery of κ_{33} is beneficial. Without the recovery of κ_{33}, the solutions are too stiff and the error is 9%, as $\nu = 0.3$.

Table 7.1 Slender cantilever. Effect of recovery of normal strains for different loads.

Recovered strains	$P_x = 1$	$M_z = 1$		$P_y = 1$	
	$u_x \times 10^4$	$u_y \times 10^2$	$\omega \times 10^3$	u_y	$\omega \times 10^2$
none	0.91	5.46	1.0920	3.6402	5.46
ε_{33}	1.00	5.46	1.0920	3.6402	5.46
κ_{33}	0.91	6.00	1.2000	4.0002	6.00
Ref.	1.00	6.00	1.2000	4.0000	6.00

7.2.2 Reduced constitutive equations for incompressibility condition

The incompressibility condition, eq. (7.26) or (7.30), can be exploited in two ways:

1. It can be appended to the potential energy, i.e. $\Pi'(\chi, p) \doteq \Pi(\chi) + \int_V p\ (\det \mathrm{Grad}\chi - 1)\ \mathrm{d}V$, where the pressure p serves as the Lagrange multiplier. Note that unless p is included as a variable, the calculated stress is determined up to the pressure, see [239], pp. 70–72. This method is generally applicable, see [218] and the literature cited therein.
2. It can be treated as an auxiliary equation to recover the normal strain for shells made of an incompressible material. This application is of interest in this section and two formulations are presented below:
   a) For membranes, the description is given in terms of principal stretches and we assume that all strains are constant over the thickness, see Sect. 7.2.2A,
   b) For arbitrary shells, we assume that all components, except the normal one, are linear polynomials of ζ. Hence, the recovered U_{33} is a rational function of ζ, and the question arises of how many terms in the expansion should be retained, see Sect. 7.2.2B.

A. Membranes. Description in principal stretches

The principal directions of the right stretching tensor $\mathbf{U}$ are defined as follows:

$$\mathbf{Q} \in \mathrm{SO}(3): \qquad \mathbf{Q}\mathbf{U}\mathbf{Q}^{\mathrm{T}} = \hat{\mathbf{U}}, \tag{7.87}$$

where $\hat{\mathbf{U}} = \mathrm{diag}\{\lambda_1, \lambda_2, \lambda_3\}$ and λ_i $(i = 1, 2, 3)$ are the principal stretches. For isotropic materials, the Biot stress $\hat{\mathbf{T}}_s^B$ is coaxial with $\hat{\mathbf{U}}$ and, hence, also $\mathbf{Q}\mathbf{T}_s^B\mathbf{Q}^T = \hat{\mathbf{T}}_s^B$ holds, where $\hat{\mathbf{T}}_s^B = \mathrm{diag}\{t_1, t_2, t_3\}$ and t_i are the principal values of the Biot stress. Note that $\mathbf{Q}$ can vary during deformation.

For membranes, the stretches λ_i are constant over the thickness. Besides, the transverse shear stresses and strains are equal to zero, so one principal direction is normal to the membrane; we designate it as λ_3. As a consequence a one-parameter rotation $\mathbf{Q}$ describes the orientation of the tangent principal axes.

To find the principal directions in the tangent plane, we note that the Cauchy–Green tensor $\mathbf{C} \doteq \mathbf{F}^T\mathbf{F} = \mathbf{U}^2$ has the same principal directions as $\mathbf{U}$, but has a simpler form. Hence, instead of eq. (7.87), we use the equation $\hat{\mathbf{C}} = \mathbf{QCQ}^T$, where

$$\hat{\mathbf{C}} \doteq \begin{bmatrix} \hat{C}_{11} & \hat{C}_{12} \\ \hat{C}_{12} & \hat{C}_{22} \end{bmatrix}, \qquad \mathbf{Q} \doteq \begin{bmatrix} \cos\theta & -\sin\theta \\ \sin\theta & \cos\theta \end{bmatrix}, \qquad \mathbf{C} \doteq \begin{bmatrix} C_{11} & C_{12} \\ C_{12} & C_{22} \end{bmatrix} \tag{7.88}$$

and θ is the angle defining the first principal direction. From the condition $\hat{C}_{12} = 0$, we find $\theta(C_{\alpha\beta})$ and, next, $\hat{C}_{11}(C_{\alpha\beta})$ and $\hat{C}_{22}(C_{\alpha\beta})$, where $\alpha, \beta = 1, 2$. Besides, we have the relations to stretches

$$\lambda_1^2 = \hat{C}_{11}(C_{\alpha\beta}), \qquad \lambda_2^2 = \hat{C}_{22}(C_{\alpha\beta}), \tag{7.89}$$

which are used to calculate the derivatives needed in constitutive equations, see eq. (7.100).

For incompressible materials, we can use the incompressibility condition of eq. (7.30) written in terms of the principal stretches, $\det \hat{\mathbf{U}} = \lambda_1\lambda_2\lambda_3 = 1$, to calculate the normal stretch

$$\lambda_3 = (\lambda_1\lambda_2)^{-1}, \tag{7.90}$$

and, next, to obtain the reduced strain energy.

Ogden's strain energy. The Ogden form of the strain energy is an isotropic function of principal stretches

$$\mathcal{W}(\lambda_i) = \sum_r \frac{\mu_r}{\alpha_r} \left[\lambda_1^{\alpha_r} + \lambda_2^{\alpha_r} + \lambda_3^{\alpha_r} - 3\right], \qquad i = 1, 2, 3, \tag{7.91}$$

where μ_r and α_r are the material constants, see [158, 159]. The number of terms r is selected to characterize a particular material, e.g. for rubber $r = 3$ is used. The principal values of the Biot stress are obtained as

$$t_i \doteq \frac{\partial \mathcal{W}(\lambda_j)}{\partial \lambda_i}, \qquad i, j = 1, 2, 3. \tag{7.92}$$

For incompressible materials, we can use λ_3 of eq. (7.90), and then the reduced strain energy depends only on two stretches

$$\mathcal{W}^*(\lambda_\alpha) = \sum_r \frac{\mu_r}{\alpha_r} \left[\lambda_1^{\alpha_r} + \lambda_2^{\alpha_r} + (\lambda_1 \lambda_2)^{-\alpha_r} - 3 \right], \qquad \alpha = 1,2 \tag{7.93}$$

and the principal values of the Biot stress are

$$t_\alpha^* \doteq \frac{\partial \mathcal{W}(\lambda_\beta)}{\partial \lambda_\alpha}, \qquad \alpha, \beta = 1,2. \tag{7.94}$$

Strain energy depending on invariants of U. Assume that the strain energy is a function of the principal invariants of $\mathbf{U}$ which, in turn, are expressed by stretches λ_i,

$$\tilde{\mathcal{W}}(I_i(\mathbf{U})) = \hat{\mathcal{W}}(I_i(\hat{\mathbf{U}})), \qquad i = 1,2,3, \tag{7.95}$$

where the principal invariants of $\hat{\mathbf{U}}$ are as follows:

$$I_1(\hat{\mathbf{U}}) = \lambda_1 + \lambda_2 + \lambda_3, \quad I_2(\hat{\mathbf{U}}) = \lambda_1\lambda_2 + \lambda_2\lambda_3 + \lambda_3\lambda_1, \quad I_3(\hat{\mathbf{U}}) = \lambda_1\lambda_2\lambda_3. \tag{7.96}$$

The constitutive equation for the principal values of the Biot stress is calculated as

$$t_i \doteq \frac{\partial \hat{\mathcal{W}}(I_j(\hat{\mathbf{U}}))}{\partial \lambda_i} = \frac{\partial \hat{\mathcal{W}}}{\partial I_j} \frac{\partial I_j}{\partial \lambda_i}, \qquad i,j = 1,2,3. \tag{7.97}$$

Using λ_3 of eq. (7.90) in the two first invariants, we obtain

$$I_1^*(\hat{\mathbf{U}}) = \lambda_1 + \lambda_2 + (\lambda_1\lambda_2)^{-1}, \qquad I_2^*(\hat{\mathbf{U}}) = \lambda_1\lambda_2 + (\lambda_1)^{-1} + (\lambda_2)^{-1}. \tag{7.98}$$

The strain energy becomes a function of λ_1 and λ_2, i.e. $\tilde{\mathcal{W}}(I_i(\mathbf{U})) = \hat{\mathcal{W}}(I_\alpha^*(\hat{\mathbf{U}}))$, $\alpha = 1,2,$ and the constitutive equation is calculated as

$$t_\alpha^* \doteq \frac{\partial \hat{\mathcal{W}}(I_\beta^*(\hat{\mathbf{U}}))}{\partial \lambda_\alpha} = \frac{\partial \hat{\mathcal{W}}}{\partial I_\beta^*} \frac{\partial I_\beta^*}{\partial \lambda_\alpha}, \qquad \alpha, \beta = 1,2. \tag{7.99}$$

Remark. The above formulas are simple, but the computational procedure for rubber-like membranes is not trivial because the relation between the stretches and strain components is complicated. For instance, the constitutive equation for the Ogden energy is

$$S^{\alpha\beta} = \frac{\partial \mathcal{W}(\lambda_\gamma)}{\partial E_{\alpha\beta}} = \frac{\partial \mathcal{W}(\lambda_\gamma)}{\partial \lambda_1}\frac{\partial \lambda_1}{\partial E_{\alpha\beta}} + \frac{\partial \mathcal{W}(\lambda_\gamma)}{\partial \lambda_2}\frac{\partial \lambda_2}{\partial E_{\alpha\beta}}, \qquad \alpha, \beta, \gamma = 1, 2, \tag{7.100}$$

where the derivatives $\partial\lambda_1/\partial E_{\alpha\beta}$ and $\partial\lambda_2/\partial E_{\alpha\beta}$ are computed by using eq. (7.89) and are complex. The computational procedure for the formulation in terms the second Piola–Kirchhoff stress and Green strain is given in [86].

Incompressibility condition for small strains. For small strains, we can obtain an alternative expression for the incompressibility condition. Consider the following linear Taylor expansions at $\lambda_1 = \lambda_2 = \lambda_3 = 1$,

$$\det \mathbf{U} \doteq \lambda_1\lambda_2\lambda_3 = 1 + \mathrm{d}\lambda_1 + \mathrm{d}\lambda_2 + \mathrm{d}\lambda_3 + \mathrm{O}(\mathrm{d}\lambda_1^2, \mathrm{d}\lambda_2^2, \mathrm{d}\lambda_3^2), \tag{7.101}$$

$$\mathrm{tr}(\mathbf{U} - \mathbf{I}) \doteq \lambda_1 + \lambda_2 + \lambda_3 - 3 = \mathrm{d}\lambda_1 + \mathrm{d}\lambda_2 + \mathrm{d}\lambda_3 + \mathrm{O}(\mathrm{d}\lambda_1^2, \mathrm{d}\lambda_2^2, \mathrm{d}\lambda_3^2). \tag{7.102}$$

Hence,

$$\det \mathbf{U} - 1 = \mathrm{tr}(\mathbf{U} - \mathbf{I}) = \mathrm{tr}\mathbf{H}, \tag{7.103}$$

with the second-order accuracy. Thus, for small strains, the incompressibility condition $\det \mathbf{U} = 1$ can be replaced by the condition $\mathrm{tr}\mathbf{H} = 0$.

B. Arbitrary shells

Consider the incompressibility condition of eq. (7.30), which is expressed in terms of the right stretching tensor. Let us write this condition as

$$\det \mathbf{U} = U_{31}D_{31} - U_{32}D_{32} + U_{33}D_{33} = 1, \tag{7.104}$$

where the minors are

$$D_{31} \doteq U_{12}U_{23} - U_{13}U_{22}, \quad D_{32} \doteq U_{11}U_{23} - U_{13}U_{12}, \quad D_{33} \doteq U_{11}U_{22} - U_{12}^2. \tag{7.105}$$

The normal component U_{33} appears in this equation only once and can be calculated as

$$U_{33} = \frac{1}{D_{33}}(1 - U_{31}D_{31} + U_{32}D_{32}), \tag{7.106}$$

where the r.h.s. depends on all components of $\mathbf{U}$ except the normal one.

We denote the $\mathbf{U}$ without the 33 component by $\mathbf{U}^*$, and assume that it is a linear polynomial of ζ, i.e. $\mathbf{U}^* = \mathbf{e}^* + \zeta\mathbf{k}^*$. Then U_{33} is a rational function of ζ with a polynomial of the third order in the nominator, and a polynomial of the second order in the denominator. Hence, unless the denominators are equal to zero, $U_{33}(\zeta)$ is infinitely times differentiable and we can perform the Taylor series expansion of it around the middle surface, retaining as many terms as necessary,

$$U_{33}(\zeta) = (U_{33})_0 + (U_{33,\zeta})_0\,\zeta + \tfrac{1}{2}\,(U_{33,\zeta\zeta})_0\,\zeta^2 + \mathrm{O}(\zeta^3), \tag{7.107}$$

where we can denote $(U_{33})_0 \doteq e_{33}$ and $(U_{33,\zeta})_0 \doteq k_{33}$ to keep the notation consistent. The so-recovered e_{33} and k_{33} can be used in $\mathbf{U}$.

A rigorous analysis of the question of how many terms of the expansion should be retained is difficult, see [202], because we must define, in advance, the class of deformation and geometry which is analyzed. Some insight provides the example given below in which we determine accuracy of the linear expansion of the normal strain.

Example. Inversion of a spherical cap. The example of an inversion of a spherical cap is solved analytically in [232]. The current position vector is assumed as $\mathbf{x}(\zeta) = \mathbf{x}_0 + \lambda(\zeta)\,\bar{\mathbf{n}}$, where $\bar{\mathbf{n}}$ is a unit vector normal to the deformed middle surface and $\lambda(\zeta)$ is the extension function. The deformed configuration also has a spherical shape, so $\lambda(\zeta)$ is obtained analytically, see eq. (6.4) therein. The obtained normal strain has the following form:

$$\varepsilon_{33}(\hat{\xi}) \doteq \frac{\partial \lambda}{\partial \zeta} = \frac{(1+\frac{h}{R}\hat{\xi})^2}{\left[1+\eta^3-(1+\frac{h}{R}\hat{\xi})^3\right]^{2/3}}, \tag{7.108}$$

where $\hat{\xi} = -(\zeta + h/2)/h$, $\hat{\xi} \in [-1, 0]$ and η is the in-plane stretch of the reference surface. We see that ε_{33} is a complicated function of $\hat{\xi}$, but a constant approximation of it is sufficient for the following limit cases:

1. Thin and/or flat shells. For $h/R \to 0$, we have $\varepsilon_{33}(\hat{\xi}) \to 1/\eta^3$.
2. Large stretches. For $\eta \to \infty$, we have $\varepsilon_{33}(\hat{\xi}) \to 0$, i.e. the normal strain vanishes. In reality, for the inflated structures made of rubber-like materials, $\eta \leq 10$.

The relative error of a linear expansion of $\varepsilon_{33}(\hat{\xi})$ at the middle surface $(\hat{\xi} = -0.5)$ is given in Table 7.2. We see that, when the 1% error at the external surface is acceptable, the linear approximation of ε_{33} can be used for a range of values of η and h/R.

Table 7.2 Relative error [in %] for linear expansion of normal strain ε_{33}.

η	h/R				
	0.2	0.1	0.05	0.01	0.001
0.5	14.95	4.98	1.53	0.070	0.00080
1.0	8.57	2.42	0.67	0.030	0.00030
1.5	3.54	0.83	0.20	0.008	0.00008
2.0	2.37	0.51	0.12	0.005	0.00005
5.0	1.61	0.32	0.07	0.003	0.00003
10.0	1.57	0.31	0.07	0.003	0.00003

7.3 Shear correction factor

The value of the shear correction factor k can be determined in several ways, see [259, 260]. Below, for the assumption that the distribution of the in-plane stresses is linear across thickness, we find that the transverse shear is parabolic and determine the value of the shear correction factor.

3D equilibrium equations and traction boundary conditions. The 3D equilibrium equations in a local Cartesian basis $\{\mathbf{t}_i\}$ at the reference surface of a shell are as follows:

$$\sigma_{\alpha\beta,\alpha} + \sigma_{3\beta,3} = 0, \qquad \sigma_{\alpha 3,\alpha} + \sigma_{33,3} = 0, \tag{7.109}$$

where σ_{ij} $(i,j = 1,2,3)$ is the stress (symmetric). We assume that the body force $b_i = 0$. The indices $\alpha, \beta = 1,2$ correspond to the tangent (in-plane) directions and the index 3 to the normal direction of the basis $\{\mathbf{t}_k\}$, see Fig. 7.2.

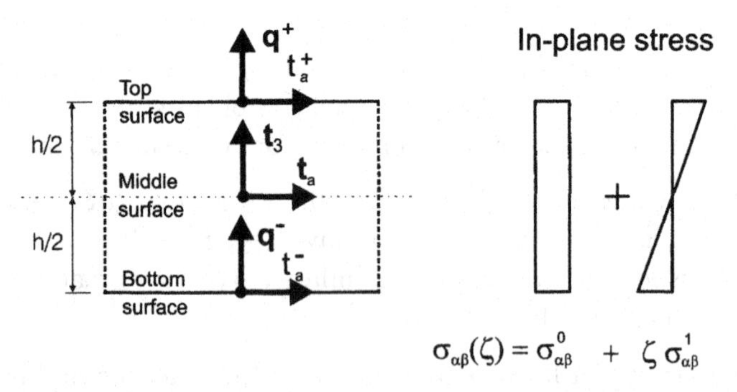

Fig. 7.2 The external loads and distribution of in-plane stress.

The transverse shear stress $\sigma_{3\beta}$ and the transverse normal stress σ_{33} have to satisfy the traction boundary conditions at surfaces bounding the shell,

$$\sigma_{3\beta}|_{\zeta=+\frac{h}{2}} = \tau_\beta^+, \qquad \sigma_{3\beta}|_{\zeta=-\frac{h}{2}} = \tau_\beta^-, \tag{7.110}$$

$$\sigma_{33}|_{\zeta=+\frac{h}{2}} = q^+, \qquad \sigma_{33}|_{\zeta=-\frac{h}{2}} = q^-, \tag{7.111}$$

where τ_β^+ and τ_β^- are the tangent components while q^+ and q^- are the normal components of the external load on the top and bottom surfaces, respectively.

Distribution of transverse shear stress. We assume that the in-plane stresses are linear over the thickness, i.e. $\sigma_{\alpha\beta}(\zeta) = \sigma^0_{\alpha\beta} + \zeta\sigma^1_{\alpha\beta}$, where $\zeta \in [-h/2, +h/2]$. Using the equilibrium equations (7.109), we determine the distribution of the transverse shear stress $\sigma_{3\beta}$ over the thickness.

Integrating eq. $(7.109)_1$ w.r.t. ζ (or 3), we have

$$\sigma_{3\beta}(\zeta) = C - \zeta\sigma^0_{\alpha\beta,\alpha} - \frac{\zeta^2}{2}\sigma^1_{\alpha\beta,\alpha}. \tag{7.112}$$

By the boundary condition at the bottom boundary, $\sigma_{3\beta}|_{\zeta=-\frac{h}{2}} = \tau_\beta^-$, we obtain

$$\sigma_{3\beta}(\zeta) = \tau_\beta^- - \left(\frac{h}{2} + \zeta\right)\sigma^0_{\alpha\beta,\alpha} + \left(\frac{h^2}{8} - \frac{\zeta^2}{2}\right)\sigma^1_{\alpha\beta,\alpha}. \tag{7.113}$$

There is no another constant to account for the condition at the top boundary $\zeta = +h/2$, but it is satisfied, as shown below. The integral of eq. $(7.109)_1$ over the thickness yields the relation

$$\int_{-\frac{h}{2}}^{+\frac{h}{2}} \sigma_{\alpha\beta,\alpha}\, d\zeta + (\tau_\beta^+ - \tau_\beta^-) = h\sigma^0_{\alpha\beta,\alpha} + (\tau_\beta^+ - \tau_\beta^-) = 0. \tag{7.114}$$

On the other hand, for the top boundary, eq. (7.113) yields

$$\sigma_{3\beta}(+h/2) = \tau_\beta^- - h\sigma^0_{\alpha\beta,\alpha} \tag{7.115}$$

and, by eq. (7.114), the r.h.s. of this equation is equal to τ_β^+. Hence, $\sigma_{3\beta}(\zeta)$ of eq. (7.113) satisfies both the boundary conditions.

We can rewrite eq. (7.113) in several equivalent forms. By using $\sigma^0_{\alpha\beta,\alpha}$ calculated from eq. (7.114), we rewrite eq. (7.113) as

$$\sigma_{3\beta}(\zeta) = \frac{1}{2}\left(1 - \frac{2\zeta}{h}\right)\tau_\beta^- + \frac{1}{2}\left(1 + \frac{2\zeta}{h}\right)\tau_\beta^+ + \left(\frac{h^2}{8} - \frac{\zeta^2}{2}\right)\sigma^1_{\alpha\beta,\alpha} \tag{7.116}$$

or, using the natural coordinate $\bar{\zeta} \doteq 2\zeta/h \in [-1, +1]$,

$$\sigma_{3\beta}(\bar{\zeta}) = S_1(\bar{\zeta})\, \tau_\beta^- + S_2(\bar{\zeta})\, \tau_\beta^+ + \frac{h^2}{8} S_3(\bar{\zeta})\, \sigma^1_{\alpha\beta,\alpha}, \tag{7.117}$$

where the component functions are

$$S_1(\bar{\zeta}) \doteq \frac{1}{2}\left(1 - \bar{\zeta}\right), \quad S_2(\bar{\zeta}) \doteq \frac{1}{2}\left(1 + \bar{\zeta}\right), \quad S_3(\bar{\zeta}) \doteq 1 - \bar{\zeta}^2, \tag{7.118}$$

see Fig. 7.3. For the zero boundary conditions, $\tau_\beta^+ = \tau_\beta^- = 0$, we obtain a very simple formula

$$\sigma_{3\beta}(\bar{\zeta}) = \frac{h^2}{8}(1 - \bar{\zeta}^2)\, \sigma^1_{\alpha\beta,\alpha}. \tag{7.119}$$

Concluding, for the in-plane stress linearly distributed over the thickness, the distribution of the transverse shear stress is parabolic in $\bar{\zeta}$.

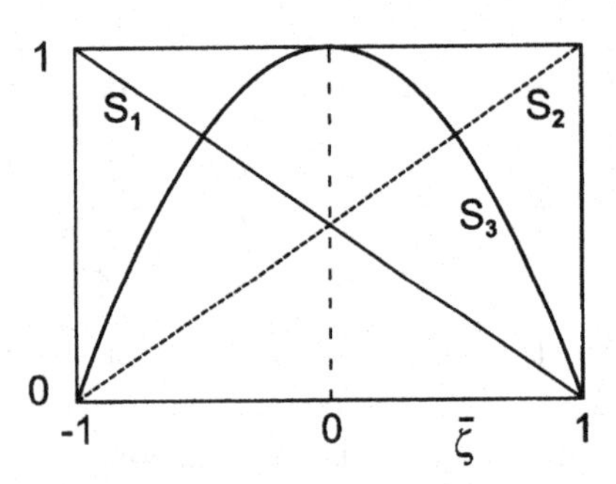

Fig. 7.3 Component functions for transverse shear stress.

Transverse shear stress in terms of shell resultants. We can express $\sigma_{3\beta}$ of eq. (7.117) in terms of the stress and couple resultants.

For the membrane stress $\sigma_{\alpha\beta}(\zeta) = \sigma^0_{\alpha\beta} + \zeta\sigma^1_{\alpha\beta}$, the in-plane stress and couple resultants are

$$N_{\alpha\beta} \doteq \int_{-\frac{h}{2}}^{+\frac{h}{2}} \sigma_{\alpha\beta}(\zeta)\, \mathrm{d}\zeta = h\, \sigma^0_{\alpha\beta}, \qquad M_{\alpha\beta} \doteq \int_{-\frac{h}{2}}^{+\frac{h}{2}} \zeta\sigma_{\alpha\beta}(\zeta)\, \mathrm{d}\zeta = \frac{h^3}{12}\, \sigma^1_{\alpha\beta}. \tag{7.120}$$

We calculate $\sigma^1_{\alpha\beta} = (12/h^3)\, M_{\alpha\beta}$ from the last formula and use it in the transverse shear stress of eq. (7.117),

$$\sigma_{3\beta}(\bar{\zeta}) = S_1(\bar{\zeta})\,\tau_\beta^- + S_2(\bar{\zeta})\,\tau_\beta^+ + \frac{3}{2h} S_3(\bar{\zeta})\,M_{\alpha\beta,\alpha}. \tag{7.121}$$

For this form of $\sigma_{3\beta}$, the transverse shear stress and couple resultants are

$$N_{3\beta} \doteq \int_{-\frac{h}{2}}^{+\frac{h}{2}} \sigma_{3\beta}(\zeta)\,\mathrm{d}\zeta = \frac{h}{2}(\tau_\beta^+ + \tau_\beta^-) + M_{\alpha\beta,\alpha}, \tag{7.122}$$

$$M_{3\beta} \doteq \int_{-\frac{h}{2}}^{+\frac{h}{2}} \zeta\,\sigma_{3\beta}(\zeta)\,\mathrm{d}\zeta = \frac{h^2}{12}(\tau_\beta^+ - \tau_\beta^-). \tag{7.123}$$

For the zero tangent loads, $\hat{\tau}_\beta^- = \hat{\tau}_\beta^+ = 0$, these resultants are reduced to

$$N_{3\beta} = M_{\alpha\beta,\alpha}, \qquad M_{3\beta} = 0, \tag{7.124}$$

where the first equation is a well-known formula linking the bending moment and the transverse shear resultant. By using it in the transverse shear stress of eq. (7.121), we obtain

$$\sigma_{3\beta}(\bar{\zeta}) = \frac{3}{2h} S_3(\bar{\zeta})\,M_{\alpha\beta,\alpha} = \frac{3}{2h} S_3(\bar{\zeta})\,N_{3\beta}, \tag{7.125}$$

which depends on the transverse shear resultant. The last form is identical to eq. $(20.5)_2$ of [153], p. 573.

Remark. Note that for the zero tangent loads, we have $M_{3\beta} = 0$ in eq. (7.124) and, hence, by the inverse constitutive equation, the first-order shell strain $\kappa_{3\beta} = 0$. Then we can omit the term with $\kappa_{3\beta}$ in the shell strain energy.

Shear correction factor. We can use the parabolic transverse shear stress to derive the shear correction factor. Note that, for the Reissner kinematics, the transverse shear strain is linear in ζ and cannot match the parabolic shear stress of eq. (7.125).

For the SVK material, the complementary energy density is

$$\begin{aligned}\mathcal{W}_c \doteq{}& \frac{1+\nu}{2E}(\sigma_{11}^2 + \sigma_{22}^2 + \sigma_{33}^2 + 2\sigma_{21}^2 + 2\sigma_{31}^2 + 2\sigma_{32}^2) \\ &- \frac{\nu}{2E}(\sigma_{11} + \sigma_{22} + \sigma_{33})^2,\end{aligned} \tag{7.126}$$

where E is Young's modulus and ν is the Poisson's ratio. For simplicity, we separate the term for the transverse shear stress,

$$\mathcal{W}_c^{3\beta} \doteq \frac{1+\nu}{E} \sigma_{3\beta}^2, \qquad \beta = 1,2. \tag{7.127}$$

For the transverse shear stress of eq. (7.125), the shell (integral) counterpart of $\mathcal{W}_c^{3\beta}$ becomes

$$\Sigma_c^{3\beta} \doteq \int_{-\frac{h}{2}}^{+\frac{h}{2}} \mathcal{W}_c^{3\beta}(\zeta)\, \mathrm{d}\zeta = \frac{1+\nu}{E} \frac{6}{5h} N_{3\beta}^2. \tag{7.128}$$

Then the inverse constitutive equations for the transverse shear strain is

$$\varepsilon_{3\beta} \doteq \frac{\partial \Sigma_c^{3\beta}}{\partial N_{3\beta}} = \frac{6}{5h} \frac{2(1+\nu)}{E} N_{3\beta}, \tag{7.129}$$

from which we can obtain the constitutive equation for the transverse shear stress resultant

$$N_{3\beta} = \frac{5}{6} \frac{E}{2(1+\nu)} h\, \varepsilon_{3\beta} = k\, Gh\, \varepsilon_{3\beta}, \tag{7.130}$$

where $G \doteq E/[2(1+\nu)]$ is the shear modulus and $k = 5/6$ is the shear correction factor. This factor accounts for the parabolic distribution of $\sigma_{3\beta}$ corresponding to the linear distribution of $\sigma_{\alpha\beta}$ over the thickness, and was obtained in [190]. Equation (7.130) corresponds to eq. $(20.12)_2$ of [153], p. 574.

Finally, we note that the shear correction factor can also be derived for the shearing moment $M_{3\beta}$ but it is rarely used, as usually the strain energy of $\kappa_{3\beta}$ is omitted in shell elements, as the second order quantity.

Summarizing, three results were obtained for shells in this section:

1. the formula for distributions of $\sigma_{3\beta}$ over the shell thickness, eq. (7.116) or (7.117),
2. the motivation for omitting the first-order shell strain $\kappa_{3\beta}$ in the strain energy, see eq. (7.124), and the remark which follows,
3. the shear correction factor for constitutive equation for $N_{3\beta}$, eq. (7.130).

Part III
FINITE ROTATIONS FOR SHELLS

8

Parametrization of finite rotations

In this chapter, we describe the basic questions related to properties and parametrization of rotations; the subject of the algorithmic treatment of rotations is addressed separately in Chap. 9.

The topic of finite rotations is very important in practice and often undertaken in the works on rigid-body dynamics, see [197, 79] and on multi-body dynamics of rigid and flexible bodies, see [258, 4, 44, 77, 76, 9]. There are also mathematical works on rotations, such as, e.g., [231, 45, 2]. This subject is also covered in the works on the Cosserat continuum and on structures with rotational degrees of freedom, such as shells and 3D beams; these works are cited in Chap. 4.

The rotations are described by a proper orthogonal tensor and its basic properties are presented in Sect. 8.1. However, in numerical implementations, we have to use some rotational parameters; several of them are in use and their properties are very different. We describe a wide, although not complete, selection of parametrizations in Sect. 8.2; some of them provide a theoretical background but are not used in computation of structures. For more details, see [5, 8, 107].

8.1 Basic properties of rotations

In this section, we provide elementary information related to rotations, such as the definition of the rotation tensor, and two basic problems: the rotation of a vector about an axis and the rotation of a Cartesian triad of vectors. Basic properties of orthogonal tensors and skew-symmetric tensors are provided.

8.1.1 Rotation tensor

Let us denote by $\mathbf{R}$ the rotation tensor belonging to the special orthogonal group defined as follows

$$\mathrm{SO}(3) := \{\mathbf{R} : \mathbb{R}^3 \rightarrow \mathbb{R}^3 \;\text{ is linear } \mid \mathbf{R}^T\mathbf{R} = \mathbf{I} \;\text{ and }\; \det \mathbf{R} = +1\}. \tag{8.1}$$

The orthogonality condition $\mathbf{R}^T\mathbf{R} = \mathbf{I}$ renders that preserved are (i) the angle between two rotated vectors $\mathbf{a}$, $\mathbf{b}$, because $(\mathbf{Ra}) \cdot (\mathbf{Rb}) = \mathbf{a} \cdot (\mathbf{R}^T\mathbf{Rb}) = \mathbf{a} \cdot \mathbf{b}$, and (ii) the length of a rotated vector, because $\sqrt{(\mathbf{Ra}) \cdot (\mathbf{Ra})} = \sqrt{\mathbf{a} \cdot (\mathbf{R}^T\mathbf{Ra})} = \sqrt{\mathbf{a} \cdot \mathbf{a}}$.

Example. The orthogonality condition $\mathbf{R}^T\mathbf{R} = \mathbf{I}$ itself does not suffice to define the rotation as the transformation representing a reflection also satisfies this condition. Consider two orthogonal matrices

$$\mathbf{R}_1 \doteq \begin{bmatrix} \cos\omega & -\sin\omega \\ \sin\omega & \cos\omega \end{bmatrix}, \qquad \mathbf{R}_2 \doteq \begin{bmatrix} \cos\omega & \sin\omega \\ \sin\omega & -\cos\omega \end{bmatrix}. \tag{8.2}$$

For the vector $\mathbf{a} \doteq [1, 1]^T$ and $\omega \doteq \pi$, we obtain

$$\mathbf{b} \doteq \mathbf{R}_1 \mathbf{a} = [-1, -1]^T, \qquad \mathbf{c} \doteq \mathbf{R}_2 \mathbf{a} = [-1, 1]^T, \tag{8.3}$$

which are shown in Fig. 8.1. Note that $\mathbf{R}_1$ rotates $\mathbf{a}$ about point "0", while $\mathbf{R}_2$ reflects it across the line 0Y. We can check that $\det \mathbf{R}_1 = +1$, while $\det \mathbf{R}_2 = -1$.

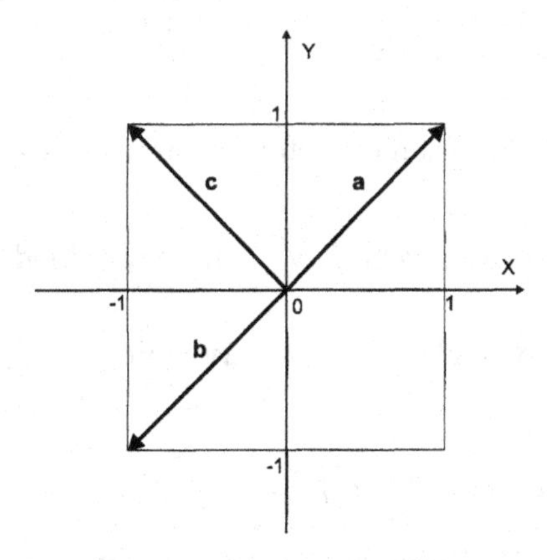

Fig. 8.1 Rotation of vector $\mathbf{a}$ yields vector $\mathbf{b}$; reflection yields vector $\mathbf{c}$.

Consider the right-handed triad of ortho-normal vectors, such that $[\mathbf{t}_1 \times \mathbf{t}_2] \cdot \mathbf{t}_3 > 0$. The rotation tensor $\mathbf{R} \in \mathrm{SO}(3)$ applied to each vector of this triad yields a triad of the same handedness, i.e. $[(\mathbf{R}\mathbf{t}_1) \times (\mathbf{R}\mathbf{t}_2)] \cdot (\mathbf{R}\mathbf{t}_3) > 0$, see Sect. 8.1.3. The parallelepipeds spanned by these triads have identical volumes, i.e. $[(\mathbf{R}\mathbf{t}_1) \times (\mathbf{R}\mathbf{t}_2)] \cdot (\mathbf{R}\mathbf{t}_3) = [\mathbf{t}_1 \times \mathbf{t}_2] \cdot \mathbf{t}_3$.

8.1.2 Rotation of vector about axis

Consider an arbitrary vector $\mathbf{v}$ and rotate it about an axis A-B to the position $\mathbf{v}'$, see Fig. 8.2. Associate with the axis of rotation A-B, a unit vector $\mathbf{e}$ and define two auxiliary vectors

$$\mathbf{e}_2 = \frac{\mathbf{e} \times \mathbf{v}}{|\mathbf{e} \times \mathbf{v}|}, \qquad \mathbf{e}_1 = \mathbf{e}_2 \times \mathbf{e}, \tag{8.4}$$

which are perpendicular to $\mathbf{e}$ and ortho-normal. The angle ω between the planes $\mathbf{e}-\mathbf{v}$ and $\mathbf{e}-\mathbf{v}'$ is called the rotation angle and it is assumed positive when it is in accord with the handedness of $\{\mathbf{e}_1, \mathbf{e}_2\}$.

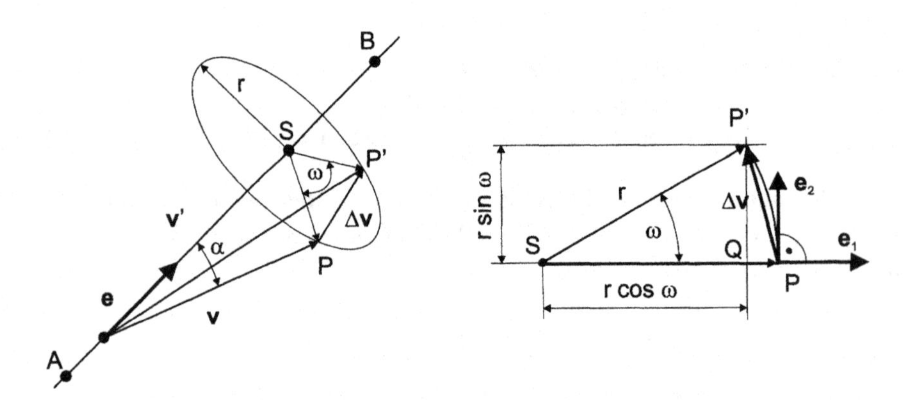

Fig. 8.2 Rotation of vector $\mathbf{v}$ around axis $\mathbf{e}$.

A position of the rotated vector $\mathbf{v}'$ can be obtained from the following formula:

$$\mathbf{v}' = \mathbf{v} + \Delta\mathbf{v}, \qquad \Delta\mathbf{v} = \overrightarrow{PQ} + \overrightarrow{QP'}, \tag{8.5}$$

where

$$\overrightarrow{PQ} = -\mathbf{e}_1\, r\,(1 - \cos\omega), \qquad \overrightarrow{QP'} = \mathbf{e}_2\, r\, \sin\omega, \qquad r = |\mathbf{v}| \sin\alpha,$$

and α is an angle between $\mathbf{e}$ and $\mathbf{v}$. After elementary algebraic transformations, we obtain

$$\mathbf{v}' = \mathbf{v} + \sin\omega\,(\mathbf{e}\times\mathbf{v}) + (1-\cos\omega)\,[\mathbf{e}\times(\mathbf{e}\times\mathbf{v})]. \tag{8.6}$$

Let us introduce a tensor $\mathbf{S} \doteq \mathbf{e}\times\mathbf{I} \in \mathrm{so}(3)$, where $\mathrm{so}(3)$ is a linear space of skew-symmetric tensors such that

$$\mathrm{so}(3) := \{\mathbf{S} :\ \mathbb{R}^3 \to \mathbb{R}^3 \ \text{ is linear } \mid \ \mathbf{S}^T = -\mathbf{S}\}. \tag{8.7}$$

Some useful properties of $\mathbf{S}$ are given in Table 8.1. Note that $\mathbf{e}$ is the axial vector of $\mathbf{S}$, i.e. $\mathbf{S}\,\mathbf{e} = \mathbf{0}$. (To emphasize this, we can use the tilde and define: $\mathbf{S} \doteq \tilde{\mathbf{e}}$.) $\mathbf{S}$ can be expressed by an arbitrary pair of ortho-normal vectors of this plane (see *Property 2* below), e.g. $\mathbf{S} \doteq \mathbf{e}_2\otimes\mathbf{e}_1 - \mathbf{e}_1\otimes\mathbf{e}_2$, where $\mathbf{e}_1$ and $\mathbf{e}_2$ are the vectors of eq. (8.4). Then, the cross-products in eq. (8.6) can be expressed as

$$\mathbf{e}\times\mathbf{v} = \mathbf{S}\mathbf{v}, \qquad \mathbf{e}\times(\mathbf{e}\times\mathbf{v}) = \mathbf{S}^2\mathbf{v}, \tag{8.8}$$

and eq. (8.6) can be rewritten as a linear mapping $\mathbf{v}' = \mathbf{R}\,\mathbf{v}$, where

$$\mathbf{R} \doteq \mathbf{I} + \sin\omega\mathbf{S} + (1-\cos\omega)\mathbf{S}^2. \tag{8.9}$$

Several properties of $\mathbf{R}$ are listed in Table 8.2. The meaning of particular terms of $\mathbf{R}$ becomes clear if we consider components of $\Delta\mathbf{v} = \mathbf{R}\mathbf{v} - \mathbf{v}$, see Fig. 8.3.

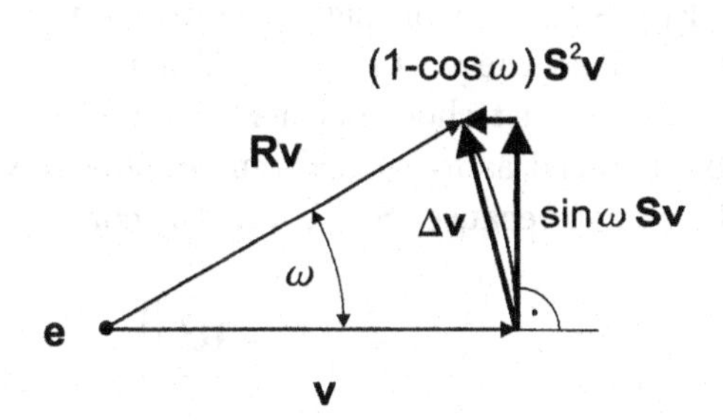

Fig. 8.3 Tangent and normal components of increment $\Delta\mathbf{v} \doteq \mathbf{R}\mathbf{v} - \mathbf{v}$.

Property 1. The coefficients of $\mathbf{R}$ of eq. (8.9) are identical for $\omega + k\,2\pi$, where k is some integer, due to periodicity of the sine and cosine functions. In particular, for $0 + k2\pi$, $\mathbf{R} = \mathbf{I}$. Hence, we can restrict the range of the rotation angle to $-\pi \leq \omega \leq +\pi$. The number of turns indicated by the integer k often is meaningless, e.g., in free rigid body motion, where only a position has significance. But for rotational springs, the number of turns is important as it changes the internal force.

Property 2. It can be seen that $\mathbf{R}$ of eq. (8.9) is an isotropic function of $\mathbf{S}$. Let us calculate

$$\mathbf{QRQ}^T = \mathbf{QQ}^T + \sin\omega \mathbf{QSQ}^T + (1 - \cos\omega)\mathbf{QS}^2\mathbf{Q}^T, \tag{8.10}$$

where $\mathbf{Q} \in \mathrm{SO}(3)$. Noting that $\mathbf{QS}^2\mathbf{Q}^T = \mathbf{QS}(\mathbf{Q}^T\mathbf{Q})\mathbf{SQ}^T = (\mathbf{QSQ}^T)^2$, we obtain $\mathbf{QR}(\mathbf{S})\mathbf{Q}^T = \mathbf{R}(\mathbf{QSQ}^T)$, which is a definition of an isotropic function, see [239] p. 23, eq. (8.7). If we denote the rotate-forward operation as $(\cdot)^* \doteq \mathbf{Q}\,(\cdot)\,\mathbf{Q}^T$, then

$$\begin{aligned}
&\mathbf{I}^* = \mathbf{Q}(\mathbf{e}_i \otimes \mathbf{e}_i)\mathbf{Q}^T = \mathbf{e}_i^* \otimes \mathbf{e}_i^*, \\
&\mathbf{S}^* = \mathbf{Q}(\mathbf{e}_2 \otimes \mathbf{e}_1 - \mathbf{e}_1 \otimes \mathbf{e}_2)\mathbf{Q}^T = \mathbf{e}_2^* \otimes \mathbf{e}_1^* - \mathbf{e}_1^* \otimes \mathbf{e}_2^*, \\
&(\mathbf{S}^*)^2 = -(\mathbf{e}_1^* \otimes \mathbf{e}_1^* + \mathbf{e}_2^* \otimes \mathbf{e}_2^*),
\end{aligned} \tag{8.11}$$

where $\mathbf{e}^* \doteq \mathbf{Q}\,\mathbf{e}$ and $\mathbf{e}_\alpha^* \doteq \mathbf{Q}\,\mathbf{e}_\alpha$. By these equations, the components of $\mathbf{R}$ are identical in all bases $\{\mathbf{e}^*, \mathbf{e}_\alpha^*\}$. In particular, if $\mathbf{Q}$ is chosen as a rotation about $\mathbf{e}$, then $\{\mathbf{e}_1, \mathbf{e}_2\}$ can be replaced by any co-planar orthonormal pair $\{\mathbf{e}_1^*, \mathbf{e}_2^*\}$.

Property 3. The relation between $\mathbf{R}$ and the set $\{\omega, \mathbf{e}\}$ is 1-to-2, because two sets, $\{\omega, \mathbf{e}\}$ and $\{-\omega, -\mathbf{e}\}$, yield the same $\mathbf{R}$ of eq. (8.9) and, as shown in Fig. 8.4, are physically equivalent, i.e. for both, the vector $\mathbf{v}$ is rotated into the vector $\mathbf{v}'$. When calculating ω and $\mathbf{e}$ from a given $\mathbf{R}$, we must choose either the sign of ω or the sense of $\mathbf{e}$. This can be deduced as follows. From eq. (8.9), we can calculate $\sin\omega\,\mathbf{S} = \frac{1}{2}(\mathbf{R}-\mathbf{R}^T)$, and because $\mathbf{S} = \mathbf{e}\times\mathbf{I}$, the corresponding vectorial equation is

$$\sin\omega\,\mathbf{e} = \frac{1}{2}\left[\mathbf{I} \times \frac{1}{2}(\mathbf{R} - \mathbf{R}^T)\right], \tag{8.12}$$

where the right-hand side vector is known. Hence, only the product $\sin\omega\,\mathbf{e}$ is defined by this equation and whether we use $\{\sin\omega, \mathbf{e}\}$ or $\{-\sin\omega, -\mathbf{e}\}$ is a matter of choice.

Skew-symmetric tensor and its axial vector. Designate the skew-symmetric tensor by $\tilde{\mathbf{v}} \in \mathrm{so}(3)$, i.e. using a tilde, and its axial vector by $\mathbf{v}$. They are linked by the relations

$$\tilde{\mathbf{v}} = \mathbf{v} \times \mathbf{I}, \qquad \mathbf{v} = \tfrac{1}{2}(\mathbf{I} \times \tilde{\mathbf{v}}), \tag{8.13}$$

where the cross-product of a vector and a tensor is defined as in [33] p. 74.

Table 8.1 Properties of $\mathbf{S} \in so(3)$.

dyadic representation:	$\mathbf{S} = \mathbf{e} \times \mathbf{I} = \mathbf{e}_2 \otimes \mathbf{e}_1 - \mathbf{e}_1 \otimes \mathbf{e}_2, \qquad \mathbf{e}_1, \mathbf{e}_2 \perp \mathbf{e}$
skew-symmetricity:	$\mathbf{S}^T = -\mathbf{S}$
powers:	odd $\mathbf{S}^{2n+1} = (-1)^n \mathbf{S}$, i.e. $\mathbf{S},\ \mathbf{S}^3 = -\mathbf{S},\ \mathbf{S}^5 = \mathbf{S},\ \mathbf{S}^7 = -\mathbf{S}, \ldots$ even $\mathbf{S}^{2n+2} = (-1)^{n+2}\mathbf{S}^2$, $n = 0, 1, \ldots$ i.e. $\mathbf{S}^2,\ \mathbf{S}^4 = -\mathbf{S}^2,\ \mathbf{S}^6 = \mathbf{S}^2,\ \mathbf{S}^8 = -\mathbf{S}^2, \ldots$ where $\mathbf{S}^2 = \mathbf{S}\mathbf{S} = -(\mathbf{e}_1 \otimes \mathbf{e}_1 + \mathbf{e}_2 \otimes \mathbf{e}_2) = \mathbf{e} \otimes \mathbf{e} - \mathbf{I}$
norms:	$\Vert\mathbf{S}\Vert_E = \sqrt{\mathbf{S} \cdot \mathbf{S}} = \sqrt{2}, \quad \Vert\mathbf{S}\Vert_2 = 1$
scalar invariants:	$\mathrm{tr}\mathbf{S} = 0, \quad \mathrm{tr}\mathbf{S}^2 = -2, \quad \det \mathbf{S} = 0$
principal invariants:	$I_1(\mathbf{S}) = 0, \ I_2(\mathbf{S}) = 1, \ I_3(\mathbf{S}) = 0$
eigenpair in $\mathbb{R} \times \mathbb{R}^3$:	$\{0, \mathbf{e}\}$, i.e. $(\mathbf{S} - 0\mathbf{I})\mathbf{e} = \mathbf{0}$
zero products:	$\mathbf{S}\mathbf{e} = \mathbf{0}, \quad \mathbf{S}(\mathbf{e} \otimes \mathbf{e}) = \mathbf{0}, \quad \mathbf{S} \cdot (\mathbf{e} \otimes \mathbf{e}) = 0, \quad \mathbf{S}^2\mathbf{e} = \mathbf{0}$
relation to rotations:	$\mathbf{S} = \frac{1}{2\sin\omega}(\mathbf{R} - \mathbf{R}^T), \quad \mathbf{S}^2 = \frac{1}{2(1-\cos\omega)}(\mathbf{R} + \mathbf{R}^T - 2\mathbf{I})$

Table 8.2 Properties of $\mathbf{R} \in SO(3)$.

dyadic representation:	$\mathbf{R} = \mathbf{I} + \sin\omega\ \mathbf{S} + (1 - \cos\omega)\ \mathbf{S}^2$
exponential representation:	$\mathbf{R} = \exp(\omega\mathbf{S}) = \mathbf{I} + \omega\mathbf{S} + \frac{1}{2!}\omega^2\mathbf{S}^2 + \ldots$
orthogonality:	$\mathbf{R}^T = \mathbf{R}^{-1}$
norms:	$\Vert\mathbf{R}\Vert_E = \sqrt{\mathbf{R} \cdot \mathbf{R}} = \sqrt{3}, \quad \Vert\mathbf{R}\Vert_2 = 1$
scalar invariants:	$\mathrm{tr}\mathbf{R} = 1 + 2\cos\omega, \quad \mathrm{tr}\mathbf{R}^2 = 1 + 2\cos 2\omega, \quad \det \mathbf{R} = 1$
principal invariants:	$I_1(\mathbf{R}) = \ I_2(\mathbf{R}) = 1 + 2\cos\omega, \ I_3(\mathbf{R}) = 1$
eigenpair in $\mathbb{R} \times \mathbb{R}^3$:	$\{1, \mathbf{e}\}$, i.e. $(\mathbf{R} - 1\mathbf{I})\mathbf{e} = \mathbf{0}$
products:	$\mathbf{R}\mathbf{e} = \mathbf{e}, \quad \mathbf{R}(\mathbf{e} \otimes \mathbf{e}) = \mathbf{e} \otimes \mathbf{e}$
invariance of length:	$\Vert\mathbf{R}\mathbf{x}\Vert_2 = \Vert\mathbf{x}\Vert_2, \quad \mathbf{x}$ - vector
invariance of angle:	$\mathbf{x} \cdot \mathbf{y} = (\mathbf{R}\mathbf{x}) \cdot (\mathbf{R}\mathbf{y}), \quad \mathbf{x}, \mathbf{y}$ - vectors

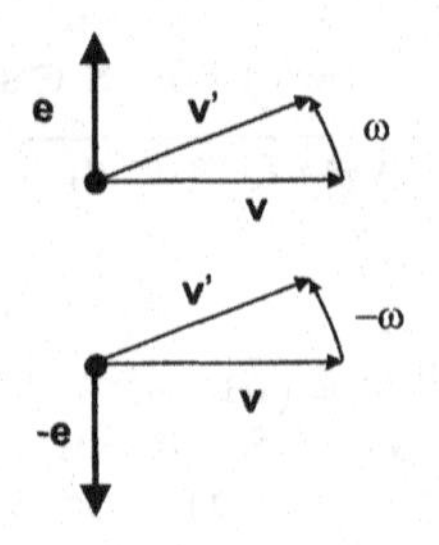

Fig. 8.4 Rotations $\{\omega, \mathbf{e}\}$ and $\{-\omega, -\mathbf{e}\}$ are physically equivalent.

These relations are quite simple in terms of components. Skew-symmetry of $\tilde{\mathbf{v}}$ (i.e. $\tilde{\mathbf{v}} = -\tilde{\mathbf{v}}^T$) implies that its representation must have the form

$$(\tilde{\mathbf{v}})_{ij} = \begin{bmatrix} 0 & -a & -b \\ a & 0 & -c \\ b & c & 0 \end{bmatrix}. \tag{8.14}$$

Note that $\{0, \mathbf{v}\}$ are the eigenpair of $\tilde{\mathbf{v}}$, see Table 8.1, and because $\det \tilde{\mathbf{v}} = 0$, the set of eigenequations $(\tilde{\mathbf{v}} - 0\,\mathbf{I})\,\mathbf{v} = \tilde{\mathbf{v}}\mathbf{v} = \mathbf{0}$ is undetermined and one solution must be chosen.

Designate the components of the vector $\mathbf{v}$ as $(\mathbf{v})_i \doteq [v_1, v_2, v_3]^T$. If we choose $a = v_3$, then the remaining two equations yield $b = -v_2$ and $c = v_1$, so, in terms of the components of $\mathbf{v}$, we obtain

$$(\tilde{\mathbf{v}})_{ij} = \begin{bmatrix} 0 & -v_3 & v_2 \\ v_3 & 0 & -v_1 \\ -v_2 & v_1 & 0 \end{bmatrix}, \qquad (\tilde{\mathbf{v}}^2)_{ij} = \begin{bmatrix} v_1^2 - v^2 & v_2 v_1 & v_3 v_1 \\ v_1 v_2 & v_2^2 - v^2 & v_3 v_2 \\ v_1 v_3 & v_2 v_3 & v_3^2 - v^2 \end{bmatrix}, \tag{8.15}$$

where $v^2 \doteq v_1^2 + v_2^2 + v_3^2$. Note that the components of $\mathbf{v}$ can be obtained by extraction of the appropriate components from the skew-symmetric $(\tilde{\mathbf{v}})_{ij}$,

$$(\mathbf{v})_i = [(\tilde{\mathbf{v}})_{32}, (\tilde{\mathbf{v}})_{13}, (\tilde{\mathbf{v}})_{21}]^T. \tag{8.16}$$

which corresponds to eq. $(8.13)_2$.

Eigenvalues of R. Let us find the eigenpairs $\{\lambda, \mathbf{z}\}$ of $\mathbf{R}$, such that $(\mathbf{R} - \lambda\mathbf{I})\mathbf{z} = \mathbf{0}$. This equation is homogeneous, and has a nontrivial solution $\mathbf{z}$ if and only if $\det(\mathbf{R} - \lambda\mathbf{I}) = 0$. Consider a matrix of components of $\mathbf{R}$ in the basis $\{\mathbf{e}_1, \mathbf{e}_2, \mathbf{e}\}$. We can select the following basis vectors:

$$\mathbf{e}_1 \doteq [1, 0, 0]^T, \qquad \mathbf{e}_2 \doteq [0, 1, 0]^T, \qquad \mathbf{e} \doteq [0, 0, 1]^T,$$

for which $\mathbf{R}$ of eq. (8.9) has the following representation:

$$\mathbf{R} = \begin{bmatrix} \cos\omega & -\sin\omega & 0 \\ \sin\omega & \cos\omega & 0 \\ 0 & 0 & 1 \end{bmatrix}. \tag{8.17}$$

Then

$$\mathbf{R} - \lambda\mathbf{I} = \begin{bmatrix} \cos\omega - \lambda & -\sin\omega & 0 \\ \sin\omega & \cos\omega - \lambda & 0 \\ 0 & 0 & 1-\lambda \end{bmatrix}, \tag{8.18}$$

and $\det(\mathbf{R} - \lambda\mathbf{I}) = \mathbf{0}$ yields the characteristic equation

$$\lambda^3 - A\lambda^2 + A\lambda - 1 = 0, \qquad A \doteq 1 + 2\cos\omega. \tag{8.19}$$

We note that $\lambda_1 = +1$ is a root of this equation, and rewrite it as

$$(\lambda - 1)\,[\,\lambda^2 + \lambda(1-A) + 1)\,] = 0. \tag{8.20}$$

For the quadratic equation in brackets, $\Delta = (1-A)^2 - 4 = -4\sin^2\omega \leq 0$, and hence it has complex roots, $\lambda_{2,3} = \cos\omega \pm i\sin\omega$. The associated eigenvectors are

$$\mathbf{z}_1 \doteq [0,0,1]^T, \qquad \mathbf{z}_2 \doteq [-i,1,0]^T, \qquad \mathbf{z}_3 \doteq [i,1,0], \tag{8.21}$$

where only $\mathbf{z}_1$ is real.

Note that for $\{\lambda_1, \mathbf{z}_1\}$, the eigenequation $(\mathbf{R} - \lambda\mathbf{I})\,\mathbf{z} = \mathbf{0}$ becomes $\mathbf{R}\,\mathbf{z}_1 = \mathbf{z}_1$. Hence, the vector $\mathbf{z}_1$ is unaffected by the rotation $\mathbf{R}$ and therefore it defines the axis of rotation, i.e. $\mathbf{z}_1 = \mathbf{e}$. We can also check the property $\mathbf{R}\,\mathbf{e} = \mathbf{e}$ using eqs. (8.9) and (8.8). Hence, $\{+1, \mathbf{e}\}$ is the real eigenpair of the rotation tensor.

Remark. To ensure a noticeable effect of any $\mathbf{R} \neq \mathbf{I}$, we must use at least two non-parallel vectors. The traditional choice is to use an ortho-normal triad of vectors, although a dyad of vectors plus the assumption that the third vector is ortho-normal to these two vectors would suffice, as shown in Sect. 8.2.1. The triads are used not only in analytic geometry of space, when a transformation of coordinates is considered, but also in mechanics of rigid bodies and shells.

8.1.3 Rotation of a triad of vectors

Consider two Cartesian triads (of ortho-normal vectors) $\{\mathbf{t}_k\}$ and $\{\mathbf{a}_k\}$, $k = 1,2,3$, shown in Fig. 8.5. Assume that the latter triad is obtained by a rotation of the first triad, i.e. $\mathbf{a}_k \doteq \mathbf{R}\mathbf{t}_k$. Hence, both triads have the same handedness. The Cartesian reference basis is denoted by $\{\mathbf{i}_k\}$.

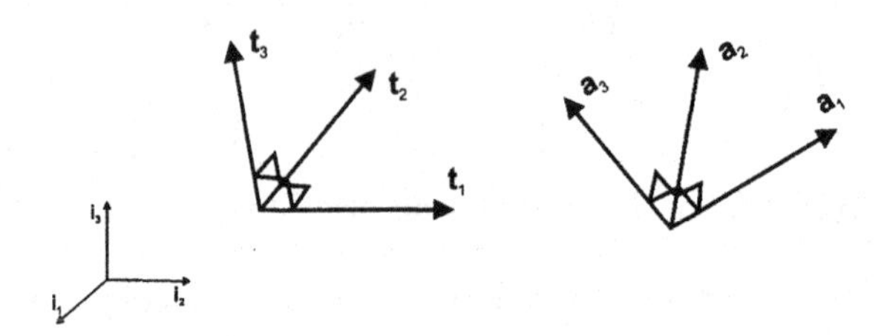

Fig. 8.5 Ortho-normal triads of vectors.

For these triads, we can calculate nine scalar products $(\mathbf{a}_i \cdot \mathbf{t}_j)$, $i,j = 1,2,3$, which can be interpreted in three ways, either (i) as coordinates of the vectors $\mathbf{a}_i$ in the basis $\{\mathbf{t}_j\}$ or (ii) as coordinates of the vectors $\mathbf{t}_i$ in the basis $\{\mathbf{a}_j\}$, or (iii) as direction cosines of angles between the vectors $\mathbf{a}_i$ and $\mathbf{t}_j$.

Let us define the rotation tensor as the following tensorial product:

$$\mathbf{R} \doteq \mathbf{a}_k \otimes \mathbf{t}_k. \tag{8.22}$$

We can easily check that $\mathbf{R}\mathbf{t}_i = \mathbf{a}_i$, as we earlier assumed. Besides, $R_{ij} \doteq (\mathbf{R}\mathbf{t}_i) \cdot \mathbf{t}_j = \mathbf{a}_i \cdot \mathbf{t}_j$, i.e. the components of $\mathbf{R}$ in the basis $\{\mathbf{t}_i\}$ are scalar products of $\mathbf{a}_i$ and $\mathbf{t}_j$. Let us check that the so-defined $\mathbf{R}$ is proper orthogonal.

1. The orthogonality can be easily checked,

$$\mathbf{R}^T\mathbf{R} = (\mathbf{t}_i \otimes \mathbf{a}_i)(\mathbf{a}_k \otimes \mathbf{t}_k) = (\mathbf{a}_i \cdot \mathbf{a}_k)(\mathbf{t}_i \otimes \mathbf{t}_k) = \mathbf{t}_k \otimes \mathbf{t}_k = \mathbf{I}. \tag{8.23}$$

2. The determinant property, $\det \mathbf{R} = +1$, is much more complicated to prove. We shall use the fact that both $\{\mathbf{t}_i\}$ and $\{\mathbf{a}_i\}$ are right-handed. Then we can write $\mathbf{t}_3 = \mathbf{t}_1 \times \mathbf{t}_2$ and $\mathbf{a}_3 = \mathbf{a}_1 \times \mathbf{a}_2$, so the representation of rotation tensor is

$$\mathbf{R} = \begin{bmatrix} (\mathbf{a}_1 \cdot \mathbf{t}_1) & (\mathbf{a}_1 \cdot \mathbf{t}_2) & \mathbf{a}_1 \cdot (\mathbf{t}_1 \times \mathbf{t}_2) \\ (\mathbf{a}_2 \cdot \mathbf{t}_1) & (\mathbf{a}_2 \cdot \mathbf{t}_2) & \mathbf{a}_2 \cdot (\mathbf{t}_1 \times \mathbf{t}_2) \\ (\mathbf{a}_1 \times \mathbf{a}_2) \cdot \mathbf{t}_1 & (\mathbf{a}_1 \times \mathbf{a}_2) \cdot \mathbf{t}_2 & (\mathbf{a}_1 \times \mathbf{a}_2) \cdot (\mathbf{t}_1 \times \mathbf{t}_2) \end{bmatrix}. \tag{8.24}$$

Next we use the results of Sect. 8.2.1, where it is shown that if (a) the first two rows of $\mathbf{R}$ are ortho-normal and (b) the third row of $\mathbf{R}$ is a vector product of the first two rows, then $\det \mathbf{R} = +1$. Hence, for

$$\mathbf{R} = \begin{bmatrix} \mathbf{a} \\ \mathbf{b} \\ \mathbf{c} \end{bmatrix}, \tag{8.25}$$

where the rows are $\mathbf{a} = [a_1, a_2, a_3]$, $\mathbf{b} = [b_1, b_2, b_3]$, and $\mathbf{c} = [c_1, c_2, c_3]$, we have to check that

$$\mathbf{c} = \mathbf{a} \times \mathbf{b} = [(a_2 b_3 - a_3 b_2),\ (a_3 b_1 - a_1 b_3),\ (a_1 b_2 - a_2 b_1)]. \tag{8.26}$$

Let us check this for the first term $c_1 = (a_2 b_3 - a_3 b_2)$, which can be transformed as follows:

$$\begin{aligned} c_1 &= (\mathbf{a}_1 \cdot \mathbf{t}_2)\,[\,\mathbf{a}_2 \cdot (\mathbf{t}_1 \times \mathbf{t}_2)\,] - [\,\mathbf{a}_1 \cdot (\mathbf{t}_1 \times \mathbf{t}_2)\,]\,(\mathbf{a}_2 \cdot \mathbf{t}_2) \\ &= (\mathbf{a}_1 \otimes \mathbf{a}_2) \cdot [\,\mathbf{t}_2 \otimes (\mathbf{t}_1 \times \mathbf{t}_2) - (\mathbf{t}_1 \times \mathbf{t}_2) \otimes \mathbf{t}_2\,], \end{aligned} \tag{8.27}$$

where in brackets we have a skew-symmetric tensor $\mathbf{t}_2 \otimes \mathbf{t}_3 - \mathbf{t}_3 \otimes \mathbf{t}_2 = -\mathbf{t}_1 \times \mathbf{I}$. Hence,

$$c_1 = (\mathbf{a}_1 \otimes \mathbf{a}_2) \cdot (-\mathbf{t}_1 \times \mathbf{I}\,) = \text{skew}(\mathbf{a}_1 \otimes \mathbf{a}_2) \cdot (-\mathbf{t}_1 \times \mathbf{I}\,) \tag{8.28}$$

or, in terms of the axial vectors,

$$c_1 = 2\tfrac{1}{2}(\mathbf{a}_2 \times \mathbf{a}_1) \cdot (-\mathbf{t}_1) = (\mathbf{a}_1 \times \mathbf{a}_2) \cdot \mathbf{t}_1, \tag{8.29}$$

which is identical as the 31-component of $\mathbf{R}$ of eq. (8.24). Similar results can be derived for c_2 and c_3, thus we conclude that the third row of $\mathbf{R}$ is a vector product of the first two rows and $\det \mathbf{R} = +1$ indeed.

Euler's theorem. If two triads are given, then we can always calculate $\mathbf{R}$ using eq. (8.24) and, as it is a proper orthogonal matrix, we can determine the vector $\pm\mathbf{e}$, which defines the axis of rotation, as in Sect. 8.1.2. This fact is stated as Euler's theorem: *If two ortho-normal triads of vectors are given in space, one can always specify a line such that* $\{\mathbf{t}_i\}$ *goes into* $\{\mathbf{a}_i\}$ *by a rotation about this line.*

Numerical example. Consider two ortho-normal triads, $\{\mathbf{t}_i\}$ and $\{\mathbf{a}_i\}$ shown in Fig. 8.5. Their components in the reference basis $\{\mathbf{i}_k\}$ are orthogonal matrices, which we denote as $\mathbf{R}_t$ and $\mathbf{R}_a$, respectively. For the known $\mathbf{R}_t, \mathbf{R}_a \in \mathrm{SO}(3)$, we would like to determine the matrix $\mathbf{R}$, such that

$$\mathbf{R}_a = \underbrace{\mathbf{R}}_{\text{unknown}} \mathbf{R}_t. \tag{8.30}$$

The matrix $\mathbf{R}$ can be obtained in two ways:

1. using orthogonality of $\mathbf{R}_t$, i.e. $\mathbf{R}_t^{-1} = \mathbf{R}_t^T$. Then, by right post-multiplication by $\mathbf{R}_t^T$, we have

$$\mathbf{R} = \mathbf{R}_a \mathbf{R}_t^{-1} = \mathbf{R}_a \mathbf{R}_t^T, \tag{8.31}$$

2. not using orthogonality of $\mathbf{R}_t$. Then, we have to solve the matrix equation corresponding to eq. (8.30). To obtain this matrix equation, we assume a general form of the unknown matrix $\mathbf{R}$, with nine unknown components R_{ij}, $(i,j = 1,2,3)$. They are arranged as a vector $\mathbf{x} \doteq [R_{11}, R_{12}, R_{13}, ..., R_{31}, R_{32}, R_{33}]^T$. Performing the multiplication in eq. (8.30) and next writing each equation separately, and extracting $\mathbf{x}$, we obtain a set of nine linear equations

$$\mathbf{A}\,\mathbf{x} = \mathbf{b}, \tag{8.32}$$

where $\mathbf{A}$ is non-singular for any $\mathbf{R}_t$ and $\mathbf{R}_a$, so the set can be uniquely solved. Re-arranging $\mathbf{x}$, we obtain $\mathbf{R}$ and, as we checked, it is proper orthogonal.

Obviously, the first way, using the orthogonality of $\mathbf{R}_t$, is much more effective.

Numerical example. Assume that positions of two Cartesian dyads of vectors are known, see Fig. 8.6. These dyads are denoted by $\{\mathbf{t}_1, \mathbf{t}_2\}$ and $\{\mathbf{a}_1, \mathbf{a}_2\}$ and belong to the plane $\mathbf{i}_1 - \mathbf{i}_2$. Our objective is to calculate the angle ω characterizing the difference in the position of these frames, such that

$$\omega: \quad \mathbf{a}_k = \mathbf{R}(\omega)\, \mathbf{t}_k, \qquad \mathbf{R} \in \mathrm{SO}(1), \qquad k = 1,2, \tag{8.33}$$

for the following parametrization of ω:

$$\omega = \sqrt{\psi^2 + \tau}, \qquad \text{where} \quad \tau = 10^{-8}. \tag{8.34}$$

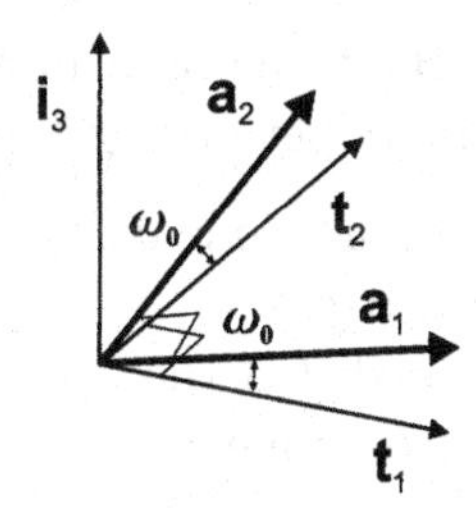

Fig. 8.6 Position of two Cartesian dyads of vectors.

(Note that we can use a similar perturbed form for the canonical rotation tensor, see eq. (8.89).) The rotational parameter ψ is determined from the minimization problem

$$\min e(\psi), \tag{8.35}$$

where

$$e(\psi) = \sum_{k=1}^{2} e_k(\psi), \qquad e_k(\psi) \doteq [\mathbf{a}_k - \mathbf{R}(\psi)\,\mathbf{t}_k] \cdot [\mathbf{a}_k - \mathbf{R}(\psi)\,\mathbf{t}_k], \tag{8.36}$$

i.e. $e(\psi)$ is the sum of squares of errors. The Newton method is used to solve this problem with the tangent operator and the residual given by eqs. (9.94)–(9.96).

The calculations are performed for a range of values of angles ω_0, where ω_0 is the angle characterizing the assumed difference in the position of the two frames. The number of iterations N in which the convergence is obtained is shown in Table 8.3 and we see that up to $\omega_0 = 175^{\circ}$, less than 10 iterations are required.

Table 8.3 Rate of convergence for various positions of two Cartesian dyads of vectors.

ω_0 [deg]	10	20	40	60	70	80	100	120	140	160	170	175	180
N [iterations]	4	4	4	5	5	5	6	6	7	7	8	9	20

8.2 Parametrization of rotations

If two ortho-normal triads $\{\mathbf{t}_i\}$ and $\{\mathbf{a}_i\}$ are given, then the rotation tensor can be easily found by using eq. (8.22). However, in real physical problems, we have a different and much more complicated situation: we must assume some form of $\mathbf{R}$ to be used in governing equations. This form must ensure that $\mathbf{R}$ is proper orthogonal.

The most obvious form is to use nine components of the rotation tensor $\mathbf{R}$, but it has two drawbacks: (i) a large number of components must be manipulated and stored, (ii) the orthogonality conditions must be additionally enforced. For components of $\mathbf{R}$ written as

$$\mathbf{R} \doteq \begin{bmatrix} \mathbf{a} \\ \mathbf{b} \\ \mathbf{c} \end{bmatrix}, \tag{8.37}$$

where the rows are $\mathbf{a} \doteq [a_1, a_2, a_3]$, $\mathbf{b} \doteq [b_1, b_2, b_3]$, and $\mathbf{c} \doteq [c_1, c_2, c_3]$, we have to enforce the following conditions:

$$\mathbf{a} \cdot \mathbf{a} = 1, \qquad \mathbf{b} \cdot \mathbf{b} = 1, \qquad \mathbf{c} \cdot \mathbf{c} = 1, \tag{8.38}$$

$$\mathbf{a} \cdot \mathbf{b} = 0, \qquad \mathbf{a} \cdot \mathbf{c} = 0, \qquad \mathbf{b} \cdot \mathbf{c} = 0, \tag{8.39}$$

$$\det \mathbf{R} = (\mathbf{a} \times \mathbf{b}) \cdot \mathbf{c} = +1. \tag{8.40}$$

Identical relations are obtained if $\mathbf{a}$, $\mathbf{b}$ and $\mathbf{c}$ denote the columns of $\mathbf{R}$. As there is no easy way to account for these conditions, other ways of dealing with them have been developed. Basically, we have three possibilities:

- to neglect the orthogonality conditions when the governing equation is solved and later fit in an orthogonal matrix. This way is tested for free body dynamics, e.g. in [90] p. 115, but its accuracy is doubtful.
- to devise such an algorithm that the calculated $\mathbf{R}$ is orthogonal, see [105], where the generalized mid-point method is applied to the equation generating rotations $\dot{\mathbf{R}}(t) = \tilde{\boldsymbol{\psi}} \mathbf{R}(t)$, where $\mathbf{R}(0) = \mathbf{I}$ and $\tilde{\boldsymbol{\psi}}$ is a skew-symmetric tensor.
- to use a certain specific parametrization of $\mathbf{R}$, which either accounts for the orthogonality conditions or allows us to correct the set of parameters in a simple way. Several types of parametrization of rotations have already been proposed; they are surveyed in [231, 224].
  We discuss them in the next sections, while here, we only note that the number of orthogonality conditions (constraints) which must be enforced increases with the number of rotational parameters, see Table 8.4.

Table 8.4 Number of orthogonality constraints for various parametrizations.

no of rotation parameters	9	6	5	4	3
no of additional constraints	7	3	2	1	0

8.2.1 Six parameters

For the (right-handed) ortho-normal triad $\{\mathbf{t}_i\}$, the vector $\mathbf{t}_3$ can be expressed in terms of $\mathbf{t}_1$ and $\mathbf{t}_2$ as $\mathbf{t}_3 = \mathbf{t}_1 \times \mathbf{t}_2$. Therefore, for $\mathbf{a}_3 \doteq \mathbf{R}\mathbf{t}_3$, we obtain $\mathbf{a}_3 = \mathbf{R}(\mathbf{t}_1 \times \mathbf{t}_2) = (\mathbf{R}\mathbf{t}_1) \times (\mathbf{R}\mathbf{t}_2) = \mathbf{a}_1 \times \mathbf{a}_2$, i.e. $\mathbf{a}_3$ can be calculated using $\mathbf{a}_1$ and $\mathbf{a}_2$. We see that we do not need to use a triad as it suffices to describe the rotation of the dyad $\{\mathbf{t}_1, \mathbf{t}_2\}$ and calculate $\mathbf{a}_3$ afterwards. Hence, instead of eq. (8.22), we can use the following form of the rotation tensor:

$$\mathbf{R} \doteq \mathbf{a}_1 \otimes \mathbf{t}_1 + \mathbf{a}_2 \otimes \mathbf{t}_2, \tag{8.41}$$

for which $\mathbf{R}\mathbf{t}_\alpha = \mathbf{a}_\alpha$ and $R_{\alpha j} \doteq (\mathbf{R}\mathbf{t}_\alpha) \cdot \mathbf{t}_j = (\mathbf{a}_\alpha \cdot \mathbf{t}_j)$ are six direction cosines of angles between $\mathbf{a}_\alpha$ and $\mathbf{t}_j$. The same can be shown for the representation of $\mathbf{R}$ in $\{\mathbf{t}_i\}$,

$$\mathbf{R} = \begin{bmatrix} \mathbf{a} \\ \mathbf{b} \\ \mathbf{x} \end{bmatrix}, \tag{8.42}$$

where the rows are $\mathbf{a} \doteq [a_1, a_2, a_3]$, $\mathbf{b} \doteq [b_1, b_2, b_3]$, and $\mathbf{x} \doteq [x_1, x_2, x_3]$. Assume that $\mathbf{a}$ and $\mathbf{b}$ are ortho-normal, i.e.

$$\mathbf{a} \cdot \mathbf{a} = 1, \qquad \mathbf{b} \cdot \mathbf{b} = 1, \qquad \mathbf{a} \cdot \mathbf{b} = 0. \tag{8.43}$$

The remaining conditions of eqs. (8.38)–(8.40), i.e. the orthogonality of rows of $\mathbf{R}$ and $\det \mathbf{R} = +1$ can be written as follows:

$$\left. \begin{aligned} \mathbf{a} \cdot \mathbf{x} &= 0 \\ \mathbf{b} \cdot \mathbf{x} &= 0 \\ (\mathbf{a} \times \mathbf{b}) \cdot \mathbf{x} &= 1 \end{aligned} \right\}. \tag{8.44}$$

This set of three equations has the solution $\mathbf{x} = \mathbf{a} \times \mathbf{b}$, where $\mathbf{x}$ is a unit vector, as $\mathbf{x} \cdot \mathbf{x} = 1$. Hence, the representation of the rotation matrix is

$$\mathbf{R} = \begin{bmatrix} \mathbf{a} \\ \mathbf{b} \\ \mathbf{a} \times \mathbf{b} \end{bmatrix} = \begin{bmatrix} a_1 & a_2 & a_3 \\ b_1 & b_2 & b_3 \\ (a_2 b_3 - a_3 b_2) & (a_3 b_1 - a_1 b_3) & (a_1 b_2 - a_2 b_1) \end{bmatrix}, \tag{8.45}$$

where six components of $\mathbf{a}$ and $\mathbf{b}$ serve as rotational parameters. Still, however, $\mathbf{a}$ and $\mathbf{b}$ must be ortho-normal, i.e. we have to enforce three conditions of eq. (8.43).

The relation between the rotation matrix and the six parameters is non-singular as $\{\mathbf{a}, \mathbf{b}\} \to \mathbf{R}$ is realized by the vector product $\mathbf{a} \times \mathbf{b}$, while $\mathbf{R} \to \{\mathbf{a}, \mathbf{b}\}$ is obtained by a selection of the first two rows of $\mathbf{R}$.

Example. We can calculate the angle between two given dyads $\{\mathbf{t}_1, \mathbf{t}_2\}$ and $\{\mathbf{a}_1, \mathbf{a}_2\}$ in 3D space, where $\mathbf{a}_\alpha = \mathbf{R}\mathbf{t}_\alpha$, which is of interest in crystallography, where dyads are associated with some specific material directions. For $R_{\alpha j} = \mathbf{a}_\alpha \cdot \mathbf{t}_j$, eq. (8.45) becomes

$$\mathbf{R} = \begin{bmatrix} (\mathbf{a}_1 \cdot \mathbf{t}_1) & (\mathbf{a}_1 \cdot \mathbf{t}_2) & \cdots \\ (\mathbf{a}_2 \cdot \mathbf{t}_1) & (\mathbf{a}_2 \cdot \mathbf{t}_2) & \cdots \\ \cdots & \cdots & (\mathbf{a}_1 \cdot \mathbf{t}_1)(\mathbf{a}_2 \cdot \mathbf{t}_2) - (\mathbf{a}_1 \cdot \mathbf{t}_2)(\mathbf{a}_2 \cdot \mathbf{t}_1) \end{bmatrix}, \tag{8.46}$$

where only the components obtained from the dyads are specified. Then, we calculate

$$\mathrm{tr}\mathbf{R} = (\mathbf{a}_1 \cdot \mathbf{t}_1) + (\mathbf{a}_2 \cdot \mathbf{t}_2) + (\mathbf{a}_1 \cdot \mathbf{t}_1)(\mathbf{a}_2 \cdot \mathbf{t}_2) - (\mathbf{a}_1 \cdot \mathbf{t}_2)(\mathbf{a}_2 \cdot \mathbf{t}_1) \tag{8.47}$$

and the rotation angle $\omega = \arccos \frac{1}{2}(\mathrm{tr}\mathbf{R} - 1) \in [0, \pi]$.

8.2.2 Five parameters

The number of independent parameters which are used in eq. (8.45) to describe a position of the dyad $\{\mathbf{t}_1, \mathbf{t}_2\}$, can be reduced to five by using the constraint equations (8.43) in several ways.

For instance, from the condition $\mathbf{a} \cdot \mathbf{a} = 1$, we can calculate $a_1 = \sqrt{1 - a_2^2 - a_3^2}$ and insert it in eq. (8.45). However, this parametrization will not be generally applicable because of the square-root problem for the Newton scheme, see the description of this problem for eq. (8.71).

Another way is to take the orthogonality condition $\mathbf{a} \cdot \mathbf{b} = 0$, calculate $a_1 = -(a_2 b_2 + a_3 b_3)/b_1$, and next use it in eq. (8.45). However, this formula is singular for $b_1 = 0$ and then we have to use either $a_2 = -(a_1 b_1 + a_3 b_3)/b_2$ or $a_3 = -(a_1 b_1 + a_2 b_2)/b_3$ instead; one of them certainly is non-singular because $b_1^2 + b_2^2 + b_3^2 = 1$. Hence, in the case of singularity, we have to replace the parametrization in terms of $\{a_2, a_3, \mathbf{b}\}$ by either $\{a_1, a_3, \mathbf{b}\}$ or $\{a_1, a_2, \mathbf{b}\}$, which complicates a solution algorithm. We note that two constrains, $\mathbf{a} \cdot \mathbf{a} = 1$ and $\mathbf{b} \cdot \mathbf{b} = 1$, still remain and must be enforced.

The above difficulties are avoided if we use the five-parameter representation based on the concept of the stereographic projection, [231].

Stereographic projection. This projection establishes a one-to-one correspondence between the points of a sphere (except the north pole) and the points of the plane passing through the equator; the historical note on it is given in [230], p. 175. Here, we consider a 2D form of the projection, with a circle instead of a sphere. The circle has the center at point $O(0,0)$, and the radius 1. A line passing through the north pole, $P(1,0)$, and an arbitrary point on the circle, $X(x_1, x_2)$, intersects OY-axis at point $Y(0, y_2)$, which is called a stereographic projection of X, see Fig. 8.7. By simple geometric calculations,

$$y_2 = \frac{x_2}{1 - x_1}, \tag{8.48}$$

which is singular at $x_1 = 1$, for X coinciding with P. The inverse relation is as follows

$$x_1 = \frac{y_2^2 - 1}{y_2^2 + 1}, \qquad x_2 = \frac{2y_2}{y_2^2 + 1}, \tag{8.49}$$

and $x_1^2 + x_2^2 = 1$, indeed. Note that (x_1, x_2) and $(0, y_2,)$ are coordinates of points X and Y in the orthonormal basis $\{\mathbf{e}_1, \mathbf{e}_2\}$, such that $\mathbf{e}_1$ locates the north pole.

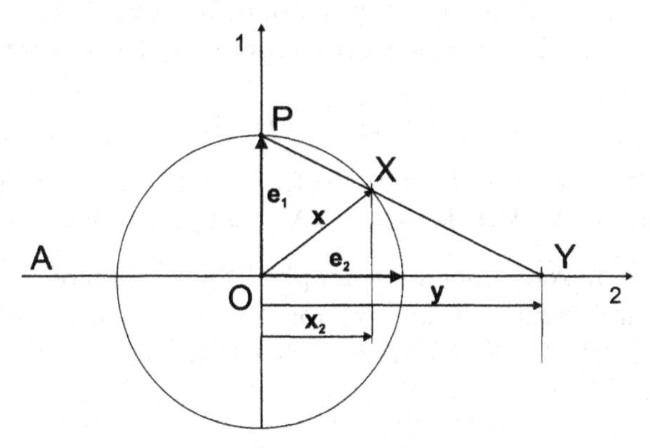

Fig. 8.7 Stereographic projection.

Five-parameter representation based on stereographic projection. Let us define a vector composed of the first two rows of the six-parameter representation, $\mathbf{x} = [\mathbf{a}/\sqrt{2}, \mathbf{b}/\sqrt{2}]$, such that

$$\mathrm{M} = \{\mathbf{x} \in E^6 \; : \quad \mathbf{x} \cdot \mathbf{x} = 1, \quad \mathbf{x} \cdot (\mathbf{J}_i \, \mathbf{x}) = 0, \quad i = 1, 2\}, \tag{8.50}$$

where

$$\mathbf{J}_1 = \begin{bmatrix} \mathbf{I}_3 & \mathbf{0} \\ \mathbf{0} & -\mathbf{I}_3 \end{bmatrix}, \qquad \mathbf{J}_2 = \begin{bmatrix} \mathbf{0} & \mathbf{I}_3 \\ \mathbf{I}_3 & \mathbf{0} \end{bmatrix}. \tag{8.51}$$

We see that the set M is on a unit sphere S^5 in E^6. The properties of eq. (8.50) can be rewritten in terms of $\mathbf{a}$ and $\mathbf{b}$ as follows

$$\mathbf{x} \cdot \mathbf{x} = (\mathbf{a} \cdot \mathbf{a})/2 + (\mathbf{b} \cdot \mathbf{b})/2 = 1, \tag{8.52}$$

$$\mathbf{x} \cdot (\mathbf{J}_1 \, \mathbf{x}) = (\mathbf{a} \cdot \mathbf{a} - \mathbf{b} \cdot \mathbf{b})/2 = 0, \qquad \mathbf{x} \cdot (\mathbf{J}_2 \, \mathbf{x}) = \mathbf{a} \cdot \mathbf{b} = 0. \tag{8.53}$$

By the first of eq. (8.53), $\mathbf{a}$ and $\mathbf{b}$ have the same length and, from eq. (8.52), $\mathbf{x} \cdot \mathbf{x} = \mathbf{a} \cdot \mathbf{a} = 1$, and $\mathbf{b} \cdot \mathbf{b} = 1$. Besides, $\mathbf{a}$ and $\mathbf{b}$ are orthogonal by the second of eq. (8.53). Hence, the properties of eq. (8.50) are an alternative form of eq. (8.43).

Using the concept of the stereographic projection, we project the set M onto the hyperplane A in E^5. The hyperplane is perpendicular to $\mathbf{e}_1$, which locates the north pole, and cannot be in M. We assume that A is spanned by one of two bases of five orthonormal vectors, either by $\{\mathbf{r}_2, ..., \mathbf{r}_6\}$ where $\mathbf{r}_i$ are of dimension 6, or by $\{\mathbf{v}_1, ..., \mathbf{v}_5\}$ where $\mathbf{v}_i$ are of dimension 5. The relation between the two bases is set up by the tensor

$$\mathbf{V} = \mathbf{v}_j \otimes \mathbf{r}_{j+1}, \qquad j = 1, .., 5, \tag{8.54}$$

where $\mathbf{v}_i = \mathbf{V}\mathbf{r}_{i+1}, i = 1, .., 5$. On the other hand, $\mathbf{r}_{i+1} = (\mathbf{r}_{j+1} \otimes \mathbf{v}_j)\, \mathbf{v}_i = \mathbf{V}^T \mathbf{v}_i$. Note that $\mathbf{V}^T\mathbf{V} = \mathbf{I}_6$ and $\mathbf{V}\mathbf{V}^T = \mathbf{I}_5$. Besides, $\mathbf{r}_1 = \mathbf{e}_1$.

Denote an orthogonal (not stereographic !) projection of $\mathbf{x} \in E^6$ onto A as $\mathbf{x}_2$, and define the unit vector $\mathbf{e}_2 \doteq \mathbf{x}_2/\|\mathbf{x}_2\|$, see Fig. 8.7. Multiplying eq. (8.48) by $\mathbf{e}_2$ and noting that $y_2 \, \mathbf{e}_2 = \mathbf{y} \in E^5$ and $x_2 = \|\mathbf{x}_2\|$, we obtain

$$\mathbf{y} = \frac{x_2 \mathbf{e}_2}{1 - x_1} = \frac{\mathbf{x}_2}{1 - \mathbf{x} \cdot \mathbf{e}_1}. \tag{8.55}$$

We can also find $\mathbf{x}$ for a given $\mathbf{y}$ by using $\mathbf{x} = x_1 \mathbf{e}_1 + x_2 \mathbf{e}_2$ with x_1 and x_2 specified in eq. (8.49),

$$\mathbf{x} = \frac{\mathbf{y}\cdot\mathbf{y}-1}{\mathbf{y}\cdot\mathbf{y}+1}\,\mathbf{e}_1 + \frac{2\|\mathbf{y}\|}{(\mathbf{y}\cdot\mathbf{y}+1)}\frac{\mathbf{y}}{\|\mathbf{y}\|} = \frac{(\mathbf{y}\cdot\mathbf{y}-1)\,\mathbf{e}_1+2\mathbf{y}}{1+\mathbf{y}\cdot\mathbf{y}}, \tag{8.56}$$

where $\mathbf{e}_2 \doteq \mathbf{y}/\|\mathbf{y}\|$, $y_2 = \|\mathbf{y}\| = \sqrt{\mathbf{y}\cdot\mathbf{y}}$ were used. Note that eqs. (8.55) and (8.56) are always non-singular. In the denominator of eq. (8.55), $\mathbf{x}\cdot\mathbf{e}_1 \neq 1$ because we assumed that the north pole cannot be in M, while in the denominator of eq. (8.56), $\mathbf{y}\cdot\mathbf{y} \neq -1$ because $\mathbf{y}\cdot\mathbf{y} \geq 0$.

For $\mathbf{x}$ specified by eq. (8.56), we can check conditions (8.50), i.e.

$$\mathbf{x}\cdot\mathbf{x} = \frac{1}{(c+2)^2}\left(4\mathbf{y}\cdot\mathbf{y} + 4c\mathbf{y}\cdot\mathbf{e}_1 + c^2\,\mathbf{e}_1\cdot\mathbf{e}_1\right) = \frac{4(c+1)+c^2}{(c+2)^2} = 1, \tag{8.57}$$

$$\mathbf{x}\cdot(\mathbf{J}_i\,\mathbf{x}) = 4\mathbf{y}\cdot(\mathbf{J}_i\,\mathbf{y}) + 2c[\mathbf{y}\cdot(\mathbf{J}_i\,\mathbf{e}_1) + \mathbf{e}_1\cdot(\mathbf{J}_i\,\mathbf{y})] + c^2\,\mathbf{e}_1\cdot(\mathbf{J}_i\,\mathbf{e}_1) = 0, \tag{8.58}$$

where $c = \mathbf{y}\cdot\mathbf{y}-1$. We see that the first condition, which is the equation of the unit sphere, is satisfied, but the next two, eq. (8.58), remain as constraints.

Five-parameter representation in terms of components. Let us now write eqs. (8.55) and (8.56) in terms of components. Components of eq. (8.55) in the basis $\{\mathbf{v}_1, ..., \mathbf{v}_5\}$ are as follows:

$$\{(\mathbf{y}\cdot\mathbf{v}_1), ..., (\mathbf{y}\cdot\mathbf{v}_5)\}^T = \frac{\{(\mathbf{x}\cdot\mathbf{v}_1), ..., (\mathbf{x}\cdot\mathbf{v}_5)\}^T}{1-\mathbf{x}\cdot\mathbf{e}_1}, \tag{8.59}$$

where $\mathbf{x}_2\cdot\mathbf{v}_k = \mathbf{x}\cdot\mathbf{v}_k$ is used. Components of eq.(8.56) in the basis $\{\mathbf{r}_1, ..., \mathbf{r}_6\}$, are

$$[(\mathbf{x}\cdot\mathbf{r}_1), (\mathbf{x}\cdot\mathbf{r}_2), ..., (\mathbf{x}\cdot\mathbf{r}_6)]^T = \frac{1}{1+\mathbf{y}\cdot\mathbf{y}}\left[(\mathbf{y}\cdot\mathbf{y}-1)\,[1,0,...,0]^T + 2[(\mathbf{y}\cdot\mathbf{r}_1), (\mathbf{y}\cdot\mathbf{r}_2), ..., (\mathbf{y}\cdot\mathbf{r}_6)]^T\right]. \tag{8.60}$$

We define the following 5D and 6D vectors of components

$$\mathsf{y} = [(\mathbf{y}\cdot\mathbf{v}_1), ..., (\mathbf{y}\cdot\mathbf{v}_5)]^T, \qquad \mathsf{r}_i = [(\mathbf{r}_i\cdot\mathbf{v}_1), ..., (\mathbf{r}_i\cdot\mathbf{v}_5)]^T, \tag{8.61}$$

$$\mathsf{x} = [(\mathbf{x}\cdot\mathbf{r}_1), ..., (\mathbf{x}\cdot\mathbf{r}_6)]^T, \qquad \mathsf{v}_i = [(\mathbf{v}_i\cdot\mathbf{r}_1), ..., (\mathbf{v}_i\cdot\mathbf{r}_6)]^T, \tag{8.62}$$

$$\mathsf{e}_1 = [(\mathbf{e}_1\cdot\mathbf{r}_1), ..., (\mathbf{e}_1\cdot\mathbf{r}_6)]^T = [1,0,0,0,0,0]^T. \tag{8.63}$$

Using v_i and r_i, we can form either a 5×6 matrix $\mathsf{V} = [\mathsf{v}_1, \ldots, \mathsf{v}_5]^T$, i.e. with rows formed by v_i of dimension 6, or a 6×5 matrix $\mathsf{V}^T = [\mathsf{r}_1, \ldots, \mathsf{r}_6]^T$, i.e. with rows formed by r_i of dimension 5. Because,

$(\mathbf{Vr}_{i+1}) \cdot \mathbf{r}_{j+1} = \mathbf{v}_i \cdot \mathbf{r}_{j+1}$, we see that $\mathbf{V}$ is a matrix of components of $\mathbf{V}$, which can be used to rewrite some terms of eqs. (8.59) and (8.60),

$$(\mathbf{x} \cdot \mathbf{v}_i) = \sum_{k=1}^{6} (\mathbf{x} \cdot \mathbf{r}_k)(\mathbf{v}_i \cdot \mathbf{r}_k) = \mathbf{v}_i \, \mathbf{x}, \quad [(\mathbf{x} \cdot \mathbf{v}_1), ..., (\mathbf{x} \cdot \mathbf{v}_5)]^T = \mathbf{V}\mathbf{x}, \tag{8.64}$$

$$(\mathbf{y} \cdot \mathbf{r}_i) = \sum_{k=1}^{5} (\mathbf{y} \cdot \mathbf{v}_k)(\mathbf{r}_i \cdot \mathbf{v}_k) = \mathbf{r}_i \, \mathbf{y}, \quad [(\mathbf{y} \cdot \mathbf{r}_1), ..., (\mathbf{y} \cdot \mathbf{r}_6)]^T = \mathbf{V}^T\mathbf{y}. \tag{8.65}$$

Then, eqs. (8.59) and (8.60) can be rewritten in terms of the vectors and matrices of components as follows:

$$\mathbf{y} = \frac{\mathbf{V}\mathbf{x}}{1 - \mathbf{e}_1^T\mathbf{x}}, \qquad \mathbf{x} = \frac{(\mathbf{y}^T\mathbf{y} - 1)\,\mathbf{e}_1 + 2\mathbf{V}^T\mathbf{y}}{1 + \mathbf{y}^T\mathbf{y}}. \tag{8.66}$$

Comparing the 5- and six-parameter representations, we see that the number of parameters and orthogonality constraints is reduced by one, as $\mathbf{x}$ is replaced by $\mathbf{y}$, and eq. (8.43) is replaced by eq. (8.58). However, the new parameters and constraint equations are more complicated and do not have such a clear interpretation as the previous ones.

Finally, we note that new formulas obtained by using the stereographic projection, are also more complicated than these derived at the beginning of the section.

8.2.3 Four parameters: Euler parameters (quaternions)

The set $\{\omega, \mathbf{e}\}$ introduced in Sect. 8.1 can be replaced by the so-called Euler parameters (or quaternions), [2, 224], which are defined as

$$q_0 \doteq \cos(\omega/2), \qquad \mathbf{q} \doteq \sin(\omega/2)\,\mathbf{e}. \tag{8.67}$$

The rotation tensor, eq. (8.9), in terms of $\{q_0, \mathbf{q}\}$ is expressed as

$$\mathbf{R} \doteq (2q_0^2 - 1)\mathbf{I} + 2q_0\mathbf{q} \times \mathbf{I} + 2\mathbf{q} \otimes \mathbf{q}, \tag{8.68}$$

or by using the skew-symmetric $\tilde{\mathbf{q}} \doteq \sin(\omega/2)\,\mathbf{S}$, as

$$\mathbf{R} \doteq \mathbf{I} + 2q_0\tilde{\mathbf{q}} + 2\tilde{\mathbf{q}}^2 \tag{8.69}$$

and both these forms are never singular. The parameters $\{q_0, \mathbf{q}\}$ must satisfy the constraint equation, $q_0^2 + \mathbf{q} \cdot \mathbf{q} = 1$, which is equivalent to $\mathbf{e} \cdot \mathbf{e} = 1$ and, in this sense, they form a unit quaternion. The constraint equation must be appended either to eq. (8.68) or (8.69), e.g.

$$\begin{cases} \mathbf{R} \doteq (2q_0^2 - 1)\mathbf{I} + 2q_0\mathbf{q} \times \mathbf{I} + 2\mathbf{q} \otimes \mathbf{q}, \\ q_0^2 + \mathbf{q} \cdot \mathbf{q} = 1. \end{cases} \tag{8.70}$$

If we eliminate the constraint equation by calculating $q_0 = \sqrt{1 - \mathbf{q} \cdot \mathbf{q}}$ and inserting it into the first equation, then the expression for $\mathbf{R}$ contains the square root, which can cause a failure of the Newton method, typically used to solve the equilibrium equations (see the example below).

Example. Consider a simple equation with a square root:

$$y \doteq \sqrt{1 - x^2} = 0, \qquad |x| \leq 1, \tag{8.71}$$

The Newton solution scheme can be written as

$$\begin{cases} (\mathrm{d}y/\mathrm{d}x)_i \, \Delta x = -y_i, \\ x_{i+1} = x_i + \Delta x, \end{cases} \tag{8.72}$$

where $y_i = \sqrt{1 - x_i^2}$ and $(\mathrm{d}y/\mathrm{d}x)_i = -x_i/\sqrt{1 - x_i^2}$. This scheme fails if $|x_i| > 1$ appears in iterations, because then the argument of the square root is negative. In fact, for eq. (8.71) such a situation will always occur because eqs. (8.72) are equivalent to one equation: $x_{i+1} = 1/x_i$. Hence, $|x_i| < 1$ yields $|x_{i+1}| > 1$, for which the argument of the square root is negative.

Note that the relation between a quaternion and the rotation tensor is 2-to-1, because $\{q_0, \mathbf{q}\}$ and $\{-q_0, -\mathbf{q}\}$ represent the same rotation similarly as $\{\omega, \mathbf{e}\}$ and $\{-\omega, -\mathbf{e}\}$.

The matrix of components of the rotation tensor, eq. (8.68), for $\mathbf{q} = [q_1, q_2, q_3]^T$, becomes

$$(\mathbf{R})_{ij} = 2 \begin{bmatrix} q_0^2 + q_1^2 - \frac{1}{2} & q_1 q_2 - q_3 q_0 & q_1 q_3 + q_2 q_0 \\ q_1 q_2 + q_3 q_0 & q_0^2 + q_2^2 - \frac{1}{2} & q_2 q_3 - q_1 q_0 \\ q_1 q_3 - q_2 q_0 & q_2 q_3 + q_1 q_0 & q_0^2 + q_3^2 - \frac{1}{2} \end{bmatrix}. \tag{8.73}$$

If $\mathbf{R}$ is given, then we can calculate

$$q_0 = \frac{1}{2}\sqrt{\mathrm{tr}\mathbf{R} + 1}, \qquad \tilde{\mathbf{q}} = \frac{1}{4q_0}(\mathbf{R} - \mathbf{R}^T). \tag{8.74}$$

Note that the formula for skew-symmetric $\tilde{\mathbf{q}}$ is singular for $q_0 = 0$, i.e. for $\mathrm{tr}\mathbf{R} = -1$ or $\omega = \arccos\frac{1}{2}(\mathrm{tr}\mathbf{R} - 1) = \arccos(-1) = \pi$. The rotation axis and angle can be extracted from the quaternion $\mathbf{q}$ as follows:

$$\mathbf{e} = \frac{\mathbf{q}}{\sqrt{\mathbf{q}\cdot\mathbf{q}}}, \qquad \begin{cases} \omega = 2\arccos q_0, & 0 \leq \omega \leq +\pi, \\ \text{or} & \\ \omega = 2\arcsin\sqrt{\mathbf{q}\cdot\mathbf{q}}, & 0 \leq \omega \leq +\pi. \end{cases} \tag{8.75}$$

Note that the arguments of the arccos and arcsin functions must be smaller than 1, i.e. $q_0 \leq 1$ and $\sqrt{\mathbf{q}\cdot\mathbf{q}} \leq 1$. To avoid arguments out of the domain due to round off errors, we can use the minimum function

$$\omega = 2\arcsin\left(\min\left(\sqrt{\mathbf{q}\cdot\mathbf{q}},\ 1\right)\right), \qquad 0 \leq \omega \leq +\pi. \tag{8.76}$$

Only $\omega \geq 0$ is yielded by this equation, while the sign of $\mathbf{q}$ is transferred to $\mathbf{e}$.

Remark 1. Note that the ω extracted from the quaternion is $\leq +\pi$, which means that for rotations exceeding this value, the rotation vector is shortened, similarly as by the operation of eq. (8.82). Recall that this operation is correct for the rotation tensor $\mathbf{R}$, but not for the tangent operator $\mathbf{T}(\boldsymbol{\psi})$ of Sect. 9.1, which is not periodic, see eq. (9.23). Hence, we can use the operation of extraction of ω from a quaternion only to the rotations $\leq +\pi$.

Remark 2. Using the rotation tensor in the form of eq. (8.68), the rotated (current) shell director can be expressed as

$$\mathbf{a}_3 \doteq \mathbf{Q}_0\mathbf{t}_3 = (2q_0^2 - 1)\,\mathbf{t}_3 + 2q_0\mathbf{q}\times\mathbf{t}_3 + 2(\mathbf{q}\cdot\mathbf{t}_3)\,\mathbf{q}, \tag{8.77}$$

where only operations on scalars and vectors are performed. As we shall see in Sect. 8.3.2, the quaternions can be conveniently used to compose (accumulate) rotations.

8.2.4 Three parameters: rotation pseudo-vectors

The rotation pseudo-vectors are used to parameterize the rotation tensor of eq. (8.9) obtained for the rotation of a vector around an axis. Several rotation pseudo-vectors were proposed in the literature; they all have the direction of the axis of rotation $\mathbf{e}$, but differ in length. The two most often used vectors are presented in detail below.

The rotation pseudo-vectors have only three components, and directly generate $\mathbf{R} \in \mathrm{SO}(3)$. Hence, there is no need to reduce the number of parameters or to additionally impose the orthogonality constraints.

The basic relations for the rotation vector are

$$\tilde{\boldsymbol{\psi}} = \boldsymbol{\psi} \times \mathbf{I}, \qquad \boldsymbol{\psi} = \tfrac{1}{2}(\mathbf{I} \times \tilde{\boldsymbol{\psi}}), \qquad \|\boldsymbol{\psi}\| = \sqrt{\boldsymbol{\psi} \cdot \boldsymbol{\psi}}, \qquad \mathbf{e} = \frac{\boldsymbol{\psi}}{\|\boldsymbol{\psi}\|}, \tag{8.78}$$

where the tilde denotes the associated skew-symmetric tensor and the cross-product of a vector and a tensor is defined as in [33] p. 74. In terms of components, the relation between the rotation vector and the associated skew-symmetric tensor can be specified as in eq. (8.15).

A. Canonical rotation vector

This vector is defined as

$$\boldsymbol{\psi} \doteq \omega\, \mathbf{e}, \tag{8.79}$$

for which we have

$$\tilde{\boldsymbol{\psi}} = \omega \mathbf{S}, \qquad \tilde{\boldsymbol{\psi}}^2 = \omega^2 \mathbf{S}^2, \qquad \boldsymbol{\psi} \cdot \boldsymbol{\psi} = \tfrac{1}{2} \tilde{\boldsymbol{\psi}} \cdot \tilde{\boldsymbol{\psi}} = \omega^2.$$

The rotation tensor, eq. (8.9), can be rewritten in terms of $\boldsymbol{\psi}$ as follows

$$\mathbf{R} \doteq \mathbf{I} + c_1\, \tilde{\boldsymbol{\psi}} + c_2\, \tilde{\boldsymbol{\psi}}^2, \qquad \|\boldsymbol{\psi}\| = \sqrt{\boldsymbol{\psi} \cdot \boldsymbol{\psi}} \geq 0, \tag{8.80}$$

where the scalar coefficients are

$$c_1 \doteq \frac{\sin \|\boldsymbol{\psi}\|}{\|\boldsymbol{\psi}\|}, \qquad c_2 \doteq \frac{1 - \cos \|\boldsymbol{\psi}\|}{\|\boldsymbol{\psi}\|^2}. \tag{8.81}$$

The representation (8.80) can be obtained from the exponential representation of the rotation tensor, $\mathbf{R} = \exp(\tilde{\boldsymbol{\psi}}) \doteq \mathbf{I} + \tilde{\boldsymbol{\psi}} + \cdots + \frac{1}{n!}\tilde{\boldsymbol{\psi}}^n + \cdots$, by using the expressions for the odd and even powers of $\mathbf{S}$ of Table 8.1.

The representation (8.80) is a periodic function of $\|\boldsymbol{\psi}\|$, which can be shown by transforming it to the form of eq. (8.9) by using $\omega \doteq \|\boldsymbol{\psi}\|$. Because $\omega > 0$, we can restrict its range to $\omega \leq 2\pi$. As a consequence, we can shorten the rotation vector for $\|\boldsymbol{\psi}\| > 2\pi$ using the following transformation:

$$k \doteq \text{Integer Part}\left(\frac{\|\boldsymbol{\psi}\|}{2\pi}\right), \qquad \boldsymbol{\psi}^* \doteq \left(1 - k\frac{2\pi}{\|\boldsymbol{\psi}\|}\right)\boldsymbol{\psi}. \tag{8.82}$$

We can check that $\mathbf{R}(\boldsymbol{\psi}^*) = \mathbf{R}(\boldsymbol{\psi})$. Note that this operation is correct for the rotation tensor $\mathbf{R}$ but not for the tangent operator $\mathbf{T}(\boldsymbol{\psi})$ of Sect. 9.1, which is not periodic, see eq. (9.23). Hence, we cannot shorten the rotation vectors, e.g., for shells.

The representation (8.80) is never singular but the coefficients must be considered for three characteristic values of $\|\boldsymbol{\psi}\|$:

1. $\|\boldsymbol{\psi}\| = 0$, at which c_1 and c_2 are numerically indeterminate. We have to extend their domain and define c_1 and c_2 at $\|\boldsymbol{\psi}\| = 0$ as the limit values, see the details below. Then, $\mathbf{R}(\boldsymbol{\psi} = \mathbf{0}) = \mathbf{I}$.
2. $\|\boldsymbol{\psi}\| = k\pi$, $(k = 1, 2, ...)$, at which $c_1 = 0$ and $c_2 \neq 0$ and then $\mathbf{R} = \mathbf{I} + c_2\, \tilde{\boldsymbol{\psi}}^2$ and is symmetric ! For these values, we cannot uniquely recover $\tilde{\boldsymbol{\psi}}$ from $\mathbf{R}$, see eq. (8.93).
3. $\|\boldsymbol{\psi}\| = 2k\pi$, $(k = 1, 2, ...)$, at which $c_1 = 0$ and $c_2 = 0$ and then $\mathbf{R} = \mathbf{I}$.

In numerical calculations, we encounter two problems:

1. Only c_1 converges correctly, while c_2 does not, see Table 8.5, due to round-off errors. Then we use the identity $1 - \cos\|\boldsymbol{\psi}\| = 2\sin^2(\|\boldsymbol{\psi}\|/2)$ to obtain
$$c_2 \doteq \frac{1 - \cos\|\boldsymbol{\psi}\|}{\|\boldsymbol{\psi}\|^2} = \frac{1}{2}\left[\frac{\sin(\|\boldsymbol{\psi}\|/2)}{(\|\boldsymbol{\psi}\|/2)}\right]^2, \tag{8.83}$$
which ensures a correct numerical behavior for a wider range of values of $\|\boldsymbol{\psi}\| \to 0$, see Table 8.5, with the results of calculations in double precision.

Table 8.5 Values yielded by two expressions of c_2 for $\|\boldsymbol{\psi}\| = 10^{-n}$.

n	5	6	7	8	9	10
$\frac{1-\cos\|\boldsymbol{\psi}\|}{\|\boldsymbol{\psi}\|^2}$	0.50000	0.50004	0.49960	0.00000	0.00000	0.00000
$\frac{1}{2}\left[\frac{\sin(\|\boldsymbol{\psi}\|/2)}{(\|\boldsymbol{\psi}\|/2)}\right]^2$	0.50000	0.50000	0.50000	0.50000	0.50000	0.50000

2. At exactly $\|\boldsymbol{\psi}\| = 0$, both coefficients are indeterminate,
$$c_1 = \left.\frac{\sin\|\boldsymbol{\psi}\|}{\|\boldsymbol{\psi}\|}\right|_{\|\boldsymbol{\psi}\|=0} = \frac{0}{0}, \qquad c_2 = \left.\frac{1}{2}\left[\frac{\sin(\|\boldsymbol{\psi}\|/2)}{(\|\boldsymbol{\psi}\|/2)}\right]^2\right|_{\|\boldsymbol{\psi}\|=0} = \frac{0}{0}, \tag{8.84}$$
which indicates that the point $\|\boldsymbol{\psi}\| = 0$ does not belong to their domains. However, for $\|\boldsymbol{\psi}\| \to 0$, limits of the coefficients are finite, i.e.
$$\lim_{\|\boldsymbol{\psi}\|\to 0} c_1 = 1, \qquad \lim_{\|\boldsymbol{\psi}\|\to 0} c_2 = \tfrac{1}{2}, \tag{8.85}$$
so we can define c_1 and c_2 at $\|\boldsymbol{\psi}\| = 0$ as the limit values. Then $\mathbf{R}(\boldsymbol{\psi} = \mathbf{0}) = \mathbf{I}$. Nonetheless, we still have the problem of derivatives of $\mathbf{R}$ at $\|\boldsymbol{\psi}\| = 0$. Simple remedies can be used to alleviate this problem; they are explained below.

Methods of avoiding numerical indeterminacy. Instead of c_1 of eq. (8.81), we consider the expression

$$c \doteq \frac{\sin\sqrt{x^2}}{\sqrt{x^2}}, \tag{8.86}$$

where x is a scalar. Because calculations of the tangent operator involve the first and second derivative, we calculate

$$\frac{\mathrm{d}c}{\mathrm{d}x} = \frac{\cos\sqrt{x^2}}{x} - \frac{\sin\sqrt{x^2}}{x\sqrt{x^2}}, \qquad \frac{\mathrm{d}^2c}{\mathrm{d}x^2} = -\frac{2\cos\sqrt{x^2}}{x^2} + \frac{2\sin\sqrt{x^2}}{(x^2)^{3/2}} - \frac{\sin\sqrt{x^2}}{(x^2)^{1/2}}$$

and they need to be considered. Their limits for $x \to 0$ are

$$\lim_{x\to 0} c = 1, \qquad \lim_{x\to 0} \frac{\mathrm{d}c}{\mathrm{d}x} = 0, \qquad \lim_{x\to 0} \frac{\mathrm{d}^2c}{\mathrm{d}x^2} = -\frac{1}{3}, \tag{8.87}$$

and the derivatives exist. However, numerically evaluating the derivatives at $x = 0$, we obtain indeterminate expressions. Two remedies to this problem are as follows:

a) *Perturbation.* We can use the perturbed value $\sqrt{x^2 + \tau}$ instead of $\sqrt{x^2}$ and consider

$$c_p \doteq \frac{\sin\sqrt{x^2+\tau}}{\sqrt{x^2+\tau}} \tag{8.88}$$

instead of c. Then, for $x = 0$, we obtain

$$c_p\big|_{x=0} = \frac{\sin\sqrt{\tau}}{\sqrt{\tau}}, \qquad \left.\frac{\mathrm{d}c_p}{\mathrm{d}x}\right|_{x=0} = 0, \qquad \left.\frac{\mathrm{d}^2c_p}{\mathrm{d}x^2}\right|_{x=0} = \frac{\cos\sqrt{\tau}}{\tau} - \frac{\sin\sqrt{\tau}}{\tau^{3/2}},$$

and for $\tau \to 0$ they yield the limits of eq. (8.87). The first derivative provides a correct limit regardless of the value of τ, but the value of τ must be selected properly because of c_p and its second derivative. Evaluating them for $\tau = 0$, we find that they are indeterminate, which indicates that the value of perturbation cannot be too small. The perturbation value $\tau = 10^{-8}$ is suitable for calculations in double precision and it yields a good approximation of the limits of eq. (8.87).

Similar considerations can be performed for c_2 of eq. (8.83), and they provide similar conclusions. We can check that the modified form of c_2 of eq. (8.83) is still needed to obtain a correct value of the second derivative.

Note that, for the canonical rotation vector $\boldsymbol{\psi}$, the perturbation is applied as follows:

$$\|\boldsymbol{\psi}\| = \sqrt{\boldsymbol{\psi} \cdot \boldsymbol{\psi} + \tau}. \tag{8.89}$$

b) *Taylor expansion.* Another way to avoid problems at $x = 0$ is to use a truncated Taylor series expansion at $x = 0$, e.g.

$$c_1 = \frac{\sin x}{x} \approx 1 - \frac{1}{6}x^2 + \frac{1}{120}x^4,$$
$$c_2 = \frac{1}{2}\left[\frac{\sin(x/2)}{(x/2)}\right]^2 \approx \frac{1}{2} - \frac{1}{24}x^2 + \frac{1}{720}x^4,$$

where $x \doteq \|\boldsymbol{\psi}\|$. The above three-term expansions preserve a good accuracy of both coefficients and their first and second derivatives for the range up to $|x| \approx 1$, see Fig. 8.8. In this figure, the three-term expansions are indicated by letter "e", and we see that the curves for them coincide with the original curves.

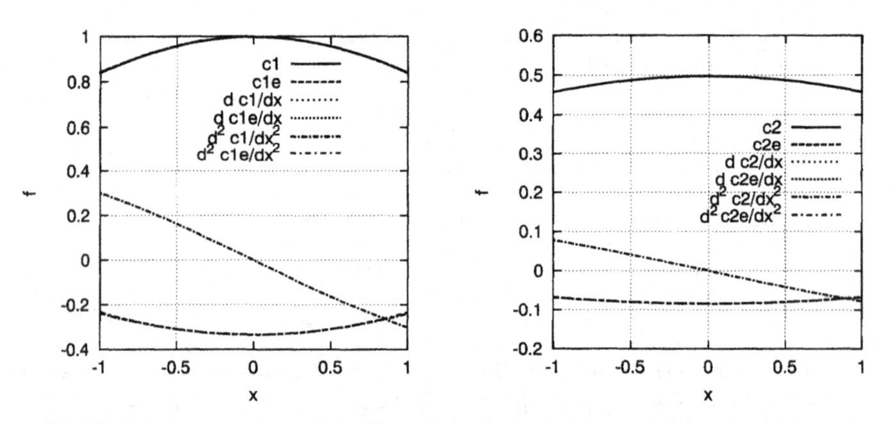

Fig. 8.8 Approximations of coefficients c_1 and c_2 and their derivatives.

Matrix of components. The matrix of components of the rotation tensor of eq. (8.80), for c_1 given by eq. (8.81) and c_2 by eq. (8.83), is as follows:

$$(\mathbf{R})_{ij} = \begin{bmatrix} \frac{\psi_1^2+(\psi_2^2+\psi_3^2)c}{\psi^2} & \frac{2\psi_1\psi_2\sin^2(\psi/2)}{\psi^2} - \frac{\psi_3 s}{\psi} & \frac{2\psi_1\psi_3\sin^2(\psi/2)}{\psi^2} + \frac{\psi_2 s}{\psi} \\ \frac{2\psi_1\psi_2\sin^2(\psi/2)}{\psi^2} + \frac{\psi_3 s}{\psi} & \frac{\psi_2^2+(\psi_1^2+\psi_3^2)c}{\psi^2} & \frac{2\psi_2\psi_3\sin^2(\psi/2)}{\psi^2} - \frac{\psi_1 s}{\psi} \\ \frac{2\psi_1\psi_3\sin^2(\psi/2)}{\psi^2} - \frac{\psi_2 s}{\psi} & \frac{2\psi_2\psi_3\sin^2(\psi/2)}{\psi^2} + \frac{\psi_1 s}{\psi} & \frac{\psi_3^2+(\psi_1^2+\psi_2^2)c}{\psi^2} \end{bmatrix}, \tag{8.90}$$

where $\boldsymbol{\psi} \doteq \{\psi_1, \psi_2, \psi_3\}$, $s \doteq \sin\psi$, $c \doteq \cos\psi$, and $\|\boldsymbol{\psi}\| \doteq \psi = \sqrt{\psi_1^2+\psi_2^2+\psi_3^2}$. Note that $\mathrm{tr}\mathbf{R} = 1+2\cos\psi$. For instance, for the rotation around the axis 0–3, we have $[\psi_1, \psi_2, \psi_3] = [0, 0, \psi_3]$ and $\psi \doteq \sqrt{\psi_3^2}$, so the matrix becomes

$$(\mathbf{R})_{ij} = \begin{bmatrix} \frac{\psi_3^2 c}{\psi^2} & -\frac{\psi_3 s}{\psi} & 0 \\ \frac{\psi_3 s}{\psi} & \frac{\psi_3^2 c}{\psi^2} & 0 \\ 0 & 0 & \frac{\psi_3^2}{\psi^2} \end{bmatrix} = \begin{bmatrix} \frac{\psi_3^2}{\psi_3^2}\cos\psi & -\frac{\psi_3}{\sqrt{\psi_3^2}}\sin\psi & 0 \\ \frac{\psi_3}{\sqrt{\psi_3^2}}\sin\psi & \frac{\psi_3^2}{\psi_3^2}\cos\psi & 0 \\ 0 & 0 & \frac{\psi_3^2}{\psi_3^2} \end{bmatrix}. \tag{8.91}$$

Note that this matrix cannot be reduced to

$$\begin{bmatrix} \cos\psi & -\sin\psi & 0 \\ \sin\psi & \cos\psi & 0 \\ 0 & 0 & 1 \end{bmatrix}, \quad \text{but can be reduced to} \quad \begin{bmatrix} \cos\psi_3 & -\sin\psi_3 & 0 \\ \sin\psi_3 & \cos\psi_3 & 0 \\ 0 & 0 & 1 \end{bmatrix}.$$

Extraction of rotation vector from rotation matrix. Given the rotation tensor $\mathbf{R}$, we can extract the skew-symmetric tensor as follows:

$$\tilde{\boldsymbol{\psi}} = \frac{1}{2}\frac{\omega}{\sin\omega}(\mathbf{R}-\mathbf{R}^T), \quad \text{where} \quad \omega = \arccos\frac{1}{2}(\mathrm{tr}\mathbf{R}-1) \in [0, \pi]. \tag{8.92}$$

If $\mathrm{tr}\mathbf{R} = 3$, then $\omega = 0$ and $\mathbf{R} = \mathbf{I}$ but, by the first of eq. (8.85), the expression for $\tilde{\boldsymbol{\psi}}$ is not singular and we obtain $\tilde{\boldsymbol{\psi}} = \mathbf{0}$. If $\mathrm{tr}\mathbf{R} = -1$ then we have $\omega = \pi$, $(\sin\omega)/\omega = 0$, $(1-\cos\omega)/\omega^2 = 2/\pi^2$, and, from eq. (8.80), we obtain

$$\tilde{\boldsymbol{\psi}} = \pm\sqrt{\frac{\pi^2}{2}(\mathbf{R}-\mathbf{I})}, \tag{8.93}$$

i.e. two solutions for one $\mathbf{R}$. Hence, the relation $\mathbf{R} \rightarrow \tilde{\boldsymbol{\psi}}$ is 1-to-2. The canonical rotation vector $\boldsymbol{\psi} = [\psi_1, \psi_2, \psi_3]^T$ is obtained by extraction of proper components of $\tilde{\boldsymbol{\psi}}$.

Using eq. (8.92), we must control that the argument of arccos is in the range $[-1, 1]$. Besides, the ω extracted from $\mathbf{R}$ is $\leq \pi$, which means that for rotations exceeding this value, the rotation vector is shortened as by the operation of eq. (8.82).

Extraction of rotation vector from quaternion. Given the quaternion $\{q_0, \mathbf{q}\}$, see eq. (8.67), we can extract $\mathbf{e}$ and ω from the quaternion using eqs. (8.75) and (8.76). The canonical rotation vector is calculated as

$$\boldsymbol{\psi} = \omega\, \mathbf{e}. \tag{8.94}$$

Quaternion for given rotation vector. From a canonical rotation vector, we can obtain

$$\omega = \|\boldsymbol{\psi}\| > 0, \qquad \mathbf{e} = \boldsymbol{\psi}/\|\boldsymbol{\psi}\| = \boldsymbol{\psi}/\omega \tag{8.95}$$

and then we can calculate the quaternion as follows:

$$q_0 \doteq \cos(\omega/2), \qquad \mathbf{q} \doteq \sin(\omega/2)\,\mathbf{e} = \frac{\sin(\omega/2)}{\omega}\,\boldsymbol{\psi}. \tag{8.96}$$

B. Semi-tangential rotation vector

This vector is defined as follows:

$$\boldsymbol{\psi} \doteq t\,\mathbf{e}, \qquad t \doteq \tan(\omega/2), \tag{8.97}$$

for which we have

$$\tilde{\boldsymbol{\psi}} = t\mathbf{S}, \qquad \tilde{\boldsymbol{\psi}}^2 = t^2\mathbf{S}^2, \qquad \boldsymbol{\psi}\cdot\boldsymbol{\psi} = \tfrac{1}{2}\tilde{\boldsymbol{\psi}}\cdot\tilde{\boldsymbol{\psi}} = t^2.$$

This vector is attributed to B.O. Rodrigues (1795-1851), [79]. Components of the semi-tangential vector are also called the Cayley parameters.

The semi-tangential vector is undetermined at $\omega = (2k+1)\pi$ for some integer k, at which $\tan(\omega/2)$ has asymptotes. By using t of eq. (8.97) and the trigonometric identities

$$\sin\omega = \frac{2t}{1+t^2}, \qquad \cos\omega = \frac{1-t^2}{1+t^2}, \tag{8.98}$$

the rotation tensor of eq. (8.9) can be represented in terms of $\boldsymbol{\psi}$ as

$$\mathbf{R} \doteq \mathbf{I} + \frac{2}{1+\|\boldsymbol{\psi}\|^2}\left(\tilde{\boldsymbol{\psi}} + \tilde{\boldsymbol{\psi}}^2\right), \qquad \|\boldsymbol{\psi}\| = \sqrt{\boldsymbol{\psi}\cdot\boldsymbol{\psi}} \geq 0. \tag{8.99}$$

This representation never is singular because the denominator $1+\|\boldsymbol{\psi}\|^2 \neq 0$ for $\|\boldsymbol{\psi}\|^2 \geq 0$. An alternative form of eq. (8.99), called the Cayley form, is

$$\mathbf{R} \doteq (\mathbf{I} + \tilde{\boldsymbol{\psi}})(\mathbf{I} - \tilde{\boldsymbol{\psi}})^{-1}. \tag{8.100}$$

To verify this formula, we must post-multiply eqs. (8.99) and (8.100) by $(\mathbf{I} - \tilde{\boldsymbol{\psi}})$ and then use $\tilde{\boldsymbol{\psi}}^3 = -\|\boldsymbol{\psi}\|^2 \tilde{\boldsymbol{\psi}}$, which directly leads to the identity of both forms.

The representation (8.99) does not contain trigonometric functions and is not a periodic function of $\|\boldsymbol{\psi}\|$, which can be shown by transforming it to the form

$$\mathbf{R} \doteq \mathbf{I} + c_1 \mathbf{S} + c_2 \mathbf{S}^2, \qquad c_1 \doteq \frac{2t}{1+t^2}, \qquad c_2 \doteq \frac{2t^2}{1+t^2}, \tag{8.101}$$

where c_1 and c_2 are plotted in Fig. 8.9 as functions of $t \doteq \|\boldsymbol{\psi}\|$. Hence, we cannot shorten the rotation vector as we could for the canonical parametrization, see eq. (8.82).

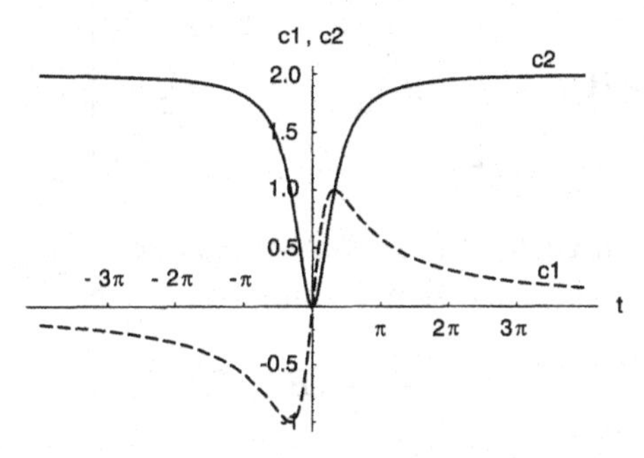

Fig. 8.9 Semi-tangential parametrization. Coefficients c_1 and c_2.

Remark. We note that even if t grows large, the rotation angle ω remains restricted because the mapping $t \rightarrow \omega = 2 \arctan t$: $(-\infty, +\infty) \rightarrow (-\pi, +\pi)$ and the points $\pm\pi$ are asymptotic. To establish how the rotation tensor behaves for a growing t, we rewrite eq. (8.99) as $\mathbf{R} \doteq \mathbf{I} + \frac{2}{1+t^2}\left(t\mathbf{S} + t^2\mathbf{S}^2\right)$. For $t \rightarrow \pm\infty$, we have

$$\frac{2t}{1+t^2} \rightarrow 0, \qquad \frac{2t^2}{1+t^2} \rightarrow 2$$

and, hence, $\mathbf{R} \rightarrow \mathbf{I} + 2\mathbf{S}^2 = -\mathbf{I} + 2\,\mathbf{e} \otimes \mathbf{e}$, where the formula for $\mathbf{S}^2$ of Table 8.1 was used. In the basis $\{\mathbf{e}, \mathbf{e}_1, \mathbf{e}_2\}$,

$$t \rightarrow \pm\infty: \quad \mathbf{R} \rightarrow \begin{bmatrix} +1 & 0 & 0 \\ 0 & -1 & 0 \\ 0 & 0 & -1 \end{bmatrix}, \qquad \det \mathbf{R} \rightarrow +1, \qquad \operatorname{tr} \mathbf{R} \rightarrow -1.$$

Matrix of components. The matrix of components of the rotation tensor of eq. (8.99) is

$$(\mathbf{R})_{ij} = \begin{bmatrix} 1 - \frac{2(\psi_2^2+\psi_3^2)}{1+\psi^2} & \frac{2(\psi_1\psi_2-\psi_3)}{1+\psi^2} & \frac{2(\psi_1\psi_3+\psi_2)}{1+\psi^2} \\ \frac{2(\psi_1\psi_2+\psi_3)}{1+\psi^2} & 1 - \frac{2(\psi_1^2+\psi_3^2)}{1+\psi^2} & \frac{2(\psi_2\psi_3-\psi_1)}{1+\psi^2} \\ \frac{2(\psi_1\psi_3-\psi_2)}{1+\psi^2} & \frac{2(\psi_2\psi_3+\psi_1)}{1+\psi^2} & 1 - \frac{2(\psi_1^2+\psi_2^2)}{1+\psi^2} \end{bmatrix}, \tag{8.102}$$

where $\boldsymbol{\psi} \doteq [\psi_1, \psi_2, \psi_3]^T$ and $\|\boldsymbol{\psi}\| \doteq \psi = \sqrt{\psi_1^2 + \psi_2^2 + \psi_3^2}$. Note that $\mathrm{tr}\mathbf{R} = (3 - \psi^2)/(1 + \psi^2) = 4/(1 + \psi^2) - 1$. For instance, for the rotation around the axis 0-3, we have $[\psi_1, \psi_2, \psi_3] = [0, 0, \psi_3]$ and $\psi \doteq \sqrt{\psi_3^2}$, so the matrix becomes

$$(\mathbf{R})_{ij} = \begin{bmatrix} 1 - \frac{2\psi_3^2}{1+\psi^2} & \frac{-2\psi_3}{1+\psi^2} & 0 \\ \frac{2\psi_3}{1+\psi^2} & 1 - \frac{2\psi_3^2}{1+\psi^2} & 0 \\ 0 & 0 & 1 \end{bmatrix}. \tag{8.103}$$

Extraction of rotation vector from rotation matrix. Given a rotation tensor $\mathbf{R}$, we can calculate the skew-symmetric tensor as

$$\tilde{\boldsymbol{\psi}} = \frac{1}{1 + \mathrm{tr}\mathbf{R}}(\mathbf{R} - \mathbf{R}^T), \tag{8.104}$$

where we used $(1 + \|\boldsymbol{\psi}\|^2)/4 = 1/(1 + \mathrm{tr}\mathbf{R})$, and the form of $\mathbf{R}$ of eq. (8.99). Equation (8.104) is singular for $\mathrm{tr}\mathbf{R} = -1$, but this value is not attained for $\psi \in [0, +\infty]$ and only $\lim_{\psi\to\infty} \mathrm{tr}\mathbf{R} = -1$. The rotation angle can be calculated from $\omega = 2\arctan\sqrt{\frac{1}{2}\tilde{\boldsymbol{\psi}} \cdot \tilde{\boldsymbol{\psi}}} \in [0, \pi]$.

Extraction of rotation vector from quaternion. Given the quaternion $\{q_0, \mathbf{q}\}$ (see eq. (8.67)), note that

$$\tan(\omega/2) = \frac{\pm\sqrt{\sin^2(\omega/2)}}{\cos(\omega/2)}, \qquad \|\mathbf{q}\| \doteq \sqrt{\mathbf{q}\cdot\mathbf{q}} = \pm\sqrt{\sin^2(\omega/2)}. \tag{8.105}$$

We can extract the semi-tangential rotation vector as follows:

$$\begin{aligned} \boldsymbol{\psi} = \tan(\omega/2)\frac{\mathbf{q}}{\|\mathbf{q}\|} &= \frac{\pm\sqrt{\sin^2(\omega/2)}}{\cos(\omega/2)} \frac{\mathbf{q}}{\pm\sqrt{\sin^2(\omega/2)}} \\ &= \frac{1}{\cos(\omega/2)}\mathbf{q} = \mathbf{q}/q_0. \end{aligned} \tag{8.106}$$

Quaternion for given rotation vector. For the semi-tangential rotation vector, eq. (8.97), we can calculate

$$\|\boldsymbol{\psi}\|^2 = \boldsymbol{\psi} \cdot \boldsymbol{\psi} = t^2. \tag{8.107}$$

Because $\tan(\omega/2) = \sin(\omega/2)/\cos(\omega/2)$ and $\sin^2(\omega/2)+\cos^2(\omega/2) = 1$, we can obtain

$$\cos(\omega/2) = \frac{1}{\sqrt{1+t^2}}, \qquad \sin(\omega/2) = \frac{\sqrt{t^2}}{\sqrt{1+t^2}}. \tag{8.108}$$

Besides, $\mathbf{e} = \boldsymbol{\psi}/\|\boldsymbol{\psi}\| = \boldsymbol{\psi}/\sqrt{t^2}$. Then we can calculate the quaternion as follows:

$$q_0 \doteq \cos(\omega/2) = \frac{1}{\sqrt{1+t^2}}, \qquad \mathbf{q} \doteq \sin(\omega/2)\,\mathbf{e} = \frac{\sqrt{t^2}}{\sqrt{1+t^2}}\frac{\boldsymbol{\psi}}{\sqrt{t^2}} = q_0\,\boldsymbol{\psi} \tag{8.109}$$

and it depends only on $\boldsymbol{\psi}$ and t^2.

Example. Extraction of rotation vectors from rotation matrix. In this example, we test the extraction formulas for the parametrizations based on rotation vectors. Given is the rotation matrix

$$(\mathbf{R})_{ij} = \begin{bmatrix} \cos\alpha & -\sin\alpha & 0 \\ \sin\alpha & \cos\alpha & 0 \\ 0 & 0 & 1 \end{bmatrix} \tag{8.110}$$

and the angle α is an independent parameter. We calculate the skew-symmetric $\tilde{\boldsymbol{\psi}}$ and the angle ω. Next, the rotation vector $\boldsymbol{\psi}$ and the axis $\mathbf{e}$ are obtained.

a. For the canonical vector, by eq. (8.92), we obtain

$$\beta = \arccos \tfrac{1}{2}(\mathrm{tr}\mathbf{R} - 1) = \arccos(\cos\alpha), \tag{8.111}$$

$$\tilde{\boldsymbol{\psi}} = \tfrac{1}{2}\frac{\beta}{\sin\beta}(\mathbf{R} - \mathbf{R}^T) = A \begin{bmatrix} 0 & -1 & 0 \\ 1 & 0 & 0 \\ 0 & 0 & 0 \end{bmatrix}, \qquad A \doteq \frac{\beta\sin\alpha}{\sin\beta}, \tag{8.112}$$

$$\omega = \sqrt{\frac{1}{2}\tilde{\boldsymbol{\psi}} \cdot \tilde{\boldsymbol{\psi}}} = |A|. \tag{8.113}$$

b. For the semi-tangential vector, by eq. (8.104) we obtain

$$\tilde{\psi} = \frac{1}{1+\mathrm{tr}\mathbf{R}}(\mathbf{R}-\mathbf{R}^T) = A\begin{bmatrix} 0 & -1 & 0 \\ 1 & 0 & 0 \\ 0 & 0 & 0 \end{bmatrix}, \qquad A \doteq \frac{\sin\alpha}{1+\cos\alpha}, \tag{8.114}$$

$$\omega = 2\arctan\sqrt{-\tfrac{1}{2}\mathrm{tr}\,\tilde{\psi}^2} = 2\arctan|A|. \tag{8.115}$$

The angle of rotation is plotted in Fig. 8.10a, and we see that the curves coincide and $\omega \in [0, \pi]$. For all parametrizations, the rotation vector and the rotation axis are as follows:

$$\psi = [0,0,A]^T, \qquad \mathbf{e} = \frac{\psi}{\|\psi\|} = [0,0,\mathrm{sign}(A)]^T \tag{8.116}$$

and their third components are shown in Figs. 8.10b and c. Note that the component ψ_3 is different for each parametrization, while the component e_3 is identical for all parametrizations and has the sign of ψ_3.

Example. Extraction of canonical rotation vector from quaternion. In this example, we test the extraction formulas for the canonical parametrization, assuming that a quaternion $\{q_0, \mathbf{q}\}$ is given. Note that in the previous example, the rotation matrix was given.

Assume that $\mathbf{e} \doteq [0,0,1]^T$ and that α is an independent parameter, so the quaternion of eq. (8.67) becomes

$$q_0 \doteq \cos(\alpha/2), \qquad \mathbf{q} \doteq \sin(\alpha/2)\,\mathbf{e}. \tag{8.117}$$

(Note that for this data, using eq. (8.69), we obtain the same rotation matrix (8.110) as in the previous example.) To extract the canonical rotation vector from the given quaternion, we use eq. (8.94). The rotation angle ω is calculated as

$$\|\mathbf{q}\| = \sqrt{\sin^2(\alpha/2)}, \qquad \omega \doteq \|\psi\| = 2\arcsin\sqrt{\sin^2(\alpha/2)} \in [0,\pi] \tag{8.118}$$

and ω is shown in Fig. 8.11a. The rotation vector and the rotation axis are as follows:

$$\psi = \frac{\|\psi\|}{\|\mathbf{q}\|}\mathbf{q} = [0,0,A]^T, \qquad \mathbf{e} = \frac{\psi}{\|\psi\|} = [0,0,\mathrm{sign}(A)]^T, \tag{8.119}$$

$$A \doteq \frac{2\arcsin\sqrt{\sin^2(\alpha/2)}\,\sin(\alpha/2)}{\sqrt{\sin^2(\alpha/2)}} \tag{8.120}$$

and their third components are shown in Figs. 8.11b and c.

Comparing Figs. 8.10a and 8.11a, we see that there is no difference in the recovered ω. The difference is in

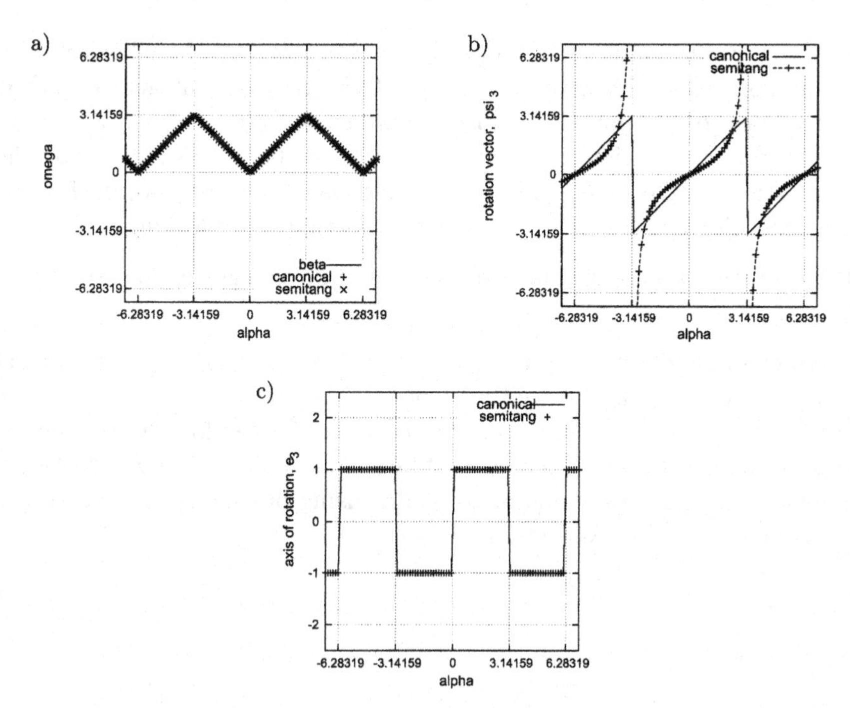

Fig. 8.10 Example. Extraction of rotation vectors from rotation matrix.

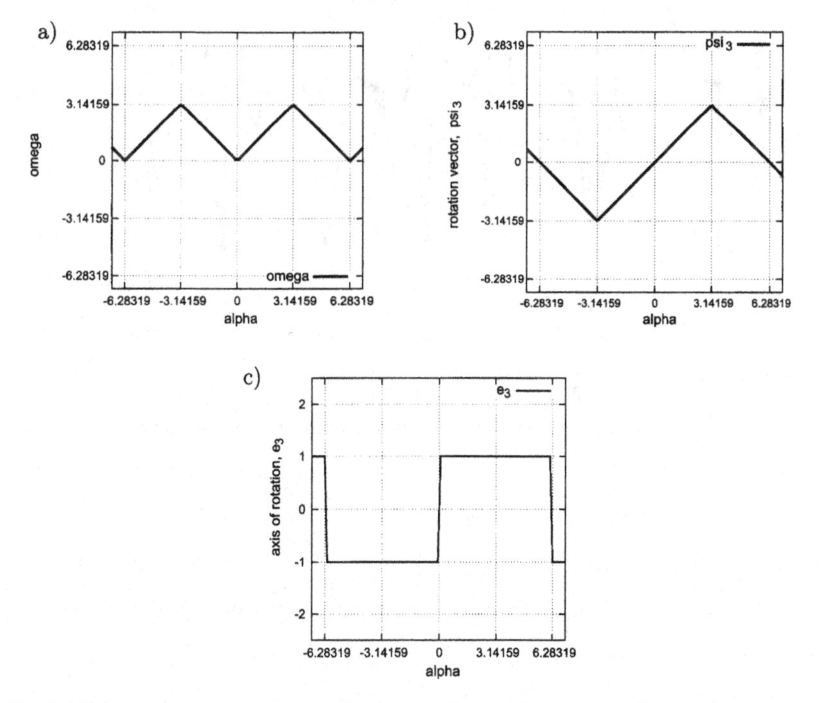

Fig. 8.11 Example. Extraction of canonical rotation vector from quaternion.

1. the rotation vector ψ (compare Figs. 8.10b and 8.11b). The rotation matrix yields a jump of ψ at $\pi + 2k\pi$ for some integer k, while there is no such jump if we use a quaternion and
2. the axis of rotation $\mathbf{e}$ (compare Figs. 8.10c and 8.11c). For the rotation matrix, $\mathbf{e}$ changes the sign at $0 + k\pi$, while for the quaternion at $0 + 2k\pi$, i.e. with the period twice as long.

Hence, using the quaternions, we avoid abrupt changes of the recovered ψ and $\mathbf{e}$.

An example of such jumps is shown in Fig. 8.12 where the canonical rotation vector is obtained either from a quaternion or from a rotation matrix, where, for the latter, the jumps occur. These plots are for the example of unstable rotations of a rigid body about the axis of intermediate moment of inertia (Example 6.1 of [221]) using our modified form of the conserving algorithm ALGO-C1.

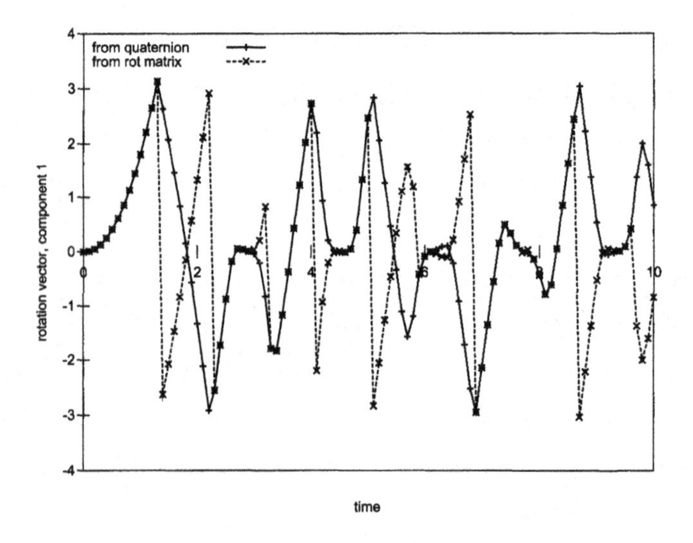

Fig. 8.12 Rotation vector obtained either from quaternion or from rotation matrix. Note vertical jumps of the line obtained from rotation matrix !

8.2.4.1 Transformation of rotation vector to another basis

Here we show that rotation vectors can be transformed from one basis to another as ordinary vectors. For the sake of simplicity, the proof is for the semi-tangential vector but the same is true for the canonical vector.

Consider two Cartesian bases; the global basis $\{\mathbf{e}_i\}$ and the local basis $\{\mathbf{t}_i\}$, with the position of the latter described by $\mathbf{R}_0 \in \mathrm{SO}(3)$, i.e. $\mathbf{t}_i = \mathbf{R}_0\,\mathbf{e}_i$. We assume that a semi-tangential rotation vector is given in $\{\mathbf{e}_i\}$, so $\boldsymbol{\psi} = \psi_i \mathbf{e}_i$, and we wish to transform it to $\{\mathbf{t}_i\}$. An ordinary vector $\mathbf{v}$ transforms from $\{\mathbf{e}_i\}$ to $\{\mathbf{t}_i\}$ as follows: $\mathbf{v}^* = \mathbf{R}_0^T \mathbf{v}$ and, as shown below, this rule is also valid for the semi-tangential $\boldsymbol{\psi}$. The proof amounts to showing that

$$\mathbf{R}(\boldsymbol{\psi}^*) = \mathbf{R}_0^T\,\mathbf{R}(\boldsymbol{\psi})\,\mathbf{R}_0, \quad \text{where} \quad \boldsymbol{\psi}^* = \mathbf{R}_0^T\,\boldsymbol{\psi}, \tag{8.121}$$

i.e. that the rotation tensor for $\boldsymbol{\psi}^*$ and the back-rotated rotation tensor for $\boldsymbol{\psi}$ are identical. This relation can be easily verified numerically.

First, we establish the transformation rule for $\boldsymbol{\psi}$. For a given $\boldsymbol{\psi}$, we calculate the rotation tensor on use of eq. (8.99) and transform this tensor from $\{\mathbf{e}_i\}$ to $\{\mathbf{t}_i\}$ using $\mathbf{R}^* = \mathbf{R}_0^T \mathbf{R}\,\mathbf{R}_0$. Next, we extract the skew-symmetric tensor $\tilde{\boldsymbol{\psi}}^* \in \mathrm{so}(3)$ from $\mathbf{R}^*$ using eq. (8.104), i.e.

$$\tilde{\boldsymbol{\psi}}^* = \frac{1}{1 + \mathrm{tr}\mathbf{R}^*}(\mathbf{R}^* - \mathbf{R}^{*T}), \tag{8.122}$$

and, finally, we calculate the rotation vector $\boldsymbol{\psi}^* = \frac{1}{2}(\mathbf{I} \times \tilde{\boldsymbol{\psi}}^*)$. As a result, $\boldsymbol{\psi}^*$ is a function of $\boldsymbol{\psi}$ which defines the transformation rule for the semi-tangential vector.

Below, we inspect the above-obtained expression for the skew-symmetric $\tilde{\boldsymbol{\psi}}^*$ (instead of its axial vector $\boldsymbol{\psi}^*$), and show that

$$\tilde{\boldsymbol{\psi}}^* = \tilde{\boldsymbol{\psi}}^R, \tag{8.123}$$

where $\tilde{\boldsymbol{\psi}}^R \doteq \mathbf{R}_0^T \tilde{\boldsymbol{\psi}}\,\mathbf{R}_0 \in \mathrm{so}(3)$ is obtained by a standard transformation rule for tensors. Using eq. (8.99), we obtain

$$\mathbf{R}^* = \mathbf{R}_0^T \mathbf{R} \mathbf{R}_0 = \mathbf{I} + A\left[\tilde{\boldsymbol{\psi}}^R + (\tilde{\boldsymbol{\psi}}^R)^2\right], \tag{8.124}$$

where $A \doteq 2/(1+t^2)$, and $t^2 \doteq \boldsymbol{\psi} \cdot \boldsymbol{\psi} = \frac{1}{2}\tilde{\boldsymbol{\psi}} \cdot \tilde{\boldsymbol{\psi}}$. Hence, $\mathbf{R}^* - \mathbf{R}^{*T} = 2A\,\tilde{\boldsymbol{\psi}}^R$, and eq. (8.122) becomes

$$\tilde{\psi}^* = \frac{2A}{1 + \mathrm{tr}\mathbf{R}^*}\, \tilde{\psi}^R. \tag{8.125}$$

To prove that $\tilde{\psi}^* = \tilde{\psi}^R$, we show that the fraction in front of $\tilde{\psi}^R$ is equal to 1. Note that

$$\mathrm{tr}\mathbf{R}^* = 3 - \mathrm{A}\tilde{\psi}^R \cdot \tilde{\psi}^R = 3 - \mathrm{A}\tilde{\psi} \cdot \tilde{\psi}, \tag{8.126}$$

using the property that the trace of a tensor is the first invariant, i.e. $\tilde{\psi}^R \cdot \tilde{\psi}^R = -\mathrm{tr}(\tilde{\psi}^R)^2 = -\mathrm{tr}(\mathbf{R}_0^{\mathrm{T}} \tilde{\psi}^2 \mathbf{R}_0) = -\mathrm{tr}\tilde{\psi}^2 = \tilde{\psi} \cdot \tilde{\psi}$. Then

$$A = \frac{4}{2 + \tilde{\psi} \cdot \tilde{\psi}}, \qquad A\tilde{\psi}^R \cdot \tilde{\psi}^R = \frac{4\tilde{\psi} \cdot \tilde{\psi}}{2 + \tilde{\psi} \cdot \tilde{\psi}},$$

$$1 + \mathrm{tr}\mathbf{R}^* = 4 - \frac{4\tilde{\psi} \cdot \tilde{\psi}}{2 + \tilde{\psi} \cdot \tilde{\psi}} = \frac{8}{2 + \tilde{\psi} \cdot \tilde{\psi}}$$

and we obtain

$$\frac{2A}{1 + \mathrm{tr}\mathbf{R}^*} = \frac{8}{2 + \tilde{\psi} \cdot \tilde{\psi}}\, \frac{2 + \tilde{\psi} \cdot \tilde{\psi}}{8} = 1, \tag{8.127}$$

which ends the proof. □

Because the skew-symmetric $\tilde{\psi}$ transforms as an ordinary tensor, we can conclude that its axial vector ψ transforms as an ordinary vector.

8.2.5 Three parameters: Euler angles

The parametrization based on Euler angles is classical and has been used in many problems of mechanics; e.g. to describe rotations of a shell director in the shell theories with two rotational parameters, see Sect. 8.2.6.

Elementary rotation. The elementary rotation is defined as a rotation about a chosen vector of an ortho-normal basis. For instance, the elementary rotation about the axis $\mathbf{t}_3$ by the angle ω_3 is defined as follows: $\mathbf{e} \doteq \mathbf{t}_3$ and $\omega \doteq \omega_3$ and denoted as $\mathbf{R}(\omega_3, \mathbf{t}_3)$. For convenience, we choose $\mathbf{e}_1 \doteq \mathbf{t}_1$ and $\mathbf{e}_2 \doteq \mathbf{t}_2$. Then the skew-symmetric tensor, eq. (8.7), and the rotation tensor, eq. (8.9), are expressed as

$$\mathbf{S} = \mathbf{t}_2 \otimes \mathbf{t}_1 - \mathbf{t}_1 \otimes \mathbf{t}_2, \tag{8.128}$$

$$\mathbf{R}(\omega_3, \mathbf{t}_3) = \cos\omega_3(\mathbf{t}_1 \otimes \mathbf{t}_1 + \mathbf{t}_2 \otimes \mathbf{t}_2) + \sin\omega_3(\mathbf{t}_2 \otimes \mathbf{t}_1 - \mathbf{t}_1 \otimes \mathbf{t}_2) + \mathbf{t}_3 \otimes \mathbf{t}_3, \tag{8.129}$$

or, in terms of components in $\{\mathbf{t}_k\}$,

$$(\mathbf{S})_{ij} = \begin{bmatrix} 0 & -1 & 0 \\ +1 & 0 & 0 \\ 0 & 0 & 0 \end{bmatrix}, \qquad (\mathbf{R})_{ij} = \begin{bmatrix} \cos\omega_3 & -\sin\omega_3 & 0 \\ +\sin\omega_3 & \cos\omega_3 & 0 \\ 0 & 0 & 1 \end{bmatrix}. \tag{8.130}$$

The sign of ω_3 is positive for the "shortest" rotation of $\mathbf{t}_1$ to $\mathbf{t}_2$. The elementary rotations about the axes $\mathbf{t}_1$ to $\mathbf{t}_2$ are defined in an analogous way, permuting the indices.

Because each elementary rotation introduces one rotation angle, the rotation matrix can be obtained as a multiplication of three elementary rotations; the most popular are the Euler angles (3–1–3 Euler angles), and Brayant or Cardan angles (1–2–3 Euler angles). A form of the resulting rotation matrix depends on the sequence in which the elementary rotations are performed and in this sense it is not invariant. A systematic list of 24 sequences of rotations is given, e.g., in [122] and further comments are provided in [90] p. 113.

This type of parametrization of rotations is well suited for motion constrained by hinges such as, e.g., of a gyroscope in gimbals. In the absence of physical hinges, the choice of the axes of rotations is arbitrary and not related to properties of the motion. For such problems, this type of parametrization is not considered as optimal, and for free rigid body rotation, it leads to singularities of the equation of motion, even for physically sound conditions (see [231], p. 427). Such difficulties can be overcome to a certain extent by an update of rotations.

Rotation tensor $\mathbf{R}$ for Euler angles. Consider the following (left) sequence of the elementary rotations

$$\mathbf{R} = \mathbf{R}_3(-\omega_3, \mathbf{t}_3'') \, \mathbf{R}_2(\omega_2, \mathbf{t}_2') \, \mathbf{R}_1(-\omega_1, \mathbf{t}_1), \tag{8.131}$$

where ω_1, ω_2, and ω_3 are three Euler angles. The rotated bases, designated by the prime and the double prime, are used in this definition and they are

$$\mathbf{t}_k' = \mathbf{R}_1(-\omega_1, \mathbf{t}_1)\, \mathbf{t}_k, \qquad \mathbf{t}_k'' = \mathbf{R}_2(\omega_2, \mathbf{t}_2')\, \mathbf{t}_k', \qquad k = 1, 2, 3. \tag{8.132}$$

In order to perform the elementary rotations about the axes of the initial basis $\{\mathbf{t}_k\}$, we can use the right sequence of rotations, see Sect. 8.3.1, which yields

$$\mathbf{R} = \mathbf{R}_1(-\omega_1, \mathbf{t}_1)\ \mathbf{R}_2^*(\omega_2, \mathbf{t}_2)\ \mathbf{R}_3^*(-\omega_3, \mathbf{t}_3). \tag{8.133}$$

We can check that the resulting matrix of components for these two sequences of rotations is identical. In terms of components in $\{\mathbf{t}_k\}$, for the right sequence, we have

$$(\mathbf{R})_{ij} = \begin{bmatrix} 1 & 0 & 0 \\ 0 & c_1 & s_1 \\ 0 & -s_1 & c_1 \end{bmatrix} \begin{bmatrix} c_2 & 0 & s_2 \\ 0 & 1 & 0 \\ -s_2 & 0 & c_2 \end{bmatrix} \begin{bmatrix} c_3 & s_3 & 0 \\ -s_3 & c_3 & 0 \\ 0 & 0 & 1 \end{bmatrix}, \tag{8.134}$$

where $s_i \doteq \sin\omega_i$ and $c_i \doteq \cos\omega_i$. After multiplication, we obtain

$$(\mathbf{R})_{ij} = \begin{bmatrix} c_2 c_3 & c_2 s_3 & s_2 \\ -c_3 s_1 s_2 - c_1 s_3 & c_1 c_3 - s_1 s_2 s_3 & c_2 s_1 \\ -c_1 c_3 s_2 + s_1 s_3 & -c_3 s_1 - c_1 s_2 s_3 & c_1 c_2 \end{bmatrix} \tag{8.135}$$

and it can be checked that indeed $(\mathbf{R}^T)_{ij}(\mathbf{R})_{ij} = \delta_{ij}$ and $\det(\mathbf{R})_{ij} = 1$.

Extraction of Euler angles from rotation matrix. The values of sine and cosine functions s_i and c_i can be extracted from R_{ij} in the following way:

1. Note that $R_{11} = c_2 c_3$ and $R_{12} = c_2 s_3$, so $R_{11}^2 + R_{12}^2 = c_2^2$. Besides, $R_{13} = s_2$. Therefore,

$$c_2 = \sqrt{R_{11}^2 + R_{12}^2}, \qquad s_2 = R_{13}. \tag{8.136}$$

2. Upon inserting the above relations into the first row of $\mathbf{R}$, we have $c_3\sqrt{R_{11}^2 + R_{12}^2} = R_{11}$ and $s_3\sqrt{R_{11}^2 + R_{12}^2} = R_{12}$, from which, if $\sqrt{R_{11}^2 + R_{12}^2} \neq 0$, we can calculate

$$c_3 = \frac{R_{11}}{\sqrt{R_{11}^2 + R_{12}^2}}, \qquad s_3 = \frac{R_{12}}{\sqrt{R_{11}^2 + R_{12}^2}}. \tag{8.137}$$

3. Similarly, for the last column of $\mathbf{R}$, we have $s_1\sqrt{R_{11}^2 + R_{12}^2} = R_{23}$ and $c_1\sqrt{R_{11}^2 + R_{12}^2} = R_{33}$, from which we calculate

$$c_1 = \frac{R_{33}}{\sqrt{R_{11}^2 + R_{12}^2}}, \qquad s_1 = \frac{R_{23}}{\sqrt{R_{11}^2 + R_{12}^2}}. \tag{8.138}$$

In the update scheme, we can directly use s_i and c_i, without the recovery of angles.

Having s_i and c_i of eqs. (8.136)–(8.138), we can uniquely determine the Euler angles ω_1, ω_2, and ω_3 in $[-\pi, +\pi]$. For instance, for ω_1, we can use

$$\omega_1 = \text{Sign}(\alpha_1)\ \alpha_2 \ \in \ [-\pi, +\pi], \tag{8.139}$$

where $\alpha_1 \doteq \arcsin s_1 \in [-\pi/2, +\pi/2]$ and $\alpha_2 \doteq \arccos c_1 \in [0, +\pi]$, see Fig. 8.13. Similar formulas can be used for ω_2 and ω_3.

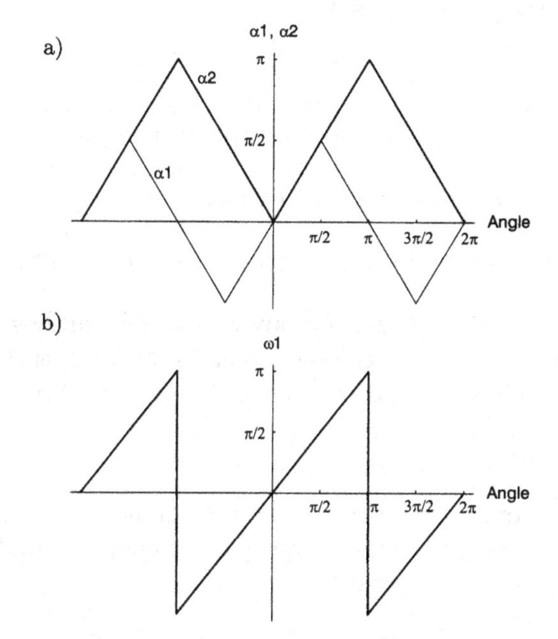

Fig. 8.13 Euler angles. a) $\alpha_1 \in [-\pi/2, +\pi/2]$, $\alpha_2 \in [0, +\pi]$. b) $\omega_1 \in [-\pi, +\pi]$.

Non-uniqueness of Euler angles for $\omega_2 = \pm\pi/2$. Note that eq. (8.137) and (8.138) are derived for the assumption that $\sqrt{R_{11}^2 + R_{12}^2} \neq 0$, which is equivalent to $c_2 \neq 0$, by eq. (8.136). Hence, we have to consider the case $c_2 = 0$ separately; we proceed similarly as in [231].

Note that $c_2 \doteq \cos\omega_2 = 0$ for $\omega_2 = \pm\pi/2$, if $|\omega_2| \leq \pi$. For $\omega_2 = +\pi/2$, we have $s_2 = 1$ and then

$$(\mathbf{R})_{ij} = \begin{bmatrix} 0 & 0 & -1 \\ \sin(\omega_1 - \omega_3) & \cos(\omega_1 - \omega_3) & 0 \\ \cos(\omega_1 - \omega_3) & -\sin(\omega_1 - \omega_3) & 0 \end{bmatrix}, \tag{8.140}$$

where we used $c_3 s_1 - c_1 s_3 = \sin(\omega_1 - \omega_3)$ and $c_3 c_1 + s_3 s_1 = \cos(\omega_1 - \omega_3)$. This form of $\mathbf{R}$ implies four equations:

$$\sin(\omega_1 - \omega_3) = R_{21}, \qquad \sin(\omega_1 - \omega_3) = -R_{32}, \tag{8.141}$$

$$\cos(\omega_1 - \omega_3) = R_{22}, \qquad \cos(\omega_1 - \omega_3) = R_{31}. \tag{8.142}$$

The above pairs of equations are not contradictory, which can be checked by considering the representation

$$(\mathbf{R})_{ij} = \begin{bmatrix} 0 & 0 & -1 \\ R_{21} & R_{22} & 0 \\ R_{31} & R_{32} & 0 \end{bmatrix}. \tag{8.143}$$

The orthogonality conditions, eq. (8.1), yield

$$R_{21}R_{22} + R_{31}R_{32} = 0, \qquad R_{21}^2 + R_{31}^2 = 1, \qquad R_{22}^2 + R_{32}^2 = 1 \tag{8.144}$$

and for R_{21} and R_{22} specified as above, these conditions yield $R_{32} = -R_{21}$ and $R_{31} = R_{22}$, i.e. the same relation between entries of $\mathbf{R}$ as in eq. (8.140). Thus, instead of eqs. (8.141) and (8.142), it suffices to consider

$$\sin(\omega_1 - \omega_3) = R_{21}, \qquad \cos(\omega_1 - \omega_3) = R_{22}. \tag{8.145}$$

From these two equations we can find the difference of angles $(\omega_1 - \omega_3)$ but ω_1 and ω_3 cannot be not uniquely determined. A similar reasoning for the case $\omega_2 = -\pi/2$, yields

$$\sin(\omega_1 + \omega_3) = -R_{21}, \qquad \cos(\omega_1 + \omega_3) = R_{22}, \tag{8.146}$$

from which we can find the sum of angles $(\omega_1 + \omega_3)$ but not ω_1 and ω_3, separately. Hence, we conclude that we cannot uniquely extract the rotation angles from $\mathbf{R}$ for $\omega_2 = \pm\pi/2$. Finally, we note that in the incremental formulations, we can use the Euler angles for the step only and limit the step size to remain below this value.

8.2.6 Two parameters: constrained rotations of shell director

Consider the case when a rotation of the shell director $\mathbf{t}_3$ is constrained in such a way that the drilling rotation (around $\mathbf{t}_3$) is excluded. This case is important for shells based on Reissner kinematics which have five dofs/node.

A. Canonical rotation vector

The constrained rotation can be parameterized using the rotation vector $\boldsymbol{\psi}$, and requiring that the axis of rotation $\mathbf{e}$ is perpendicular to the shell director, i.e. $\mathbf{e} \cdot \mathbf{t}_3 = 0$, see Fig. 8.14.

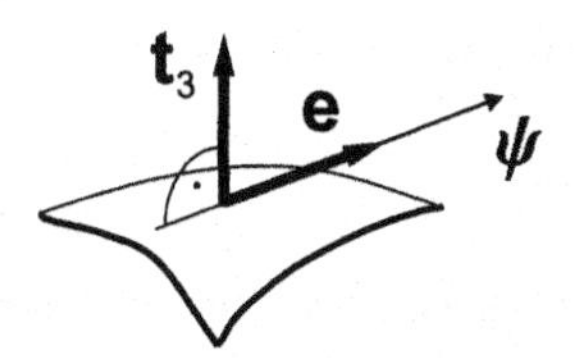

Fig. 8.14 Constrained rotation: the axis of rotation is perpendicular to the director.

For the constrained rotation we have $\psi_3 \doteq \boldsymbol{\psi} \cdot \mathbf{t}_3 = \omega \, \mathbf{e} \cdot \mathbf{t}_3 = 0$, so $\boldsymbol{\psi} = [\psi_1, \psi_2, 0]^T$. Hence, only two components of the rotation vector remain nonzero in the basis $\{\mathbf{t}_i\}$. The rotation tensor of eq. (8.9) can be rewritten as

$$\mathbf{R} = \cos\omega \, \mathbf{I} + \sin\omega \, (\mathbf{e} \times \mathbf{I}) + (1 - \cos\omega) \, \mathbf{e} \otimes \mathbf{e}, \tag{8.147}$$

where $\mathbf{S} = \tilde{\mathbf{e}} \doteq \mathbf{e} \times \mathbf{I}$ and $\mathbf{S}^2 = \mathbf{e} \otimes \mathbf{e} - \mathbf{I}$, see Table 8.1. The position of the rotated shell director is given as

$$\mathbf{a}_3 \doteq \mathbf{R} \, \mathbf{t}_3 = \cos\omega \, \mathbf{t}_3 + \sin\omega (\mathbf{e} \times \mathbf{t}_3), \tag{8.148}$$

where the term $(\mathbf{e} \otimes \mathbf{e}) \, \mathbf{t}_3 = \mathbf{0}$, because $\mathbf{e} \cdot \mathbf{t}_3 = 0$. For the canonical rotation vector, $\boldsymbol{\psi} \doteq \omega \, \mathbf{e}$, we obtain

$$\mathbf{a}_3 = \cos \|\boldsymbol{\psi}\| \, \mathbf{t}_3 + \frac{\sin \|\boldsymbol{\psi}\|}{\|\boldsymbol{\psi}\|} (\boldsymbol{\psi} \times \mathbf{t}_3). \tag{8.149}$$

To grasp the geometrical meaning of this equation, we can take $\boldsymbol{\psi}/\|\boldsymbol{\psi}\| = \mathbf{t}_2$, for which $\mathbf{a}_3 = \cos \|\boldsymbol{\psi}\| \, \mathbf{t}_3 + \sin \|\boldsymbol{\psi}\| \, \mathbf{t}_1$, see Fig. 8.15a, where $\omega \doteq \|\boldsymbol{\psi}\|$.

Note that $\|\boldsymbol{\psi} \times \mathbf{t}_3\| = \|\boldsymbol{\psi}\|$, e.g. by the identity (4.9.11) of [33]. Then, eq. (8.149) becomes

$$\mathbf{a}_3 = \cos \|\boldsymbol{\psi} \times \mathbf{t}_3\| \, \mathbf{t}_3 + \frac{\sin \|\boldsymbol{\psi} \times \mathbf{t}_3\|}{\|\boldsymbol{\psi} \times \mathbf{t}_3\|} (\boldsymbol{\psi} \times \mathbf{t}_3), \tag{8.150}$$

in which we have only one argument, $\boldsymbol{\psi} \times \mathbf{t}_3$. We stress that this simple form can be obtained only for two-parameter rotations!

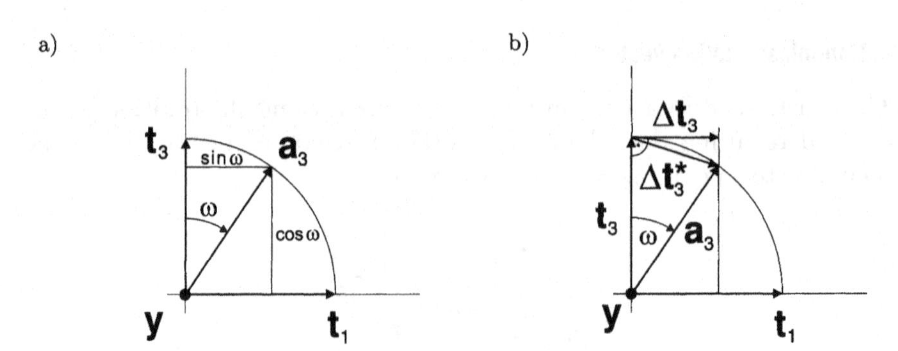

Fig. 8.15 Rotated director for two-parameter rotations.
a) Components of $\mathbf{a}_3$, b) Increment $\Delta\mathbf{t}_3^*$ and its projection $\Delta\mathbf{t}_3$.

Director update scheme. Denote $\boldsymbol{\psi} \times \mathbf{t}_3 \doteq \Delta\mathbf{t}_3$. Note that $\Delta\mathbf{t}_3$ is not the full increment of the director, $\Delta\mathbf{t}_3^*$, but only its projection on the $\mathbf{t}_1 - \mathbf{t}_2$ plane, see Fig. 8.15b. In the basis $\{\mathbf{t}_i\}$, only two first components of $\Delta\mathbf{t}_3$ are nonzero. For the so-defined $\Delta\mathbf{t}_3$, eq. (8.150) can be used as the director update formula

$$\mathbf{t}_3^{k+1} = \cos\|\Delta\mathbf{t}_3\|\,\mathbf{t}_3^k + \frac{\sin\|\Delta\mathbf{t}_3\|}{\|\Delta\mathbf{t}_3\|}\Delta\mathbf{t}_3, \tag{8.151}$$

where $\|\boldsymbol{\psi} \times \mathbf{t}_3\| = \|\Delta\mathbf{t}_3\|$ and k and $k+1$ are the iteration indices. Besides, by $\omega \doteq \|\boldsymbol{\psi}\| = \|\Delta\mathbf{t}_3\|$, eq. (8.147) can be used to define the increment of rotation tensor,

$$\Delta\mathbf{R} = \cos\|\Delta\mathbf{t}_3\|\,\mathbf{I} + \sin\|\Delta\mathbf{t}_3\|\,(\mathbf{e}\times\mathbf{I}) + (1-\cos\|\Delta\mathbf{t}_3\|)\,\mathbf{e}\otimes\mathbf{e}, \tag{8.152}$$

where

$$\mathbf{e} \doteq \frac{\boldsymbol{\psi}}{\|\boldsymbol{\psi}\|} = \mathbf{t}_3^k \times \frac{\Delta\mathbf{t}_3}{\|\Delta\mathbf{t}_3\|}.$$

Hence, $\mathbf{t}_3^{k+1}$ and $\Delta\mathbf{R}$ are expressed in terms of $\Delta\mathbf{t}_3$ and the known $\mathbf{t}_3^k$. The current rotation is computed by the left rule as $\mathbf{R}^{k+1} = \Delta\mathbf{R}\,\mathbf{R}^k$.

Equations (8.151) and (8.152) are used in [211], see BOX 2 therein. In this paper, the directors are updated at nodes and then are interpolated within an element, which implies that at the Gauss points we have $\|\mathbf{t}_3^{k+1}\| \neq 1$. But no loss of accuracy due to this feature was noted in the computed examples, see Example 4.1 therein.

Rotation tensor for given $\mathbf{t}_3$ and $\mathbf{a}_3$. The vectors $\boldsymbol{\psi}$ and $\mathbf{t}_3$ are perpendicular and the resulting vector $\mathbf{a}_3 = \mathbf{R}\mathbf{t}_3$ is perpendicular to $\boldsymbol{\psi}$. Hence, for $\mathbf{t}_3$ and $\mathbf{a}_3$ known, we have

$$\mathbf{t}_3 \cdot \mathbf{a}_3 = \cos\omega, \qquad \mathbf{t}_3 \times \mathbf{a}_3 = \sin\omega\, \mathbf{e}, \tag{8.153}$$

where ω is the angle of rotation. To avoid the problem with signs, we choose the axis of rotation as follows:

$$\mathbf{e} = \frac{\mathbf{t}_3 \times \mathbf{a}_3}{\|\mathbf{t}_3 \times \mathbf{a}_3\|}. \tag{8.154}$$

The rotation tensor $\mathbf{R}$ for this constrained problem can be found from eq. (8.147) by noting that the particular terms are now as follows:

$$\sin\omega(\mathbf{e} \times \mathbf{I}) = (\mathbf{t}_3 \times \mathbf{a}_3) \times \mathbf{I}, \quad (1 - \cos\omega)\, \mathbf{e} \otimes \mathbf{e} = A\, (\mathbf{t}_3 \times \mathbf{a}_3) \otimes (\mathbf{t}_3 \times \mathbf{a}_3), \tag{8.155}$$

where

$$A \doteq \frac{1 - \cos\omega}{\|\mathbf{t}_3 \times \mathbf{a}_3\|^2} = \frac{1 - \cos\omega}{\sin^2\omega} = \frac{1 - \cos\omega}{1 - \cos^2\omega} = \frac{1}{1 + \cos\omega} = \frac{1}{1 + (\mathbf{t}_3 \cdot \mathbf{a}_3)}. \tag{8.156}$$

Inserting the above formulas into eq. (8.147), we obtain

$$\mathbf{R} = (\mathbf{t}_3 \cdot \mathbf{a}_3)\mathbf{I} + (\mathbf{t}_3 \times \mathbf{a}_3) \times \mathbf{I} + \frac{1}{1 + (\mathbf{t}_3 \cdot \mathbf{a}_3)} (\mathbf{t}_3 \times \mathbf{a}_3) \otimes (\mathbf{t}_3 \times \mathbf{a}_3), \tag{8.157}$$

which is singular for $(\mathbf{t}_3 \cdot \mathbf{a}_3) = -1$, i.e. for $\mathbf{a}_3 = -\mathbf{t}_3$ or $\omega = \pi$. Such an expression for the constrained rotation is obtained, e.g., in [213], eq. (2.15).

Components of $\mathbf{R}$ *in basis* $\{\mathbf{t}_i\}$. Denote $c_i \doteq \mathbf{a}_3 \cdot \mathbf{t}_i$ and specify particular terms of eq. (8.157) as follows:

$$\mathbf{t}_3 \times \mathbf{a}_3 = \mathbf{t}_3 \times (c_i \mathbf{t}_i) = c_1 \mathbf{t}_3 \times \mathbf{t}_1 + c_2 \mathbf{t}_3 \times \mathbf{t}_2 = c_1 \mathbf{t}_2 - c_2 \mathbf{t}_1,$$
$$(\mathbf{t}_3 \times \mathbf{a}_3) \otimes (\mathbf{t}_3 \times \mathbf{a}_3) = c_2^2 \mathbf{t}_1 \otimes \mathbf{t}_1 + c_1^2 \mathbf{t}_2 \otimes \mathbf{t}_2 - c_1 c_2 (\mathbf{t}_1 \otimes \mathbf{t}_2 + \mathbf{t}_2 \otimes \mathbf{t}_1).$$

Besides, noting that for $\boldsymbol{\psi} = \psi_i\, \mathbf{t}_i$ we have

$$\tilde{\boldsymbol{\psi}} = \boldsymbol{\psi} \times \mathbf{I} = \psi_1(\mathbf{t}_3 \otimes \mathbf{t}_2 - \mathbf{t}_2 \otimes \mathbf{t}_3) + \psi_2(\mathbf{t}_1 \otimes \mathbf{t}_3 - \mathbf{t}_3 \otimes \mathbf{t}_1) + \psi_3(\mathbf{t}_2 \otimes \mathbf{t}_1 - \mathbf{t}_1 \otimes \mathbf{t}_2), \tag{8.158}$$

we can write

$$(\mathbf{t}_3 \times \mathbf{a}_3) \times \mathbf{I} = (-c_2 \mathbf{t}_1 + c_1 \mathbf{t}_2) \times \mathbf{I} = -c_2(\mathbf{t}_3 \otimes \mathbf{t}_2 - \mathbf{t}_2 \otimes \mathbf{t}_3) + c_1(\mathbf{t}_1 \otimes \mathbf{t}_3 - \mathbf{t}_3 \otimes \mathbf{t}_1). \tag{8.159}$$

Hence,

$$(\mathbf{R})_{ij} = \begin{bmatrix} c_3 + A c_2^2 & -c_1 c_2 & c_1 \\ c_1 c_2 & c_3 + A c_1^2 & c_2 \\ -c_1 & -c_2 & c_3 \end{bmatrix}, \qquad A = \frac{1}{1 + c_3}, \tag{8.160}$$

where in the last column of the matrix we have the components of $\mathbf{a}_3$ in the basis $\{\mathbf{t}_i\}$.

Example. Numerical solution for rotation around known axis. Let us assume that the vectors $\mathbf{a}$ and $\mathbf{b} = \mathbf{Ra}$ are known. For simplicity, we take the axis of rotation $\mathbf{e} \doteq \mathbf{t}_3$, and $\mathbf{a}$ and $\mathbf{b}$ which belong to the plane spanned by $\mathbf{t}_1$ and $\mathbf{t}_2$. The governing equation can be written as

$$\mathbf{r} \doteq \mathbf{Ra} - \mathbf{b} = \mathbf{0}, \qquad \mathbf{R} = \mathbf{R}(\omega, \mathbf{t}_3), \tag{8.161}$$

where $\mathbf{r}$ is a residual. This equation constitutes a set of two nonlinear equations for one unknown, ω. Note that, due to periodicity of the sine and cosine functions, each of the equations has multiple solutions; of interest are only those which satisfy both equations. The analytical solution of this problem is

$$\omega = \arccos(\mathbf{a} \cdot \mathbf{b}) + 2k\pi, \tag{8.162}$$

but we want to obtain it by using the Newton method combined with the penalty method. In terms of components, we have

$$\mathbf{a} = a_1\mathbf{t}_1 + a_2\mathbf{t}_2, \qquad \mathbf{b} = b_1\mathbf{t}_1 + b_2\mathbf{t}_2, \tag{8.163}$$

$$\mathbf{S} = \mathbf{t}_2 \otimes \mathbf{t}_1 - \mathbf{t}_1 \otimes \mathbf{t}_2, \qquad \mathbf{S}^2 = -\mathbf{I}, \qquad \mathbf{R} = \cos\omega\mathbf{I} + \sin\omega\mathbf{S}. \tag{8.164}$$

The Newton scheme can be written as

$$D\mathbf{g} \cdot \Delta\omega = -\mathbf{g}, \qquad \omega = \bar{\omega} + \Delta\omega, \tag{8.165}$$

where the differential of $\mathbf{r}$ is $D\mathbf{r} \cdot \Delta\omega = (D\mathbf{R} \cdot \Delta\omega)\mathbf{a}$. For

$$D\mathbf{R} \cdot \Delta\omega = \frac{\mathrm{d}\mathbf{R}}{\mathrm{d}\omega}\Delta\omega, \qquad \frac{\mathrm{d}\mathbf{R}}{\mathrm{d}\omega} = -\sin\omega\mathbf{I} + \cos\omega\mathbf{S}, \tag{8.166}$$

we have

$$D\mathbf{r} \cdot \Delta\omega = (-\sin\omega\,\mathbf{I} + \cos\omega\,\mathbf{S})\mathbf{a}\,\Delta\omega = \mathbf{SRa}\,\Delta\omega, \tag{8.167}$$

where the identity $\mathbf{S}^2 = -\mathbf{I}$ and the definition of $\mathbf{R}$ are used to obtain the last form. Note that $\mathbf{SR}$ defines the tangent plane at $\mathbf{R}$ and $\mathbf{SRa} = \mathbf{S}\bar{\mathbf{a}}$, where $\bar{\mathbf{a}} = \mathbf{Ra}$ is the updated (rotated) $\mathbf{a}$.

We use the penalty method with $\alpha = 10^{-10}$ and the strategy of choosing the sequence in which the equations are solved, which is based on a comparison of residuals. To establish the convergence of this method, the vector $\mathbf{a}$ is taken at 360 uniformly spaced initial locations, differing from $\mathbf{b}$ by the angle from 0° to 360°.

For $\mathbf{b} = [0, 1]^T$ and the tolerance $\tau = 10^{-10}$, the convergence is attained on average in about 2.45 iteration per one starting point, with the maximum of seven iterations. Such a fast convergence confirms that the method has been properly implemented. The same strategy but with $\alpha = 0$ yields divergence for several starting points.

B. Euler angles

A survey of the literature indicates that the Euler angles dominate in the class of two-parameter rotations. Then the rotation tensor is assumed as a composition of two elementary rotations, each defined as in eq. (8.129).

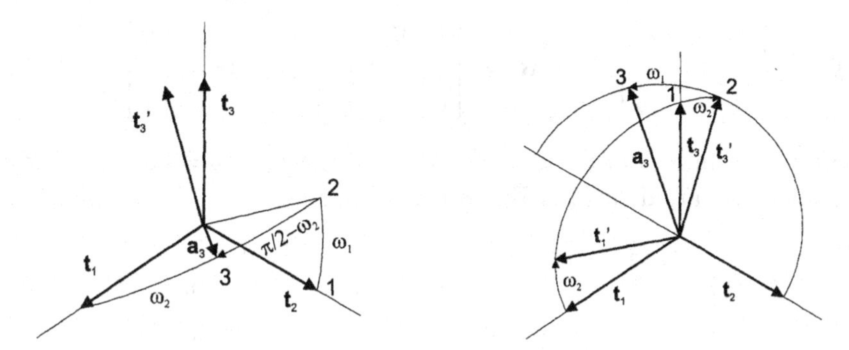

Fig. 8.16 Rotation of shell director: a) in [186], b) in [262].

a. In [186], the following formula is used to describe a position of the rotated shell director

$$\mathbf{a}_3 = \mathbf{R}_2(\pi/2 - \omega_2, \mathbf{t}_3') \, \mathbf{R}_1(\omega_1, \mathbf{t}_1) \, \mathbf{t}_2 = \mathbf{R}_1(\omega_1, \mathbf{t}_1) \, \mathbf{R}_2^*(\pi/2 - \omega_2, \mathbf{t}_3) \, \mathbf{t}_2, \tag{8.168}$$

where $\mathbf{t}_3' = \mathbf{R}_1(\omega_1, \mathbf{t}_1) \, \mathbf{t}_3$ and the angles ω_1 and ω_2 are shown in Fig. 8.16a. In the basis $\{\mathbf{t}_i\}$, the components are

$$\mathbf{R}_1 = \begin{bmatrix} 1 & 0 & 0 \\ 0 & c_1 & -s_1 \\ 0 & s_1 & c_1 \end{bmatrix}, \qquad \mathbf{R}_2^* = \begin{bmatrix} c_2 & s_2 & 0 \\ -s_2 & c_2 & 0 \\ 0 & 0 & 1 \end{bmatrix}, \qquad \mathbf{t}_2 = \begin{bmatrix} 0 \\ 1 \\ 0 \end{bmatrix}, \tag{8.169}$$

where $s_1 \doteq \sin \omega_1$, $c_1 \doteq \cos \omega_1$, $s_2 \doteq \sin(\pi/2 - \omega_2)$, $c_2 \doteq \cos(\pi/2 - \omega_2)$. On use of the trigonometric identities $\sin(\pi/2 - \omega_2) = \cos \omega_2$ and $\cos(\pi/2 - \omega_2) = \sin \omega_2$, and by a multiplication, we obtain

$$\mathbf{R}_1\mathbf{R}_2^* = \begin{bmatrix} s_2 & c_2 & 0 \\ -c_1c_2 & c_1s_2 & -s_1 \\ -s_1c_2 & s_1s_2 & c_1 \end{bmatrix}, \qquad \mathbf{a}_3 = \begin{bmatrix} c_2 \\ c_1s_2 \\ s_1s_2 \end{bmatrix}, \tag{8.170}$$

where $s_2 \doteq \sin\omega_2$ and $c_2 \doteq \cos\omega_2$.

b. Alternatively, the director can be related to the normal vector $\mathbf{t}_3$, see, e.g., in [262],

$$\mathbf{a}_3 = \mathbf{R}_2(\omega_1, \mathbf{t}_1')\,\mathbf{R}_1(\omega_2, \mathbf{t}_2)\,\mathbf{t}_3 = \mathbf{R}_1(\omega_2, \mathbf{t}_2)\,\mathbf{R}_2^*(\omega_1, \mathbf{t}_1)\,\mathbf{t}_3, \tag{8.171}$$

where $\mathbf{t}_1' = \mathbf{R}_1(\omega_2, \mathbf{t}_2)\,\mathbf{t}_1$. In the basis $\{\mathbf{t}_i\}$, the components are

$$\mathbf{R}_1 = \begin{bmatrix} c_2 & 0 & -s_2 \\ 0 & 1 & 0 \\ s_2 & 0 & c_2 \end{bmatrix}, \qquad \mathbf{R}_2^* = \begin{bmatrix} 1 & 0 & 0 \\ 0 & c_1 & -s_1 \\ 0 & s_1 & c_1 \end{bmatrix}, \qquad \mathbf{t}_3 = \begin{bmatrix} 0 \\ 0 \\ 1 \end{bmatrix}. \tag{8.172}$$

Here $s_1 \doteq \sin\omega_1$, $c_1 \doteq \cos\omega_1$, and $s_2 \doteq \sin\omega_2$, $c_2 \doteq \cos\omega_2$, where ω_1 and ω_2 are defined in Fig. 8.16b. Then

$$\mathbf{R}_1\mathbf{R}_2^* = \begin{bmatrix} c_2 & -s_1s_2 & -c_1s_2 \\ 0 & c_1 & -s_1 \\ s_2 & s_1c_2 & c_1c_2 \end{bmatrix}, \qquad \mathbf{a}_3 = \begin{bmatrix} -c_1s_2 \\ -s_1 \\ c_1c_2 \end{bmatrix}. \tag{8.173}$$

8.3 Composition of rotations

In this section, we consider the composition of rotation tensors and the left and right sequences of rotations are introduced. Then the corresponding formulas are derived for rotational parameters, i.e. the quaternions and the semi-tangential rotation vector. For the canonical rotation vector, the procedure based on quaternions is described instead.

8.3.1 Composition of rotation tensors

Composition of orthogonal tensors yields an orthogonal tensor. If $\mathbf{R}_1, \mathbf{R}_2 \in$ SO(3), then $\mathbf{R}_3 \doteq \mathbf{R}_2\,\mathbf{R}_1 \in$ SO(3). The proof is immediate:

$$\mathbf{R}_3^T\mathbf{R}_3 = (\mathbf{R}_2\mathbf{R}_1)^T(\mathbf{R}_2\mathbf{R}_1) = \mathbf{R}_1^T(\mathbf{R}_2^T\mathbf{R}_2)\mathbf{R}_1 = \mathbf{R}_1^T\mathbf{I}\,\mathbf{R}_1 = \mathbf{I}, \tag{8.174}$$

$$\det\mathbf{R}_3 = \det(\mathbf{R}_2\mathbf{R}_1) = (\det\mathbf{R}_2)(\det\mathbf{R}_1) = +1, \tag{8.175}$$

where, for the determinant of a product of tensors, we used eq. (4.9.98) of [33].

Left and right sequence of rotations. Let us rotate a vector from the position "0" to the position "1", $\mathbf{v}_1 = \mathbf{R}_1\mathbf{v}$, and then from the position "1" to the position "2", $\mathbf{v}_2 = \mathbf{R}_2^*\mathbf{v}_1$, where $\mathbf{R}_1, \mathbf{R}_2^* \in \mathrm{SO}(3)$. The final position is $\mathbf{v}_2 = \mathbf{R}_2^*\mathbf{v}_1 = \mathbf{R}_2^*(\mathbf{R}_1\mathbf{v}) = (\mathbf{R}_2^*\mathbf{R}_1)\mathbf{v}$. Hence, we can define the total rotation tensor as a multiplicative left composition of the component rotation tensors, i.e.

$$\mathbf{R}_t = \mathbf{R}_2^*\,\mathbf{R}_1. \tag{8.176}$$

This composition is not commutative, i.e. $\mathbf{R}_2^*\mathbf{R}_1 \neq \mathbf{R}_1\mathbf{R}_2^*$, except for the case of rotations performed around the same axis when $\mathbf{R}_1$ and $\mathbf{R}_2^*$ have the same eigenpair $\{+1, \mathbf{e}\}$.

To have the right composition rule, we define another rotation tensor $\mathbf{R}_2$, such that it yields the same total rotation when applied to $\mathbf{R}_1$ from the right, i.e.

$$\mathbf{R}_2 \in \mathrm{SO}(3) \quad | \quad \mathbf{R}_2^*\,\mathbf{R}_1 = \mathbf{R}_1\,\mathbf{R}_2. \tag{8.177}$$

This implies

$$\mathbf{R}_2 = \mathbf{R}_1^T\,\mathbf{R}_2^*\,\mathbf{R}_1, \tag{8.178}$$

i.e. $\mathbf{R}_2$ is obtained by a back-rotation of $\mathbf{R}_2^*$, and their properties are linked as follows:

1. The eigenvalues of $\mathbf{R}_2$ and $\mathbf{R}_2^*$ are identical while their eigenvectors satisfy the relation $\boldsymbol{\lambda} = \mathbf{R}_1^T\boldsymbol{\lambda}^*$.
   For the eigenpair $\{\mu, \boldsymbol{\lambda}^*\}$ of $\mathbf{R}_2^*$, satisfying the equation $\mathbf{R}_2^*\boldsymbol{\lambda}^* = \mu\boldsymbol{\lambda}^*$, and the back-rotated eigenvector, $\boldsymbol{\lambda} \doteq \mathbf{R}_1^T\boldsymbol{\lambda}^*$, we check that

$$\mathbf{R}_2\boldsymbol{\lambda} = (\mathbf{R}_1^T\mathbf{R}_2^*\mathbf{R}_1)(\mathbf{R}_1^T\boldsymbol{\lambda}^*) = \mathbf{R}_1^T(\mathbf{R}_2^*\boldsymbol{\lambda}^*) = \mathbf{R}_1^T(\mu\boldsymbol{\lambda}^*) = \mu\boldsymbol{\lambda},$$

   i.e. $\{\mu, \boldsymbol{\lambda}\}$ is the eigenpair of $\mathbf{R}_2$.
2. The angles of rotation ω_2^* and ω_2 corresponding to $\mathbf{R}_2^*$ and $\mathbf{R}_2$ are equal.
   Note that $\mathrm{tr}\mathbf{R}_2 = \mathrm{tr}(\mathbf{R}_1^T\mathbf{R}_2^*\mathbf{R}_1) = \mathrm{tr}(\mathbf{R}_1\mathbf{R}_1^T\mathbf{R}_2^*) = \mathrm{tr}\mathbf{R}_2^*$ and, hence, $\cos\omega_2^* = \frac{1}{2}(\mathrm{tr}\mathbf{R}_2^* - 1)$ and $\cos\omega_2 = \frac{1}{2}(\mathrm{tr}\mathbf{R}_2 - 1)$ are equal.

Analogy with polar decomposition of deformation gradient. The left and right composition rules introduced above for rotations are also used in continuum mechanics in conjunction with the polar decomposition of the deformation gradient

$$\mathbf{F} = \mathbf{V}\mathbf{R} = \mathbf{R}\mathbf{U}, \tag{8.179}$$

where $\mathbf{F}$ is the deformation gradient and $\mathbf{V}$, $\mathbf{U}$ are the left and right stretching tensors. Equation (8.179) implies $\mathbf{U} = \mathbf{R}^T\mathbf{V}\,\mathbf{R}$, i.e. $\mathbf{U}$ is obtained by a back-rotation of $\mathbf{V}$. Hence, the properties of $\mathbf{V}$ and $\mathbf{U}$ are linked identically as the properties of $\mathbf{R}_2$ and $\mathbf{R}_2^*$.

Components of $\mathbf{R}_2^*$ in co-rotational basis. Consider two bases, the reference basis $\{\mathbf{t}_i\}$ and the co-rotational basis $\{\mathbf{a}_i\}$, where $\mathbf{a}_i \doteq \mathbf{R}_1\,\mathbf{t}_i$. For the representations

$$\mathbf{R}_2 = R_{2ij}\,\mathbf{t}_i \otimes \mathbf{t}_j, \qquad \mathbf{R}_2^* = R_{2ij}^*\,\mathbf{a}_i \otimes \mathbf{a}_j, \tag{8.180}$$

eq. (8.178) implies

$$\begin{aligned}\mathbf{R}_2 = \mathbf{R}_1^T\,\mathbf{R}_2^*\,\mathbf{R}_1 &= \mathbf{R}_1^T(R_{2ij}^*\,\mathbf{a}_i \otimes \mathbf{a}_j)\,\mathbf{R}_1 \\ &= R_{2ij}^*\,(\mathbf{R}_1^T\mathbf{a}_i \otimes \mathbf{R}_1^T\mathbf{a}_j) = R_{2ij}^*\,\mathbf{t}_i \otimes \mathbf{t}_j. \end{aligned} \tag{8.181}$$

Hence, $R_{2ij} = R_{2ij}^*$, i.e. the components of $\mathbf{R}_2$ in $\{\mathbf{t}_i\}$ and the components of $\mathbf{R}_2^*$ in the co-rotational basis $\{\mathbf{a}_i\}$ are equal. Therefore, $\mathbf{R}_2$ can be considered as $\mathbf{R}_2^*$ parallel transported from the co-rotational basis to the reference basis.

In incrementally formulated problems, $\mathbf{R}_1$ is given and the total rotation $\mathbf{R}_t$ is sought. Then either R_{2ij}^* or R_{2ij} can be used as the unknown.

Rotation of a shell director. Let us consider a rotation of the shell director, which we can write as $\mathbf{a}_3 = \mathbf{R}_2^*\,\mathbf{t}_3$. Using $\mathbf{t}_3 \doteq \mathbf{R}_1\mathbf{i}_3$ and eq. (8.178), we can write

$$\mathbf{R}_2^*\,\mathbf{t}_3 = (\mathbf{R}_1\,\mathbf{R}_2\,\mathbf{R}_1^T)(\mathbf{R}_1\mathbf{i}_3) = \mathbf{R}_1\,\mathbf{R}_2\,\mathbf{i}_3. \tag{8.182}$$

Hence, either the representation of $\mathbf{R}_2^*$ in $\{\mathbf{t}_k\}$ or the representations of $\mathbf{R}_1$ and $\mathbf{R}_2$ in $\{\mathbf{i}_k\}$ can be used.

8.3.2 Composition of Euler parameters (quaternions)

The Euler parameters $\{q_0, \mathbf{q}\}$ defined in eq. (8.67) are a very convenient tool for composing rotations. Consider a composition of two rotation tensors, $\mathbf{R}_t = \mathbf{R}_2\mathbf{R}_1$, and for each of them use the representation (8.68):

$$\begin{aligned}\mathbf{R}_1 &\doteq (2q_0^2 - 1)\mathbf{I} + 2q_0\mathbf{q} \times \mathbf{I} + 2\mathbf{q} \otimes \mathbf{q}, \\ \mathbf{R}_2 &\doteq (2p_0^2 - 1)\mathbf{I} + 2p_0\mathbf{p} \times \mathbf{I} + 2\mathbf{p} \otimes \mathbf{p}, \end{aligned} \tag{8.183}$$

in which $\mathbf{R}_1$ is represented by the quaternion $\{q_0, \mathbf{q}\}$ and $\mathbf{R}_2$ by the quaternion $\{p_0, \mathbf{p}\}$. The total rotation must have a form analogous to (8.68):

$$\mathbf{R}_t \doteq (2r_0^2 - 1)\mathbf{I} + 2r_0\mathbf{r} \times \mathbf{I} + 2\mathbf{r} \otimes \mathbf{r}. \tag{8.184}$$

To satisfy $\mathbf{R}_t = \mathbf{R}_2\mathbf{R}_1$ for the representations (8.183) and (8.184), the quaternion corresponding to $\mathbf{R}_t$ is obtained as

$$\{r_0, \mathbf{r}\} = \{p_0, \mathbf{p}\} \circ \{q_0, \mathbf{q}\}, \tag{8.185}$$

where the composition of two quaternions is defined as follows:

$$\{p_0, \mathbf{p}\} \circ \{q_0, \mathbf{q}\} \doteq \{p_0 q_0 - \mathbf{p} \cdot \mathbf{q}, \quad p_0\mathbf{q} + q_0\mathbf{p} + \mathbf{p} \times \mathbf{q}\}. \tag{8.186}$$

This formula is non-singular and contains only simple scalar and vector operations. Note that in the cross-product, the vectors $\mathbf{p}$ and $\mathbf{q}$ appear in the same order as the associated rotation tensors in $\mathbf{R}_t = \mathbf{R}_2\mathbf{R}_1$.

Remark. If we compose rotation matrices many times, then the product loses orthogonality and its re-orthogonalization is complicated. The advantage of quaternions is that they can be easily re-normalized

$$r_0 = r_0 / \sqrt{r_0^2 + \mathbf{r} \cdot \mathbf{r}}, \qquad \mathbf{r} = \mathbf{r} / \sqrt{r_0^2 + \mathbf{r} \cdot \mathbf{r}}, \tag{8.187}$$

which yields a unit quaternion.

8.3.3 Composition of rotation pseudo-vectors

Below, we derive the composition formula for the semi-tangential vector and for the canonical vector.

Composition of semi-tangential vectors. The rule of composition of the semi-tangential vectors is derived below from the rule for quaternions. The quaternion parameters involved in the composition (8.185) are defined as:

$$\begin{aligned} q_0 &= \cos(\omega_1/2), & \mathbf{q} &= \sin(\omega_1/2)\,\mathbf{c}_1, \\ p_0 &= \cos(\omega_2/2), & \mathbf{p} &= \sin(\omega_2/2)\,\mathbf{c}_2, \\ r_0 &= \cos(\omega_3/2), & \mathbf{r} &= \sin(\omega_3/2)\,\mathbf{c}_3, \end{aligned} \tag{8.188}$$

where $\mathbf{c}_1$, $\mathbf{c}_2$, and $\mathbf{c}_3$ are the unit axes of rotation (corresponding with $\mathbf{e}$ of eq. (8.67)). If the first two of the above definitions are used in eq. (8.186), then we obtain

$$r_0 = c_1c_2 - s_1s_2(\mathbf{c}_1 \cdot \mathbf{c}_2), \qquad \mathbf{r} = c_2s_1\mathbf{c}_1 + c_1s_2\mathbf{c}_2 + s_1s_2(\mathbf{c}_2 \times \mathbf{c}_1),$$

where $s_i \doteq \sin(\omega_i/2)$ and $c_i \doteq \cos(\omega_i/2)$, $(i = 1,2,3)$. By the definitions of the semi-tangential rotation vectors, $\boldsymbol{\psi}_i \doteq t_i\,\mathbf{c}_i$, where $t_i \doteq \tan(\omega_i/2)$, the above equations can be rewritten as

$$r_0 = c_1c_2\left[1 - t_1t_2(\mathbf{c}_1 \cdot \mathbf{c}_2)\right] = c_1c_2\left[1 - \boldsymbol{\psi}_1 \cdot \boldsymbol{\psi}_2\right],$$
$$\mathbf{r} = c_1c_2\left[t_1\mathbf{c}_1 + t_2\mathbf{c}_2 + t_1t_2(\mathbf{c}_1 \times \mathbf{c}_2)\right] = c_1c_2\left[\boldsymbol{\psi}_1 + \boldsymbol{\psi}_2 + \boldsymbol{\psi}_2 \times \boldsymbol{\psi}_1\right].$$

We can easily obtain a semi-tangential vector from the quaternion and, therefore by eq. (8.188), $\boldsymbol{\psi}_3 \doteq \tan(\omega_3/2)\,\mathbf{c}_3 = \mathbf{r}/r_0$. Using the above expressions for r_0 and $\mathbf{r}$, we obtain the formula for the **left** composition of semi-tangential rotation vectors

$$\boldsymbol{\psi}_3 = \frac{1}{1 - \boldsymbol{\psi}_1 \cdot \boldsymbol{\psi}_2}(\boldsymbol{\psi}_1 + \boldsymbol{\psi}_2 + \boldsymbol{\psi}_2 \times \boldsymbol{\psi}_1). \tag{8.189}$$

Note that in the cross-product, the vectors $\boldsymbol{\psi}_2$ and $\boldsymbol{\psi}_1$ appear in the same order as the rotation tensors in the product $\mathbf{R}_t = \mathbf{R}_2\mathbf{R}_1$. The rotation tensor $\mathbf{R}$ for the semi-tangential vector is given by eq. (8.99).

Remark 1. Formula (8.189) is singular for $\boldsymbol{\psi}_1 \cdot \boldsymbol{\psi}_2 = 1$, for which the denominator is equal to zero. For instance, for two equal co-axial rotations, ω, this condition becomes $\tan^2(\omega/2) = 1$, from which we calculate $\omega = \pm\pi/2$, at which the singularity occurs. Because of the singularity, the semi-tangential rotation vectors are less convenient in compounding rotations than quaternions, but can still be used for parametrization of increments. Besides, due to its simplicity, eq. (8.189) is useful in analytical derivations; see below, and Sect. 9.1.

Remark 2. For $\boldsymbol{\psi}_3$ of eq. (8.189), the relation

$$\mathbf{R}(\boldsymbol{\psi}_3) = \mathbf{R}(\boldsymbol{\psi}_2)\,\mathbf{R}(\boldsymbol{\psi}_1) \tag{8.190}$$

is satisfied. If we take $\boldsymbol{\psi}_3 = \boldsymbol{\psi}_1 + \boldsymbol{\psi}_2$, i.e. only a part of eq. (8.189) instead of the full formula, then, in general,

$$\mathbf{R}(\boldsymbol{\psi}_1 + \boldsymbol{\psi}_2) \neq \mathbf{R}(\boldsymbol{\psi}_2)\,\mathbf{R}(\boldsymbol{\psi}_1), \tag{8.191}$$

i.e. a rotation tensor for the sum of two rotation vectors is not equal to a product of the rotation tensors for each vector taken separately.

Composition of canonical vectors. For the canonical vectors, the procedure analogous to the one used for the semi-tangential vectors yields very complicated formulas. Hence, instead of a single formula, we can use the procedure outlined below:

1. the canonical vectors $\boldsymbol{\psi}_1$ and $\boldsymbol{\psi}_2$ are converted to quaternions $\{q_0, \mathbf{q}\}$ and $\{p_0, \mathbf{p}\}$, respectively, as follows:

$$\omega_1 \doteq \|\boldsymbol{\psi}_1\|, \qquad q_0 = \cos(\omega_1/2), \qquad \mathbf{q} = [\sin(\omega_1/2)/\omega_1]\, \boldsymbol{\psi}_1,$$
$$\omega_2 \doteq \|\boldsymbol{\psi}_2\|, \qquad p_0 = \cos(\omega_2/2), \qquad \mathbf{p} = [\sin(\omega_2/2)/\omega_2]\, \boldsymbol{\psi}_2,$$

2. the quaternions are composed using eq. (8.185), which yields the total quaternion $\{r_0, \mathbf{r}\}$,
3. the total canonical rotation vector $\boldsymbol{\psi}_3$ is extracted from the total quaternion $\{r_0, \mathbf{r}\}$ using

$$\boldsymbol{\psi}_3 = \frac{2 \arcsin \|\mathbf{r}\|}{\|\mathbf{r}\|}\, \mathbf{r}. \tag{8.192}$$

The above formula was obtained by noting that the final quaternion depends on the final canonical vector as follows:

$$r_0 \doteq \cos(\|\boldsymbol{\psi}_3\|/2), \qquad \mathbf{r} \doteq \sin(\|\boldsymbol{\psi}_3\|/2)\, \mathbf{e},$$

which is in accord with eq. (8.67). Hence, $\|\mathbf{r}\| = \sin(\|\boldsymbol{\psi}_3\|/2)$ and we can calculate

$$\|\boldsymbol{\psi}_3\| = 2 \arcsin \|\mathbf{r}\|, \qquad \mathbf{c}_3 = \mathbf{r}/\|\mathbf{r}\|, \qquad \boldsymbol{\psi}_3 = \|\boldsymbol{\psi}_3\|\, \mathbf{c}_3,$$

which can be rewritten as eq. (8.192).

Coaxial vectors of the same sense. Let us convert two coaxial canonical vectors of the same sense $\boldsymbol{\psi}_1$ and $\boldsymbol{\psi}_2$ to quaternions

$$\omega_1 \doteq \|\boldsymbol{\psi}_1\|, \qquad q_0 \doteq \cos(\omega_1/2), \qquad \mathbf{q} \doteq \sin(\omega_1/2)\, \mathbf{e}, \tag{8.193}$$

$$\omega_2 \doteq \|\boldsymbol{\psi}_2\|, \qquad p_0 \doteq \cos(\omega_2/2), \qquad \mathbf{p} \doteq \sin(\omega_2/2)\, \mathbf{e}. \tag{8.194}$$

The composition formula of eq. (8.186) yields

$$\begin{aligned} r_0 = p_0 q_0 - \mathbf{p} \cdot \mathbf{q} &= \cos(\omega_1/2)\cos(\omega_2/2) - \sin(\omega_1/2)\sin(\omega_2/2) \\ &= \cos[(\omega_1 + \omega_2)/2], \\ \mathbf{r} = p_0 \mathbf{q} + q_0 \mathbf{p} + \mathbf{p} \times \mathbf{q} &= [\cos(\omega_2/2)\sin(\omega_1/2) + \cos(\omega_1/2)\sin(\omega_2/2)]\, \mathbf{e} \\ &= \sin[(\omega_1 + \omega_2)/2]\, \mathbf{e}, \end{aligned}$$

using trigonometric identities. Because, $\omega_1 + \omega_2 = \|\boldsymbol{\psi}_1\| + \|\boldsymbol{\psi}_2\|$, and for coaxial vectors of the same sense $\|\boldsymbol{\psi}_1\| + \|\boldsymbol{\psi}_2\| = \|\boldsymbol{\psi}_1 + \boldsymbol{\psi}_2\|$, the obtained quaternion $\{r_0, \mathbf{r}\}$ is identical to the quaternion for a sum of vectors $\boldsymbol{\psi}_1 + \boldsymbol{\psi}_2$. This implies

$$\mathbf{R}(\boldsymbol{\psi}_2)\,\mathbf{R}(\boldsymbol{\psi}_1) = \mathbf{R}(\boldsymbol{\psi}_1 + \boldsymbol{\psi}_2), \tag{8.195}$$

i.e. the additive composition of coaxial rotation vectors of the same sense is exact.

Composition of tangent rotation and drilling rotation. Let us examine the composition formula (8.189) for $\boldsymbol{\psi}_1$ being tangent to the reference surface, and $\boldsymbol{\psi}_2$ normal to this surface. The local ortho-normal basis $\{\mathbf{t}_i\}$ on the shell reference surface is used.

Semi-tangential rotation vector. Assume the semi-tangential vectors as $\boldsymbol{\psi}_1 = A\,\mathbf{t}_1 + B\,\mathbf{t}_2$ and $\boldsymbol{\psi}_2 = C\,\mathbf{t}_3$. Because $\boldsymbol{\psi}_1 \cdot \boldsymbol{\psi}_2 = 0$ and $\boldsymbol{\psi}_2 \times \boldsymbol{\psi}_1 = CA\mathbf{t}_2 - CB\mathbf{t}_1$, eq. (8.189) yields the composed rotation

$$\boldsymbol{\psi}_3 = (A - CB)\,\mathbf{t}_1 + (B + CA)\,\mathbf{t}_2 + C\,\mathbf{t}_3, \tag{8.196}$$

i.e. the normal component C also contributes to the tangent components $(A - CB)$ and $(B + CA)$. For given $\boldsymbol{\psi}_3 = \psi_1\mathbf{t}_1 + \psi_2\mathbf{t}_2 + \psi_3\mathbf{t}_3$, we can readily find the inverse relations

$$A = \frac{\psi_1 + \psi_2\psi_3}{1 + \psi_3^2}, \qquad B = \frac{\psi_2 - \psi_1\psi_3}{1 + \psi_3^2}, \qquad C = \psi_3. \tag{8.197}$$

We see that for semi-tangential rotation vectors, the relation between the components $\{A, B, C\}$ and $\{\psi_1, \psi_2, \psi_3\}$ are nonlinear but non-singular.

Canonical rotation vector. Assume the canonical rotation vectors in the form $\boldsymbol{\psi}_1 = a\,\mathbf{t}_1 + b\,\mathbf{t}_2$ and $\boldsymbol{\psi}_2 = c\,\mathbf{t}_3$. We can define the corresponding semi-tangential vectors:

$$\bar{\boldsymbol{\psi}}_1 \doteq \tan\left(\frac{\|\boldsymbol{\psi}_1\|}{2}\right)\frac{\boldsymbol{\psi}_1}{\|\boldsymbol{\psi}_1\|} = \frac{\tan\left(\|\boldsymbol{\psi}_1\|/2\right)}{\|\boldsymbol{\psi}_1\|}\boldsymbol{\psi}_1 = A\,\mathbf{t}_1 + B\,\mathbf{t}_2, \tag{8.198}$$

$$\bar{\boldsymbol{\psi}}_2 \doteq \tan\left(\frac{\|\boldsymbol{\psi}_2\|}{2}\right)\frac{\boldsymbol{\psi}_2}{\|\boldsymbol{\psi}_2\|} = \frac{\tan\left(\|\boldsymbol{\psi}_2\|/2\right)}{\|\boldsymbol{\psi}_2\|}\boldsymbol{\psi}_2 = C\,\mathbf{t}_3, \tag{8.199}$$

where

$$A \doteq \frac{\tan(\|\boldsymbol{\psi}_1\|/2)}{\|\boldsymbol{\psi}_1\|} a, \quad B \doteq \frac{\tan(\|\boldsymbol{\psi}_2\|/2)}{\|\boldsymbol{\psi}_2\|} b, \quad C \doteq \frac{\tan(\|\boldsymbol{\psi}_3\|/2)}{\|\boldsymbol{\psi}_3\|} c. \tag{8.200}$$

Then, using the formula for composition of semi-tangential rotation vectors, eq. (8.189), we find $\overline{\boldsymbol{\psi}}_3$ in a form given by eq. (8.196). Using the semi-tangential $\overline{\boldsymbol{\psi}}_3$, we can calculate the resulting canonical rotation vector as follows:

$$\boldsymbol{\psi}_3 = \omega\, \mathbf{e}, \qquad \omega = 2 \arctan \sqrt{\tfrac{1}{2}\overline{\boldsymbol{\psi}}_3 \cdot \overline{\boldsymbol{\psi}}_3}, \qquad \mathbf{e} = \frac{\overline{\boldsymbol{\psi}}_3}{\|\overline{\boldsymbol{\psi}}_3\|}. \tag{8.201}$$

We see that the formula for the canonical vector will be much more complicated than for the semi-tangential one and for this reason, it is not given here explicitly.

8.3.4 Composition of Euler angles

The parametrization in terms of Euler angles consists of a sequence of three elementary rotations around known vectors of a Cartesian basis, hence it inherently utilizes a composition of rotations. We can use either the left sequence of eq. (8.131) or the right sequence of eq. (8.133), where in the latter, the elementary rotations are performed around the vectors of the initial Cartesian basis.

9

Algorithmic schemes for finite rotations

This chapter presents the topics related to the algorithmic treatment of finite rotations. Mostly static (time-independent) problems are considered, although angular velocity and acceleration also are defined in Sect. 9.4.

We assume that the Newton method is used to solve the non-linear equilibrium equations. To generate the tangent matrix and the residual vector, the total rotation and the increment of rotation are needed.

1. For the total rotation, we use either the rotations matrices or quaternions. Both are used in the most general Scheme 2, when the total rotation is composed of the part which is a result of the update and the part which is parameterized.
2. For increments of rotations, we use the rotation vector to avoid additional orthogonality constraints.

The respective formulas are provided for the parametrization by the canonical rotation vector and by the semi-tangential vector. For both, it is essential use the formulae which are free of numerical indeterminacy; this applies not only to the rotation tensor but also to its first and second differentials.

Besides, the formulae to convert the parameters used for the increment to those used for the total rotation are required; we show that the aforementioned rotation vectors and quaternions can be conveniently matched together. Note that the rotation vector and its increment can belong to different tangent planes to SO(3), which is a matter of choice and can affect the effectiveness and stability of computations.

Several update schemes of rotational parameters can be considered but not all of them perform equally well. This cannot be fully predicted theoretically and a numerical verification is always needed. The answer to the question which type of update is optimal, multiplicative or additive, is quite convoluted, see [30].

Note that the situation in dynamics is more complicated, due to the presence of the angular velocity and acceleration. The time-stepping (e.g. Newmark) scheme must be extended to incorporate the rotational dofs, which can be done in various ways. This is illustrated by the examples for the rigid-body dynamics in Sect. 9.4 but the dynamics of shells remains beyond the scope of this work.

9.1 Increments of rotation vectors in two tangent planes

In this section, we consider the tangent spaces at two different rotations, $\mathbf{R}_A$ and $\mathbf{R}_B$, and establish the relation between the infinitesimal rotation vectors belonging to these spaces, using either a left or right composition rule. The tangent operators $\mathbf{T}$ and their inverses are given for the semi-tangential and canonical rotation vectors. Finally, the differentials $\chi\mathbf{T}$, which are needed in the second variation of the rotation tensor, are obtained.

Tangent plane. The set of all infinitesimal rotations $\tilde{\boldsymbol{\theta}}$ superposed onto the finite rotation $\mathbf{R}$ is referred to as the plane tangent to SO(3) at $\mathbf{R}$, and denoted by $T_R\mathrm{SO}(3) \doteq \{\tilde{\boldsymbol{\theta}}\mathbf{R} \mid \text{for } \tilde{\boldsymbol{\theta}} \in \mathrm{so}(3)\}$. The plane tangent at $\mathbf{R} = \mathbf{I}$ is called the initial tangent plane and denoted by $T_I\mathrm{SO}(3) \doteq \{\tilde{\boldsymbol{\theta}} \mid \text{for } \tilde{\boldsymbol{\theta}} \in \mathrm{so}(3)\}$.

The definitions are analogous for the right composition rule; the tangent plane at $\mathbf{R}$ is defined as $T_R\mathrm{SO}(3) \doteq \{\mathbf{R}\tilde{\boldsymbol{\Theta}} \mid \text{for } \tilde{\boldsymbol{\Theta}} \in \mathrm{so}(3)\}$, and the initial tangent plane as $T_I\mathrm{SO}(3) \doteq \{\tilde{\boldsymbol{\Theta}} \mid \text{for } \tilde{\boldsymbol{\Theta}} \in \mathrm{so}(3)\}$.

9.1.1 Operator T

Generally, in this section we use the notation similar to that of [44], with the exception of the tangent operator $\mathbf{T}$, which we associate with the left composition. In [44], the operator for the right composition rule is designated by $\mathbf{T}$, see eq. (38) therein, while we denote it by $\mathbf{T}^T$. Our $\mathbf{T}$ is identical to $\boldsymbol{\Gamma}$ of [9], eq. (8.31), and $\mathbf{T}$ of [40], eq. (68). The operator $\mathbf{T}$ used in [219] and [221], eq. (88), is equivalent to our $\mathbf{T}^{-T}$. Note that, in these papers, the tangent operators are given only for the canonical rotation vector, while we also provide the operators for the semi-tangential parametrization.

A. Left composition rule

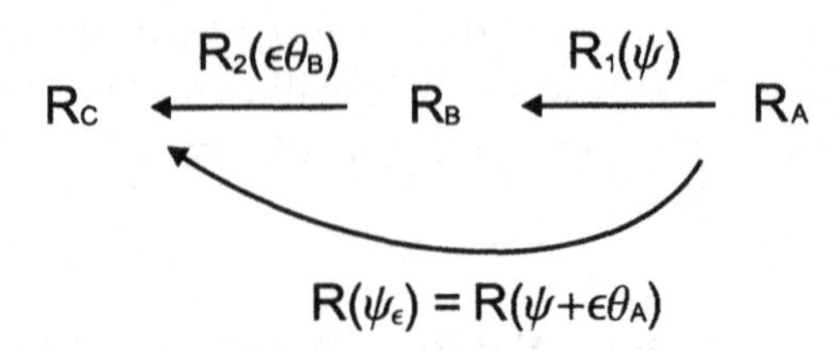

Fig. 9.1 Scheme of increments of rotations for the left composition rule.

We adopt the left composition rule and express $\mathbf{R}_B = \mathbf{R}_1(\boldsymbol{\psi})\,\mathbf{R}_A$, where $\boldsymbol{\psi}$ is the rotation vector, see Fig. 9.1. The perturbed rotation $\mathbf{R}_C$ can be related either to $\mathbf{R}_A$ or to $\mathbf{R}_B$,

$$\mathbf{R}_C = \mathbf{R}(\boldsymbol{\psi}_\epsilon)\,\mathbf{R}_A, \qquad \mathbf{R}_C = \mathbf{R}_2(\epsilon\boldsymbol{\theta}_B)\,\mathbf{R}_B, \tag{9.1}$$

where $\boldsymbol{\psi}_\epsilon \doteq \boldsymbol{\psi} + \epsilon\boldsymbol{\theta}_A$ and ϵ is a scalar parameter. Note that, using the notation established in mechanics, we can also designate $\boldsymbol{\theta}_A$ as $\Delta\boldsymbol{\psi}$. Besides, $\boldsymbol{\theta}_A$ and $\boldsymbol{\theta}_B$ are infinitesimal rotation vectors, and

$$\tilde{\boldsymbol{\psi}}_\epsilon\,\mathbf{R}_A = \left(\tilde{\boldsymbol{\psi}} + \epsilon\tilde{\boldsymbol{\theta}}_A\right)\mathbf{R}_A \in T_{R_A}\mathrm{SO}(3), \qquad \epsilon\tilde{\boldsymbol{\theta}}_B\,\mathbf{R}_B \in T_{R_B}\mathrm{SO}(3), \tag{9.2}$$

i.e. the perturbations $\epsilon\tilde{\boldsymbol{\theta}}_A$ and $\epsilon\tilde{\boldsymbol{\theta}}_B$ belong to different tangent planes, see Fig. 9.2.
Because both relations (9.1) must yield the same $\mathbf{R}_C$, we obtain

$$\mathbf{R}_2(\epsilon\boldsymbol{\theta}_B)\,\mathbf{R}_B = \mathbf{R}(\boldsymbol{\psi}_\epsilon)\,\mathbf{R}_A, \tag{9.3}$$

which, using $\mathbf{R}_B = \mathbf{R}_1(\boldsymbol{\psi})\,\mathbf{R}_A$, reduces to

$$\mathbf{R}_2(\epsilon\boldsymbol{\theta}_B) = \mathbf{R}(\boldsymbol{\psi}_\epsilon)\,\mathbf{R}_1^T(\boldsymbol{\psi}). \tag{9.4}$$

This is a non-linear equation of $\boldsymbol{\theta}_A$ and $\boldsymbol{\theta}_B$, and to find the relation between $\boldsymbol{\theta}_A$ and $\boldsymbol{\theta}_B$, we have to

1. select a specific parametrization of $\mathbf{R}$,
2. convert the tensorial equation to the vectorial form, and
3. locally linearize it using the scheme defined below to obtain the tangent operator.

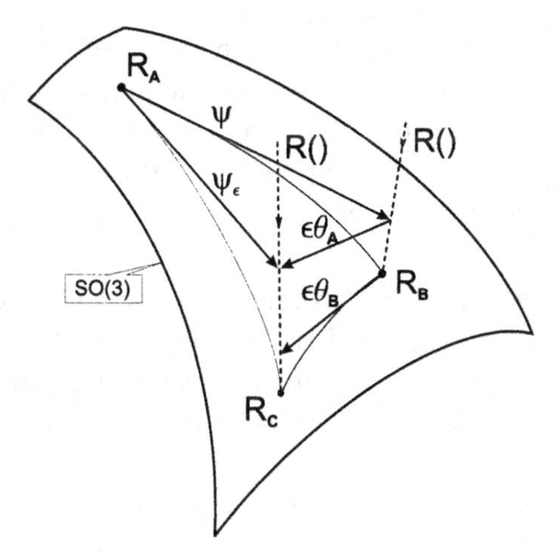

Fig. 9.2 Geometrical interpretation of SO(3) and increments of rotation. $\boldsymbol{\theta}_A$ and $\boldsymbol{\theta}_B$ are not parallel!

Scheme of calculation of operator T. Assume that the l.h.s. of an equation, such as, e.g., eq. (9.8), depends on $\boldsymbol{\theta}_B$ while the r.h.s. depends on $\boldsymbol{\theta}_A$. Hence, the equation can be rewritten as $\mathbf{l}(\boldsymbol{\theta}_B) = \mathbf{r}(\boldsymbol{\theta}_A)$, where $\mathbf{l}$ and $\mathbf{r}$ are, in general, non-linear functions. The differentials of both sides of the considered equation must be equal,

$$D\mathbf{l} \cdot \boldsymbol{\theta}_B = D\mathbf{r} \cdot \boldsymbol{\theta}_A. \tag{9.5}$$

Hence, we calculate two directional derivatives and obtain two tangent operators,

$$\mathbf{T}_l: \quad D\mathbf{l} \cdot \boldsymbol{\theta}_B = \mathbf{T}_l\, \boldsymbol{\theta}_B, \qquad \mathbf{T}_r: \quad D\mathbf{r} \cdot \boldsymbol{\theta}_B = \mathbf{T}_r\, \boldsymbol{\theta}_A, \tag{9.6}$$

where $\mathbf{T}_l$ is the left tangent operator, while $\mathbf{T}_r$ is the right tangent operator. By using them, from eq. (9.5), we obtain

$$\boldsymbol{\theta}_B = \mathbf{T}\, \boldsymbol{\theta}_A, \quad \text{where} \qquad \mathbf{T} \doteq \mathbf{T}_l^{-1}\mathbf{T}_r. \tag{9.7}$$

The directional derivative is calculated in a standard manner, i.e. we differentiate the perturbed expression with respect to the perturbation parameter ϵ and evaluate it for $\epsilon = 0$.

Remark. The hand derivation of tangent operators is, in general, quite tedious work and prone to errors. The same operators can be obtained in a relatively easier way using a symbolic manipulation program, such as *Mathematica* or *Maple.* Using such programs, we can differentiate a vectorial expression w.r.t. a scalar variable and evaluate it for $\epsilon = 0$, so they are well suited to calculate the directional derivative. Then, it remains to recast the obtained expressions into a concise tensorial form and extract the tangent operator.

Semi-tangential parametrization. Assume that $\boldsymbol{\theta}_A$, $\boldsymbol{\theta}_B$, $\boldsymbol{\psi}$ and $\boldsymbol{\psi}_\epsilon$ are semi-tangential rotation vectors. For the semi-tangential parametrization, we can apply the composition formula (8.189), use the identity $\mathbf{R}_1^T(\boldsymbol{\psi}) = \mathbf{R}_1(-\boldsymbol{\psi})$ in eq. (9.4), and write its vectorial counterpart as follows:

$$\epsilon\boldsymbol{\theta}_B = \frac{1}{1+\boldsymbol{\psi}_\epsilon \cdot \boldsymbol{\psi}} \left[\boldsymbol{\psi}_\epsilon - \boldsymbol{\psi} - \boldsymbol{\psi}_\epsilon \times \boldsymbol{\psi}\right]. \tag{9.8}$$

This is an equation non-linear in $\boldsymbol{\theta}_A$ which we can linearize by using the scheme explained in eq. (9.7). Then

$$\boldsymbol{\theta}_B = \mathbf{T}(\boldsymbol{\psi})\,\boldsymbol{\theta}_A, \qquad \mathbf{T} \doteq \frac{\partial \boldsymbol{\theta}_B}{\partial \boldsymbol{\theta}_A}, \tag{9.9}$$

where the tangent operator is

$$\mathbf{T}(\boldsymbol{\psi}) = \frac{1}{1+\boldsymbol{\psi}\cdot\boldsymbol{\psi}}(\mathbf{I} + \tilde{\boldsymbol{\psi}}). \tag{9.10}$$

The operator $\mathbf{T}$ has the following properties:

1. It is non-singular for arbitrary $\boldsymbol{\psi}$, as $\det \mathbf{T} = 1/(1+\|\boldsymbol{\psi}\|^2)^2$.
2. If $\boldsymbol{\psi} = \mathbf{0}$, then $\mathbf{T}(\boldsymbol{\psi}) = \mathbf{I}$. Then we obtain $\boldsymbol{\theta}_B = \boldsymbol{\theta}_A$, as expected.
3. If $\boldsymbol{\psi}$ and $\boldsymbol{\theta}_A$ are coaxial, i.e. $\boldsymbol{\theta}_A = \alpha\,\boldsymbol{\psi}$, where α is an arbitrary scalar, then

$$\boldsymbol{\theta}_B = \mathbf{T}(\boldsymbol{\psi})\,\boldsymbol{\theta}_A = \alpha\,\mathbf{T}(\boldsymbol{\psi})\,\boldsymbol{\psi} = \frac{1}{1+\boldsymbol{\psi}\cdot\boldsymbol{\psi}}\boldsymbol{\theta}_A, \tag{9.11}$$

   i.e. $\mathbf{T}(\boldsymbol{\psi})$ only shortens $\boldsymbol{\theta}_A$. In the proof, we used $\tilde{\boldsymbol{\psi}}\boldsymbol{\psi} = \mathbf{0}$.
4. For $\|\boldsymbol{\psi}\| \to \infty$, $\mathbf{T}(\boldsymbol{\psi}) \to \mathbf{0}$. Hence, we cannot use very long rotation vectors, the norms of which are big numbers.
5. $\mathbf{T}(\boldsymbol{\psi})$ is not a periodic function of $\|\boldsymbol{\psi}\|$. To show this, we use $\tilde{\boldsymbol{\psi}} = \|\boldsymbol{\psi}\|\,\tilde{\mathbf{e}}$ in eq. (9.10), where $\tilde{\mathbf{e}} = \mathbf{S}$, to obtain

$$\mathbf{T}(\boldsymbol{\psi}) = \frac{1}{1+\|\boldsymbol{\psi}\|^2}\mathbf{I} + \frac{\|\boldsymbol{\psi}\|}{1+\|\boldsymbol{\psi}\|^2}\,\tilde{\mathbf{e}}, \tag{9.12}$$

where $\tilde{\mathbf{e}}$ does not depend on $\|\boldsymbol{\psi}\|$. For example, for $\mathbf{e} = [0,0,1]^T$, we obtain the following representation:

$$(\mathbf{T})_{ij} = \begin{bmatrix} \frac{1}{1+t^2} & -\frac{t}{1+t^2} & 0 \\ \frac{t}{1+t^2} & \frac{1}{1+t^2} & 0 \\ 0 & 0 & \frac{1}{1+t^2} \end{bmatrix}, \tag{9.13}$$

where its components are not periodical functions of $t \doteq \|\boldsymbol{\psi}\|$. (Note that the coefficients in this matrix are equal to c_1 and c_2 of eq. (8.101) divided by $2t$.) In consequence, we cannot use the shortened rotation vector.

6. The inverse operator is as follows:

$$\mathbf{T}^{-1}(\boldsymbol{\psi}) = \mathbf{I} - \tilde{\boldsymbol{\psi}} + \boldsymbol{\psi} \otimes \boldsymbol{\psi}. \tag{9.14}$$

Note that the derivation of $\mathbf{T}$ is relatively simple for semi-tangential vectors due to the vectorial form of eq. (9.8).

Finally, our $\mathbf{T}$ of eq. (9.10) is different from the $\mathbf{T}$ operator of [76], eq. (4.32), in which the additional multiplier 2 appears. Both operators are correct and their various forms result from different definitions.

Canonical parametrization. Assume that $\boldsymbol{\theta}_A$, $\boldsymbol{\theta}_B$, $\boldsymbol{\psi}$, and $\boldsymbol{\psi}_\epsilon$ are canonical rotation vectors. For the canonical parametrization, we do not have a vectorial composition formula, such as for the semi-tangential parametrization. Therefore, in order to use eq. (8.189), we shall first introduce auxiliary semi-tangential vectors as functions of the canonical vectors

$$\overline{\boldsymbol{\psi}} \doteq \tan(\|\boldsymbol{\psi}\|/2)\, \frac{\boldsymbol{\psi}}{\|\boldsymbol{\psi}\|}, \qquad \overline{\boldsymbol{\psi}_\epsilon} \doteq \tan(\|\boldsymbol{\psi}_\epsilon\|/2)\, \frac{\boldsymbol{\psi}_\epsilon}{\|\boldsymbol{\psi}_\epsilon\|},$$

$$\overline{\epsilon\boldsymbol{\theta}_B} \doteq \tan(\|\epsilon\boldsymbol{\theta}_B\|/2)\, \frac{\epsilon\boldsymbol{\theta}_B}{\|\epsilon\boldsymbol{\theta}_B\|}, \tag{9.15}$$

marked by a horizontal overbar. Using these auxiliary vectors and exploiting $\mathbf{R}_1^T(\overline{\boldsymbol{\psi}}) = \mathbf{R}_1(-\overline{\boldsymbol{\psi}})$ in eq. (9.4), we can write its vectorial counterpart as

$$\overline{\epsilon\boldsymbol{\theta}_B} = \frac{1}{1 + \overline{\boldsymbol{\psi}_\epsilon} \cdot \overline{\boldsymbol{\psi}}} \left[\overline{\boldsymbol{\psi}_\epsilon} - \overline{\boldsymbol{\psi}} - \overline{\boldsymbol{\psi}_\epsilon} \times \overline{\boldsymbol{\psi}} \right]. \tag{9.16}$$

Analogously, as in eq. (9.9), the operator $\mathbf{T}$ is defined as

$$\boldsymbol{\theta}_B = \mathbf{T}(\boldsymbol{\psi})\, \boldsymbol{\theta}_A, \qquad \mathbf{T} \doteq \frac{\partial \boldsymbol{\theta}_B}{\partial \boldsymbol{\theta}_A}. \tag{9.17}$$

To obtain the operator $\mathbf{T}$, we use the earlier-described scheme of eq. (9.7), which yields

$$\mathbf{T}(\boldsymbol{\psi}) = c_1\,\mathbf{I} + (1-c_1)\,\mathbf{e}\otimes\mathbf{e} + c_2\,\tilde{\boldsymbol{\psi}}, \tag{9.18}$$

where $\mathbf{e} = \boldsymbol{\psi}/\|\boldsymbol{\psi}\|$ and the scalar coefficients c_1 and c_2 are the same as in the rotation tensor of eqs. (8.81) and (8.83). Another, equivalent form of this operator is obtained by using $\tilde{\boldsymbol{\psi}}^2 = \|\boldsymbol{\psi}\|^2\mathbf{S}^2 = \|\boldsymbol{\psi}\|^2(\mathbf{e}\otimes\mathbf{e}-\mathbf{I})$, and is as follows:

$$\mathbf{T}(\boldsymbol{\psi}) = \mathbf{I} + c_2\,\tilde{\boldsymbol{\psi}} + c_3\,\tilde{\boldsymbol{\psi}}^2, \tag{9.19}$$

where $c_3 \doteq (1-c_1)/\|\boldsymbol{\psi}\|^2$. The operator $\mathbf{T}$ has the following properties:

1. If $\boldsymbol{\psi}\to\mathbf{0}$, then $\mathbf{T}(\boldsymbol{\psi})\to\mathbf{I}$, i.e. the operator tends to the identity operator.
2. At $\|\boldsymbol{\psi}\|=0$, the coefficients of $\mathbf{T}$ are numerically indeterminate. This problem also appeared for the rotation tensor, see eq. (8.85), and was already solved for c_1 and c_2. For c_3, the problem can be solved in the same way, by defining it for $\|\boldsymbol{\psi}\|=0$ as the limit value, i.e.

$$\lim_{\|\boldsymbol{\psi}\|\to 0} c_3 = \frac{1}{6}. \tag{9.20}$$

   We have to consider the first derivative of c_3, which is used in the tangent operator. Again, we can use either the perturbation, $\|\boldsymbol{\psi}\| = \sqrt{\boldsymbol{\psi}\cdot\boldsymbol{\psi}+\tau}$, where $\tau = 10^{-8}$, or the truncated Taylor series expansion at $x=0$, e.g.

$$c_3 = \frac{1-(\sin x/x)}{x^2} \approx \frac{1}{6} - \frac{1}{120}x^2 + \frac{1}{5040}x^4,$$

   where $x \doteq \|\boldsymbol{\psi}\|$. The above three-term expansion preserves a good accuracy of c_3 and its first derivative for the range exceeding $|x|=1$. In consequence, we have $\mathbf{T}(\boldsymbol{\psi}=\mathbf{0})=\mathbf{I}$.
3. $\mathbf{T}(\boldsymbol{\psi})$ is singular at $\|\boldsymbol{\psi}\| = 2k\pi$, $(k=1,2,...)$, at which the determinant $\det\mathbf{T} = 2(1-\cos\|\boldsymbol{\psi}\|)/\|\boldsymbol{\psi}\|^2$ is equal to zero. At $\|\boldsymbol{\psi}\| = 0$, $\det\mathbf{T}$ is indeterminate although $\lim_{\|\boldsymbol{\psi}\|\to 0}\det\mathbf{T} = 1$.
4. If $\boldsymbol{\psi}$ and $\boldsymbol{\theta}_A$ are coaxial, i.e. $\boldsymbol{\theta}_A = \alpha\,\boldsymbol{\psi}$, where α is an arbitrary scalar, then

$$\boldsymbol{\theta}_B = \mathbf{T}(\boldsymbol{\psi})\,\boldsymbol{\theta}_A = \alpha\,\mathbf{T}(\boldsymbol{\psi})\,\boldsymbol{\psi} = \boldsymbol{\theta}_A, \tag{9.21}$$

   i.e. $\mathbf{T}(\boldsymbol{\psi})$ acts as the identity operator. In the proof, we used $\tilde{\boldsymbol{\psi}}\boldsymbol{\psi}=\mathbf{0}$ and $\mathbf{e}(\mathbf{e}\cdot\boldsymbol{\psi})=\boldsymbol{\psi}$.

5. $\mathbf{T}(\boldsymbol{\psi})$ is not a periodic function of $\|\boldsymbol{\psi}\|$. To show this, we use $\tilde{\boldsymbol{\psi}} = \|\boldsymbol{\psi}\|\,\tilde{\mathbf{e}}$ in eq. (9.18), where $\tilde{\mathbf{e}} = \mathbf{S}$, to obtain

$$\mathbf{T}(\boldsymbol{\psi}) = c_1\,\mathbf{I} + (1 - c_1)\;\mathbf{e}\otimes\mathbf{e} + c_2\|\boldsymbol{\psi}\|\,\tilde{\mathbf{e}}, \tag{9.22}$$

where $\mathbf{e}$ and $\tilde{\mathbf{e}}$ do not depend on $\|\boldsymbol{\psi}\|$. For example, for $\mathbf{e} = [0,0,1]^T$, we obtain the following representation:

$$(\mathbf{T})_{ij} = \begin{bmatrix} \frac{\sin\omega}{\omega} & -\frac{\sin^2(\omega/2)}{(\omega/2)} & 0 \\ \frac{\sin^2(\omega/2)}{(\omega/2)} & \frac{\sin\omega}{\omega} & 0 \\ 0 & 0 & 1 \end{bmatrix}, \tag{9.23}$$

and we see that its components are not periodical functions of $\omega \doteq \|\boldsymbol{\psi}\|$. In consequence, we cannot use the shortened canonical rotation vector $\boldsymbol{\psi}^*$ of eq. (8.82) because $\mathbf{T}(\boldsymbol{\psi}) \neq \mathbf{T}(\boldsymbol{\psi}^*)$.

6. $\mathbf{T}(\boldsymbol{\psi})$ is singular for $\|\boldsymbol{\psi}\| \to \infty$. Note that for $\omega \doteq \|\boldsymbol{\psi}\| \to \infty$, eq. (9.22) yields

$$\mathbf{T}(\boldsymbol{\psi}) \to \mathbf{e}\otimes\mathbf{e}, \tag{9.24}$$

where $\det(\mathbf{e}\otimes\mathbf{e}) = 0$, so the representation of $\mathbf{T}(\boldsymbol{\psi})$ is singular. Hence, it is not advisable to use very long rotation vectors as their norms are big numbers.

7. $\mathbf{T}(\boldsymbol{\psi})$ can be represented as the following series:

$$\mathbf{T}(\boldsymbol{\psi}) = \mathbf{I} + \frac{1}{2!}\tilde{\boldsymbol{\psi}} + \frac{1}{3!}\tilde{\boldsymbol{\psi}}^2 + \dots + \frac{1}{(n+1)!}\tilde{\boldsymbol{\psi}}^n \dots, \tag{9.25}$$

which can be truncated for small $\boldsymbol{\psi}$.

8. The inverse operator is

$$\mathbf{T}^{-1}(\boldsymbol{\psi}) = c_3\mathbf{I} + c_4\,\boldsymbol{\psi}\otimes\boldsymbol{\psi} - \tfrac{1}{2}\tilde{\boldsymbol{\psi}}, \tag{9.26}$$

where

$$c_3 \doteq \frac{\|\boldsymbol{\psi}\|/2}{\tan(\|\boldsymbol{\psi}\|/2)}, \qquad c_4 \doteq \frac{1 - c_3}{\|\boldsymbol{\psi}\|^2}.$$

In terms of components of $\boldsymbol{\psi} \doteq [\psi_1, \psi_2, \psi_3]^T$, the operator $\mathbf{T}$ is as follows:

$$(\mathbf{T})_{ij} = \begin{bmatrix} c_1 + A\psi_1^2 & -c_2\psi_3 + A\psi_1\psi_2 & c_2\psi_2 + A\psi_1\psi_3 \\ c_2\psi_3 + A\psi_2\psi_1 & c_1 + A\psi_2^2 & -c_2\psi_1 + A\psi_2\psi_3 \\ -c_2\psi_2 + A\psi_3\psi_1 & c_2\psi_1 + A\psi_3\psi_2 & c_1 + A\psi_3^2 \end{bmatrix}, \tag{9.27}$$

where $A \doteq (1 - c_1)/\|\boldsymbol{\psi}\|^2 = (1 - c_1)/(\psi_1^2 + \psi_2^2 + \psi_3^2)$.

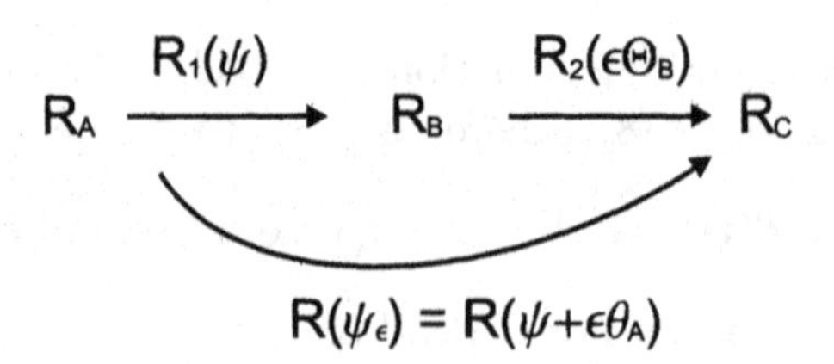

Fig. 9.3 Scheme of increments of rotations for the right composition rule.

B. Right composition rule

We adopt the right composition rule and express $\mathbf{R}_B = \mathbf{R}_A \mathbf{R}_1(\boldsymbol{\psi})$, where $\boldsymbol{\psi}$ is the rotation vector, see Fig. 9.3. The perturbed rotation $\mathbf{R}_C$ can be related either to $\mathbf{R}_A$ or to $\mathbf{R}_B$,

$$\mathbf{R}_C = \mathbf{R}_A \mathbf{R}(\boldsymbol{\psi}_\epsilon), \qquad \mathbf{R}_C = \mathbf{R}_B \mathbf{R}_2(\epsilon \boldsymbol{\Theta}_B), \tag{9.28}$$

where $\boldsymbol{\psi}_\epsilon \doteq \boldsymbol{\psi} + \epsilon \boldsymbol{\theta}_A$ and ϵ is a scalar parameter. Besides, $\boldsymbol{\theta}_A$ and $\boldsymbol{\Theta}_B$ are infinitesimal rotation vectors, and

$$\mathbf{R}_A \tilde{\boldsymbol{\psi}}_\epsilon = \mathbf{R}_A \left(\tilde{\boldsymbol{\psi}} + \epsilon \tilde{\boldsymbol{\theta}}_A \right) \in T_{R_A}\mathrm{SO}(3), \quad \mathbf{R}_B \, \epsilon \tilde{\boldsymbol{\Theta}}_B \in T_{R_B}\mathrm{SO}(3), \tag{9.29}$$

i.e. the perturbations $\epsilon \tilde{\boldsymbol{\theta}}_A$ and $\epsilon \tilde{\boldsymbol{\Theta}}_B$ belong to different tangent planes. Because both relations (9.28) must yield the same $\mathbf{R}_C$, we obtain

$$\mathbf{R}_B \mathbf{R}_2(\epsilon \boldsymbol{\Theta}_B) = \mathbf{R}_A \mathbf{R}(\boldsymbol{\psi}_\epsilon), \tag{9.30}$$

which, using $\mathbf{R}_B = \mathbf{R}_A \mathbf{R}_1(\boldsymbol{\psi})$, is reduced to

$$\mathbf{R}_2(\epsilon \boldsymbol{\Theta}_B) = \mathbf{R}_1^T(\boldsymbol{\psi}) \, \mathbf{R}(\boldsymbol{\psi}_\epsilon). \tag{9.31}$$

To find the relation between $\boldsymbol{\theta}_A$ and $\boldsymbol{\Theta}_B$ from this non-linear equation, we have to use the same steps as for the left composition rule, outlined below eq. (9.4).

Semi-tangential parametrization. Assume that $\boldsymbol{\theta}_A$, $\boldsymbol{\Theta}_B$, $\boldsymbol{\psi}$, and $\boldsymbol{\psi}_\epsilon$ are semi-tangential rotation vectors. For the semi-tangential parametrization, we use $\mathbf{R}_1^T(\boldsymbol{\psi}) = \mathbf{R}_1(-\boldsymbol{\psi})$, apply the composition formula (8.189), and directly write a vectorial counterpart of eq. (9.31) as follows:

$$\epsilon \boldsymbol{\Theta}_B = \frac{1}{1 + \boldsymbol{\psi}_\epsilon \cdot \boldsymbol{\psi}} \left[\boldsymbol{\psi}_\epsilon - \boldsymbol{\psi} - \boldsymbol{\psi} \times \boldsymbol{\psi}_\epsilon \right]. \tag{9.32}$$

Comparing with eq. (9.8), we note that $\boldsymbol{\psi}_\epsilon$ and $\boldsymbol{\psi}$ in the cross-product are interchanged. To obtain the relation between $\boldsymbol{\theta}_A$ and $\boldsymbol{\Theta}_B$, we use the earlier-described scheme of eq. (9.7), which yields

$$\boldsymbol{\Theta}_B = \mathbf{T}^T(\boldsymbol{\psi})\,\boldsymbol{\theta}_A, \tag{9.33}$$

where $\mathbf{T}$ was defined in eq. (9.10). Finally, we note that transposition of $\mathbf{T}$ changes the sign of the skew-symmetric $\tilde{\boldsymbol{\psi}}$ so this term needs special attention.

Canonical parametrization. Assume that $\boldsymbol{\theta}_A$, $\boldsymbol{\Theta}_B$, $\boldsymbol{\psi}$, and $\boldsymbol{\psi}_\epsilon$ are canonical rotation vectors. For the canonical parametrization, we do not have a vectorial composition formula as for the semi-tangential one. Therefore, in order to use eq. (8.189), we shall first introduce auxiliary semi-tangential vectors as functions of the canonical vectors

$$\overline{\boldsymbol{\psi}} \doteq \tan(\|\boldsymbol{\psi}\|/2)\,\frac{\boldsymbol{\psi}}{\|\boldsymbol{\psi}\|}, \qquad \overline{\boldsymbol{\psi}_\epsilon} \doteq \tan(\|\boldsymbol{\psi}_\epsilon\|/2)\,\frac{\boldsymbol{\psi}_\epsilon}{\|\boldsymbol{\psi}_\epsilon\|},$$

$$\overline{\epsilon\boldsymbol{\Theta}_B} \doteq \tan(\|\epsilon\boldsymbol{\Theta}_B\|/2)\,\frac{\epsilon\boldsymbol{\Theta}_B}{\|\epsilon\boldsymbol{\Theta}_B\|}, \tag{9.34}$$

marked by a horizontal overbar. On use of the auxiliary vectors, and $\mathbf{R}_1^T(\overline{\boldsymbol{\psi}}) = \mathbf{R}_1(-\overline{\boldsymbol{\psi}})$, we can write a vectorial counterpart of eq. (9.31), as follows:

$$\overline{\epsilon\boldsymbol{\Theta}_B} = \frac{1}{1+\overline{\boldsymbol{\psi}_\epsilon}\cdot\overline{\boldsymbol{\psi}}}\left[\overline{\boldsymbol{\psi}_\epsilon} - \overline{\boldsymbol{\psi}} - \overline{\boldsymbol{\psi}}\times\overline{\boldsymbol{\psi}_\epsilon}\right]. \tag{9.35}$$

Comparing with eq. (9.16), we note that $\overline{\boldsymbol{\psi}_\epsilon}$ and $\overline{\boldsymbol{\psi}}$ in the cross-product are interchanged. To obtain the relation between $\boldsymbol{\theta}_A$ and $\boldsymbol{\Theta}_B$, we use the earlier-described scheme of eq. (9.7), which yields

$$\boldsymbol{\Theta}_B = \mathbf{T}^T(\boldsymbol{\psi})\,\boldsymbol{\theta}_A, \tag{9.36}$$

where $\mathbf{T}$ was defined in eq. (9.18).

Remark. Note that the multiplicative right update of rotations and the relation of eq. (9.36) are used, e.g., in the energy and momentum conserving algorithm for rigid-body dynamics, see Sect. 9.4.3.

Numerical verification of $\mathbf{T}$. To verify correctness of the form of $\mathbf{T}$, in particular of the sign in front of the skew-symmetric term, we can numerically calculate the value of $\boldsymbol{\Theta}_B$ in two different ways. Let us assume, e.g., that $\boldsymbol{\psi} = [0,0,1]^T$, $\boldsymbol{\theta}_A = [1,0,0]^T$ and $\epsilon = 10^{-8}$. Then

1. Using $\mathbf{T}$, we obtain $\epsilon\boldsymbol{\Theta}_B = \mathbf{T}^T \epsilon\boldsymbol{\theta}_A \approx [8.41, \; -4.59, \; 0]^T \times 10^{-9}$.
2. For $\boldsymbol{\psi}_\epsilon = \boldsymbol{\psi} + \epsilon\boldsymbol{\theta}_A = [10^{-8}, 0, 1]^T$, we calculate $\overline{\epsilon\boldsymbol{\Theta}_B}$ using the vectorial eq. (9.35) and, next, by the approximation

$$\overline{\epsilon\boldsymbol{\Theta}_B} \doteq \tan(\|\epsilon\boldsymbol{\Theta}_B\|/2)\,\frac{\epsilon\boldsymbol{\Theta}_B}{\|\epsilon\boldsymbol{\Theta}_B\|} \approx \frac{\|\epsilon\boldsymbol{\Theta}_B\|}{2}\,\frac{\epsilon\boldsymbol{\Theta}_B}{\|\epsilon\boldsymbol{\Theta}_B\|} = \frac{\epsilon\boldsymbol{\Theta}_B}{2}, \tag{9.37}$$

we obtain $(\epsilon\boldsymbol{\Theta}_B) \approx 2\,(\overline{\epsilon\boldsymbol{\Theta}_B}) \approx [8.41, \; -4.59, \; 0]^T \times 10^{-9}$.

Hence, both methods yield the same value. If we change the sign at the skew-symmetric term in $\mathbf{T}$, then we obtain: $\epsilon\boldsymbol{\Theta}_B = \mathbf{T}^T \epsilon\boldsymbol{\theta}_A \approx [8.41, \; +4.59, \; 0]^T \times 10^{-9}$, with a plus at the second component.

9.1.2 Differential $\chi\mathbf{T}$

In calculations of the second variation of the rotation tensor in Sect. 9.2.4, we will need the directional derivative of the tangent operator $\mathbf{T}$, which is defined as

$$\chi\mathbf{T} \doteq D\mathbf{T}(\boldsymbol{\psi}) \cdot \boldsymbol{\theta}^+, \tag{9.38}$$

where the direction $\boldsymbol{\theta}^+$ is defined in eq. (9.73).

A. For the semi-tangential parametrization, and the operator $\mathbf{T}$ of eq. (9.10), we obtain

$$\chi\mathbf{T}(\boldsymbol{\psi}, \boldsymbol{\theta}^+) = -2a_1^2\,(\boldsymbol{\theta}^+ \cdot \boldsymbol{\psi})\,(\mathbf{I} + \tilde{\boldsymbol{\psi}}) + a_1\,\tilde{\boldsymbol{\theta}}^+, \tag{9.39}$$

where $a_1 \doteq 1/(1 + \boldsymbol{\psi} \cdot \boldsymbol{\psi})$. For $\boldsymbol{\psi} \to \mathbf{0}$, $\chi\mathbf{T} \to \tilde{\boldsymbol{\theta}}^+ \in \mathrm{so}(3)$.

B. For the canonical rotation vector, and the operator $\mathbf{T}$ of eq. (9.18), we obtain

$$\begin{aligned}\chi\mathbf{T}(\boldsymbol{\psi}, \boldsymbol{\theta}^+) &= a_1\,(\mathbf{e} \cdot \boldsymbol{\theta}^+)\,\mathbf{I} + a_2\,(\boldsymbol{\theta}^+ \otimes \mathbf{e} + \mathbf{e} \otimes \boldsymbol{\theta}^+) \\ &\quad + a_3\,(\mathbf{e} \cdot \boldsymbol{\theta}^+)(\mathbf{e} \otimes \mathbf{e}) + a_4\,(\mathbf{e} \cdot \boldsymbol{\theta}^+)\tilde{\boldsymbol{\psi}} + a_5\,\tilde{\boldsymbol{\theta}}^+, \end{aligned} \tag{9.40}$$

where $\mathbf{e} = \boldsymbol{\psi}/\|\boldsymbol{\psi}\|$, $\omega = \|\boldsymbol{\psi}\| = \sqrt{\boldsymbol{\psi} \cdot \boldsymbol{\psi}}$, and the scalar coefficients are

$$a_1 = b_2 - b_1, \qquad a_2 = b_3 - b_1,$$
$$a_3 = 3b_1 - b_2 - 2b_3, \qquad a_4 = -b_3 b_4 + b_1, \qquad a_5 = \tfrac{1}{2} b_4,$$
$$b_1 = \frac{\sin\omega}{\omega^2}, \quad b_2 = \frac{\cos\omega}{\omega}, \quad b_3 = \frac{1}{\omega}, \quad b_4 = \left[\frac{\sin(\omega/2)}{(\omega/2)}\right]^2. \qquad (9.41)$$

We see that the coefficients $b_i \ (i = 1, \ldots, 4)$ are numerically indeterminate at $\omega = 0$. For $\boldsymbol{\psi} \to \mathbf{0}$, $\chi\mathbf{T} \to \frac{1}{2}\tilde{\boldsymbol{\theta}}^+ \in \mathrm{so}(3)$.

Remark. The derivation of formula (9.40) is cumbersome, see [44], but its correctness can be verified easier. For instance, we can assume $(\boldsymbol{\psi})_i = \{0, 0, 1\}$ and $(\boldsymbol{\theta}^+)_i = \{0, 0, \tau\}$, where τ is a small value chosen in the way established for the finite difference operators, see [65]. Then we can calculate $D\mathbf{T} \cdot \boldsymbol{\theta}^+$ in two ways: first, using eq. (9.40) and next, by an approximate difference formula $D\mathbf{T} \cdot \boldsymbol{\theta}^+ \approx \mathbf{T}(\boldsymbol{\psi} + \boldsymbol{\theta}^+) - \mathbf{T}(\boldsymbol{\psi})$.

A summary of the tangent operators is provided in Table 9.1 where the variations of the rotation tensor in which they are used are indicated.

Table 9.1 Tangent operators and variations of rotation tensor.

Tangent operator	Semi-tangential parametrization	Canonical parametrization	Used in variations of rotation
$\mathbf{T}(\boldsymbol{\psi})$	eq. (9.10)	eq. (9.18)	first, second
$\chi\mathbf{T}(\boldsymbol{\psi}, \boldsymbol{\theta}^+)$	eq. (9.39)	eq. (9.40)	second

9.2 Variation of rotation tensor

In this section we derive the formulae for the variation of the rotation tensor assuming the additive and multiplicative (left and right) compositions of the semi-tangential and canonical rotation vectors. Then we can relate the obtained variation to each other using the tangent operators derived in the preceding section.

The notation used below is the same as in the preceding section, see Figs. 9.1 and 9.3.

9.2.1 Variation of rotation tensor for additive composition

For the additive composition of the rotation parameters, $\boldsymbol{\psi}_\epsilon = \boldsymbol{\psi} + \epsilon\boldsymbol{\theta}_A \in T_I\mathrm{SO}(3)$, we define the variation as the following directional derivative:

$$\delta_{\tilde{\theta}_A}\mathbf{R}(\boldsymbol{\psi}) \doteq D\mathbf{R}(\boldsymbol{\psi})\cdot\tilde{\boldsymbol{\theta}}_A = \frac{\mathrm{d}}{\mathrm{d}\epsilon}[\mathbf{R}(\boldsymbol{\psi}_\epsilon)]_{\epsilon=0}, \tag{9.42}$$

where ϵ is a scalar perturbation parameter. Note that we can also denote $\boldsymbol{\theta}_A \doteq \delta\boldsymbol{\psi}$, following the established convention in mechanics. Because the function $\mathbf{R}(\boldsymbol{\psi})$ is different for each parametrization, the variation must be derived separately for the semi-tangential and canonical parametrization.

The above directional derivative can be calculated using a symbolic manipulation program, such as *Mathematica* or *Maple*, but then we obtain long and complicated formulas. On the other hand, concise forms can be derived using the multiplicative composition of rotation tensors, as we show below.

9.2.2 Variation of rotation tensor for multiplicative composition

We derive the formulae for the variation of the rotation tensor w.r.t. the rotation vectors (semi-tangential and canonical), at two characteristic points: (A) at $\mathbf{R}_B = \mathbf{I}$ (or for $\mathbf{R}_1 = \mathbf{I}$) and (B) at arbitrary $\mathbf{R}_B$.

A. Variation of rotation tensor $\mathbf{R}_B = \mathbf{I}$

Define the variation of $\mathbf{R}_B$ w.r.t. the skew-symmetric $\tilde{\boldsymbol{\theta}}_B \in \mathrm{so}(3)$, as the directional derivative of $\mathbf{R}_2$ in the direction $\tilde{\boldsymbol{\theta}}_B$,

$$\delta_{\tilde{\psi}}\mathbf{R}_B \doteq D\mathbf{R}_B\cdot\tilde{\boldsymbol{\theta}}_B = \frac{\mathrm{d}}{\mathrm{d}\epsilon}[\mathbf{R}_2(\epsilon\tilde{\boldsymbol{\theta}}_B)]_{\epsilon=0}, \tag{9.43}$$

where ϵ is a scalar perturbation parameter.

Semi-tangential rotation vector. For the skew-symmetric $\tilde{\boldsymbol{\theta}}_B \in \mathrm{so}(3)$ associated with the semi-tangential rotation vector of eq. (8.97), we obtain

$$\delta_{\tilde{\theta}_B}\mathbf{R}_B = 2\,\tilde{\boldsymbol{\theta}}_B, \tag{9.44}$$

where the form of $\mathbf{R}_B$ for the semi-tangential vector was used. The proof is as follows. For $\epsilon\tilde{\boldsymbol{\theta}}_B$, eq. (8.99) becomes

$$\mathbf{R}(\epsilon\tilde{\boldsymbol{\theta}}_B) = \mathbf{I} + \frac{2}{1+\|\epsilon\boldsymbol{\theta}_B\|^2}\left(\epsilon\tilde{\boldsymbol{\theta}}_B + \epsilon^2\tilde{\boldsymbol{\theta}}_B^2\right), \qquad \|\epsilon\boldsymbol{\theta}_B\|^2 = \tfrac{1}{2}\epsilon^2\tilde{\boldsymbol{\theta}}_B\cdot\tilde{\boldsymbol{\theta}}_B \geq 0. \tag{9.45}$$

Denoting the nominator by $N \doteq 2\left(\epsilon\tilde{\boldsymbol{\theta}}_B + \epsilon^2\tilde{\boldsymbol{\theta}}_B^2\right)$ and the denominator by $D \doteq 1 + \frac{1}{2}\epsilon^2\tilde{\boldsymbol{\theta}}_B\cdot\tilde{\boldsymbol{\theta}}_B$, we calculate the derivative

$$\frac{\mathrm{d}}{\mathrm{d}\epsilon}\left[\mathbf{R}_B(\epsilon\tilde{\boldsymbol{\theta}}_B)\right] = \frac{1}{D^2}\left(\frac{\mathrm{dN}}{\mathrm{d}\epsilon}D - \frac{\mathrm{dD}}{\mathrm{d}\epsilon}N\right), \tag{9.46}$$

where $\frac{\mathrm{dN}}{\mathrm{d}\epsilon} = 2\left(\tilde{\boldsymbol{\theta}}_B + 2\epsilon\tilde{\boldsymbol{\theta}}_B^2\right)$, and $\frac{\mathrm{dD}}{\mathrm{d}\epsilon} = \epsilon\boldsymbol{\theta}_B \cdot \boldsymbol{\theta}_B$. For $\epsilon = 0$, we obtain

$$[N]_{\epsilon=0} = 0, \qquad [D]_{\epsilon=0} = 1, \qquad \left[\frac{\mathrm{dN}}{\mathrm{d}\epsilon}\right]_{\epsilon=0} = 2\tilde{\boldsymbol{\theta}}_B, \qquad \left[\frac{\mathrm{dD}}{\mathrm{d}\epsilon}\right]_{\epsilon=0} = 0,$$

and eq. (9.46) yields the r.h.s. of eq. (9.44). □

Canonical rotation vector. For the skew-symmetric $\tilde{\boldsymbol{\theta}}_B \in \mathrm{so}(3)$ associated with the canonical rotation vector of eq. (8.79), we obtain

$$\delta_{\tilde{\theta}_B}\mathbf{R}_B = \tilde{\boldsymbol{\theta}}_B. \tag{9.47}$$

The proof is immediate, as for $\mathbf{R}_B$ we may use the exponential representation, i.e. $\mathbf{R}_{B\epsilon} = \exp(\epsilon\tilde{\boldsymbol{\theta}}_B) = \mathbf{I} + \epsilon\tilde{\boldsymbol{\theta}}_B + ... + \frac{1}{n!}(\epsilon\tilde{\boldsymbol{\theta}}_B)^n + ...$. Then

$$\frac{\mathrm{d}}{\mathrm{d}\epsilon}[\exp(\epsilon\tilde{\boldsymbol{\theta}}_B)] = \left[\mathbf{I} + \epsilon\tilde{\boldsymbol{\theta}}_B + ... + \frac{1}{(n-1)!}(\epsilon\tilde{\boldsymbol{\theta}}_B)^{n-1} + ...\right]\tilde{\boldsymbol{\theta}}_B, \tag{9.48}$$

and, by setting $\epsilon = 0$, we obtain the r.h.s. of eq. (9.47). □

B. Variation of arbitrary rotation tensor $\mathbf{R}_B$

To calculate the variation of an arbitrary rotation $\mathbf{R}_B$, we use the composition rules for the rotation tensors of eqs. (8.176) and (8.177) with $\mathbf{R}_1 = \mathbf{R}(\boldsymbol{\psi})$, and the variations of the rotation tensor $\mathbf{R}_B = \mathbf{I}$ of eqs. (9.44) and (9.47).

Left (or spatial) variation. For the left composition rule, the perturbed rotation is defined as $\mathbf{R}_{B\epsilon} \doteq \mathbf{R}_2(\epsilon\tilde{\boldsymbol{\theta}}_B)\,\mathbf{R}_B$, where $\tilde{\boldsymbol{\theta}}_B \in \mathrm{so}(3)$. The variation of $\mathbf{R}_B$ w.r.t. $\tilde{\boldsymbol{\theta}}_B$ is defined as the derivative of $\mathbf{R}_B$ in the direction $\tilde{\boldsymbol{\theta}}_B$,

$$\delta_{\tilde{\theta}_B}\mathbf{R}_B \doteq D\mathbf{R}_B \cdot \tilde{\boldsymbol{\theta}}_B = \frac{\mathrm{d}}{\mathrm{d}\epsilon}[\mathbf{R}_2(\epsilon\tilde{\boldsymbol{\theta}}_B)\,\mathbf{R}_B]_{\epsilon=0} = \frac{\mathrm{d}}{\mathrm{d}\epsilon}[\mathbf{R}_2(\epsilon\tilde{\boldsymbol{\theta}}_B)]_{\epsilon=0}\,\mathbf{R}_B. \tag{9.49}$$

Right (Lagrangian, or material) variation. For the right composition rule, the perturbed rotation is defined as $\mathbf{R}_{B\epsilon} \doteq \mathbf{R}_B \mathbf{R}_2(\epsilon \tilde{\boldsymbol{\Theta}}_B)$, where $\tilde{\boldsymbol{\Theta}}_B \in \mathrm{so}(3)$. The variation of $\mathbf{R}_B$ w.r.t. $\tilde{\boldsymbol{\Theta}}_B$ is defined as the derivative of $\mathbf{R}_B$ in the direction $\tilde{\boldsymbol{\Theta}}_B$,

$$\delta_{\tilde{\Theta}_B} \mathbf{R}_B \doteq D\mathbf{R} \cdot \tilde{\boldsymbol{\Theta}}_B = \frac{\mathrm{d}}{\mathrm{d}\epsilon} [\mathbf{R}_B \mathbf{R}_2(\epsilon \tilde{\boldsymbol{\Theta}}_B)]_{\epsilon=0} = \mathbf{R}_B \frac{\mathrm{d}}{\mathrm{d}\epsilon} [\mathbf{R}_2(\epsilon \tilde{\boldsymbol{\Theta}}_B)]_{\epsilon=0}. \tag{9.50}$$

In the above two definitions, we should use the directional derivative of $\mathbf{R}_2$ given for the semi-tangential vector by eq. (9.44) and for the canonical vector by eq. (9.47).

9.2.3 Relations between variations for various composition rules

In this section, we establish the relations between variations for the additive composition and the multiplicative composition of the canonical rotational parameters. For the semi-tangential vector, the procedure is analogous and the results are in eq. (9.60).

The composition equations for the rotation tensors, eqs. (8.176) and (8.177), can be rewritten together as

$$\mathbf{R}_t = \mathbf{R}_2^* \, \mathbf{R}_1 = \mathbf{R}_1 \, \mathbf{R}_2. \tag{9.51}$$

Let us define $\mathbf{R}_t \doteq \mathbf{R}(\boldsymbol{\psi} + \epsilon \boldsymbol{\theta}_A)$, $\mathbf{R}_1 \doteq \mathbf{R}(\boldsymbol{\psi})$, $\mathbf{R}_2^* \doteq \mathbf{R}_2(\epsilon \tilde{\boldsymbol{\theta}}_B)$, and $\mathbf{R}_2 \doteq \mathbf{R}_2(\epsilon \tilde{\boldsymbol{\Theta}}_B)$, where $\boldsymbol{\theta}_A$, $\boldsymbol{\theta}_B$, and $\boldsymbol{\Theta}_B$ are the infinitesimal rotation vectors, shown in Figs. 9.1 and 9.3. Then eq. (9.51) becomes

$$\underbrace{\mathbf{R}\,(\boldsymbol{\psi} + \epsilon \boldsymbol{\theta}_A)}_{\text{additive}} = \underbrace{\mathbf{R}_2(\epsilon \tilde{\boldsymbol{\theta}}_B)\, \mathbf{R}(\boldsymbol{\psi})}_{\text{multiplicative, left}} = \underbrace{\mathbf{R}(\boldsymbol{\psi})\, \mathbf{R}_2(\epsilon \tilde{\boldsymbol{\Theta}}_B)}_{\text{multiplicative, right}}, \tag{9.52}$$

where the additive and multiplicative compositions of rotational parameters were used.

Below, we calculate three variations (directional derivatives) of eq. (9.52): (1) in the direction $\tilde{\boldsymbol{\theta}}_A \in T_I\mathrm{SO}(3)$, (2) in the direction $\tilde{\boldsymbol{\theta}}_B \mathbf{R} \in T_R\mathrm{SO}(3)$, and (3) in the direction $\mathbf{R}\tilde{\boldsymbol{\Theta}}_B \in T_R\mathrm{SO}(3)$. The relations are derived for the canonical rotation vector.

Variation in direction $\tilde{\boldsymbol{\theta}}_A \in T_I\mathrm{SO}(3)$. We can calculate the derivative of eq. (9.52) in the direction $\tilde{\boldsymbol{\theta}}_A$ in a standard manner and the derivatives of particular parts are as follows:

a) For the additive composition, $\mathbf{R}(\boldsymbol{\psi}+\epsilon\boldsymbol{\theta}_A)$, we obtain the derivative of eq. (9.42).

b) For the left multiplicative composition, $\mathbf{R}_2(\epsilon\tilde{\boldsymbol{\theta}}_B)\,\mathbf{R}(\boldsymbol{\psi})$,

$$\delta_{\tilde{\theta}_A}\left[\mathbf{R}_2(\epsilon\tilde{\boldsymbol{\theta}}_B)\,\mathbf{R}(\boldsymbol{\psi})\right] = \frac{\mathrm{d}}{\mathrm{d}\epsilon}\left[\mathbf{R}_2(\epsilon\tilde{\boldsymbol{\theta}}_B)\,\mathbf{R}(\boldsymbol{\psi})\right]_{\epsilon=0}, \tag{9.53}$$

in which we must express $\tilde{\boldsymbol{\theta}}_B$ as a function of the perturbation $\tilde{\boldsymbol{\theta}}_A$. Note that $\boldsymbol{\theta}_B = \mathbf{T}(\boldsymbol{\psi})\,\boldsymbol{\theta}_A$, where the tangent operator $\mathbf{T}$ is defined in eqs. (9.10) and (9.18) and we can write $\epsilon\tilde{\boldsymbol{\theta}}_B = [\mathbf{T}(\boldsymbol{\psi})\,\epsilon\boldsymbol{\theta}_A]\times\mathbf{I}$ to obtain

$$\delta_{\tilde{\theta}_A}\left[\mathbf{R}_2(\epsilon\tilde{\boldsymbol{\theta}}_B)\,\mathbf{R}(\boldsymbol{\psi})\right] = \underbrace{\{[\mathbf{T}(\boldsymbol{\psi})\,\boldsymbol{\theta}_A]\times\mathbf{I}\}}_{\text{skew-symm.}}\,\mathbf{R}(\boldsymbol{\psi}). \tag{9.54}$$

c) For the right multiplicative composition, $\mathbf{R}(\boldsymbol{\psi})\,\mathbf{R}_2(\epsilon\tilde{\boldsymbol{\Theta}}_B)$, we obtain

$$\delta_{\tilde{\theta}_A}\left[\mathbf{R}(\boldsymbol{\psi})\,\mathbf{R}_2(\epsilon\tilde{\boldsymbol{\Theta}}_B)\right] = \frac{\mathrm{d}}{\mathrm{d}\epsilon}\left[\mathbf{R}(\boldsymbol{\psi})\,\mathbf{R}_2(\epsilon\tilde{\boldsymbol{\Theta}}_B)\right]_{\epsilon=0}, \tag{9.55}$$

in which we must express $\epsilon\tilde{\boldsymbol{\Theta}}_B$ as a function of the perturbation $\epsilon\tilde{\boldsymbol{\theta}}_A$. Note that $\boldsymbol{\Theta}_B = \mathbf{T}^T(\boldsymbol{\psi})\,\boldsymbol{\theta}_A$, by eqs. (9.33) and (9.36), and we can write $\epsilon\tilde{\boldsymbol{\Theta}}_B = [\mathbf{T}^T(\boldsymbol{\psi})\,\epsilon\boldsymbol{\theta}_A]\times\mathbf{I}$ to obtain

$$\delta_{\tilde{\theta}_A}\left[\mathbf{R}(\boldsymbol{\psi})\mathbf{R}_2(\epsilon\tilde{\boldsymbol{\Theta}}_B)\right] = \mathbf{R}(\boldsymbol{\psi})\,\underbrace{\{[\mathbf{T}^T(\boldsymbol{\psi})\,\boldsymbol{\theta}_A]\times\mathbf{I}\}}_{\text{skew-symm.}}. \tag{9.56}$$

Writing the above results together, we have the relation linking the variations for various compositions of rotations

$$\underbrace{\delta_{\tilde{\theta}_A}\mathbf{R}(\boldsymbol{\psi})}_{\text{additive}} = \underbrace{\{[\mathbf{T}(\boldsymbol{\psi})\,\boldsymbol{\theta}_A]\times\mathbf{I}\}\,\mathbf{R}(\boldsymbol{\psi})}_{\text{left, multiplicative}} = \underbrace{\mathbf{R}(\boldsymbol{\psi})\,\{[\mathbf{T}^T(\boldsymbol{\psi})\,\boldsymbol{\theta}_A]\times\mathbf{I}\}}_{\text{right, multiplicative}}. \tag{9.57}$$

Note that this relation allows us to express the variation for the additive composition of eq. (9.42) in the concise forms of variations for multiplicative compositions to avoid long and complicated formulas.

Remark 1. Note that writing the above relation in the short form as

$$\underbrace{\delta_{\tilde{\theta}_A}\mathbf{R}}_{\text{additive}} = \underbrace{\delta_{\tilde{\theta}_B}\mathbf{R}}_{\text{left}} = \underbrace{\delta_{\tilde{\Theta}_B}\mathbf{R}}_{\text{right}}, \tag{9.58}$$

we must remember that it holds only if $\boldsymbol{\theta}_B = \mathbf{T}(\boldsymbol{\psi})\,\boldsymbol{\theta}_A$ and $\boldsymbol{\Theta}_B = \mathbf{T}^T(\boldsymbol{\psi})\,\boldsymbol{\theta}_A$.

Remark 2. The forms of the variations for the multiplicative rules specified above are not suitable for numerical implementations. However, if these variations are multiplied by a vector, e.g. the shell director $\mathbf{t}_3$, then $\boldsymbol{\theta}_A$ can be separated. For instance, for the variation of a shell director, we can perform the following transformations:

$$\begin{aligned}\delta\mathbf{a}_3 &= \delta_{\tilde{\theta}_A}\mathbf{R}\,\mathbf{t}_3 = \mathbf{R}(\boldsymbol{\psi})\,\{[\mathbf{T}^T(\boldsymbol{\psi})\,\boldsymbol{\theta}_A]\times\mathbf{I}\}\,\mathbf{t}_3 = \mathbf{R}(\boldsymbol{\psi})\,\{[\mathbf{T}^T(\boldsymbol{\psi})\,\boldsymbol{\theta}_A]\times\mathbf{t}_3\} \\ &= -\mathbf{R}(\boldsymbol{\psi})\,\{\mathbf{t}_3\times[\mathbf{T}^T(\boldsymbol{\psi})\,\boldsymbol{\theta}_A]\} = -\underbrace{\mathbf{R}(\boldsymbol{\psi})\,(\mathbf{t}_3\times\mathbf{I})\,\mathbf{T}^T(\boldsymbol{\psi})}_{3\times 3\text{ matrix}}\ \underbrace{\boldsymbol{\theta}_A}_{\text{vector}}\,,\end{aligned} \tag{9.59}$$

where, in the final form, we have a product of the matrix and $\boldsymbol{\theta}_A$.

Variations for the semi-tangential rotation vector. For the semi-tangential rotation vector, the procedure is analogous and eq. (9.57) linking the variations for various compositions of rotations, becomes

$$\underbrace{\delta_{\tilde{\theta}_A}\mathbf{R}(\boldsymbol{\psi})}_{\text{additive}} = \underbrace{\{[2\,\mathbf{T}(\boldsymbol{\psi})\,\boldsymbol{\theta}_A]\times\mathbf{I}\}\,\mathbf{R}(\boldsymbol{\psi})}_{\text{left, multiplicative}} = \underbrace{\mathbf{R}(\boldsymbol{\psi})\,\{[2\,\mathbf{T}^T(\boldsymbol{\psi})\,\boldsymbol{\theta}_A]\times\mathbf{I}\}}_{\text{right, multiplicative}}. \tag{9.60}$$

Note the multiplier 2; it appeared earlier in eq. (9.44), as compared to eq. (9.47).

Variation in direction $\tilde{\boldsymbol{\theta}}_B\mathbf{R} \in T_R SO(3)$. We calculate a derivative of eq. (9.52) in the direction $\tilde{\boldsymbol{\theta}}_B\mathbf{R} \in T_R SO(3)$ in a standard manner and obtain the following derivatives of each side:

a) For the additive composition, $\mathbf{R}(\boldsymbol{\psi}+\epsilon\boldsymbol{\theta}_A)$, we must express $\tilde{\boldsymbol{\theta}}_A$ as a function of the perturbation $\tilde{\boldsymbol{\theta}}_B$. Note that $\boldsymbol{\theta}_A = \mathbf{T}^{-1}(\boldsymbol{\psi})\,\boldsymbol{\theta}_B$ and we can write

$$\delta_{\tilde{\theta}_B}\mathbf{R}(\boldsymbol{\psi}) = \frac{\mathrm{d}}{\mathrm{d}\epsilon}[\mathbf{R}(\boldsymbol{\psi}+\mathbf{T}^{-1}\epsilon\boldsymbol{\theta}_B)]_{\epsilon=0}, \tag{9.61}$$

b) For the left multiplicative composition, $\mathbf{R}_2(\epsilon\tilde{\boldsymbol{\theta}}_B)\,\mathbf{R}(\boldsymbol{\psi})$, we obtain eq. (9.49), i.e.

$$\delta_{\tilde{\theta}_B}\mathbf{R} = \frac{\mathrm{d}}{\mathrm{d}\epsilon}[\mathbf{R}_2(\epsilon\tilde{\boldsymbol{\theta}}_B)]_{\epsilon=0}\,\mathbf{R}, \tag{9.62}$$

where the directional derivative of $\mathbf{R}_2$ is given either by eq. (9.44) or by eq. (9.47), but with $\tilde{\boldsymbol{\psi}}$ replaced by $\tilde{\boldsymbol{\theta}}_B$.

Hence, the variation for the additive composition and the left multiplicative composition are mutually related as follows:

$$\underbrace{\delta_{\tilde{\theta}_B} \mathbf{R}(\boldsymbol{\psi})}_{\text{additive}} = \underbrace{a\,\tilde{\boldsymbol{\theta}}_B\, \mathbf{R}(\boldsymbol{\psi})}_{\text{left}}, \tag{9.63}$$

where $a = 2$ for the semi-tangential vector and $a = 1$ for the canonical vector. Note that we could have also calculated a variation for the right multiplicative composition $\mathbf{R}(\boldsymbol{\psi})\,\mathbf{R}_2(\epsilon\tilde{\boldsymbol{\Theta}}_B)$, using $\boldsymbol{\Theta}_B = \mathbf{T}^T\mathbf{T}^{-1}\boldsymbol{\theta}_B$, but it is not used in subsequent calculations.

Variation in direction $\mathbf{R}\tilde{\boldsymbol{\Theta}}_B \in T_R\mathrm{SO}(3)$. We calculate a derivative of eq. (9.52) in the direction $\mathbf{R}\tilde{\boldsymbol{\Theta}}_B \in T_R\mathrm{SO}(3)$ in a standard manner, and obtain the following derivatives of each side.

a) For the additive composition, $\mathbf{R}(\boldsymbol{\psi}+\epsilon\boldsymbol{\theta}_A)$, we must express $\tilde{\boldsymbol{\theta}}_A$ as a function of the perturbation $\tilde{\boldsymbol{\Theta}}_B$. Note that $\boldsymbol{\theta}_A = \mathbf{T}^{-T}(\boldsymbol{\psi})\,\boldsymbol{\Theta}_B$ and we can write

$$\delta_{\tilde{\Theta}_B} \mathbf{R}(\boldsymbol{\psi}) = \frac{\mathrm{d}}{\mathrm{d}\epsilon}[\mathbf{R}(\boldsymbol{\psi} + \mathbf{T}^{-T}\epsilon\boldsymbol{\Theta}_B)]_{\epsilon=0}. \tag{9.64}$$

b) For the right multiplicative composition, $\mathbf{R}(\boldsymbol{\psi})\,\mathbf{R}_2(\epsilon\tilde{\boldsymbol{\Theta}}_B)$, we obtain eq. (9.50), i.e.

$$\delta_{\tilde{\Theta}_B} \mathbf{R} = \mathbf{R}\,\frac{\mathrm{d}}{\mathrm{d}\epsilon}[\mathbf{R}_2(\epsilon\tilde{\boldsymbol{\Theta}}_B)]_{\epsilon=0}, \tag{9.65}$$

where the directional derivative of $\mathbf{R}_2$ is given either by eq. (9.44) or by eq. (9.47) but with $\tilde{\boldsymbol{\psi}}$ replaced by $\tilde{\boldsymbol{\Theta}}_B$.

Hence, the variation for the additive composition and the right multiplicative composition are mutually related as follows:

$$\underbrace{\delta_{\tilde{\Theta}_B} \mathbf{R}(\boldsymbol{\psi})}_{\text{additive}} = \underbrace{a\,\mathbf{R}(\boldsymbol{\psi})\,\tilde{\boldsymbol{\Theta}}_B}_{\text{right}}, \tag{9.66}$$

where $a = 2$ for the semi-tangential vector and $a = 1$ for the canonical vector. Note that we could also calculate a variation for the left multiplicative composition $\mathbf{R}_2(\epsilon\tilde{\boldsymbol{\theta}}_B)\,\mathbf{R}(\boldsymbol{\psi})$, using $\boldsymbol{\theta}_B = \mathbf{T}\mathbf{T}^{-T}\boldsymbol{\Theta}_B$, but it is not used in calculations.

Remark. By using $\boldsymbol{\theta}_B = \mathbf{T}(\boldsymbol{\psi})\,\boldsymbol{\theta}_A$ and $\boldsymbol{\Theta}_B = \mathbf{T}^T(\boldsymbol{\psi})\,\boldsymbol{\theta}_A$, eq. (9.57) yields,

$$\tilde{\boldsymbol{\theta}}_B = \mathbf{R}\tilde{\boldsymbol{\Theta}}_B\mathbf{R}^T, \qquad \boldsymbol{\theta}_B = \mathbf{R}\boldsymbol{\Theta}_B, \tag{9.67}$$

where $\boldsymbol{\Theta}_B$ and $\boldsymbol{\theta}_B$ are the axial vectors of $\tilde{\boldsymbol{\Theta}}_B$ and $\tilde{\boldsymbol{\theta}}_B$, respectively. We see that $\tilde{\boldsymbol{\theta}}_B$ is the forward-rotated $\tilde{\boldsymbol{\Theta}}_B$, hence their properties are linked, as discussed in Sect. 8.3.1. In particular, if we assume two Cartesian bases, the reference basis $\{\mathbf{i}_i\}$ and the rotated basis $\{\mathbf{t}_i\}$, where $\mathbf{t}_i \doteq \mathbf{R}\,\mathbf{i}_i$, and the representations,

$$\tilde{\boldsymbol{\Theta}} = \tilde{\Theta}_{ij}\,\mathbf{i}_i \otimes \mathbf{i}_j, \qquad \tilde{\boldsymbol{\theta}} = \tilde{\theta}_{ij}\,\mathbf{t}_i \otimes \mathbf{t}_j, \tag{9.68}$$

then eq. (9.67) implies

$$\tilde{\boldsymbol{\Theta}} = \mathbf{R}^T(\tilde{\theta}_{ij}\,\mathbf{t}_i \otimes \mathbf{t}_j)\,\mathbf{R} = \tilde{\theta}_{ij}\,(\mathbf{R}^T\mathbf{t}_i \otimes \mathbf{R}^T\mathbf{t}_j) = \tilde{\theta}_{ij}\,\mathbf{i}_i \otimes \mathbf{i}_j. \tag{9.69}$$

Hence, $\tilde{\Theta}_{ij} = \tilde{\theta}_{ij}$, i.e. the components of $\tilde{\boldsymbol{\Theta}}$ in the basis $\{\mathbf{i}_i\}$ and the components of $\tilde{\boldsymbol{\theta}}$ in the basis $\{\mathbf{t}_i\}$ are identical. In other words, $\tilde{\boldsymbol{\Theta}}$ can be considered as $\tilde{\boldsymbol{\theta}}$ parallel transported from the rotated basis to the reference basis.

Example. Consider the case when the rotation and the variations are performed around the axis $\mathbf{t}_3$ of the basis $\{\mathbf{t}_i\}$. Assume the following representations of two canonical rotation vectors $\boldsymbol{\psi} = [0, 0, \psi_3]^T$ and $\boldsymbol{\theta}_A = [0, 0, \theta_3]^T$, which yield

$$\mathbf{R}(\psi_3) = \begin{bmatrix} \cos\psi_3 & -\sin\psi_3 & 0 \\ +\sin\psi_3 & \cos\psi_3 & 0 \\ 0 & 0 & 1 \end{bmatrix}, \qquad \tilde{\boldsymbol{\theta}}_A = \theta_3 \begin{bmatrix} 0 & -1 & 0 \\ +1 & 0 & 0 \\ 0 & 0 & 0 \end{bmatrix}.$$

For the additive variation, on use of eq. (9.42), we can calculate,

$$\delta_{\tilde{\theta}_A}\mathbf{R}(\boldsymbol{\psi}) \doteq D\mathbf{R}(\psi_3)\cdot\tilde{\boldsymbol{\theta}}_A = \frac{\mathrm{d}}{\mathrm{d}\epsilon}[\mathbf{R}(\psi_3 + \epsilon\theta_3)]_{\epsilon=0} = \frac{\mathrm{d}\mathbf{R}}{\mathrm{d}\psi_3}\,\theta_3, \tag{9.70}$$

where

$$\frac{\mathrm{d}\mathbf{R}}{\mathrm{d}\psi_3} = \begin{bmatrix} -\sin\psi_3 & -\cos\psi_3 & 0 \\ \cos\psi_3 & -\sin\psi_3 & 0 \\ 0 & 0 & 0 \end{bmatrix}.$$

For the left variation of $\mathbf{R}$, we only perform a multiplication, which yields

$$\delta_{\tilde{\theta}_B}\mathbf{R} = \tilde{\boldsymbol{\theta}}_A\,\mathbf{R} = \theta_3 \begin{bmatrix} -\sin\psi_3 & -\cos\psi_3 & 0 \\ \cos\psi_3 & -\sin\psi_3 & 0 \\ 0 & 0 & 0 \end{bmatrix}. \tag{9.71}$$

Similarly for the right variation of $\mathbf{R}$,

$$\delta_{\tilde{\Theta}_B}\mathbf{R} = \mathbf{R}\,\tilde{\boldsymbol{\theta}}_A = \theta_3 \begin{bmatrix} -\sin\psi_3 & -\cos\psi_3 & 0 \\ \cos\psi_3 & -\sin\psi_3 & 0 \\ 0 & 0 & 0 \end{bmatrix}. \tag{9.72}$$

Note that $\delta_{\tilde{\Theta}_B}\mathbf{R} = \delta_{\tilde{\theta}_B}\mathbf{R}$, indeed. This shows that the difference between the left and right variation vanishes for the rotation and the variations around one axis. Then the tangent operator $\mathbf{T}(\boldsymbol{\psi})$ acts as the identity operator, see eq. (9.21).

Example. In this example, we calculate the variations of the rotation tensor in a slightly different way; directly using the composition formula of eq. (8.189) and the rotation tensor of eq. (8.99) for the semi-tangential parametrization.

Assume that $\tilde{\boldsymbol{\psi}}, \tilde{\boldsymbol{\theta}}_A \in T_I SO(3)$, and their axial vectors are: $\boldsymbol{\psi} = [9, 5, -1]^T$ and $\boldsymbol{\theta}_A = [-0.8, -0.1, 0.4]^T$. The variations are defined as follows:

1. for the additive composition, we use $\boldsymbol{\psi}_A = \boldsymbol{\psi} + \epsilon\boldsymbol{\theta}_A$, and calculate

$$\delta_{\tilde{\theta}_A}\mathbf{R}(\boldsymbol{\psi}) = \frac{\mathrm{d}}{\mathrm{d}\epsilon}[\mathbf{R}(\boldsymbol{\psi}_A)]_{\epsilon=0} = \begin{bmatrix} -0.0388889 & 0.0296296 & 0.0685185 \\ 0.0388889 & 0.0537037 & 0.0148148 \\ 0.0444444 & 0.0351852 & -0.00925926 \end{bmatrix}.$$

2. for the left composition, we use $\boldsymbol{\psi}_B = \frac{1}{1-\boldsymbol{\psi}\cdot\epsilon\boldsymbol{\theta}_B}(\boldsymbol{\psi} + \epsilon\boldsymbol{\theta}_B + \epsilon\boldsymbol{\theta}_B \times \boldsymbol{\psi})$, and calculate

$$\delta_{\tilde{\theta}_B}\mathbf{R}(\boldsymbol{\psi}) = \frac{\mathrm{d}}{\mathrm{d}\epsilon}[\mathbf{R}(\boldsymbol{\psi}_B)]_{\epsilon=0} = \begin{bmatrix} -0.0388889 & 0.0296296 & 0.0685185 \\ 0.0388889 & 0.0537037 & 0.0148148 \\ 0.0444444 & 0.0351852 & -0.00925926 \end{bmatrix},$$

where $\boldsymbol{\theta}_B = \mathbf{T}(\boldsymbol{\psi})\ \boldsymbol{\theta}_A = [0.0101852, -0.0268519, 0.0324074]^T$. We checked that $\delta_{\tilde{\theta}_B}\mathbf{R} = (2\tilde{\boldsymbol{\theta}}_B)\ \mathbf{R}(\boldsymbol{\psi})$ yields exactly the same matrix.

3. for the right composition, we use $\boldsymbol{\psi}_B = \frac{1}{1-\boldsymbol{\psi}\cdot\epsilon\boldsymbol{\Theta}_B}(\boldsymbol{\psi}+\epsilon\boldsymbol{\Theta}_B-\epsilon\boldsymbol{\Theta}_B\times\boldsymbol{\psi})$, and calculate

$$\delta_{\tilde{\Theta}_B}\mathbf{R}(\boldsymbol{\psi}) = \frac{\mathrm{d}}{\mathrm{d}\epsilon}[\mathbf{R}(\boldsymbol{\psi}_B)]_{\epsilon=0} = \begin{bmatrix} -0.0388889 & 0.0296296 & 0.0685185 \\ 0.0388889 & 0.0537037 & 0.0148148 \\ 0.0444444 & 0.0351852 & -0.00925926 \end{bmatrix},$$

where $\boldsymbol{\Theta}_B = \mathbf{T}^T(\boldsymbol{\psi})\ \boldsymbol{\theta}_A = [-0.025, 0.025, -0.025]^T$. We check that $\delta_{\tilde{\Theta}_B}\mathbf{R} = \mathbf{R}(\boldsymbol{\psi})\ (2\tilde{\boldsymbol{\Theta}}_B)$ yields exactly the same matrix.

The above results confirm that, as derived in eq. (9.58), the variations are equal indeed.

Now, we calculate the rotation matrices $\mathbf{R}$ for $\boldsymbol{\psi}_A$ and $\boldsymbol{\psi}_B$ obtained by taking $\epsilon = 1$, i.e. for the finite increment of rotation vectors. Note that for $\boldsymbol{\psi} = \mathbf{0}$ we obtain $\boldsymbol{\psi}_A = \boldsymbol{\psi}_B$, because then $\boldsymbol{\psi}_A = \boldsymbol{\theta}_A$ and $\mathbf{T} = \mathbf{I}$ so $\boldsymbol{\theta}_B = \boldsymbol{\theta}_A$ and $\boldsymbol{\psi}_B = \boldsymbol{\theta}_B$, for which we finally obtain

$\boldsymbol{\psi}_B = \boldsymbol{\theta}_A = \boldsymbol{\psi}_A$. However, for $\boldsymbol{\psi} \neq \mathbf{0}$, a difference between $\boldsymbol{\psi}_A$ and $\boldsymbol{\psi}_B$ exists so the rotation matrices are different, i.e.

$$\mathbf{R}(\boldsymbol{\psi}_A) = \begin{bmatrix} 0.473707 & 0.880682 & -0.000431 \\ 0.854767 & -0.459886 & -0.240579 \\ -0.212072 & 0.113595 & -0.970630 \end{bmatrix},$$

$$\mathbf{R}(\boldsymbol{\psi}_B) = \begin{bmatrix} 0.477253 & 0.878746 & -0.005914 \\ 0.851921 & -0.464315 & -0.242162 \\ -0.215545 & 0.110533 & -0.970218 \end{bmatrix}.$$

For both multiplicative composition rules, we obtain the same $\mathbf{R}(\boldsymbol{\psi}_B)$, which is in agreement with eq. (8.177).

9.2.4 Second variation of rotation tensor

Definition of second variation. To define the second variation, we must extend the notation used earlier.

For instance, for the additive composition of rotational parameters, we used $\boldsymbol{\psi}_\epsilon = \boldsymbol{\psi} + \epsilon \boldsymbol{\theta}_A \in T_I SO(3)$ to define the first variation in eq. (9.42). Now, we need two perturbed vectors

$$\boldsymbol{\psi}^- = \boldsymbol{\psi} + \epsilon \boldsymbol{\theta}^- \in T_I SO(3), \qquad \boldsymbol{\psi}^+ = \boldsymbol{\psi} + \epsilon \boldsymbol{\theta}^+ \in T_I SO(3), \tag{9.73}$$

so we use two superscripts, "–" and "+", and omit the subscript "ϵ", to simplify the notation. (Note that we could also denote $\boldsymbol{\theta}^- \doteq \delta\boldsymbol{\psi}$ and $\boldsymbol{\theta}^+ \doteq \Delta\boldsymbol{\psi}$, using the notation typical in mechanics.) We define two variations of some function f as the following directional derivatives:

$$\delta f \doteq \frac{\mathrm{d}}{\mathrm{d}\epsilon} f(\boldsymbol{\psi} + \epsilon \boldsymbol{\theta}^-)\Big|_{\epsilon=0}, \qquad \chi f \doteq \frac{\mathrm{d}}{\mathrm{d}\epsilon} f(\boldsymbol{\psi} + \epsilon \boldsymbol{\theta}^+)\Big|_{\epsilon=0}, \tag{9.74}$$

where "δ" and "χ" are associated with the directions $\boldsymbol{\theta}^-$ and $\boldsymbol{\theta}^+$, respectively. The second variation is defined as the directional derivative of the first variation,

$$\chi(\delta f) \doteq \frac{\mathrm{d}}{\mathrm{d}\epsilon} \delta f(\boldsymbol{\psi} + \epsilon \boldsymbol{\theta}^+)\Big|_{\epsilon=0}. \tag{9.75}$$

Analogous expressions will be used for multiplicative composition of rotations.

Below, the second variations are derived for the canonical rotation vector; for the semi-tangential vector, the procedure is analogous and the obtained results are provided in eqs. (9.87) and (9.88).

A. Second variation of rotation tensor for additive composition

For the additive composition of the rotation parameters (in the notation introduced above), the first variation is defined as

$$\delta\mathbf{R} \doteq D\mathbf{R}(\boldsymbol{\psi}) \cdot \tilde{\boldsymbol{\theta}}^{-} = \frac{\mathrm{d}}{\mathrm{d}\epsilon}[\mathbf{R}(\boldsymbol{\psi}^{-})]_{\epsilon=0}, \tag{9.76}$$

while the second variation is defined as

$$\chi\delta\mathbf{R} \doteq D\,[\delta\mathbf{R}] \cdot \tilde{\boldsymbol{\theta}}^{+} = \frac{\mathrm{d}}{\mathrm{d}\epsilon}[\delta\mathbf{R}(\boldsymbol{\psi}^{+})]_{\epsilon=0}. \tag{9.77}$$

The above directional derivatives can be calculated using automatic differentiation of a symbolic manipulation program. The concise forms can be also derived using the multiplicative composition of rotation tensors, as shown below.

B. Second variation of rotation tensor for multiplicative composition

Recall eq. (9.57) linking the variations for the additive composition and the multiplicative (left and right) compositions of canonical rotation parameters, which can be rewritten as follows:

$$\underbrace{\delta\mathbf{R}}_{\text{additive}} = \underbrace{\{[\mathbf{T}(\boldsymbol{\psi})\,\boldsymbol{\theta}^{-}] \times \mathbf{I}\}\,\mathbf{R}(\boldsymbol{\psi})}_{\text{left, multiplicative}} = \underbrace{\mathbf{R}(\boldsymbol{\psi})\,\{[\mathbf{T}^{T}(\boldsymbol{\psi})\,\boldsymbol{\theta}^{-}] \times \mathbf{I}\}}_{\text{right, multiplicative}}, \tag{9.78}$$

where the tangent operator $\mathbf{T}$ is defined in eq. (9.17). Below, we only consider the canonical parametrization for which $\mathbf{T}$ has the form given in eq. (9.18).

Left multiplicative rule. The second differential of $\mathbf{R}$ is defined as the directional derivative of the respective first variation of eq. (9.78) in direction $\boldsymbol{\theta}^{+}$,

$$\chi(\delta\,\mathbf{R}) \doteq \frac{\mathrm{d}}{\mathrm{d}\epsilon}\left.\{[\mathbf{T}(\boldsymbol{\psi} + \epsilon\boldsymbol{\theta}^{+})\;\boldsymbol{\theta}^{-}] \times \mathbf{I}\}\,\mathbf{R}(\boldsymbol{\psi} + \epsilon\boldsymbol{\theta}^{+})\right|_{\epsilon=0}. \tag{9.79}$$

The derivative of a cross-product of a vector $\mathbf{a}$ and a tensor $\mathbf{A}$ with respect to the scalar ϵ is $(\mathbf{a} \times \mathbf{A})^{'} = \mathbf{a}^{'} \times \mathbf{A} + \mathbf{a} \times \mathbf{A}^{'}$ and, hence,

$$\chi(\delta\,\mathbf{R}) = \{(\chi\mathbf{T}\,\boldsymbol{\theta}^{-}) \times \mathbf{I}\}\,\mathbf{R} \;+\; \{(\mathbf{T}\,\boldsymbol{\theta}^{-}) \times \mathbf{I}\}\,\chi\mathbf{R}. \tag{9.80}$$

Using $\chi\mathbf{R} = [(\mathbf{T}\boldsymbol{\theta}^{+}) \times \mathbf{I}]\,\mathbf{R},$ this becomes

$$\chi(\delta\,\mathbf{R}) = \{[(\chi\mathbf{T}\,\boldsymbol{\theta}^{-})\times\mathbf{I}] + [(\mathbf{T}\boldsymbol{\theta}^{-})\times\mathbf{I}][(\mathbf{T}\boldsymbol{\theta}^{+})\times\mathbf{I}]\}\;\mathbf{R}, \tag{9.81}$$

where $\mathbf{R}$ is factored out of the braces.

The second component can be directly evaluated. Regarding the first component, the differential $\chi\mathbf{T}$ is defined in eq. (9.40) and we can calculate the product

$$\begin{aligned}\chi\mathbf{T}(\boldsymbol{\psi},\boldsymbol{\theta}^{+})\,\boldsymbol{\theta}^{-} &= a_1\,(\mathbf{e}\cdot\boldsymbol{\theta}^{+})\,\boldsymbol{\theta}^{-} + a_2\,(\mathbf{e}\cdot\boldsymbol{\theta}^{-})\,\boldsymbol{\theta}^{+} + a_2\,(\boldsymbol{\theta}^{+}\cdot\boldsymbol{\theta}^{-})\,\mathbf{e}\\ &+ a_3\,(\mathbf{e}\cdot\boldsymbol{\theta}^{+})(\mathbf{e}\cdot\boldsymbol{\theta}^{-})\,\mathbf{e} + a_4\,(\mathbf{e}\cdot\boldsymbol{\theta}^{+})\,\tilde{\boldsymbol{\psi}}\,\boldsymbol{\theta}^{-} + a_5\,(\boldsymbol{\theta}^{+}\times\boldsymbol{\theta}^{-}),\end{aligned} \tag{9.82}$$

which is a vector, so the term $(\chi\mathbf{T}\,\boldsymbol{\theta}^{-})\times\mathbf{I}$ is the associated skew-symmetric tensor. We see that only two terms (third and fourth) are symmetric with respect to $\boldsymbol{\theta}^{-}$ and $\boldsymbol{\theta}^{+}$.

Right multiplicative rule. The second differential of $\mathbf{R}$ is defined as the directional derivative of the respective first variation of eq. (9.78) in the direction $\boldsymbol{\theta}^{+}$,

$$\chi(\delta\,\mathbf{R}) \doteq \frac{\mathrm{d}}{\mathrm{d}\epsilon}\left.\{\mathbf{R}(\boldsymbol{\psi}+\epsilon\boldsymbol{\theta}^{+})\,[\mathbf{T}^{T}(\boldsymbol{\psi}+\epsilon\boldsymbol{\theta}^{+})\;\boldsymbol{\theta}^{-}\times\mathbf{I}]\}\right|_{\epsilon=0}. \tag{9.83}$$

Then we obtain

$$\chi(\delta\,\mathbf{R}) = \chi\mathbf{R}\,[(\mathbf{T}^{T}\boldsymbol{\theta}^{-})\times\mathbf{I}] + \mathbf{R}\,[(\chi\mathbf{T}^{T}\boldsymbol{\theta}^{-})\times\mathbf{I}], \tag{9.84}$$

which, using $\chi\mathbf{R} = \mathbf{R}\,[(\mathbf{T}^{T}\boldsymbol{\theta}^{+})\times\mathbf{I}]$, becomes

$$\chi(\delta\,\mathbf{R}) = \mathbf{R}\,\{[(\chi\mathbf{T}^{T}\boldsymbol{\theta}^{-})\times\mathbf{I}] + [(\mathbf{T}^{T}\boldsymbol{\theta}^{+})\times\mathbf{I}][(\mathbf{T}^{T}\boldsymbol{\theta}^{-})\times\mathbf{I}]\}, \tag{9.85}$$

where $\mathbf{R}$ pre-multiplies the term in braces, and in the last term the position of $\boldsymbol{\theta}^{+}$ and $\boldsymbol{\theta}^{-}$ is interchanged, comparing to eq. (9.81) for the left composition.

The second component can be directly evaluated. In the first component, the differential $\chi\mathbf{T}$ is defined in eq. (9.40), and its transposition changes the sign of skew-symmetric terms at a_4 and a_5. Then the product becomes

$$\begin{aligned}\chi\mathbf{T}^{T}(\boldsymbol{\psi},\boldsymbol{\theta}^{+})\,\boldsymbol{\theta}^{-} &= a_1\,(\mathbf{e}\cdot\boldsymbol{\theta}^{+})\,\boldsymbol{\theta}^{-} + a_2\,(\mathbf{e}\cdot\boldsymbol{\theta}^{-})\,\boldsymbol{\theta}^{+} + a_2\,(\boldsymbol{\theta}^{+}\cdot\boldsymbol{\theta}^{-})\,\mathbf{e}\\ &+ a_3\,(\mathbf{e}\cdot\boldsymbol{\theta}^{+})(\mathbf{e}\cdot\boldsymbol{\theta}^{-})\,\mathbf{e} - a_4\,(\mathbf{e}\cdot\boldsymbol{\theta}^{+})\,\tilde{\boldsymbol{\psi}}\,\boldsymbol{\theta}^{-} - a_5\,(\boldsymbol{\theta}^{+}\times\boldsymbol{\theta}^{-}),\end{aligned} \tag{9.86}$$

where only the third and fourth terms are symmetric with respect to $\boldsymbol{\theta}^{-}$ and $\boldsymbol{\theta}^{+}$.

Second variations for semi-tangential rotation vector. The second differential of $\mathbf{R}$ is defined as the directional derivative of the respective first variation of eq. (9.60) in direction $\boldsymbol{\theta}^+$. Then, for the left composition of rotations, we obtain

$$\chi(\delta\mathbf{R}) = \left\{2\left[(\chi\mathbf{T}\boldsymbol{\theta}^-)\times\mathbf{I}\right] + 4\left[(\mathbf{T}\boldsymbol{\theta}^-)\times\mathbf{I}\right]\left[(\mathbf{T}\boldsymbol{\theta}^+)\times\mathbf{I}\right]\right\}\mathbf{R}, \tag{9.87}$$

while for the right composition of rotations, we obtain

$$\chi(\delta\mathbf{R}) = \mathbf{R}\left\{2\left[(\chi\mathbf{T}^T\boldsymbol{\theta}^-)\times\mathbf{I}\right] + 4\left[(\mathbf{T}^T\boldsymbol{\theta}^+)\times\mathbf{I}\right]\left[(\mathbf{T}^T\boldsymbol{\theta}^-)\times\mathbf{I}\right]\right\}. \tag{9.88}$$

Note that the multipliers 2 and 4 have appeared in these formulas, in comparison with eqs. (9.81) and (9.85).

Special case: co-axial rotation vectors. Consider the case when $\boldsymbol{\psi}, \boldsymbol{\theta}^+, \boldsymbol{\theta}^-$ are co-axial vectors, i.e. the rotations are performed about one axis, $\mathbf{e}$. Let $\boldsymbol{\psi} = \psi\,\mathbf{e}$, $\boldsymbol{\theta}^- = \alpha\,\mathbf{e}$, and $\boldsymbol{\theta}^+ = \beta\,\mathbf{e}$, where ψ, α, β denote the angles of rotation. Then, by the property of eq. (9.21), we have

$$\mathbf{T}(\boldsymbol{\psi})\,\boldsymbol{\theta}^- = \alpha\mathbf{e}, \qquad \mathbf{T}(\boldsymbol{\psi})\,\boldsymbol{\theta}^+ = \beta\mathbf{e}, \qquad \chi\mathbf{T}\,\boldsymbol{\theta}^- = \alpha\beta\,A\,\mathbf{e}, \tag{9.89}$$

where $A \doteq (-2b_1 + a_3 + a_2 + b_2 + b_3) = -[2(\cos\omega - 1) + \omega]/\omega^2$, and $\omega \doteq \|\boldsymbol{\psi}\| = \sqrt{\boldsymbol{\psi}^2}$. The plot of A is presented in Fig. 9.4. For $\omega \to \infty$, A tends to zero.

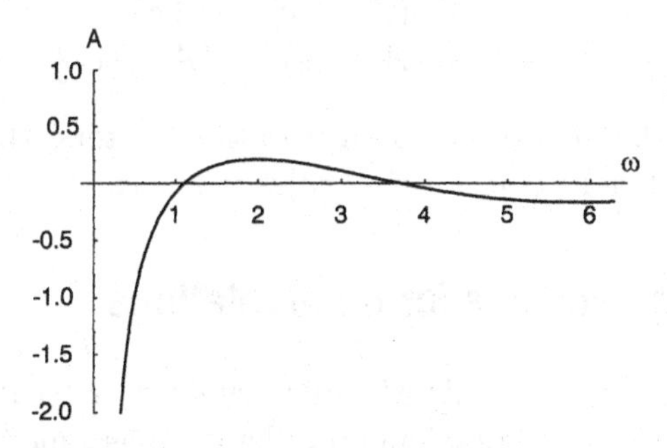

Fig. 9.4 Coefficient A as a function of ω.

The first differential of $\mathbf{R}$ of eq. (9.78) becomes

$$\delta\mathbf{R} = \left\{[\mathbf{T}(\boldsymbol{\psi})\,\boldsymbol{\theta}^-]\times\mathbf{I}\right\}\mathbf{R}(\boldsymbol{\psi}) = \alpha\,(\mathbf{e}\times\mathbf{I})\,\mathbf{R}(\boldsymbol{\psi}), \tag{9.90}$$

while the second differential of eq. (9.81) becomes

$$\chi(\delta \mathbf{R}) = \alpha\beta\,(A\,\mathbf{e}\times\mathbf{I} + \,\mathbf{e}\otimes\mathbf{e} - \mathbf{I})\,\mathbf{R}. \tag{9.91}$$

Next, we can use the relations $\mathbf{S} \doteq \mathbf{e}\times\mathbf{I}$ and $\mathbf{S}^2 = \mathbf{e}\otimes\mathbf{e} - \mathbf{I}$ from Table 8.1, and the rotation tensor in the form of eq. (8.9), $\mathbf{R} \doteq \mathbf{I} + s\,\mathbf{S} + (1-c)\,\mathbf{S}^2$, where $s \doteq \sin\omega$ and $c \doteq \cos\omega$. Then the differentials of the rotation tensors are as follows:

$$\delta\mathbf{R} = \alpha\mathbf{S}\,\mathbf{R}(\psi) = \alpha[\mathbf{S} + s\,\mathbf{S}^2 + (1-c)\,\mathbf{S}^3] = \alpha(c\,\mathbf{S} + s\,\mathbf{S}^2), \tag{9.92}$$

$$\chi(\delta\mathbf{R}) = \alpha\beta\,(A\,\mathbf{S} \;+\; \mathbf{S}^2)\,\mathbf{R} = \alpha\beta\left[(A\,c - s)\,\mathbf{S} + (A\,s + c)\,\mathbf{S}^2\right]. \tag{9.93}$$

For simplicity assume that the rotations are performed around the reference axis $\mathbf{i}_3$, i.e. $\mathbf{e} \doteq \mathbf{i}_3$. Then

$$\mathbf{S} = \mathbf{i}_2\otimes\mathbf{i}_1 - \mathbf{i}_1\otimes\mathbf{i}_2, \qquad \mathbf{S}^2 = -(\mathbf{i}_1\otimes\mathbf{i}_1 + \mathbf{i}_2\otimes\mathbf{i}_2),$$

see Table 8.1, and representations of the tensors are

$$\mathbf{R} = \mathbf{I} + s\,\mathbf{S} + (1-c)\,\mathbf{S}^2 = \begin{bmatrix} c & -s \\ s & c \end{bmatrix}, \tag{9.94}$$

$$\delta\mathbf{R} = \alpha(c\,\mathbf{S} + s\,\mathbf{S}^2) = \alpha\begin{bmatrix} -s & -c \\ c & -s \end{bmatrix}, \tag{9.95}$$

$$\begin{aligned} \chi(\delta\mathbf{R}) &= \alpha\beta\left[(A\,c - s)\,\mathbf{S} + (A\,s + c)\,\mathbf{S}^2\right] \\ &= \alpha\beta\begin{bmatrix} -(A\,s + c) & -(A\,c - s) \\ (A\,c - s) & -(A\,s + c) \end{bmatrix}. \end{aligned} \tag{9.96}$$

Finally, we note that for the co-axial rotation vectors, the difference between the left and right composition rules vanishes.

9.3 Algorithmic schemes for finite rotations

In this section we consider algorithmic schemes of treating finite rotations for a static (time-independent) problem and assume that the Newton method is used to solve the non-linear equilibrium equations.

The tangent matrix and residual for the Newton method can be obtained from any of the three forms of variations which were presented earlier as follows:

1. For the canonical parametrization: the first variation in eq. (9.57) and the second variation in eqs. (9.77), (9.81), and (9.85).

2. For the semi-tangential parametrization: the first variation in eq. (9.60) and the second variation in eqs. (9.77), (9.87), and (9.88).

The tangent matrix and residual obtained for these three forms are fully equivalent.

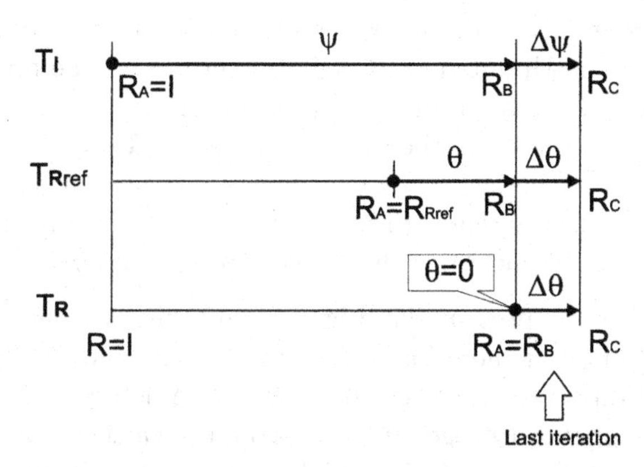

Fig. 9.5 Three schemes of treating finite rotations.

In the algorithmic treatment of rotations, we have to make several choices, regarding (a) the rotation vector used for the increment of rotations (e.g. canonical or semi-tangential vector), (b) parameters used to store nodal rotations (e.g. quaternions or rotation matrices), and (c) an approximation of nodal rotation parameters over the element.

To select the best algorithmic procedure, we have to consider the theoretical properties of each combination of choices, implement them, and subject them to rigorous testing. This is not only laborious but also requires accounting for certain limitations. For instance, in testing of rotations, we must avoid too complex examples to be able to run them automatically without the user's intervention. Non-linear solutions can be very complex and possess extremum and turning points, as well as bifurcation points. To obtain some solutions, not only is an arc-length procedure required, but also additional advanced capabilities, enabling localization of bifurcation points and branch switching. Such capabilities require the user's assistance and, for this reason, are not currently available in commercial FE codes. When testing schemes for rotations we should avoid examples which require them.

Below, we describe three algorithmic schemes of treating finite rotations, formulated in various tangent planes to SO(3):

Scheme 1 is formulated in the initial tangent plane at $\mathbf{R}_A = \mathbf{I}$, i.e. in T_I, and corresponds to the Total Lagrangian description. The global rotation vector $\boldsymbol{\psi}$ is used and updated throughout the whole solution process.

Scheme 2 is formulated in the tangent plane at $\mathbf{R}_A = \mathbf{R}_{\text{ref}}$, i.e. in $T_{R_{\text{ref}}}$, where $\mathbf{R}_{\text{ref}}$ is the last converged solution. This scheme corresponds to the Updated Lagrangian description. The rotation vector $\boldsymbol{\theta}$ is used and updated for each step (increment).

Scheme 3 is formulated in the tangent plane at $\mathbf{R}_A = \mathbf{R}_B$, i.e. in T_R, where $\mathbf{R}_B$ is the last available solution, in general, non-converged. This scheme corresponds to the Eulerian description. The rotation vector $\boldsymbol{\theta} = \mathbf{0}$ throughout the whole solution process.

All these schemes are presented in Fig. 9.5 and the increment of a rotation vector is used in all of them for an iteration of the Newton method. Note that the rotation vector involves only three parameters and, hence, there is no need to append orthogonality constraints, which is convenient. The increment of a rotation vector can also belong to various tangent planes, so we use either $\Delta\boldsymbol{\psi}$ or $\Delta\boldsymbol{\theta}$. Additional questions must also be considered, such as

1. how to update the rotation vector $\boldsymbol{\psi}$ used by Scheme 1 and $\boldsymbol{\theta}$ used by Scheme 2. We can use either a multiplicative update scheme or an additive update scheme.
   In the multiplicative update, we can use either the rotation matrices or quaternions. The quaternions give the advantage that they can be easily renormalized, so they always yield orthogonal rotation matrices. It is complicated to recover the orthogonality for the rotation matrices.
2. How to update the rotation matrix $\mathbf{R}_A$, which is used by Schemes 2 and 3. If, instead of the rotation matrices, we use quaternions, then the update can be achieved via the composition of quaternions given by eq. (8.185).

For both updates, the increment of the rotation vector must be converted to a quaternion and composed with the known quaternion; either for the previous step (in Scheme 2) or for the previous iteration (in Scheme 3).

9.3.1 Scheme 1: formulation in $\mathrm{T}_I\mathrm{SO}(3)$

In this scheme we use as the rotational unknown, the rotation vector $\boldsymbol{\psi}$ related to $\mathbf{R}_A = \mathbf{I}$, i.e. $\tilde{\boldsymbol{\psi}} \in T_I\mathrm{SO}(3)$, where T_I is the initial tangent plane, see Fig. 9.5. The total rotation is represented by $\mathbf{R}_B \doteq \mathbf{R}(\boldsymbol{\psi})$.

The additive update procedure for the rotation vector and its increment belonging to the initial tangent plane, i.e. $\tilde{\psi}, \Delta\tilde{\psi} \in T_I\mathrm{SO}(3)$, is presented in Table 9.2.

Table 9.2 Scheme 1. Additive update. $\tilde{\psi}, \Delta\tilde{\psi} \in T_I\mathrm{SO}(3)$.

Initialize: $\psi = \mathbf{0}$	$\leftarrow$	total
Step		
Newton loop		
Form equilibrium equations using ψ, solve for $\Delta\psi$		
Update $\psi = \psi + \Delta\psi$	$\leftarrow$	total (additive)
End of Newton loop		

The multiplicative update of rotations is exact, see Sect. 8.3, but the additive update of Scheme 1 also yields very accurate results for shells, even in examples involving finite rotations. This can be explained as follows:

1. The additive update is exact if ψ and $\Delta\psi$ are parallel and have the same sense, i.e.
$$\mathbf{R}_C = \underbrace{\mathbf{R}(\Delta\psi)\,\mathbf{R}(\psi)}_{\text{left}} = \mathbf{R}(\psi + \Delta\psi), \tag{9.97}$$
see the proof of eq. (8.195). Such, or almost such, rotations and their increments are characteristic for the most common one-parameter external loads, i.e. with the magnitude varied by one parameter.
2. Even if the formula $\psi + \Delta\psi$ is not exact, still the operation $(\psi + \Delta\psi) \to \mathbf{R}$ is exact.
3. In problems involving large strains and non-linear materials, the rotations are not the only source of non-linearity and, hence, we do not have to be too exact when updating rotational parameters, as long as the Newton scheme converges.

However, we must be aware that there are limits of this scheme and, e.g. in the twisted ring example of Sect. 9.3.5, it yields a wrong solution at the rotations close to 2π.

9.3.2 Scheme 2: formulation in $\mathrm{T}_{R_{\mathrm{ref}}}\mathrm{SO}(3)$

In this scheme, we use as the rotational unknown, the rotation vector $\boldsymbol{\theta}$ (or $\boldsymbol{\Theta}$) related to $\mathbf{R}_{\mathrm{ref}} \doteq \mathbf{R}_A$ where $\mathbf{R}_A$ is the last converged solution.

Hence, $\tilde{\boldsymbol{\theta}}\,\mathbf{R}_A$ (or $\mathbf{R}_A\,\tilde{\boldsymbol{\Theta}}$) $\in T_{R_{\text{ref}}}\text{SO}(3)$, where $T_{R_{\text{ref}}}$ is the reference tangent plane, see Fig. 9.5.

The total rotation $\mathbf{R}_C$ is related to the known rotation $\mathbf{R}_B$ with the help of the rotation for the step by either the left or the right composition rule as follows:

$$\mathbf{R}_C = \underbrace{\mathbf{R}(\Delta\boldsymbol{\theta})\;\mathbf{R}(\boldsymbol{\theta})}_{\text{left, for step}}\;\mathbf{R}_B, \qquad \mathbf{R}_C = \mathbf{R}_B\;\underbrace{\mathbf{R}(\boldsymbol{\Theta})\;\mathbf{R}(\Delta\boldsymbol{\Theta})}_{\text{right, for step}}. \tag{9.98}$$

The notation used above is similar to that in eqs. (9.1) and (9.28).

For this formulation, we use two schemes based on quaternions which are presented below; the multiplicative update in Table 9.3 and the multiplicative/additive update in Table 9.4. They have the following features:

1. In both schemes, the total quaternion is used. In the first scheme, the total quaternion $\mathbf{X}$ is updated in each iteration,

$$\mathbf{R}_B = \mathbf{R}(\Delta\boldsymbol{\theta})\;\underbrace{\mathbf{R}(\boldsymbol{\theta})\;\mathbf{R}_A}_{\mathbf{X}}, \qquad \mathbf{R}_B = \underbrace{\mathbf{R}_A\;\mathbf{R}(\boldsymbol{\Theta})}_{\mathbf{X}}\;\mathbf{R}(\Delta\boldsymbol{\Theta}), \tag{9.99}$$

while in the second scheme, the total quaternion $\mathbf{X}_n$ is updated when the Newton iterations have converged,

$$\mathbf{R}_B = \mathbf{R}(\Delta\boldsymbol{\theta})\;\mathbf{R}(\boldsymbol{\theta})\;\underbrace{\mathbf{R}_A}_{\mathbf{X}_n}, \qquad \mathbf{R}_B = \underbrace{\mathbf{R}_A}_{\mathbf{X}_n}\;\mathbf{R}(\boldsymbol{\Theta})\;\mathbf{R}(\Delta\boldsymbol{\Theta}). \tag{9.100}$$

In both schemes, the updates of the total quaternion are multiplicative, via a composition of quaternions, as in eq. (8.185). The calculations for the step are different in each scheme.

2. In the multiplicative scheme, the increment of the rotation vector $\Delta\boldsymbol{\Theta}$ is converted to the quaternion $\Delta\mathbf{q}$ using eq. (8.96). The previous quaternion for the step $\mathbf{q}$ and the quaternion for the increment $\Delta\mathbf{q}$ are composed as in eq. (8.185) and then the rotation vector for the step $\boldsymbol{\Theta}$ is extracted from $\mathbf{q}$ by using eq. (8.94).
3. In the multiplicative/additive scheme, the increment of rotation vector $\Delta\boldsymbol{\Theta}$ is added to the rotation vector for the step $\boldsymbol{\Theta}$, converted to the quaternion for the step $\mathbf{q}$, and composed with the quaternion $\mathbf{X}_n$ for the previous converged solution. Note that $\mathbf{X}_n$ is updated only when the Newton iterations for the step have converged.

4. In both schemes, the rotation vector for the step, $\boldsymbol{\Theta}$ belongs to the tangent plane at the converged rotation for the previous increment, $\mathbf{R}^0_{n+1}\tilde{\boldsymbol{\Theta}} \in T_{R^0_{n+1}}SO(3)$. However, the increments $\Delta\boldsymbol{\Theta}$ belong in the multiplicative scheme to $T_{R^i_{n+1}}SO(3)$, which is the tangent plane at the last available rotation, not necessarily converged, and to $T_{R^0_{n+1}}SO(3)$ in the multiplicative/additive scheme.
5. After the computations shown in Tables 9.3 and 9.4, the total rotation vector χ is extracted from the total quaternion $\mathbf{X}$ by using eq. (8.94).

Note that the presented update schemes are extended is Sect. 9.4.3 for the rigid body dynamics, see Tables 9.6 and 9.8.

Table 9.3 Scheme 2. Multiplicative updates. $\mathbf{R}^i_{n+1}\Delta\tilde{\boldsymbol{\Theta}} \in T_{R^i_{n+1}}SO(3)$.

Initialize $\mathbf{X}$	$\leftarrow$	total
Step		
$\quad\boldsymbol{\Theta} = \mathbf{0}$, initialize $\mathbf{q}$		
$\quad$**Newton loop**		
$\quad$Form equilibrium equations using $(\mathbf{X}, \boldsymbol{\Theta})$, solve for $\Delta\boldsymbol{\Theta}$		
$\quad$Update		
$\qquad\Delta\boldsymbol{\Theta} \to \Delta\mathbf{q} \to \mathbf{X} = \mathbf{X} \circ \Delta\mathbf{q}$	$\leftarrow$	total (multiplicative)
$\qquad\to \mathbf{q} = \mathbf{q} \circ \Delta\mathbf{q} \to \boldsymbol{\Theta}$	$\leftarrow$	for increment (multiplicative)
$\quad$**End of Newton loop**		

Table 9.4 Scheme 2. Multiplicative/additive updates. $\mathbf{R}^0_{n+1}\Delta\tilde{\boldsymbol{\Theta}} \in T_{R^0_{n+1}}SO(3)$.

Initialize $\mathbf{X}_n$	$\leftarrow$	total
Step		
$\quad\boldsymbol{\Theta} = \mathbf{0}$		
$\quad$**Newton loop**		
$\quad$Form equilibrium equations using $(\mathbf{X}, \boldsymbol{\Theta})$, solve for $\Delta\boldsymbol{\Theta}$		
$\quad$Update		
$\qquad\boldsymbol{\Theta} = \boldsymbol{\Theta} + \Delta\boldsymbol{\Theta}$	$\leftarrow$	for increment (additive)
$\qquad\boldsymbol{\Theta} \to \mathbf{q} \to \mathbf{X} = \mathbf{X}_n \circ \mathbf{q}$	$\leftarrow$	total (multiplicative)
$\quad$**End of Newton loop**		
$\quad$**Update**		
$\qquad\mathbf{X}_n = \mathbf{X}$	$\leftarrow$	total

9.3.3 Scheme 3: formulation in $\mathrm{T}_R\mathrm{SO}(3)$

In this scheme we use as the rotational unknown, the rotation vector $\boldsymbol{\theta}$ related to $\mathbf{R}_B$, where $\mathbf{R}_B$ is the last available solution, in general, non-converged. Hence, $\tilde{\boldsymbol{\theta}}\,\mathbf{R}_B$ (or $\mathbf{R}_B\,\tilde{\boldsymbol{\Theta}}$) $\in T_{R_B}\mathrm{SO}(3)$, where T_{R_B} is the current tangent plane, see Fig. 9.5.

We use the left composition rule and we assume $\boldsymbol{\psi} = \mathbf{0}$ in the scheme of Fig. 9.1, for which $\mathbf{R}_1(\boldsymbol{\psi}) = \mathbf{I}$ and $\mathbf{R}_B = \mathbf{R}_A$. Hence, $\mathbf{R}(\epsilon\boldsymbol{\theta}_A) = \mathbf{R}_2(\epsilon\boldsymbol{\theta}_B)$ and both forms of the perturbed rotation of eq. (9.1) become identical,

$$\mathbf{R}_C = \mathbf{R}_2(\epsilon\boldsymbol{\theta})\,\mathbf{R}_B, \tag{9.101}$$

where we denoted $\boldsymbol{\theta}$ instead of $\boldsymbol{\theta}_B$, and $\epsilon\tilde{\boldsymbol{\theta}}\,\mathbf{R}_B \in T_{R_B}\mathrm{SO}(3)$. Similarly, for the right composition of eq. (9.28), see Fig. 9.3, which yields

$$\mathbf{R}_C = \mathbf{R}_B\,\mathbf{R}_2(\epsilon\boldsymbol{\Theta}), \tag{9.102}$$

where we denoted $\boldsymbol{\Theta}$ instead of $\boldsymbol{\Theta}_B$, and $\mathbf{R}_B\,\epsilon\tilde{\boldsymbol{\Theta}} \in T_{R_B}\mathrm{SO}(3)$.

For this formulation, we use the multiplicative update scheme which is presented in Table 9.5, and has the following features:

1. Only one quaternion is used in this scheme. The quaternion $\mathbf{X}$ is used for the total rotation and is updated in every iteration,

$$\mathbf{R}_B = \underbrace{\mathbf{R}(\Delta\boldsymbol{\theta})}_{\Delta\mathbf{X}}\ \underbrace{\mathbf{R}(\boldsymbol{\theta})\,\mathbf{R}_A}_{\mathbf{X}}, \qquad \mathbf{R}_B = \underbrace{\mathbf{R}_A\,\mathbf{R}(\boldsymbol{\Theta})}_{\mathbf{X}}\ \underbrace{\mathbf{R}(\Delta\boldsymbol{\Theta})}_{\Delta\mathbf{X}}. \tag{9.103}$$

   The previous quaternion $\mathbf{X}$, and the quaternion for the iteration, $\Delta\mathbf{X}$, are composed multiplicatively, as in eq. (8.185).
2. The increment $\Delta\boldsymbol{\Theta}$ belongs to $T_{R^i_{n+1}}\mathrm{SO}(3)$, which is the tangent plane at the last available rotation, not necessarily converged. The increment $\Delta\boldsymbol{\Theta}$ is converted to the quaternion $\Delta\mathbf{X}$ using eq. (8.96).

Table 9.5 Scheme 3. Multiplicative update. $\mathbf{R}^i_{n+1}\Delta\tilde{\boldsymbol{\Theta}} \in T_{R^i_{n+1}}\mathrm{SO}(3)$.

Initialize $\mathbf{X}$	$\leftarrow$	total
Step		
Newton loop		
Form equilibrium equations using $(\mathbf{X}, \boldsymbol{\Theta} = \mathbf{0})$, solve for $\Delta\boldsymbol{\Theta}$		
Update		
$\Delta\boldsymbol{\Theta} \rightarrow \Delta\mathbf{X} \rightarrow \mathbf{X} = \mathbf{X} \circ \Delta\mathbf{X}$	$\leftarrow$	total (multiplicative)
End of Newton loop		

9.3.4 Symmetry of tangent operator for structures with rotational dofs

The question of symmetry of the tangent operator (stiffness matrix) for the Newton method is very important in numerical implementations and, for structures with rotational dofs, was considered, e.g., in [220, 44, 206, 40, 148].

Consider a conservative system, e.g. a shell or a beam made of a hyperelastic material and deformation-independent loads, for which the potential energy exists. The potential energy of the whole body is $\Pi = \int_V \pi \, \mathrm{d}V$, where π is the potential energy density. In general, π depends on displacements and rotations but, for the sake of simplicity, the displacements are disregarded below.

Below, the notation is the same as in Sect. 9.1, where we considered increments of rotation vectors in two tangent planes. The variations are defined as in Sect. 9.2, see eqs. (9.74) and (9.75). If we designate $\boldsymbol{\theta}^- \doteq \delta\boldsymbol{\psi}$ and $\boldsymbol{\theta}^+ \doteq \Delta\boldsymbol{\psi}$ in these equations, then this notation becomes suitable for incremental formulations.

Consider the potential energy density $\pi(\mathbf{R})$ at some $\mathbf{R} \in \mathrm{SO}(3)$. The first and second differentials of the potential energy are

$$\delta\pi = \frac{\partial \pi}{\partial \mathbf{R}} \cdot \delta\mathbf{R}, \qquad \chi(\delta\pi) = \chi\left(\frac{\partial \pi}{\partial \mathbf{R}}\right) \cdot \delta\mathbf{R} + \frac{\partial \pi}{\partial \mathbf{R}} \cdot \chi(\delta\mathbf{R}). \tag{9.104}$$

The second differential $\chi(\delta\pi)$ yields the tangent operator (stiffness matrix), therefore its symmetry is of interest and is examined below.

1. For the first component of $\chi(\delta\pi)$ of eq. (9.104), we have

$$\chi(\frac{\partial \pi}{\partial \mathbf{R}})\cdot\delta\mathbf{R} = \left(\left[\frac{\partial^2 \pi}{\partial \mathbf{R}\partial \mathbf{R}}\right] \chi\mathbf{R}\right)\cdot\delta\mathbf{R} = \left(\left[\frac{\partial^2 \pi}{\partial \mathbf{R}\partial \mathbf{R}}\right]^T \delta\mathbf{R}\right)\cdot\chi\mathbf{R}, \tag{9.105}$$

   where the first differentials of the rotation tensor, e.g. for the left composition rule and the canonical parametrization, by eq. (9.78) are

$$\delta\mathbf{R} = (\mathbf{T}\boldsymbol{\theta}^-) \times \mathbf{R}, \qquad \chi\mathbf{R} = (\mathbf{T}\boldsymbol{\theta}^+) \times \mathbf{R}. \tag{9.106}$$

   The last form of eq. (9.105) is obtained by the identity (K8) from [33], p. 62, and the term in brackets is a fourth-rank tensor. Symmetry of this term implies symmetry of the whole component w.r.t. $\boldsymbol{\theta}^-$ and $\boldsymbol{\theta}^+$.

2. In the second component of $\chi(\delta\pi)$ of eq. (9.104), we have a scalar product of $\partial\pi/\partial\mathbf{R}$ and the second differential of the rotation tensor, $\chi(\delta\mathbf{R})$. At the equilibrium configuration $\partial\pi/\partial\mathbf{R} = \mathbf{0}$, so the whole component vanishes, but otherwise its contribution is non-trivial. The second differential $\chi(\delta\mathbf{R})$, e.g. for the left composition rule and the canonical parametrization, is defined by eq. (9.81).

We see that the second differential of π has a complicated form and non-symmetric components and it is not easy to resolve the question of symmetry of the corresponding tangent matrix by inspection of the above formulas. Hence, it is advisable to verify numerically whether the tangent matrix of a newly developed finite element is symmetric.

9.3.5 Example: twisted ring by 3D beam element

The twisted ring example is highly non-linear and demanding, due to the presence of finite rotations. It is described in detail and computed using the shell elements in Sect. 15.3.15. Here, we use our two-node 3D beam element; it is relatively simple and, therefore, convenient to test various schemes of treating finite rotations. Below, the results obtained for the canonical rotation vector are presented.

The ring is twisted by a moment applied at one point and fixed at the opposite point, both on the same axis. The whole ring is computed using the arc-length method for the initial $M_x^{\text{ref}} = 50$. Below, we report the rotation r_x and the displacement u_x obtained by the schemes of the preceding sections at point A of Fig. 15.42.

Two solutions by Scheme 1 are shown in Fig. 9.6 for the mesh with 124 and 1000 elements. The curves coincide in the almost whole range for both meshes, but differ when the rotation r_x approaches 2π. It seems that at this value the curves for the displacement and the rotation have vertical asymptotes, which are not physically correct. Note that the curves for the 1000-element mesh are closer to them than the curves for the 124-element mesh. Other examples of erroneous behavior of Scheme 1 at rotations equal to 2π are given in [107].

The solutions from the two schemes, Schemes 1 and 2, are shown in Fig. 9.7, and are obtained for the 124-element mesh. Scheme 1 yields rotations for up to $r_x = 2\pi$ but Scheme 2 allows to perform several turns and we plotted the rotation for up to $r_x \approx 30$. Note that in Scheme 2 we do not use the total rotation vector ψ but, to visualize the

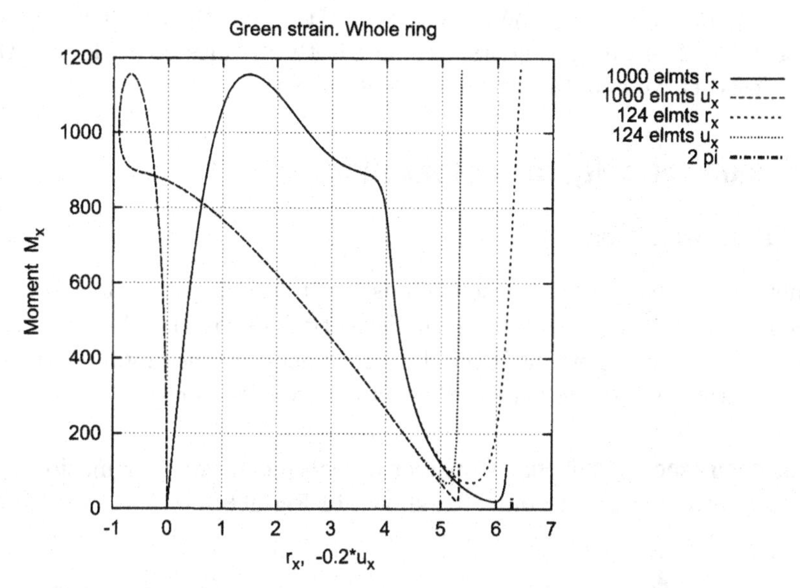

Fig. 9.6 Twisted ring: vertical asymptote at $r_x = 2\pi$ for Scheme 1.
$E = 2.1 \cdot 10^6, \quad \nu = 0.3, \quad w = 1, \quad h = 1/3, \quad r = 20.$

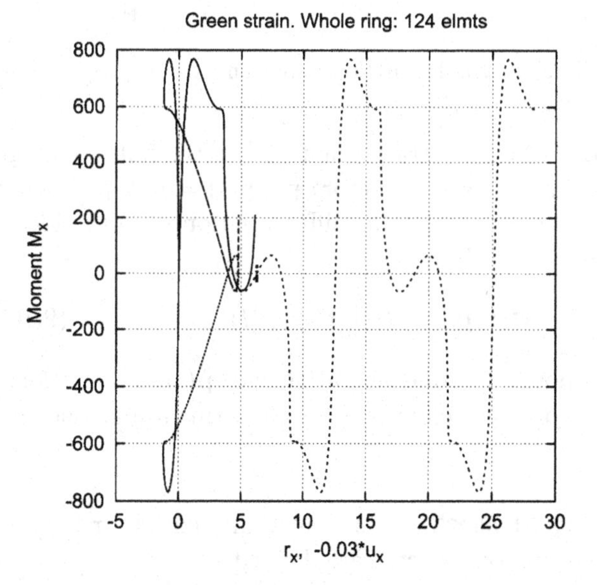

Fig. 9.7 Twisted ring: Scheme 1 and Scheme 2.
$E = 2 \cdot 10^5, \quad \nu = 0.3, \quad w = 6, \quad h = 0.6, \quad r = 120.$

rotations, we summed up the increments $\Delta\psi$, and the result is plotted in Fig. 9.7. The displacements u_x for both schemes coincide and the same closed curve is obtained for multiple turns.

9.4 Angular velocity and acceleration

9.4.1 Basic definitions

In this section, we present basic notions and relations pertaining to rotations in dynamics, i.e. involving time derivatives of rotational tensors and parameters. Very important is the difference between the angular velocity and acceleration and the time derivatives of rotation vectors.

Instantaneous angular motion. Consider instantaneous angular motion of a rigid body about a fixed point, 0, shown in Fig. 9.8.

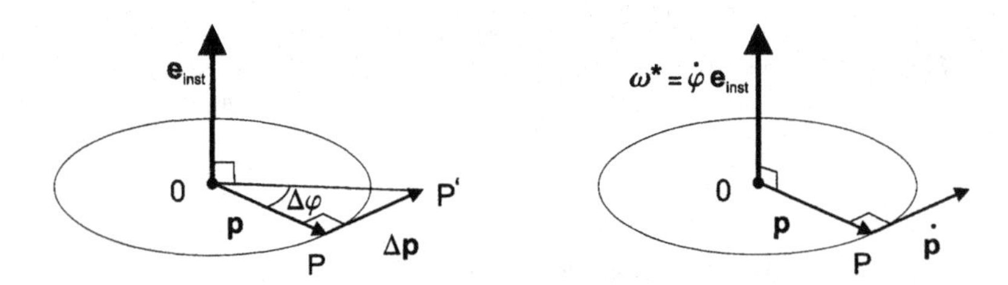

Fig. 9.8 Instantaneous angular motion about an axis $\mathbf{e}_{\text{inst}}$.

The position vector $\mathbf{p}_0$ of an arbitrary point P in the initial configuration is mapped smoothly into a new vector $\mathbf{p}$: $\mathbf{p}(t) = \mathbf{R}(t)\,\mathbf{p}_0$, where $\mathbf{R} \in \mathrm{SO}(3)$ and time $t \in [0, t_{\max}]$. By differentiation of both sides of this relation w.r.t. time t, we obtain

$$\dot{\mathbf{p}}(t) = \dot{\mathbf{R}}(t)\,\mathbf{p}_0 = \dot{\mathbf{R}}(t)\,\mathbf{R}^T(t)\,\mathbf{p}(t), \tag{9.107}$$

where $\dot{(\cdot)} \doteq \mathrm{d}(\cdot)/\mathrm{d}t$ denotes the time derivative and $\dot{\mathbf{p}}(t)$ is the velocity of point P. Note that $\mathbf{p}_0$ was eliminated so all terms are at one time instant, t.

Eulerian (spatial, or left) angular velocity. The Eulerian (spatial, or left) angular velocity tensor and its axial vector are defined as

$$\tilde{\boldsymbol{\omega}}^* \doteq \dot{\mathbf{R}}\mathbf{R}^T \in \mathrm{so}(3), \qquad \boldsymbol{\omega}^* \doteq \tfrac{1}{2}(\mathbf{I} \times \tilde{\boldsymbol{\omega}}^*). \tag{9.108}$$

Note that $\tilde{\omega}^*$ is a skew-symmetric tensor, which can be shown as follows: From the time differentiation of the orthogonality condition $\mathbf{R}\mathbf{R}^T = \mathbf{I}$, we obtain $\dot{\mathbf{R}}\mathbf{R}^T + \mathbf{R}\dot{\mathbf{R}}^T = \mathbf{0}$, which, by eq. (9.108), is equivalent to $\tilde{\omega}^* = -(\tilde{\omega}^*)^T$, which is the definition of skew-symmetry.

Using $\tilde{\omega}^*$, eq. (9.107) can be rewritten in two characteristic ways:

1. $\dot{\mathbf{p}} = \boldsymbol{\omega}^* \times \mathbf{p}$, from which we see that the velocity $\dot{\mathbf{p}}$ is perpendicular to the angular velocity vector $\boldsymbol{\omega}^*$ and to the position vector $\mathbf{p}$, see Fig. 9.8b.
2. $\dot{\mathbf{R}} - \tilde{\omega}^*\mathbf{R} = \mathbf{0}$, obtained by using $\mathbf{p}(t) = \mathbf{R}(t)\,\mathbf{p}_0$. Given the $\tilde{\omega}^*$, this is the ODE generating rotations $\mathbf{R}$, for the initial condition $\mathbf{R}(t=0) = \mathbf{R}_0$.

Direction of $\boldsymbol{\omega}^*$. The direction of the angular velocity vector $\boldsymbol{\omega}^*$ can be established as follows. Denote the instantaneous axis of rotation by $\mathbf{e}_{\text{inst}}$, and the rotation angle by $\Delta\psi$. Then we can write a simple geometrical formula,

$$\Delta\mathbf{p} = (\Delta\psi\,\mathbf{e}_{\text{inst}}) \times \mathbf{p}, \tag{9.109}$$

see Fig. 9.8a. Dividing by Δt, and taking the limit $\Delta t \to 0$, we obtain

$$\dot{\mathbf{p}} = (\dot{\psi}\,\mathbf{e}_{\text{inst}}) \times \mathbf{p}, \tag{9.110}$$

where $\dot{\mathbf{p}} \doteq \lim_{\Delta t \to 0}(\Delta\mathbf{p}/\Delta t)$ and $\dot{\psi} \doteq \lim_{\Delta t \to 0}(\Delta\psi/\Delta t)$. By comparison with the earlier derived formula, $\dot{\mathbf{p}} = \boldsymbol{\omega}^* \times \mathbf{p}$, we see that

$$\boldsymbol{\omega}^* = \dot{\psi}\,\mathbf{e}_{\text{inst}}, \tag{9.111}$$

i.e. $\boldsymbol{\omega}^*$ has direction of the instantaneous axis $\mathbf{e}_{\text{inst}}$, see Fig. 9.8b. We stress that $\mathbf{e}_{\text{inst}}$ is instantaneous, i.e. is valid only for an infinitesimal Δt, and is usually different from the axis of rotation $\mathbf{e}$ for a finite time period !

Remark. The above relations are typical for the rigid-body mechanics, but can be also used for shells. We can just replace the position vector $\mathbf{p}$ by the current director $\mathbf{a}_3$, to obtain the relation $\dot{\mathbf{a}}_3 = \boldsymbol{\omega}^* \times \mathbf{a}_3$, see the study on shell intersections in [207].

Lagrangian (material, or right) angular velocity. The Lagrangian (material, or right) angular velocity tensor and its axial vector are defined as

$$\tilde{\omega} \doteq \mathbf{R}^T \dot{\mathbf{R}} \in \mathrm{so}(3), \qquad \omega \doteq \tfrac{1}{2}(\mathbf{I} \times \tilde{\omega}). \tag{9.112}$$

They can be obtained by back-rotation of the left angular velocity, i.e.

$$\tilde{\omega} = \mathbf{R}^T \tilde{\omega}^* \mathbf{R}, \qquad \omega = \mathbf{R}^T \omega^*. \tag{9.113}$$

Using the above relation, the equation generating rotations becomes

$$\dot{\mathbf{R}} - \tilde{\omega}^*\mathbf{R} = \dot{\mathbf{R}} - (\mathbf{R}\tilde{\omega}\mathbf{R}^T)\mathbf{R} = \dot{\mathbf{R}} - \mathbf{R}\tilde{\omega} = \mathbf{0}, \tag{9.114}$$

where $\tilde{\omega}$ multiplies $\mathbf{R}$ from the right.

Angular acceleration. The angular acceleration vectors are defined as time derivatives of the left and right velocity vectors

$$\mathbf{a}_a^* \doteq \dot{\omega}^*, \qquad \mathbf{a}_a \doteq \dot{\omega}, \tag{9.115}$$

where $\mathbf{a}_a^*$ is the Eulerian (spatial, or left) angular acceleration and $\mathbf{a}_a$ is the Lagrangian (material, or right) angular acceleration.

To find the relation between these accelerations and time derivatives of the rotation tensor, we have to introduce skew-symmetric tensors associated with the accelerations and use the time-differentiated eqs. (9.108) and (9.112). More useful, however, are vectorial formulas obtained for particular parametrizations of rotations.

9.4.2 Angular velocity and acceleration for parametrizations

Below, we derive the relations between the earlier-defined angular velocity and acceleration vectors and the time derivatives of the rotation vector for the semi-tangential and canonical parametrization.

A. Left angular velocity

a. For the semi-tangential parametrization, we rewrite the variation for the left composition rule of eq. (9.60) as follows:

$$\underbrace{\delta_{\tilde{\theta}_A}\mathbf{R}(\boldsymbol{\psi})}_{\text{additive}} = \underbrace{\{[2\,\mathbf{T}(\boldsymbol{\psi})\,\boldsymbol{\theta}_A] \times \mathbf{I}\}\,\mathbf{R}(\boldsymbol{\psi})}_{\text{left, multiplicative}} \tag{9.116}$$

in which the variation on both sides are calculated in the direction $\tilde{\boldsymbol{\theta}}_A \in T_I\mathrm{SO}(3)$. We can link the variations to the time derivatives by using $\delta\mathbf{R} = \dot{\mathbf{R}}\,\delta t$ and $\boldsymbol{\theta}_A = \dot{\boldsymbol{\psi}}\,\delta t$, from which the above equation becomes

$$\dot{\mathbf{R}}(\boldsymbol{\psi}) = \{[2\,\mathbf{T}(\boldsymbol{\psi})\,\dot{\boldsymbol{\psi}}] \times \mathbf{I}\}\,\mathbf{R}(\boldsymbol{\psi}). \tag{9.117}$$

By the post-multiplication by $\mathbf{R}^T$, we obtain

$$\underbrace{\dot{\mathbf{R}}(\boldsymbol{\psi})\,\mathbf{R}^T(\boldsymbol{\psi})}_{=\tilde{\boldsymbol{\omega}}^*} = [2\,\mathbf{T}(\boldsymbol{\psi})\,\dot{\boldsymbol{\psi}}] \times \mathbf{I}, \tag{9.118}$$

where $\tilde{\boldsymbol{\omega}}^* = \boldsymbol{\omega}^* \times \mathbf{I}$ is the skew-symmetric tensor of the left angular velocity of eq. (9.108). Hence, in terms of the axial vectors, we have

$$\boldsymbol{\omega}^* = \mathbf{T}(\boldsymbol{\psi})\,\dot{\boldsymbol{\psi}} = \frac{2}{1+\boldsymbol{\psi}\cdot\boldsymbol{\psi}}(\dot{\boldsymbol{\psi}} + \boldsymbol{\psi} \times \dot{\boldsymbol{\psi}}), \tag{9.119}$$

where $\mathbf{T}$ of eq. (9.10) was used.

b. For the canonical parametrization, we begin from eq. (9.57):

$$\underbrace{\delta\mathbf{R}(\boldsymbol{\psi})}_{\text{additive}} = \underbrace{\{[\mathbf{T}(\boldsymbol{\psi})\,\boldsymbol{\theta}_A] \times \mathbf{I}\}\,\mathbf{R}(\boldsymbol{\psi})}_{\text{left, multiplicative}}, \tag{9.120}$$

and, in the same way as for the semi-tangential parametrization, we obtain

$$\boldsymbol{\omega}^* = \mathbf{T}(\boldsymbol{\psi})\,\dot{\boldsymbol{\psi}} = c_1\,\dot{\boldsymbol{\psi}} + A\,\boldsymbol{\psi} + c_2\,\boldsymbol{\psi} \times \dot{\boldsymbol{\psi}}, \tag{9.121}$$

where $A \doteq (1-c_1)\,(\boldsymbol{\psi}\cdot\dot{\boldsymbol{\psi}})/(\boldsymbol{\psi}\cdot\boldsymbol{\psi})$, and $\mathbf{T}$ of eq. (9.18) was used.

Note that eqs. (9.119) and (9.121) link the left angular velocity vector $\boldsymbol{\omega}^*$ and the time derivative of the rotation vector $\boldsymbol{\psi}$.

B. Right angular velocity

a. For the semi-tangential parametrization, we rewrite the variation for the right composition rule of eq. (9.60) as follows:

$$\underbrace{\delta_{\tilde{\boldsymbol{\theta}}_A}\mathbf{R}(\boldsymbol{\psi})}_{\text{additive}} = \underbrace{\mathbf{R}(\boldsymbol{\psi})\,\left\{\left[2\,\mathbf{T}^T(\boldsymbol{\psi})\,\boldsymbol{\theta}_A\right] \times \mathbf{I}\right\}}_{\text{right, multiplicative}}. \tag{9.122}$$

Using $\delta\mathbf{R} = \dot{\mathbf{R}}\,\delta t$ and $\boldsymbol{\theta}_A = \dot{\boldsymbol{\psi}}\,\delta t$, we obtain

$$\dot{\mathbf{R}}(\boldsymbol{\psi}) = \mathbf{R}(\boldsymbol{\psi})\,\left\{\left[2\,\mathbf{T}^T(\boldsymbol{\psi})\,\dot{\boldsymbol{\psi}}\right] \times \mathbf{I}\right\}. \tag{9.123}$$

By the left multiplication by $\mathbf{R}^T$, we obtain

$$\underbrace{\mathbf{R}^T(\boldsymbol{\psi})\,\dot{\mathbf{R}}(\boldsymbol{\psi})}_{=\tilde{\boldsymbol{\omega}}} = \left[2\,\mathbf{T}^T(\boldsymbol{\psi})\,\dot{\boldsymbol{\psi}}\right] \times \mathbf{I}, \tag{9.124}$$

where $\tilde{\boldsymbol{\omega}} = \boldsymbol{\omega} \times \mathbf{I}$ is the skew-symmetric tensor of the right angular velocity of eq. (9.112). Hence, for the axial vectors we have

$$\boldsymbol{\omega} = 2\,\mathbf{T}^T(\boldsymbol{\psi})\,\dot{\boldsymbol{\psi}} = \frac{2}{1+\boldsymbol{\psi}\cdot\boldsymbol{\psi}}(\dot{\boldsymbol{\psi}} - \boldsymbol{\psi}\times\dot{\boldsymbol{\psi}}), \tag{9.125}$$

where $\mathbf{T}$ of eq. (9.10) was used.

b. For the canonical parametrization, we rewrite the variation for the right composition rule of eq. (9.57) as

$$\underbrace{\delta\mathbf{R}(\boldsymbol{\psi})}_{\text{additive}} = \underbrace{\mathbf{R}(\boldsymbol{\psi})\,\left\{\left[\mathbf{T}^T(\boldsymbol{\psi})\,\boldsymbol{\theta}_A\right] \times \mathbf{I}\right\}}_{\text{right, multiplicative}} \tag{9.126}$$

and, in the same way as for the semi-tangential parametrization, we obtain

$$\boldsymbol{\omega} = \mathbf{T}^T(\boldsymbol{\psi})\,\dot{\boldsymbol{\psi}} = c_1\,\dot{\boldsymbol{\psi}} + A\,\boldsymbol{\psi} - c_2\,\boldsymbol{\psi}\times\dot{\boldsymbol{\psi}}, \tag{9.127}$$

where $A \doteq (1-c_1)\,(\boldsymbol{\psi}\cdot\dot{\boldsymbol{\psi}})/(\boldsymbol{\psi}\cdot\boldsymbol{\psi})$, and $\mathbf{T}$ of eq. (9.18) was used.

Note that eqs. (9.125) and (9.127) link the right angular velocity vector $\boldsymbol{\omega}^*$ and the time derivative of the rotation vector $\boldsymbol{\psi}$.

Remark 1. Recall the properties of $\mathbf{T}$ for the canonical rotation vector of Sect. 9.1, where we found that if $\boldsymbol{\psi}$ and $\boldsymbol{\theta}_A$ are coaxial, then $\mathbf{T}(\boldsymbol{\psi})$ acts as the identity operator. (For the semi-tangential vector, $\mathbf{T}(\boldsymbol{\psi})$ shortens $\boldsymbol{\theta}_A$.) The same property holds if we replace $\mathbf{T}$ by $\mathbf{T}^T$, and $\boldsymbol{\theta}_A$ by $\dot{\boldsymbol{\psi}}$. If $\boldsymbol{\psi}$ and $\dot{\boldsymbol{\psi}}$, are coaxial, then $\mathbf{e}_{\text{inst}} = \mathbf{e}$, i.e. the instantaneous axis of rotation $\mathbf{e}_{\text{inst}}$ coincides with the axis of rotation $\mathbf{e}$ for a finite time period. Hence, for the angular motion about a fixed axis, we have $\boldsymbol{\omega} = \dot{\boldsymbol{\psi}}$.

Remark 2. Let us rewrite eq. (9.121), linking the left angular velocity $\boldsymbol{\omega}^*$ and the time derivative of the canonical rotation vector $\boldsymbol{\psi}$, in the form

$$\dot{\boldsymbol{\psi}} - \mathbf{T}^{-1}(\boldsymbol{\psi})\,\boldsymbol{\omega}^* = \mathbf{0}. \tag{9.128}$$

If $\boldsymbol{\omega}^*$ is known, then this is the ODE which generates $\boldsymbol{\psi}$, given the initial condition $\boldsymbol{\psi}(t=0) = \boldsymbol{\psi}_0$. This equation is an analogue of the equation generating rotations $\mathbf{R} \in \mathrm{SO}(3)$,

$$\dot{\mathbf{R}} - \tilde{\boldsymbol{\omega}}^* \mathbf{R} = \mathbf{0}, \tag{9.129}$$

where the skew-symmetric $\tilde{\boldsymbol{\omega}}^* \in \mathrm{so}(3)$ is known. Note that $\boldsymbol{\omega}^*$ of eq. (9.128) is the axial vector of $\tilde{\boldsymbol{\omega}}^*$ of eq. (9.129).

Update of angular velocity. Consider the right angular velocities at two time instants, t_n and t_{n+1}, which, by eq. (9.112), are defined as

$$\tilde{\boldsymbol{\omega}}_{n+1} \doteq \mathbf{R}_{n+1}^T \dot{\mathbf{R}}_{n+1}, \qquad \tilde{\boldsymbol{\omega}}_n \doteq \mathbf{R}_n^T \dot{\mathbf{R}}_n. \tag{9.130}$$

Using the incremental rotation $\Delta\mathbf{R}$, we have

$$\mathbf{R}_{n+1} = \mathbf{R}_n \Delta\mathbf{R}, \qquad \dot{\mathbf{R}}_{n+1} = \dot{\mathbf{R}}_n \Delta\mathbf{R} + \mathbf{R}_n \dot{\widehat{\Delta\mathbf{R}}} \tag{9.131}$$

and, using them in eq. $(9.130)_1$, we obtain

$$\tilde{\boldsymbol{\omega}}_{n+1} = \Delta\mathbf{R}^T \, \tilde{\boldsymbol{\omega}}_n \, \Delta\mathbf{R} + \Delta\tilde{\mathbf{w}}, \tag{9.132}$$

where $\Delta\tilde{\mathbf{w}} \doteq \Delta\mathbf{R}^T \, \dot{\widehat{\Delta\mathbf{R}}} \in \mathrm{so}(3)$ is the (right) angular velocity tensor for the incremental rotation. In terms of the axial vectors, we can write

$$\boldsymbol{\omega}_{n+1} = \Delta\mathbf{R}^T \boldsymbol{\omega}_n + \Delta\mathbf{w}, \tag{9.133}$$

in which $\Delta\mathbf{R}$ and $\Delta\mathbf{w}$ are associated with the increment.

The above formula can be simplified for small increments of the rotation vector and its time derivatives. For $\Delta\mathbf{R} \approx \mathbf{I} + \Delta\tilde{\boldsymbol{\psi}}$, we have $\dot{\widehat{\Delta\mathbf{R}}} = \Delta\dot{\tilde{\boldsymbol{\psi}}}$ and $\Delta\tilde{\mathbf{w}} = (\mathbf{I} - \Delta\tilde{\boldsymbol{\psi}})\Delta\dot{\tilde{\boldsymbol{\psi}}}$, which yield

$$\tilde{\boldsymbol{\omega}}_{n+1} \approx (\mathbf{I} - \Delta\tilde{\boldsymbol{\psi}})(\tilde{\boldsymbol{\omega}}_n + \Delta\dot{\tilde{\boldsymbol{\psi}}}) \approx \tilde{\boldsymbol{\omega}}_n + \Delta\dot{\tilde{\boldsymbol{\psi}}}. \tag{9.134}$$

Generally, the update of the angular velocity should be consistent with the time-stepping algorithm.

C. Angular acceleration, left and right

a. For the semi-tangential parametrization, eqs. (9.119) and (9.125) link the angular velocity vectors $\boldsymbol{\omega}^*$ and $\boldsymbol{\omega}$ and the time derivative of the rotation vector $\boldsymbol{\psi}$. By time differentiation of them, we obtain

$$\mathbf{a}_a^* \doteq \dot{\boldsymbol{\omega}}^* = 2\,\mathbf{T}(\boldsymbol{\psi})\,\ddot{\boldsymbol{\psi}} + 2\,\dot{\mathbf{T}}(\boldsymbol{\psi})\,\dot{\boldsymbol{\psi}}, \quad \mathbf{a}_a \doteq \dot{\boldsymbol{\omega}} = 2\,\mathbf{T}^T(\boldsymbol{\psi})\,\ddot{\boldsymbol{\psi}} + 2\,\dot{\mathbf{T}}^T(\boldsymbol{\psi})\,\dot{\boldsymbol{\psi}}. \tag{9.135}$$

For $\mathbf{T}$ given by eq. (9.10), we obtain

$$\dot{\mathbf{T}}(\boldsymbol{\psi}) = a_1(\mathbf{I} + \tilde{\boldsymbol{\psi}}) + \frac{1}{1 + \boldsymbol{\psi} \cdot \boldsymbol{\psi}} \dot{\tilde{\boldsymbol{\psi}}}, \tag{9.136}$$

where $a_1 \doteq -2(\dot{\boldsymbol{\psi}} \cdot \boldsymbol{\psi})/(1 + \boldsymbol{\psi} \cdot \boldsymbol{\psi})^2$. Note that $\dot{\mathbf{T}}$ is analogous to $\chi\mathbf{T}$ of eq. (9.39).

b. For the canonical parametrization, eqs. (9.121) and (9.127) link the angular velocity vectors $\boldsymbol{\omega}^*$ and $\boldsymbol{\omega}$ and the time derivative of the rotation vector $\boldsymbol{\psi}$. By time differentiation of them, we obtain

$$\mathbf{a}_a^* \doteq \dot{\boldsymbol{\omega}}^* = \mathbf{T}(\boldsymbol{\psi})\,\ddot{\boldsymbol{\psi}} + \dot{\mathbf{T}}(\boldsymbol{\psi})\,\dot{\boldsymbol{\psi}}, \qquad \mathbf{a}_a \doteq \dot{\boldsymbol{\omega}} = \mathbf{T}^T(\boldsymbol{\psi})\,\ddot{\boldsymbol{\psi}} + \dot{\mathbf{T}}^T(\boldsymbol{\psi})\,\dot{\boldsymbol{\psi}}. \tag{9.137}$$

For $\mathbf{T}$ given by eq. (9.18), we obtain

$$\begin{aligned} \dot{\mathbf{T}}(\boldsymbol{\psi}) = {} & a_1\,(\dot{\boldsymbol{\psi}} \cdot \mathbf{e})\,\mathbf{I} + a_2\,(\dot{\boldsymbol{\psi}} \otimes \mathbf{e} + \mathbf{e} \otimes \dot{\boldsymbol{\psi}}) \\ & + a_3\,(\dot{\boldsymbol{\psi}} \cdot \mathbf{e})\,(\mathbf{e} \otimes \mathbf{e}) + a_4\,(\dot{\boldsymbol{\psi}} \cdot \mathbf{e})\,\tilde{\boldsymbol{\psi}} + a_5\,(\dot{\boldsymbol{\psi}} \times \mathbf{I}), \end{aligned} \tag{9.138}$$

where the scalar coefficients are defined by eq. (9.41). Note that $\dot{\mathbf{T}}$ is analogous to $\chi\mathbf{T}$ of eq. (9.40). For $\boldsymbol{\psi} \to \mathbf{0}$: $\dot{\mathbf{T}}^T(\boldsymbol{\psi}) \to -\frac{1}{2}(\dot{\boldsymbol{\psi}} \times \mathbf{I})$ and $\dot{\boldsymbol{\omega}} \to \ddot{\boldsymbol{\psi}}$, as $\dot{\boldsymbol{\psi}} \times \dot{\boldsymbol{\psi}} = \mathbf{0}$.

Finally, we note that using the above-derived relations between the angular velocities and accelerations and the time derivatives of rotation vectors, we can formulate various algorithms of dynamics in terms of $\{\boldsymbol{\psi}, \dot{\boldsymbol{\psi}}, \ddot{\boldsymbol{\psi}}\}$, as in [250].

9.4.3 Examples of updates for rigid body motion

The updates of rotational parameters can be conveniently presented for the equations of angular motion of a rigid body, which are relatively simple. We base on the algorithm ALGO-C1 of [221], which conserves the angular momentum and the kinetic energy and develop our algorithms as modifications of ALGO-C1.

In the formulation presented below, the right composition rule of rotations is used, i.e. $\boldsymbol{\Lambda}_{n+1} = \boldsymbol{\Lambda}_n \exp\tilde{\boldsymbol{\Theta}}$, where $\boldsymbol{\Lambda} \in \mathrm{SO}(3)$ is the rotation tensor and $\boldsymbol{\Theta}$ is the canonical rotation vector for the time step. This vector belongs to the tangent plane at the converged rotation for the previous time step, i.e. $\boldsymbol{\Lambda}^0_{n+1}\tilde{\boldsymbol{\Theta}} \in T_{\boldsymbol{\Lambda}^0_{n+1}}\mathrm{SO}(3)$, where $\boldsymbol{\Lambda}^0_{n+1} = \boldsymbol{\Lambda}^{\mathrm{conv}}_n$. Hence, we use the formulation in $T_{R_{\mathrm{ref}}}\mathrm{SO}(3)$ of Sect. 9.3.

The notation used below is similar to that in [221]. Hence, $\mathbf{W}$ is the material (right) angular velocity vector (equal to $\boldsymbol{\omega}$ of eq. (9.112)) and $\mathbf{A} \doteq \dot{\mathbf{W}}$ is the angular acceleration (equal to $\mathbf{a}$ of eq. (9.115)). The spatial angular momentum is $\boldsymbol{\pi}(t) \doteq \boldsymbol{\Lambda}(t)\,\mathbb{J}\,\mathbf{W}(t)$ and the kinetic energy of the angular motion is $E_k(t) \doteq \frac{1}{2}\mathbf{W}(t) \cdot [\mathbb{J}\mathbf{W}(t)]$, both relative to the center of mass, where $\mathbb{J}$ is the material (time-independent) inertia tensor.

The basic idea underlying the ALGO-C1 algorithm follows that proposed earlier for the dynamics with translational dofs in [270]. The second Newton law for the angular motion, $\mathrm{d}\boldsymbol{\pi}/\mathrm{d}t = \mathbf{m}$, where $\mathbf{m}$ is the external torque, is integrated w.r.t. time in the interval $[t_n, t_{n+1}]$, which yields

$$\boldsymbol{\pi}_{n+1} - \boldsymbol{\pi}_n = \int_{t_n}^{t_{n+1}} \mathbf{m}(t)\,\mathrm{d}t. \tag{9.139}$$

This eliminates acceleration from the governing equations; still, however, it can be recovered using angular velocities. Finally, the equation of motion is

$$\boldsymbol{\Lambda}_{n+1}\,\mathbb{J}\,\mathbf{W}_{n+1} - \boldsymbol{\Lambda}_n\,\mathbb{J}\,\mathbf{W}_n - h\mathbf{m}_{n+\alpha} = \mathbf{0}, \tag{9.140}$$

where $h \doteq t_{n+1} - t_n$, and $\alpha \in [0,1]$. The Newmark algorithm for the rotational parameters used in [221] is as follows:

$$\boldsymbol{\Theta} = h\mathbf{W}_n + h^2\left[\left(\frac{1}{2} - \beta\right)\mathbf{A}_n + \beta\mathbf{A}_{n+1}\right], \qquad \beta \in \left[0, \frac{1}{2}\right], \tag{9.141}$$

$$\mathbf{W}_{n+1} = \mathbf{W}_n + h[(1-\gamma)\mathbf{A}_n + \gamma\mathbf{A}_{n+1}], \qquad \gamma \in [0,1], \tag{9.142}$$

and involves $\{\boldsymbol{\Theta}, \mathbf{W}, \dot{\mathbf{W}}\}$. Note that $\{\boldsymbol{\Theta}, \dot{\boldsymbol{\Theta}}, \ddot{\boldsymbol{\Theta}}\}$ are used, e.g., in [44, 250]. In [146], the latter set of variables is treated as more correct, but our experience indicates that its use diminishes the radius of convergence of the Newton method and the time steps must be smaller.

The motion is free when either the integral in eq. (9.139) is equal to zero or $\mathbf{m}_{n+\alpha} = \mathbf{0}$ in eq. (9.140) and then $\boldsymbol{\pi}_{n+1} = \boldsymbol{\pi}_n$, i.e. the angular momentum is preserved. It is shown in [221] that the kinetic energy E_k is conserved for any $\gamma = 2\beta$.

The algorithms which are defined below have the following features:

1. In all presented algorithms, the converged results for the step are combined in the same way with the results for previous time steps. It is done via a composition of two quaternions: the total quaternion $\mathbf{X}$, and the quaternion for time step $\mathbf{q}$. Hence, the difference between the algorithms is confined to the computations within the time step.
2. The canonical rotation vector for the time step $\boldsymbol{\Theta}$ belongs to the tangent plane at the converged rotation for the previous time step, i.e. $\boldsymbol{\Lambda}^0_{n+1}\tilde{\boldsymbol{\Theta}} \in T_{\Lambda^0_{n+1}}\mathrm{SO}(3)$. However, its increments $\Delta\boldsymbol{\Theta}$ can belong either to the same plane or to $T_{\Lambda^i_{n+1}}\mathrm{SO}(3)$, which is the tangent plane at the last available rotation, not necessarily converged. This constitutes the difference between Algorithm 1 and Algorithms 2 and 3.
3. The update of the angular velocity is as follows:

$$\mathbf{W}^{i+1}_{n+1} = \mathbf{W}^i_{n+1} + \frac{\gamma}{\beta h}\left[\boldsymbol{\Theta}^{(i+1)} - \boldsymbol{\Theta}^{(i)}\right], \tag{9.143}$$

   where i and $i+1$ designated iterations. In Algorithm 1 of Table 9.6, $\boldsymbol{\Theta}^{(i+1)}$ is recovered from the updated quaternion $\mathbf{q}$ for the time step, while in the algorithms of Tables 9.7 and 9.8, $\boldsymbol{\Theta}^{(i+1)}$ is obtained by the additive update. Hence, $\tilde{\boldsymbol{\Theta}}$ used in the update of angular velocity belongs to the initial tangent plane, i.e. $\boldsymbol{\Lambda}^0_{n+1}\tilde{\boldsymbol{\Theta}} \in T_{\Lambda^0_{n+1}}\mathrm{SO}(3)$, in all algorithms.
4. After the computations shown in Tables 9.6, 9.7 and 9.8, the total rotation vector $\boldsymbol{\chi}$ is extracted from the total quaternion $\mathbf{X}$ and the angular acceleration $\mathbf{A}$ is computed from the angular velocity $\mathbf{W}$.

Algorithm 1. Multiplicative updates. In this algorithm $\Delta\tilde{\boldsymbol{\Theta}}$ belongs to the plane tangent at $\boldsymbol{\Lambda}^i_{n+1}$, which is the last available solution, not necessarily converged. This is basically the ALGO-C1 algorithm of [221].

We assume that the external torque $\mathbf{m}$ is independent of $\boldsymbol{\Theta}$. Then the tangent operator is

$$\mathbf{K}(\boldsymbol{\Lambda}_{n+1}, \boldsymbol{\Theta}) = \boldsymbol{\Lambda}_{n+1}\left[\frac{\gamma}{\beta h}\mathbb{J}\mathbf{T}(\boldsymbol{\Theta}) - (\widetilde{\mathbf{J}\mathbf{W}_{n+1}})\right], \tag{9.144}$$

where the operator $\mathbf{T}$, which is used here, is defined as in [221], i.e. it is equal to our $\mathbf{T}^{-T}$, where our $\mathbf{T}$ is given in eq. (9.18). To derive the above form of $\mathbf{K}$, the directional derivative of rotation $\boldsymbol{\Lambda}_{n+1}^{i}$ is calculated as

$$\frac{\mathrm{d}}{\mathrm{d}\epsilon}[\underbrace{\boldsymbol{\Lambda}_n \exp \tilde{\boldsymbol{\Theta}}^i}_{=\boldsymbol{\Lambda}_{n+1}^{i}} \exp(\epsilon \Delta\tilde{\boldsymbol{\Theta}})]_{\epsilon=0} = \boldsymbol{\Lambda}_{n+1}^{i} \Delta\tilde{\boldsymbol{\Theta}}, \tag{9.145}$$

where $\boldsymbol{\Lambda}_{n+1}^{i} \Delta\tilde{\boldsymbol{\Theta}} \in T_{\Lambda_{n+1}^{i}}\mathrm{SO}(3)$, i.e. the increment belongs to a different tangent plane than the total rotation vector $\tilde{\boldsymbol{\Theta}}$.

On the other hand, the directional derivative of angular velocity, $\mathbf{W}_{n+1}^{i} = (\gamma/\beta h)\, \boldsymbol{\Theta}^i + (\cdot)_n$, where $(\cdot)_n$ denotes the terms for t_n, is calculated as follows:

$$\frac{\mathrm{d}}{\mathrm{d}\epsilon}[\frac{\gamma}{\beta h}(\boldsymbol{\Theta}^i + \epsilon \Delta\tilde{\boldsymbol{\Theta}})]_{\epsilon=0} = \frac{\gamma}{\beta h} \Delta\tilde{\boldsymbol{\Theta}}, \tag{9.146}$$

where $\boldsymbol{\Lambda}_{n+1}^{0} \Delta\tilde{\boldsymbol{\Theta}} \in T_{\Lambda_{n+1}^{0}}\mathrm{SO}(3)$, i.e. it belongs to the same tangent plane as $\boldsymbol{\Lambda}_{n+1}^{0}\tilde{\boldsymbol{\Theta}}$. The transformation to the plane $T_{\Lambda^i}\mathrm{SO}(3)$ is

$$\underbrace{\Delta\boldsymbol{\Theta}}_{\in T_{\Lambda_{n+1}^{i}}} = \mathbf{T} \underbrace{\Delta\boldsymbol{\Theta}}_{\in T_{\Lambda_{n+1}^{0}}} . \tag{9.147}$$

The multiplicative updates of rotational parameters for the algorithm in $T_R\mathrm{SO}(3)$ are presented in Table 9.6. Note that two quaternions are used: $\mathbf{X}$ is the total quaternion, while $\mathbf{q}$ is the quaternion for the time step.

Algorithms 2 and 3. Multiplicative/additive updates. In these algorithms, $\Delta\tilde{\boldsymbol{\Theta}}$ belongs to the plane tangent at $\boldsymbol{\Lambda}_{n+1}^{0}$, which is the last converged solution for the previous time step, i.e. $\boldsymbol{\Lambda}_{n+1}^{0} = \boldsymbol{\Lambda}_{n}^{\mathrm{conv}}$. These algorithms are our modifications of ALGO-C1.

We assume that the external moment $\mathbf{m}$ is independent of $\boldsymbol{\Theta}$. Then the tangent operator is

$$\mathbf{K}(\boldsymbol{\Lambda}_{n+1}, \boldsymbol{\Theta}) = \boldsymbol{\Lambda}_{n+1} \left[\frac{\gamma}{\beta h} \mathbb{J} - (\widetilde{\mathbf{J}\mathbf{W}_{n+1}}) \mathbf{T}^{-1}(\boldsymbol{\Theta}) \right]. \tag{9.148}$$

This form of $\mathbf{K}$ is obtained when we use the inverse relation

$$\underbrace{\Delta\boldsymbol{\Theta}}_{\in T_{\Lambda_{n+1}^{0}}} = \mathbf{T}^{-1} \underbrace{\Delta\boldsymbol{\Theta}}_{\in T_{\Lambda_{n+1}^{i}}} \tag{9.149}$$

Table 9.6 Multiplicative updates of Algorithm 1. Quaternions.

Initialize		
$\mathbf{X}$	$\leftarrow$	total
Time step		
$\mathbf{W}$	$\leftarrow$	predict for step
$\mathbf{W} \to \boldsymbol{\Theta} \to \mathbf{q}$	$\leftarrow$	for step
$\mathbf{X} = \mathbf{X} \circ \mathbf{q} \to \boldsymbol{\Lambda}$	$\leftarrow$	total (multiplicative)
Newton loop		
Form governing equations using $\mathbf{K}(\boldsymbol{\Lambda}, \boldsymbol{\Theta})$ of eq. (9.144)		
Solve for $\Delta\boldsymbol{\Theta}$		
Update		
$\Delta\boldsymbol{\Theta} \to \Delta\mathbf{q} \to \mathbf{q} = \mathbf{q} \circ \Delta\mathbf{q} \to \boldsymbol{\Theta}$	$\leftarrow$	for step (multiplicative)
$\Delta\mathbf{q} \to \mathbf{X} = \mathbf{X} \circ \Delta\mathbf{q} \to \boldsymbol{\Lambda}$	$\leftarrow$	total (multiplicative)
$\boldsymbol{\Theta} \to \mathbf{W}$	$\leftarrow$	for step
End of Newton loop		

to transform $\Delta\tilde{\boldsymbol{\Theta}} = \Delta\boldsymbol{\Theta} \times \mathbf{I}$ in the directional derivative of rotation $\boldsymbol{\Lambda}^i_{n+1}$ of eq. (9.145), but when we leave the derivative of the angular velocity in eq. (9.146) unchanged. Here, $\boldsymbol{\Lambda}^0_{n+1}\Delta\tilde{\boldsymbol{\Theta}} \in T_{\boldsymbol{\Lambda}^0_{n+1}}\mathrm{SO}(3)$.

The multiplicative/additive updates of rotational parameters are presented in two versions: Algorithm 2 uses the rotation matrices, see Table 9.7, while Algorithm 3 uses two quaternions, the total quaternion $\mathbf{X}$ and the quaternion for the time step $\mathbf{q}$, see Table 9.8.

The rotation matrix $\boldsymbol{\Lambda}_n$ in Table 9.7 and the total quaternion $\mathbf{X}$ in Table 9.8 are not updated until the Newton iterations for the time step have converged. The additive update of $\boldsymbol{\Theta}$ affects $\mathbf{T}$ and either the rotation matrix $\boldsymbol{\Lambda}^{(i)}$ or the quaternion $\mathbf{q}$ for the time step.

In both schemes, after the convergence of the Newton method, the total rotation vector $\boldsymbol{\chi}$ is recovered from the quaternion $\mathbf{X}$, which is essential to obtain $\boldsymbol{\chi}$ without jumps.

Example. Unstable rotations. In this example, unstable rotations about the axis of intermediate moment of inertia are simulated, see [221], and we compare different update schemes of rotation parameters.

The motion consists of three phases: (1) unstable rotations about the axis of intermediate moment of inertia, (2) small disturbance acts for a duration of one time step, and (3) free unstable motion. In the third phase, the kinetic energy and the angular momentum should be preserved. The

Table 9.7 Multiplicative/additive updates of Algorithm 2. Rotation matrices.

Initialize		
$\mathbf{X}, \boldsymbol{\Lambda}$	$\leftarrow$	total
Time step		
$\mathbf{W}$	$\leftarrow$	predict for step
$\mathbf{W} \to \boldsymbol{\Theta} \to \boldsymbol{\Lambda}^{(0)}$	$\leftarrow$	for step
$\boldsymbol{\Lambda} = \boldsymbol{\Lambda}_n \boldsymbol{\Lambda}^{(0)}$	$\leftarrow$	total (multiplicative)
Newton loop		
Form governing equations using $\mathbf{K}(\boldsymbol{\Lambda}, \boldsymbol{\Theta})$ of eq. (9.148)		
Solve for $\Delta\boldsymbol{\Theta}$		
Update		
$\boldsymbol{\Theta} = \boldsymbol{\Theta} + \Delta\boldsymbol{\Theta}$	$\leftarrow$	for step (additive)
$\boldsymbol{\Theta} \to \boldsymbol{\Lambda}^{(i)}$	$\leftarrow$	for step
$\boldsymbol{\Lambda} = \boldsymbol{\Lambda}_n \boldsymbol{\Lambda}^{(i)}$	$\leftarrow$	total (multiplicative)
$\boldsymbol{\Theta} \to \mathbf{W}$	$\leftarrow$	for step
End of Newton loop		
Update		
$\boldsymbol{\Theta} \to \mathbf{q} \to \mathbf{X} = \mathbf{X} \circ \mathbf{q}$	$\leftarrow$	total (multiplicative)
$\boldsymbol{\Lambda}_n = \boldsymbol{\Lambda}$	$\leftarrow$	total

external torque $\mathbf{m}$ is defined as

$$\mathbf{m} = \begin{cases} C_1\,\mathbf{e}_1 & 0 \leq t \leq t_z \\ C_2\,\mathbf{e}_2 & t_z \leq t \leq t_z + h\,, \qquad t_z + h = 2, \quad C_1 = 20, \quad h\,C_2 = 0.2, \\ \mathbf{0} & t > t_z + h \end{cases}$$

the moment of inertia $\mathbb{J} = \text{diag}[5, 10, 1]$, and the initial conditions $\chi(0) = \mathbf{0}$, $\mathbf{W}(0) = \mathbf{0}$, $(\mathbf{A}(0) = \mathbf{0})$. The parameters for the Newmark algorithm are $\beta = 1/2$, $\gamma = 1$. The time step $h = 0.1$ is used to show large convergence radius of the algorithms used; nonetheless, it is too large to yield good accuracy of results.

The results for the multiplicative update (A1) and the multiplicative/additive updates (A1 and A2) are compared in Fig. 9.9 for $t_{\max}$ =10 sec. The results for A2 and A3 are identical. We see that all algorithms yield exactly the same solution and conserve the kinetic energy and the angular momentum during free motion. For a longer simulation, up to $t_{\max}$ =1000 sec, the results were also identical.

Table 9.8 Multiplicative/additive updates of Algorithm 3. Quaternions.

Initialize		
$\mathbf{X}_n$	$\leftarrow$	total
Time step		
$\mathbf{W}$	$\leftarrow$	predict for step
$\mathbf{W} \to \boldsymbol{\Theta} \to \mathbf{q}^{(0)}$	$\leftarrow$	for step ($\mathbf{q}^{(0)}$ - intermediate)
$\mathbf{X} = \mathbf{X}_n \circ \mathbf{q}^{(0)} \to \boldsymbol{\Lambda}$	$\leftarrow$	total (multiplicative)
Newton loop		
Form governing equations using $\mathbf{K}(\boldsymbol{\Lambda}, \boldsymbol{\Theta})$ of eq. (9.148)		
Solve for $\Delta\boldsymbol{\Theta}$		
Update		
$\boldsymbol{\Theta} = \boldsymbol{\Theta} + \Delta\boldsymbol{\Theta}$	$\leftarrow$	for step (additive)
$\boldsymbol{\Theta} \to \mathbf{q}^{(i)} \to \mathbf{X}_n \circ \mathbf{q}^{(i)} \to \boldsymbol{\Lambda}$	$\leftarrow$	total (multiplicative) ($\mathbf{X}$-not updated)
$\boldsymbol{\Theta} \to \mathbf{W}$	$\leftarrow$	for step
End of Newton loop		
Update		
$\boldsymbol{\Theta} \to \mathbf{q} \to \mathbf{X}_n = \mathbf{X}_n \circ \mathbf{q}$	$\leftarrow$	total (multiplicative)

The total number of iterations in the whole simulation is given in Table 9.9, and we see that the differences are minor, up to about 1.2% for the longer run.

Table 9.9 Number of iterations for particular updates. Unstable rotations. $h = 0.1$.

Algorithm	Updates	Table	Number of iterations in 10 sec	1000 sec
A1	multiplicative	9.6	359	41020
A2,A3	multiplicative/additive	9.7, 9.8	344	41485
(A1/A2) × 100%			104.4	98.9

Example. Fast spinning top. In this example, the motion of a symmetrical top in a uniform gravitational field is simulated, see [221] for more details. The top is not rotating freely so the energy and angular momentum are not to be conserved, but still we can compare the performance of the developed algorithms. The external torque is rendered by gravitation and is defined as follows:

$$\mathbf{m}_{n+\alpha} = -Mg\,l\,(\boldsymbol{\Lambda}_{n+\alpha}\,\mathbf{e}_3) \times \mathbf{e}_3, \tag{9.150}$$

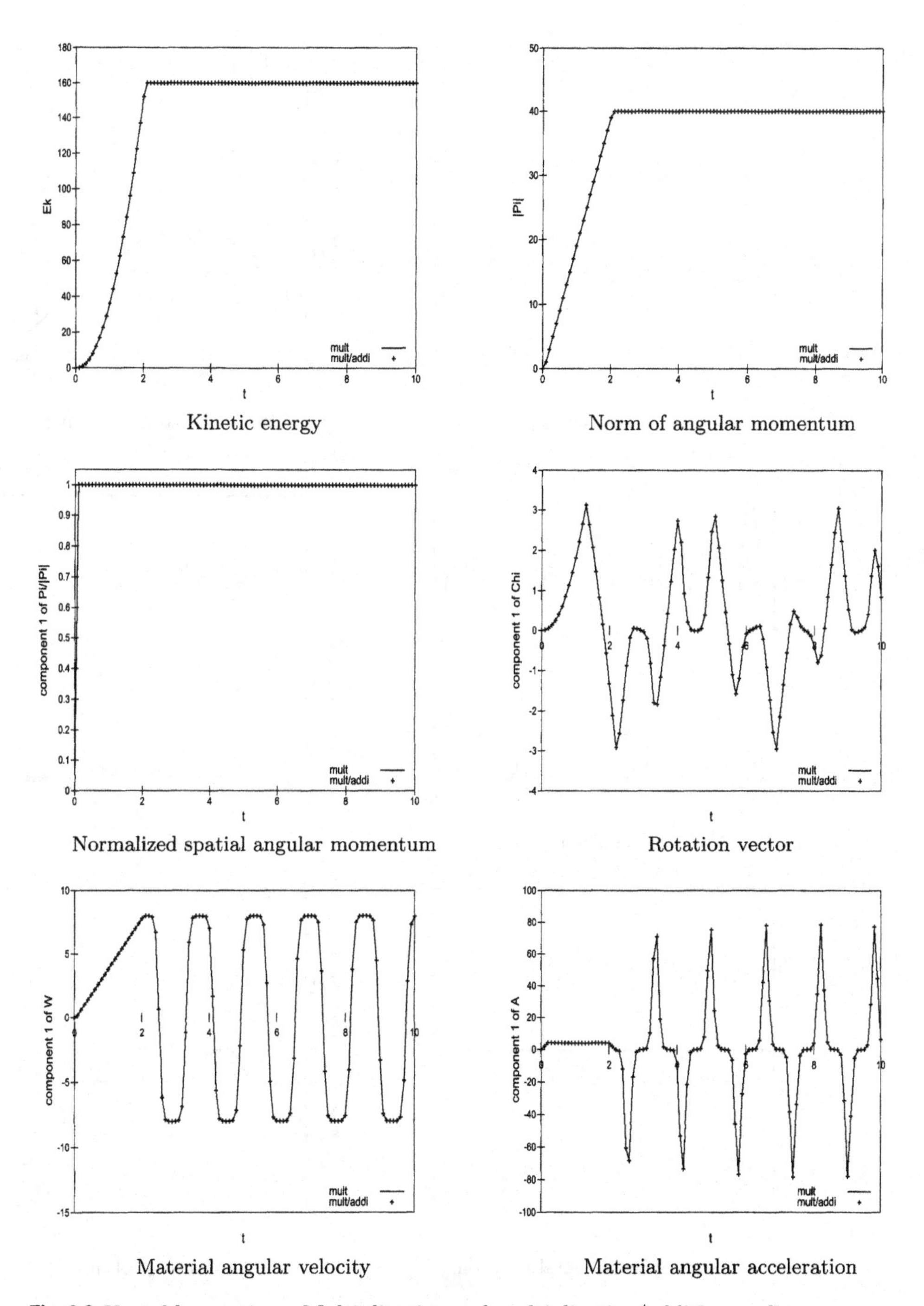

Fig. 9.9 Unstable rotations. Multiplicative and multiplicative/additive updates.

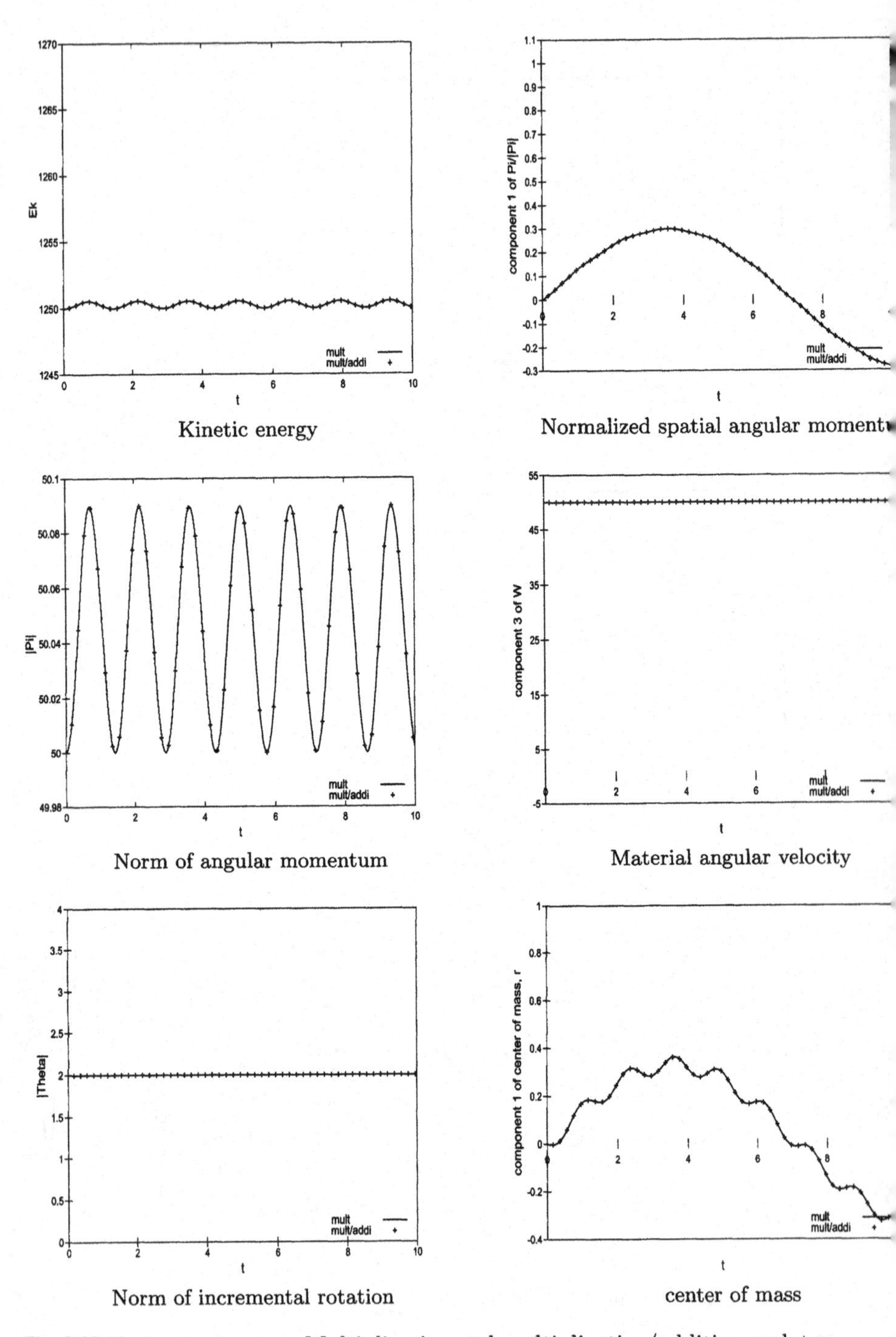

Fig. 9.10 Fast spinning top. Multiplicative and multiplicative/additive updates.

where $\boldsymbol{\Lambda}_{n+\alpha} = \boldsymbol{\Lambda}_n \exp(\alpha\tilde{\boldsymbol{\Theta}})$. Besides, M is mass, g is the gravitational acceleration, l is the distance from the center of mass to the fixed contact point, which is in the origin of the global frame $\{\mathbf{e}_i\}$. The basis vector $\mathbf{e}_3 \doteq \{0, 0, 1\}$.

The data is as follows: $Mg = 20$, $l = 1$, $\mathbb{J} = \text{diag}[5, 5, 1]$ and the initial conditions are $\boldsymbol{\chi}(0) = [0.3, 0, 0]^T$, $\mathbf{W}(0) = [0, 0, 50]^T$, $\mathbf{A}(0) = \mathbf{0}$, where $\boldsymbol{\chi}$ is the canonical rotation vector parameterizing $\boldsymbol{\Lambda}$. At $t = 0$, $\boldsymbol{\Lambda}(0) = \exp\tilde{\boldsymbol{\chi}}(0)$. Besides, $\alpha = 1/2$.

The torque $\mathbf{m}_{n+\alpha}$ depends on $\boldsymbol{\Theta}$ and contributes to the tangent operator in the following way:

$$\mathbf{K}_B(\boldsymbol{\Lambda}_{n+1}, \boldsymbol{\Theta}) = \mathbf{K}(\boldsymbol{\Lambda}_{n+1}, \boldsymbol{\Theta}) - h\,\mathbf{K}_{m\alpha}(\boldsymbol{\Theta}),$$

$$\mathbf{K}_{m\alpha}(\boldsymbol{\Theta}) = -Mg\,l\,\alpha\,\tilde{\mathbf{e}}_3\,\boldsymbol{\Lambda}_{n+\alpha}\,\tilde{\mathbf{e}}_3, \tag{9.151}$$

where $\mathbf{K}(\boldsymbol{\Lambda}_{n+1}, \boldsymbol{\Theta})$ is defined either by eq. (9.144) or by (9.148).

The results for the multiplicative update (A1) and the multiplicative/additive updates (A2 and A3) are compared in Fig. 9.10 for $t_{\max} = 10$ sec. The results for A2 and A3 are identical. For $h = 0.04$, all algorithms give identical results, and needed the same number of iterations (1000 iterations).

Finally, we can conclude, that for the incremental formulation with the iterative (Newton) solution within the time step, the additive update of rotation vectors provides the same accuracy and a similar effectiveness as the multiplicative update.

Part IV

FOUR-NODE SHELL ELEMENTS

10

Basic relations for four-node shell elements

In this chapter we describe the basic relations for four-node shell elements related to the FE approximations, numerical integration, and derivation of the tangent matrix and residual vector. The literature on four-node shell elements is vast, see, e.g., [123, 101, 162, 117, 118, 192, 235, 209, 85, 213, 165, 73, 217, 106, 242, 41, 108, 32, 68, 201, 243, 53], and many others.

The finite element method has achieved remarkable sophistication, but also great complexity, see the classical textbooks on FEs, such as [36, 98, 125, 61, 62, 16, 268, 58] and the new ones [160] and [161]. The requirements which new shell elements have to satisfy, are better defined and more demanding than they were some years ago.

10.1 Bilinear isoparametric approximations

Bilinear shape functions. Consider a bilinear function,

$$f(\xi,\eta) \doteq a_0 + a_1\xi + a_2\eta + a_3\xi\eta, \tag{10.1}$$

where $\xi,\eta \in [-1,+1]$ are the *natural coordinates*. Note that the domain is a bi-unit square, spanned by the corner nodes of coordinates $\{\xi_I,\eta_I\} = \{\pm 1, \pm 1\}$, see Fig. 10.1.

The coefficients a_i can be expressed in terms of values of f at corner nodes, using the conditions (i) $f = 1$ at one of the corner nodes, say I, and (ii) $f = 0$ at all other nodes. This yields a set of four equations, from which we can determine a_i for a selected I. The f, with so-determined a_i, is denoted as N_I, and designated as the *shape function*. Repeating this procedure for all corner nodes, we obtain

$$N_1(\xi,\eta) \doteq \frac{1}{4}(1-\xi)(1-\eta), \qquad N_2(\xi,\eta) \doteq \frac{1}{4}(1+\xi)(1-\eta),$$
$$N_3(\xi,\eta) \doteq \frac{1}{4}(1+\xi)(1+\eta), \qquad N_4(\xi,\eta) \doteq \frac{1}{4}(1-\xi)(1+\eta) \quad (10.2)$$

or, in concise form,

$$N_I(\xi,\eta) \doteq \frac{1}{4}(1+\xi_I\,\xi)\,(1+\eta_I\,\eta), \qquad I=1,2,3,4, \qquad (10.3)$$

where $(\xi_I,\eta_I) = (\pm 1, \pm 1)$ are coordinates of node I in the bi-unit domain. Note that N_I is a hyperbolic paraboloid (saddle) surface of ξ,η.

Vector of shape functions. Let us define the following vector of shape functions

$$\mathbf{N}(\xi,\eta) \doteq [N_1(\xi,\eta),\, N_2(\xi,\eta),\, N_3(\xi,\eta),\, N_4(\xi,\eta)]\,, \qquad (10.4)$$

where N_I are defined in eq. (10.2). This vector can be re-arranged as follows:

$$\mathbf{N}(\xi,\eta) = \frac{1}{4}\,(\mathbf{s} + \boldsymbol{\xi}\,\xi + \boldsymbol{\eta}\,\eta + \mathbf{h}\,\xi\eta)\,, \qquad (10.5)$$

where the auxiliary vectors are defined as

$$\mathbf{s} \doteq [1,1,1,1], \qquad \boldsymbol{\xi} \doteq [-1,1,1,-1],$$
$$\boldsymbol{\eta} \doteq [-1,-1,1,1], \qquad \mathbf{h} \doteq [1,-1,1,-1], \qquad (10.6)$$

and the subsequent entries correspond to the consecutive nodes. It is easy to check that the vectors $\mathbf{s}$, $\boldsymbol{\xi}$, $\boldsymbol{\eta}$, $\mathbf{h}$ are mutually orthogonal. The vector $\mathbf{s}$ is the translation vector, while the vectors $\boldsymbol{\xi}$ and $\boldsymbol{\eta}$ define ξ and η positions of consecutive nodes. The hourglass vector $\mathbf{h}$ multiplies the bilinear term $\xi\eta$.

Isoparametric approximations for shell elements. In an isoparametric shell FE, the initial position vector $\mathbf{y}_0$, the displacement vector $\mathbf{u}$ and the rotation vector $\boldsymbol{\psi}$, all for the reference surface, are approximated by the same shape functions $N_I(\xi^\alpha)$ as follows:

$$\mathbf{y}_0(\xi^\alpha) = \sum_{I=1}^{nel} N_I(\xi^\alpha)\,\mathbf{y}_{0I}, \qquad \mathbf{u}(\xi^\alpha) = \sum_{I=1}^{nel} N_I(\xi^\alpha)\,\mathbf{u}_I,$$
$$\boldsymbol{\psi}(\xi^\alpha) = \sum_{I=1}^{nel} N_I(\xi^\alpha)\,\boldsymbol{\psi}_I, \qquad \alpha = 1,2, \qquad (10.7)$$

where $(\cdot)_I$ denotes a value at node I, and nel is the number of nodes on an element; for four-node elements $nel = 4$. In the sequel, the natural coordinates are designated in two ways,

$$\xi^k \doteq \{\xi^\alpha, \xi^3\} = \{\xi, \eta, \xi^3\}, \qquad \xi^\alpha, \xi^3, \xi, \eta \in [-1, +1], \tag{10.8}$$

where $\alpha = 1,2$ and $k = 1,2,3$.

10.2 Geometry and bases of shell element

The initial geometry of a four-node shell element is defined by (i) positions of four corner nodes and (ii) bilinear shape functions. Using them, we can define the initial position of the reference surface and construct the local vector normal to this surface. The shell as a 3D body is generated by assuming some thickness in this normal direction.

Parametrization of reference surface. The position vector $\mathbf{y}_0$, which defines the reference surface in the initial configuration, see Fig. 10.1, is parameterized in terms of the natural coordinates $\xi, \eta \in [-1, +1]$ in the following way:

$$\mathbf{y}_0(\xi, \eta) = \sum_{I=1}^{4} N_I(\xi, \eta)\, \mathbf{y}_{0I}, \tag{10.9}$$

where $\mathbf{y}_{0I}$ is a position vector of node I, and $N_I(\xi, \eta)$ are shape functions of eq. (10.3). Alternatively, we can write this expression for each component k separately,

$$y_{0k}(\xi, \eta) = \sum_{I=1}^{4} N_I(\xi, \eta)\, y_{0kI}, \qquad k = 1,2,3, \tag{10.10}$$

where $y_{0k} \doteq \mathbf{y}_0 \cdot \mathbf{i}_k$ and $y_{0kI} \doteq \mathbf{y}_{0I} \cdot \mathbf{i}_k$. We can use the vector of shape functions of eq. (10.4), to avoid the summation sign,

$$y_{0k}(\xi, \eta) = \mathbf{N}(\xi, \eta)\, \mathbf{y}_{0kI}, \tag{10.11}$$

where $\mathbf{y}_{0kI} \doteq [y_{0k1}, y_{0k2}, y_{0k3}, y_{0k4}]$ is the vector of kth components of the position vectors of all nodes.

The parametrization of the reference surface defined by eq. (10.10) spans either a planar element or a warped element when one of the nodes is shifted out of the plane spanned by the other three nodes. The latter case is illustrated by an example below. More information on warped elements is provided in Sect. 14.

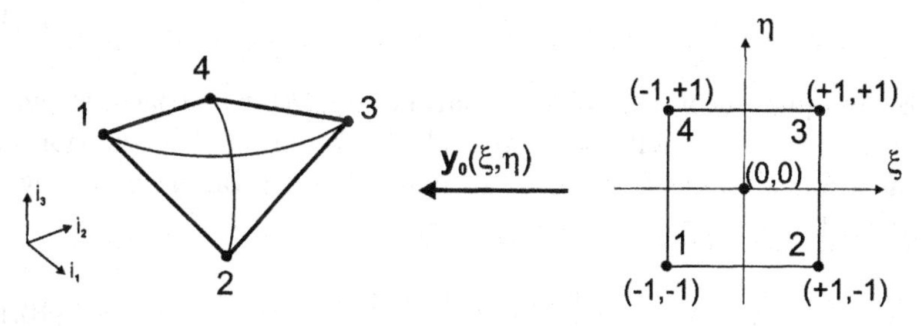

Fig. 10.1 Physical and reference (parent) bi-unit domain of a four-node element.

Example. Consider a square 2×2 element, with nodes 1, 2, 4 located in the X0Y plane, and node 3 elevated in the z-direction by w, see Fig. 10.2a.

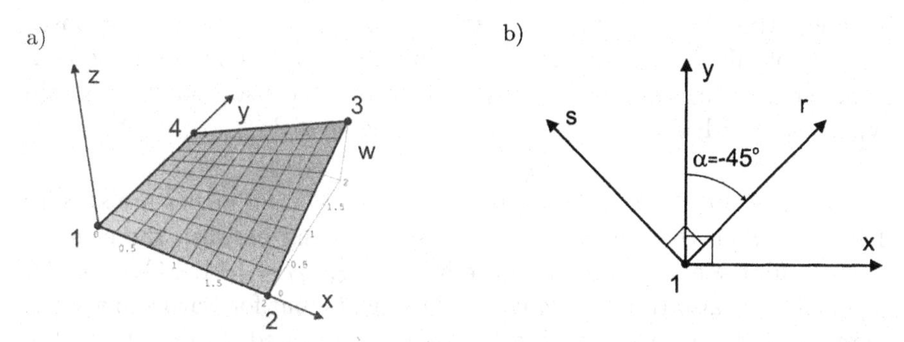

Fig. 10.2 a) The hyperbolic paraboloidal surface spanned by four nodes. b) Two coordinate systems.

Denote the components of the position vector as follows: $y_{01} \doteq x$, $y_{02} \doteq y$, $y_{03} \doteq z$. The Cartesian coordinates (x, y, z) of corner nodes are assumed to be $p_1 = (0,0,0)$, $p_2 = (2,0,0)$, $p_3 = (2,2,w)$, and $p_4 = (0,2,0)$. Grouping the x, y, and z components of all nodes as follows::

$$x_I = [0,2,2,0], \qquad y_I = [0,0,2,2], \qquad z_I = [0,0,w,0],$$

and using eq. (10.10) for each component, we obtain

$$x = 1 + \xi, \qquad y = 1 + \eta, \qquad z = \frac{w}{4}(1+\xi)(1+\eta).$$

From the first two equations, we have $\xi = x - 1$ and $\eta = y - 1$, and the third equation can be expressed solely in Cartesian coordinates,

$$z = \frac{w}{4} xy. \tag{10.12}$$

This is an equation of a hyperbolic paraboloidal (h-p) surface, shown in Fig. 10.2a, having a saddle point at node 1. To obtain this equation in a classical form, we introduce the coordinates $r, s,$ rotated by 45° w.r.t. $[x, y]$ coordinates, see Fig. 10.2b. Then

$$\begin{bmatrix} x \\ y \end{bmatrix} = \begin{bmatrix} c & s \\ -s & c \end{bmatrix} \begin{bmatrix} r \\ s \end{bmatrix}, \tag{10.13}$$

where $s \doteq \sin\alpha$, $c \doteq \cos\alpha$ and $\alpha = -45^o$. Using this relation in eq. (10.12), we obtain

$$z = \frac{w}{8}(r^2 - s^2), \tag{10.14}$$

which has the standard form of the h-p surface equation, see [150] p. 545. If we cut this surface using the vertical planes r=const. or s=const., then we obtain parabolas with either minimum or maximum at node 1. If we cut this surface using horizontal planes $z =$ const., we then obtain hyperbolas, with the asymptotes intersecting at node 1.

Remark 1. For planar (2D) four-node elements, if all angles are smaller than π, then there exists a one-to-one mapping between the element and a bi-unit square spanned on the nodes $\{\xi, \eta\}_I = \{\pm 1, \pm 1\}$, see [98] p. 116. This is also true for planar shell elements, but for warped ones, the question of the inverse mapping becomes complicated. In fact, using the FE method and a numerical integration, we do not need this information, and the Jacobian matrix and its inverse at integration points suffice.

Remark 2. The four-node quadrilateral shell element can also be defined in another way by additionally using the normal vectors at nodes. These vectors must be either computed, e.g. using normals of all elements connected to the node, or be provided as input data, which can be cumbersome. Another possibility is to use a CAD program, in which we can define typical shapes of regular surfaces and directly obtain a normal vector at a selected point.

Natural tangent vectors. The natural vectors tangent to the reference surface are defined as

$$\mathbf{g}_1(\xi, \eta) \doteq \frac{\partial \mathbf{y}_0(\xi, \eta)}{\partial \xi}, \qquad \mathbf{g}_2(\xi, \eta) \doteq \frac{\partial \mathbf{y}_0(\xi, \eta)}{\partial \eta}. \tag{10.15}$$

In general, these vectors are neither unit nor mutually orthogonal. For the bilinear approximation and $\mathbf{y}_0$ in the form of eq. (10.9), we obtain

$$\mathbf{g}_\alpha(\xi,\eta) = \sum_{I=1}^{4} N_{I,\alpha}\,\mathbf{y}_{0I}, \qquad \alpha = 1,2, \tag{10.16}$$

where the derivatives of the shape functions are $N_{I,1} = \frac{1}{4}\xi_I\,(1+\eta_I\,\eta)$ and $N_{I,2} = \frac{1}{4}(1+\xi_I\,\xi)\,\eta_I$. Using the vector of shape functions $\mathbf{N}(\xi,\eta)$ of eq. (10.11), we have

$$(\mathbf{g}_\alpha)_k(\xi,\eta) = \mathbf{N}_{,\alpha}\,\mathbf{y}_{0kI}, \tag{10.17}$$

where $\mathbf{N}_{,1} = \frac{1}{4}\,(\boldsymbol{\xi} + \mathbf{h}\,\eta)$ and $\mathbf{N}_{,2} = \frac{1}{4}\,(\boldsymbol{\eta} + \mathbf{h}\,\xi)$. We see that the tangent vectors vary over the element; $\mathbf{g}_1$ is constant in ξ and linear in η, while $\mathbf{g}_2$ is the opposite way round. At the element's center, $\xi = \eta = 0$, the vectors $\mathbf{g}_\alpha$ are equal to 1/2 of vectors connecting the opposite middle-edge points.

Normal vector. The vector normal to the reference surface is defined as a cross-product of the natural tangent vectors

$$\bar{\mathbf{g}}_3(\xi,\eta) \doteq \mathbf{g}_1(\xi,\eta) \times \mathbf{g}_2(\xi,\eta). \tag{10.18}$$

Note that $\bar{\mathbf{g}}_3$ is perpendicular to the tangent vectors $\mathbf{g}_\alpha$, but is not of unit length. This vector is not associated with the coordinate $\xi^3 \in [-1,+1]$; see eq. (10.30).

Local Cartesian basis. For an irregular geometry of an element, the basis $\{\mathbf{g}_\alpha, \mathbf{g}_3\}$ is normal but skew, which is not convenient, e.g., to define the constitutive relations for non-isotropic materials. Hence, a local Cartesian basis is introduced as described below.

Designate the local Cartesian basis by $\{\mathbf{t}_k\}$ $(k = 1,2,3)$. Define the normal vector as

$$\mathbf{t}_3 \doteq \frac{\bar{\mathbf{g}}_3}{\|\bar{\mathbf{g}}_3\|}, \tag{10.19}$$

where $\bar{\mathbf{g}}_3$ is defined by eq. (10.18). The tangent vectors of the local Cartesian basis can be constructed in several ways; we define them in terms of the auxiliary normalized natural vectors

$$\tilde{\mathbf{g}}_\alpha \doteq \frac{\mathbf{g}_\alpha}{\|\mathbf{g}_\alpha\|}, \tag{10.20}$$

designated by a tilde. Below three types of bases used in shell elements are presented.

Basis 1. *One vector is parallel to the vector of the natural basis*

$$\mathbf{t}_1 \doteq \tilde{\mathbf{g}}_1, \qquad \mathbf{t}_2 \doteq \mathbf{t}_3 \times \mathbf{t}_1, \tag{10.21}$$

where $\mathbf{t}_1$ is identical as $\tilde{\mathbf{g}}_1$, see Fig. 10.3a. This basis was used, e.g., in DYNA3D, see [89], eq. (35).

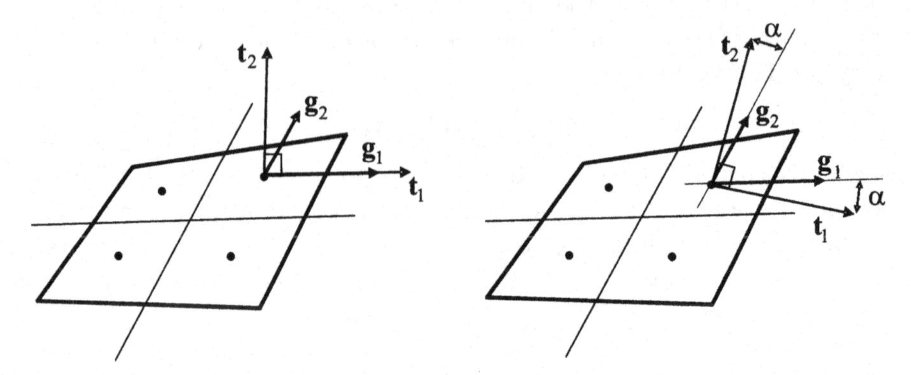

Fig. 10.3 Ortho-normal frames at a Gauss point. a) **Basis 1** and b) **Basis 2**.

Basis 2. *Vectors are equally distant from vectors of the natural basis*

$$\mathbf{t}_1 \doteq \frac{1}{\sqrt{2}}(\tilde{\mathbf{t}}_1 - \tilde{\mathbf{t}}_2), \qquad \mathbf{t}_2 \doteq \frac{1}{\sqrt{2}}(\tilde{\mathbf{t}}_1 + \tilde{\mathbf{t}}_2), \tag{10.22}$$

where the auxiliary vectors are

$$\tilde{\mathbf{t}}_1 \doteq \frac{\tilde{\mathbf{g}}_1 + \tilde{\mathbf{g}}_2}{\|\tilde{\mathbf{g}}_1 + \tilde{\mathbf{g}}_2\|}, \qquad \tilde{\mathbf{t}}_2 \doteq \mathbf{t}_3 \times \tilde{\mathbf{t}}_1. \tag{10.23}$$

This is the most popular basis, see in [98] p. 386 or [50] p. 111 and is shown in Fig. 10.3b.

For this basis, we can show that the average of $\mathbf{t}_1$ and $\mathbf{t}_2$ and the average of $\tilde{\mathbf{g}}_1$ and $\tilde{\mathbf{g}}_2$ are parallel, but have different lengths, i.e.

$$\tfrac{1}{2}(\mathbf{t}_1 + \mathbf{t}_2) = \frac{1}{2\sqrt{2}}\tilde{\mathbf{t}}_1 = a\,\tfrac{1}{2}(\tilde{\mathbf{g}}_1 + \tilde{\mathbf{g}}_2),$$

where $a = 1/(\sqrt{2}\,\|\tilde{\mathbf{g}}_1 + \tilde{\mathbf{g}}_2\|)$.

Besides, the term "equally distant" means that the angle between $\tilde{\mathbf{g}}_1$ and $\mathbf{t}_1$ is equal to the angle between $\tilde{\mathbf{g}}_2$ and $\mathbf{t}_2$. This can be checked in the following way:

$$\begin{aligned} \mathbf{t}_1 \cdot \tilde{\mathbf{g}}_1 &= a\left\{(\tilde{\mathbf{g}}_1 + \tilde{\mathbf{g}}_2) \cdot \tilde{\mathbf{g}}_1 - [\mathbf{t}_3 \times (\tilde{\mathbf{g}}_1 + \tilde{\mathbf{g}}_2)] \cdot \tilde{\mathbf{g}}_1\right\} \\ &= a\left\{\tilde{\mathbf{g}}_1 \cdot \tilde{\mathbf{g}}_1 - [\mathbf{t}_3 \times \tilde{\mathbf{g}}_2] \cdot \tilde{\mathbf{g}}_1\right\}, \\ \mathbf{t}_2 \cdot \tilde{\mathbf{g}}_2 &= a\left\{(\tilde{\mathbf{g}}_1 + \tilde{\mathbf{g}}_2) \cdot \tilde{\mathbf{g}}_2 + [\mathbf{t}_3 \times (\tilde{\mathbf{g}}_1 + \tilde{\mathbf{g}}_2)] \cdot \tilde{\mathbf{g}}_2\right\} \\ &= a\left\{\tilde{\mathbf{g}}_2 \cdot \tilde{\mathbf{g}}_2 + [\mathbf{t}_3 \times \tilde{\mathbf{g}}_1] \cdot \tilde{\mathbf{g}}_2\right\}. \end{aligned}$$

These two scalar products are equal, as $\tilde{\mathbf{g}}_1 \cdot \tilde{\mathbf{g}}_1 = \tilde{\mathbf{g}}_2 \cdot \tilde{\mathbf{g}}_2 = 1$ by eq. (10.20).

Finally, we note that for 2D problems formulated in the $\{\mathbf{g}_1, \mathbf{g}_2\}$-plane, we can obtain the components of $\tilde{\mathbf{t}}_2$ as follows::

$$\tilde{\mathbf{t}}_2 = \mathbf{t}_3 \times \tilde{\mathbf{t}}_1 = [0, 0, 1]^T \times [t_1,\, t_2,\, 0]^T = -[t_2,\, t_1,\, 0]^T,$$

where the components of $\tilde{\mathbf{t}}_1 = [t_1, t_2, 0]^T$ are known.

Basis 2. Version 2. *Vectors equally distant from vectors of natural basis*

$$\mathbf{t}_1 \doteq \cos(-\beta/2)\, \tilde{\mathbf{t}}_1 + \sin(-\beta/2)\, \tilde{\mathbf{t}}_2, \quad \mathbf{t}_2 \doteq -\sin(-\beta/2)\, \tilde{\mathbf{t}}_1 + \cos(-\beta/2)\, \tilde{\mathbf{t}}_2, \tag{10.24}$$

where

$$\beta = \arctan \frac{(\mathbf{t}_1 \cdot \mathbf{g}_2)}{(\mathbf{t}_2 \cdot \mathbf{g}_2)}, \qquad \tilde{\mathbf{t}}_1 = \tilde{\mathbf{g}}_1, \qquad \tilde{\mathbf{t}}_2 = \mathbf{t}_3 \times \tilde{\mathbf{t}}_1. \tag{10.25}$$

Derivation. To obtain the same angle between $\mathbf{g}_1$ and $\mathbf{t}_1$ and between $\mathbf{g}_2$ and $\mathbf{t}_2$, we generate an orthonormal basis and then rotate it around the normal vector $\mathbf{t}_3$. We shall use **Basis 1** of eq. (10.21) as the orthonormal basis to start from and denote its vectors as follows:

$$\tilde{\mathbf{t}}_1 = \tilde{\mathbf{g}}_1, \qquad \tilde{\mathbf{t}}_2 = \mathbf{t}_3 \times \tilde{\mathbf{t}}_1.$$

The angle between $\mathbf{t}_2$ and $\mathbf{g}_2$, is denoted by β and we assume that $|\beta| < \pi/2$. To determine β, we can use the following formulas:

$$\sin\beta = \frac{(\tilde{\mathbf{t}}_1 \cdot \mathbf{g}_2)}{\|\mathbf{g}_2\|}, \qquad \cos\beta = \frac{(\tilde{\mathbf{t}}_2 \cdot \mathbf{g}_2)}{\|\mathbf{g}_2\|}, \qquad \tan\beta = \frac{\sin\beta}{\cos\beta} = \frac{(\tilde{\mathbf{t}}_1 \cdot \mathbf{g}_2)}{(\tilde{\mathbf{t}}_2 \cdot \mathbf{g}_2)},$$

from which we obtain

$$\beta = \arctan \frac{(\tilde{\mathbf{t}}_1 \cdot \mathbf{g}_2)}{(\tilde{\mathbf{t}}_2 \cdot \mathbf{g}_2)}.$$

Then we rotate the frame $\{\tilde{\mathbf{t}}_1, \tilde{\mathbf{t}}_2\}$ by the angle $-\beta/2$ around the normal vector $\mathbf{t}_3$. For the rotation tensor defined as

$$\mathbf{R} = \cos(-\beta/2)(\tilde{\mathbf{t}}_1 \otimes \tilde{\mathbf{t}}_1 + \tilde{\mathbf{t}}_2 \otimes \tilde{\mathbf{t}}_2) + \sin(-\beta/2)(\tilde{\mathbf{t}}_2 \otimes \tilde{\mathbf{t}}_1 - \tilde{\mathbf{t}}_1 \otimes \tilde{\mathbf{t}}_2) + \mathbf{t}_3 \otimes \mathbf{t}_3,$$

from $\mathbf{t}_1 = \mathbf{R}\tilde{\mathbf{t}}_1$ and $\mathbf{t}_2 = \mathbf{R}\tilde{\mathbf{t}}_2$, we obtain eq. (10.24).

Note that we can obtain $\sin(-\beta/2)$ and $\cos(-\beta/2)$ in another way, without the use of the arctan function. Then, first we calculate

$$\sin\beta = \tilde{\mathbf{t}}_1 \cdot \tilde{\mathbf{g}}_2, \qquad \cos\beta = \tilde{\mathbf{t}}_2 \cdot \tilde{\mathbf{g}}_2$$

and then, using the half-angle formulas, we obtain

$$\sin(-\beta/2) = s\sqrt{\tfrac{1}{2}(1-\cos\beta)}, \qquad \cos(-\beta/2) = +\sqrt{\tfrac{1}{2}(1+\cos\beta)},$$

where $s = \text{sign}(-\sin\beta)$.

Basis 3. *Vectors related to the reference basis.*

$$\text{If } (\mathbf{t}_3 \cdot \mathbf{i}_1) < (1-\tau) \text{ then } \mathbf{t}_2 = \frac{\mathbf{t}_3 \times \mathbf{i}_1}{\|\mathbf{t}_3 \times \mathbf{i}_1\|} = \frac{1}{\sqrt{t_2^2 + t_3^2}} \begin{bmatrix} 0 \\ t_3 \\ -t_2 \end{bmatrix}, \tag{10.26}$$

$$\text{otherwise } \mathbf{t}_2 = \frac{\mathbf{t}_3 \times \mathbf{i}_2}{\|\mathbf{t}_3 \times \mathbf{i}_2\|} = \frac{1}{\sqrt{t_1^2 + t_2^2}} \begin{bmatrix} t_2 \\ -t_1 \\ 0 \end{bmatrix}, \tag{10.27}$$

and

$$\mathbf{t}_1 = \mathbf{t}_2 \times \mathbf{t}_3, \tag{10.28}$$

where $\mathbf{t}_3 = [t_1, t_2, t_3]^T$ denotes the components of the normal vector in the reference (global) basis $\{\mathbf{i}_k\}$ and τ is a small parameter. The advantage of this basis is that it provides easy identification of directions for complicated curved structures.

The second formula, eq. (10.27), is used when $\mathbf{t}_3$ and $\mathbf{i}_1$ are almost or exactly parallel and the cross-product $\mathbf{t}_3 \times \mathbf{i}_1$ is not well conditioned. Consider the result of the above definitions for two limit cases.

1. If $\mathbf{t}_3 \| \mathbf{i}_1$, then $\mathbf{t}_2 = \mathbf{t}_3 \times \mathbf{i}_2 \| \mathbf{i}_3$ and $\mathbf{t}_1 = \mathbf{t}_2 \times \mathbf{t}_3 \| \mathbf{i}_2$.
2. If $\mathbf{t}_3 \| \mathbf{i}_2$, then $\mathbf{t}_2 = \mathbf{t}_3 \times \mathbf{i}_1 \| -\mathbf{i}_3$ and $\mathbf{t}_1 = \mathbf{t}_2 \times \mathbf{t}_3 \| \mathbf{i}_1$.

Both these cases are shown in Fig. 10.4. Another basis related to the reference basis $\{\mathbf{i}_k\}$ is defined in [71] p. 242.

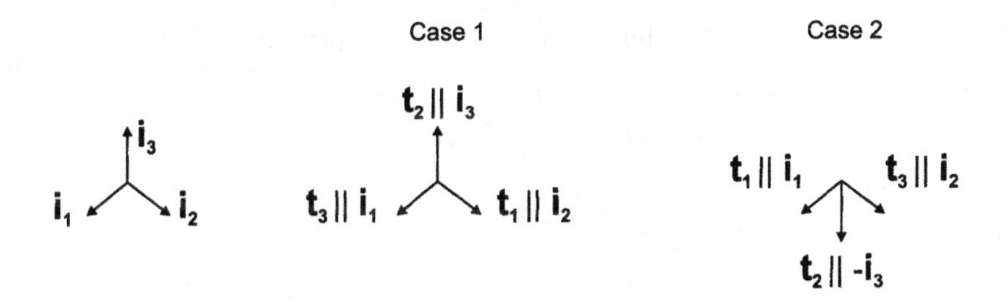

Fig. 10.4 Basis 3. Two limit cases of its position.

Normal vector associated with ξ^3. Note that the vector $\bar{\mathbf{g}}_3$ of eq. (10.18) is not associated with the coordinate $\xi^3 \in [-1, +1]$ and below we derive the proper vector.

The position vector of a shell lamina $\zeta = \text{const.}$ relative to the middle surface is $(\mathbf{y} - \mathbf{y}_0) = \zeta \mathbf{t}_3$, where $\zeta \in [-h/2, +h/2]$. We can parameterize ζ in terms of $\xi^3 \in [-1, +1]$, as $\zeta = (h/2)\, \xi^3$, and define the normal vector as a derivative w.r.t. ξ^3, i.e.

$$\mathbf{g}_3 \doteq \frac{\mathrm{d}(\mathbf{y} - \mathbf{y}_0)}{\mathrm{d}\xi^3} = \frac{h}{2} \mathbf{t}_3. \tag{10.29}$$

This vector stretches from the middle surface to the top surface of a shell, see Fig. 10.5.

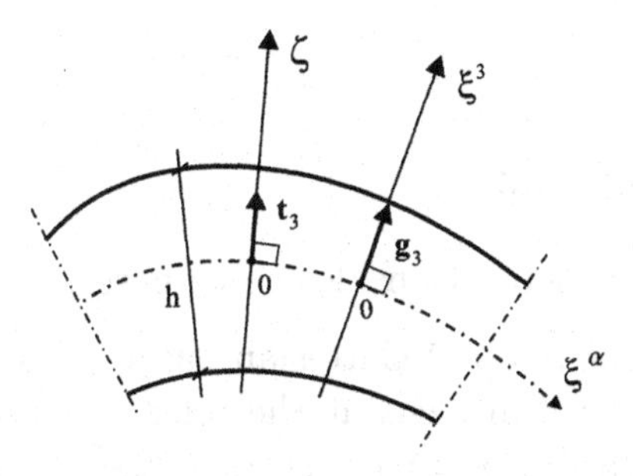

Fig. 10.5 Associated pairs: $(\xi^3, \mathbf{g}_3)$ and $(\zeta, \mathbf{t}_3)$.

Note that the vector $\mathbf{g}_3$ is different from $\bar{\mathbf{g}}_3$ of eq. (10.18) and is associated with ξ^3 because

$$\xi^3 \mathbf{g}_3 = \zeta\, \mathbf{t}_3. \tag{10.30}$$

Hence, we can use either $(\xi^3, \mathbf{g}_3)$ or $(\zeta, \mathbf{t}_3)$, but certainly not $(\xi^3, \bar{\mathbf{g}}_3)$.

Remark. In some works, the normal vector is approximated as

$$\mathbf{t}_3(\xi, \eta) \doteq \sum_{I=1}^{4} N_I(\xi, \eta) \, \mathbf{n}_I, \tag{10.31}$$

where $\mathbf{n}_I$ is the normal unit vector at node I defined as

$$\mathbf{n}_I = \frac{\mathbf{a}_I \times \mathbf{b}_I}{\|\mathbf{a}_I \times \mathbf{b}_I\|}$$

and $\mathbf{a}_I$ and $\mathbf{b}_I$ are the vectors connecting node I with the adjacent corners, see Fig. 10.6. When the element is planar, then this definition is equivalent to eq. (10.19). However, when the element is warped, vectors $\mathbf{n}_I$ are not parallel and $\mathbf{t}_3$ is neither unit nor perpendicular to the local tangent vectors $\mathbf{g}_\alpha$.

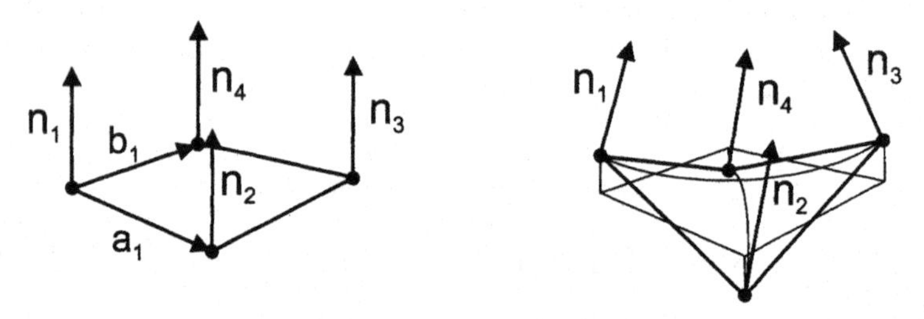

Fig. 10.6 Normal corner vectors for a planar element and a warped element.

10.3 Jacobian matrices

Bases for initial configuration. Consider three bases:

1. $\{\mathbf{i}_k\}$ - the reference (global) Cartesian basis, $k = 1, 2, 3$,
2. $\{\mathbf{g}_k\}$ - the local natural basis at the reference surface for the initial configuration with the tangent vectors defined by eq. (10.15) and the normal vector by eq. (10.29).
3. $\{\mathbf{t}_k\}$ - the local Cartesian basis at the reference surface for the initial configuration with the normal vector defined by eq. (10.19) and the tangent vectors generated in one of the three ways described earlier. For simplicity, we denote $S^3 = \zeta$.

The coordinates associated with these bases are designated as y^k, ξ^k, S^k, respectively. Note that the zero of the natural coordinates ξ^α is at the element's center, while the zero of the Cartesian coordinates S^α is at any considered Gauss point g at which we define the local bases.

The initial position vector $\mathbf{y}$ relative to the position vector of the Gauss point, $\mathbf{y}_g$, can be expressed in the following three ways:

$$\mathbf{y} - \mathbf{y}_g = (y^k - y^k_g)\,\mathbf{i}_k = (\xi^k - \xi^k_g)\,\mathbf{g}_k = S^k\,\mathbf{t}_k. \tag{10.32}$$

This equation links the above-defined bases and coordinates on the tangent plane spanned at a Gauss point.

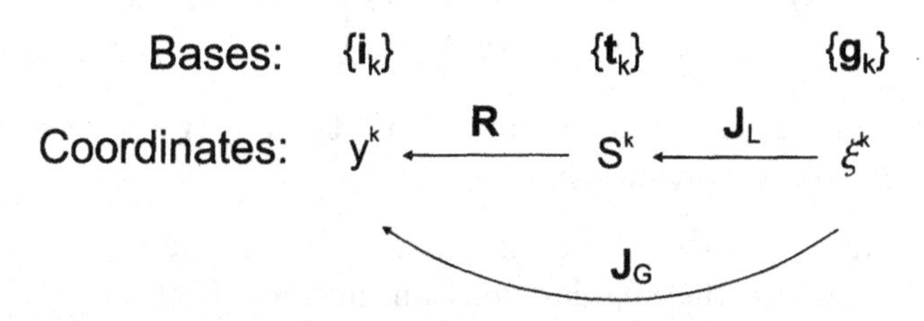

Fig. 10.7 Mappings and Jacobian matrices for the initial configuration.

Jacobian matrices for initial configuration. Let us define the following three types of mappings of coordinates and the Jacobian matrices, see Fig. 10.7:

$$\begin{aligned}
\xi^k &\mapsto y^l : &\quad \mathbf{J}_G &\doteq \left[\frac{\partial y^l}{\partial \xi^k}\right], \\
S^l &\mapsto y^k : &\quad \mathbf{R} &\doteq \left[\frac{\partial y^k}{\partial S^l}\right], \\
\xi^k &\mapsto S^l : &\quad \mathbf{J}_L &\doteq \left[\frac{\partial S^l}{\partial \xi^k}\right],
\end{aligned} \tag{10.33}$$

where the components of gradients are the matrices arranged as in eq. (2.41).

To obtain the equation linking the above gradients, we use the chain rule of differentiation

$$\frac{\partial y^k}{\partial \xi^l} = \frac{\partial y^k}{\partial S^m}\frac{\partial S^m}{\partial \xi^l}, \qquad \text{which yields} \quad \mathbf{J}_G = \mathbf{R}\,\mathbf{J}_L, \tag{10.34}$$

where $k, l, m = 1, 2, 3$.

Rotation matrix $\mathbf{R}$. The angular position of the local basis $\{\mathbf{t}_k\}$ relative to the reference $\{\mathbf{i}_k\}$ is described by the rotation tensor $\mathbf{R} \doteq \mathbf{t}_l \otimes \mathbf{i}_l \in$ SO(3). This definition implies that $\mathbf{R}\,\mathbf{i}_k = \mathbf{t}_k$, i.e. $\mathbf{t}_k$ is a forward-rotated $\mathbf{i}_k$. The components of $\mathbf{R}$ are

$$R_{jk} = \mathbf{i}_j \cdot (\mathbf{R}\mathbf{i}_k) = \mathbf{i}_j \cdot \mathbf{t}_k \tag{10.35}$$

and, in matrix form,

$$[R_{jk}] = \begin{bmatrix} \mathbf{i}_1 \cdot \mathbf{t}_1 & \mathbf{i}_1 \cdot \mathbf{t}_2 & \mathbf{i}_1 \cdot \mathbf{t}_3 \\ \mathbf{i}_2 \cdot \mathbf{t}_1 & \mathbf{i}_2 \cdot \mathbf{t}_2 & \mathbf{i}_2 \cdot \mathbf{t}_3 \\ \mathbf{i}_3 \cdot \mathbf{t}_1 & \mathbf{i}_3 \cdot \mathbf{t}_2 & \mathbf{i}_3 \cdot \mathbf{t}_3 \end{bmatrix} = [\mathsf{t}_1 \,|\, \mathsf{t}_2 \,|\, \mathsf{t}_3], \tag{10.36}$$

where the columns contain components of $\mathbf{t}_k$ in $\{\mathbf{i}_k\}$. The vectors of these components we denote as t_k.

We can show that the Jacobian matrix of the mapping of coordinates $S^l \mapsto y^k$ is equal to the angular position matrix $[R_{jk}]$, i.e.

$$\left[\frac{\partial y^k}{\partial S^j}\right] = [R_{jk}]. \tag{10.37}$$

By eq. (10.32), $(y^j - y^j_g)\mathbf{i}_j = S^k \mathbf{t}_k$, from which we obtain $(y^j - y^j_g) = S^k\,(\mathbf{t}_k \cdot \mathbf{i}_j)$, and the differentiation of both sides w.r.t. S^k yields $\partial y^j / \partial S^k = \mathbf{i}_j \cdot \mathbf{t}_k$, where the r.h.s. is identical as the r.h.s. of eq. (10.35), which ends the proof. □

Global Jacobian matrix $\mathbf{J}_G$. For the mapping of coordinates $\xi^k \mapsto y^i$ of eq. (10.33), the Jacobian matrix is defined as

$$\mathbf{J}_G \doteq \left[\frac{\partial y^i}{\partial \xi^k}\right] = \begin{bmatrix} \frac{\partial y^1}{\partial \xi^1} & \frac{\partial y^1}{\partial \xi^2} & \frac{\partial y^1}{\partial \xi^3} \\ \frac{\partial y^2}{\partial \xi^1} & \frac{\partial y^2}{\partial \xi^2} & \frac{\partial y^2}{\partial \xi^3} \\ \frac{\partial y^3}{\partial \xi^1} & \frac{\partial y^3}{\partial \xi^2} & \frac{\partial y^3}{\partial \xi^3} \end{bmatrix} = [\mathsf{g}_1 \,|\, \mathsf{g}_2 \,|\, \mathsf{g}_3], \tag{10.38}$$

where the columns contain components of $\mathbf{g}_k$ in $\{\mathbf{i}_k\}$. We designate this Jacobian as "global" because the global (reference) Cartesian coordinates y^i are differentiated.

Local Jacobian matrices $\mathbf{J}_L$. For the mapping of coordinates $\xi^k \mapsto S^l$ of eq. (10.33), we define the Jacobian matrix

$$\mathbf{J}_L \doteq \left[\frac{\partial S^i}{\partial \xi^k}\right] = \left[\begin{array}{cc|c} \frac{\partial S^1}{\partial \xi^1} & \frac{\partial S^1}{\partial \xi^2} & 0 \\ \frac{\partial S^2}{\partial \xi^1} & \frac{\partial S^2}{\partial \xi^2} & 0 \\ \hline 0 & 0 & \frac{h}{2} \end{array}\right], \tag{10.39}$$

where the last form is obtained for

$$\frac{\partial S^3}{\partial \xi} = 0, \quad \frac{\partial S^3}{\partial \eta} = 0, \quad \frac{\partial S^\alpha}{\partial \xi^3} = 0, \quad \frac{\partial S^3}{\partial \xi^3} = \frac{\partial(\xi^3 h/2)}{\partial \xi^3} = \frac{h}{2}. \tag{10.40}$$

We designate this Jacobian as "local", because the local Cartesian coordinates S^i are differentiated.

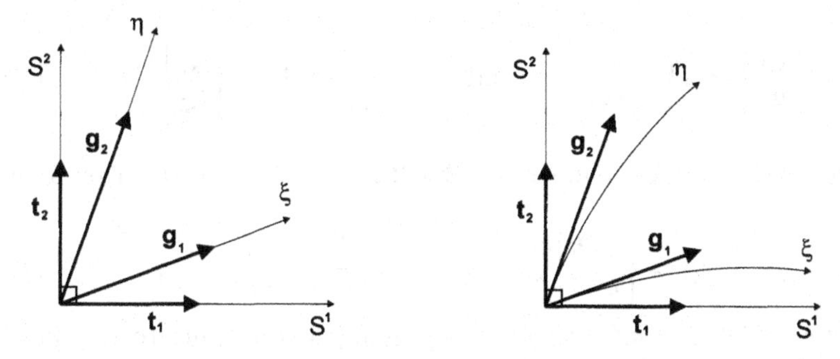

Fig. 10.8 Local bases and coordinates. a) for four-node element, b) for 9-node element with curved boundaries.

Consider only this part of the local mapping $\xi^\beta \mapsto S^\alpha$ $(\alpha, \beta = 1, 2)$ which is related to the tangent plane, see Fig. 10.8. To obtain a Jacobian matrix for this part, we extract the upper 2×2 part of $\mathbf{J}_L$, and denote it as $\mathbf{J}$,

$$\mathbf{J} \doteq \left[\frac{\partial S^\alpha}{\partial \xi^\beta}\right] = \begin{bmatrix} \frac{\partial S^1}{\partial \xi^1} & \frac{\partial S^1}{\partial \xi^2} \\ \frac{\partial S^2}{\partial \xi^1} & \frac{\partial S^2}{\partial \xi^2} \end{bmatrix} = \begin{bmatrix} \mathbf{g}_1 \cdot \mathbf{t}_1 & \mathbf{g}_2 \cdot \mathbf{t}_1 \\ \mathbf{g}_1 \cdot \mathbf{t}_2 & \mathbf{g}_2 \cdot \mathbf{t}_2 \end{bmatrix}, \tag{10.41}$$

where columns contain components of $\mathbf{g}_\beta$ in $\{\mathbf{t}_\alpha\}$.

The last form of $\mathbf{J}$ is obtained as follows. By eq. (10.32), $S^k = (\mathbf{y} - \mathbf{y}_g) \cdot \mathbf{t}_k$, in which $\mathbf{y}$ and $\mathbf{t}_k$ are functions of ξ^α, and the differentiation yields

$$\frac{\partial S^{\alpha}}{\partial \xi^{\beta}} = \mathbf{g}_{\beta} \cdot \mathbf{t}_{\alpha} + \left(\mathbf{y} - \mathbf{y}_{g}\right) \cdot \frac{\partial \mathbf{t}_{\alpha}}{\partial \xi^{\beta}} \quad \overset{\text{at GP}}{=} \quad \mathbf{g}_{\beta} \cdot \mathbf{t}_{\alpha}, \tag{10.42}$$

where $\mathbf{g}_{\beta} \doteq \partial \mathbf{y}/\partial \xi^{\beta}$ by eq. (10.15). The last form is valid only for the Gauss point, when $\left(\mathbf{y} - \mathbf{y}_{g}\right) = \mathbf{0}$, and the second term drops out so we obtain eq. (10.41). The Jacobian at the element center is denoted as $\mathbf{J}_{c} \doteq \mathbf{J}|_{\xi=\eta=0}$.

Relation between $\mathbf{g}_{\alpha}$ and $\mathbf{t}_{\alpha}$. The coordinate gradients imply relations linking the bases with which these coordinates are associated. The natural basis vectors $\mathbf{g}_{\alpha}$ can be decomposed in the ortho-normal $\{\mathbf{t}_{\alpha}\}$ as follows:

$$\mathbf{g}_{\alpha} = (\mathbf{g}_{\alpha} \cdot \mathbf{t}_{1})\, \mathbf{t}_{1} + (\mathbf{g}_{\alpha} \cdot \mathbf{t}_{2})\, \mathbf{t}_{2}, \tag{10.43}$$

in which $(\mathbf{g}_{\alpha} \cdot \mathbf{t}_{\beta})$ are components of $\mathbf{J}$ of eq. (10.41). Hence, we can rewrite

$$\begin{bmatrix} \mathbf{g}_1 \\ \mathbf{g}_2 \end{bmatrix} = \mathbf{J}^{T} \begin{bmatrix} \mathbf{t}_1 \\ \mathbf{t}_2 \end{bmatrix} \quad \text{and} \quad \begin{bmatrix} \mathbf{t}_1 \\ \mathbf{t}_2 \end{bmatrix} = \mathbf{J}^{-T} \begin{bmatrix} \mathbf{g}_1 \\ \mathbf{g}_2 \end{bmatrix}. \tag{10.44}$$

Inverse Jacobian. An inverse of a 2×2 matrix A is given by a simple formula,

$$A = \begin{bmatrix} a & b \\ c & d \end{bmatrix}, \qquad A^{-1} = \frac{1}{\det A} \begin{bmatrix} d & -b \\ -c & a \end{bmatrix}, \tag{10.45}$$

providing $\det A = ad - bc \neq 0$. Applying this formula to the Jacobian matrix $\mathbf{J}$ of eq. (10.41), we obtain

$$\mathbf{J}^{-1} = \frac{1}{\det \mathbf{J}} \begin{bmatrix} \mathbf{g}_2 \cdot \mathbf{t}_2 & -\mathbf{g}_2 \cdot \mathbf{t}_1 \\ -\mathbf{g}_1 \cdot \mathbf{t}_2 & \mathbf{g}_1 \cdot \mathbf{t}_1 \end{bmatrix}, \tag{10.46}$$

where $\det \mathbf{J} \doteq (\mathbf{g}_1 \cdot \mathbf{t}_1)(\mathbf{g}_2 \cdot \mathbf{t}_2) - (\mathbf{g}_1 \cdot \mathbf{t}_2)(\mathbf{g}_1 \cdot \mathbf{t}_1)$.

Another form of the inverse of Jacobian can be obtained with the help of the co-basis $\{\mathbf{g}^{\alpha}, \mathbf{t}_3\}$. The co-basis vectors $\mathbf{g}^{\alpha}$ are defined as follows:

$$\mathbf{g}^{\alpha}: \qquad \mathbf{g}^{\alpha} \cdot \mathbf{g}_{\beta} = \delta^{\alpha}_{\beta} \quad \text{and} \quad \mathbf{g}^{\alpha} \cdot \mathbf{t}_3 = 0 \tag{10.47}$$

and by analogy with eq. (5.8), they can be calculated as

$$\mathbf{g}^{1} = \frac{(\mathbf{g}_2 \times \mathbf{t}_3)}{(\mathbf{g}_2 \times \mathbf{t}_3) \cdot \mathbf{g}_1}, \qquad \mathbf{g}^{2} = \frac{(\mathbf{t}_3 \times \mathbf{g}_1)}{(\mathbf{t}_3 \times \mathbf{g}_1) \cdot \mathbf{g}_2}. \tag{10.48}$$

In terms of the co-basis vectors $\mathbf{g}^{\alpha}$, the inverse of Jacobian is

$$\mathbf{J}^{-1} = \begin{bmatrix} \mathbf{t}_1 \cdot \mathbf{g}^1 & \mathbf{t}_2 \cdot \mathbf{g}^1 \\ \mathbf{t}_1 \cdot \mathbf{g}^2 & \mathbf{t}_2 \cdot \mathbf{g}^2 \end{bmatrix}, \tag{10.49}$$

where columns of $\mathbf{J}^{-1}$ contain components of $\mathbf{t}_\beta$ in $\{\mathbf{g}^\alpha\}$.

Check. We can check that the inverse matrices of eqs. (10.46) and (10.49) are identical. Let us transform the 12-component of the matrix of eq. (10.49) as follows:

$$\mathbf{t}_2 \cdot \mathbf{g}^1 = \frac{\mathbf{t}_2 \cdot (\mathbf{g}_2 \times \mathbf{t}_3)}{(\mathbf{g}_2 \times \mathbf{t}_3) \cdot \mathbf{g}_1} = \frac{\mathbf{g}_2 \cdot (\mathbf{t}_3 \times \mathbf{t}_2)}{(\mathbf{g}_1 \times \mathbf{g}_2) \cdot \mathbf{t}_3} = \frac{-\mathbf{g}_2 \cdot \mathbf{t}_1}{\det \mathbf{J}}, \tag{10.50}$$

where we used $\mathbf{g}_1 \times \mathbf{g}_2 = \mathbf{t}_3 \det \mathbf{J}$, see eq. (10.106) for details. Hence, the obtained expression is identical to the 12-component of eq. (10.46). For the other components of $\mathbf{J}^{-1}$, we can proceed similarly.

Relation between $\mathbf{g}^\alpha$ and $\mathbf{t}_\alpha$. As previously in the derivation of eq. (10.44), we can use the fact that the coordinate gradients imply relations linking the bases with which these coordinates are associated. The co-basis vectors $\mathbf{g}^\alpha$ can be decomposed in the ortho-normal $\{\mathbf{t}_\alpha\}$ as follows:

$$\mathbf{g}^\alpha = (\mathbf{g}^\alpha \cdot \mathbf{t}_1)\, \mathbf{t}_1 + (\mathbf{g}^\alpha \cdot \mathbf{t}_2)\, \mathbf{t}_2, \tag{10.51}$$

in which $(\mathbf{g}^\alpha \cdot \mathbf{t}_\beta)$ are components of the inverse Jacobian matrix $\mathbf{J}^{-1}$ of eq. (10.49). Hence, we can rewrite

$$\begin{bmatrix} \mathbf{g}^1 \\ \mathbf{g}^2 \end{bmatrix} = \mathbf{J}^{-1} \begin{bmatrix} \mathbf{t}_1 \\ \mathbf{t}_2 \end{bmatrix} \quad \text{and} \quad \begin{bmatrix} \mathbf{t}_1 \\ \mathbf{t}_2 \end{bmatrix} = \mathbf{J} \begin{bmatrix} \mathbf{g}^1 \\ \mathbf{g}^2 \end{bmatrix}. \tag{10.52}$$

Co-basis definition expressed by Jacobian matrices. The condition defining the co-basis $\mathbf{g}^\alpha \cdot \mathbf{g}_\beta = \delta^\alpha_\beta$ can be rewritten as

$$(\mathbf{t}_i \cdot \mathbf{g}^\alpha)(\mathbf{g}_\beta \cdot \mathbf{t}_i) = \delta^\alpha_\beta, \qquad i = 1, 2, \tag{10.53}$$

where $\mathbf{g}^\alpha = (\mathbf{g}^\alpha \cdot \mathbf{t}_i)\, \mathbf{t}_i$ and $\mathbf{g}_\beta = (\mathbf{g}_\beta \cdot \mathbf{t}_k)\, \mathbf{t}_k$ $(i, k = 1, 2)$, and we can calculate

$$\mathbf{g}^\alpha \cdot \mathbf{g}_\beta = (\mathbf{g}^\alpha \cdot \mathbf{t}_i)(\mathbf{g}_\beta \cdot \mathbf{t}_k)\, \mathbf{t}_i \cdot \mathbf{t}_k = (\mathbf{t}_i \cdot \mathbf{g}^\alpha)(\mathbf{g}_\beta \cdot \mathbf{t}_i). \tag{10.54}$$

We note that $\mathbf{g}_\beta \cdot \mathbf{t}_i = \partial S^i / \partial \xi^\beta$ by eq. (10.42) and, hence, on the basis of eq. (10.53), we can define the gradient

$$\frac{\partial \xi^\alpha}{\partial S^i} \doteq \mathbf{t}_i \cdot \mathbf{g}^\alpha. \tag{10.55}$$

On the other hand, by eq. (10.49), $\mathbf{g}^\alpha \cdot \mathbf{t}_i = \left[\mathbf{J}^{-1}\right]_{\alpha i}$ and, hence,

$$\mathbf{J}^{-1} = \begin{bmatrix} \mathbf{t}_1 \cdot \mathbf{g}^1 & \mathbf{t}_2 \cdot \mathbf{g}^1 \\ \mathbf{t}_1 \cdot \mathbf{g}^2 & \mathbf{t}_2 \cdot \mathbf{g}^2 \end{bmatrix} = \begin{bmatrix} \frac{\partial \xi^1}{\partial S^1} & \frac{\partial \xi^1}{\partial S^2} \\ \frac{\partial \xi^2}{\partial S^1} & \frac{\partial \xi^2}{\partial S^2} \end{bmatrix} \doteq \begin{bmatrix} \frac{\partial \xi^\alpha}{\partial S^i} \end{bmatrix}. \tag{10.56}$$

Therefore, eq. (10.53) can be rewritten simply as $\mathbf{J}^{-1}\mathbf{J} = \mathbf{I}$, where $\mathbf{I}$ is the identity matrix.

Example. Note that the procedure of calculation of $\mathbf{J}^{-1}$ allows us to avoid expressing explicitly the natural coordinates ξ^α in terms of the ortho-normal coordinates S^α. This is an advantage because such relations can be quite complicated. For instance, for **Basis 1** attached at the element center, these relations are as follows:

$$\mathbf{g}_1 \xi^1 = \mathbf{t}_1 (S^1 - S^2 \tan\beta), \qquad \mathbf{g}_2 \xi^2 = \mathbf{t}_1 S^2 \tan\beta + \mathbf{t}_2 S^2, \tag{10.57}$$

where $\beta < \pi/2$ is the angle between $\mathbf{t}_2$ and $\mathbf{g}_2$ and $\tan\beta = (\mathbf{t}_1 \cdot \mathbf{g}_2)/(\mathbf{t}_2 \cdot \mathbf{g}_2)$. Then

$$\mathbf{J}^{-1} = \begin{bmatrix} \frac{\partial \xi^1}{\partial S^1} & \frac{\partial \xi^2}{\partial S^1} \\ \frac{\partial \xi^1}{\partial S^2} & \frac{\partial \xi^2}{\partial S^2} \end{bmatrix} = \begin{bmatrix} 1/\sqrt{g_{11}} & 0 \\ -\tan\beta/\sqrt{g_{11}} & 1/(\cos\beta\sqrt{g_{22}}) \end{bmatrix}, \tag{10.58}$$

where $g_{\alpha\alpha} \doteq \mathbf{g}_\alpha \cdot \mathbf{g}_\alpha$ and $\cos\beta = (\mathbf{t}_2 \cdot \mathbf{g}_2)/\sqrt{g_{22}}$.

Local Jacobian and its inverse for the vector of shape functions. We can approximate the relative vector $(\mathbf{y} - \mathbf{y}_g)$, see eq. (10.32), using a vector of shape functions $\mathbf{N}(\xi^1, \xi^2)$ of eq. (10.5), and write

$$S^\alpha = \frac{1}{4}\left[(\mathbf{s}\mathbf{S}^\alpha) + (\boldsymbol{\xi}\mathbf{S}^\alpha)\,\xi + (\boldsymbol{\eta}\mathbf{S}^\alpha)\,\eta + (\mathbf{h}\mathbf{S}^\alpha)\,\xi\eta\right], \tag{10.59}$$

where $\mathbf{S}^\alpha \doteq (\mathbf{y}_I - \mathbf{y}_g) \cdot \mathbf{t}_\alpha = [S_1^\alpha, S_2^\alpha, S_3^\alpha, S_4^\alpha]^T$ is the vector of projections of nodal relative position vectors on $\mathbf{t}_\alpha$. Differentiating eq. (10.59) w.r.t. the natural coordinates, we obtain

$$\frac{\partial S^\alpha}{\partial \xi} = \frac{1}{4}\left[(\boldsymbol{\xi}\mathbf{S}^\alpha) + (\mathbf{h}\mathbf{S}^\alpha)\,\eta\right], \qquad \frac{\partial S^\alpha}{\partial \eta} = \frac{1}{4}\left[(\boldsymbol{\eta}\mathbf{S}^\alpha) + (\mathbf{h}\mathbf{S}^\alpha)\,\xi\right] \tag{10.60}$$

and, hence, the Jacobian matrix of eq. (10.41) is

$$\mathbf{J} = \begin{bmatrix} \frac{1}{4}\left[(\boldsymbol{\xi}\mathbf{S}^1) + (\mathbf{h}\mathbf{S}^1)\,\eta\right] & \frac{1}{4}\left[(\boldsymbol{\eta}\mathbf{S}^1) + (\mathbf{h}\mathbf{S}^1)\,\xi\right] \\ \frac{1}{4}\left[(\boldsymbol{\xi}\mathbf{S}^2) + (\mathbf{h}\mathbf{S}^2)\,\eta\right] & \frac{1}{4}\left[(\boldsymbol{\eta}\mathbf{S}^2) + (\mathbf{h}\mathbf{S}^2)\,\xi\right] \end{bmatrix}. \tag{10.61}$$

Note that column 1 varies linearly with η, while column 2 varies linearly with ξ.

An inverse of the Jacobian can be obtained by eq. (10.45) and it is as follows:

$$\mathbf{J}^{-1} = \frac{1}{\det \mathbf{J}} \begin{bmatrix} \frac{1}{4}\left[(\boldsymbol{\eta}\mathbf{S}^2) + (\mathbf{h}\mathbf{S}^2)\,\xi\right] & -\frac{1}{4}\left[(\boldsymbol{\eta}\mathbf{S}^1) + (\mathbf{h}\mathbf{S}^1)\,\xi\right] \\ -\frac{1}{4}\left[(\boldsymbol{\xi}\mathbf{S}^2) + (\mathbf{h}\mathbf{S}^2)\,\eta\right] & \frac{1}{4}\left[(\boldsymbol{\xi}\mathbf{S}^1) + (\mathbf{h}\mathbf{S}^1)\,\eta\right] \end{bmatrix}, \tag{10.62}$$

where

$$\det \mathbf{J} = J(\xi, \eta) = J_0 + J_1 \xi + J_2 \eta, \tag{10.63}$$

and its components are

$$J_0 \doteq \frac{1}{16}\left[(\boldsymbol{\xi}\mathbf{S}^1)\,(\boldsymbol{\eta}\mathbf{S}^2) - (\boldsymbol{\eta}\mathbf{S}^1)\,(\boldsymbol{\xi}\mathbf{S}^2)\right],$$

$$J_1 \doteq \frac{1}{16}\left[(\boldsymbol{\xi}\mathbf{S}^1)\,(\mathbf{h}\mathbf{S}^2) - (\mathbf{h}\mathbf{S}^1)\,(\boldsymbol{\xi}\mathbf{S}^2)\right],$$

$$J_2 \doteq \frac{1}{16}\left[(\mathbf{h}\mathbf{S}^1)\,(\boldsymbol{\eta}\mathbf{S}^2) - (\boldsymbol{\eta}\mathbf{S}^1)\,(\mathbf{h}\mathbf{S}^2)\right].$$

The bilinear term $J_{12}\,\xi\eta$ is not present in the expansion eq. (10.63) because

$$J_{12} \doteq \frac{1}{16}\left[(\mathbf{h}\mathbf{S}^1)\,(\mathbf{h}\mathbf{S}^2) - (\mathbf{h}\mathbf{S}^1)\,(\mathbf{h}\mathbf{S}^2)\right] = 0.$$

It can be checked for parallelograms that only $J_0 \neq 0$, while $J_1 = J_2 = 0$.

Example. Jacobian matrices for basic shapes of element. The Jacobian matrix contains information about the initial shape of the element. In Fig. 10.9, we show several basic shapes of a planar four-node element. The Jacobian matrix and its determinant for these shapes are as follows:

a) square 2×2: $\mathbf{J} = \begin{bmatrix} 1 & 0 \\ 0 & 1 \end{bmatrix}$, $\quad \det \mathbf{J} = 1$,

b) rectangle: $\mathbf{J} = \begin{bmatrix} 5/2 & 0 \\ 0 & 1 \end{bmatrix}$, $\quad \det \mathbf{J} = 5/2$,

c) parallelogram: $\mathbf{J} = \begin{bmatrix} 5/2 & 0 \\ 1 & 1 \end{bmatrix}$, $\quad \det \mathbf{J} = 5/2$,

$$\text{d) trapezoid: } \mathbf{J} = \begin{bmatrix} 5/2+\eta & 0 \\ \xi & 1 \end{bmatrix}, \qquad \det \mathbf{J} = 5/2+\eta,$$

$$\text{e) trapezoid: } \mathbf{J} = \begin{bmatrix} (5+\eta)/2 & 0 \\ (1+\xi)/2 & 1 \end{bmatrix}, \qquad \det \mathbf{J} = (5+\eta)/2,$$

$$\text{f) irregular: } \mathbf{J} = \begin{bmatrix} (5+\eta)/2 & (1+\eta)/4 \\ (1+\xi)/2 & (5+\xi)/4 \end{bmatrix}, \det \mathbf{J} = (6+\xi+\eta)/2.$$

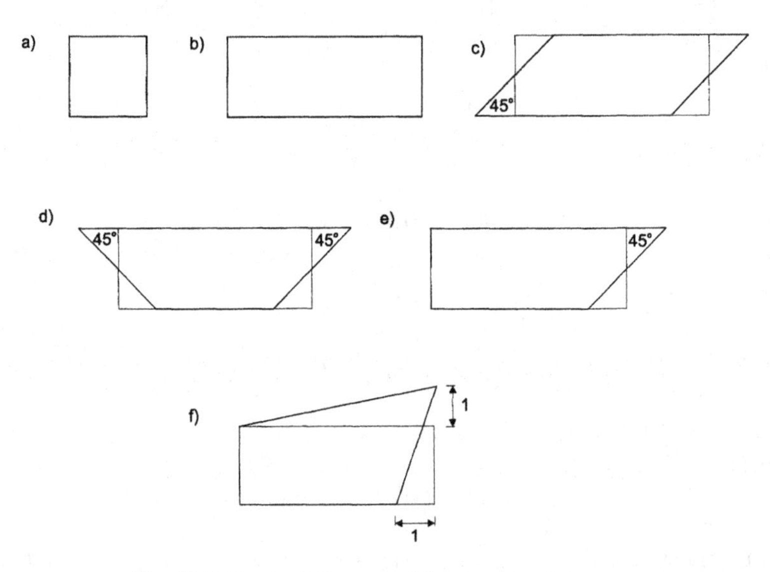

Fig. 10.9 Typical shapes of four-node elements.
Elements c), d), e), f) are obtained from the rectangle of size 5×2.

We see that $\mathbf{J}$ is a diagonal matrix only for the square and the rectangle. For the square, the rectangle, and the parallelogram, $\mathbf{J}$ is constant over the element.

For trapezoids and irregular elements, $\mathbf{J}$ is non-diagonal and non-constant and linearly depends on ξ, η. Note that for non-rectangular shapes, especially when the element aspect ratio is far from 1, the accuracy of four-node elements substantially decreases, see tests of Sects. 15.2.4 and 15.2.6.

10.4 Deformation gradient, $\mathbf{F}^T\mathbf{F}$ and $\mathbf{Q}^T\mathbf{F}$ products

Using various bases in FE computations. To enable linking of finite elements of various spatial orientation, the displacement vector $\mathbf{u}$ and the rotation vector $\boldsymbol{\psi}$ are represented in the reference Cartesian basis $\{\mathbf{i}_k\}$. However, in computations on the level of an element, we have a choice and one of the following two bases can be used:

1. the reference Cartesian basis $\{\mathbf{i}_k\}$. In order to perform the local operations, we have to transform strain components and components of the $\mathbf{Q}^T\mathbf{F}$ product to the local basis $\{\mathbf{t}_k\}$ at a Gauss point.
2. the elemental Cartesian basis $\{\mathbf{t}_k^c\}$ at the element center. Then, first, the displacement and rotation components must be transformed from the reference basis $\{\mathbf{i}_k\}$ to this basis. In order to perform the local operations, we transform the strain components and components of the $\mathbf{Q}^T\mathbf{F}$ product to the local basis $\{\mathbf{t}_k\}$ at a Gauss point. Afterwards, the tangent stiffness matrix at the residual vector must be transformed to the reference basis $\{\mathbf{i}_k\}$.
   The use of this basis requires additional transformations but, contrary to expectations, can significantly improve the efficiency of a four-node element if zero values are accounted for in the implementation.
   Note that the use of the elemental basis $\{\mathbf{t}_k^c\}$ is indispensable in the case of warped four-node elements if the substitute flat element and the warpage correction are used, see Sect. 14.3 for details.

Local operations for shell element. The shell assumptions and several techniques related to the formulation of a finite element require a local definition of directions, e.g.

- the Reissner hypothesis,
- imposition of the zero normal stress (ZNS) condition,
- modifications of transverse shear strains (using the ANS method),
- formulation of the drilling RC equation,
- integration of the strain energy, when it is separated into the integration in the normal (fiber) direction and the integration in the tangent (lamina) direction.

Hence, no matter whether the basis $\{\mathbf{i}_k\}$ or $\{\mathbf{t}_k^c\}$ is used on the element's level, we have to transform the strain components and components of the $\mathbf{Q}^T\mathbf{F}$ product to the local basis $\{\mathbf{t}_k\}$ at a Gauss point. Note that instead of transforming components from one basis to another, we can operate on the backward-rotated objects, as described in Sect. 2.

Remarks on use of a skew basis at element's center. The skew basis at the element center, $\{\mathbf{g}^c_\alpha, \mathbf{t}^c_3\}$, is used in mixed elements based on the Hellinger–Reissner functional and the Hu–Washizu functional, see Sect. 11.5. Representations of stress are assumed in this basis, while representations of strain for the Hu–Washizu functional are assumed either in this basis or in its co-basis. These representations are next transformed to the local orthonormal basis at the element center $\{\mathbf{t}^c_k\}$.

The formulas for a transformation between a non-orthogonal basis and a Cartesian basis are derived in Sect. 2. In the case of the in-plane $(\alpha\beta)$ components, for non-symmetric tensors we use eqs. (2.21) and (2.25), while for symmetric tensors we use eq. (2.29) with $\mathbf{T}^*$ replaced by $\mathbf{T}^{**}$ of eq. (2.30) or eq. (2.31). For the transverse $(\alpha 3)$ components, we use eqs. (2.26) and (2.27).

Deformation gradient, $\mathbf{F}^T\mathbf{F}$ and $\mathbf{Q}^T\mathbf{F}$ products. Below, we derive matrices of components for the deformation gradient $\mathbf{F}$, the Cauchy–Green deformation tensor $\mathbf{C} \doteq \mathbf{F}^T\mathbf{F}$, and the $\mathbf{Q}^T\mathbf{F}$ product. They are derived for the reference basis $\{\mathbf{i}_k\}$ and, subsequently, the latter two matrices are transformed to the local bases $\{\mathbf{t}_k\}$. As mentioned earlier, instead of the reference basis $\{\mathbf{i}_k\}$, the elemental basis $\{\mathbf{t}^c_k\}$ can be used as well. Two ways of derivation are presented below in which the position vectors are treated as functions of either (A) the natural coordinates, or (B) the local Cartesian coordinates.

(A) Natural coordinates. For the coordinates $\{\xi^\alpha, \zeta\}$, we take $\zeta = \frac{h}{2}\xi^3$, express $\zeta \in [-h/2, +h/2]$ in terms of $\xi^3 \in [-1, +1]$, and use the natural coordinates $\{\xi^k\}$ $(k = 1, 2, 3)$.

Then the position vector in the initial configuration of eq. (5.1) is as follows:

$$\mathbf{y}(\xi^k) = \mathbf{y}_0(\xi^\alpha) + \frac{h}{2}\xi^3\, \mathbf{t}_3(\xi^\alpha), \qquad \alpha = 1, 2, \tag{10.64}$$

and the current position vector is $\mathbf{x} = \mathbf{x}(\xi^k(\mathbf{y}))$. The deformation gradient of eq. (5.15) can be written simply as

$$\mathbf{F} \doteq \frac{\partial \mathbf{x}}{\partial \mathbf{y}} = \frac{\partial \mathbf{x}}{\partial \xi^k} \otimes \frac{\partial \xi^k}{\partial \mathbf{y}}, \tag{10.65}$$

with ξ^k serving as intermediate variables. Let us use the components of $\mathbf{y}$ and $\mathbf{x}$ in the reference basis $\{\mathbf{i}_k\}$. Then $\mathbf{y} = y^m\, \mathbf{i}_m$ and $\mathbf{x} = x^l\, \mathbf{i}_l$, $(m, l = 1, 2, 3)$ and we differentiate only the components,

$$\frac{\partial \mathbf{x}}{\partial \xi^k} = \frac{\partial x^l}{\partial \xi^k}\,\mathbf{i}_l, \qquad \frac{\partial \mathbf{y}}{\partial \xi^k} = \frac{\partial y^m}{\partial \xi^k}\,\mathbf{i}_m. \tag{10.66}$$

Hence, the inverse derivative is

$$\frac{\partial \xi^k}{\partial \mathbf{y}} = \frac{\partial \xi^k}{\partial y^m}\,\mathbf{i}_m \tag{10.67}$$

and the deformation gradient becomes

$$\mathbf{F} = \frac{\partial \mathbf{x}}{\partial \xi^k} \otimes \frac{\partial \xi^k}{\partial \mathbf{y}} = \frac{\partial x^l}{\partial \xi^k}\frac{\partial \xi^k}{\partial y^m}\,\mathbf{i}_l \otimes \mathbf{i}_m = F_{lm}\,\mathbf{i}_l \otimes \mathbf{i}_m, \tag{10.68}$$

where

$$F_{lm} \doteq \frac{\partial x^l}{\partial \xi^k}\frac{\partial \xi^k}{\partial y^m}. \tag{10.69}$$

Let us introduce the matrices of components

$$\mathbf{F} \doteq [F_{lm}], \qquad \mathbf{J}_G^{\mathrm{curr}} \doteq \left[\frac{\partial x^l}{\partial \xi^k}\right], \qquad \mathbf{J}_G \doteq \left[\frac{\partial y^m}{\partial \xi^k}\right], \tag{10.70}$$

where $\mathbf{J}_G$ is the global Jacobian of eq. (10.33). Then the deformation gradient matrix can be computed as

$$\mathbf{F} = \mathbf{J}_G^{\mathrm{curr}}\,\mathbf{J}_G^{-1}. \tag{10.71}$$

Now we can compute the components of the Cauchy–Green tensor $\mathbf{C} \doteq \mathbf{F}^T\mathbf{F}$ and of the $\mathbf{Q}_0^T\mathbf{F}$ product, and transform them to the local orthonormal basis $\{\mathbf{t}_k\}$. This can be done, as derived in Sect. 2, by the transformation of components $(\cdot)_{\mathrm{local}} = \mathbf{R}^T\,(\cdot)_{\mathrm{global}}\,\mathbf{R}$, see eq. (2.13). Note that $\mathbf{R}$ is the rotation matrix of eq. (10.33).

1. Components of the Cauchy–Green deformation tensor $\mathbf{C} \doteq \mathbf{F}^T\,\mathbf{F}$,

$$\mathbf{C}_* = \mathbf{R}^T\,\mathbf{C}\,\mathbf{R} = \bar{\mathbf{F}}^T\bar{\mathbf{F}}, \tag{10.72}$$

2. Components of the $\mathbf{Q}^T\mathbf{F}$ product, i.e. $\mathbf{Q}^T\mathbf{F}$,

$$(\mathbf{Q}^T\mathbf{F})_* = \mathbf{R}^T\,(\mathbf{Q}^T\mathbf{F})\,\mathbf{R} = (\mathbf{Q}\,\mathbf{R})^T\,\bar{\mathbf{F}}, \tag{10.73}$$

where

$$\bar{\mathbf{F}} \doteq \mathbf{F}\mathbf{R} = \mathbf{J}_G^{\mathrm{curr}}\mathbf{J}_G^{-1}\,\mathbf{R} = \mathbf{J}_G^{\mathrm{curr}}\mathbf{J}_L^{-1}, \tag{10.74}$$

with the last form obtained on use of eq. (10.34).

(B) Orthonormal local coordinates. We can express $\zeta \in [-h/2, +h/2]$ in terms of $\xi^3 \in [-1, +1]$ as $\zeta = (h/2)\,\xi^3$, and define $S^3 \doteq \xi^3$. Then we can use the Cartesian coordinates $\{S^k\}$ $(k = 1, 2, 3)$ instead of $\{S^\alpha, \zeta\}$.

The position vector in the initial configuration of eq. (5.1) is now as follows:

$$\mathbf{y}(S^k) = \mathbf{y}_0(S^\alpha) + \frac{h}{2} S^3\, \mathbf{t}_3(S^\alpha), \qquad \alpha = 1, 2, \tag{10.75}$$

and the current position vector is $\mathbf{x} = \mathbf{x}(S^k(\mathbf{y}))$. The deformation gradient of eq. (5.15) can be written simply as

$$\mathbf{F} \doteq \frac{\partial \mathbf{x}}{\partial \mathbf{y}} = \frac{\partial \mathbf{x}}{\partial S^k} \otimes \frac{\partial S^k}{\partial \mathbf{y}}, \tag{10.76}$$

with S^k being intermediate variables. Let us use the components of $\mathbf{y}$ and $\mathbf{x}$ in the reference basis $\{\mathbf{i}_k\}$. Then $\mathbf{y} = y^m\, \mathbf{i}_m$ and $\mathbf{x} = x^l\, \mathbf{i}_l$ and we can differentiate the components

$$\frac{\partial \mathbf{x}}{\partial S^k} = \frac{\partial x^l}{\partial S^k}\, \mathbf{i}_l, \qquad \frac{\partial \mathbf{y}}{\partial S^k} = \frac{\partial y^m}{\partial S^k}\, \mathbf{i}_m. \tag{10.77}$$

Hence, the inverse derivative is

$$\frac{\partial S^k}{\partial \mathbf{y}} = \frac{\partial S^k}{\partial y^m}\, \mathbf{i}_m \tag{10.78}$$

and the deformation gradient becomes

$$\mathbf{F} = \frac{\partial \mathbf{x}}{\partial S^k} \otimes \frac{\partial S^k}{\partial \mathbf{y}} = \frac{\partial x^l}{\partial S^k} \frac{\partial S^k}{\partial y^m}\, \mathbf{i}_l \otimes \mathbf{i}_m = F_{lm}\, \mathbf{i}_l \otimes \mathbf{i}_m, \tag{10.79}$$

where

$$F_{lm} \doteq \frac{\partial x^l}{\partial S^k} \frac{\partial S^k}{\partial y^m}. \tag{10.80}$$

Let us define the following matrices of components

$$\mathbf{F} \doteq [F_{lm}], \qquad \nabla \mathbf{x} \doteq \left[\frac{\partial x^l}{\partial S^k}\right], \qquad \mathbf{R} \doteq \left[\frac{\partial y^m}{\partial S^k}\right], \tag{10.81}$$

where $\mathbf{R}$ is the rotation matrix of eq. (10.33). Then the deformation gradient matrix can be computed as

$$\mathbf{F} = \nabla \mathbf{x}\, \mathbf{R}^T. \tag{10.82}$$

Now we can compute the components of the Cauchy–Green tensor $\mathbf{C} \doteq \mathbf{F}^T\mathbf{F}$ and of the $\mathbf{Q}_0^T\mathbf{F}$ product and transform them to the local orthonormal basis $\{\mathbf{t}_k\}$. This can be done, as derived in Sect. 2, by the transformation of components $(\cdot)_\mathrm{L} = \mathbf{R}^T\,(\cdot)_\mathrm{G}\,\mathbf{R}$, see eq. (2.13).

1. Components of the Cauchy–Green deformation tensor $\mathbf{C} \doteq \mathbf{F}^T\,\mathbf{F}$,

$$\mathbf{C}_* = \mathbf{R}^T\,\mathbf{C}\,\mathbf{R} = (\nabla\mathbf{x})^T\nabla\mathbf{x}, \tag{10.83}$$

2. Components of the $\mathbf{Q}^T\mathbf{F}$ product, i.e. $\mathbf{Q}^T\mathbf{F}$,

$$(\mathbf{Q}^T\mathbf{F})_* = \mathbf{R}^T\,(\mathbf{Q}^T\mathbf{F})\,\mathbf{R} = (\mathbf{Q}\,\mathbf{R})^T\,\nabla\mathbf{x}. \tag{10.84}$$

Note that, formally, $\nabla\mathbf{x}$ plays the same role as $\bar{\mathbf{F}}$ in case (A).

The formulation based on the coordinates S^k makes sense when the derivatives of shape functions are expressed in terms of ortho-normal S^α, as, e.g., for the one-integration point element. Besides, it is an analogue of the formulation used in the analytical studies in Sect. 6.

Increment of Green strain. Version 1. For $\mathbf{x} = \mathbf{x}_n + \Delta\mathbf{u}$, where n refers to the last known configuration and Δ to the increment from the last known configuration to the current one, the deformation gradient can be multiplicatively decomposed as follows:

$$\mathbf{F} \doteq \frac{\partial\mathbf{x}}{\partial\mathbf{y}} = \frac{\partial\mathbf{x}}{\partial\mathbf{x}_n}\frac{\partial\mathbf{x}_n}{\partial\mathbf{y}} = \Delta\mathbf{F}\,\mathbf{F}_n. \tag{10.85}$$

Then, the Green strain can be rewritten as

$$\mathbf{E} \doteq \tfrac{1}{2}(\mathbf{F}^T\mathbf{F} - \mathbf{I}) = \tfrac{1}{2}\left[\mathbf{F}_n^T\,(\Delta\mathbf{F}^T\Delta\mathbf{F})\,\mathbf{F}_n - \mathbf{I}\right], \tag{10.86}$$

where

$$\Delta\mathbf{F} \doteq \frac{\partial\mathbf{x}}{\partial\mathbf{x}_n} = \nabla_n\mathbf{x} = \mathbf{I} + \nabla_n(\Delta\mathbf{u}) \tag{10.87}$$

and $\nabla_n(\cdot) \doteq \partial(\cdot)/\partial\mathbf{x}_n$ denotes the gradient w.r.t. the known position vector. We can linearize the $\Delta\mathbf{F}^T\Delta\mathbf{F}$ product w.r.t. $\Delta\mathbf{u}$, which yields

$$\begin{aligned}(\Delta\mathbf{F}^T\Delta\mathbf{F}) &= \mathbf{I} + \nabla_n(\Delta\mathbf{u}) + \nabla_n^T(\Delta\mathbf{u}) + \underbrace{\nabla_n^T(\Delta\mathbf{u})\nabla_n(\Delta\mathbf{u})}_{\text{neglected}} \\ &\approx \mathbf{I} + 2\Delta\boldsymbol{\varepsilon},\end{aligned} \tag{10.88}$$

where $\Delta\boldsymbol{\varepsilon} \doteq \mathrm{sym}\nabla_n(\Delta\mathbf{u})$ is the infinitesimal strain increment. Hence, the Green strain can be expressed as

$$\mathbf{E} = \tfrac{1}{2}\left[\mathbf{F}_n^T (\mathbf{I} + 2\Delta\varepsilon)\,\mathbf{F}_n - \mathbf{I}\right] = \mathbf{E}_n + \mathbf{F}_n^T\,\Delta\varepsilon\,\mathbf{F}_n, \tag{10.89}$$

where $\mathbf{E}_n \doteq \frac{1}{2}(\mathbf{F}_n^T\mathbf{F}_n - \mathbf{I})$, and the increment of the Green strain can be obtained as the pull-back of the infinitesimal strain increment, $\Delta\mathbf{E} \doteq \mathbf{E} - \mathbf{E}_n = \mathbf{F}_n^T\,\Delta\varepsilon\,\mathbf{F}_n$.

Increment of Green strain. Version 2. The increment of the Green strain can be defined as $\Delta\mathbf{E} \doteq \mathbf{E}_{n+1} - \mathbf{E}_n$ and expressed as

$$\Delta\mathbf{E} = \tfrac{1}{2}(\mathbf{F}_{n+1}^T\mathbf{F}_{n+1} - \mathbf{F}_n^T\mathbf{F}_n). \tag{10.90}$$

The deformation gradients at t^n and t^{n+1} can be expressed by the mid-point deformation gradient $\mathbf{F}_{n+1/2} \doteq \partial\mathbf{x}_{n+1/2}/\partial\mathbf{y}$ as follows:

$$\mathbf{F}_{n+1} \doteq \frac{\partial\mathbf{x}_{n+1}}{\partial\mathbf{X}} = \frac{\partial\mathbf{x}_{n+1}}{\partial\mathbf{x}_{n+1/2}}\frac{\partial\mathbf{x}_{n+1/2}}{\partial\mathbf{y}} = \left[\frac{\partial\mathbf{x}_{n+1}}{\partial\mathbf{x}_{n+1/2}}\right]\mathbf{F}_{n+1/2}, \tag{10.91}$$

$$\mathbf{F}_{n} \doteq \frac{\partial\mathbf{x}_{n}}{\partial\mathbf{X}} = \frac{\partial\mathbf{x}_{n}}{\partial\mathbf{x}_{n+1/2}}\frac{\partial\mathbf{x}_{n+1/2}}{\partial\mathbf{y}} = \left[\frac{\partial\mathbf{x}_{n+1/2}}{\partial\mathbf{x}_{n}}\right]^{-1}\mathbf{F}_{n+1/2}. \tag{10.92}$$

Then, the increment of the Green strain can be written as

$$\Delta\mathbf{E} = \mathbf{F}_{n+1/2}^T\,\Delta\varepsilon\,\mathbf{F}_{n+1/2}, \tag{10.93}$$

where the part which is pushed-forward to the mid-point position is

$$2\Delta\varepsilon = \left[\frac{\partial\mathbf{x}_{n+1}}{\partial\mathbf{x}_{n+1/2}}\right]^T\left[\frac{\partial\mathbf{x}_{n+1}}{\partial\mathbf{x}_{n+1/2}}\right] - \left[\frac{\partial\mathbf{x}_{n}}{\partial\mathbf{x}_{n+1/2}}\right]^T\left[\frac{\partial\mathbf{x}_{n}}{\partial\mathbf{x}_{n+1/2}}\right]. \tag{10.94}$$

We note that

$$\frac{\partial\mathbf{x}_{n+1}}{\partial\mathbf{x}_{n+1/2}} = \frac{\partial(\mathbf{x}_{n+1/2} + \frac{1}{2}\Delta\mathbf{u})}{\partial\mathbf{x}_{n+1/2}} = \mathbf{I} + \frac{1}{2}\frac{\partial\Delta\mathbf{u}}{\partial\mathbf{x}_{n+1/2}} \tag{10.95}$$

and

$$\frac{\partial\mathbf{x}_{n}}{\partial\mathbf{x}_{n+1/2}} = \frac{\partial(\mathbf{x}_{n+1/2} - \frac{1}{2}\Delta\mathbf{u})}{\partial\mathbf{x}_{n+1/2}} = \mathbf{I} - \frac{1}{2}\frac{\partial\Delta\mathbf{u}}{\partial\mathbf{x}_{n+1/2}}. \tag{10.96}$$

Using the above relations in eq. (10.94), we obtain

$$\Delta\varepsilon = \frac{1}{2}\left[\frac{\partial\Delta\mathbf{u}}{\partial\mathbf{x}_{n+1/2}} + \left(\frac{\partial\Delta\mathbf{u}}{\partial\mathbf{x}_{n+1/2}}\right)^T\right]. \tag{10.97}$$

Formula (10.93) is used in finite strain plasticity, e.g., in [97, 250].

Rate of Green strain. Differentiation of the Green strain $\mathbf{E} \doteq \frac{1}{2}(\mathbf{F}^T\mathbf{F} - \mathbf{I})$ w.r.t. time t, yields

$$\begin{aligned} 2\dot{\mathbf{E}} &= \dot{\mathbf{F}}^T\mathbf{F} + \mathbf{F}^T\dot{\mathbf{F}} = \mathbf{F}^T(\mathbf{F}^{-T}\dot{\mathbf{F}}^T + \dot{\mathbf{F}}\mathbf{F}^{-1})\mathbf{F} \\ &= \mathbf{F}^T(\nabla\mathbf{v}^T + \nabla\mathbf{v})\mathbf{F} = 2\mathbf{F}^T\mathbf{d}\,\mathbf{F}, \end{aligned} \tag{10.98}$$

where the spatial velocity gradient $\nabla\mathbf{v} \doteq \dot{\mathbf{F}}\mathbf{F}^{-1}$, and the rate of deformation $\mathbf{d} \doteq \frac{1}{2}(\nabla\mathbf{v}^T + \nabla\mathbf{v})$. Using this formula, we can obtain an interpretation of the above two forms of increment of the Green strain.

Writing eq. (10.98) at time instant t^n and multiplying by Δt, we obtain

$$\Delta\mathbf{E} = \mathbf{F}_n^T\,\Delta\boldsymbol{\varepsilon}\,\mathbf{F}_n, \tag{10.99}$$

where $\Delta\mathbf{E} \doteq \dot{\mathbf{E}}_n\Delta t$ and $\Delta\boldsymbol{\varepsilon} \doteq \mathbf{d}_n\Delta t$ by the forward Euler finite-difference scheme, which is first-order accurate. This equation corresponds to eq. (10.89).

Writing eq. (10.98) at time instant $t^{n+1/2}$ and multiplying by Δt, we obtain

$$\Delta\mathbf{E} = \mathbf{F}_{n+1/2}^T\,\Delta\boldsymbol{\varepsilon}\,\mathbf{F}_{n+1/2}, \tag{10.100}$$

where $\Delta\mathbf{E} \doteq \dot{\mathbf{E}}_{n+1/2}\Delta t$ and $\Delta\boldsymbol{\varepsilon} \doteq \mathbf{d}_{n+1/2}\Delta t$ by the central finite-difference scheme, which is second-order accurate. This equation corresponds to eq. (10.93).

10.5 Numerical integration of shell elements

Infinitesimal volume and area of shell element. Below we consider the formulas suitable for (i) the ortho-normal coordinates S^k and (ii) the natural coordinates ξ^k. The latter are actually used in our computations.

(i) Ortho-normal coordinates S^k. The differential of the position vector $\mathbf{y}$ can be written as

$$\mathrm{d}\mathbf{y} \doteq \mathbf{t}_1\,\mathrm{d}S^1 + \mathbf{t}_2\,\mathrm{d}S^2 + \mathbf{t}_3\,\mathrm{d}S^3, \tag{10.101}$$

where $\mathrm{d}S^3 \doteq \zeta \in [-h/2, +h/2]$. An infinitesimal volume of the rectangular parallelepiped spanned by vectors $(\mathrm{d}\mathbf{y})_i = \mathbf{t}_i\,\mathrm{d}S^i$ (no summation) is as follows:

$$\mathrm{d}V \doteq (\mathbf{t}_1\,\mathrm{d}S^1 \times \mathbf{t}_2\,\mathrm{d}S^2)\cdot(\mathbf{t}_3\,\mathrm{d}S^3) = (\mathbf{t}_1 \times \mathbf{t}_2)\cdot\mathbf{t}_3\,\mathrm{d}S^1\mathrm{d}S^2\mathrm{d}S^3 = \mathrm{d}S^1\mathrm{d}S^2\mathrm{d}S^3, \tag{10.102}$$

as $\mathbf{t}_1 \times \mathbf{t}_2 = \mathbf{t}_3$. The infinitesimal area of a rectangle spanned by the tangent vectors $(\mathrm{d}\mathbf{y})_\alpha$ $(\alpha = 1,2)$ is

$$\mathrm{d}A \doteq (\mathbf{t}_1\,\mathrm{d}S^1 \times \mathbf{t}_2\,\mathrm{d}S^2)\cdot\mathbf{t}_3 = \mathrm{d}S^1\mathrm{d}S^2. \tag{10.103}$$

(ii) Natural coordinates ξ^k. The differential of the position vector $\mathbf{y}$ can be written as

$$\mathrm{d}\mathbf{y} \doteq \frac{\partial \mathbf{y}_0}{\partial \xi^1}\,\mathrm{d}\xi^1 + \frac{\partial \mathbf{y}_0}{\partial \xi^2}\,\mathrm{d}\xi^2 + \mathbf{t}_3\frac{h}{2}\,\mathrm{d}\xi^3 = \mathbf{g}_1\,\mathrm{d}\xi^1 + \mathbf{g}_2\,\mathrm{d}\xi^2 + \mathbf{t}_3\frac{h}{2}\,\mathrm{d}\xi^3, \tag{10.104}$$

where $\mathbf{g}_\alpha \doteq \partial\mathbf{y}_0/\partial\xi^\alpha$, $\xi^3 \doteq 2\zeta/h$, and $\xi^k \in [-1,+1]$. An infinitesimal volume of the parallelepiped spanned by the component vectors is

$$\mathrm{d}V \doteq (\mathbf{g}_1\,\mathrm{d}\xi^1 \times \mathbf{g}_2\,\mathrm{d}\xi^2)\cdot\left(\mathbf{t}_3\frac{h}{2}\,\mathrm{d}\xi^3\right) = \frac{h}{2}(\mathbf{g}_1 \times \mathbf{g}_2)\cdot\mathbf{t}_3\;\mathrm{d}\xi^1\mathrm{d}\xi^2\mathrm{d}\xi^3. \tag{10.105}$$

Note that

$$\mathbf{g}_1 \times \mathbf{g}_2 = J\,\mathbf{t}_3, \qquad J = (\mathbf{g}_1\cdot\mathbf{t}_1)(\mathbf{g}_2\cdot\mathbf{t}_2) - (\mathbf{g}_2\cdot\mathbf{t}_1)(\mathbf{g}_1\cdot\mathbf{t}_2), \tag{10.106}$$

for $\mathbf{g}_\alpha$ decomposed in $\{\mathbf{t}_i\}$ as $\mathbf{g}_\alpha = (\mathbf{g}_\alpha\cdot\mathbf{t}_1)\,\mathbf{t}_1 + (\mathbf{g}_\alpha\cdot\mathbf{t}_2)\,\mathbf{t}_2$ and $\mathbf{t}_1 \times \mathbf{t}_2 = \mathbf{t}_3$. Besides, $J \doteq \det\mathbf{J}$, where $\mathbf{J}$ is the Jacobian matrix of eq. (10.41). Using the above relation $(\mathbf{g}_1 \times \mathbf{g}_2)\cdot\mathbf{t}_3 = J\,\mathbf{t}_3\cdot\mathbf{t}_3 = J$ and

$$\mathrm{d}V = \frac{h}{2}\,J\;\mathrm{d}\xi^1\mathrm{d}\xi^2\mathrm{d}\xi^3. \tag{10.107}$$

We note that $(h/2)\,J = (h/2)\,\det\mathbf{J} = \det\mathbf{J}_L$, for $\mathbf{J}_L$ of eq. (10.39). Besides, $\mathbf{J}_L = \mathbf{R}^T\mathbf{J}_G$, by eq. (10.34), and hence $\det\mathbf{J}_L = \det\mathbf{J}_G$, as $\det\mathbf{R} = 1$. The infinitesimal area of the parallelogram spanned by vectors $(\mathrm{d}\mathbf{y})_\alpha$ $(\alpha = 1,2)$ is

$$\mathrm{d}A \doteq (\mathbf{g}_1\,\mathrm{d}\xi^1 \times \mathbf{g}_2\,\mathrm{d}\xi^2)\cdot\mathbf{t}_3 = (\mathbf{g}_1 \times \mathbf{g}_2)\cdot\mathbf{t}_3\,\mathrm{d}\xi^1\mathrm{d}\xi^2 = J\;\mathrm{d}\xi^1\mathrm{d}\xi^2. \tag{10.108}$$

Remark. In the above derivations, we assumed that the element's geometry is approximately flat, see the restriction of eq. (5.19). To account for curvature, the infinitesimal parallelepiped should be spanned by the vectors $\hat{\mathbf{g}}_\alpha(\zeta)$ of eq. (5.5) and integrated over the thickness.

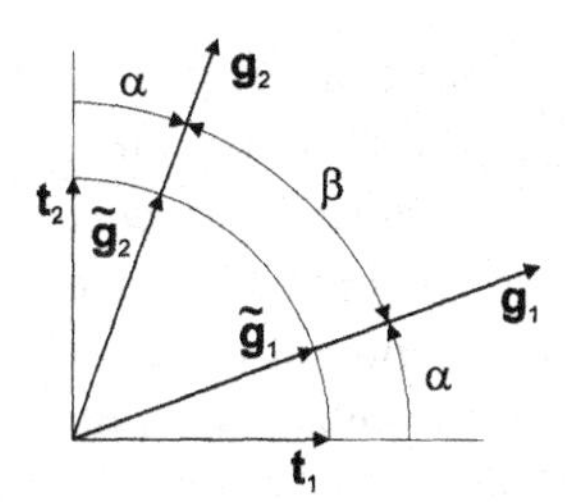

Fig. 10.10 Sign of the Jacobian determinant depends on α or β.

Example. We show that the sign of the Jacobian determinant J is a function of the angle between $\mathbf{t}_1$ and $\mathbf{g}_1$, denoted as α in Fig. 10.10. We assume that the basis $\{\mathbf{t}_i\}$ is constructed as **Basis 2**, see eq. (10.22).

The natural basis vectors $\mathbf{g}_\alpha$ can be expressed as $\mathbf{g}_\alpha = \tilde{\mathbf{g}}_\alpha \|\mathbf{g}_\alpha\|$, where $\tilde{\mathbf{g}}_\alpha$ are unit vectors given in $\{\mathbf{t}_\alpha\}$ as follows:

$$\tilde{\mathbf{g}}_1 = \cos\alpha\, \mathbf{t}_1 - \sin\alpha\, \mathbf{t}_2, \qquad \tilde{\mathbf{g}}_2 = -\sin\alpha\, \mathbf{t}_1 + \cos\alpha\, \mathbf{t}_2.$$

Then the Jacobian determinant becomes

$$\begin{aligned} J &= (\mathbf{g}_1 \cdot \mathbf{t}_1)(\mathbf{g}_2 \cdot \mathbf{t}_2) - (\mathbf{g}_2 \cdot \mathbf{t}_1)(\mathbf{g}_1 \cdot \mathbf{t}_2) \\ &= \|\mathbf{g}_1\|\|\mathbf{g}_2\|(\cos^2\alpha - \sin^2\alpha) = \|\mathbf{g}_1\|\|\mathbf{g}_2\|(1 - 2\sin^2\alpha). \end{aligned} \tag{10.109}$$

Because $\|\mathbf{g}_1\|\|\mathbf{g}_2\| > 0$, the sign of J depends on the angle α, i.e.

$$\begin{aligned} &J > 0, \text{ for } |\sin\alpha| < 1/\sqrt{2}, \quad \text{or } |\alpha| < 45^\circ, \\ &J = 0, \text{ for } |\sin\alpha| = \pm 1/\sqrt{2}, \text{ or } |\alpha| = 45^\circ, \\ &J < 0, \text{ for } |\sin\alpha| > 1/\sqrt{2}, \quad \text{or } |\alpha| > 45^\circ. \end{aligned} \tag{10.110}$$

We can rewrite these conditions in terms of the angle between $\mathbf{g}_1$ and $\mathbf{g}_2$, defined as $\beta \doteq 90^\circ - 2\alpha$, see Fig. 10.10, as follows:

$$\begin{aligned} &J > 0, \text{ for } 0^\circ < \beta < 180^\circ, \\ &J = 0, \text{ for } \beta = 0^\circ \text{ or } \beta = 180^\circ, \\ &J < 0, \text{ for } \beta > 180^\circ\,. \end{aligned} \tag{10.111}$$

We see that J is singular when $\mathbf{g}_1$ and $\mathbf{g}_2$ are co-linear and is negative if they are inclined at the angle greater than 180°.

Finally, we note that the Jacobian determinant J is computed at the Gauss points and it should be positive to avoid negative volumes which are non-physical.

Remark. Note that the condition requiring that the internal angles between adjacent edges of a four-node element be within the range $[45^\circ, 135^\circ]$ is motivated by accuracy concerns, as it is more restrictive than necessary to avoid $J < 0$. For instance, if node 3 is placed on the line linking node 2 and 4, then the angle at node 3 is 180°, far beyond the above range. Still, $J > 0$ everywhere except at node 3, where $J = 0$.

Volume and area of shell element. Below we consider the formulas for integration suitable for the natural coordinates ξ^k.

The volume of a shell element is defined as an integral,

$$V \doteq \int_V \mathrm{d}V = \int_{-1}^{+1}\int_{-1}^{+1}\int_{-1}^{+1} \frac{h}{2} J \, \mathrm{d}\xi^1 \mathrm{d}\xi^2 \mathrm{d}\xi^3, \tag{10.112}$$

where we used $\mathrm{d}V$ of eq. (10.107). If the thickness h is constant in the element, then we obtain $V = hA$, where the shell element area is defined as

$$A \doteq \int_A \mathrm{d}A = \int_{-1}^{+1}\int_{-1}^{+1} J \, \mathrm{d}\xi^1 \mathrm{d}\xi^2, \tag{10.113}$$

where $\mathrm{d}A$ is defined in eq. (10.108). For a four-node element and the bilinear approximation, $\det \mathbf{J} = J_0 + J_1 \xi^1 + J_2 \xi^2$, see eq. (10.63), and the area $A = 4J_0$.

Equation (10.112) is general, can be applied to elements of arbitrary shape, also to the warped ones, while a simpler expression can be found for a flat four-node element. We can divide the four-node elements into two triangles, e.g., by the diagonal 1-3, and calculate its area as follows:

$$A \doteq \tfrac{1}{2} \left(\mathbf{y}_{32} \times \mathbf{y}_{12} - \mathbf{y}_{34} \times \mathbf{y}_{14} \right) \cdot \mathbf{t}_3, \tag{10.114}$$

where $\mathbf{t}_3$ is a unit normal vector and $\mathbf{y}_{KL} \doteq \mathbf{y}_K - \mathbf{y}_L$, i.e. the vector connecting nodes K and L $(K, L \in \{1,2,3,4\})$. Then $\mathbf{y}_{KL} = S^\alpha_{KL} \mathbf{t}_\alpha$, where $S^\alpha_{KL} \doteq S^\alpha_K - S^\alpha_L$ and $S^\alpha_I \doteq (\mathbf{y}_I - \mathbf{y}_c) \cdot \mathbf{t}_\alpha$, where $\mathbf{y}_c$ is the position of the element center. Finally, the area can be expressed as

$$A = \tfrac{1}{2} \left(S^1_{31} S^2_{42} - S^2_{31} S^1_{24} \right) \tag{10.115}$$

or

$$A = \frac{1}{4} \left[(\boldsymbol{\xi} \mathbf{S}^1)(\boldsymbol{\eta} \mathbf{S}^2) - (\boldsymbol{\xi} \mathbf{S}^2)(\boldsymbol{\eta} \mathbf{S}^1) \right]. \tag{10.116}$$

To prove the correctness of the last form, we have to perform multiplications and introduce the differences of coordinates S^α_{KL}, which yields eq. (10.115). Comparing eq. (10.116) with the definition of J_0 of eq. (10.63), we obtain $A = 4J_0$.

Integration over the element volume. The volume of the shell element is mapped onto a unit cube and a numerical integration is performed as follows:

$$\int_V F\, \mathrm{d}V = \int_{-1}^{+1}\int_{-1}^{+1}\int_{-1}^{+1} \bar{F}(\xi^k)\, \frac{h}{2}\, J\, \mathrm{d}\xi^1 \mathrm{d}\xi^2 \mathrm{d}\xi^3 = \sum_{n=1}^{N_{IP}} w_n\, \bar{F}(\xi^k_n), \tag{10.117}$$

where $\mathrm{d}V$ is defined in eq. (10.107), $\bar{F}(\xi^k) \doteq \frac{1}{2} f(\xi^k)\, h(\xi^\alpha)\, J(\xi^\alpha)$, and w_n denotes the weighing factor for the integration point n.

We can separately specify the integration rule for the reference lamina (l) and for the fiber (f), see Fig. 5.2, as follows:

$$\int_V F\, \mathrm{d}V = \sum_{l=1}^{N^l_{IP}} w_l \sum_{f=1}^{N^f_{IP}} w_f\, \bar{F}(\xi^\alpha_l, \xi^3_f). \tag{10.118}$$

The order in which the above summations are performed can have a significant effect on the speed of the FE code and the effect can be contrary to our expectations; this issue is discussed in [89]. Note that, generally, it is better to write two integration loops instead of one, as usually it makes a difference to the compiler's optimizer.

Integration of strain energy over thickness. For the Reissner kinematics, the deformation gradient $\mathbf{F}$ is a linear function of the normal coordinate $\zeta \in [-h/2, +h/2]$. Hence, the strain $\mathbf{E}$ and the strain energy density function $\mathcal{W}$ are polynomials of ζ,

$$\mathbf{E}(\zeta) = \sum_{n=0}^{N} \frac{1}{n!} \mathbf{E}^{(n)}\, \zeta^n, \qquad \mathcal{W}(\zeta) = \sum_{n=0}^{N} \mathcal{W}_n\, \zeta^n, \tag{10.119}$$

where $\mathbf{E}^{(n)}$ denotes the nth derivative w.r.t ζ at the middle surface, $\zeta = 0$. The shell-type strain energy is defined as $\mathcal{W}_{\mathrm{sh}} \doteq \int_{-h/2}^{+h/2} \mathcal{W}(\zeta)\, \mathrm{d}\zeta$, and involves the integration through-the-thickness.

In Table 10.1 are (1) the analytically integrated $\mathcal{W}_{\mathrm{sh}}$, and (2) the minimum number of integration points to obtain exact $\mathcal{W}_{\mathrm{sh}}$ for two types of numerical quadratures, Gauss and Newton–Cotes (NC). Various forms of strain are assumed. Note that $\mathcal{W}_2^*$ depends on all derivatives of the strain, while $\mathcal{W}_2$ only on $\mathbf{E}^{(1)}$.

Table 10.1 Integration of SVK strain energy over ζ for various forms of strain. MNIP=minimum no. of integration points

Form of strain $\mathbf{E}(\zeta)$	Coefficients of strain energy $\mathcal{W}(\zeta)$	analytically integr. $\mathcal{W}_{sh}$	MNIP Gauss	NC
$\mathbf{E}^{(0)}$	$\mathcal{W}_0(\mathbf{E}^{(0)})$	$h\mathcal{W}_0$	1	1
$\mathbf{E}^{(0)} + \zeta\mathbf{E}^{(1)}$	$\mathcal{W}_0(\mathbf{E}^{(0)}), \mathcal{W}_1(\mathbf{E}^{(0)}, \mathbf{E}^{(1)}),$ $\mathcal{W}_2(\mathbf{E}^{(1)})$	$h\mathcal{W}_0 + \frac{h^3}{12}\mathcal{W}_2$	2	3
$\mathbf{E}^{(0)} + \zeta\mathbf{E}^{(1)} + \frac{\zeta^2}{2}\mathbf{E}^{(2)}$	$\mathcal{W}_0(\mathbf{E}^{(0)}), \mathcal{W}_1(\mathbf{E}^{(0)}, \mathbf{E}^{(1)}),$ $\mathcal{W}_2^*(\mathbf{E}^{(0)}, \mathbf{E}^{(1)}, \mathbf{E}^{(2)}),$ $\mathcal{W}_3(\mathbf{E}^{(1)}, \mathbf{E}^{(2)}), \mathcal{W}_4(\mathbf{E}^{(2)})$	$h\mathcal{W}_0 + \frac{h^3}{12}\mathcal{W}_2^* + \frac{h^5}{80}\mathcal{W}_4$	3	5

Numerical integration of four-node shell elements over thickness. In the four-node shell element, we use a 2×2 Gauss rule for integration over the reference lamina. (The analytical integration over the lamina is used, e.g., in the so-called one-integration point element.) The integration over the fiber is performed either analytically or one of the following 1D integration rules is used:

1. the 2-point Gauss rule. The locations of sampling points for the interval $\xi \in [-1, +1]$ and weighing factors for the 2-point Gauss rule are given in Table 10.2.

Table 10.2 2-point Gauss integration rule.

m	1	2
ξ_m	$-1/\sqrt{3}$	$1/\sqrt{3}$
w_m	1	1

2. the 5-point Simpson rule. In the Simpson method, the interval is divided into an even number of intervals and within each pair of intervals the function is approximated by a parabola. The method is exact for polynomials of degree at most 3. The locations of sampling points for the interval $\xi \in [-1, +1]$ and weighing factors for the 5-point Simpson rule are given in Table 10.3. In the context of shells, the Simpson rule has the advantage that the sampling points are also located at the ends of the interval, $\xi_m = \pm 1$, i.e. at the most external laminas.

Table 10.3 5-point Simpson integration rule.

m	1	2	3	4	5
ξ_m	-1	-1/2	0	1/2	1
w_m	1/6	4/6	3/6	4/6	1/6

10.6 Newton method and tangent operator

Newton method. Consider the potential energy functional defined as

$$F(\mathbf{z}) \doteq \int_V \mathcal{W}(\mathbf{z})\, \mathrm{d}V - F_{\text{ext}}, \tag{10.120}$$

where $\mathcal{W}(\mathbf{z})$ is the strain energy expressed by $\mathbf{z} \doteq \{\mathbf{u}, \boldsymbol{\psi}\}$ and F_{ext} is a functional of external forces. The below-described procedure is analogous for other governing functionals.

We write the stationarity condition of F as

$$\delta F \doteq DF(\bar{\mathbf{z}}) \cdot \delta\mathbf{z} = \frac{\mathrm{d}}{\mathrm{d}t} F(\bar{\mathbf{z}} + t\,\delta\mathbf{z})\Big|_{t=0} = 0, \tag{10.121}$$

where $DF(\bar{\mathbf{z}}) \cdot \delta\mathbf{z}$ is the directional derivative of F at $\bar{\mathbf{z}}$ in the direction $\delta\mathbf{z}$, t is a scalar parameter, and $\bar{\mathbf{z}}$ denotes the known (last computed) solution. This is the virtual work (VW) equation.

We designate $G \doteq \delta F$ and rewrite the VW equation as $G(\mathbf{z}) = 0$. It can be linearized and solved iteratively, e.g., using the Newton scheme defined as follows:

$$\left.\begin{aligned} & DG(\bar{\mathbf{z}}) \cdot \Delta\mathbf{z} = -G(\bar{\mathbf{z}}) \\ & \mathbf{z} = \bar{\mathbf{z}} + \Delta\mathbf{z} \end{aligned}\right\}, \tag{10.122}$$

where

$$DG(\bar{\mathbf{z}}) \cdot \Delta\mathbf{z} \doteq \frac{\mathrm{d}}{\mathrm{d}t} G(\bar{\mathbf{z}} + t\,\Delta\mathbf{z})\Big|_{t=0} \tag{10.123}$$

is the directional derivative of G at $\bar{\mathbf{z}}$ in the direction $\Delta\mathbf{z}$. This derivative provides the tangent operator $\mathbf{K}$.

Tangent operator for the linear material. Consider the strain energy $\mathcal{W}(\mathbf{z}) \doteq \frac{1}{2}\boldsymbol{\varepsilon} \cdot (\mathbb{C}\,\boldsymbol{\varepsilon})$, where $\mathbb{C} \doteq \partial\boldsymbol{\sigma}/\partial\boldsymbol{\varepsilon}$ is the tangent constitutive matrix. For simplicity, we omit the integral and F_{ext} in eq. (10.120).

Then $F(\mathbf{z}) \doteq \mathcal{W}(\mathbf{z})$ and its variation is $G \doteq \delta F = \delta\mathcal{W} = \boldsymbol{\sigma} \cdot \delta\boldsymbol{\varepsilon}$, where the constitutive equation $\boldsymbol{\sigma} \doteq \partial\mathcal{W}/\partial\boldsymbol{\varepsilon} = \mathbb{C}\,\boldsymbol{\varepsilon}$. The directional derivative is calculated as

$$DG(\bar{\mathbf{z}}) \cdot \Delta\mathbf{z} = \frac{\partial G}{\partial \mathbf{z}} \Delta\mathbf{z} = \Delta\boldsymbol{\sigma} \cdot \delta\boldsymbol{\varepsilon} + \boldsymbol{\sigma} \cdot \Delta(\delta\boldsymbol{\varepsilon}). \tag{10.124}$$

The first component of eq. (10.124) can be transformed using the constitutive relation in the incremental form $\Delta\boldsymbol{\sigma} = \mathbb{C}\,\Delta\boldsymbol{\varepsilon}$, where $\mathbf{B} \doteq \partial\boldsymbol{\varepsilon}/\partial\mathbf{z}$ is the kinematical strain-displacement matrix, to obtain

$$\Delta\boldsymbol{\sigma} \cdot \delta\boldsymbol{\varepsilon} = (\mathbb{C}\,\Delta\boldsymbol{\varepsilon}) \cdot \delta\boldsymbol{\varepsilon} = (\mathbb{C}\,\mathbf{B}\,\Delta\mathbf{z}) \cdot (\mathbf{B}\,\delta\mathbf{z}) = \delta\mathbf{z} \cdot (\mathbf{B}^T\mathbb{C}\,\mathbf{B}\,\Delta\mathbf{z}). \tag{10.125}$$

The second component of eq. (10.124) can be rewritten as

$$\boldsymbol{\sigma} \cdot \Delta(\delta\boldsymbol{\varepsilon}) = \Delta(\boldsymbol{\sigma}^* \cdot \delta\boldsymbol{\varepsilon}) = \Delta\delta(\boldsymbol{\sigma}^* \cdot \boldsymbol{\varepsilon}), \tag{10.126}$$

provided that, in differentiation, $\boldsymbol{\sigma}$ is treated as independent of $\mathbf{z}$, which is indicated by the asterisk, i.e. $\boldsymbol{\sigma}^*$. Note that the scalar $(\boldsymbol{\sigma}^* \cdot \boldsymbol{\varepsilon})$ is differentiated in the last form of this equation. Then the second component becomes

$$\Delta\delta(\boldsymbol{\sigma}^* \cdot \boldsymbol{\varepsilon}) = \delta\mathbf{z} \cdot \left[\frac{\partial^2(\boldsymbol{\varepsilon} \cdot \boldsymbol{\sigma}^*)}{\partial \mathbf{z}^2}\,\Delta\mathbf{z}\right]. \tag{10.127}$$

Finally,

$$DG(\bar{\mathbf{z}}) \cdot \Delta\mathbf{z} = \delta\mathbf{z} \cdot (\mathbf{K}\,\Delta\mathbf{z}), \qquad \mathbf{K} \doteq \mathbf{B}^T\mathbf{C}\mathbf{B} + \frac{\partial^2(\boldsymbol{\varepsilon} \cdot \boldsymbol{\sigma}^*)}{\partial \mathbf{z}^2}, \tag{10.128}$$

where $\mathbf{K}$ is the tangent stiffness operator.

Computation of tangent matrix. The major part of computation of the tangent matrix is the computation of derivatives and for non-linear functions these derivatives are more complicated than the function itself. There are three ways to compute the tangent matrix:

1. Analytic, i.e. by hand or using one of the symbolic manipulators such as *Mathematica*, *Maple*, and others, which can be used for manipulating equations and obtaining expressions for partial derivatives. This way is exact but laborious.
2. Numerical, i.e. by finite difference (FD) approximations and either two-sided or one-sided differences can be used. This yields an inefficient code, which can be also inaccurate.
3. Automatic (or algorithmic) differentiation of the computer program. The automatic differentiation (AD) programs can deal with constructs such as branches and loops and derivatives are correct up to the machine precision. The AD has strong theoretical foundations and is a mature computational technology, which can be used with confidence, see, e.g., [185, 82, 84, 83].

Below, we discuss these three ways of generating the tangent matrix.

1. Stiffness tangent matrix derived analytically. Again, consider the VW equation in the simplified form, i.e. $G(\mathbf{z}) \doteq \delta \mathcal{W} \doteq \boldsymbol{\sigma} \cdot \delta \boldsymbol{\varepsilon}$, where $\boldsymbol{\sigma} \doteq \partial \mathcal{W} / \partial \boldsymbol{\varepsilon}$. We can split

$$\delta \boldsymbol{\varepsilon} = [\delta_1 \boldsymbol{\varepsilon}, ..., \delta_N \boldsymbol{\varepsilon}], \qquad \delta_i \boldsymbol{\varepsilon} = \frac{\partial \boldsymbol{\varepsilon}}{\partial z_i} \delta q_i, \qquad i = 1, ..., N, \tag{10.129}$$

where $z_i \in \mathbf{z}_I$ denotes a nodal variable of a discrete FE model and, for simplicity, we take, only one of its components, i.e. $G_i(\mathbf{z}) \doteq \boldsymbol{\sigma} \cdot \delta_i \, \boldsymbol{\varepsilon}$. We shall calculate a derivative of this component w.r.t. one component, $z_j \in \mathbf{z}_I$,

$$\frac{\partial G_i(\mathbf{z}_I)}{\partial z_j} = \frac{\partial}{\partial z_j} (\boldsymbol{\sigma} \cdot \delta_i \boldsymbol{\varepsilon}) = \boldsymbol{\sigma}_{,j} \cdot (\mathbf{B}_i \, \delta z_i) + \boldsymbol{\sigma} \cdot (\mathbf{B}_{i,j} \, \delta z_i), \tag{10.130}$$

where

$$\boldsymbol{\sigma}_{,j} = \frac{\partial \boldsymbol{\sigma}}{\partial \boldsymbol{\varepsilon}} \frac{\partial \boldsymbol{\varepsilon}}{\partial z_j} = \mathbb{C} \, \mathbf{B}_j, \quad \mathbf{B}_i = \frac{\partial \boldsymbol{\varepsilon}}{\partial z_i}, \quad \mathbf{B}_{i,j} = \frac{\partial \mathbf{B}_i}{\partial z_j} = \frac{\partial^2 \boldsymbol{\varepsilon}}{\partial z_i \, \partial z_j}, \tag{10.131}$$

and $\mathbb{C} \doteq \partial \boldsymbol{\sigma} / \partial \boldsymbol{\varepsilon}$ is the constitutive tangent operator. Note that

1. If components of $\boldsymbol{\sigma}$ and $\boldsymbol{\varepsilon}$ are written as matrices, then $\mathbf{B}_i$ and $\mathbf{B}_{i,j}$ are also matrices but $\mathbb{C}$ must be a 4D matrix. Using the identity $\mathbf{T}_1 \cdot \mathbf{T}_2 = \mathrm{tr}(\mathbf{T}_2^T \mathbf{T}_1)$, we can rewrite eq. (10.130) as

$$\frac{\partial G_i(\mathbf{z}_I)}{\partial z_j} = \mathrm{tr} \, (\mathbf{B}_i^T \, \mathbb{C} \, \mathbf{B}_j + \mathbf{B}_{i,j}^T \, \boldsymbol{\sigma}) \, \delta z_i. \tag{10.132}$$

   This form is not used in computations because of the inconvenient form of $\mathbb{C}$.

2. If components of $\boldsymbol{\sigma}$ and $\boldsymbol{\varepsilon}$ are written as vectors, then $\mathbf{B}_i$ and $\mathbf{B}_{i,j}$ are also vectors, while $\mathbb{C}$ can be written as a 2D matrix; this is the so-called Voigt's notation. Then we can rewrite eq. (10.130) as follows:

$$\frac{\partial G_i(\mathbf{z}_I)}{\partial z_j} = (\mathbf{B}_i \, \delta z_i) \cdot (\mathbb{C} \, \mathbf{B}_j) + (\mathbf{B}_{i,j} \, \delta z_i) \cdot \boldsymbol{\sigma}. \tag{10.133}$$

   For all components taken into account, we have $i, j = 1, ..., N$ and the vectors $\mathbf{B}_i$ can be arranged as a matrix $\mathbf{B} = [\mathbf{B}_1, ..., \mathbf{B}_N]$. Then

$$\frac{\partial G(\mathbf{z}_I)}{\partial \mathbf{z}} \doteq \left[\frac{\partial G_i(\mathbf{z})}{\partial z_j} \right]_{i,j=1,...,N} = \delta \mathbf{B}^T \mathbb{C} \mathbf{B} \; + \; \left(\frac{\mathrm{d} \delta \mathbf{B}}{\mathrm{d} \mathbf{z}} \right)^T \boldsymbol{\sigma}, \tag{10.134}$$

where $\delta \mathbf{B} = [\mathbf{B}_1\, \delta z_1, ..., \mathbf{B}_N\, \delta z_N]$ and

$$\left(\frac{\mathrm{d}\delta \mathbf{B}}{\mathrm{d}\mathbf{z}}\right)^T = \begin{bmatrix} \mathbf{B}_{1,1}\, \delta z_1 & ... & \mathbf{B}_{1,N}\, \delta z_1 \\ ... & ... & ... \\ \mathbf{B}_{N,1}\, \delta z_N & ... & \mathbf{B}_{N,N}\, \delta z_N \end{bmatrix}. \tag{10.135}$$

Equation (10.134) is used to generate the tangent stiffness matrix which is defined as follows:

$$\mathbf{K} \doteq \int_A \left[\mathbf{B}^T \mathbb{C} \mathbf{B} + \left(\frac{\partial \mathbf{B}}{\partial \mathbf{z}}\right)^T \boldsymbol{\sigma} \right] \mathrm{d}V, \tag{10.136}$$

where the displacement matrix and the initial stress matrix are defined as

$$\mathbf{K}_0 + \mathbf{K}_L \doteq \int_A \mathbf{B}^T \mathbb{C} \mathbf{B}\, \mathrm{d}V, \qquad \mathbf{K}_\sigma \doteq \int_A \left(\frac{\partial \mathbf{B}}{\partial \mathbf{z}}\right)^T \boldsymbol{\sigma}\, \mathrm{d}V. \tag{10.137}$$

Here, $\mathbf{K}_0$ is the infinitesimal (linear) stiffness matrix, while $\mathbf{K}_u$ and $\mathbf{K}_\sigma$ are the parts which appear for non-linear strains and/or non-linear constitutive relations, see the classical textbooks on FEs.

2. Stiffness tangent matrix derived by finite difference method. The stiffness matrix can be approximated by a secant operator obtained using the Finite Difference (FD) method. This is a very inefficient method which can be accelerated by deriving $\mathbf{B}$ analytically and using it to compute the initial stress matrix $\mathbf{K}_\sigma$ by the FD method. This is obtained as follows:

1. First, the analytical formula for $\mathbf{B}$ is derived and compared with the FD approximation

$$\mathbf{B}_i^{\mathrm{FD}} = \frac{\boldsymbol{\varepsilon}(z_i + \tau) - \boldsymbol{\varepsilon}(z_i + \tau)}{2\tau}, \qquad z_i \in \mathbf{z}_I, \tag{10.138}$$

   where $\tau = 10^{-8}$ for double precision and $\mathbf{B}_i^{\mathrm{FD}}$ is the ith column of $\mathbf{B}^{\mathrm{FD}}$. This verification should be done for $z_i \neq 0$.
2. The so-verified analytical $\mathbf{B}$ is used to compute the initial stress matrix $\mathbf{K}_\sigma$ by the FD method,

$$(\mathbf{K}_\sigma)_i^{\mathrm{FD}} = \frac{\mathbf{B}^T(z_i + \tau)\, \boldsymbol{\sigma} - \mathbf{B}^T(z_i - \tau)\, \boldsymbol{\sigma}}{2\tau}, \tag{10.139}$$

   where $(\mathbf{K}_\sigma)_i^{\mathrm{FD}}$ is the ith column of $(\mathbf{K}_\sigma)^{\mathrm{FD}}$. If components of $\boldsymbol{\sigma}$ are written as a vector, then $\mathbf{B}$ is a matrix and, in the nominator, we obtain a difference of two vectors.

Remark. Note that the use of the analytically derived $\mathbf{B}$ is very important for efficiency. Otherwise we have to compute the second derivative of strain because

$$\mathbf{K}_\sigma = \left(\frac{\mathrm{d}\mathbf{B}}{\mathrm{d}\mathbf{z}}\right)^T \boldsymbol{\sigma} = \left(\frac{\mathrm{d}^2\boldsymbol{\varepsilon}}{\mathrm{d}\mathbf{z}^2}\right)^T \boldsymbol{\sigma}. \tag{10.140}$$

Computation of the second derivative is time-consuming for multidimensional $\mathbf{z}$ because, first, a second-order hyper-surface must be spanned. For instance, for each strain component ε_{kl}, we have to span

$$\varepsilon_{kl} = a_0 + \sum_{i=1}^{N} a_i\, z_i + \sum_{i=1}^{N-1} \sum_{j=1+i}^{N} a_{ij}\, z_i\, z_j + \sum_{i=1}^{N} a_{ii}\, z_i^2, \tag{10.141}$$

where the base functions are polynomials of up to the second order. The number of coefficients a_0, a_i, a_{ij}, a_{ii} which have to be calculated is

$$p = \underbrace{1}_{\text{constant}} + \underbrace{N}_{\text{linear}} + \underbrace{(N^2 - N)/2}_{\text{mixed quadratic}} + \underbrace{N}_{\text{pure quadratic}} = (N^2 + 3N + 2)/2, \tag{10.142}$$

and for $N = 8$ (four-node element $\times$ 2 dofs/node), we obtain $p = 45$. The semi-analytical method is more efficient but even this method can be used only when the efficiency of the FE is not important.

Finally, we note that the formulas given above may also be used to verify the correctness of $\mathbf{B}$ and $\mathbf{K}_\sigma$ derived either analytically or by an automatic differentiation program.

3. Stiffness tangent matrix by automatic differentiation. Even for a single shell element, we have to use many independent variables, e.g. in case of shell elements with six dofs/node, we use 24 variables in the four-node element and 54 in the 9-node element. This means that the functional F must be differentiated w.r.t. this number of variables, which produces thousands of formulas.

In consequence, the process of derivation of this matrix and coding is time-consuming and error prone. Controlling and modifications of a code become difficult because of its size. For this reason, the programs in which we can write operations in a very compact way and perform automatically differentiation are very useful.

In the automatic differentiation (AD) programs, there are several options which can be used to calculate the residual vector and the stiffness tangent matrix. The simplest possibility is

1. Apply the FE approximations and integrate the functional F, which yields an algebraic function $\hat{F}(\mathbf{z}_I)$ of the set of nodal variables $\mathbf{z}_I$,

$$F \stackrel{\text{FE}}{\longrightarrow} \hat{F}(\mathbf{z}_I). \tag{10.143}$$

2. The residual vector and the stiffness tangent matrix are calculated as derivatives of $\hat{F}(\mathbf{z}_I)$ w.r.t. nodal variables

$$\mathbf{r} \doteq \frac{\mathrm{d}\hat{F}(\mathbf{z}_I)}{\mathrm{d}\mathbf{z}_I}, \qquad \mathbf{K} \doteq \frac{\mathrm{d}\mathbf{r}}{\mathrm{d}\mathbf{z}_I}. \tag{10.144}$$

   These operations of differentiation can be coded in a few lines, and the form of results depends on the features of a particular AD program.

We develop finite elements using two programs: *FEAP* [183] and *AceGen* [181] and they are combined in the following way. In *AceGen*, we derive the algebraic function $\hat{F}(\mathbf{z}_I)$ and we code the automatic differentiation operations of eq. (10.144), to obtain the tangent matrix and the residual for an element. The resulting subroutine is in Fortran and is included into *FEAP* to build an executable program which is the subject of tests.

- *AceGen* is a fully reliable system enabling automatic derivation of formulae for numerical procedures developed by J. Korelc[1]. It is written as an add-on package for *Mathematica* and uses the symbolic language of *Mathematica*. The approach implemented in *AceGen* combines several techniques such as: (1) symbolic and algebraic capabilities of *Mathematica*, (2) automatic differentiation (forward and backward mode), (3) automatic code generation, (4) simultaneous optimization of expressions, and many other techniques. For details, see [130, 131] and the manuals.
- *FEAP* is a research finite element environment developed by R.L. Taylor[2], and its source is distributed by the University of California at Berkeley. *FEAP* has an open architecture which allows us to connect user subroutines through a pre-defined interface, see [268]. This program is used in many universities as an excellent environment for developing new finite elements.

[1] Prof. Jože Korelc, University of Ljubljana, Ravnikova 4, SI-1000, Ljubljana, Slovenia. E-mail: AceProducts@fgg.uni-lj.si (http://www.fgg.uni-lj.si/Symech/).

[2] Prof. Robert L. Taylor, Department of Civil Engineering, University of California at Berkeley, Berkeley, CA 94720. E-mail: rlt@ce.berkeley.edu (http://www.ce.berkeley.edu/~rlt/feap/).

Example: forward and backward automatic differentiation. Automatic differentiation can be performed in two ways, designated as *forward* and *backward.* The basic idea is well illustrated by the simple example used, e.g., in the manual of *AceGen* which shows two ways of differentiation of a composite function, $f_3(z_i, f_1(z_i), f_2(z_i, f_1(z_i)))$, depending on the variables z_i, $i = 1, .., n$.

The *forward* differentiation of the function f_3 yields the following formulas:

$$\begin{aligned}
&v_1 = f_1(z_i), && \frac{\partial v_1}{\partial z_i} = \frac{\partial f_1}{\partial z_i}, && i = 1, ..., n, \\
&v_2 = f_2(z_i, v_1), && \frac{\partial v_2}{\partial z_i} = \frac{\partial f_2}{\partial z_i} + \frac{\partial f_2}{\partial v_1}\frac{\partial v_1}{\partial z_i}, && i = 1, ..., n, \\
&v_3 = f_3(z_i, v_1, v_2), && \frac{\partial v_3}{\partial z_i} = \frac{\partial f_3}{\partial z_i} + \frac{\partial f_3}{\partial v_1}\frac{\partial v_1}{\partial z_i} + \frac{\partial f_3}{\partial v_2}\frac{\partial v_2}{\partial z_i}, && i = 1, ..., n,
\end{aligned} \tag{10.145}$$

where v_1, v_2, v_3 are the intermediate variables generated during differentiation.

The *backward* differentiation looks like

$$\begin{aligned}
&v_3 = f_3(z_i, v_1, v_2), && \overline{v_3} = \frac{\partial v_3}{\partial v_3} = 1, \\
&v_2 = f_2(z_i, v_1), && \overline{v_2} = \frac{\partial v_3}{\partial v_2} = \frac{\partial f_3}{\partial v_2}\overline{v_3}, \\
&v_1 = f_1(z_i), && \overline{v_1} = \frac{\partial v_3}{\partial v_1} = \frac{\partial f_3}{\partial v_1}\overline{v_3} + \frac{\partial f_2}{\partial v_1}\overline{v_2}, \\
&z_i && \frac{\partial v_3}{\partial z_i} = \frac{\partial f_3}{\partial z_i}\overline{v_3} + \frac{\partial f_2}{\partial z_i}\overline{v_2} + \frac{\partial f_1}{\partial z_i}\overline{v_1}, \quad i = 1, ..., n.
\end{aligned} \tag{10.146}$$

The *backward* differentiation yields more effective algorithms for large n, although is more time consuming.

11

Plane four-node elements (without drilling rotation)

In this chapter, we describe techniques used to derive plane (2D) four-node elements with translational degrees of freedom, but without drilling rotations. Such elements are relatively simple so can be used to test the concepts which are later incorporated into either the 3D or shell elements.

The 2D elements can be directly used as a membrane part of the shell elements without drilling rotations, i.e. either in the shell elements with five dofs/node or in the "solid-shell" elements (without rotational dofs). However, the 2D elements are flat so for the warped shell elements, the formulation must be generalized as described in Sect. 14.

11.1 Basic equations

Consider the classical configuration space of the non-polar Cauchy continuum defined as $\mathcal{C} \doteq \{\boldsymbol{\chi}: \ B \rightarrow R^3\}$, where $\boldsymbol{\chi}$ is the deformation function defined on the reference configuration of the body B.

Basic functionals. The following functionals are used in this chapter:

1. The three-field Hu–Washizu (HW) functional.

   A. For linear elastic materials, we can use the classical form of the HW functional

$$F_{\mathrm{HW}}(\mathbf{u}, \boldsymbol{\sigma}, \boldsymbol{\varepsilon}) \doteq \int_B \{\mathcal{W}(\boldsymbol{\varepsilon}) + \boldsymbol{\sigma} \cdot [\mathbf{E}(\nabla \mathbf{u}) - \boldsymbol{\varepsilon}]\} \, \mathrm{d}V - F_{\mathrm{ext}}, \qquad (11.1)$$

   where $\mathcal{W}(\boldsymbol{\varepsilon})$ is the strain energy expressed by the independent strain $\boldsymbol{\varepsilon}$ and the stress $\boldsymbol{\sigma}$ plays the role of the Lagrange multiplier of the relation involving the independent strain $\boldsymbol{\varepsilon}$ and the Green strain

$\mathbf{E}(\mathbf{u})$, which is a function of the displacement $\mathbf{u}$. At the solution, we have $\boldsymbol{\varepsilon} = \mathbf{E}(\nabla \mathbf{u})$ and $\boldsymbol{\sigma} = \mathbf{S}$, where $\mathbf{S}$ is the second Piola–Kirchhoff stress tensor. F_{ext} is the potential of the body force, the external loads, and the displacement boundary conditions.

B. For non-linear materials, the constitutive operator $\mathbb{C}(\boldsymbol{\varepsilon}) \doteq \partial^2 \mathcal{W}(\boldsymbol{\varepsilon})/(\partial \boldsymbol{\varepsilon})^2$ depends on strain $\boldsymbol{\varepsilon}$ and we can use it only with increments. We write the displacements, stress, and strain in the incremental form

$$\mathbf{u}^i = \mathbf{u}^{i-1} + \Delta \mathbf{u}, \qquad \boldsymbol{\sigma}^i = \boldsymbol{\sigma}^{i-1} + \Delta \boldsymbol{\sigma}, \qquad \boldsymbol{\varepsilon}^i = \boldsymbol{\varepsilon}^{i-1} + \Delta \boldsymbol{\varepsilon}, \tag{11.2}$$

where i, $i-1$ are iteration indices and the increment $\Delta(\cdot) \doteq (\cdot)^i - (\cdot)^{i-1}$. Inserting these formulas into eq. (11.1), we obtain an incremental HW functional

$$\begin{aligned} F^*_{\text{HW}}(\Delta \mathbf{u}, \Delta \boldsymbol{\sigma}, \Delta \boldsymbol{\varepsilon}) \doteq \int_B \{ & \mathcal{W}(\boldsymbol{\varepsilon} + \Delta \boldsymbol{\varepsilon}) \\ & + (\boldsymbol{\sigma} + \Delta \boldsymbol{\sigma}) \cdot [\mathbf{E}(\nabla(\mathbf{u} + \Delta \mathbf{u})) - (\boldsymbol{\varepsilon} + \Delta \boldsymbol{\varepsilon})]\} \, \mathrm{d}V - F_{\text{ext}}, \end{aligned} \tag{11.3}$$

where the index $(i-1)$ was omitted for clarity.

2. The two-field Hellinger–Reissner (HR) functional. A. For linear elastic materials, we can use the classical form of the HR functional

$$F_{\text{HR}}(\mathbf{u}, \boldsymbol{\sigma}) \doteq \int_B \left[-\tfrac{1}{2} \boldsymbol{\sigma} \cdot (\mathbb{C}^{-1} \boldsymbol{\sigma}) + \boldsymbol{\sigma} \cdot \mathbf{E}(\mathbf{u}) \right] \mathrm{d}V - F_{\text{ext}}. \tag{11.4}$$

This functional is obtained as follows. For the linear elastic material, the strain energy is $\mathcal{W} \doteq \frac{1}{2} \boldsymbol{\varepsilon} \cdot (\mathbb{C} \boldsymbol{\varepsilon}) = \frac{1}{2} \boldsymbol{\varepsilon} \cdot \boldsymbol{\sigma}$, where $\mathbb{C}$ is the constitutive operator. Then, using $\boldsymbol{\varepsilon} = \mathbb{C}^{-1} \boldsymbol{\sigma}$, we obtain

$$\mathcal{W} - \boldsymbol{\sigma} \cdot \boldsymbol{\varepsilon} = -\tfrac{1}{2} \boldsymbol{\sigma} \cdot \boldsymbol{\varepsilon} = -\tfrac{1}{2} \boldsymbol{\sigma} \cdot (\mathbb{C}^{-1} \boldsymbol{\sigma}).$$

By using this expression in the HW functional of eq. (11.1), we obtain the classical form of the HR functional of eq. (11.4).

B. For non-linear materials, the constitutive operator $\mathbb{C}(\boldsymbol{\varepsilon}) \doteq \partial^2 \mathcal{W}(\boldsymbol{\varepsilon})/(\partial \boldsymbol{\varepsilon})^2$ depends on strain $\boldsymbol{\varepsilon}$ and we can use it only with increments. We write the displacements, stress, and strain in the incremental form

$$\mathbf{u}^i = \mathbf{u}^{i-1} + \Delta \mathbf{u}, \qquad \boldsymbol{\sigma}^i = \boldsymbol{\sigma}^{i-1} + \Delta \boldsymbol{\sigma}, \qquad \boldsymbol{\varepsilon}^i = \boldsymbol{\varepsilon}^{i-1} + \Delta \boldsymbol{\varepsilon}, \tag{11.5}$$

where i, $i-1$ are iteration indices and the increment $\Delta(\cdot) \doteq (\cdot)^i - (\cdot)^{i-1}$. The strain increment is expressed by an inverse constitutive equation

$$\Delta\varepsilon = (\mathbb{C}^{i-1})^{-1}\, \Delta\sigma, \qquad \mathbb{C}^{i-1} \doteq \mathbb{C}(\varepsilon^{i-1}). \tag{11.6}$$

Inserting these formulas into the Hu–Washizu functional of eq. (11.1), we obtain an incremental HR functional

$$F^*_{\text{HR}}(\Delta\mathbf{u}, \Delta\sigma) \doteq \int_B \{\mathcal{W}\left(\varepsilon + \mathbb{C}^{-1}\Delta\sigma\right) \\ - (\sigma + \Delta\sigma)\cdot\left[\varepsilon + \mathbb{C}^{-1}\Delta\sigma - \mathbf{E}(\mathbf{u} + \Delta\mathbf{u})\right]\}\, \mathrm{d}V - F_{\text{ext}}, \tag{11.7}$$

where the index $(i-1)$ was omitted for clarity. This functional depends on two fields, similarly as in the classical HR functional of eq. (11.4). The values from the previous $(i-1)$th iteration, i.e. $\mathbf{u}$, σ and ε, must be stored as history variables.

3. The potential energy (PE) functional.

$$F_{\text{PE}}(\mathbf{u}) \doteq \int_B \mathcal{W}(\mathbf{u})\, \mathrm{d}V - F_{\text{ext}}, \tag{11.8}$$

where $\mathcal{W}(\mathbf{u})$ is the strain energy expressed by displacements $\mathbf{u}$. This functional is obtained from eq. (11.1) assuming that $\varepsilon = \mathbf{E}(\mathbf{u})$, for which the term with stress vanishes. Then $\mathcal{W}(\varepsilon) = \mathcal{W}(\mathbf{E}(\mathbf{u})) = \mathcal{W}(\mathbf{u})$.

These three functionals form the basis of the elements developed in the next sections.

Strain energy and constitutive equation. Assume that the strain energy density $\mathcal{W}$, defined per unit non-deformed volume, is a function of the right Cauchy–Green deformation tensor $\mathbf{C} \doteq \mathbf{F}^T\mathbf{F}$, where $\mathbf{F}$ is the deformation gradient, so that the objectivity requirement is satisfied. The constitutive law for the second Piola–Kirchhoff stress $\mathbf{S}$ is as follows:

$$\mathbf{S} = 2\,\partial_C \mathcal{W}(\mathbf{C}). \tag{11.9}$$

The work conjugate to $\mathbf{S}$ is the Green strain $\mathbf{E} \doteq \frac{1}{2}(\mathbf{C} - \mathbf{I})$. The constitutive tangent operator is defined as $\mathbb{C} \doteq \partial\mathbf{S}/\partial\mathbf{E} = \partial^2\mathcal{W}(\mathbf{E})/(\partial\mathbf{E})^2$.

The two-dimensional (2D) incremental constitutive equations and the constitutive operator can be obtained by applying the plane stress condition to the incremental constitutive equation written for 3D strains and stresses, see Sect. 7.2.1.

Natural basis at the element's center. The position vector in the initial configuration is approximated as

$$\mathbf{y}(\xi,\eta) = \sum_{I=1}^{4} N_I(\xi,\eta)\, \mathbf{y}_I, \tag{11.10}$$

where $N_I(\xi,\eta)$ are the bilinear shape functions of eq. (10.3) and $\xi,\eta \in [-1,+1]$ are natural coordinates.

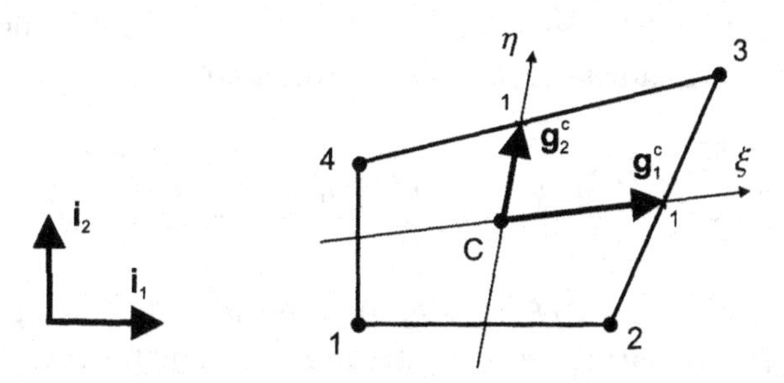

Fig. 11.1 Natural basis at the element's center $\{\mathbf{g}_k^c\}$ and the reference basis $\{\mathbf{i}_k\}$.

The vectors of the natural basis are defined as in eq. (10.15),

$$\mathbf{g}_1(\xi,\eta) \doteq \frac{\partial \mathbf{y}(\xi,\eta)}{\partial \xi}, \qquad \mathbf{g}_2(\xi,\eta) \doteq \frac{\partial \mathbf{y}(\xi,\eta)}{\partial \eta}, \tag{11.11}$$

and the vectors of the natural basis at the element's center, i.e. $\{\mathbf{g}_k^c\}$ $(k=1,2)$, are defined as

$$\mathbf{g}_1^c \doteq \mathbf{g}_1|_{\xi,\eta=0}, \qquad \mathbf{g}_2^c \doteq \mathbf{g}_2|_{\xi,\eta=0}. \tag{11.12}$$

In general, $\mathbf{g}_1^c$ and $\mathbf{g}_2^c$ are neither unit nor orthogonal, see Fig. 11.1. The co-basis vectors $\mathbf{g}_c^k$ are defined as in eq. (10.47), by the relation $\mathbf{g}_c^k \cdot \mathbf{g}_l^c = \delta_l^k \quad (l=1,2)$.

In the reference Cartesian basis $\{\mathbf{i}_k\}$, we have $\mathbf{y} = x\mathbf{i}_1 + y\mathbf{i}_2$, and the global Jacobian matrix is

$$\mathbf{J} \doteq \begin{bmatrix} \frac{\partial x}{\partial \xi} & \frac{\partial x}{\partial \eta} \\ \frac{\partial y}{\partial \xi} & \frac{\partial y}{\partial \eta} \end{bmatrix} = \begin{bmatrix} \mathbf{g}_1 \cdot \mathbf{i}_1 & \mathbf{g}_2 \cdot \mathbf{i}_1 \\ \mathbf{g}_1 \cdot \mathbf{i}_2 & \mathbf{g}_2 \cdot \mathbf{i}_2 \end{bmatrix}, \tag{11.13}$$

where $\mathbf{g}_1, \mathbf{g}_2$ are the vectors of the natural basis of eq. (11.11).

11.2 Displacement element Q4

The basic four-node element derived from the PE functional for displacements approximated by bilinear shape functions, is designated as Q4. The displacements preserve inter-element continuity, i.e. are compatible, and the neighboring elements are congruent (conform). However, accuracy of Q4 is so poor that is of no practical importance.

Compatible displacements, deformation gradient and Green strain. The position vector in the initial configuration and the compatible displacements for the four-node quadrilateral are approximated as

$$\mathbf{y}(\xi,\eta)=\sum_{I=1}^{4} N_I(\xi,\eta)\,\mathbf{y}_I, \qquad \mathbf{u}^c(\xi,\eta)=\sum_{I=1}^{4} N_I(\xi,\eta)\,\mathbf{u}_I, \tag{11.14}$$

where $N_I(\xi,\eta)\doteq\frac{1}{4}(1+\xi_I\,\xi)\,(1+\eta_I\,\eta)$ are the bilinear shape functions, $\xi,\eta\in[-1,+1]$ are natural coordinates and I designates the corner nodes. The deformation gradient is defined as

$$\mathbf{F}^c=\frac{\partial(\mathbf{y}+\mathbf{u}^c)}{\partial\boldsymbol{\xi}}\,\frac{\partial\boldsymbol{\xi}}{\partial\mathbf{y}}=\mathbf{F}_\xi\,\mathbf{J}^{-1}, \tag{11.15}$$

where $\mathbf{F}_\xi\doteq\partial(\mathbf{y}+\mathbf{u}^c)/\partial\boldsymbol{\xi}$, $\mathbf{J}\doteq\partial\mathbf{y}/\partial\boldsymbol{\xi}$ is the Jacobian matrix and $\boldsymbol{\xi}\doteq[\xi,\eta]^T$. Then the compatible Green strain in the global frame is

$$\mathbf{E}^c=\tfrac{1}{2}(\mathbf{F}^{cT}\mathbf{F}^c-\mathbf{I})=\tfrac{1}{2}\left[\mathbf{J}^{-T}(\mathbf{F}_\xi^T\,\mathbf{F}_\xi)\,\mathbf{J}^{-1}-\mathbf{I}\right]. \tag{11.16}$$

The vectors and matrices of components used above are expressed in the global reference basis $\{\mathbf{i}_k\}$.

Global and local forms of deformation gradient and Green strain. Below, vectors and matrices of components are considered and the index "G" indicates that the components are in the global reference basis, while "L" indicates that they are in the local Cartesian basis at the element's center. Define the local position vectors and the local displacements as follows:

$$\mathbf{y}_L\doteq\mathbf{R}_{0c}^T\,\mathbf{y}_G, \qquad \mathbf{u}_L^c\doteq\mathbf{R}_{0c}^T\,\mathbf{u}_G^c, \tag{11.17}$$

where $\mathbf{R}_{0c}\in SO(3)$ defines the position of the local frame in the global reference frame. Then,

$$\mathbf{J}_G \doteq \frac{\partial \mathbf{y}_G}{\partial \boldsymbol{\xi}} = \frac{\partial(\mathbf{R}_{0c}\mathbf{y}_L)}{\partial \boldsymbol{\xi}} = \mathbf{R}_{0c}\frac{\partial \mathbf{y}_L}{\partial \boldsymbol{\xi}} = \mathbf{R}_{0c}\mathbf{J}_L,$$

$$\frac{\partial(\mathbf{y}_G + \mathbf{u}_G^c)}{\partial \boldsymbol{\xi}} = \mathbf{R}_{0c}\frac{\partial(\mathbf{y}_L + \mathbf{u}_L^c)}{\partial \boldsymbol{\xi}}$$

and the deformation gradient can be expressed as

$$\mathbf{F}_G^c = \frac{\partial(\mathbf{y}_G + \mathbf{u}_G^c)}{\partial \boldsymbol{\xi}} \frac{\partial \boldsymbol{\xi}}{\partial \mathbf{y}_G} = \mathbf{R}_{0c}\,\mathbf{F}_L^c\,\mathbf{R}_{0c}^T, \tag{11.18}$$

where

$$\mathbf{F}_L^c \doteq \frac{\partial(\mathbf{y}_L + \mathbf{u}_L^c)}{\partial \boldsymbol{\xi}}\mathbf{J}_L^{-1} \tag{11.19}$$

is the local form of the deformation gradient. In a similar manner, the Green strain can be expressed as

$$\mathbf{E}_G^c \doteq \tfrac{1}{2}[\mathbf{F}_G^{cT}\mathbf{F}_G^c - \mathbf{I}] = \tfrac{1}{2}\left[\mathbf{R}_{0c}^T\,(\mathbf{F}_L^c)^T\,\mathbf{F}_L^c\,\mathbf{R}_{0c} - \mathbf{I}\right] = \mathbf{R}_{0c}^T\,\mathbf{E}_L^c\,\mathbf{R}_{0c}, \tag{11.20}$$

where

$$\mathbf{E}_L^c \doteq \tfrac{1}{2}\left[(\mathbf{F}_L^c)^T\,\mathbf{F}_L^c - \mathbf{I}\,\right] \tag{11.21}$$

is the local form of the Green strain. The local $\mathbf{F}_L^c$ and $\mathbf{E}_L^c$ can be used to derive the local tangent matrix and the residual vector, which is more convenient. Afterwards, the matrix and the vector must be rotated to the global basis.

Approximation of strains in Q4. The bilinear approximations of displacement components can be written as

$$u(\xi,\eta) = u_0+\xi\,u_1+\eta\,u_2+\xi\eta\,u_3, \quad v(\xi,\eta) = v_0+\xi\,v_1+\eta\,v_2+\xi\eta\,v_3, \tag{11.22}$$

where $\xi,\eta \in [-1,+1]$ and the coefficients u_i and v_i $(i = 0,1,2,3)$ are functions of the nodal displacement components.

Consider a bi-unit (2×2) square element, for which the position vector components are $x = \xi$ and $y = \eta$, and the Jacobian matrix is the identity matrix. Then we have the following approximations:

- the displacement gradient,

$$\nabla\mathbf{u} \doteq \begin{bmatrix} u_{,\xi} & u_{,\eta} \\ v_{,\xi} & v_{,\eta} \end{bmatrix} = \begin{bmatrix} u_1 + \eta\,u_3 & u_2 + \xi\,u_3 \\ v_1 + \eta\,v_3 & v_2 + \xi\,v_3 \end{bmatrix}, \tag{11.23}$$

- the linear strain,

$$\boldsymbol{\varepsilon} \doteq \operatorname{sym}\nabla\mathbf{u} = \begin{bmatrix} u_1 + \eta\, u_3 & \frac{1}{2}[(u_2 + v_1) + \xi\, u_3 + \eta\, v_3] \\ \text{sym.} & v_2 + \xi\, v_3 \end{bmatrix}. \tag{11.24}$$

We see that ε_{11} and ε_{22} are incomplete linear polynomials of ξ and η, while the shear strain ε_{12} is a complete linear polynomial.

Despite the completeness of ε_{12}, the Q4 element performs poorly in tests involving shear strains. When ε_{12} is calculated (sampled) only at the element center and this value is used to approximate the whole field within the element, i.e.

$$\varepsilon_{12}(\xi, \eta) \approx (\varepsilon_{12})_c, \tag{11.25}$$

then the element still has a correct rank and the results are improved, see numerical results for the AS12 element in [256]. This feature leads to the concept of the "one-integration point" elements. The accuracy of the AS12 element is worse than of the elements discussed in the next sections.

Fig. 11.2 Pure bending: a) exact deformation, b) deformation of Q4 element.

Another observation made in [248] is that the quadratic terms are missing in eq. (11.22), so pure bending of the element cannot be properly represented, see Fig. 11.2.

11.3 Solution of FE equations for problems with additional variables

Improved formulations of four-node element. A lot of research has been devoted to improving the formulation of a four-node element and two directions were taken:

1. approximations of strains were enhanced, leading to the *enhanced strain* methods, see Sect. 11.4. The stress was eliminated from these formulations.

2. Mixed HR and HW functionals were applied instead of the PE functional, leading to the *mixed* methods, see Sect. 11.5. The stress was retained in these formulations.

Both these directions are combined by the mixed/enhanced methods.

Set of equations for problems with additional variables. For the considered classes of methods, the governing functional F depends on two sets of variables: the nodal displacements $\mathbf{u}_I$ and the elemental multipliers $\mathbf{q}$.

For kinematically non-linear problems, the stationarity condition of $F(\mathbf{u}_I, \mathbf{q})$ yields a system of equilibrium equations for an element,

$$\mathbf{r}_u \doteq \frac{\partial F(\mathbf{u}_I, \mathbf{q})}{\partial \mathbf{u}_I} = \mathbf{0}, \qquad \mathbf{r}_q \doteq \frac{\partial F(\mathbf{u}_I, \mathbf{q})}{\partial \mathbf{q}} = \mathbf{0}. \tag{11.26}$$

The linearized (Newton) form of these equations is as follows:

$$\begin{bmatrix} \mathbf{K} & \mathbf{L} \\ \mathbf{L}^T & \mathbf{K}_{qq} \end{bmatrix} \begin{bmatrix} \Delta \mathbf{u}_I \\ \Delta \mathbf{q} \end{bmatrix} = - \begin{bmatrix} \mathbf{r}_u \\ \mathbf{r}_q \end{bmatrix}, \tag{11.27}$$

where

$$\mathbf{K} \doteq \frac{\partial \mathbf{r}_u}{\partial \mathbf{u}_I}, \qquad \mathbf{L} \doteq \frac{\partial \mathbf{r}_u}{\partial \mathbf{q}}, \qquad \mathbf{K}_{qq} \doteq \frac{\partial \mathbf{r}_q}{\partial \mathbf{q}}. \tag{11.28}$$

To eliminate $\Delta \mathbf{q}$ at the element level, we calculate it from the second of eq. (11.27) as follows:

$$\Delta \mathbf{q} = -\mathbf{K}_{qq}^{-1} (\mathbf{r}_q + \mathbf{L}^T \Delta \mathbf{u}_I) \tag{11.29}$$

and, next, we use it in the first equation, which yields

$$\mathbf{K}^* \, \Delta \mathbf{u}_I = -\mathbf{r}^*, \tag{11.30}$$

where

$$\mathbf{K}^* \doteq \mathbf{K} - \mathbf{L}\mathbf{K}_{qq}^{-1}\mathbf{L}^T, \qquad \mathbf{r}^* \doteq \mathbf{r}_u - \mathbf{L}\mathbf{K}_{qq}^{-1}\mathbf{r}_q. \tag{11.31}$$

Subsequently, $\mathbf{K}^*$ and $\mathbf{r}^*$ are aggregated for all elements and the global set of equations is solved for $\Delta \mathbf{u}_I$. Then we update the nodal displacements $\mathbf{u}_I = \mathbf{u}_I + \Delta \mathbf{u}_I$. The elemental multipliers $\mathbf{q}$ are treated as described below.

Remark. Note that the system of equations (11.27) written for all elements is solvable if the matrix $\mathbf{K}_{qq}$ for each element and the matrix $\mathbf{K}^*$ for all elements additionally modified by boundary conditions, are invertible. Note that eigenvalues of the non-reduced tangent matrix of eq. (11.27) are different from those of matrix $\mathbf{K}^*$ of eq. (11.30).

Schemes of update of multipliers. Several update schemes of the vector of multipliers can be developed and each requires specific storage and implementation. We have implemented and tested two update schemes; in both, the update $\mathbf{q} = \mathbf{q} + \Delta\mathbf{q}$ is local, i.e. it is performed in each element separately.

Scheme U1. In this scheme, the storage space is minimal, as we store only one vector $\mathbf{q}$. For the iteration i, we use eq. (11.29) in the following form:

$$(\Delta\mathbf{q})^i = -\mathbf{K}_{qq}^{-1}\left(\mathbf{r}_q + \mathbf{L}^T\Delta\mathbf{u}_I^{i-1}\right), \qquad \mathbf{q}^i = \mathbf{q}^{i-1} + (\Delta\mathbf{q})^i, \tag{11.32}$$

where $\mathbf{K}_{qq}$, $\mathbf{r}_q$ and $\mathbf{L}$ are calculated for $(\mathbf{u}_I^{i-1}, \mathbf{q}^{i-1})$. This update is performed just after the local matrices have been generated and the updated $\mathbf{q}^i$ is stored. Note that so-updated multipliers are not used until the next iteration, $i+1$, and $\mathbf{q}^i$ is obtained for $\Delta\mathbf{u}_I^{i-1}$, so the difference is of two iterations!

We tested that this scheme performs better (less often causes divergence) if the update is performed only in the first iteration of each step. (For the convergent solution of the previous step, $\Delta\mathbf{u}_I^{i-1} \approx \mathbf{0}$, and then we can omit the last term in the above equation.)

Scheme U2. This scheme is more exact than the previous one because it uses the last increment $\Delta\mathbf{u}_I^i$, but requires more storage. It is equivalent to the one globally treating $\mathbf{q}$, in the same way as $\mathbf{u}_I$, instead of eliminating it at the element's level. The advantage of the scheme U2 over such a global treatment is that to invert $\mathbf{K}_{qq}$, we can use a specialized solver and retain an effective band-profile solver for the global equations. Let us rewrite eq. (11.29) as follows:

$$\Delta\mathbf{q} = -\underbrace{\mathbf{K}_{qq}^{-1}\mathbf{r}_q}_{\doteq\mathbf{v}} + \underbrace{\mathbf{K}_{qq}^{-1}\mathbf{L}^T}_{\doteq\mathbf{A}}\Delta\mathbf{u}_I, \tag{11.33}$$

where $\mathbf{v}$ and $\mathbf{A}$ are updated and stored. Hence, we have to store vectors $\mathbf{q}$ and $\mathbf{v}$ of dimension nM, and a matrix $\mathbf{A}$ of dimension $nM \times nst$, where nM is a number of additional modes and nst is a number of dofs/element ($nst = 8$ for four-node element with two dofs/node). The update is performed before the local matrices are generated, as follows:

1. retrieve $\mathbf{v}$ and $\mathbf{A}$,
2. calculate

$$(\Delta \mathbf{q})^i = -\mathbf{v} + \mathbf{A}\, \Delta \mathbf{u}_I^{i-1}, \qquad \mathbf{q}^i = \mathbf{q}^{i-1} + (\Delta \mathbf{q})^i, \tag{11.34}$$

   for the last available $\Delta \mathbf{u}_I^{i-1}$,
3. generate the local elemental matrices and vectors.

As a by-product of step 3, we obtain the updated $\mathbf{v}$ and $\mathbf{A}$, which we store for the next step.

The convergence rate for both these schemes is compared in [256, 257], for the slender cantilever example of Sect.15.3.1. The scheme U1 performs reasonably well only for the enhanced strain elements; the use of scheme U2 appears to be crucial for mixed elements.

Remark. Finally, we mention that the matrix $\mathbf{K}_{qq}$ is symmetric and sparse and we can use these properties to effectively compute its inverse. Besides, we note that to find $\mathbf{q}_i$, we can use another approach and directly solve the set of equations $\mathbf{r}_q = \mathbf{0}$ for fixed $\mathbf{u}_I$. This can be a useful approach, e.g., in dynamics and an explicit time-integration scheme.

11.4 Enhanced strain elements based on potential energy

The class of the enhanced strain methods is based on the technique of adding additional terms either to displacements, or strains, or the displacement gradient, with the purpose of improving the element's performance.

In all the methods described below, the multipliers $\mathbf{q}$ are associated with the element (not with the nodes) and are eliminated (condensed out) on the element level. They are discontinuous across the element boundaries.

11.4.1 ID4 element

The incompatible displacements (ID) method was proposed in [248] to improve the behavior of the Q4 element in pure bending, see Fig. 11.2. Later, it was discovered that the ID element does not pass the patch test for distorted meshes and a correction was proposed in [234]. From today's perspective, the idea of the ID method was ingenious and the whole class of the enhanced strain methods stems from it.

In the ID method, the assumed incompatible displacements are added to the compatible ones as follows:

$$\underbrace{\mathbf{u}(\xi,\eta)}_{\text{enhanced}} \doteq \underbrace{\mathbf{u}^{c}(\xi,\eta)}_{\text{compatible}} + \underbrace{\mathbf{u}^{\text{inc}}(\xi,\eta)}_{\text{incompatible}} . \tag{11.35}$$

Original formulation. In the original paper, the incompatible displacements are assumed in the following form:

$$\mathbf{u}^{\text{inc}}(\xi,\eta) \doteq \mathbf{i}_1\, u^{\text{inc}}(\xi,\eta) + \mathbf{i}_2\, v^{\text{inc}}(\xi,\eta), \tag{11.36}$$

where

$$\begin{bmatrix} u^{\text{inc}}(\xi,\eta) \\ v^{\text{inc}}(\xi,\eta) \end{bmatrix} \doteq \begin{bmatrix} q_1\, P_1(\xi) + q_3\, P_2(\eta) \\ q_2\, P_1(\xi) + q_4\, P_2(\eta) \end{bmatrix}. \tag{11.37}$$

and the quadratic (bubble) modes $P_1(\xi) \doteq 1-\xi^2$ and $P_2(\eta) \doteq 1-\eta^2$. Four multipliers q_i are used, see Fig. 11.3. The incompatible displacements are assumed in the Cartesian basis $\{\mathbf{i}_k\}$, similarly to the compatible displacements.

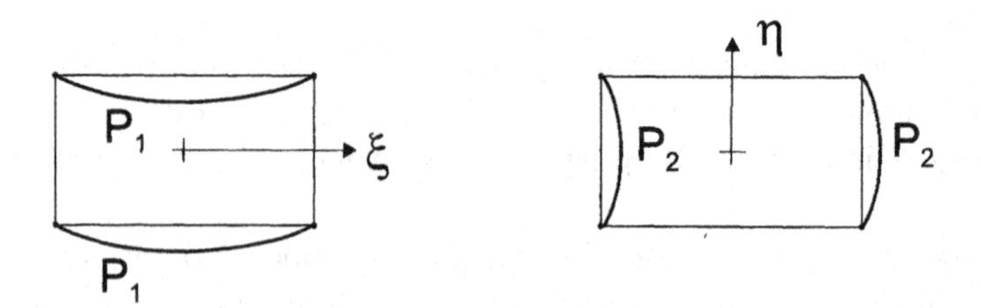

Fig. 11.3 Incompatible modes P_1 and P_2.

The effect of introducing the incompatible displacements can be shown as follows. Consider a bi-unit (2×2) square element, for which the position vector components are $x = \xi$ and $y = \eta$ and the Jacobian matrix is the identity matrix. Then, for the incompatible displacements, we obtain

- the "incompatible" displacement gradient,

$$\nabla \mathbf{u}^{\text{inc}} \doteq \begin{bmatrix} u^{\text{inc}}_{,\xi} & u^{\text{inc}}_{,\eta} \\ v^{\text{inc}}_{,\xi} & v^{\text{inc}}_{,\eta} \end{bmatrix} = -2 \begin{bmatrix} q_1\xi & q_3\eta \\ q_2\xi & q_4\eta \end{bmatrix}, \tag{11.38}$$

- the "incompatible" linear strain,

$$\boldsymbol{\varepsilon}^{\text{inc}} \doteq \text{sym}\nabla \mathbf{u}^{\text{inc}} = -2 \begin{bmatrix} q_1\xi & \frac{1}{2}(q_2\xi + q_3\eta) \\ \text{sym.} & q_4\eta \end{bmatrix}. \tag{11.39}$$

Note that $\varepsilon^{\text{inc}}_{11}$ and $\varepsilon^{\text{inc}}_{22}$ of (11.39) enhance the compatible ε_{11} and ε_{22} of eq. (11.24), so these components become complete linear polynomials. The role of the shear component $\varepsilon^{\text{inc}}_{12}$ is different; it rather de-enhances the compatible ε_{12}, in which the ξ and η terms were already present. Nonetheless, this de-enhancement is beneficial and significantly improves accuracy in tests involving the in-plane shear. It is more effective than the sampling of ε_{12} at the element's center of eq. (11.25).

Modified formulation. Element ID4. We can define the incompatible displacements in the natural basis at the element's center $\{\mathbf{g}^c_k\}$ of Fig. 11.1,

$$\mathbf{u}^{\text{inc}}(\xi, \eta) \doteq \mathbf{g}^c_1 \, u^{\text{inc}}(\xi, \eta) + \mathbf{g}^c_2 \, v^{\text{inc}}(\xi, \eta), \tag{11.40}$$

which can be rewritten as

$$\begin{bmatrix} u^{\text{inc}}_C \\ v^{\text{inc}}_C \end{bmatrix} = \mathbf{J}_c \begin{bmatrix} u^{\text{inc}} \\ v^{\text{inc}} \end{bmatrix}, \tag{11.41}$$

where $\mathbf{J}_c$ is the Jacobian matrix of eq. (11.13) at the element's center, and $u^{\text{inc}}_C, v^{\text{inc}}_C$ are the components in the Cartesian basis $\{\mathbf{i}_k\}$. The last form is obtained by using $\mathbf{g}^c_1 = (\mathbf{g}^c_1 \cdot \mathbf{i}_1)\mathbf{i}_1 + (\mathbf{g}^c_1 \cdot \mathbf{i}_2)\mathbf{i}_2$ and $\mathbf{g}^c_2 = (\mathbf{g}^c_2 \cdot \mathbf{i}_1)\mathbf{i}_1 + (\mathbf{g}^c_2 \cdot \mathbf{i}_2)\mathbf{i}_2$, by separation of terms multiplied by $\mathbf{i}_1$ and $\mathbf{i}_2$.

The discrete F_{PE} functional depends on two sets of variables: the nodal displacements $\mathbf{u}_I$ and the elemental multipliers $\mathbf{q}$ of the incompatible displacement modes. The obtained set of FE equations is given by eq. (11.27) and to update the stress multipliers, the scheme U2 of eq. (11.34) should be used.

Variational basis of the ID method. We can write the PE functional of eq. (11.8) for the enhanced displacements,

$$F_{\mathrm{PE}}(\mathbf{u}^{\mathrm{enh}}) \doteq \int_B \mathcal{W}(\mathbf{u}^{\mathrm{enh}})\,\mathrm{d}V - F_{\mathrm{ext}}, \tag{11.42}$$

on use of eqs. (11.35) and (11.40), which furnishes a general formula. The original ID method was developed for small strains and the SVK material, for which the strain energy can be written as

$$\mathcal{W}(\mathbf{u}^{\mathrm{enh}}) \doteq \tfrac{1}{2}\int_B (\boldsymbol{\varepsilon}_v^c + \boldsymbol{\varepsilon}_v^{\mathrm{inc}})^T\,\mathbb{C}\,(\boldsymbol{\varepsilon}_v^c + \boldsymbol{\varepsilon}_v^{\mathrm{inc}})\,\mathrm{d}V, \tag{11.43}$$

where $\boldsymbol{\varepsilon}_v^c \doteq \boldsymbol{\varepsilon}_v(\mathbf{u}^c)$, $\boldsymbol{\varepsilon}_v^{\mathrm{inc}} \doteq \boldsymbol{\varepsilon}_v(\mathbf{u}^{\mathrm{inc}})$. Here, $(\cdot)_v$ denotes a vector of tensorial components. The obtained set of FE equations has the structure given by eq. (11.27), where

$$\mathbf{K} \doteq \int_B \mathbf{B}^T\mathbb{C}\,\mathbf{B}\,\mathrm{d}V, \qquad \mathbf{L} \doteq \int_B \mathbf{B}^T\mathbb{C}\,\mathbf{G}\,\mathrm{d}V, \qquad \mathbf{K}_{qq} \doteq \int_B \mathbf{G}^T\mathbb{C}\,\mathbf{G}\,\mathrm{d}V, \tag{11.44}$$

$$\mathbf{r}_u = -\mathbf{p}, \qquad \mathbf{r}_q = \mathbf{0}.$$

The tangent operators are defined as

$$\mathbf{B} \doteq \partial\boldsymbol{\varepsilon}_v^c/\partial\mathbf{u}_I, \qquad \mathbf{G} \doteq \partial\boldsymbol{\varepsilon}_v^{\mathrm{inc}}/\partial\mathbf{q}, \tag{11.45}$$

where $\mathbf{B}$ is for compatible strains, while $\mathbf{G}$ for incompatible strains. Besides, $\mathbf{p}$ is the vector of external loads.

The stress for the linear material is obtained in the ID method as follows:

$$\boldsymbol{\sigma}_v^{\mathrm{enh}} \doteq \partial\mathcal{W}(\mathbf{u}^{\mathrm{enh}})/\partial\boldsymbol{\varepsilon}_v^{\mathrm{enh}} = \mathbb{C}\,(\boldsymbol{\varepsilon}_v^c + \boldsymbol{\varepsilon}_v^{\mathrm{inc}}). \tag{11.46}$$

Sufficient condition to pass the patch test. The incompatible modes are quadratic functions and yield the incompatible strains which are linear, see eq. (11.39). Hence, these modes should not be activated in the patch test, in which the strains are constant. This leads to the requirement that the formulation of the ID element should yield $\mathbf{q} = \mathbf{0}$ in this test and, generally, for any nodal displacements $\mathbf{u}_I$ generating constant strains.

It can be shown that to obtain $\mathbf{q} = \mathbf{0}$ in the constant strain patch test, it suffices to satisfy the condition

$$\int_B \mathbf{G}\,\mathrm{d}V = \mathbf{0}, \tag{11.47}$$

by the reasoning of [234], which we outline below. From the second of eq. (11.27), we can calculate

$$\mathbf{q} = -\mathbf{K}_{qq}^{-1}\, \mathbf{L}^T\, \mathbf{u}_I. \tag{11.48}$$

To obtain $\mathbf{q} = \mathbf{0}$, we may require the condition $\mathbf{L}^T \mathbf{u}_I = \mathbf{0}$ to be satisfied. By the definition of $\mathbf{L}$ of eq. (11.44),

$$\mathbf{L}^T \mathbf{u}_I = \int_B \mathbf{G}^T (\mathbb{C}\mathbf{B}\mathbf{u}_I)\, \mathrm{d}V = \int_B \mathbf{G}^T \boldsymbol{\sigma}\, \mathrm{d}V, \tag{11.49}$$

where $\boldsymbol{\sigma} \doteq \mathbb{C}\,(\mathbf{B}\mathbf{u}_I)$ is the stress, as $\mathbf{B}\mathbf{u}_I$ is the strain in a kinematically linear problem. In the patch test, the nodal displacements yield a constant strain and, hence, for a constant $\mathbb{C}$, also the stress $\boldsymbol{\sigma}$ is constant and can be taken away from under the integral, which yields the condition of eq. (11.47). This condition is enough to yield $\mathbf{L}^T \mathbf{u}_I = \mathbf{0}$, $\mathbf{q} = -\mathbf{K}_{qq}^{-1}\mathbf{L}^T \mathbf{u}_I = \mathbf{0}$, and to pass the constant strain patch test.

The original version of the ID method of [248], based on eq. (11.36), did not pass the patch test for elements of distorted initial geometry (non-parallelograms) and was subsequently corrected in [234] by using the Jacobian matrix at the element's center and the modified Jacobian inverse, see eq. (11.50). In the modified version of the ID method of eq. (11.41), only the second of these corrections is necessary because the $\mathbf{J}_c$ is present as a natural consequence of the use of $\{\mathbf{g}_k^c\}$.

Modification of the Jacobian inverse. The Jacobian inverse $\mathbf{J}^{-1}$ varies over the element area. To eliminate the dependence of it on ξ, η, we may define

$$(\mathbf{J}^{-1})^* \doteq \mathbf{J}_c^{-1} \left(\frac{j_c}{j} \right), \tag{11.50}$$

where $j \doteq \det \mathbf{J}$, and the subscript c indicates the value at the element's center. The 2×2-point Gauss integration of $(\mathbf{J}^{-1})^*$ yields

$$\int_A (\mathbf{J}^{-1})^*\, \mathrm{d}A = \sum_{g=1}^{4} (\mathbf{J}^{-1})_g^*\, j_g = 4\, \mathbf{J}_c^{-1}\, j_c, \tag{11.51}$$

which is exactly the result of the 1-point integration of $\mathbf{J}^{-1}$. Here, g is the index of integration points.

Displacement gradient for incompatible displacements. The displacement gradient can be split into a compatible and incompatible part,

$$\nabla \mathbf{u} = \left[\frac{\partial(\mathbf{u}^c + \mathbf{u}^{\text{inc}})}{\partial \xi} \right] \mathbf{J}^{-1} = \left[\frac{\partial \mathbf{u}^c}{\partial \xi} \right] \mathbf{J}^{-1} + \underline{\left[\frac{\partial \mathbf{u}^{\text{inc}}}{\partial \xi} \right] \mathbf{J}^{-1}}. \tag{11.52}$$

The incompatible (underlined) part is evaluated at the Gauss point g by using the incompatible displacements of eq. (11.41) and the modified inverse Jacobian of eq. (11.50),

$$\nabla \mathbf{u}_g^{\text{inc}} = \mathbf{J}_c \begin{bmatrix} \frac{\partial u^{\text{inc}}}{\partial \xi} & \frac{\partial u^{\text{inc}}}{\partial \eta} \\ \frac{\partial v^{\text{inc}}}{\partial \xi} & \frac{\partial v^{\text{inc}}}{\partial \eta} \end{bmatrix}_g \mathbf{J}_c^{-1} \left(\frac{j_c}{j_g} \right), \tag{11.53}$$

where

$$\begin{bmatrix} \frac{\partial u^{\text{inc}}}{\partial \xi} & \frac{\partial u^{\text{inc}}}{\partial \eta} \\ \frac{\partial v^{\text{inc}}}{\partial \xi} & \frac{\partial v^{\text{inc}}}{\partial \eta} \end{bmatrix} = -2 \begin{bmatrix} q_1 \xi & q_3 \eta \\ q_2 \xi & q_4 \eta \end{bmatrix}.$$

The deformation gradient, $\mathbf{F} = \mathbf{I} + \nabla \mathbf{u}$, is the sum of the compatible deformation gradient $\mathbf{F}_g^c$ and the incompatible displacement gradient

$$\mathbf{F}_g = \mathbf{F}_g^c + \nabla \mathbf{u}_g^{\text{inc}}. \tag{11.54}$$

Recall that in the back-rotated $\mathbf{C}_*$ of eq. (10.72) and $(\mathbf{Q}_0^T \mathbf{F})_*$ of eq. (10.73), we use the product $\bar{\mathbf{F}} \doteq \mathbf{F}\mathbf{R}_0$ of eq. (10.74), which is now calculated as follows:

$$\bar{\mathbf{F}}_g \doteq \mathbf{F}_g \mathbf{R}_{0g} = \mathbf{F}_g^c \mathbf{R}_{0g} + \nabla \mathbf{u}_g^{\text{inc}} \mathbf{R}_{0g}. \tag{11.55}$$

In the above equation, we still use $\mathbf{R}_{0g}$ at the Gauss point because, as we have verified, if it is replaced by $\mathbf{R}_{0c}$ for the element's center, then the patch test is not satisfied. The compatible term $\mathbf{F}_g^c \mathbf{R}_{0k}$ can be transformed as shown in eq. (10.74).

The functional F_{PE} depends on two sets of variables: the nodal displacements $\mathbf{u}_I$ and the elemental multipliers of incompatible modes $\mathbf{q}$. The obtained set of FE equations is given by eq. (11.27) and to update the stress and strain multipliers, the scheme U2 of eq. (11.34) should be used.

The finite element for the incompatible displacement gradient of eq. (11.53) is designated as ID4. It is invariant, has a correct rank, and passes the patch test. Its accuracy and robustness is much better than that of the Q4 element; in linear tests it performs identically as the EAS4 and EADG4 elements.

Note that we can also use only two modes, q_2 and q_3, and enhance only the shear strain, see eq. (11.39). Such an element (designated as ID2) is particularly useful for shells, for which it performs in a very stable way in nonlinear tests.

11.4.2 EAS4 element

Introduction. The Enhanced Assumed Strain (EAS) method was introduced in [216] and it embodies the following modifications of the ID method:

1. Not displacements but strains are enhanced. The enhancing modes are directly introduced on the level of strains without resorting to displacements.
2. The HW functional is used instead of the PE functional. This change strengthens the variational background and shows the importance of orthogonality of the assumed strain to stress. The crucial result pertaining to the ID method, see eq. (11.47), is fully adopted.

Within the EAS method, the strain for the compatible displacements $\mathbf{E}^c \doteq \mathbf{E}(\mathbf{u}^c)$ is enhanced additively by the strain $\boldsymbol{\varepsilon}^{\text{enh}}$ as follows

$$\underbrace{\boldsymbol{\varepsilon}(\xi,\eta)}_{\text{enhanced}} \doteq \underbrace{\mathbf{E}^c(\xi,\eta)}_{\text{compatible}} + \underbrace{\boldsymbol{\varepsilon}^{\text{enh}}(\xi,\eta)}_{\text{enhancing}}. \tag{11.56}$$

Variational basis of the EAS method. We take the three-field HW functional of eq. (11.1), and use eq. (11.56), which yields

$$F(\mathbf{u},\boldsymbol{\sigma},\boldsymbol{\varepsilon}^{\text{enh}}) = \int_B \left[\mathcal{W}(\mathbf{E}^c + \boldsymbol{\varepsilon}^{\text{enh}}) - \boldsymbol{\sigma}\cdot\boldsymbol{\varepsilon}^{\text{enh}} \right] \mathrm{d}V \;-\; F_{\text{ext}}. \tag{11.57}$$

We wish to eliminate the stress $\boldsymbol{\sigma}$ from this functional, thus we require the enhancing strain to be orthogonal to the stress, i.e.

$$\int_B \boldsymbol{\sigma}\cdot\boldsymbol{\varepsilon}^{\text{enh}}\,\mathrm{d}V = 0, \tag{11.58}$$

for which the term with $\boldsymbol{\sigma}$ in eq. (11.57) vanishes and F becomes the two-field potential energy functional of eq. (11.8) in the following form:

$$F_{\mathrm{PE}}(\mathbf{u}, \boldsymbol{\varepsilon}^{\mathrm{enh}}) = \int_B \mathcal{W}(\mathbf{E}^c + \boldsymbol{\varepsilon}^{\mathrm{enh}})\, \mathrm{d}V \; - \; F_{\mathrm{ext}}. \tag{11.59}$$

We note that the orthogonality condition plays an important role in the above derivation, because (1) it allows us to reduce the number of independent fields, (2) it establishes the relation with the elements explicitly using the assumed stress, and (3) it defines the admissible strain enhancements, as such for which the integral (11.58) vanishes for the assumed stress.

Kinematically linear problems. For small strains, i.e. when $\boldsymbol{\varepsilon}_v^c$ is a linear function of $\mathbf{u}^c$, we proceed as follows. A quadratic Taylor's expansion of the strain energy at some $\boldsymbol{\varepsilon}_v = \boldsymbol{\varepsilon}_v^+$ is as follows

$$\mathcal{W}(\boldsymbol{\varepsilon}_v) \approx \mathcal{W}^+ + \boldsymbol{\sigma}_v^+ \cdot \Delta\boldsymbol{\varepsilon}_v + \frac{1}{2} \Delta\boldsymbol{\varepsilon}_v^T\, \mathbb{C}^+\, \Delta\boldsymbol{\varepsilon}_v, \tag{11.60}$$

where the symbols with "+" are evaluated at $\boldsymbol{\varepsilon}_v^+$, and $(\cdot)_v$ denotes a vector of tensorial components. Besides, $\boldsymbol{\sigma}_v \doteq \partial\mathcal{W}/\partial\boldsymbol{\varepsilon}_v$ is the stress and $\mathbb{C}_{vv} \doteq \partial^2\mathcal{W}/\partial\boldsymbol{\varepsilon}_v^2$ is the constitutive matrix. For kinematically linear problems, we have $\boldsymbol{\varepsilon}_v^+ = \mathbf{0}$, $\mathcal{W}^+ = 0$, $\boldsymbol{\sigma}_v^+ = \mathbf{0}$, $\Delta\boldsymbol{\varepsilon}_v = \boldsymbol{\varepsilon}_v$, for which we obtain $\mathcal{W}(\boldsymbol{\varepsilon}_v) = \frac{1}{2}\, \boldsymbol{\varepsilon}_v \cdot (\mathbb{C}_{vv}\, \boldsymbol{\varepsilon}_v)$. Hence, the strain energy of eq. (11.59) becomes

$$\mathcal{W}(\boldsymbol{\varepsilon}_v^c + \boldsymbol{\varepsilon}_v^{\mathrm{enh}}) = \tfrac{1}{2}\, (\boldsymbol{\varepsilon}_v^c + \boldsymbol{\varepsilon}_v^{\mathrm{enh}})^T\, \mathbb{C}_{vv}\, (\boldsymbol{\varepsilon}_v^c + \boldsymbol{\varepsilon}_v^{\mathrm{enh}}), \tag{11.61}$$

which can be compared with eq. (11.43) for the ID method. We see that $\boldsymbol{\varepsilon}_v^{\mathrm{enh}}$ plays an analogous role as $\boldsymbol{\varepsilon}_v^{\mathrm{inc}}$ in the ID method.

Enhancing strain. The enhancing strain is constructed as follows:

$$\boldsymbol{\varepsilon}^{\mathrm{enh}} = \mathbf{J}_c^{-T}\, \boldsymbol{\varepsilon}_\xi\, \mathbf{J}_c^{-1}, \tag{11.62}$$

which is the transformation rule for covariant components $\boldsymbol{\varepsilon}_\xi$ of a second-rank tensor, from the natural basis at the element's center $\{\mathbf{g}_k^c\}$ to the reference Cartesian basis. We note that the modification of [234], where the Jacobian matrix at the element's center is used to enable passing the patch test by the ID element, is naturally present in eq. (11.62), as a consequence of the use of the basis at element's center. At the Gauss integration point g, we write

$$\boldsymbol{\varepsilon}_g^{\mathrm{enh}} = \mathbf{J}_c^{-T}\; \boldsymbol{\varepsilon}_{\xi g}\; \mathbf{J}_c^{-1} \left(\frac{j_c}{j_g}\right), \qquad \boldsymbol{\varepsilon}_\xi \doteq \begin{bmatrix} q_1\, \xi & q_3\, \xi + q_4\, \eta \\ q_3\, \xi + q_4\, \eta & q_2\, \eta \end{bmatrix}, \tag{11.63}$$

where ε_ξ is a matrix of the assumed strain, and $j \doteq \det \mathbf{J}$. Note that (j_c/j_g) is added, and that the 2×2-point Gauss integration of it yields $4j_c$, which is the result of the 1-point integration of j. This modification can be compared with that of eq. (11.50) for $\mathbf{J}_c^{-1}$. The matrix ε_ξ involves four parameters q_i, and two modes $\{\xi, \eta\}$.

The discrete F_{PE} functional depends on two sets of variables: the nodal displacements $\mathbf{u}_I$ and the elemental multipliers $\mathbf{q}$ of the assumed strain modes. The obtained set of FE equations is given by eq. (11.27) and to update the stress multipliers, the scheme U2 of eq. (11.34) should be used.

The finite element for the assumed strains of eq. (11.62) is designated as EAS4 and, currently, it is a standard in the class of four-node EAS elements. It is invariant, has the correct rank, and passes the patch test. Its accuracy and robustness is much better than that of the Q4 element.

Remark 1. Other representations. Several other forms of ε_ξ were tested in the literature. The representation with seven parameters (EAS7), obtained from EAS4 by adding the bilinear term $\xi\eta$ to each component, also gained some popularity, but it turned out that it does not satisfy the compatibility condition. The same is true about the five-parameter representation (EAS5), obtained from EAS4 by adding the bilinear term $\xi\eta$ to the shear component only. The EAS2 representation, which uses two parameters for the shear strain enhancement, is particularly stable in non-linear shell applications, but the response is slightly stiffer, which renders that more elements must be used.

Remark 2. Enhancement of Cauchy–Green tensor. In eq. (11.15), the deformation gradient for compatible displacements is written down as $\mathbf{F}^c = \mathbf{F}_\xi \mathbf{J}^{-1}$, for which the Cauchy–Green tensor becomes $\mathbf{C}^c \doteq (\mathbf{F}^c)^T\mathbf{F}^c = \mathbf{J}^{-T}(\mathbf{F}_\xi^T\mathbf{F}_\xi)\,\mathbf{J}^{-1}$ and involves the transformation $\mathbf{J}^{-T}(\,\cdot\,)\,\mathbf{J}^{-1}$. The same transformation, but with $\mathbf{J}$ replaced by $\mathbf{J}_c$, is used in eq. (11.62). Hence, when we use the Green strain, we can interpret the EAS method as the enhancement of the Cauchy–Green tensor.

Analytical verification of orthogonality condition for constant stress. Assume that the stress $\boldsymbol{\sigma}$ is constant over the element domain. Then, in eq. (11.64), $\int_B \boldsymbol{\sigma} \cdot \boldsymbol{\varepsilon}^{\text{enh}}\, \mathrm{d}V = \boldsymbol{\sigma} \cdot \int_B \boldsymbol{\varepsilon}^{\text{enh}}\, \mathrm{d}V$, and the orthogonality condition is reduced to

$$\int_B \varepsilon^{\mathrm{enh}} \, dV = 0, \tag{11.64}$$

which is analogous to eq. (11.47) for the ID method and suffices to pass the patch test. On use of the 2 × 2-point Gauss integration, the integral of the enhancing strain becomes

$$\int_B \varepsilon^{\mathrm{enh}} \, dV = \sum_{g=1}^{4} \mathbf{J}_c^{-T} \, \varepsilon_{\xi g} \, \mathbf{J}_c^{-1} \left(\frac{j_c}{j_g} \right) j_g = \mathbf{J}_c^{-T} \left(\sum_{g=1}^{4} \varepsilon_{\xi g} \right) \mathbf{J}_c^{-1} j_c, \tag{11.65}$$

where eq. (11.62) was used and g is a Gauss point. To satisfy eq. (11.64), it suffices that

$$\sum_{g=1}^{4} \varepsilon_{\xi g} = \mathbf{0}. \tag{11.66}$$

We can check that this condition is satisfied for the EAS4 element, because

$$\sum_{g=1}^{4} \begin{bmatrix} q_1 \, \xi_g & q_3 \, \xi_g + q_4 \, \eta_g \\ q_3 \, \xi_g + q_4 \, \eta_g & q_2 \, \eta_g \end{bmatrix} = \mathbf{0}, \tag{11.67}$$

for $\xi_g, \eta_g = \pm 1/\sqrt{3}$. This element passes the patch test of Sect. 15.2.3.

Verification of orthogonality condition for non-constant stress. The orthogonality condition is checked for the non-constant five- and seven-parameter stress representations in [256], Appendix B. The stress is assumed as $\boldsymbol{\sigma} = \mathbf{J}_c \, \boldsymbol{\sigma}^{\xi} \, \mathbf{J}_c^T$, which is the transformation rule of the contra-variant tensor components from the $\{\mathbf{g}_k^c\}$ basis at the element center to the reference Cartesian basis. The enhancing strain $\varepsilon^{\mathrm{enh}}$ is taken in the form given by eq. (11.62). Then, the orthogonality condition becomes

$$\int_B \boldsymbol{\sigma} \cdot \varepsilon^{\mathrm{enh}} \, dV = h \int_{-1}^{+1} \int_{-1}^{+1} \mathrm{tr}[(\mathbf{J}_c \, \boldsymbol{\sigma}^{\xi} \, \mathbf{J}_c^{\mathrm{T}})^{\mathrm{T}} (\mathbf{J}_c^{-\mathrm{T}} \, \varepsilon_{\xi} \, \mathbf{J}_c^{-1})] \, j \, d\xi \, d\eta. \tag{11.68}$$

Evaluating this integral for various forms of $\boldsymbol{\sigma}^{\xi}$ and ε_{ξ}, we can test the orthogonality of the involved fields. As $\boldsymbol{\sigma}^{\xi}$, we take the five-parameter stress of eq. (11.125), or the seven-parameter stress of eq. (12.96), and we use ε_{ξ} of eq. (11.62), both assumed either in the natural coordinates $\{\xi, \eta\}$ or in the skew coordinates of eq. (11.81).

We have verified, using a symbolic manipulator, that the orthogonality condition is not satisfied for these representations for irregular elements but is satisfied for parallelograms. Hence, for irregular elements, the PE functional (11.59) is not fully equivalent to the HW functional (11.57), but only approximates it.

Verification of compatibility of enhancing strains. The compatibility condition for 2D strains is as follows,

$$\frac{\partial^2 \varepsilon_{xx}}{\partial y^2} + \frac{\partial^2 \varepsilon_{yy}}{\partial x^2} = 2\frac{\partial^2 \varepsilon_{xy}}{\partial x\, \partial y}. \tag{11.69}$$

It was evaluated for the following specifications of the assumed strain and its derivatives:

1. The enhancing strain in the reference basis is obtained from the assumed representation $\boldsymbol{\varepsilon}_\xi$ using eq. (11.62).
2. The first and the second derivatives w.r.t. x, y are expressed by the derivatives w.r.t. ξ, η as specified in eqs. (11.132) and (11.133).

This condition is satisfied in the case of the EAS4 and EAS2 enhancement, only for parallelograms. Because the strain enhancement is added to the compatible strain, eq. (11.56), the total strain has the same property.

The compatibility condition is not satisfied by the EAS5 and EAS7 representations, even for parallelograms, which is caused by the term $\xi\eta$ in ε_{12}. The use of them is therefore not advisable.

Couplings of $\mathbf{u}_I$ and $\mathbf{q}$ in matrix $\mathbf{K}$. For kinematically nonlinear problems, the tangent matrix $\mathbf{K}$ of eq. (11.28) can be a function of multipliers $\mathbf{q}$. This is a consequence of couplings of the compatible strain $\boldsymbol{\varepsilon}^c$ and the enhancing strain $\boldsymbol{\varepsilon}^{\text{inc}}$ in the strain energy.

Consider the SVK strain energy, $\mathcal{W}(\boldsymbol{\varepsilon}) \doteq \frac{1}{2}\lambda\,(\text{tr}\boldsymbol{\varepsilon})^2 + \mu\,\text{tr}\boldsymbol{\varepsilon}^2$, where λ and μ are Lamé constants. For $\boldsymbol{\varepsilon} = \boldsymbol{\varepsilon}^c + \boldsymbol{\varepsilon}^{\text{enh}}$, we obtain

$$\begin{aligned}
&\text{tr}\boldsymbol{\varepsilon} = \text{tr}\boldsymbol{\varepsilon}^c + \text{tr}\boldsymbol{\varepsilon}^{\text{enh}},\\
&(\text{tr}\boldsymbol{\varepsilon})^2 = (\text{tr}\boldsymbol{\varepsilon}^c)^2 + \underline{2(\text{tr}\boldsymbol{\varepsilon}^c)(\text{tr}\boldsymbol{\varepsilon}^{\text{enh}})} + (\text{tr}\boldsymbol{\varepsilon}^{\text{enh}})^2,\\
&\boldsymbol{\varepsilon}^2 = (\boldsymbol{\varepsilon}^c)^2 + (\boldsymbol{\varepsilon}^c\boldsymbol{\varepsilon}^{\text{enh}} + \boldsymbol{\varepsilon}^{\text{enh}}\boldsymbol{\varepsilon}^c) + (\boldsymbol{\varepsilon}^{\text{enh}})^2,\\
&\text{tr}(\boldsymbol{\varepsilon})^2 = \text{tr}(\boldsymbol{\varepsilon}^c)^2 + \underline{2\text{tr}(\boldsymbol{\varepsilon}^c\boldsymbol{\varepsilon}^{\text{enh}})} + \text{tr}(\boldsymbol{\varepsilon}^{\text{enh}})^2.
\end{aligned}$$

Hence, $\mathcal{W}(\boldsymbol{\varepsilon}) \neq \mathcal{W}(\boldsymbol{\varepsilon}^c) + \mathcal{W}(\boldsymbol{\varepsilon}^{\text{enh}})$, i.e. the contribution of $\boldsymbol{\varepsilon}^c$ and $\boldsymbol{\varepsilon}^{\text{enh}}$ to the strain energy is not additive, due to the coupling (underlined) terms. Due to these terms, the tangent matrix $\mathbf{K} \doteq \partial^2\mathcal{W}/\partial\mathbf{u}_I\partial\mathbf{u}_J$ can be a function of multipliers $\mathbf{q}$ and this depends on the type of strain used.

a) The compatible strain $\boldsymbol{\varepsilon}^c$ is a linear function of $\mathbf{u}_I$ for (i) a kinematically linear problem when $\boldsymbol{\varepsilon} \doteq \frac{1}{2}(\nabla\mathbf{u}+\nabla^T\mathbf{u})$ and (ii) for a nonlinear problem when we use the right stretch strain $\mathbf{H} \doteq \mathrm{sym}[\mathbf{Q}^T(\mathbf{I}+\nabla\mathbf{u})]$. Then the coupling terms do not affect $\mathbf{K}$.
b) The compatible strain $\boldsymbol{\varepsilon}^c$ is a quadratic function of $\mathbf{u}_I$ for the Green strain $\mathbf{E} \doteq \frac{1}{2}(\nabla\mathbf{u}+\nabla^T\mathbf{u}+\nabla^T\mathbf{u}\nabla\mathbf{u})$. Then $\mathbf{K}$ for the coupling terms is non-zero and depends on $\mathbf{q}$.

Summarizing, we obtain additional couplings of $\mathbf{u}_I$ and $\mathbf{q}$ in matrix $\mathbf{K}$ for the Green strain but neither for the infinitesimal strain nor for the right stretch strain.

11.4.3 EADG4 element

The method of Enhanced Assumed Displacement Gradient (EADG) was proposed in [208] and, in fact, its basic concept is deeper rooted in the ID method than the concept of the EAS method which was published two years earlier.

Within the EADG method, the gradient of compatible displacements $\mathbf{u}^c$ is additively enhanced by the enhancing matrix $\tilde{\mathbf{H}}$ as follows:

$$\underbrace{\mathbf{F}(\xi,\eta)}_{\text{enhanced}} \doteq \mathbf{I} + \underbrace{\nabla\mathbf{u}^c(\xi,\eta)}_{\text{compatible}} + \underbrace{\tilde{\mathbf{H}}(\xi,\eta)}_{\text{enhancing}} . \tag{11.70}$$

Construction of $\tilde{\mathbf{H}}$. In eq. (11.53) for the ID method, the incompatible displacements were differentiated to calculate the matrix

$$\begin{bmatrix} \frac{\partial u^{\mathrm{inc}}}{\partial \xi} & \frac{\partial u^{\mathrm{inc}}}{\partial \eta} \\ \frac{\partial v^{\mathrm{inc}}}{\partial \xi} & \frac{\partial v^{\mathrm{inc}}}{\partial \eta} \end{bmatrix} = -2 \begin{bmatrix} q_1\xi & q_3\eta \\ q_2\xi & q_4\eta \end{bmatrix}.$$

In the EADG method, we directly assume the form of this matrix, designated as $\mathbf{G}^\xi$, without resorting to the concept of incompatible displacements and without differentiation. Equation (11.53) of the ID method is rewritten for the EADG method as follows:

$$\tilde{\mathbf{H}}_g \doteq \mathbf{J}_c\, \mathbf{G}^\xi_g\, \mathbf{J}_c^{-1} \left(\frac{j_c}{j_g}\right), \qquad \mathbf{G}^\xi \doteq \begin{bmatrix} q_1\xi & q_3\eta \\ q_2\xi & q_4\eta \end{bmatrix}, \tag{11.71}$$

where the factor (-2) was omitted in $\mathbf{G}^\xi$ and g is a Gauss point. Other representations can also be used in $\mathbf{G}^\xi$ so the EADG and EAS methods are equally versatile.

Variational basis of the EADG method. The EADG method is based on the three-field HW functional, although involving not strains but the deformation gradient

$$F(\mathbf{u},\mathbf{F},\mathbf{P}) \doteq \int_B \{ \mathcal{W}(\mathbf{F}^T\mathbf{F}) + \mathbf{P}\cdot[(\mathbf{I}+\nabla\mathbf{u})-\mathbf{F}] \}\, \mathrm{d}V - F_{\mathrm{ext}}, \quad (11.72)$$

where $\mathbf{P}$ is the nominal stress, $\mathbf{F}$ is an independent field, and F_{ext} is a potential of the body force, the external loads, and the displacement boundary conditions. Note that $\mathbf{P}$ serves as a Lagrange multiplier for the relation $(\mathbf{I}+\nabla\mathbf{u})-\mathbf{F}$.

Using eq. (11.70), we obtain

$$F(\mathbf{u},\tilde{\mathbf{H}},\mathbf{P}) = \int_B \left\{ \mathcal{W}[(\mathbf{I}+\nabla\mathbf{u}+\tilde{\mathbf{H}})^T(\mathbf{I}+\nabla\mathbf{u}+\tilde{\mathbf{H}})] - \mathbf{P}\cdot\tilde{\mathbf{H}} \right\} \mathrm{d}V - F_{\mathrm{ext}}, \quad (11.73)$$

in which we do not have $\mathbf{F}$ but the enhancing $\tilde{\mathbf{H}}$. If the assumed $\tilde{\mathbf{H}}$ is orthogonal to the stress, i.e. $\int_B \mathbf{P}\cdot\tilde{\mathbf{H}}\, \mathrm{d}V = 0$, then the last term of the above functional vanishes and we obtain a two-field enhanced PE functional

$$F_{\mathrm{PE}}(\mathbf{u},\tilde{\mathbf{H}}) = \int_B \mathcal{W}[(\mathbf{I}+\nabla\mathbf{u}+\tilde{\mathbf{H}})^T(\mathbf{I}+\nabla\mathbf{u}+\tilde{\mathbf{H}})]\, \mathrm{d}V - F_{\mathrm{ext}}, \quad (11.74)$$

which does not depend on the stress $\mathbf{P}$.

The discrete F_{PE} functional depends on two types of variables: the nodal displacements $\mathbf{u}_I$ and the elemental multipliers $\mathbf{q}$ of assumed displacement gradient modes. The obtained set of FE equations is given by eq. (11.27), and the scheme U2 of eq. (11.34) should be used to update the stress multipliers.

The finite element for the representation of eq. (11.71) is designated as EADG4 and, currently, it is a standard in the class of four-node EADG elements. It is invariant, has a correct rank, and passes the patch test. Its accuracy and robustness are much better than those of the Q4 element. In linear tests, it performs identically as ID4 and EAS4 elements, but is superior to them in the case of elements with a drilling rotation, see Sect. 12.

Remark 1. Relation to EAS method. The EADG and ID method use the $\mathbf{J}_c(\cdot)\,\mathbf{J}_c^{-1}$ transformation, see eqs. (11.53) and (11.71), while the EAS method is based on the $\mathbf{J}_c^{-T}(\cdot)\,\mathbf{J}_c^{-1}$ transformation, see eq. (11.62).

These transformations are identical only when $\mathbf{J}_c = \mathbf{J}_c^{-T}$, e.g. when $\mathbf{J}_c \in \mathrm{SO}(3)$. Hence, in general, these methods are different although based on the same concept and perform similarly in some tests. The variational foundations of the assumed strain methods are revised in [215].

Remark 2. Spatial formulation. Using the deformation function, $\chi : \mathbf{x} = \chi(\mathbf{y})$, the approximation of eq. (11.70) can be rewritten as $\mathbf{F} = \nabla\chi + \tilde{\mathbf{H}}$. Defining the spatial enhanced displacement gradient $\tilde{\mathbf{h}} \doteq \tilde{\mathbf{H}}\, \nabla\chi^{-1}$, we obtain $\mathbf{F} = (\mathbf{I} + \tilde{\mathbf{h}})\, \nabla\chi$, in which the enhanced deformation gradient $(\mathbf{I} + \tilde{\mathbf{h}})$ is superposed multiplicatively on the standard deformation gradient $\nabla\chi$. This form of $\mathbf{F}$ and the variational problem in the spatial setting, i.e. $\mathbf{P} \cdot \delta\mathbf{F} = \boldsymbol{\tau} \cdot [\nabla(\delta\mathbf{u})\mathbf{F}^{-1}]$, where $\boldsymbol{\tau}$ is the Kirchhoff stress, are used in [208].

Finally, we note that some enhanced strain elements can experience problems in the range of large compressive strains. This problem was detected in [63] and studied in [264], where a single square element was compressed by two equal forces and the solution was obtained for the compressible neo-Hookean material. At the first zero eigenvalue, the non-symmetric bifurcation point was obtained. This test can also be performed for a block of elements, as in [263] where the eigenvector at the bifurcation point is checked for the presence of hourglassing. This topic is also addressed in [154].

11.5 Mixed Hellinger–Reissner and Hu–Washizu elements

Definition of mixed formulations. To improve the performance of early elements, several non-standard formulations were tested, including the mixed formulation in [168], and the hybrid mixed formulation in [121]. A lot of work has been done since these pioneering papers to improve mixed methods; the elements and their theoretical foundations.

Various definitions exist of the *mixed* formulation in the literature; we adopt the one referring to the features of the governing functional:

1. the governing functional must depend on several types of variables,
2. some of the variables must be Lagrange multipliers. Hence, the governing functional attains a saddle point, not a minimum, at a solution.

This definition implies that the Hellinger–Reissner (HR) functional and the Hu–Washizu (HW) functional are mixed, but the potential energy

(PE) functional is not. For shells, the formulations with rotations of Sect. 4 are mixed but the use of the Reissner hypothesis does not yield a mixed formulation, although it introduces the rotational dofs. Note that the formulation remains mixed, even if the multipliers are eliminated by a local regularization of the functional.

Compared with the standard elements, the mixed elements have the following features:

1. the inter-element continuity of certain fields is relaxed,
2. the level of non-linearity for finite strains is reduced,
3. the non-zero eigenvalues of the non-reduced tangent matrix of eq. (11.27) for mixed elements are either positive or negative because the discrete HR and HW functionals have a saddle point at $(\mathbf{u} = \mathbf{0}, \mathbf{q} = \mathbf{0})$. The number of negative eigenvalues is identical to the number of stress parameters, i.e. five in Fig. 11.4.

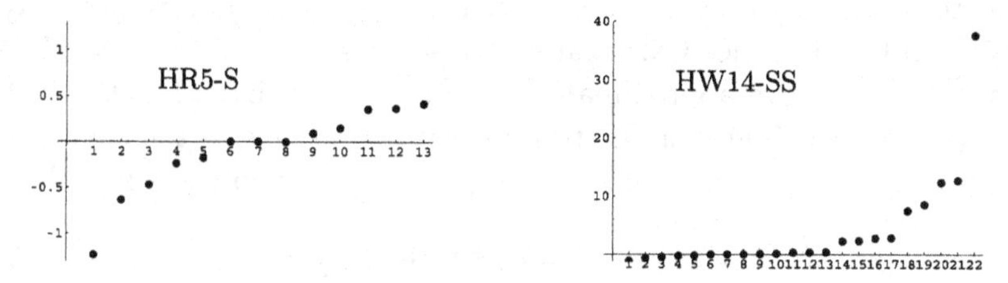

Fig. 11.4 Eigenvalues of non-reduced matrix of mixed elements.

The mixed finite elements show (i) a slightly higher accuracy of displacements and stresses for coarse distorted meshes, (ii) a better convergence rate in non-linear problems than elements based on other formulations. They can be cast in a similar form to the standard elements by eliminating the additional variables on the element level.

In this section, we describe 2D mixed elements based on the HR and the HW functionals. We also provide comments on the mixed/enhanced elements.

Skew coordinates. To define the representation of stress (and strain) in mixed elements, we use the skew coordinates instead of the natural coordinates as proposed in [256, 257]. This modification improves the accuracy of mixed elements.

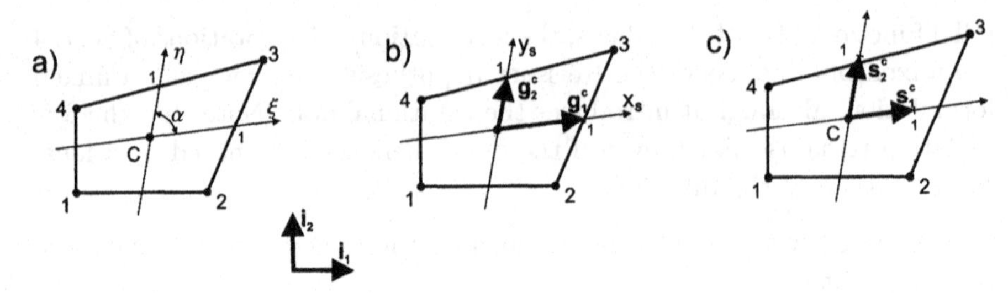

Fig. 11.5 Bases at element's center. a) Natural coordinates $\{\xi, \eta\}$. b) Natural basis $\{\mathbf{g}_k^c\}$ and skew coordinates $\{x_S, y_S\}$. c) Oblique basis $\{\mathbf{s}_k^c\}$, which is not used here!

The skew coordinates relative to the natural basis at the element's center $\{\mathbf{g}_k^c\}$ are designated by $\{x_S, y_S\}$. They can be defined in relation to the Cartesian coordinates $\{x, y\}$ associated with the reference basis $\{\mathbf{i}_k\}$ as follows:

The position vector of a particle in the initial configuration can be expressed in the reference Cartesian basis, see Fig. 11.5A, as $\mathbf{y} = x\mathbf{i}_1 + y\mathbf{i}_2$, where x, y are approximated by the bilinear shape functions of ξ, η of eq. (11.10). Consider the position vector relative to the element's center, i.e. $\bar{\mathbf{y}} = \mathbf{y} - \mathbf{y}_c$, and write it relative to these two bases as follows

$$\bar{\mathbf{y}} = \bar{x}\,\mathbf{i}_1 + \bar{y}\,\mathbf{i}_2 = x_S\,\mathbf{g}_1^c + y_S\,\mathbf{g}_2^c. \tag{11.75}$$

Taking the scalar product of this equation with the vectors $\mathbf{i}_1$ and $\mathbf{i}_2$, we obtain two equations which can be written in the following form:

$$\begin{bmatrix} \bar{x} \\ \bar{y} \end{bmatrix} = \mathbf{J}_c \begin{bmatrix} x_S \\ y_S \end{bmatrix}, \qquad \mathbf{J}_c = \begin{bmatrix} \mathbf{g}_1^c \cdot \mathbf{i}_1 & \mathbf{g}_2^c \cdot \mathbf{i}_1 \\ \mathbf{g}_1^c \cdot \mathbf{i}_2 & \mathbf{g}_2^c \cdot \mathbf{i}_2 \end{bmatrix}, \tag{11.76}$$

where $\mathbf{J}_c$ is the Jacobian of eq. (11.13) at the element's center. Then the skew coordinates are calculated as

$$\begin{bmatrix} x_S \\ y_S \end{bmatrix} = \mathbf{J}_c^{-1} \begin{bmatrix} \bar{x} \\ \bar{y} \end{bmatrix}. \tag{11.77}$$

This relation implies

$$\mathbf{J}_c^{-1} = \begin{bmatrix} \frac{\partial x_S}{\partial \bar{x}} & \frac{\partial x_S}{\partial \bar{y}} \\ \frac{\partial y_S}{\partial \bar{x}} & \frac{\partial y_S}{\partial \bar{y}} \end{bmatrix}. \tag{11.78}$$

For the position vector of eq. (11.10) rewritten as

$$\begin{bmatrix} x \\ y \end{bmatrix} = \begin{bmatrix} a_0 + a_1\xi + a_2\eta + a_3\xi\eta \\ b_0 + b_1\xi + b_2\eta + b_3\xi\eta \end{bmatrix}, \tag{11.79}$$

where the coefficients a_i, b_i are functions of the positions of nodes, we have

$$\begin{bmatrix} \bar{x} \\ \bar{y} \end{bmatrix} \doteq \begin{bmatrix} x - a_0 \\ y - b_0 \end{bmatrix} = \begin{bmatrix} a_1\xi + a_2\eta + a_3\xi\eta \\ b_1\xi + b_2\eta + b_3\xi\eta \end{bmatrix}, \tag{11.80}$$

where a_0, b_0 are coordinates of the element's center. Using this relation in eq. (11.77), the skew coordinates become the following functions of the natural coordinates:

$$\begin{bmatrix} x_S \\ y_S \end{bmatrix} = \begin{bmatrix} \xi + A\,\xi\eta \\ \eta + B\,\xi\eta \end{bmatrix}, \tag{11.81}$$

where

$$A \doteq \frac{a_3b_2 - a_2b_3}{a_1b_2 - a_2b_1}, \qquad B \doteq \frac{a_1b_3 - a_3b_1}{a_1b_2 - a_2b_1}.$$

The coefficients A and B can be expressed using the determinant of the Jacobian $\mathbf{J}$ of eq. (11.13). This Jacobian, using eq. (11.79), becomes

$$\mathbf{J} \doteq \begin{bmatrix} \mathbf{g}_1 \cdot \mathbf{i}_1 & \mathbf{g}_2 \cdot \mathbf{i}_1 \\ \mathbf{g}_1 \cdot \mathbf{i}_2 & \mathbf{g}_2 \cdot \mathbf{i}_2 \end{bmatrix} = \begin{bmatrix} a_1 + a_3\eta & a_2 + a_3\xi \\ b_1 + b_3\eta & b_2 + b_2\xi \end{bmatrix}, \tag{11.82}$$

where $\mathbf{g}_1, \mathbf{g}_2$ are defined in eq. (11.11). Note that this Jacobian is not associated with the element's center, differently from the Jacobian of eq. (11.76). We can expand the determinant of this Jacobian as follows:

$$\det \mathbf{J} = j_c + (j_{,\xi})_c\,\xi + (j_{,\eta})_c\,\eta, \tag{11.83}$$

where $j_c = a_1b_2 - a_2b_1$, $(j_{,\xi})_c = a_1b_3 - a_3b_1$, and $(j_{,\eta})_c = a_3b_2 - a_2b_3$. Hence, an alternative form of the coefficients is

$$A = \frac{(j_{,\eta})_c}{j_c}, \qquad B = \frac{(j_{,\xi})_c}{j_c}. \tag{11.84}$$

For the elements of a parallelogram shape, $(j_{,\xi})_c = (j_{,\eta})_c = 0$, so $A = B = 0$ and, by eq. (11.81), the skew coordinates $\{x_S, y_S\}$ are equal to the natural coordinates $\{\xi, \eta\}$.

Remark 1. It is a common error that the natural coordinates are treated as being associated with the natural basis at the element's center. To prove that it is incorrect, it suffices to define the position vector not as $\bar{\mathbf{y}} = x_S\,\mathbf{g}_1^c + y_S\,\mathbf{g}_2^c$, which is the correct form, but using the natural coordinates,

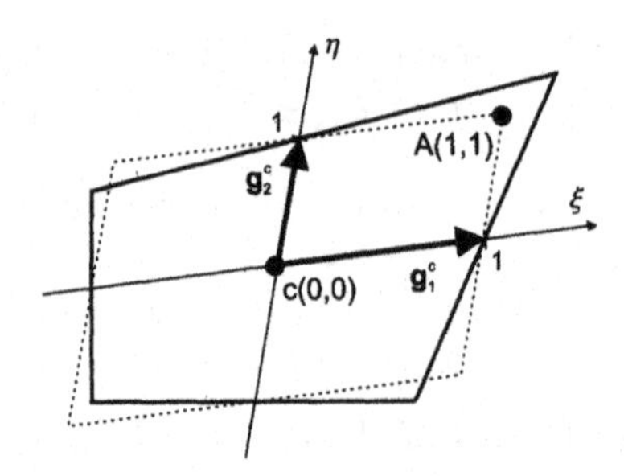

Fig. 11.6 "Fictitious" parallelogram yielded by the use of natural coordinates.

i.e. as $\bar{\mathbf{y}} = \xi\,\mathbf{g}_1^c + \eta\,\mathbf{g}_2^c$. For the latter form and $\xi, \eta \in [-1, +1]$, we obtain a "fictitious" parallelogram element as shown in Fig. 11.6 by the dotted line. The difference between the two forms of $\bar{\mathbf{y}}$ vanishes for parallelograms, as then the skew and natural coordinates are identical.

Remark 2. In literature, the idea to replace a trapezoidal element by an "equivalent" parallelogram element, identical with the "fictitious" parallelogram of Fig. 11.6, is put forward. Note, however, that the "equivalent" element does not pass the patch test!

Remark 3. Note that we can also define another basis at the element's center, the so-called oblique basis, as follows:

$$\mathbf{s}_1^c \doteq \frac{\mathbf{g}_1^c}{\|\mathbf{g}_1^c\|}, \qquad \mathbf{s}_2^c \doteq \frac{\mathbf{g}_2^c}{\|\mathbf{g}_2^c\|}, \tag{11.85}$$

where $\mathbf{s}_k^c$ are unit vectors, co-linear with the natural basis, see Fig. 11.5C. The oblique basis and the corresponding oblique coordinates are described in [151] where they are used to skew membranes and plates. They were also applied in several elements, e.g., in [184]. The advantage of using the oblique stresses is that the bi-harmonic equation retains a simple form. Therefore, the Airy stress function can be easily found and the homogenous equilibrium equation and the strain compatibility equation are satisfied. The disadvantage is that the oblique stresses are different from the real stresses, for which the constitutive equation is written, see [151], p. 25. The oblique basis and the associated coordinates are not used in our work.

***Inf-sup* (LBB) condition.** In mixed formulations, the full (non-reduced) tangent matrix of eq. (11.27) is not positive definite, which can cause problems with the well-posedness of the equations, i.e. with solvability and stability. The requirement to safely solve the system of equations is called the *inf-sup* condition and there is a vast mathematical literature related to it, see e.g. [11, 12, 38, 37, 193]. This condition depends on the FE discretization and, hence, analytical expressions are difficult and beyond the scope of this work.

On the other hand, we can much easier check a numerical counterpart of this condition, called the LBB condition, where the LBB is the acronym for the names Ladyzhenskaya–Babuška–Brezzi. Below, we consider the problem which is kinematically and materially linear, using a procedure similar to that presented in [38].

For the purely displacement formulation, the FE equilibrium equations have the form

$$\mathbf{K}_0\, \mathbf{u} = \mathbf{p}, \tag{11.86}$$

where $\mathbf{K}_0$ is the tangent matrix, $\mathbf{u}$ is the vector of nodal values of displacements, and $\mathbf{p}$ is the vector of external nodal loads. Equation (11.86) is well-posed if the following condition of positive definiteness (ellipticity) is satisfied

$$\exists \beta > 0 \qquad \mathbf{u}^T \mathbf{K}_0\, \mathbf{u} \geq \beta\, \|\mathbf{u}\|^2 \tag{11.87}$$

for an arbitrary non-zero vector $\mathbf{u}$ and some norm $\|\cdot\|$ for the space of $\mathbf{u}$. Usually, the energy norm is used, i.e. $\|\mathbf{u}\|^2 = \mathbf{u}^T\mathbf{K}_0\mathbf{u}$, and then $\beta = 1$. Below, we consider the mixed formulations and procedures for obtaining their reduced displacement form.

***Inf-sup* (LBB) for two-field mixed formulation.** For a mixed two-field formulation, the equilibrium equations have the form

$$\begin{bmatrix} \mathbf{0} & \mathbf{L} \\ \mathbf{L}^T & \mathbf{K} \end{bmatrix} \begin{bmatrix} \mathbf{u} \\ \mathbf{q} \end{bmatrix} = \begin{bmatrix} \mathbf{p} \\ \mathbf{0} \end{bmatrix}, \tag{11.88}$$

where $\mathbf{q}$ is the vector of additional variables, e.g. the stress parameters for the HR functional. The matrix of eq. (11.88) is symmetric, but indefinite i.e. has positive and negative eigenvalues. The sub-matrix $\mathbf{K}$ is symmetric and positive definite, $\mathbf{L}$ can be rectangular.

To obtain the reduced displacement form of the mixed equations, we calculate $\mathbf{q} = \mathbf{K}^{-1}\mathbf{L}^T\mathbf{u}$ from the second equation of the system (11.88) and use it in the first equation,

$$\mathbf{K}^* \mathbf{u} = \mathbf{p}, \qquad \text{where} \quad \mathbf{K}^* \doteq \mathbf{L}\,\mathbf{K}^{-1}\mathbf{L}^T. \tag{11.89}$$

This equation is well-posed if

$$\exists \beta > 0 \qquad \mathbf{u}^T \mathbf{K}^* \mathbf{u} \geq \beta \, \|\mathbf{u}\|^2, \tag{11.90}$$

for an arbitrary non-zero vector $\mathbf{u}$. We can use the energy norm, $\|\mathbf{u}\|^2 \doteq \mathbf{u}^T\mathbf{K}_0\mathbf{u}$, where $\mathbf{K}_0$ is the matrix of the purely displacement eq. (11.86), so the above condition becomes

$$\exists \beta > 0 \qquad \mathbf{u}^T \mathbf{K}^* \mathbf{u} \geq \beta\, \mathbf{u}^T\mathbf{K}_0\mathbf{u}. \tag{11.91}$$

Thus, we have to find

$$\beta \doteq \inf_{\mathbf{u}} \frac{\mathbf{u}^T \mathbf{K}^* \mathbf{u}}{\mathbf{u}^T \mathbf{K}_0 \mathbf{u}} \tag{11.92}$$

and check whether $\beta > 0$. The fraction on the r.h.s. is the Rayleigh quotient, hence β is the smallest eigenvalue of the generalized eigenvalue problem

$$\mathbf{K}^* \mathbf{u} = \gamma\, \mathbf{K}_0\, \mathbf{u}. \tag{11.93}$$

A more general form of eq. (11.92) is obtained if we note that $\mathbf{u}^T \mathbf{K}^* \mathbf{u} = \mathbf{u}^T \mathbf{L}\,\mathbf{K}^{-1}\mathbf{L}^T \mathbf{u}$ and use the following *equivalence*:

$$\mathbf{u}^T \mathbf{L}\,\mathbf{K}^{-1}\mathbf{L}^T \mathbf{u} = \sup_{\mathbf{q}} \frac{(\mathbf{q}^T \mathbf{L}^T \mathbf{u})^2}{\mathbf{q}^T \mathbf{K}\mathbf{q}}, \tag{11.94}$$

the proof of which is given below. Then we obtain the *inf-sup* condition for the system (11.88),

$$\beta \doteq \inf_{\mathbf{u}} \sup_{\mathbf{q}} \frac{(\mathbf{q}^T \mathbf{L}^T \mathbf{u})^2}{(\mathbf{q}^T \mathbf{K}\,\mathbf{q})\,(\mathbf{u}^T \mathbf{K}_0\,\mathbf{u})} > 0. \tag{11.95}$$

The advantage of this condition is that it does not contain inverse matrices, i.e. we don't have to solve the problem to see if it is solvable.

Proof of equivalence, eq. (11.94). ([12], Sect. 7) The crucial fact is that $\mathbf{K}$ is symmetric and positive definite, so there exists a symmetric and positive definite $\mathbf{K}^{1/2}$ such that $\mathbf{K}^{1/2}\mathbf{K}^{1/2} = \mathbf{K}$. Let us denote $\mathbf{w} \doteq \mathbf{K}^{1/2}\mathbf{q}$, so $\mathbf{q} \doteq \mathbf{K}^{-1/2}\mathbf{w}$. By substituting $\mathbf{q}^T$, we have

$$\sup_{\mathbf{q}} \frac{(\mathbf{q}^T \mathbf{L}^T \mathbf{u})^2}{\mathbf{q}^T \mathbf{K}\,\mathbf{q}} = \sup_{\mathbf{w}} \frac{(\mathbf{w}^T \mathbf{K}^{-1/2} \mathbf{L}^T \mathbf{u})^2}{\mathbf{w}^T \mathbf{w}} \tag{11.96}$$

and we shall prove that

$$\mathbf{u}^T \mathbf{L} \mathbf{K}^{-1} \mathbf{L}^T \mathbf{u} = \sup_{\mathbf{w}} \frac{(\mathbf{w}^T \mathbf{K}^{-1/2} \mathbf{L}^T \mathbf{u})^2}{\mathbf{w}^T \mathbf{w}}, \tag{11.97}$$

instead of eq. (11.94). The proof is divided into two parts.

(i) The Schwartz inequality, $(\mathbf{a}^T\mathbf{b})^2 \leq (\mathbf{a}^T\mathbf{a})\,(\mathbf{b}^T\mathbf{b})$, with vectors $\mathbf{a} \doteq \mathbf{w}$ and $\mathbf{b} \doteq \mathbf{K}^{-1/2} \mathbf{L}^T \mathbf{u}$, yields

$$(\mathbf{w}^T\mathbf{K}^{-1/2} \mathbf{L}^T \mathbf{u})^2 \leq (\mathbf{w}^T\mathbf{w})\,(\mathbf{u}^T \mathbf{L} \mathbf{K}^{-1/2} \mathbf{K}^{-1/2} \mathbf{L}^T \mathbf{u}), \tag{11.98}$$

and dividing both sides by $\mathbf{w}^T\mathbf{w} = \mathbf{q}^T \mathbf{K} \mathbf{q} \neq 0$, we obtain

$$\sup_{\mathbf{w}} \frac{(\mathbf{w}^T \mathbf{K}^{-1/2} \mathbf{L}^T \mathbf{u})^2}{\mathbf{w}^T\mathbf{w}} \leq \mathbf{u}^T \mathbf{L} \mathbf{K}^{-1} \mathbf{L}^T \mathbf{u}. \tag{11.99}$$

(ii) Selecting $\mathbf{w} = \mathbf{K}^{-1/2} \mathbf{L}^T \mathbf{u}$ and using it in the r.h.s. of eq. (11.97), we obtain

$$\sup_{\mathbf{w}} \frac{(\mathbf{w}^T \mathbf{K}^{-1/2} \mathbf{L}^T \mathbf{u})^2}{\mathbf{w}^T \mathbf{w}} \geq \frac{(\mathbf{u}^T \mathbf{L} \mathbf{K}^{-1/2} \mathbf{K}^{-1/2} \mathbf{L}^T \mathbf{u})^2}{\mathbf{u}^T \mathbf{L} \mathbf{K}^{-1/2} \mathbf{K}^{-1/2} \mathbf{L}^T \mathbf{u}} = \mathbf{u}^T \mathbf{L} \mathbf{K}^{-1} \mathbf{L}^T \mathbf{u}. \tag{11.100}$$

The inequalities (11.100) and (11.99) imply eq. (11.97) and, in turn, the *equivalence* of eq. (11.94). □

***Inf-sup* (LBB) for three-field mixed formulation.** For a mixed three-field formulation, the equilibrium equations have the form

$$\begin{bmatrix} \mathbf{0} & \mathbf{L}_1 & \mathbf{0} \\ \mathbf{L}_1^T & \mathbf{0} & \mathbf{K}_{12} \\ \mathbf{0} & \mathbf{K}_{12}^T & \mathbf{K}_{22} \end{bmatrix} \begin{bmatrix} \mathbf{u} \\ \mathbf{q}_1 \\ \mathbf{q}_2 \end{bmatrix} = \begin{bmatrix} \mathbf{p} \\ \mathbf{0} \\ \mathbf{0} \end{bmatrix}, \tag{11.101}$$

where $\mathbf{q}_1$ and $\mathbf{q}_2$ are vectors of additional variables. For instance, for the HW functional, $\mathbf{q}_1$ is the vector of stress parameters and $\mathbf{q}_2$ is the vector of strain parameters. The matrix in eq. (11.101) is symmetric but indefinite i.e. has positive and negative eigenvalues. The sub-matrix $\mathbf{K}_{22}$ is symmetric and positive definite, $\mathbf{K}_{12}$ and $\mathbf{L}_1$ can be rectangular. The above set is solved as a sequence of two problems, each for two fields only.

Problem 1. The first problem is intermediate, i.e. needed to solve *Problem 2,* and is defined by the set of equations

$$\begin{bmatrix} \mathbf{0} & \mathbf{K}_{12} \\ \mathbf{K}_{12}^T & \mathbf{K}_{22} \end{bmatrix} \begin{bmatrix} \mathbf{q}_1 \\ \mathbf{q}_2 \end{bmatrix} = \begin{bmatrix} -\mathbf{L}_1^T \mathbf{u} \\ \mathbf{0} \end{bmatrix}, \tag{11.102}$$

which is analogous to eq. (11.88) for two-field mixed formulation. (Note that for $\mathbf{u}$, we use the energy norm $\|\mathbf{u}\|^2 \doteq \mathbf{u}^T \mathbf{K}_0 \mathbf{u}$, while for $\mathbf{q}_1$, we shall use the Euclidean norm $\|\mathbf{q}_1\|^2 = \mathbf{q}_1^T \mathbf{q}_1$.) To solve this set, first, from the second equation, we calculate $\mathbf{q}_2 = -\mathbf{K}_{22}^{-1} \mathbf{K}_{12}^T \mathbf{q}_1$, which is possible because $\mathbf{K}_{22}$ is invertible. Next we use $\mathbf{q}_2$ in the first equation to obtain

$$\bar{\mathbf{K}}\, \mathbf{q}_1 = \mathbf{L}_1\, \mathbf{u}, \qquad \text{where} \quad \bar{\mathbf{K}} \doteq \mathbf{K}_{12}\, \mathbf{K}_{22}^{-1}\, \mathbf{K}_{12}^T. \tag{11.103}$$

This equation is solvable if $\bar{\mathbf{K}}$ is positive definite, i.e.

$$\exists \beta_1 > 0 \qquad \mathbf{q}_1^T\, \bar{\mathbf{K}}\, \mathbf{q}_1 \geq \beta_1\, \|\mathbf{q}_1\|^2, \tag{11.104}$$

or, for the Euclidean norm $\|\mathbf{q}_1\|^2 = \mathbf{q}_1^T \mathbf{q}_1$,

$$\exists \beta_1 > 0 \qquad \mathbf{q}_1^T\, \bar{\mathbf{K}}\, \mathbf{q}_1 \geq \beta_1\, \mathbf{q}_1^T \mathbf{q}_1, \tag{11.105}$$

for any non-zero vector $\mathbf{q}_1$. Thus, we have to find

$$\beta_1 \doteq \inf_{\mathbf{q}_1} \frac{\mathbf{q}_1^T\, \bar{\mathbf{K}}\, \mathbf{q}_1}{\mathbf{q}_1^T \mathbf{q}_1} \tag{11.106}$$

and check that $\beta_1 > 0$. The fraction on the r.h.s. is the Rayleigh quotient, so β_1 is the smallest eigenvalue of the standard eigenvalue problem

$$\bar{\mathbf{K}}\, \mathbf{q}_1 = \gamma_1\, \mathbf{q}_1, \tag{11.107}$$

which can be used to verify numerically the well-posedness of *Problem 1.* On use of the *equivalence* of eq. (11.94)

$$\mathbf{q}_1^T\, \bar{\mathbf{K}}\, \mathbf{q}_1 = \mathbf{q}_1^T\, \mathbf{K}_{12} \mathbf{K}_{22}^{-1} \mathbf{K}_{12}^T\, \mathbf{q}_1 = \sup_{\mathbf{q}_2} \frac{(\mathbf{q}_2^T\, \mathbf{K}_{12}^T\, \mathbf{q}_1)^2}{\mathbf{q}_2^T\, \mathbf{K}_{22}\, \mathbf{q}_2}, \tag{11.108}$$

so the *inf-sup* condition for *Problem 1* is analogous to eq. (11.95),

$$\beta_1 \doteq \inf_{\mathbf{q}_1} \sup_{\mathbf{q}_2} \frac{(\mathbf{q}_2^T\, \mathbf{K}_{12}^T\, \mathbf{q}_1)^2}{(\mathbf{q}_2^T\, \mathbf{K}_{22}\, \mathbf{q}_2)\; (\mathbf{q}_1^T \mathbf{q}_1)} > 0. \tag{11.109}$$

This condition does not require calculation of the inverse $\mathbf{K}_{22}^{-1}$ and can be written in an alternative form,

$$\forall \mathbf{q}_1 \ \exists \mathbf{q}_2 \qquad (\mathbf{q}_2^T \mathbf{K}_{12}^T \mathbf{q}_1)^2 > \beta_1 (\mathbf{q}_2^T \mathbf{K}_{22} \mathbf{q}_2)(\mathbf{q}_1^T \mathbf{q}_1) \qquad \text{for some} \quad \beta_1 > 0, \tag{11.110}$$

allowing us to deduce that $\mathbf{q}_1$ cannot belong to the null space of $\mathbf{K}_{12}^T$, i.e. $\mathbf{K}_{12}^T$ must have the rank equal to the number of columns and that $\mathbf{K}_{12}\,\mathbf{q}_2$ cannot be orthogonal to the space of $\mathbf{q}_1$'s.

Problem 2. Using $\mathbf{q}_2 = -\mathbf{K}_{22}^{-1}\mathbf{K}_{12}^T\mathbf{q}_1$ in the second of the full set of equation (11.101), the first two equations form the set

$$\begin{bmatrix} \mathbf{0} & \mathbf{L}_1 \\ \mathbf{L}_1^T & -\bar{\mathbf{K}} \end{bmatrix} \begin{bmatrix} \mathbf{u} \\ \mathbf{q}_1 \end{bmatrix} = \begin{bmatrix} \mathbf{p} \\ \mathbf{0} \end{bmatrix}, \tag{11.111}$$

which is analogous to eq. (11.88) for two-field mixed formulation. The matrix $\bar{\mathbf{K}}$ is symmetric and, if eq. (11.109) is satisfied for *Problem 1*, then it is also positive definite. From the second equation of (11.111), we can calculate: $\mathbf{q}_1 = \bar{\mathbf{K}}^{-1}\mathbf{L}_1^T\mathbf{u}$, and use it in the first equation to obtain the reduced displacement form of the mixed equations

$$\mathbf{K}^*\,\mathbf{u} = \mathbf{p}, \qquad \text{where} \quad \mathbf{K}^* \doteq \mathbf{L}_1\,\bar{\mathbf{K}}^{-1}\mathbf{L}_1^T. \tag{11.112}$$

This equation is solvable if $\mathbf{K}^*$ is positive definite, i.e.

$$\exists \beta_2 > 0 \qquad\qquad \mathbf{u}^T\,\mathbf{K}^*\,\mathbf{u} \geq \beta_2\,\|\mathbf{u}\|^2, \tag{11.113}$$

for an arbitrary non-zero vector $\mathbf{u}$. We can use the energy norm $\|\mathbf{u}\|^2 \doteq \mathbf{u}^T\mathbf{K}_0\mathbf{u}$, where $\mathbf{K}_0$ is the matrix of the purely displacement eq. (11.86), so the above condition becomes

$$\exists \beta_2 > 0 \qquad\qquad \mathbf{u}^T\,\mathbf{K}^*\,\mathbf{u} \geq \beta_2\,\mathbf{u}^T\mathbf{K}_0\mathbf{u}. \tag{11.114}$$

Thus, we have to find

$$\beta_2 \doteq \inf_{\mathbf{u}} \frac{\mathbf{u}^T\,\mathbf{K}^*\,\mathbf{u}}{\mathbf{u}^T\mathbf{K}_0\mathbf{u}} \tag{11.115}$$

and check whether $\beta_2 > 0$. The fraction on the r.h.s. is the Rayleigh quotient, so β_2 is the smallest eigenvalue of the generalized eigenvalue problem

$$\mathbf{K}^*\,\mathbf{u} = \gamma_2\,\mathbf{K}_0\,\mathbf{u}. \tag{11.116}$$

If we write $\mathbf{u}^T\,\mathbf{K}^*\,\mathbf{u} = \mathbf{u}^T\,\mathbf{L}_1\bar{\mathbf{K}}^{-1}\mathbf{L}_1^T\,\mathbf{u}$ and use the *equivalence* eq. (11.94), we obtain the *inf-sup* form of eq. (11.115),

$$\beta_2 \doteq \inf_{\mathbf{u}} \sup_{\mathbf{q}_1} \frac{(\mathbf{q}_1^T \mathbf{L}_1^T \mathbf{u})^2}{(\mathbf{q}_1^T \bar{\mathbf{K}} \mathbf{q}_1)\,(\mathbf{u}^T \mathbf{K}_0 \mathbf{u})} > 0. \tag{11.117}$$

Note that $\bar{\mathbf{K}}$ depends on the inverse $\mathbf{K}_{22}^{-1}$, which we eliminate as follows. The property $[\sup_{\mathbf{x}} F(\mathbf{x})]^{-1} = \inf_{\mathbf{x}} F^{-1}(\mathbf{x})$ holds for a scalar continuous function $F(\mathbf{x}) > 0$. We take

$$\mathbf{x} \doteq \mathbf{q}_2, \qquad F(\mathbf{q}_2) \doteq \frac{(\mathbf{q}_2^T \mathbf{K}_{12}^T \mathbf{q}_1)^2}{\mathbf{q}_2^T \mathbf{K}_{22} \mathbf{q}_2}, \tag{11.118}$$

where $F(\mathbf{q}_2) > 0$ by eq. (11.109). Then the inverse of eq. (11.108) is

$$\frac{1}{\mathbf{q}_1^T \bar{\mathbf{K}} \mathbf{q}_1} = \inf_{\mathbf{q}_2} \frac{\mathbf{q}_2^T \mathbf{K}_{22} \mathbf{q}_2}{(\mathbf{q}_2^T \mathbf{K}_{12}^T \mathbf{q}_1)^2} \tag{11.119}$$

and we use it in eq. (11.117), obtaining the *inf-sup* condition for the three-field mixed problem,

$$\beta_2 \doteq \inf_{\mathbf{u}} \sup_{\mathbf{q}_1} \inf_{\mathbf{q}_2} \frac{(\mathbf{q}_2^T \mathbf{K}_{22} \mathbf{q}_2)\,(\mathbf{q}_1^T \mathbf{L}_1^T \mathbf{u})^2}{(\mathbf{q}_2^T \mathbf{K}_{12}^T \mathbf{q}_1)^2\,(\mathbf{u}^T \mathbf{K}_0 \mathbf{u})} > 0. \tag{11.120}$$

This condition can be written in an alternative form as

$$\forall \mathbf{u}\ \exists \mathbf{q}_1\ \forall \mathbf{q}_2 \qquad (\mathbf{q}_2^T \mathbf{K}_{22} \mathbf{q}_2)(\mathbf{q}_1^T \mathbf{L}_1^T \mathbf{u})^2 > \beta_2\,(\mathbf{q}_2^T \mathbf{K}_{12}^T \mathbf{q}_1)^2\,(\mathbf{u}^T \mathbf{K}_0 \mathbf{u}) \tag{11.121}$$

for some $\beta_2 > 0$ and we see that it does not imply that $\mathbf{q}_1 \neq \mathbf{0}$ cannot belong to the null space of $\mathbf{K}_{12}^T$ and, thus, does not guarantee that eq. (11.109) is fulfilled. Hence, both the conditions of eqs. (11.109) and (11.120) are required.

Summary. To ensure the solvability of the mixed problem the following conditions should be verified:

- For the two-field problem of eq. (11.88), we have to verify either (i) the *inf-sup* condition of eq. (11.95) or (ii) that the smallest eigenvalue for the eigenvalue problem of eq. (11.93) is greater than zero, and for the mesh size going to zero, it is still greater than zero.
- For the three-field problem of eq. (11.101), we have to verify either (i) the *inf-sup* conditions of eqs. (11.109) and (11.120), or (ii) that the smallest eigenvalues for the eigenvalue problems of eqs. (11.107) and (11.116) are greater than zero.

Moreover, we have to check that the constants in the *inf-sup* conditions, or the smallest eigenvalues, do not tend to zero for the diminishing element size.

Numerical *inf-sup* test. Two meshes were used; a regular mesh and a distorted mesh, of 2×2, 4×4, and 8×8 elements, see Fig. 11.7. Besides, two values of the Poisson ratio were used: $\nu = 0.3$ for a compressible material and $\nu = 0.4999$ for a nearly incompressible material.

For the HR5-S element, we solve the eigenvalue problem of eq. (11.93). For the HW14-SS, we solve the eigenvalue problem of eq. (11.116) and, instead of solving eq. (11.109), the pivots are controlled when calculating the inverse of

$$\begin{bmatrix} \mathbf{0} & \mathbf{K}_{12} \\ \mathbf{K}_{12}^T & \mathbf{K}_{22} \end{bmatrix} \tag{11.122}$$

and they are non-zero, which indicates that *Problem 1* is solvable.

The smallest eigenvalues γ for the HW14-SS element are shown in Fig. 11.8, where $N = 2, 4, 8$ is the number of subdivisions in one direction. The curves indicate that the discrete form of the *inf-sup* test is passed, thus the condition (11.114) is met. Note that identical curves were obtained for the HR5-S element.

For the regular meshes, the obtained curves are horizontal, similarly as for the 9/3 element shown in [17], Fig. 1, and for the MINI element shown in [46], Fig. 6. Both these elements have the property that there exists an analytical proof that they pass the *inf-sup* test and the corresponding numerical test is also passed. Hence, it is likely that the analytical *inf-sup* condition can also be verified for the HW14-SS element.

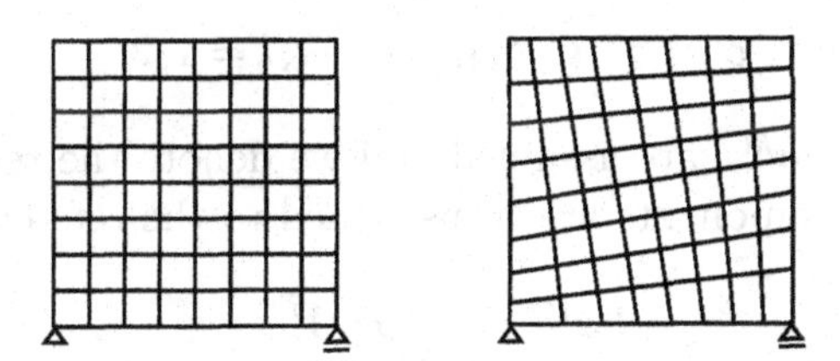

Fig. 11.7 Inf-sup test. Regular and distorted mesh of 8×8 elements.

11.5.1 Assumed stress HR elements: PS and HR5-S

In the class of the elements based on the HR functional, the PS element of [170] is standard. Currently, however, several other elements exist in the literature which perform slightly better for coarse distorted meshes; among them, the HR5-S element of [256]. Both these elements use the

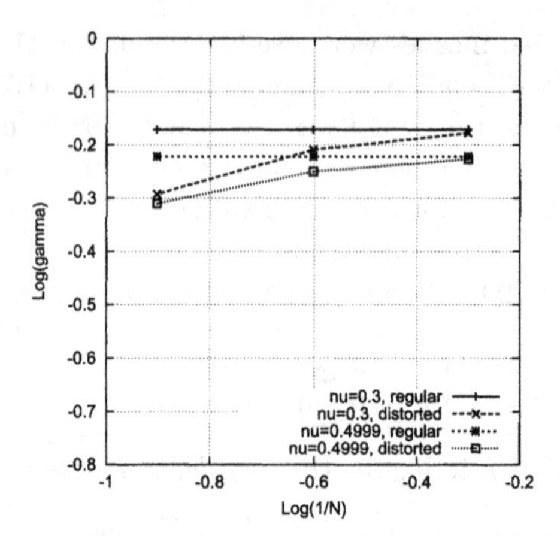

Fig. 11.8 Inf-sup test. Results for HW14-SS element. $E = 1$, $\nu = 0.3$ or 0.4999

same five-parameter representation of stress but in the PS element it is written using the natural coordinates, while in the HR5-S element the skew coordinates are used.

The early works on the HR elements, up to 1981, are reviewed in [223], while the more recent ones are reviewed in [256].

Assumed representation of stress. We use the contra-variant stress components σ^{kl} in the natural basis at the element's center $\{\mathbf{g}_k^c\}$, i.e.

$$\boldsymbol{\sigma} = \sigma^{kl}\, \mathbf{g}_k^c \otimes \mathbf{g}_l^c, \qquad k,l = 1,2. \tag{11.123}$$

The components σ^{kl} are assumed and we denote the respective matrix as $\boldsymbol{\sigma}^{\xi}$. These components are transformed to the reference basis using

$$\boldsymbol{\sigma}^{\text{ref}} = \mathbf{J}_c\, \boldsymbol{\sigma}^{\xi}\, \mathbf{J}_c^T, \tag{11.124}$$

where $\mathbf{J}_c$ is the Jacobian matrix evaluated at the element's center of eq. (11.76).

The five-parameter representation of stress was already used by Pian in 1964 in [168] in Cartesian coordinates and later in [170] in the natural coordinates

$$\boldsymbol{\sigma}^{\xi} \doteq \begin{bmatrix} q_1 + q_2\,\eta & q_5 \\ \text{sym.} & q_3 + q_4\,\xi \end{bmatrix}. \tag{11.125}$$

This representation is symmetric, and includes the modes $\{1, \xi, \eta\}$ multiplied by five parameters q_i.

In [256], the above five-parameter representation of stress was written in the skew coordinates, i.e.

$$\boldsymbol{\sigma}^{\xi} \doteq \begin{bmatrix} q_1 + q_2\, y_S & q_5 \\ \text{symm.} & q_3 + q_4\, x_S \end{bmatrix} = \begin{bmatrix} q_1 + q_2\,\eta + \underline{q_2 B\,\xi\eta} & q_5 \\ \text{sym.} & q_3 + q_4\,\xi + \underline{q_4\,A\,\xi\eta} \end{bmatrix}, \tag{11.126}$$

where A and B are defined below eq. (11.81). The bilinear (underlined) terms are non-zero only for irregular trapezoidal shapes, while they vanish for parallelograms. Still only five parameters q_i are used!

Verification of equilibrium equation for the assumed stress. For a single element, we can check whether the assumed representations of stresses satisfy the homogenous equilibrium equations. This property is not used in the construction of our elements, but it can be logically linked with their performance for characteristic shapes of the elements.

Note that in several papers, including [169, 170, 265], the satisfaction of the homogenous equilibrium equations is pivotal as they are appended to the HR functional via the Lagrange multiplier method. Then, however, the problem becomes more complicated, as the question of a suitable form of the Lagrange multiplier field arises (typically the incompatible displacement modes are exploited for this purpose). We stress that we do not use this approach.

We can check the equilibrium equations for some characteristic shapes of an element using a symbolic algebra. The homogenous equilibrium equations in the reference Cartesian coordinates, for a symmetric stress, are as follows

$$\frac{\partial \sigma_{xx}}{\partial x} + \frac{\partial \sigma_{xy}}{\partial y} = 0, \qquad \frac{\partial \sigma_{xy}}{\partial x} + \frac{\partial \sigma_{yy}}{\partial y} = 0. \tag{11.127}$$

They are checked for the following specification of the stress components and their derivatives:

1. The stresses in the reference basis are obtained from the assumed representation $\boldsymbol{\sigma}^{\xi}$ using the transformation formula (11.124),

$$\boldsymbol{\sigma}^{a} \doteq \begin{bmatrix} \sigma_{xx} & \sigma_{xy} \\ \sigma_{xy} & \sigma_{yy} \end{bmatrix} = \mathbf{J}_c \begin{bmatrix} \sigma^{\xi\xi} & \sigma^{\xi\eta} \\ \sigma^{\xi\eta} & \sigma^{\eta\eta} \end{bmatrix} \mathbf{J}_c^T = \mathbf{J}_c\, \boldsymbol{\sigma}^{\xi}\, \mathbf{J}_c^T. \tag{11.128}$$

2. When the matrix $\boldsymbol{\sigma}^{\xi}$ is assumed in terms of the skew coordinates x_S, y_S, then, to enable numerical integration of the element, x_S, y_S are treated as functions of the natural coordinates ξ, η. Hence, we

can either use the chain rule of differentiation or directly express the derivatives w.r.t. x, y in terms of derivatives w.r.t. ξ, η as follows:

$$\begin{bmatrix} \frac{\partial \sigma}{\partial x} \\ \frac{\partial \sigma}{\partial y} \end{bmatrix} = \mathbf{J}^{-T} \begin{bmatrix} \frac{\partial \sigma}{\partial \xi} \\ \frac{\partial \sigma}{\partial \eta} \end{bmatrix}, \tag{11.129}$$

where $\sigma \in [\sigma_{xx}, \sigma_{yy}, \sigma_{xy}]^T$ is an arbitrary stress component in the form of eq. (11.126). Note that here $\mathbf{J}$ is used, not $\mathbf{J}_c$.

The results of a verification of the equilibrium equation for the assumed stress are presented in Table 11.1, where "+" indicates that the equations are satisfied for an irregular shape of an element.

We see that, for the skew coordinates (HR5-S element), the equilibrium equations are satisfied point-wise, even for an irregular element. For the natural coordinates (PS element), they are satisfied point-wise only for parallelograms, while for irregular elements, only at the element's center.

Table 11.1 Verification of equilibrium equation for the assumed stress.

$\boldsymbol{\sigma}^\xi$ assumed in	At arbitrary point	At center	Integral of eq. (11.127)
skew coordinates	+	+	+
natural coordinates	–(*)	+	–(*)

(*) satisfied only for parallelograms.

Verification of compatibility of the strains for assumed stresses. The compatibility condition for 2D strains is as follows:

$$\frac{\partial^2 \varepsilon_{xx}}{\partial y^2} + \frac{\partial^2 \varepsilon_{yy}}{\partial x^2} = 2\frac{\partial^2 \varepsilon_{xy}}{\partial x \, \partial y}, \tag{11.130}$$

and we evaluate it for the strains calculated using the inverse constitutive matrix for the assumed stresses. We emphasize that we do not check the compatibility condition for the compatible strain but for the strains corresponding to (induced by) the assumed stress. They are obtained in the following steps:

1. The stresses in the reference basis are obtained as in eq. (11.128).
2. The strains corresponding to the assumed stresses are obtained from the inverse constitutive equation

$$\boldsymbol{\varepsilon}_v = \mathbb{C}_{vv}^{-1} \boldsymbol{\sigma}_v^a, \tag{11.131}$$

where $(\cdot)_v$ denotes a vector of components of a tensor $(\cdot)$, arranged in the order $\{xx, yy, xy\}$.

3. The skew coordinates x_S, y_S are treated as functions of the natural coordinates ξ, η. Hence, we can either use the chain rule of differentiation or directly express the first derivatives of strains w.r.t. x, y in terms of derivatives w.r.t. ξ, η as follows:

$$\begin{bmatrix} \frac{\partial \varepsilon}{\partial x} \\ \frac{\partial \varepsilon}{\partial y} \end{bmatrix} = \mathbf{J}^{-T} \begin{bmatrix} \frac{\partial \varepsilon}{\partial \xi} \\ \frac{\partial \varepsilon}{\partial \eta} \end{bmatrix}, \tag{11.132}$$

where $\varepsilon \in \{\varepsilon_{xx}, \varepsilon_{yy}, \varepsilon_{xy}\}$ is an arbitrary strain component. For the first derivatives of an arbitrary strain component, $\gamma \in \{\partial\varepsilon/\partial x,\ \partial\varepsilon/\partial y\}$, we similarly calculate the second derivatives,

$$\begin{bmatrix} \frac{\partial \gamma}{\partial x} \\ \frac{\partial \gamma}{\partial y} \end{bmatrix} = \mathbf{J}^{-T} \begin{bmatrix} \frac{\partial \gamma}{\partial \xi} \\ \frac{\partial \gamma}{\partial \eta} \end{bmatrix}, \tag{11.133}$$

where

$$\frac{\partial \gamma}{\partial x} = \left\{ \frac{\partial^2 \varepsilon}{\partial x^2}, \frac{\partial^2 \varepsilon}{\partial y \partial x} \right\}, \qquad \frac{\partial \gamma}{\partial y} = \left\{ \frac{\partial^2 \varepsilon}{\partial x \partial y}, \frac{\partial^2 \varepsilon}{\partial y^2} \right\}, \tag{11.134}$$

and they contain all the second derivatives needed in eq. (11.130). Note that $\mathbf{J}$ is used here not $\mathbf{J}_c$.

The results of the verification of the compatibility condition are presented in Table 11.2, where "+" indicates that the equations are satisfied for an irregular shape of an element.

We see that, for the skew coordinates (HR5-S element), the compatibility condition is satisfied, even for irregular elements, while for the natural coordinates (PS element), the compatibility condition is satisfied only for parallelograms.

Table 11.2 Verification of the compatibility condition for the assumed stress.

σ^ξ assumed in	At arbitrary point	At center	Integral of eq. (11.130)
skew coordinates	+	+	+
natural coordinates	–(*)	–(*)	–(*)

(*) satisfied only for parallelograms.

Remark. We have earlier shown that the natural coordinates cannot be treated as being associated with the natural basis at the element's center, as this leads to the "fictitious" parallelogram element shown in Fig. 11.6. The above tests of the homogenous equilibrium equation and of the compatibility of strains equation provide another argument that it is more rational to assume the representation of stress in terms of the skew coordinates than in the natural coordinates.

We do not exploit this property in the elements' formulation in any particular way. Nonetheless, the numerical results indicate that the accuracy of elements depends on the coordinates used for the stress representation.

Assumed stress elements: PS and HR5-S. The assumed stress elements are developed from the two-field HR functionals in the basic non-enhanced form of eqs. (11.4) and (11.7). In these functionals, $\mathbf{u}$ is the compatible field while $\boldsymbol{\sigma}$ is the assumed field of the form

$$\boldsymbol{\sigma}^a = \mathbf{J}_c \, \boldsymbol{\sigma}^\xi \, \mathbf{J}_c^T, \tag{11.135}$$

which is the transformation rule for the contra-variant components of a tensor of eq. (11.124). Besides, in $\boldsymbol{\sigma}^\xi$ we use the 5-parameter stress of eq. (11.125) for the PS elements, or of eq. (11.126) for the HR5-S element. The increment of the assumed stress has the analogous form, where $\Delta\boldsymbol{\sigma}^\xi$ has the structure of $\boldsymbol{\sigma}^\xi$ of eq. (11.125), but the multipliers q_i are replaced by Δq_i.

In the HR functionals, we use the reduced constitutive operator for the plane stress condition $\mathbb{C}^*$ of eq. (7.64).

The PS element is a standard in the class of mixed HR elements, but the HR5-S element performs slightly better for coarse distorted meshes. Its formulation is very simple and it yields results similar to these by the 5β-A,B,C elements of [265] and the QE2 element of [177], which are more complicated and use more parameters.

Remark. The discrete HR functional depends on two sets of variables: the nodal displacements $\mathbf{u}_I$ and the elemental stress multipliers $\mathbf{q}$. The obtained set of FE equations is given by eq. (11.27) and the scheme U2 of eq. (11.34) should be used to update the stress multipliers. Consider the non-reduced tangent matrix of eq. (11.27). At $(\mathbf{u} = \mathbf{0}, \mathbf{q} = \mathbf{0})$, the sub-matrix $\mathbf{K}$ is equal to zero and we obtain

$$\begin{bmatrix} \mathbf{0} & \mathbf{L} \\ \mathbf{L}^T & \mathbf{K}_{qq} \end{bmatrix}, \tag{11.136}$$

for which the linear element is very efficient.

Remark. Assumed stress/enhanced strain elements. The HR element can also be developed for the seven-parameter representation of stresses, but this element is too stiff, no matter in which coordinates the stresses are written. Hence, the HR functional must be enhanced and two additional EAS or EADG modes were used in the HR9 element in [256]. The HR9 element performs identically as the HR5-S element, but is less efficient. However, it still can be used in 2D and shell elements with drilling rotations, for which this type of enhancement is beneficial, see Sect. 12.

11.5.2 Assumed stress and strain HW14-SS element

The main difference between the HR elements and the HW elements is that strains are retained in the latter and we have to select their representation.

Generally, the strain representation analogous to that used for stress is too poor. A better one is implied by the inverse constitutive equation

$$\begin{bmatrix} \varepsilon_{11} \\ \varepsilon_{22} \\ \varepsilon_{12} \end{bmatrix} = \begin{bmatrix} c_1 & c_2 & 0 \\ c_2 & c_1 & 0 \\ 0 & 0 & c_3 \end{bmatrix} \begin{bmatrix} q_1 + q_2\eta \\ q_3 + q_4\xi \\ q_5 \end{bmatrix} = \begin{bmatrix} (c_1q_1 + c_2q_3) + c_1q_2\eta + c_2q_4\xi \\ (c_2q_1 + c_1q_3) + c_2q_2\eta + c_1q_4\xi \\ c_3q_5 \end{bmatrix}, \tag{11.137}$$

where the five-parameter representation of stress of eq. (11.125) and a typical structure of the inverse constitutive matrix are used. This suggests that a seven-parameter representation of strain should be used; constant representation for ε_{12} and linear representations for ε_{11} and ε_{22}. However, if ε_{12} is additionally enhanced by two modes, then the accuracy for coarse distorted meshes improves. Further improvement is obtained if this representation of strain is assumed in terms of the skew coordinates of eq. (11.81).

Assumed representation of strain. The covariant components of strain are assumed in the co-basis $\{\mathbf{g}_c^k\}$, i.e.

$$\boldsymbol{\varepsilon} = \varepsilon_{kl}\, \mathbf{g}_c^k \otimes \mathbf{g}_c^l. \tag{11.138}$$

The matrix of components $\varepsilon_{\alpha\beta}$ can be designated as $\boldsymbol{\varepsilon}_\xi$ and transformed to the ortho-normal reference basis by using

$$\boldsymbol{\varepsilon}^{\text{ref}} = \mathbf{J}_c^{-T}\, \boldsymbol{\varepsilon}_\xi\, \mathbf{J}_c^{-1}. \tag{11.139}$$

The scalar product of the assumed representations of stress and strain, eqs. (11.124) and (11.139) is invariant, which implies invariance of the derived elements.

The assumed nine-parameter representation of strain is

$$\boldsymbol{\varepsilon}_\xi \doteq \begin{bmatrix} q_6 + q_7\, y_S + q_8\, x_S & q_{12} + q_{13}\, x_S + q_{14}\, y_S \\ \text{sym.} & q_9 + q_{10}\, x_S + q_{11}\, y_S \end{bmatrix}, \tag{11.140}$$

where each component is a linear polynomial of x_S and y_S. We see that this representation consists of two parts,

$$\boldsymbol{\varepsilon}_\xi = \begin{bmatrix} q_6 + q_7\, y_S & q_{10} \\ \text{sym.} & q_8 + q_9\, x_S \end{bmatrix} + \begin{bmatrix} q_{11}\, x_S & q_{13}\, x_S + q_{14}\, y_S \\ \text{sym.} & q_{12}\, y_S \end{bmatrix}, \tag{11.141}$$

where the first part is analogous to the five-parameter representation of eq. (11.126) used for stress, while the second part is analogous to the four-parameter representation of the EAS method, see eq. (11.62), but written in the skew coordinates.

Compatibility of assumed strains. The compatibility condition for 2D strains is given by eq. (11.130). Note that, for the HR element, we verified the compatibility of the strains calculated for assumed stresses, while here we verify the compatibility of the assumed strain. Hence, we can skip point 1 of the previously defined procedure. Results of the test of the compatibility condition for the strain of eq. (11.140) are presented in Table 11.3, where "+" indicates that the condition is satisfied for an element of an arbitrary irregular shape.

Table 11.3 Verification of the compatibility condition for the assumed strain.

$\boldsymbol{\varepsilon}_\xi$ assumed in	At arbitrary point	At center
skew coordinates	+	+
natural coordinates	–(*)	–(*)

(*) satisfied only for parallelograms.

We see that for the representation in the skew coordinates, the compatibility condition is satisfied point-wise, even for irregular elements. When the strain is written in natural coordinates, then this equation is only satisfied for parallelograms. This provides the argument that it is more rational to assume the representation of strain in the skew coordinates than in the natural coordinates.

Element HW14-SS. The assumed stress/assumed strain element is developed from the three-field HW functionals in the basic non-enhanced form of eqs. (11.1) and (11.3). The compatible displacement $\mathbf{u}^c$ is defined in eq. (11.14). The independent stress $\boldsymbol{\sigma}$ and the independent strain $\boldsymbol{\varepsilon}$ are constructed as the assumed fields. The assumed fields are constructed as follows:

1. The assumed stress is constructed similarly as for the HR5-S element,
$$\boldsymbol{\sigma}^a = \mathbf{J}_c\, \boldsymbol{\sigma}^\xi\, \mathbf{J}_c^T, \tag{11.142}$$
using the transformation rule of eq. (11.124) and the five-parameter representation of $\boldsymbol{\sigma}^\xi$ of eq. (11.126). Recall that for this representation, the equilibrium equations are satisfied point-wise, even for an irregular element, see Table 11.1.
2. The assumed strain is constructed as
$$\boldsymbol{\varepsilon}^a = \mathbf{J}_c^{-T}\, \boldsymbol{\varepsilon}_\xi\, \mathbf{J}_c^{-1}, \tag{11.143}$$
using the transformation rule of eq. (11.139) for the covariant components of a tensor. The nine-parameter strain representation of $\boldsymbol{\varepsilon}_\xi$ is given by eq. (11.140) and it satisfies the compatibility condition.

We designate this element as HW14-SS because it has 14 modes and both the stress and strain representations are assumed in skew coordinates.

In numerical tests, the HW14-SS element performs identically as the HR5-S element, i.e. is slightly more accurate and less sensitive to mesh distortion than the PS element and the enhanced strain elements (ID4, EAS4, EADG4).

The HW14-SS element uses a smaller number of modes than other HW elements described in the literature, such as the QE2 element of [177] with 16 modes, and the elements with 22 modes $\bar{B}$-QE4 of [178] and $\bar{B}(x,y)$-QE4 and $\bar{B}(\xi,\eta)$-QE4 of [176], but its accuracy is identical.

Remark 1. If we use less parameters for the assumed strain, e.g., seven instead of nine, then it is beneficial to use the covariant instead of contravariant representation of strain. The results for the element based on the nine-parameter representation of strain are not altered by this change.

Remark 2. Assumed stress and strain/enhanced strain elements. The HW element can also be developed for the seven-parameter representation of stresses and the nine-parameter representation of strain, but must be enhanced; two additional EADG modes are used in the HW18 element in [257]. This element performs identically to the HR14-SS element, but is less efficient. However, it can still be used in 2D+drill and shell elements, for which the EAGD enhancement is particularly beneficial, see Sect. 12.

Remark 3. The discrete HW functional depends on two sets of variables: the nodal displacements $\mathbf{u}_I$ and the elemental stress and strain multipliers $\mathbf{q}$. The obtained set of FE equations is given by eq. (11.27) and to update the stress and strain multipliers, the scheme U2 of eq. (11.34) should be used. Consider the non-reduced tangent matrix of eq. (11.27). Several sub-matrices of it are equal to zero at $(\mathbf{u} = \mathbf{0}, \mathbf{q} = \mathbf{0})$, and we obtain

$$\begin{bmatrix} \mathbf{K} & \mathbf{L} \\ \mathbf{L}^T & \mathbf{K}_{qq} \end{bmatrix} = \begin{bmatrix} \mathbf{K} & \mathbf{L}_1 & \mathbf{L}_2 \\ \mathbf{L}_1^T & \mathbf{K}_{11} & \mathbf{K}_{12} \\ \mathbf{L}_2^T & \mathbf{K}_{12}^T & \mathbf{K}_{22} \end{bmatrix} \quad \rightarrow \quad \begin{bmatrix} \mathbf{0} & \mathbf{L}_1 & \mathbf{0} \\ \mathbf{L}_1^T & \mathbf{0} & \mathbf{K}_{12} \\ \mathbf{0} & \mathbf{K}_{12}^T & \mathbf{K}_{22} \end{bmatrix}, \tag{11.144}$$

where 1 designates the q_i parameters for stress, and 2 designates the q_i parameters for strain. The presence of zero sub-matrices can be used to obtain a very efficient linear version of this element.

11.6 Modification of $\mathbf{F}^T\mathbf{F}$ product

We can modify the $\mathbf{F}^T\mathbf{F}$ product in the Green strain in the way which preserves a correct rank of the elements and improves their coarse mesh accuracy. The deformation gradient $\mathbf{F}$ is expanded in the Taylor series w.r.t. the natural coordinates at the element's center, and the $\mathbf{F}^T\mathbf{F}$ product is approximated as follows:

$$\mathbf{F}^T\mathbf{F} \approx \mathbf{F}_c^T\mathbf{F}_c + \mathbf{A} + \mathbf{A}^T, \tag{11.145}$$

where

$$\mathbf{A} \doteq \mathbf{F}_c^T \left[\underline{\xi(\mathbf{F}_{,\xi})_c + \eta(\mathbf{F}_{,\eta})_c + \xi\eta(\mathbf{F}_{,\xi\eta})_c} + \frac{1}{2}\xi^2(\mathbf{F}_{,\xi\xi})_c + \frac{1}{2}\eta^2(\mathbf{F}_{,\eta\eta})_c \right]. \tag{11.146}$$

In other words, the Taylor expansion is combined with a selection of meaningful terms in the product. A correct rank of the reduced tangent matrix

$\mathbf{K}^*$ in eq. (11.30) is yielded by the first three (underlined) linear and bi-linear terms of $\mathbf{A}$, while the last two quadratic terms of $\mathbf{A}$ are needed to pass the patch test.

The concept of expansion was proposed in [135] and was later used in several papers, including [132], but the terms selected in these works are different from these in eq. (11.145).

1. The under-integrated and gamma-stabilized elements were developed in [135] and the following expansion was used:

$$\boldsymbol{\varepsilon}(\xi,\eta) = \mathbf{B}(\xi,\eta)\,\mathbf{u}_I, \qquad \mathbf{B}(\xi,\eta) \approx \mathbf{B}_c + \xi(\mathbf{B}_{,\xi})_c + \eta(\mathbf{B}_{,\eta})_c, \quad (11.147)$$

   where $\mathbf{B}$ is the strain-displacement matrix, see eq. (2.5a) therein. This formula corresponds to the first two of the three underlined terms in eq. (11.146).
2. In [132], the following terms of the Taylor series were selected,

$$\begin{aligned} \bar{T}(f) &\doteq \xi(f_{,\xi})_c + \eta(f_{,\eta})_c + \xi\eta(f_{,\xi\eta})_c \\ &+\frac{1}{6}\left[\xi^3(f_{,\xi\xi\xi})_c + \eta^3(f_{,\eta\eta\eta})_c + 3\xi^2\eta(f_{,\xi\xi\eta})_c + 3\eta^2\xi(f_{,\eta\eta\xi})_c\right], \quad (11.148) \end{aligned}$$

   and applied to the "stabilizing" strain and the "enhancing" strain field, see eqs. (23) and (24) therein. We see that the expansions of eqs. (11.148) and (11.145) are different in the higher-order terms.

Another difference is that small strains are used in both of the cited papers, so the term $\mathbf{F} + \mathbf{F}^T$ was modified, while we modify the product $\mathbf{F}^T\mathbf{F}$ as we use the Green strain.

A full set of tests for the EADG4, HR5-S, and HW14-S elements is given in [257]. For the mixed elements, HW14-S and HR5-S, the expansion was applied to $\mathbf{F}$, as given by eq. (11.145), while for the EADG4 element, we expanded the whole enhanced deformation gradient, $\mathbf{F} + \tilde{\mathbf{H}}$, where $\tilde{\mathbf{H}}$ is defined by eq. (11.71). The modification of the $\mathbf{F}^T\mathbf{F}$ product was proved to be beneficial in the case of coarse distorted meshes.

12

Plane four-node elements with drilling rotation

The drilling rotation, as a degree of freedom, is particularly important for shell elements, but the 2D elements with drilling rotations are much simpler and, hence, very useful in developing and testing specific ways of incorporating the drilling rotation terms. We designate the 2D elements with drilling rotations as "2D+drill".

The drilling rotation is defined as the rotation vector normal to the tangent plane of the element. However, for 2D elements, the normal direction is defined by one vector $\mathbf{t}_3$, normal to the plane of element, so it suffices to consider the angle of drilling rotation, ω. The nodal drilling rotations of a 2D+drill element are shown in Fig. 12.1.

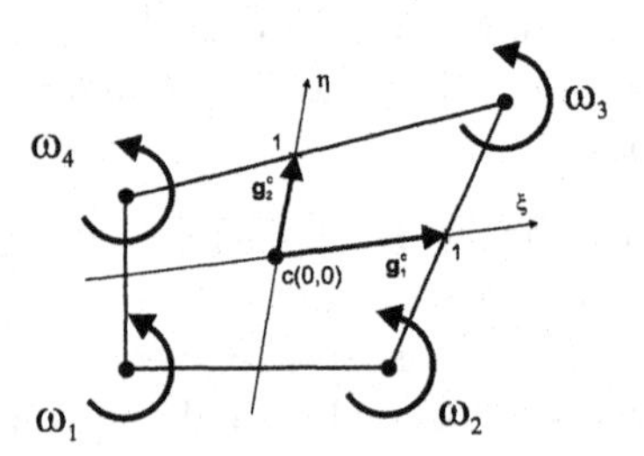

Fig. 12.1 Nodal drilling rotation angles $\omega_I \, (I = 1,2,3,4)$ of plane four-node element.

The drilling rotation can be incorporated into the 2D+drill elements, and shell elements as well, in two ways:

1. Using the so-called Allman shape functions which approximate the element's displacements in terms of nodal displacements $\mathbf{u}_I$ and nodal drilling rotations ω_I. The classical approach based on Allman

shape functions uses the potential energy functional and is valid only for small drilling rotations. The generalized version, valid for finite rotations, is described in Sect. 12.7.

2. Using bilinear shape functions and the drill RC equation, which is extracted from the RC equation, $\text{skew}(\mathbf{Q}^T\mathbf{F}) = \mathbf{0}$, as discussed in detail in Sect. 3. This equation can be implemented in a finite element in several ways, see Sect. 12.3. Some of them use the weak forms, and correspond to the 3D mixed functionals incorporating rotations of Sect. 4. The drill RC for shells was derived in Sect. 6.2. The 2D+drill elements based on bilinear shape functions are described in Sect. 12.6.

Chronologically, the approach based on the Allman shape functions was first, but tests indicate that the elements based on the drill RC equation perform slightly better.

Crucial for a good performance of 2D+drill elements is the use of EADG enhancement, which also affects the RC equation and, for this reason, is more suitable than the EAS enhancement. The EADG method was discussed in Sect. 11.4.3; its extension to formulations with rotations is given in Sect. 12.4.

All the four-node elements described in sequel are developed for finite (unrestricted) drilling rotations.

12.1 Basic relations for drill RC equation

Drill RC for shells. In the RC of eq. (3.8), we neglect the terms which do not depend on the drilling rotation. Then, only the components 12 and 21 of this equation remain and we denote $\left[\text{skew}(\mathbf{Q}^T\mathbf{F})\right]_{12} \doteq r_\omega$, where

$$r_\omega \doteq \tfrac{1}{2}(\mathbf{x}_{0,1} \cdot \mathbf{a}_2 - \mathbf{x}_{0,2} \cdot \mathbf{a}_1) \tag{12.1}$$

or, in terms of tangent displacement components u, v and the drilling rotation ω,

$$2r_\omega \doteq -(v_{,2} + u_{,1} + 2)\,\sin\omega + (v_{,1} - u_{,2})\,\cos\omega. \tag{12.2}$$

In this way, the tensorial RC, $\text{skew}(\mathbf{Q}^T\mathbf{F}) = \mathbf{0}$, is reduced to the scalar drill RC, $r_\omega = 0$.

Three forms of drill RC for 2D problem. We can obtain an alternative but equivalent form of the drill RC equation considering a 2D problem, for which we have

$$\mathbf{F} = \begin{bmatrix} F_{11} & F_{12} \\ F_{21} & F_{22} \end{bmatrix}, \qquad \mathbf{Q} = \begin{bmatrix} \cos\omega & -\sin\omega \\ \sin\omega & \cos\omega \end{bmatrix}, \tag{12.3}$$

where ω is the drilling rotation angle, and we obtain

$$\text{skew}(\mathbf{Q}^T\mathbf{F}) = \begin{bmatrix} 0 & -r_\omega \\ r_\omega & 0 \end{bmatrix}, \tag{12.4}$$

where $r_\omega \doteq \frac{1}{2}(A\sin\omega + B\cos\omega)$, $A \doteq F_{11} + F_{22} = u_{1,1} + u_{2,2} + 2$, and $B \doteq F_{12} - F_{21} = u_{1,2} - u_{2,1}$. Hence, the RC equation, $\text{skew}(\mathbf{Q}^T\mathbf{F}) = \mathbf{0}$, is reduced to one scalar equation,

$$r_\omega \doteq \tfrac{1}{2}(A\sin\omega + B\cos\omega) = 0. \tag{12.5}$$

Using this equation, we can formulate the constraint for the drilling rotation in one of the following forms:

1. For rotations $|\omega| < \pi/2$, we can divide eq. (12.5) by $\cos\omega$ to obtain

   $$\omega^* = -\arctan\frac{B}{A}. \tag{12.6}$$

   Hence, the first form of the drill RC is defined as

   $$c \doteq \omega - \omega^* = 0. \tag{12.7}$$

2. For large rotations, the constraint can be written for an increment. For $\omega \doteq \omega_n + \Delta\omega^*$, using trigonometric identities, we obtain

   $$\sin\omega = s_n\cos\Delta\omega^* + c_n\sin\Delta\omega^*, \qquad \cos\omega = c_n\cos\Delta\omega^* - s_n\sin\Delta\omega^*,$$

   where $s_n \doteq \sin\omega_n$ and $c_n \doteq \cos\omega_n$. For $|\Delta\omega^*| < \pi/2$, we can divide by $\cos\Delta\omega^*$ and, from eq. (12.5), we obtain

   $$\Delta\omega^* = -\arctan\frac{As_n + Bc_n}{Ac_n - Bs_n}. \tag{12.8}$$

   Hence, the second form of the drill RC is defined as

   $$c \doteq \Delta\omega - \Delta\omega^* = 0. \tag{12.9}$$

3. We can directly use eq. (12.5) for large rotations and define the third form of the drill RC as follows:

   $$c \doteq r_\omega = \tfrac{1}{2}\,(A\sin\omega + B\cos\omega) = 0, \tag{12.10}$$

   where A and B depend on $\mathbf{u}$. This form of the drill constraint can be linearized using symbolic differentiation.

Rotational invariance of drill RC equation for 2D problems. Consider components of displacement and rotation tensors in the reference Cartesian basis $\{\mathbf{i}_k\}$.

Let a 2D body be located in the $\{\mathbf{i}_1, \mathbf{i}_2\}$-plane and its deformation, except for the thickness change, also takes place in this plane. The orientation of the local ortho-normal basis $\{\mathbf{t}_k\}$ is defined by the rotation tensor $\mathbf{R}$, i.e. $\mathbf{t}_k = \mathbf{R}\,\mathbf{i}_k$. The normal vector $\mathbf{t}_3$ coincides with $\mathbf{i}_3$. For $\mathbf{Q}$ and $\mathbf{F}$, we use the representations of eq. (12.3) and, additionally, we define

$$\mathbf{R} \doteq \begin{bmatrix} \cos\alpha & -\sin\alpha \\ \sin\alpha & \cos\alpha \end{bmatrix}. \tag{12.11}$$

Define the following back-rotated matrices:

$$\mathbf{Q}_* \doteq \mathbf{R}^T\mathbf{Q}\,\mathbf{R}, \qquad \mathbf{F}_* \doteq \mathbf{R}^T\mathbf{F}\,\mathbf{R}, \qquad (\mathbf{Q}^T\mathbf{F})_* \doteq \mathbf{R}^T(\mathbf{Q}^T\mathbf{F})\,\mathbf{R}, \tag{12.12}$$

where $(\cdot)_*$ designates a back-rotated object. We can check that $\mathbf{Q}_*^T\mathbf{F}_* = (\mathbf{Q}^T\mathbf{F})_*$ and that

$$\text{skew}(\mathbf{Q}_*^T\mathbf{F}_*) = \text{skew}(\mathbf{Q}^T\mathbf{F})_* = \text{skew}(\mathbf{Q}^T\mathbf{F}). \tag{12.13}$$

As a consequence, in the local basis $\{\mathbf{t}_k\}$, we can use the drill RC equation in terms of components in the global basis $\{\mathbf{i}_k\}$.

Remark. The above property does not hold for 3D problems, for which $\mathbf{F}$, $\mathbf{Q}$, and $\mathbf{R}$ are 3×3 matrices. We checked this for the canonical parametrization of the rotation tensor. For 3D problems, only the property $\mathbf{Q}_*^T\mathbf{F}_* = (\mathbf{Q}^T\mathbf{F})_*$ holds and, hence, only

$$\text{skew}(\mathbf{Q}_*^T\mathbf{F}_*) = \text{skew}(\mathbf{Q}^T\mathbf{F})_* \tag{12.14}$$

can be used in the implementation of the element.

Calculation of drilling rotation for given displacement. Calculation of the drilling rotation for the given displacement $\mathbf{u}$ is a post-processing operation, but is not trivial because the equation involved is non-linear w.r.t. drilling rotation angle.

Assume that the displacement $\mathbf{u}$ is given and we wish to calculate the drilling rotation, ω. The above-defined three forms of the drill RC can be used as follows:

M1. Equation (12.6) is used, so the rotation is restricted, i.e. $\omega < |\pi/2|$.

M2. Equation (12.8) is used with the update formula $\omega = \omega_n + \Delta\omega$, where ω_n is known. Besides, A and B depend on the known $\mathbf{u}$. No iterations are needed. The increment is restricted, i.e. $\Delta\omega < |\pi/2|$, but the total rotation ω is not.

M3. Equation (12.10) is used with the Newton method,

$$\Delta\omega = -r_\omega / r_{\omega,\omega}, \qquad \omega = \omega_n + \Delta\omega. \tag{12.15}$$

Iterations are needed. Formally, $\Delta\omega$ is not restricted but the radius of convergence of the Newton method is.

Summarizing, M2 and M3 are incremental and can be used to obtain arbitrarily large drilling rotations.

Example. Drilling rotation for rigid body rotation. The above defined methods can be compared for a rigid rotation of a body, for which

$$\mathbf{F} \doteq \frac{\partial \mathbf{x}}{\partial \mathbf{y}} = \begin{bmatrix} \cos\alpha & -\sin\alpha \\ \sin\alpha & \cos\alpha \end{bmatrix}, \tag{12.16}$$

where α is the angle of a rigid rotation. Then, $A \doteq F_{11} + F_{22} = 2\cos\alpha$ and $B \doteq F_{12} - F_{21} = -2\sin\alpha$ (see the definitions following eq. (12.4)), and we can calculate ω for increasing values of α using the methods defined above. The solutions are shown in Fig. 12.2. The solutions by

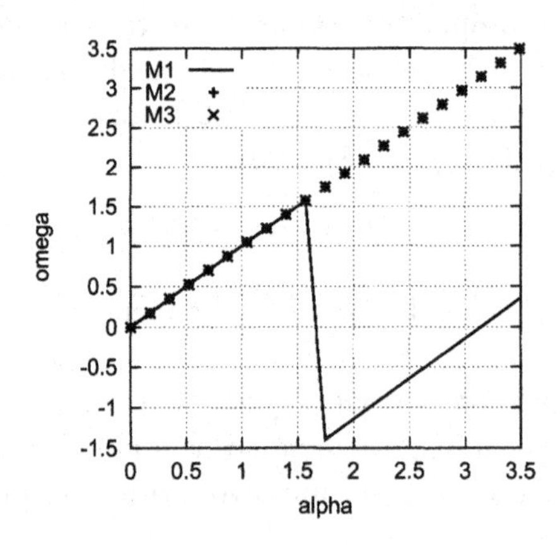

Fig. 12.2 Drilling rotation calculated by three formulas. $\Delta\alpha = 10^\circ$.

M2 and M3 coincide; they are unrestricted and $\omega = \alpha$, as required. The solution by M1 is restricted, i.e. $\omega < |\pi/2|$.

The solutions were obtained for $\Delta\alpha = 10^{\text{o}}$, but larger steps were also tested. M2 performs correctly for up to $\Delta\alpha = 89^{\text{o}}$, due to the restricted domain of arctan, while M3 for up to $\Delta\alpha = 69^{\text{o}}$; for larger steps, it converges to some shifted (incorrect) values.

Note that even $\Delta\alpha = 69^{\text{o}}$ is above the capabilities of current algorithms, e.g. the energy and momentum conserving algorithm ALGO-C1 for rotational rigid body dynamics can perform similar steps, but then the accuracy is poor, see Sect. 9.4.3.

12.2 Difficulties in approximation of drill RC

Approximation of drilling rotation. The drilling rotation is approximated by the bilinear shape functions $N_I(\xi, \eta)$ of eq. (10.3) as follows:

$$\omega(\xi, \eta) = \sum_{I=1}^{4} N_I(\xi, \eta)\, \omega_I, \tag{12.17}$$

where ω_I are the drilling rotations at the corner nodes, see Fig. 12.1, and the natural coordinates $\xi, \eta \in [-1, +1]$. Hence, the drilling rotation is analogously approximated as displacements, see eq. (10.7).

Difficulties in approximation of the drill RC equation. For the equal-order bilinear the approximations of displacements and the drilling rotation, the drill RC equation of the four-node element is incorrectly approximated.

To illustrate the problem, we consider a 2×2 square element with the center located at the origin of the Cartesian coordinate system. Then the Cartesian coordinates are equal to the natural coordinates, i.e. $x = \xi$ and $y = \eta$, and the Jacobian matrix is an identity matrix.

The bilinear approximation functions for displacements and drilling rotations can be written as

$$u(\xi, \eta) = u_0 + u_1\,\xi + u_2\,\eta + u_3\,\xi\eta, \qquad v(\xi, \eta) = v_0 + v_1\,\xi + v_2\,\eta + v_3\,\xi\eta,$$

$$\omega(\xi, \eta) = \omega_0 + \omega_1\,\xi + \omega_2\,\eta + \omega_3\,\xi\eta, \tag{12.18}$$

where u_i, v_i, ω_i $(i = 0, \ldots, 3)$ are functions of nodal values of the respective components. We consider the linearized form of the drill RC of eq. (12.10), i.e.

$$c \doteq \omega + \frac{1}{2}(u_{,\eta} - v_{,\xi}) = 0, \tag{12.19}$$

where $u_{,\eta} = u_{1,2}$ and $v_{,\xi} = u_{2,1}$. Using the above approximation functions, grouping the terms, we obtain

$$\left[\omega_0 + \frac{1}{2}(u_2 - v_1)\right] + \left(\omega_1 + \frac{1}{2}u_3\right)\xi + \left(\omega_2 - \frac{1}{2}v_3\right)\eta + \underline{\underline{\omega_3\,\xi\eta}} = 0. \quad (12.20)$$

The constant and linear terms do link the displacement and rotational parameters, indeed, but the bilinear (underlined) term contains only the rotational parameter ω_3. This last term may lead to wrong solutions in certain situations for the reason explained below.

Let us rewrite the constraint in the form $c(\xi,\eta) \doteq c_0 + c_1\xi + c_2\eta + \omega_3\xi\eta$. If we use the penalty method, then the weak form of this constraint is

$$\frac{1}{2}\int_{-1}^{+1}\int_{-1}^{+1} c(\xi,\eta)^2 \, \mathrm{d}\xi \mathrm{d}\eta = 2c_0^2 + \frac{2}{3}(c_1^2 + c_2^2) + \underline{\underline{\frac{2}{9}\omega_3^2}}, \quad (12.21)$$

i.e. ω_3 does not vanish upon integration. As a consequence, the penalty method enforces the condition $\omega_3 = 0$ which, generally, is incorrect and can yield an over-stiffened solution.

To alleviate this problem, we remove the bilinear term from eq. (12.20), and use the equation which is only linear in ξ and η, i.e.

$$\left[\omega_0 + \frac{1}{2}(u_2 - v_1)\right] + \left(\omega_1 + \frac{1}{2}u_3\right)\xi + \left(\omega_2 - \frac{1}{2}v_3\right)\eta = 0. \quad (12.22)$$

In a symbolic derivation of an element, the bilinear term can be removed in one of the following ways:

1. Using the linear expansion of eq. (12.20) at the element's center,

$$c(\xi,\eta) \doteq c_c + \xi(c_{,\xi})_c + \eta(c_{,\eta})_c, \quad (12.23)$$

where the subscript c denotes the element's center, see also eq. (12.28).

2. Evaluating this equation at the mid-points of the element's edges,

$$(\xi,\eta) = (0,\pm 1), \qquad (\xi,\eta) = (\pm 1, 0), \quad (12.24)$$

where either ξ or η is zero, so the bilinear term in eq. (12.20) is always zero.

The lack of an equation for ω_3 means that the tangent matrix for the drill RC has one spurious zero eigenvalue; the associated eigenvector Θ_2 is shown in Fig. 12.9b. The simplest way of treating this deficiency is to apply eq. (12.143), which provides the stabilization matrix $\mathbf{K}_{\omega\omega}^{\text{stab}}$ of eq. (12.144), so we have

$$\mathbf{K}_{\omega\omega} + \mathbf{K}_{\omega\omega}^{\text{stab}}, \tag{12.25}$$

where $\mathbf{K}_{\omega\omega}$ is the rank-deficient matrix obtained by differentiating twice the drill RC term modified as given either in eq. (12.23) or in eq. (12.24).

Expansion of $\mathbf{Q}^T\mathbf{F}$ product. The linear expansion of $c(\xi,\eta)$ of eq. (12.23) can be obtained in the following way. First, we expand $\mathbf{Q}$ and $\mathbf{F}$ at the element's center,

$$\mathbf{Q}(\xi,\eta) \doteq \mathbf{Q}_c + \xi\,\mathbf{Q}_{,\xi c} + \eta\,\mathbf{Q}_{,\eta c}, \qquad \mathbf{F}(\xi,\eta) \doteq \mathbf{F}_c + \xi\,\mathbf{F}_{,\xi c} + \eta\,\mathbf{F}_{,\eta c}, \tag{12.26}$$

where $(\,)_{,\xi c} \doteq (\,)_{,\xi}|_c$ and $(\,)_{,\eta c} \doteq (\,)_{,\eta}|_c$. Then, we calculate the $\mathbf{Q}^T\mathbf{F}$ product, in which we retain only the constant and linear terms,

$$\mathbf{Q}^T(\xi,\eta)\,\mathbf{F}(\xi,\eta) \approx \mathbf{Q}_c^T\mathbf{F}_c + \xi\left(\mathbf{Q}_c^T\mathbf{F}_{,\xi c} + \mathbf{Q}_{,\xi c}^T\mathbf{F}_c\right) + \eta\left(\mathbf{Q}_c^T\mathbf{F}_{,\eta c} + \mathbf{Q}_{,\eta c}^T\mathbf{F}_c\right), \tag{12.27}$$

while bilinear and quadratic terms are omitted. Finally,

$$c(\xi,\eta) \doteq \left[\text{skew}(\mathbf{Q}^T\mathbf{F})\right]_{12}, \tag{12.28}$$

i.e. we calculate the skew-symmetric part of the matrix and, provided that the matrix is given in the local Cartesian basis, we use the 12 component.

Enhancement resulting from bi-quadratic approximations of displacements. Another way of addressing the problem with the bilinear term in eq. (12.20) is to enhance displacements in such a way that ω_3 is linked with enhancing parameters.

The enhancing modes can be selected upon analysis of bi-quadratic approximations of displacements of the nine-node Lagrangian element shown in Fig. 12.3. For the bi-quadratic approximations, the vector of shape functions is defined as

$$\mathbf{N} \doteq \{P_1Q_1, P_3Q_1, P_3Q_3, P_1Q_3, P_2Q_1, P_3Q_2, P_2Q_3, P_1Q_2, P_2Q_2\}, \tag{12.29}$$

where each component is a product of P_i and Q_j, $i,j \in \{1,2,3\}$, defined as

$$P_1 \doteq \tfrac{1}{2}(\xi^2 - \xi), \qquad P_2 \doteq 1 - \xi^2, \qquad P_3 \doteq \tfrac{1}{2}(\xi^2 + \xi),$$

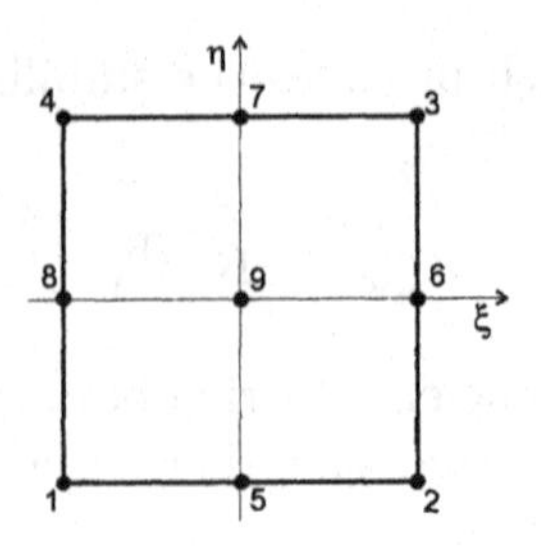

Fig. 12.3 Numeration of nodes on a nine-node Lagrangian element.

$$Q_1 \doteq \tfrac{1}{2}(\eta^2 - \eta), \qquad Q_2 \doteq 1 - \eta^2, \qquad Q_3 \doteq \tfrac{1}{2}(\eta^2 + \eta).$$

Using $\mathbf{N}$ for each displacement component separately, we have

$$u = u_0 + u_1\,\xi + u_2\,\eta + u_3\,\xi\eta + u_4\,\xi^2 + u_5\,\eta^2 + u_6\,\xi\eta^2 + u_7\,\xi^2\eta + u_8\,\xi^2\eta^2,$$
$$v = v_0 + v_1\,\xi + v_2\,\eta + v_3\,\xi\eta + v_4\,\xi^2 + v_5\,\eta^2 + v_6\,\xi\eta^2 + v_7\,\xi^2\eta + v_8\,\xi^2\eta^2,$$

where $u_i,\, v_i \quad (i = 0, \ldots, 8)$ are functions of nodal values of respective displacement components.

For simplicity, we consider a 2×2 square element with the center located at the origin of the Cartesian coordinate system. Then the bilinear term of the drill RC of eq. (12.19) yields the equation

$$u_7 - v_6 + \omega_3 = 0. \tag{12.30}$$

Using this equation, the $\xi\eta$ mode of drilling rotation is linked with the $\xi^2\eta$ and $\xi\eta^2$ modes of displacements. These modes are not available in a four-node bilinear element, but can be included as the EADG enhancement.

Let us assume the incompatible displacements in the form

$$u^{\text{inc}} = q_5\,\xi^2\eta, \qquad v^{\text{inc}} = q_6\,\xi\eta^2, \tag{12.31}$$

where q_5, q_6 are unknown multipliers. Then

$$\mathbf{G}_2^{\xi} \doteq \begin{bmatrix} \frac{\partial u^{\text{inc}}}{\partial \xi} & \frac{\partial u^{\text{inc}}}{\partial \eta} \\ \frac{\partial v^{\text{inc}}}{\partial \xi} & \frac{\partial v^{\text{inc}}}{\partial \eta} \end{bmatrix} = \begin{bmatrix} 2q_5\,\xi\eta & q_5\,\xi^2 \\ q_6\,\eta^2 & 2q_6\,\xi\eta \end{bmatrix},$$

where we have the $\xi\eta$ term on the diagonal. This matrix should be added to $\mathbf{G}^{\xi}$ in eq. (11.71) for the EADG enhancement.

We implemented the element based on the potential energy, the EADG4 enhancement, and the above-defined enhancement, but the displacements and rotations in Cook's membrane example of Sect. 15.2.7 were excessive.

12.3 Implementation of drill RC in finite elements

Overview. The drilling rotations are included in shell elements with the purpose of having three rotational degrees of freedom at each node and to facilitate linking the elements of various spatial orientation. Several methods can be applied to develop the finite element with the drilling rotation.

1. Basic method. The basic method amounts to appending the drill RCs evaluated at some points to the elemental set of equations. The method is simple but the tangent matrix is non-symmetric.

2. Methods of constrained optimization. The methods of constrained optimization, see [136, 72, 29], yield a symmetric tangent matrix, but are more complicated than the basic method. The optimization problem is defined as follows:

$$\min_{(\mathbf{u},\mathbf{Q})} F(\mathbf{u}) \qquad \text{subject to} \quad \mathbf{c}(\mathbf{u},\mathbf{Q}) = \mathbf{0}, \tag{12.32}$$

where F is the governing functional and $\mathbf{c}(\mathbf{u},\mathbf{Q}) = \mathbf{0}$ is the set of constraints related to the drill RC equation. Note that

1. various functional can be used as F, including the potential energy F_{PE}, the HR functional F_{HR}, and the HW functional F_{HW} of Sect. 11.1. However, for each functional, an optimal finite element must be developed separately.
2. Several forms of the constraint can be formulated for the drill RC equation, including strong and weak (integral) forms,
3. Several methods can be used to solve this problem of constrained optimization; in our elements we use either the penalty method or the Perturbed Lagrange method.

Below, for simplicity, we consider the potential energy $F_{\text{PE}}(\mathbf{u}) \doteq \int_B \mathcal{W}(\mathbf{u})\, dV - F_{\text{ext}}$. The extended functional which is constructed for the above-constrained optimization problem includes the part related to the drill RC equation, which can be written in two forms:

A. Strong form. Let us write the constraint related to drill RC equation as $c = 0$. We can evaluate $c(\xi, \eta)$ at four selected points within an element and form a vector, $\mathbf{c} \doteq \{c_1, c_2, c_3, c_4\}$, which is used by the strong forms discussed below.

The penalty method is based on the following extended functional,

$$F'_{\mathrm{PE}}(\mathbf{u},\mathbf{Q}) \doteq F_{\mathrm{PE}}(\mathbf{u}) + \frac{\gamma}{2}\mathbf{c}\cdot\mathbf{c}, \tag{12.33}$$

where $\gamma > 0$ is the penalty parameter. Minimization is performed w.r.t. the nodal values of $(\mathbf{u},\mathbf{Q})$. The definition of γ depends on material coefficients and the element's volume, to preserve the same degree of penalization for various volumes.

The perturbed Lagrange method is based on the following extended functional

$$F'_{\mathrm{PE}}(\mathbf{u},\mathbf{Q},\boldsymbol{\lambda}) \doteq F_{\mathrm{PE}}(\mathbf{u}) + \boldsymbol{\lambda}\cdot\mathbf{c} + \frac{1}{2\gamma}\boldsymbol{\lambda}\cdot\boldsymbol{\lambda}, \tag{12.34}$$

where $\boldsymbol{\lambda} \doteq \{\lambda_1,\lambda_2,\lambda_3,\lambda_4\}$ is a vector of Lagrange multipliers, with each multiplier for $c = 0$ written at a point within an element. Minimization is performed w.r.t. nodal values of $(\mathbf{u},\mathbf{Q})$ and the elemental vector $\boldsymbol{\lambda}$.

B. Weak (integral) form. The part related to the drill RC can also be formulated in an integral form, resembling the form of the strain energy, which is an integral over the element volume, $\int_B \mathcal{W}(\mathbf{u})\, \mathrm{d}V$. Let the drill RC have the form $c(\xi,\eta) = 0$.

The penalty method is based on the following extended functional:

$$F'_{\mathrm{PE}}(\mathbf{u},\mathbf{Q}) \doteq F_{\mathrm{PE}}(\mathbf{u}) + \int_B \frac{\gamma}{2} c^2\, \mathrm{d}V, \tag{12.35}$$

where $\gamma > 0$ is the penalty parameter. Minimization is performed w.r.t. the nodal values of $(\mathbf{u},\mathbf{Q})$. The volume of the element is automatically accounted for by the integral formulation, so it suffices to relate γ to material coefficients.

The perturbed Lagrange method is based on the following extended functional:

$$F'_{\mathrm{PE}}(\mathbf{u},\mathbf{Q},\boldsymbol{\lambda}) \doteq F_{\mathrm{PE}}(\mathbf{u}) + \int_B \left(\lambda c + \frac{1}{2\gamma}\lambda^2\right) \mathrm{d}V, \tag{12.36}$$

where λ is the Lagrange multiplier which must be approximated (assumed) over the element. Minimization is performed w.r.t. the nodal values of $(\mathbf{u},\mathbf{Q})$ and the elemental parameters of λ.

The weak forms correspond to the variational formulations of Sect. 4, and were used in implementation of our elements.

12.3.1 Selected methods to include the drill RC

Below we discuss selected methods used to include the drilling rotations in FE equations, such as the basic method, the penalty method, and the Perturbed Lagrange method.

1. Basic method

The basic method consists of two steps:

1. The drill RC equation is expanded as specified by eq. (12.23), and evaluated at four selected points within an element, which yields the equation $\mathbf{r}_\omega = \mathbf{0}$, where $\mathbf{r}_\omega \doteq \{r_{\omega 1}, r_{\omega 2}, r_{\omega 3}, r_{\omega 4}\}$. The linearized (Newton) form of this equation is

$$\mathbf{K}_{\omega u} \Delta \mathbf{u}_I + \mathbf{K}_{\omega\omega} \Delta \boldsymbol{\omega}_I = -\mathbf{r}_\omega, \tag{12.37}$$

   where $\mathbf{u}_I \doteq \{\mathbf{u}_1, \mathbf{u}_2, \mathbf{u}_3, \mathbf{u}_4\}$ and $\boldsymbol{\omega}_I \doteq \{\omega_1, \omega_2, \omega_3, \omega_4\}$ are vectors of displacements and drilling rotations at nodes, and the matrices are

$$\mathbf{K}_{\omega\omega} \doteq \frac{\partial \mathbf{r}_\omega}{\partial \boldsymbol{\omega}_I}, \qquad \mathbf{K}_{\omega u} \doteq \frac{\partial \mathbf{r}_\omega}{\partial \mathbf{u}_I}.$$

   The drill RC equation can be used in one of the forms specified in eqs. (12.7), (12.9), and (12.10); in computations, we used the last one.
2. Equation (12.37) is appended to the set of FE equations for a purely displacement problem, $\mathbf{K}\, \Delta \mathbf{u}_I = -\mathbf{r}$, where

$$\mathbf{r} \doteq \frac{\partial F_{\mathrm{PE}}(\mathbf{u}_I)}{\partial \mathbf{u}_I}, \qquad \mathbf{K} \doteq \frac{\partial \mathbf{r}}{\partial \mathbf{u}_I}.$$

   This yields

$$\begin{bmatrix} \mathbf{K} & \mathbf{0} \\ \mathbf{K}_{\omega u} & \mathbf{K}_{\omega\omega} \end{bmatrix} \begin{bmatrix} \Delta \mathbf{u}_I \\ \Delta \boldsymbol{\omega}_I \end{bmatrix} = - \begin{bmatrix} \mathbf{r} \\ \mathbf{r}_\omega \end{bmatrix}. \tag{12.38}$$

   This is a set of equations for an element. By aggregation of such sets for all elements, we obtain the global tangent matrix, which must be non-singular to provide a unique solution. The increments of displacements and drilling rotations at nodes are computed together.

Note that the matrix in eq. (12.38) is non-symmetric, which is a disadvantage, as symmetric solvers are faster. If a non-symmetric solver is used for other reasons, then this formulation also is suitable.

Consider stability of the basic formulation. We assume that the boundary conditions are accounted for in the set (12.38). From the first equation of (12.38), we calculate $\Delta \mathbf{u}_I = -\mathbf{K}^{-1}\mathbf{r}$ and using it in the second equation, we obtain

$$\Delta \omega_I = -\mathbf{K}_{\omega\omega}^{-1}(\mathbf{r}_\omega - \mathbf{K}_{\omega u}\mathbf{K}^{-1}\mathbf{r}). \tag{12.39}$$

Let us write this equation at $\mathbf{u}_I = \mathbf{0}$ and $\boldsymbol{\omega}_I = \mathbf{0}$. Then, the residuals $\mathbf{r}_u = \mathbf{0}$, $\mathbf{r}_\omega = \mathbf{0}$, and $\mathbf{r} = -\mathbf{p}$, where $\mathbf{p}$ is a vector of external loads for translational dofs, and we obtain

$$\Delta \omega_I = -(\mathbf{K}_{\omega\omega}^{-1}\mathbf{K}_{\omega u}\mathbf{K}^{-1})\ \mathbf{p}. \tag{12.40}$$

Hence, to uniquely compute the solution, $\mathbf{K}$ and $\mathbf{K}_{\omega\omega}$ must be invertible. Note that

1. elimination of ω_3 from eq. (12.20) means that $\mathbf{K}_{\omega\omega}$ becomes singular and must be stabilized, as given by eq. (12.25).
2. The drill RC cannot be evaluated at mid-side edge points, as then $\mathbf{K}_{\omega\omega}$ has complex eigenvalues, see the example below. The Gauss points or the corner nodes are good locations.

Example. Consider the single trapezoidal element of Fig. 15.1b, obtained for $d = 0.5$, and $E = 10^6$, $\nu = 0.3$, $h = 0.1$. The standard element Q4 and the basic method for the drill RC were used.

The eigenvalues of $\mathbf{K}_{\omega\omega}$ obtained for various locations of the evaluation points are shown in Table 12.1 and we see that they differ and do not depend on the element's shape, a specific property of $\mathbf{K}_{\omega\omega}$! For the mid-side nodes, we obtain two complex eigenvalues!

Table 12.1 Basic method. Eigenvalues of $\mathbf{K}_{\omega\omega}$.

Drill RC evaluated at	Eigenvalues (truncated)			
trapezoidal element				
Gauss points	2	1.15	1.15	0.66
corner nodes	2	2	2	2
mid-side nodes	2	1+i	1-i	0
square element				
Gauss points	2	1.15	1.15	0.66
corner nodes	2	2	2	2
mid-side nodes	2	1+i	1-i	0

Besides, we consider stretching the element in a vertical direction. Two parallel and equal forces are applied at the top nodes, while the boundary conditions eliminating rigid body modes are applied to displacements at the bottom nodes. The calculated drilling rotations are shown in Table 12.2, and we see that they are different for various location of the evaluation points. These differences vanish for a square element, $d = 0$.

Table 12.2 Basic method. Drilling rotations for stretched element.

Drill RC evaluated at	Drilling rotation at nodes			
trapezoidal element				
Gauss points	0.310	0.210	0.581	0.771
corner nodes	0.280	0.195	0.574	0.756
square element				
Gauss points	0	0	0	0
corner nodes	0	0	0	0

2. Penalty method

The penalty method is a classical method of solving problems of constrained optimization, [72, 29]. Generally, it is defined as a sequence of unconstrained optimization problems, which are solved for selected increasing values of the penalty parameter, [72] eq. (12.1.4). However, for efficiency reasons, the shortcut method is used in practice and not a sequence of problems, but a single unconstrained optimization problem is solved for a largish value of the penalty parameter. Hence, some errors are inevitable and we try to minimize them by selecting a suitable value of the penalty parameter; this issue is discussed in Sect. 12.3.2.

The penalty method can be used with the drill RC term in one of the previously mentioned two forms:

1. the strong form of eq. (12.33), for which we can consider several locations to evaluate the drill RC, similarly as for the basic method. For the penalty method, however, the matrix $\mathbf{K}_{\omega\omega}$ has no complex eigenvalues, for any of the considered locations.
2. the weak form of eq. (12.35), for which the drill RC is evaluated at Gauss points. Because of the integral form, the weak form automatically accounts for the element volume, so it suffices to relate the penalty coefficient γ to material coefficients.

Note that if the strong form is evaluated at Gauss points, then the only difference between these two forms are the determinants of the Jacobian used in the weak form.

The scalar drill RC equation can be used in one of the three forms specified in eqs. (12.7), (12.9), and (12.10); in computations we used the last one.

The weak form is discussed in detail below.

Relation of weak form of eq. (12.35) to variational formulation of Sect. 4. Recall the formulation based on the second Piola-Kirchhoff stress and the 3-F functional of eq. (4.65). This functional was regularized in $\mathbf{T}_a$, which is the Lagrange multiplier for the RC equation (3.8). Then, using the Euler–Lagrange equation for $\delta\mathbf{T}_a$, i.e. $\mathbf{T}_a = \gamma \,\text{skew}(\mathbf{Q}^T\mathbf{F})$, we obtained the 2-F functional of eq. (4.73), which we repeat it here as

$$\tilde{F}_2^{2\text{PK}}(\chi, \mathbf{Q}) \doteq \int_B \left[\mathcal{W}(\mathbf{F}^T\mathbf{F}) + F_{\text{RC}}^{\text{P}}(\chi, \mathbf{Q})\right] \, \text{d}V + F_{\text{ext}}, \tag{12.41}$$

where the penalty term for the RC equation is

$$F_{\text{RC}}^{\text{P}}(\chi, \mathbf{Q}) \doteq \frac{\gamma}{2} \text{skew}(\mathbf{Q}^T\mathbf{F}) \cdot \text{skew}(\mathbf{Q}^T\mathbf{F}). \tag{12.42}$$

If we restrict our considerations to a 2D problem, then the RC equation is reduced to the drill RC equation. For $\mathbf{Q}$ and $\mathbf{F}$ of eq. (12.3), we obtain $\text{skew}(\mathbf{Q}^T\mathbf{F})$ of eq. (12.4), for which

$$F_{\text{RC}}^{\text{P}}(\chi, \mathbf{Q}) = \frac{\gamma}{2} 2\, r_\omega^2. \tag{12.43}$$

Comparing this expression for F_{RC} with the drill term in the weak form, eq. (12.35), which is $(\gamma/2)\, r_\omega^2$, we see that the difference between them is the multiplier 2, which is a result of two identical (except for the sign) terms of the skew-symmetric matrix. Hence, functional (12.41) fully corresponds to the weak form of eq. (12.35).

Linearized equations. The standard procedure of consistent linearization of the functionals for the penalty method yields the following linearized (Newton) equations:

$$\left(\begin{bmatrix} \mathbf{K} & \mathbf{0} \\ \mathbf{0} & \mathbf{0} \end{bmatrix} + \gamma \begin{bmatrix} \mathbf{K}_{uu} & \mathbf{K}_{u\omega} \\ \mathbf{K}_{\omega u} & \mathbf{K}_{\omega\omega} \end{bmatrix} \right) \begin{Bmatrix} \Delta\mathbf{u}_I \\ \Delta\boldsymbol{\omega}_I \end{Bmatrix} = - \begin{Bmatrix} \mathbf{r} + \gamma\mathbf{r}_u \\ \gamma\mathbf{r}_\omega \end{Bmatrix}, \tag{12.44}$$

where

$$\mathbf{r}_u \doteq \frac{\partial F_{\text{RC}}}{\partial \mathbf{u}_I}, \qquad \mathbf{K}_{uu} \doteq \frac{\partial \mathbf{r}_u}{\partial \mathbf{u}_I}, \qquad \mathbf{K}_{u\omega} \doteq \frac{\partial \mathbf{r}_u}{\partial \boldsymbol{\omega}_I},$$

$$\mathbf{r}_\omega \doteq \frac{\partial F_{\text{RC}}}{\partial \boldsymbol{\omega}_I}, \qquad \mathbf{K}_{\omega\omega} \doteq \frac{\partial \mathbf{r}_\omega}{\partial \boldsymbol{\omega}_I}, \qquad \mathbf{K}_{\omega u} \doteq \frac{\partial \mathbf{r}_\omega}{\partial \mathbf{u}_I}.$$

The tangent matrix is symmetric, unlike the one for the basic method of eq. (12.38). The above-defined vectors and matrices do not depend on the penalty parameter γ; its value is selected as described in the next section.

Elimination of ω_3 from eq. (12.20) means that $\mathbf{K}_{\omega\omega}$ becomes singular and must be stabilized, as given by eq. (12.25), to ensure its invertibility.

Let us write eq. (12.44) at $\mathbf{u}_I = \mathbf{0}$ and $\boldsymbol{\omega}_I = \mathbf{0}$. Then the residuals $\mathbf{r}_u = \mathbf{0}$, $\mathbf{r}_\omega = \mathbf{0}$, and $\mathbf{r} = -\mathbf{p}$, where $\mathbf{p}$ is a vector of external loads for translational dofs, and we obtain

$$\left(\begin{bmatrix} \mathbf{K} & \mathbf{0} \\ \mathbf{0} & \mathbf{0} \end{bmatrix} + \gamma \begin{bmatrix} \mathbf{K}_{uu} & \mathbf{K}_{u\omega} \\ \mathbf{K}_{\omega u} & \mathbf{K}_{\omega\omega} \end{bmatrix} \right) \begin{bmatrix} \Delta \mathbf{u}_I \\ \Delta \boldsymbol{\omega}_I \end{bmatrix} = \begin{bmatrix} \mathbf{p} \\ \mathbf{0} \end{bmatrix}. \tag{12.45}$$

We assume that the boundary conditions are accounted for in this set and consider the stability requirements for the following two particular solution processes.

A. From the second equation of the set (12.45), we calculate

$$\Delta \boldsymbol{\omega}_I = -\mathbf{K}_{\omega\omega}^{-1} \mathbf{K}_{\omega u} \Delta \mathbf{u}_I \tag{12.46}$$

and use it in the first equation, which yields

$$(\mathbf{K} + \gamma \mathbf{K}_1)\, \Delta \mathbf{u}_I = \mathbf{p}, \tag{12.47}$$

where $\mathbf{K}_1 \doteq \mathbf{K}_{uu} - \mathbf{K}_{u\omega} \mathbf{K}_{\omega\omega}^{-1} \mathbf{K}_{\omega u}$. Hence, the stability requires invertibility of $\mathbf{K}_{\omega\omega}$ and $\mathbf{K} + \gamma \mathbf{K}_1$.

B. Much more complicated stability conditions are obtained if we change the order in which $\Delta \boldsymbol{\omega}_I$ and $\Delta \mathbf{u}_I$ are calculated. Then, from the first equation of the set (12.44), we calculate

$$\Delta \mathbf{u}_I = \mathbf{K}_*^{-1} (\mathbf{p} - \gamma \mathbf{K}_{u\omega} \Delta \boldsymbol{\omega}_I), \tag{12.48}$$

where $\mathbf{K}_* \doteq \mathbf{K} + \gamma \mathbf{K}_{uu}$, and using it in the second equation, we obtain

$$\Delta \boldsymbol{\omega}_I = -\mathbf{K}_{**}^{-1} \mathbf{K}_{\omega u} \mathbf{K}_*^{-1} \mathbf{p}, \tag{12.49}$$

where $\mathbf{K}_{**} \doteq \mathbf{K}_{\omega\omega} - \gamma \mathbf{K}_{\omega u} \mathbf{K}_{*}^{-1} \mathbf{K}_{u\omega}$. To obtain a stable formulation, we need the invertibility of $\mathbf{K}_{*}$ and $\mathbf{K}_{**}$, so we have to satisfy two conditions,

$$\det \mathbf{K}_{*}(\gamma) \neq 0, \qquad \det \mathbf{K}_{**}(\gamma) \neq 0. \tag{12.50}$$

They are too complicated to determine analytically which values of γ are not admissible.

Neither one of the solution processes **A** and **B** is applicable to a general problem, involving many elements and boundary conditions for displacements and drilling rotations. Then we have to consider the whole set (12.44) and invertibility of the global tangent matrix is required.

Remark on the Augmented Lagrange method. We have also implemented the Augmented Lagrange (AuL) method as an extension of the penalty method, requiring only minor modifications of the code. The update formula for the Lagrange multiplier of [179] was applied and several approximations of the Lagrange multiplier were tested. In linear tests, the AuL method performs identically to the penalty method, but in nonlinear tests, e.g. in the pinched hemisphere with a hole of Sect. 15.3.8, the performance was worse than that of the penalty method.

Example. Single element. Consider a single element of Fig. 15.1b, with $E = 10^6$, $\nu = 0.3$, $h = 0.1$. The non-enhanced element Q4 and the penalty method for the drill RC were used.

The matrix $\mathbf{K}_1$ of eq. (12.47) was calculated for two element shapes, trapezoid ($d = 0.5$) and square ($d = 0$), and the boundary conditions were either applied or not applied. In all these cases, $\mathbf{K}_1 = \mathbf{0}$. Hence, the correct solution,

$$\Delta \mathbf{u}_I = \mathbf{K}^{-1} \mathbf{p}, \qquad \Delta \omega_I = -\mathbf{K}_{\omega\omega}^{-1} \mathbf{K}_{\omega u} \Delta \mathbf{u}_I, \tag{12.51}$$

is obtained for any value of $\gamma > 0$; this is a feature of the so-called *exact* penalty method. A similar result is obtained for the formulation with the drilling rotation being a local variable, discontinuous across the element's boundaries.

The determinants of matrices $\mathbf{K}_{*}$ and $\mathbf{K}_{**}$ for selected values of γ are given in Table 12.3. We see that they are all non-zero, as required.

Table 12.3 Determinants of $\mathbf{K}_*$ and $\mathbf{K}_{**}$ for the penalty method.

Multiplier γ	det $\mathbf{K}_*$	det $\mathbf{K}_{**}$
trapezoidal element		
1	10^{22}	10^{-6}
G/1000	10^{23}	10^{4}
G	10^{38}	10^{1}
1000 G	10^{56}	10^{28}

3. Perturbed Lagrange method

The Perturbed Lagrange method belongs to the class of the Lagrange–Newton methods of the constrained optimization, see [136, 72, 29], to which the popular SQP (Sequential Quadratic Programming) method also belongs. For this class, the Newton method is used to find the stationary point of the Lagrange function w.r.t. the basic variables and the Lagrange multipliers.

In the Perturbed Lagrange method, a small perturbation term is defined in terms of the Lagrange multipliers and added to the standard Lagrange function. In computational contact mechanics, which involves inequality constraints, this method was used, e.g., in [157, 222]. For contact problems, the role of the perturbation component is to fill in the zero sub-matrix when the gap is open, see [261], eqs. (5.58) and (9.75).

The Perturbed Lagrange method can also be applied to the drill RC problem and it enables us to treat the Lagrange multipliers as local variables and to eliminate them on the element's level. Besides, we can use a simple symmetric solver on the element's level because there is no zero diagonal blocks.

The Perturbed Lagrange method can be used with the drill RC term in two forms: either the strong form of eq. (12.34) or the weak form of eq. (12.36). For both, we can use the scalar drill RC equation in one of the three forms specified in eqs. (12.7), (12.9), and (12.10); in computations we use the last one.

Relation of weak form of eq. (12.36) to variational formulation of Sect. 4. Recall the formulation based on the second Piola-Kirchhoff stress, and the three-field functional of eq. (4.65). This functional was regularized in $\mathbf{T}_a$, where $\mathbf{T}_a$ is the Lagrange multiplier for the RC equation (3.8), which yielded the 3-F functional of eq. (4.71), which we repeat here in the following form:

$$\tilde{F}_3^{\rm 2PK}(\chi, \mathbf{Q}, \mathbf{T}_a) = \int_B [\mathcal{W}(\mathbf{F}^T\mathbf{F}) + F_{\rm RC}(\chi, \mathbf{Q}, \mathbf{T}_a)] \, \mathrm{d}V - F_{\rm ext}, \quad (12.52)$$

where

$$F_{\rm RC}^{\rm PL}(\chi, \mathbf{Q}, \mathbf{T}_a) \doteq \mathbf{T}_a \cdot \mathrm{skew}(\mathbf{Q}^T\mathbf{F}) - \frac{1}{2\gamma}\mathbf{T}_a \cdot \mathbf{T}_a \quad (12.53)$$

and $\gamma \in (0, \infty)$ is the regularization parameter. If we restrict our considerations to a 2D problem, then the RC equation is reduced to the drill RC equation. For $\mathbf{Q}$ and $\mathbf{F}$ of eq. (12.3), from $\mathrm{skew}(\mathbf{Q}^T\mathbf{F})$ we obtain r_ω, as in eq. (12.4). Besides, the Lagrange multiplier is assumed to be in the following form:

$$\mathbf{T}_a \doteq \begin{bmatrix} 0 & -T \\ T & 0 \end{bmatrix}. \quad (12.54)$$

see eq. (12.56) for more details. For these representations, we obtain $\mathbf{T}_a \cdot \mathrm{skew}(\mathbf{Q}^T\mathbf{F}) = 2T r_\omega$ and $\mathbf{T}_a \cdot \mathbf{T}_a = 2T^2$, and eq. (12.53) becomes

$$F_{\rm RC}^{\rm PL}(\chi, \mathbf{Q}, \mathbf{T}_a) = 2\left(T\, r_\omega - \frac{1}{2\gamma}T^2\right). \quad (12.55)$$

Comparing this expression with the drill term in the weak form eq. (12.36), which is $\lambda\, c + (1/2\gamma)\lambda^2$, we see that the difference between them is the multiplier 2, which is a result of two identical (except for the sign) terms of the skew-symmetric matrices. Hence, functional (12.41) fully corresponds to the weak form of eq. (12.36).

Approximation of $F_{\rm RC}^{\rm PL}$. Various approximations of the functional $F_{\rm RC}$ of eq. (12.55) can be considered for a four-node element, and we selected the following ones:

1. the Lagrange multiplier is assumed as a contravariant matrix in the basis $\{\mathbf{g}_k^c\}$ and is transformed to the local orthonormal basis $\{\mathbf{t}_k^c\}$ as follows:

$$\mathbf{T}_a = \mathbf{J}_{Lc} \begin{bmatrix} 0 & T_a(\xi, \eta) \\ -T_a(\xi, \eta) & 0 \end{bmatrix} \mathbf{J}_{Lc}^T = \begin{bmatrix} 0 & T \\ -T & 0 \end{bmatrix}, \quad (12.56)$$

where $T_a(\xi, \eta)$ is the assumed representation of the Lagrange multiplier, and $T \doteq (\det \mathbf{J}_{Lc})\, T_a(\xi, \eta)$, as the $\mathbf{J}_{Lc}(\cdot)\,\mathbf{J}_{Lc}^T$ operation on a skew-symmetric matrix yields a skew-symmetric matrix. This T was used in eq. (12.54). The Jacobian is local, as both bases are located at the element's center.

2. A linear approximation of the assumed representation of the Lagrange multiplier,
$$T_a(\xi,\eta) \doteq q_0 + \xi\, q_1 + \eta\, q_2, \tag{12.57}$$
where q_0, q_1, q_2 are local multipliers eliminated on the element's level.
3. A linear expansion of the drill rotation constraint, $c = 0$, at the element center of eq. (12.28), for which the bilinear term of the drill RC of eq. (12.20) is eliminated.

Linearized equations. The standard procedure of consistent linearization of the functionals for the Perturbed Lagrange method yields the following linearized (Newton) equations:

$$\begin{bmatrix} \mathbf{K} & \mathbf{K}_{u\omega} & \mathbf{K}_{uT} \\ \mathbf{K}_{u\omega}^T & \mathbf{K}_{\omega\omega} & \mathbf{K}_{\omega T} \\ \mathbf{K}_{uT}^T & \mathbf{K}_{\omega T}^T & -\frac{1}{\gamma}\mathbf{K}_{TT} \end{bmatrix} \begin{bmatrix} \Delta\mathbf{u}_I \\ \Delta\boldsymbol{\omega}_I \\ \Delta\mathbf{T}_a \end{bmatrix} = - \begin{bmatrix} \mathbf{r}+\mathbf{r}_u \\ \mathbf{r}_\omega \\ \mathbf{r}_T \end{bmatrix}, \tag{12.58}$$

where the vectors and matrices obtained from $F_{\text{RC}}^{\text{PL}}$ of eq. (12.55) are as follows:

$$\mathbf{r}_u \doteq \frac{\partial F_{\text{RC}}}{\partial \mathbf{u}_I}, \qquad \mathbf{K}_{uu} \doteq \frac{\partial \mathbf{r}_u}{\partial \mathbf{u}_I}, \qquad \mathbf{K}_{u\omega} \doteq \frac{\partial \mathbf{r}_u}{\partial \boldsymbol{\omega}_I}, \qquad \mathbf{K}_{uT} \doteq \frac{\partial \mathbf{r}_u}{\partial \mathbf{T}_a},$$

$$\mathbf{r}_\omega \doteq \frac{\partial F_{\text{RC}}}{\partial \boldsymbol{\omega}_I}, \qquad \mathbf{K}_{\omega\omega} \doteq \frac{\partial \mathbf{r}_\omega}{\partial \boldsymbol{\omega}_I}, \qquad \mathbf{K}_{\omega u} \doteq \frac{\partial \mathbf{r}_\omega}{\partial \mathbf{u}_I}, \qquad \mathbf{K}_{\omega T} \doteq \frac{\partial \mathbf{r}_\omega}{\partial \mathbf{T}_a},$$

$$\mathbf{r}_T \doteq \frac{\partial F_{\text{RC}}}{\partial \mathbf{T}_a}, \qquad \mathbf{K}_{T\omega} \doteq \frac{\partial \mathbf{r}_T}{\partial \boldsymbol{\omega}_I}, \qquad \mathbf{K}_{Tu} \doteq \frac{\partial \mathbf{r}_T}{\partial \mathbf{u}_I}, \qquad \mathbf{K}_{TT} \doteq \frac{\partial \mathbf{r}_T}{\partial \mathbf{T}_a}.$$

Note that $\mathbf{K}_{uu} = \mathbf{0}$. The total matrix is symmetric because $\mathbf{K} = \mathbf{K}^T$, $\mathbf{K}_{\omega\omega} = \mathbf{K}_{\omega\omega}^T$, $\mathbf{K}_{TT} = \mathbf{K}_{TT}^T$, as well as $\mathbf{K}_{\omega u} = \mathbf{K}_{u\omega}^T$, $\mathbf{K}_{Tu} = \mathbf{K}_{uT}^T$, and $\mathbf{K}_{T\omega} = \mathbf{K}_{\omega T}^T$.

Let us write the set of eq. (12.58) for $\mathbf{u}_I = \mathbf{0}$ and $\boldsymbol{\omega}_I = \mathbf{0}$. Then the residuals $\mathbf{r}_u = \mathbf{0}$, $\mathbf{r}_\omega = \mathbf{0}$, $\mathbf{r}_T = \mathbf{0}$, and $\mathbf{r} = -\mathbf{p}$, where $\mathbf{p}$ is a vector of external loads for translational dofs. For the applied approximations of displacements and drilling rotations, we obtain

$$\begin{bmatrix} \mathbf{K} & \mathbf{0} & \mathbf{K}_{uT} \\ \mathbf{0} & \mathbf{0} & \mathbf{K}_{\omega T} \\ \mathbf{K}_{uT}^T & \mathbf{K}_{\omega T}^T & -\frac{1}{\gamma}\mathbf{K}_{TT} \end{bmatrix} \begin{bmatrix} \Delta\mathbf{u}_I \\ \Delta\boldsymbol{\omega}_I \\ \Delta\mathbf{T}_a \end{bmatrix} = \begin{bmatrix} \mathbf{p} \\ \mathbf{0} \\ \mathbf{0} \end{bmatrix}. \tag{12.59}$$

In the *standard* Lagrange multiplier method, the perturbation term is neglected in eq. (12.55) and, then in the above matrix, the perturbation matrix $\mathbf{K}_{TT}$ is a zero matrix.

Because, the bilinear terms are omitted in the representations of eqs. (12.57) and (12.23), the element has one spurious zero eigenvalue and can be stabilized, as given by eq. (12.25). Then, eq. (12.59) becomes

$$\begin{bmatrix} \mathbf{K} & \mathbf{0} & \mathbf{K}_{uT} \\ \mathbf{0} & \mathbf{K}_{\omega\omega}^{\text{stab}} & \mathbf{K}_{\omega T} \\ \mathbf{K}_{uT}^T & \mathbf{K}_{\omega T}^T & -\frac{1}{\gamma}\mathbf{K}_{TT} \end{bmatrix} \begin{bmatrix} \Delta\mathbf{u}_I \\ \Delta\boldsymbol{\omega}_I \\ \Delta\mathbf{T}_a \end{bmatrix} = \begin{bmatrix} \mathbf{p} \\ \mathbf{0} \\ \mathbf{0} \end{bmatrix}, \tag{12.60}$$

where the matrix $\mathbf{K}_{\omega\omega}^{\text{stab}}$ is provided by stabilization, and the number of zero eigenvalues is three, as required.

Stability of solution for local Lagrange multiplier. Assume that the parameters q_i of the Lagrange multiplier $\mathbf{T}_a$ are *local* variables, which are discontinues across the element boundaries and are eliminated on the element level. Then, first we calculate $\Delta\mathbf{T}_a$ from the third equation of (12.60),

$$\Delta\mathbf{T}_a = \gamma\,\mathbf{K}_{TT}^{-1}\,(\mathbf{K}_{uT}^T\Delta\mathbf{u}_I + \mathbf{K}_{\omega T}^T\Delta\boldsymbol{\omega}_I), \tag{12.61}$$

which is feasible because the perturbation matrix $-(1/\gamma)\mathbf{K}_{TT}$ is nonsingular. Then, we use this $\Delta\mathbf{T}_a$ in the other two equations of eq. (12.60), which yields

$$\left(\begin{bmatrix} \mathbf{K} & \mathbf{0} \\ \mathbf{0} & \mathbf{0} \end{bmatrix} + \gamma \begin{bmatrix} \mathbf{K}_{uu} & \mathbf{K}_{u\omega} \\ \mathbf{K}_{\omega u} & \mathbf{K}_{\omega\omega} \end{bmatrix} \right) \begin{bmatrix} \Delta\mathbf{u}_I \\ \Delta\boldsymbol{\omega}_I \end{bmatrix} = \begin{bmatrix} \mathbf{p} \\ \mathbf{0} \end{bmatrix}, \tag{12.62}$$

where

$$\begin{bmatrix} \mathbf{K}_{uu} & \mathbf{K}_{u\omega} \\ \mathbf{K}_{\omega u} & \mathbf{K}_{\omega\omega} \end{bmatrix} \doteq \begin{bmatrix} \mathbf{K}_{uT}\mathbf{K}_{TT}^{-1}\mathbf{K}_{uT}^T & \mathbf{K}_{uT}\mathbf{K}_{TT}^{-1}\mathbf{K}_{\omega T}^T \\ \mathbf{K}_{\omega T}\mathbf{K}_{TT}^{-1}\mathbf{K}_{uT}^T & \mathbf{K}_{\omega T}\mathbf{K}_{TT}^{-1}\mathbf{K}_{\omega T}^T + \mathbf{K}_{\omega\omega}^{\text{stab}} \end{bmatrix}.$$

Note the similarity of this set of equations to eq. (12.45) obtained for the penalty method. Stability requires invertibility of the whole above tangent matrix.

The local Lagrange multipliers are used in our elements based on the Perturbed Lagrange method.

12.3.2 Selection of regularization parameter for drill RC

Introduction. The regularization parameter γ is used by the two methods discussed earlier, the penalty method and the Perturbed Lagrange method. The value of γ affects the solution and should ensure satisfaction of the following requirements:

1. Displacements yielded by the element with drill rotations should be identical to those yielded by an analogous element without drill rotations; this requirement is incorporated in the definition of the *extended* configuration space, see eq. (4.3).
2. For coarse meshes, displacements for the formulations with and without drill rotations are similar but not identical; the former are slightly stiffer. However, in the mesh limit, i.e. for the element's size tending to zero, the displacements should converge to the displacements of an element without drill rotations.
3. Drill rotations should converge in the mesh limit from the same side as the displacements, i.e. either from below for the fully integrated (FI) elements or from above for the reduced integrated (RI) elements.

Extreme values of γ can cause the following problems:

1. Too large values of γ can yield an ill-conditioned tangent matrix; this is a characteristic deficiency of the penalty method. The ill-conditioning is typically cured by using, instead of the penalty method, the Augmented Lagrangian method, in which the Lagrange multiplier is updated iteratively, and smaller values of γ can be used. This approach is beneficial for non-linear problems which are solved iteratively, but not for the linear ones which are solved without iterations.

   Though γ cannot be too large, it should still ensure a correct transfer of drilling rotations, see the numerical example "Bending of slender cantilever by end drilling rotation" of Sect. 12.8.2. This is an important issue but often forgotten when attention is focused on avoiding over-stiffening (locking).
2. Too small values of γ cause the tangent matrix $\mathbf{K}$ to be rank deficient. In particular, $\gamma = 0$ yields four spurious zero eigenvalues for a four-node bilinear element.

Numerical examples of Sect. 12.8.2 show that a wide range of values of γ exists for which solutions are almost constant and accurate.

Selection of the penalty value in contact mechanics. The methods of constrained optimization are widely used and tested in contact mechanics, [261, 133]. The penalty method is a basic method in this area and is used in the shortened form consisting of a single unconstrained optimization problem. Hence, selection of an optimal value of the penalty parameter requires an error analysis, which takes into account the roundoff errors and the perturbation errors due to the penalty method, see [69, 70, 157, 155].

Typically, such an analysis is limited to linear constrains such that one equation constrains displacements at some points. Note that the drill RC is more complicated, as it is a nonlinear equation involving several variables of a different type, including tangent displacements and drill rotations.

Currently, the Augmented Lagrangian method is very popular in contact mechanics and in this method the question of an optimal penalty value is less acute because smaller values of the penalty can be used, see e.g. [222].

Upper bound on the penalty parameter γ for drill RC. The theoretical considerations, which provide the bounding value of the penalty parameter γ, are given in [99], where equations of linear elasticity with a non-symmetric stress tensor are considered. For the formulation based on the potential energy, the variational problem is written in the form

$$B_\gamma(\mathbf{u}, \tilde{\boldsymbol{\psi}}; \mathbf{v}, \boldsymbol{\omega}) = \mathbf{f}(\{\mathbf{v}, \boldsymbol{\omega}\}), \qquad \forall \{\mathbf{v}, \boldsymbol{\omega}\} \in U, \tag{12.63}$$

where $\mathbf{u}$ is the displacement vector and $\tilde{\boldsymbol{\psi}}$ a skew symmetric tensor for an infinitesimal drilling rotation. The corresponding trial fields are denoted as $\mathbf{v}$ and $\boldsymbol{\omega}$. Besides,

$$\begin{aligned} B_\gamma(\mathbf{u}, \tilde{\boldsymbol{\psi}}; \mathbf{v}, \boldsymbol{\omega}) = & \int_\Omega (\operatorname{sym} \nabla \mathbf{v}) \cdot [\mathbb{C}\, (\operatorname{sym} \nabla \mathbf{u})] \, \mathrm{d}\Omega \\ & + \int_\Omega (\operatorname{skew} \nabla \mathbf{v} - \boldsymbol{\omega}) \cdot \gamma (\operatorname{skew} \nabla \mathbf{u} - \tilde{\boldsymbol{\psi}}) \, \mathrm{d}\Omega \end{aligned} \tag{12.64}$$

is the symmetric bilinear form and

$$\mathbf{f}(\{\mathbf{v}, \boldsymbol{\omega}\}) = \int_\Omega \mathbf{v} \cdot \mathbf{f} \, \mathrm{d}\Omega \tag{12.65}$$

is continuous. Besides, $\mathbb{C}$ is the (rank 4) constitutive tensor.

Well-posedness of a discrete variational problem depends, among the other things, on the U-ellipticity of B_γ. It requires that a constant $\eta > 0$ exists such that

$$B_\gamma(\mathbf{u}, \boldsymbol{\psi}; \mathbf{v}, \boldsymbol{\omega}) \geq \eta \, \|\{\mathbf{v}, \boldsymbol{\omega}\}\|_U^2, \qquad \forall \{\mathbf{v}, \boldsymbol{\omega}\} \in U, \tag{12.66}$$

where

$$\|\{\mathbf{v}, \boldsymbol{\omega}\}\|_U = \|\mathbf{v}\|_V^2 + \|\boldsymbol{\omega}\|_W^2, \qquad \forall \{\mathbf{v}, \boldsymbol{\omega}\} \in U,$$

$$\|\mathbf{v}\|_V^2 = \int_\Omega \|\nabla \mathbf{v}\|^2 \, \mathrm{d}\Omega, \quad \forall \mathbf{v} \in V, \qquad \|\boldsymbol{\omega}\|_W^2 = \int_\Omega \|\boldsymbol{\omega}\|^2 \, \mathrm{d}\Omega, \quad \forall \boldsymbol{\omega} \in W.$$

Note that $U = V \times W$, where V and W are the spaces relevant to the BVP. By using the estimation of the minimum eigenvalue of $\mathbb{C}$ for an isotropic material,

$$\min_{\varepsilon=\varepsilon^T, \varepsilon\neq 0} \frac{\varepsilon \cdot (\mathbb{C}\,\varepsilon)}{\|\varepsilon\|^2} = 2G, \tag{12.67}$$

and Korn's inequality,

$$\|\mathrm{sym}\, \nabla \mathbf{v}\|^2 \geq c_k \, \|\nabla \mathbf{v}\|^2, \tag{12.68}$$

where the constant $c_k = 1/2$ for the Dirichlet problem, we obtain

$$B_\gamma(\mathbf{v}, \boldsymbol{\omega}; \mathbf{v}, \boldsymbol{\omega}) \geq \frac{G}{2} \|\nabla \mathbf{v}\|^2 + (G - \gamma) \|\mathrm{skew}\, \nabla \mathbf{v}\|^2 + \frac{\gamma}{2} \|\boldsymbol{\omega}\|^2. \tag{12.69}$$

Any $0 \leq \gamma \leq G$ is appropriate but the second term in the estimate vanishes for $\gamma = G$, so we obtain

$$B_\gamma(\mathbf{v}, \boldsymbol{\omega}; \mathbf{v}, \boldsymbol{\omega}) \geq \frac{G}{2} \left(\|\nabla \mathbf{v}\|^2 + \|\boldsymbol{\omega}\|^2 \right), \tag{12.70}$$

which is in accord with eq. (12.66). The value $\gamma = G$ was subsequently numerically tested in [102].

Numerical tests of our elements of Sect. 12.8.2 confirm that, generally, the value $\gamma = G$ is a good choice. However, in several situations, a modification of this value is beneficial.

Selection of value of the penalty parameter γ for drill RC. The shell finite elements are very complicated as they (i) involve a large number of variables, (ii) are non-linear, which means that problems are solved iteratively and the number of terms is very large, and (iii) are generated using a complex methodology which cannot easily be accounted for in theoretical considerations.

For this reason, theoretical predictions of the value of γ provide only a general guidance, while a reliable and practically meaningful value of γ must be the result of proper testing. To avoid repeating the process of selection of γ for each BVP, a set of suitable benchmark tests must be used. These problems are solved for a range of values of γ and for each, the segment in which the solution is accurate and almost constant is identified. The final single value of γ should be problem-independent, so we must choose the value which is correct for all benchmark tests. In this task, plots of the obtained displacements and drilling rotations vs. γ are particularly useful.

Generally, the value of γ should account for the element's volume and the material characteristics.

1. In the weak formulation, eq. (12.35), the element volume is accounted for by integration, so the penalty number does not have to include it. But for the strong formulation of eq. (12.33), it must be explicitly included in γ.
2. The material characteristics are accounted for by linking γ to one of the eigenvalues of the constitutive matrix. For instance, for the SVK material and the plane stress conditions, the eigenvalues of the constitutive matrix $\mathbf{C}$ are given by eq. (7.78) and the smallest eigenvalue is $E/(1+\nu) = 2G$. Hence, it is reasonable to relate the value of γ to the shear modulus G.

Hence, in the weak formulation, we use the definition

$$\gamma = \epsilon G, \tag{12.71}$$

where ϵ is a scaling factor. In the numerical tests in Sect. 12.8.2, we select the value of ϵ.

Method of calculating γ for shells of [189]. In this work, the stiffness matrix for a single shell element is divided into parts related to displacements $\mathbf{u}$ and rotation parameters $\boldsymbol{\psi}$ as follows:

$$\mathbf{K} = \begin{bmatrix} \mathbf{K}_{uu} & \mathbf{K}_{u\psi} \\ \mathbf{K}_{\psi u} & \mathbf{K}_{\psi\psi} \end{bmatrix} \tag{12.72}$$

and only diagonal sub-matrices $\mathbf{K}_{uu}$ and $\mathbf{K}_{\psi\psi}$ are considered. They consist of the classical part (C) and the part for the drill RC (D),

$$\mathbf{K}_{uu} = \mathbf{K}^{\mathrm{C}}_{uu} + \gamma\, \mathbf{K}^{\mathrm{D}}_{uu}, \qquad \mathbf{K}_{\psi\psi} = \mathbf{K}^{\mathrm{C}}_{\psi\psi} + \gamma\, \mathbf{K}^{\mathrm{D}}_{\psi\psi}. \tag{12.73}$$

Each of the sub-matrices is considered separately and maximum absolute values of their diagonal terms are compared

$$\gamma_u = \frac{\max|\mathrm{diag}\, \mathbf{K}^{\mathrm{C}}_{uu}|}{\max|\mathrm{diag}\, \mathbf{K}^{\mathrm{D}}_{uu}|}, \qquad \gamma_\psi = \frac{\max|\mathrm{diag}\, \mathbf{K}^{\mathrm{C}}_{\psi\psi}|}{\max|\mathrm{diag}\, \mathbf{K}^{\mathrm{D}}_{\psi\psi}|}. \tag{12.74}$$

The penalty parameter is defined as

$$\gamma = \frac{1}{100} \min(\gamma_u, \gamma_\psi), \tag{12.75}$$

where the value $1/100$ was selected by numerical experiments. The value of γ is calculated only once per analysis, at the beginning, for the linear stiffness matrix.

This method relies on the fact that the diagonal terms of $\mathbf{K}_{uu}$ and $\mathbf{K}_{\psi\psi}$ are much larger than the off-diagonal ones, which allows us to avoid costly eigenvalue analyses.

12.4 EADG method for formulations with rotations

The Enhanced Assumed Displacement Gradient (EADG) method for 2D elements was discussed in Sect. 11.4.3; below it is extended to the 2D+drill elements.

Consider the two-field (2-F) functionals with rotations of Sect. 4 and denote them as $F_2(\chi, \mathbf{Q})$. Let us rewrite eq. (11.70), defining the EADG method, as

$$\mathbf{F} \doteq \nabla\chi + \tilde{\mathbf{H}}, \tag{12.76}$$

where $\nabla\chi = \mathbf{I} + \nabla\mathbf{u}^c$. In the EADG method, we add two independent fields to $F_2(\chi, \mathbf{Q})$: the nominal stress $\mathbf{P}$ and the field $\mathbf{F}$, and construct the following 4-F functional

$$F_4(\chi, \mathbf{Q}, \mathbf{F}, \mathbf{P}) \doteq F_2(\mathbf{Q}, \mathbf{F}) + \int_B \mathbf{P} \cdot (\nabla\chi - \mathbf{F}) \, \mathrm{d}V, \tag{12.77}$$

where $\mathbf{P}$ is a Lagrange multiplier for the formula linking $\nabla\chi$ and the independent $\mathbf{F}$. Note that now $F_2(\mathbf{Q}, \mathbf{F})$ involves the independent $\mathbf{F}$. By using eq. (12.76), this functional becomes

$$F_4(\chi, \mathbf{Q}, \tilde{\mathbf{H}}, \mathbf{P}) = F_3(\chi, \mathbf{Q}, \tilde{\mathbf{H}}) + \int_B \mathbf{P} \cdot \tilde{\mathbf{H}} \, \mathrm{d}V, \tag{12.78}$$

in which we have the enhancing $\tilde{\mathbf{H}}$. If the enhancing $\tilde{\mathbf{H}}$ is orthogonal to the stress, i.e. $\int_B \mathbf{P} \cdot \tilde{\mathbf{H}} \, \mathrm{d}V = 0$, then the last term of eq. (12.78) vanishes and we obtain the 3-F functional

$$F_3(\chi, \mathbf{Q}, \tilde{\mathbf{H}}), \tag{12.79}$$

which does not depend on $\mathbf{P}$ and is used in the element's implementation.

In this way we can obtain the 3-F enhanced functionals for particular forms of $F_2(\chi, \mathbf{Q})$. For instance, for the functionals $\tilde{F}_2^{2PK}(\chi, \mathbf{Q})$ of eq. (4.73) and $F_2^{**}(\chi, \mathbf{Q})$ of eq. (4.77), we obtain

$$\tilde{F}_3^{2PK}(\chi, \tilde{\mathbf{H}}, \mathbf{Q}) \doteq$$
$$\int_B \left\{ \mathcal{W}\left[(\nabla\chi + \tilde{\mathbf{H}})^T (\nabla\chi + \tilde{\mathbf{H}}) \right] + F_{RC}(\chi, \tilde{\mathbf{H}}, \mathbf{Q}) \right\} \, dV - F_{ext}, \tag{12.80}$$
$$F_3^{**}(\chi, \tilde{\mathbf{H}}, \mathbf{Q}) \doteq$$
$$\int_B \left\{ \mathcal{W}\left[\mathbf{Q}^T (\nabla\chi + \tilde{\mathbf{H}}) \right] + F_{RC}(\chi, \tilde{\mathbf{H}}, \mathbf{Q}) \right\} \, dV - F_{ext}, \tag{12.81}$$

where the RC term has the penalty form

$$F_{RC}(\chi, \tilde{\mathbf{H}}, \mathbf{Q}) \doteq \frac{\gamma}{2} \text{skew}[\mathbf{Q}^T (\nabla\chi + \tilde{\mathbf{H}})] \cdot \text{skew}[\mathbf{Q}^T (\nabla\chi + \tilde{\mathbf{H}})]. \tag{12.82}$$

The RC term is also enhanced by $\tilde{\mathbf{H}}$, which is not possible within the EAS method. These functionals were used in [255].

Modification of the EADG method motivated by the EAS method. The EAS method has a certain advantage over the EADG method in non-linear 2D problems, i.e. is slightly faster and converges better. The enhancement of $\mathbf{F}$ is needed in the drill RC equation, so it should be retained in its original form, but we can simplify the Cauchy–Green deformation tensor $\mathbf{C}$ to a form which is similar to that implied by the EAS method.

1. For the EAS method, see eq. (11.62), the enhanced Cauchy–Green tensor is

$$\mathbf{C} \doteq \mathbf{F}^T\mathbf{F} + \mathbf{G}^\xi, \qquad \mathbf{G}^\xi \doteq \begin{bmatrix} q_1\xi & q_3\xi + q_4\eta \\ q_3\xi + q_4\eta & q_2\eta \end{bmatrix}, \tag{12.83}$$

where, for simplicity, we omitted the Jacobians.

2. In the EADG method, the enhancing modes are added to the deformation gradient, see eq. (12.76), and the Cauchy–Green deformation tensor is

$$\mathbf{C} = \mathbf{F}^T\mathbf{F} + \mathbf{F}^T\tilde{\mathbf{H}} + \tilde{\mathbf{H}}^T\mathbf{F} + \tilde{\mathbf{H}}^T\tilde{\mathbf{H}}. \tag{12.84}$$

Let us use the EADG4 enhancement of eq. (11.71) in which, for simplicity, we omit the Jacobians. Then, we have

$$\tilde{\mathbf{H}} = \begin{bmatrix} \xi\, q_1 & \eta\, q_3 \\ \xi\, q_4 & \eta\, q_2 \end{bmatrix}, \qquad \mathbf{F} = \begin{bmatrix} F_{11} & F_{12} \\ F_{21} & F_{22} \end{bmatrix} \tag{12.85}$$

and the last three components of eq. (12.84) involve the enhancement and are

$$\mathbf{F}^T\tilde{\mathbf{H}} = \begin{bmatrix} r_1\,\xi & r_3\,\eta \\ r_4\,\xi & r_2\,\eta \end{bmatrix}, \quad \tilde{\mathbf{H}}^T\mathbf{F} = \begin{bmatrix} r_1\,\xi & r_4\,\xi \\ r_3\,\eta & r_2\,\eta \end{bmatrix}, \quad \tilde{\mathbf{H}}^T\tilde{\mathbf{H}} = \begin{bmatrix} r_5\,\xi^2 & r_7\,\eta\xi \\ \text{sym.} & r_6\,\eta^2 \end{bmatrix},$$

where the coefficients r_i $(i = 1, \ldots, 7)$ do not depend on ξ and η. The structure of these three components can be compared with the structure of $\mathbf{G}^\xi$ for the EAS method of eq. (12.83):

a) the sum

$$\mathbf{F}^T\tilde{\mathbf{H}} + \tilde{\mathbf{H}}^T\mathbf{F} = \begin{bmatrix} 2r_1\xi & r_3\xi + r_4\eta \\ r_3\xi + r_4\eta & 2r_2\eta \end{bmatrix} \tag{12.86}$$

has a similar structure as the $\mathbf{G}^\xi$ enhancement of the EAS method, i.e. the diagonal terms are linear (and incomplete) in either ξ or η, while the off-diagonal terms are sums of linear terms in ξ and η.

b) the component $\tilde{\mathbf{H}}^T\tilde{\mathbf{H}}$ contains terms of a higher order than those in eq. (12.83) for the EAS method and, hence, this term can be safely omitted from eq. (12.84).

The above modifications make the element slightly faster, and slightly stiffer, but the difference is small. For instance, in Cook's tapered panel example of Sect. 15.2.7, the difference in the displacement and drill rotation of the tip is $< 0.2\%$.

Finally, note that we can also consider the following simplification:

$$\mathbf{F}^T\tilde{\mathbf{H}} + \tilde{\mathbf{H}}^T\mathbf{F} \approx \tilde{\mathbf{H}} + \tilde{\mathbf{H}}^T = \begin{bmatrix} 2q_1\xi & q_3\xi + q_4\eta \\ q_3\xi + q_4\eta & 2q_2\eta \end{bmatrix}, \tag{12.87}$$

and then similarity to the $\mathbf{G}^\xi$ enhancement of the EAS method is even closer. However, this version does not work well in the twisted ring example of Sect. 15.3.15.

12.5 Mixed HW and HR functionals with rotations

HW functionals with rotations. Consider the classical form of the 3-F Hu–Washizu (HW) functional of eq. (11.1). To obtain the HW functional with rotations, the Lagrange multiplier method is applied to eq. (11.1), which yields the five-field functional

$$F_{\mathrm{HW5}}(\mathbf{u}, \mathbf{Q}, \boldsymbol{\sigma}, \boldsymbol{\varepsilon}, \mathbf{T}_a) \doteq \int_B \{\mathcal{W}(\boldsymbol{\varepsilon}) + \boldsymbol{\sigma} \cdot [\mathbf{E}(\nabla\mathbf{u}) - \boldsymbol{\varepsilon}] + \mathbf{T}_a \cdot \mathrm{skew}(\mathbf{Q}^T\mathbf{F})\} \; dV - F_{\mathrm{ext}}, \quad (12.88)$$

where $\mathbf{T}_a \doteq \mathrm{skew}(\mathbf{Q}^T\mathbf{FS})$ is the Lagrange multiplier for the RC equation. Two functionals derived from eq. (12.88) are particularly useful.

A. the 4-F functional, obtained by regularization of eq. (12.88) in $\mathbf{T}_a$, and elimination of $\mathbf{T}_a$,

$$\tilde{F}_{\mathrm{HW4}}(\mathbf{u}, \boldsymbol{\sigma}, \boldsymbol{\varepsilon}, \mathbf{Q}) \doteq \int_B \{\mathcal{W}(\boldsymbol{\varepsilon}) + \boldsymbol{\sigma} \cdot [\mathbf{E}(\nabla\mathbf{u}) - \boldsymbol{\varepsilon}\} + F_{\mathrm{RC}}^{\mathrm{P}}(\nabla\mathbf{u}, \mathbf{Q})] \; dV - F_{\mathrm{ext}}, \quad (12.89)$$

where the RC term has the penalty (P) form of eqs. (12.42) and (12.43).

B. the 5-F functional obtained by regularization of eq. (12.88) in $\mathbf{T}_a$,

$$\tilde{F}_{\mathrm{HW5}}(\mathbf{u}, \mathbf{Q}, \boldsymbol{\sigma}, \boldsymbol{\varepsilon}, \mathbf{T}_a) \doteq \int_B \{\mathcal{W}(\boldsymbol{\varepsilon}) + \boldsymbol{\sigma} \cdot [\mathbf{E}(\nabla\mathbf{u}) - \boldsymbol{\varepsilon}] + F_{\mathrm{RC}}^{\mathrm{PL}}(\nabla\mathbf{u}, \mathbf{Q}, \mathbf{T}_a)\} \; dV - F_{\mathrm{ext}}, \quad (12.90)$$

where the RC term has the perturbed Lagrange (PL) form of eqs. (12.53) and (12.55).

HR functionals with rotations. Let us take the above HW functionals with rotations and apply the same procedure which was used to obtain the Hellinger–Reissner functionals of eqs. (11.4) and (11.7). Then we obtain the HR functionals with rotations applicable to linear elastic materials.

A. From the 4-F functional of eq. (12.89), we obtain

$$\tilde{F}_{\mathrm{HR3}}(\mathbf{u}, \mathbf{Q}, \boldsymbol{\sigma}) \doteq \int_B [-\tfrac{1}{2}\boldsymbol{\sigma} \cdot (\mathbb{C}^{-1}\boldsymbol{\sigma}) + \boldsymbol{\sigma} \cdot \mathbf{E}(\nabla\mathbf{u}) + F_{\mathrm{RC}}^{\mathrm{P}}(\nabla\mathbf{u}, \mathbf{Q})] \; dV - F_{\mathrm{ext}}, \quad (12.91)$$

where the penalty (P) form of the RC is given by eqs. (12.42) and (12.43).

B. From the 5-F functional of eq. (12.90), we obtain

$$\tilde{F}_{\mathrm{HR4}}(\mathbf{u}, \mathbf{Q}, \boldsymbol{\sigma}, \mathbf{T}_a) \doteq \int_B [-\tfrac{1}{2}\boldsymbol{\sigma} \cdot (\mathbb{C}^{-1}\boldsymbol{\sigma}) + \boldsymbol{\sigma} \cdot \mathbf{E}(\nabla\mathbf{u}) + F_{\mathrm{RC}}^{\mathrm{PL}}(\nabla\mathbf{u}, \mathbf{Q}, \mathbf{T}_a)] \; dV - F_{\mathrm{ext}}, \quad (12.92)$$

where the perturbed Lagrange (PL) form of the RC term is given by eqs. (12.53) and (12.55).

For non-linear materials, we use the incremental forms of the above HW and HR functionals, obtained for the increments of displacements, stress and strain of eq. (11.5).

EADG method for HW and HR functionals. In Sect. 12.4, we described the EADG method for the 3D potential energy functionals with rotations derived in Sect. 4; for the HW and HR functionals with rotations, the procedure is simpler.

For the HW and HR functionals, the EADG method can be incorporated without using additional independent fields $\mathbf{P}$ and $\mathbf{F}$ and the term $\int_B \mathbf{P} \cdot (\nabla\chi - \mathbf{F})\, \mathrm{d}V$, which was used in eq. (12.77). This is because we already have the independent fields $\boldsymbol{\sigma}$ and $\boldsymbol{\varepsilon}$ and the term $\int_B \boldsymbol{\sigma} \cdot (\mathbf{E}(\nabla\mathbf{u}) - \boldsymbol{\varepsilon})\, \mathrm{d}V$, which can be used instead. Hence, it suffices to replace $\nabla\mathbf{u}$ by $\nabla\mathbf{u} + \tilde{\mathbf{H}}$ in $\mathbf{E}(\nabla\mathbf{u})$, which is in accord with eq. (12.76). We note that the orthogonality of stress and the enhancing field is not required.

The EADG enhancement is applied to the HW functionals with rotations of eqs. (12.89) and (12.90), and the HR functionals with rotations of eqs. (12.91) and (12.92).

12.6 2D+drill elements for bilinear shape functions

The characteristics of each 2D+drill four-node element (with the drilling rotation), which are presented below, consists of three parts:

1. the designation of the plane (2D) four-node element (without the drilling rotation) of Sect. 11, which is being extended by inclusion of the drilling rotation,
2. the specification of the mixed functional on which the element is based,
3. the description of the treatment of the functional for the drill RC, F_{RC}. The weak (integral) forms of the drill RC was used; the penalty method is implemented as specified in Table 12.4, while the Perturbed Lagrange method as specified in Table 12.5.

12.6.1 EADG4 elements based on potential energy

The elements characterized below have two features: (i) they extend the EADG4 element without the drilling rotation of Sect. 11.4.3 and (ii) are

Table 12.4 Implementation of the penalty method (**P**) for drill RC.

1. weak (integral) form of drill RC, as in eq. (12.35), functional F_{RC} of eq. (12.42),
2. expansion of $\mathbf{Q}^T\mathbf{F}$ product of eq. (12.27), for which drill RC is given by eq. (12.28),
3. stabilization of spurious mode of eq. (12.25).

Table 12.5 Implementation of the Perturbed Lagrange method (**PL**) for drill RC.

1. weak (integral) form of drill RC, as in eq. (12.36), functional F_{RC} of eq. (12.53), Lagrange multiplier tensor of eq. (12.56) and representation of eq. (12.57).
2. expansion of $\mathbf{Q}^T\mathbf{F}$ product of eq. (12.27), for which drill RC is given by eq. (12.28),
3. stabilization of spurious mode of eq. (12.25).

based on the potential energy functional with rotations of eq. (4.73). Two elements were selected:

1. **Element EADG4+P**, which has the following features:
   a) the penalty method is implemented as specified in Table 12.4,
   b) it uses four additional parameters, the multipliers of the EADG modes.
2. **Element EADG4+PL**, which has the following features:
   a) the Perturbed Lagrange method is implemented as specified in Table 12.5,
   b) it uses seven additional parameters: four multipliers of the EADG modes and three parameters of the Lagrange multiplier.

As show the numerical tests, these 2D+drill elements perform very well for coarse distorted meshes, despite a small number of parameters. In non-linear tests, the second (PL) element has a larger radius of convergence.

12.6.2 Assumed stress HR5-S elements

The elements characterized below are based on the HR functionals with rotations of eqs. (12.91) and (12.92). Their 2D counterparts were described in Sect. 11.5.1. The same five-parameter representation of eq. (11.126) is used for stress. No enhancement is applied.

1. **Element HR5-S+P** which has the following features:
   a) the penalty method is implemented as specified in Table 12.4,
   b) it uses five additional parameters, which are multipliers of the stress modes.
2. **Element HR5-S+PL** which has the following features:
   a) the Perturbed Lagrange method is implemented as specified in Table 12.5,
   b) it uses eight additional parameters: five multipliers of the stress modes and three parameters of the Lagrange multiplier.

As show the numerical tests, these 2D+drill elements are worse for coarse distorted meshes than, e.g., the EADG4 element. They show a substantial decrease of accuracy, comparing to their 2D counterparts.

12.6.3 Assumed stress/enhanced strain HR7-S elements

A poor performance of the 2D+drill HR elements based on the five-parameter representation of stresses caused that we considered the seven-parameter representation of stresses and the strain enhancement. No such 2D elements are described in Sect. 11.5 because they perform identically to the HR5-S element but use more modes, so are less effective.

The assumed stress/enhanced strain elements are based on the HR functionals of eqs. (12.91) and (12.92) additionally enhanced, by replacing $\nabla \mathbf{u}$ by $\nabla \mathbf{u} + \tilde{\mathbf{H}}$, as described in Sect. 12.5. In these functionals, $\mathbf{u}$, $\mathbf{Q}$ are the compatible fields, while $\boldsymbol{\sigma}$, $\boldsymbol{\varepsilon}$, $\mathbf{T}_a$, $\tilde{\mathbf{H}}$ are the assumed fields. Besides,

1. the assumed stress and the increment of the assumed stress are constructed as follows:

$$\boldsymbol{\sigma}^a = \mathbf{J}_c \, \boldsymbol{\sigma}^\xi \, \mathbf{J}_c^T, \qquad \Delta\boldsymbol{\sigma}^a = \mathbf{J}_c \, \Delta\boldsymbol{\sigma}^\xi \, \mathbf{J}_c^T, \tag{12.93}$$

   which is the transformation rule for the contravariant components of a tensor of eq. (11.124), and $\boldsymbol{\sigma}^\xi$ contains the seven-parameter representation in terms of the skew coordinates

$$\boldsymbol{\sigma}^\xi \doteq \begin{bmatrix} q_1 + q_2\, y_S & q_5 + q_6\, x_S + q_7\, y_S \\ \text{sym.} & q_3 + q_4\, x_S \end{bmatrix}. \tag{12.94}$$

   Besides, $\Delta\boldsymbol{\sigma}^\xi$ has a structure of $\boldsymbol{\sigma}^\xi$, with the multipliers q_i replaced by Δq_i.

2. The EADG enhancement of eq. (11.71) is used in the form

$$\tilde{\mathbf{H}}_g \doteq \mathbf{J}_c \, \mathbf{G}_g^{\xi} \, \mathbf{J}_c^{-1} \left(\frac{j_c}{j_g} \right), \qquad \mathbf{G}^{\xi} \doteq \begin{bmatrix} 0 & \eta \, q_8 \\ \xi \, q_9 & 0 \end{bmatrix}, \tag{12.95}$$

where the EADG2 representation involves two parameters.

Two elements were developed:

1. **Element HR7+EADG2+P** which has the following features:
   a) the penalty method is implemented as specified in Table 12.4,
   b) it uses nine additional parameters: seven multipliers of stress modes and two multipliers of the EADG enhancement.
2. **Element HR7+EADG2+PL** which has the following features:
   a) the Perturbed Lagrange method is implemented as specified in Table 12.5,
   b) it uses 12 additional parameters; nine parameters identical as in the previous element, and three parameters of the Lagrange multiplier.

As show the numerical tests, the above elements perform very well for coarse distorted meshes, better than the EADG4 element. In non-linear tests, the second (PL) element has a much larger radius of convergence.

Remark. Note that we can also consider a different seven-parameter representation

$$\boldsymbol{\sigma}^{\xi} \doteq \begin{bmatrix} q_1 + q_2 \, y_S + q_6 \, x_S & q_5 - q_7 \, x_S - q_6 \, y_S \\ \text{sym.} & q_3 + q_4 \, x_S + q_7 \, y_S \end{bmatrix}, \tag{12.96}$$

where some parameters are repeated in the diagonal and off-diagonal terms. This representation was used in several earlier papers on mixed (or hybrid) methods; (i) in Cartesian coordinates in [223] and (ii) in oblique coordinates in [184]. More recently, it was also used in [265, 177], but in different forms; the relation between these forms was established in [256], eqs. (38) and (41).

Both these seven-parameter representations are equally good for the 2D elements, but for the 2D+drill elements, the representation (12.96) is slightly worse and, for this reason, is not used here.

12.6.4 Assumed stress and strain HW14-SS elements

The elements characterized below are based on the non-enhanced HW functionals with rotations of eq. (12.89) or eq. (12.90). Their 2D counterparts were described in Sect. 11.5.2.

1. **Element HW14-SS+P** which has the following features:
   a) the penalty method is implemented as specified in Table 12.4,
   b) it uses 14 additional parameters: five multipliers of the stress modes of eq. (11.126), and nine multipliers of the strain modes of eq. (11.140).
2. **Element HW14-SS+PL** which has the following features:
   a) the Perturbed Lagrange method is implemented as specified in Table 12.5.
   b) it uses 17 additional parameters: fourteen parameters identical as in the previous element, and three parameters of the Lagrange multiplier.

As show the numerical tests, these 2D+drill elements are worse for coarse distorted meshes than, e.g., the EADG4 element. They show a substantial decrease of accuracy compared to their 2D counterparts.

12.6.5 Assumed stress and strain/enhanced strain HW18-SS elements

A poor performance of the 2D+drill HW elements based on five-parameter representation of stress caused that we considered the seven-parameter representation of stresses and the strain enhancement. No 2D elements of this type are described in Sect. 11.5 because they perform identically to the HW14-S element but are less effective, as they use more modes.

The assumed stress and strain/enhanced strain elements characterized below are based on the HW functionals with rotations of eqs. (12.89) or (12.90) additionally enhanced, by replacing $\nabla \mathbf{u}$ by $\nabla \mathbf{u} + \tilde{\mathbf{H}}$, as described in Sect. 12.5. In these functionals, $\mathbf{u}$, $\mathbf{Q}$ are the compatible fields, while $\boldsymbol{\sigma}$, $\boldsymbol{\varepsilon}$, $\mathbf{T}_a$, $\tilde{\mathbf{H}}$ are the assumed fields. Besides,

1. the assumed stress and the increment of the assumed stress are constructed as follows:

$$\boldsymbol{\sigma}^a = \mathbf{J}_c\, \boldsymbol{\sigma}^\xi\, \mathbf{J}_c^T, \qquad \Delta\boldsymbol{\sigma}^a = \mathbf{J}_c\, \Delta\boldsymbol{\sigma}^\xi\, \mathbf{J}_c^T, \tag{12.97}$$

which is the transformation rule for the contravariant components of a tensor of eq. (11.124) and $\boldsymbol{\sigma}^\xi$ contains the seven-parameter representation in terms of the skew coordinates

$$\boldsymbol{\sigma}^\xi \doteq \begin{bmatrix} q_1 + q_2\, y_S & q_5 + q_6\, x_S + q_7\, y_S \\ \text{sym.} & q_3 + q_4\, x_S \end{bmatrix}. \tag{12.98}$$

Besides, $\Delta\boldsymbol{\sigma}^\xi$ has a structure of $\boldsymbol{\sigma}^\xi$ but with q_i replaced by Δq_i.

2. The assumed strain is constructed as

$$\boldsymbol{\varepsilon}^a = \mathbf{J}_c^{-T} \, \boldsymbol{\varepsilon}_\xi \, \mathbf{J}_c^{-1}, \tag{12.99}$$

using the transformation rule of eq. (11.139) for the covariant components of a tensor. The nine-parameter strain representation of $\boldsymbol{\varepsilon}_\xi$ is given by eq. (11.140), i.e.

$$\boldsymbol{\varepsilon}_\xi \doteq \begin{bmatrix} q_8 + q_9 \, y_S + q_{10} \, x_S & q_{14} + q_{15} \, x_S + q_{16} \, y_S \\ \text{sym.} & q_{11} + q_{12} \, x_S + q_{13} \, y_S \end{bmatrix}. \tag{12.100}$$

3. The gradient of displacements is enhanced as follows:

$$\nabla \mathbf{u} \doteq \nabla \mathbf{u}^c + \tilde{\mathbf{H}}, \tag{12.101}$$

where $\nabla \mathbf{u}^c$ is the gradient of compatible displacements and $\tilde{\mathbf{H}}$ is the assumed enhancing field constructed as follows using the EADG method:

$$\tilde{\mathbf{H}}_g = \mathbf{J}_c \, \mathbf{G}_g \, \mathbf{J}_c^{-1} \left(\frac{j_c}{j_g} \right), \qquad \mathbf{G} = \begin{bmatrix} 0 & y_S \, q_{17} \\ x_S \, q_{18} & 0 \end{bmatrix}, \tag{12.102}$$

where g indicates the Gauss point.

Two elements were developed:

1. **Element HW18-SS+EADG2+P** which has the following features:
   a) the penalty method is implemented as specified in Table 12.4,
   b) it uses 18 additional parameters: seven multipliers of the stress modes, nine multipliers of the strain modes, and two parameters of the EADG2 enhancement.
2. **Element HW18-SS+EADG2+PL** which has the following features:
   a) the Perturbed Lagrange method is implemented as specified in Table 12.5.
   b) it uses 21 additional parameters: 18 parameters identical as in the previous element, and three parameters of the Lagrange multiplier.

As show the numerical results of linear tests, e.g. of Table 12.6, the above elements with drilling rotation

1. perform identically to the HR7-S+EADG2 elements with the drilling rotation. Thus, the equivalence of linear HR and HW 2D elements established in [257] is maintained by the present 2D+drill formulation.
2. Perform very well for coarse distorted meshes, better than the EADG4 element.

12.7 2D+drill elements for Allman shape functions

Historical note. The Allman shape functions were first successfully applied to 2D triangles in [1, 27] and later extrapolated to 2D quadrilaterals in [55], where a procedure of transforming an eight-node serendipity element to a four-node element with nodal drilling rotations was proposed. This procedure is commonly used in Allman-type quadrilaterals, although it needs to be modified for large drilling rotations.

At first, the Allman shape functions were treated as a way to improve accuracy of low-order elements. Soon their ability to incorporate the drilling rotation was appreciated; this was before the role of the RC equation was recognized. The Allman shape functions can be applied in two types of four-node elements:

1. 2D+drill elements. In [144] Table 1, it is stressed that such elements are eight times faster than, e.g., eight-node elements without drilling dofs, with only slightly less accuracy in small strain problems.
2. Shell elements, where the presence of the drilling rotation in the membrane part is an advantage, as it allows us to use a three-parameter representation of rotations and to treat all rotational dofs in the same way.

An overview of the works on four-node quadrilaterals based on the Allman shape functions is given in [255] and it includes such papers as [117, 113, 235, 144, 109, 112, 87, 203]. This overview provided the motivation for the formulation which generalizes the Allman shape functions to handle large rotations and uses the EADG enhancement.

12.7.1 Allman-type shape functions

The Allman-type shape functions for a quadrilateral element are obtained by the procedure, which has two characteristic features:

1. the hierarchical shape functions are used for displacements of an eight-node 2D element,
2. the hierarchical mid-side displacements are expressed by corner drilling rotations,

which means that the element displacements become functions of corner displacements and corner drilling rotations, i.e.

$$\underbrace{\mathbf{u}(\xi, \eta,\ \mathbf{u}_I)}_{\text{8-node, 2D}} = \underbrace{\mathbf{u}(\xi, \eta,\ \mathbf{u}_I, \omega_I)}_{\text{4-node, 2D+drill}}, \qquad I = 1, 2, 3, 4. \tag{12.103}$$

Hierarchical shape functions for displacements of 2D quadrilateral. The displacements of the eight-node 2D quadrilateral of Fig. 12.4 can be approximated as follows:

$$\mathbf{u}(\xi,\eta) = \sum_{I=1}^{4} N_I(\xi,\eta)\ \mathbf{u}_I + \sum_{H=5}^{8} N_H(\xi,\eta)\ \Delta\mathbf{u}_H, \tag{12.104}$$

where $N_I(\xi,\eta)$ are the standard bilinear shape functions of eq. (10.3), and $N_H(\xi,\eta)$ are the hierarchical shape functions, defined as

$$N_5(\xi,\eta) = \tfrac{1}{2}(1-\xi^2)(1-\eta), \qquad N_7(\xi,\eta) = \tfrac{1}{2}(1-\xi^2)(1+\eta),$$
$$N_6(\xi,\eta) = \tfrac{1}{2}(1-\eta^2)(1+\xi), \qquad N_8(\xi,\eta) = \tfrac{1}{2}(1-\eta^2)(1-\xi). \tag{12.105}$$

Note that $\mathbf{u}_I$ are the nodal displacement vectors, while $\Delta\mathbf{u}_H$ are the hierarchical displacement vectors at mid-points of the element boundaries, see Fig. 12.7.

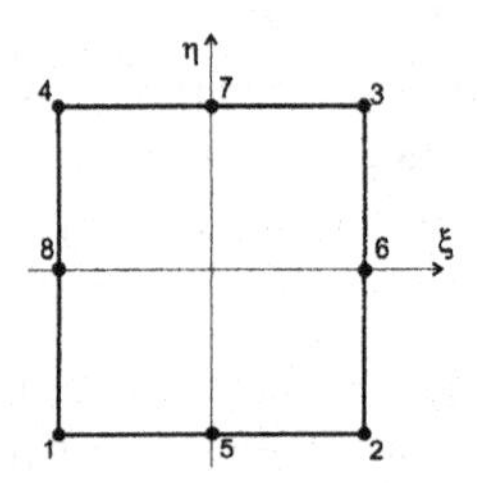

Fig. 12.4 Numeration of nodes of eight-node element.

The shape functions for a selected mid-side node of the hierarchical eight-node element and the Lagrange nine-node element are shown in Fig. 12.5. More details on the differences between these two families of shape functions can be found, e.g., in [268] Chap. 8.

To express the hierarchical displacement $\Delta\mathbf{u}_H$ in eq. (12.104) in terms of nodal drilling rotations ω_I, we consider a single boundary of a quadrilateral element and treat it as a beam. Note that

1. In the classical Allman formula, only one component of $\Delta\mathbf{u}_H$ is linked with the nodal drilling rotations, this one which is normal to the beam. Hence, this formula is valid only for small drilling rotations.
2. For large drilling rotations, the form of the Allman approximations involving two components of $\Delta\mathbf{u}_H$ must be used and was derived in [255]. Another possibility is to use the incremental formulation, but this precludes the straightforward use of automatic differentiation.

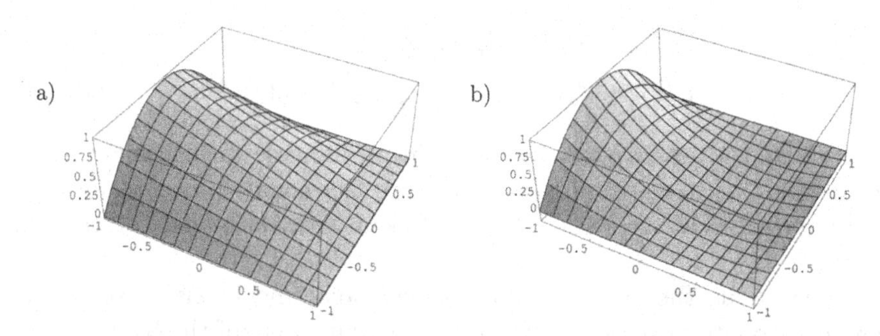

Fig. 12.5 A shape function for a mid-side node of: a) hierarchical eight-node element, b) Lagrangian nine-node element.

Classical Allman shape functions. To determine the mid-side hierarchical displacement $\Delta \mathbf{u}_I$ in eq. (12.104), we select one boundary of a quadrilateral, e.g. defined by nodes 1-5-2, and further consider a planar beam along it, see Fig. 12.6.

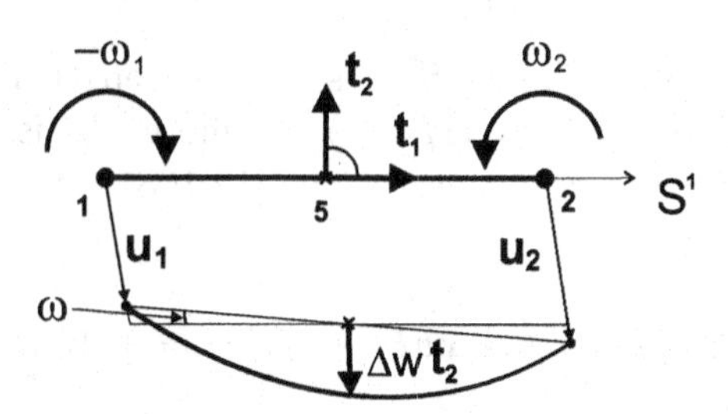

Fig. 12.6 Classical Allman shape functions: small rotation of a selected boundary.

When the rotation ω is small, then the normal displacement component u_2 is much bigger than the tangent component u_1 and we may write

$$\mathbf{u} \approx w\, \mathbf{t}_2, \tag{12.106}$$

where $w \doteq u_2$ is the normal displacement and $\mathbf{t}_2$ is a vector normal to the boundary, and a director of the beam.

Let us define the shape functions for the beam,

$$M_1(\xi) \doteq \tfrac{1}{2}(1-\xi), \quad M_2(\xi) \doteq \tfrac{1}{2}(1+\xi), \quad M_0(\xi) \doteq 1-\xi^2, \tag{12.107}$$

where $M_1(\xi)$ and $M_2(\xi)$ are linear functions and $M_0(\xi)$ is a bubble function. The rotation and the normal displacement are approximated as

$$\omega(\xi) = M_1(\xi)\,\omega_1 + M_2(\xi)\,\omega_2,$$
$$w(\xi) = M_1(\xi)\,w_1 + M_2(\xi)\,w_2 + M_0(\xi)\,\Delta w, \qquad (12.108)$$

where $M_0(\xi)$ multiplies the hierarchical mid-side displacement Δw. Note that ω_1, ω_2 and w_1, w_2 are the nodal values, while Δw is unknown. However, Δw can be linked with the nodal rotations as follows.

To calculate the hierarchical mid-side displacement Δw, we use the condition related to the transverse shear strain ε_{12} of the beam,

$$(\varepsilon_{12,\xi})|_{\xi=0} = 0, \qquad (12.109)$$

i.e. we set to zero the first derivative of the shear strain at the mid-point of the edge.

The transverse shear strain of the beam undergoing small rotations is defined as

$$\varepsilon_{12} = -\omega + w_{,1}, \qquad (12.110)$$

where $(\ )_{,1} \doteq \partial/\partial S^1$, and S^1 is the arc-length coordinate in the direction $\mathbf{t}_1$. Hence, $(\ )_{,1} = (1/L)(\ )_{,\xi}$, where L is the length of the boundary. For the approximations of eq. (12.108), the transverse shear strain becomes

$$\varepsilon_{12}(\xi,\eta) = \frac{1}{L}(w_2 - w_1 - 4\Delta w\,\xi) + M_1(\xi)\,\omega_1 + M_2(\xi)\,\omega_2, \qquad (12.111)$$

from which, using the condition (12.109), we obtain

$$\Delta w = -\frac{L}{8}\,(\omega_2 - \omega_1). \qquad (12.112)$$

This is the classical formula for the hierarchical mid-side normal displacement. Then we can write the vector of hierarchical mid-side displacement for all boundaries as follows:

$$\Delta \mathbf{u}_H \doteq -\frac{L_{JK}}{8}\,(\omega_K - \omega_J)\,\mathbf{n}_{JK}, \qquad H = 5,6,7,8, \qquad (12.113)$$

where $J = H - 4$, $K = \mathrm{mod}(H,4) + 1$, and L_{JK} is the length of the boundary JK. Here, $\mathbf{n}_{JK}$ is the vector normal to the **initial** element boundary. This formula can be directly used in eq. (12.104), which then depends only on nodal displacements and on nodal drilling rotations.

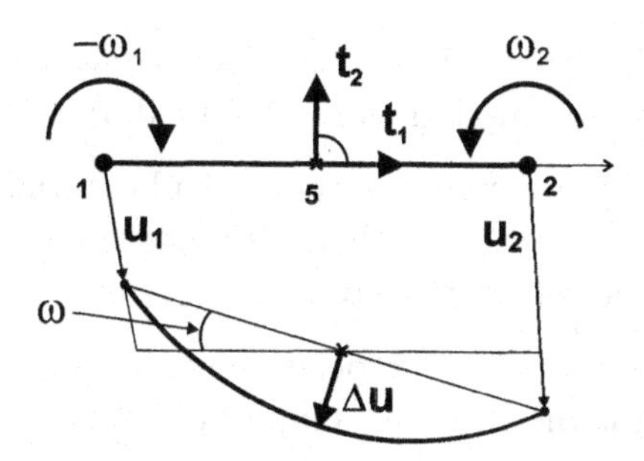

Fig. 12.7 Allman shape functions for large rotations of a selected boundary.

Allman-type shape functions for finite drilling rotation. To determine the mid-side hierarchical displacement $\Delta \mathbf{u}_I$ in eq. (12.104), we select one boundary of a quadrilateral, e.g. defined by nodes 1-5-2, and consider a planar beam along it, see Fig. 12.7.

When the rotation ω is large we have to account for both components of the displacement vectors,

$$\mathbf{u} = u_1 \, \mathbf{t}_1 + u_2 \, \mathbf{t}_2. \tag{12.114}$$

To calculate the hierarchical mid-side displacement $\Delta \mathbf{u}$, we can use two conditions related to the transverse shear strain ε_{12} of the beam,

$$(\varepsilon_{12,\xi})|_{\xi=0} = 0, \qquad (\varepsilon_{12,\xi\xi})|_{\xi=0} = 0, \tag{12.115}$$

i.e. we set to zero the first and the second derivatives of the shear strain at the mid-point of the edge. These conditions were proposed in [255].

The transverse shear strain of a beam undergoing large rotations is defined as

$$\varepsilon_{12} \doteq \tfrac{1}{2} \mathbf{x}_{,1} \cdot \mathbf{a}_2, \tag{12.116}$$

where $\mathbf{x}$ is the current position vector and the current director is $\mathbf{a}_2 \doteq \mathbf{Q} \, \mathbf{t}_2$, where $\mathbf{Q}$ is the drilling rotation tensor. For small rotations and $\mathbf{t}_{1,1} = \mathbf{t}_{2,1} \approx \mathbf{0}$, eq. (12.116) yields the transverse shear strain of eq. (12.110). However, in the derivation which follows, the magnitude of rotations is not restricted.

For this reason, instead of the approximation of the normal displacement component w we approximate the whole displacement vector

$$\mathbf{u}(\xi) = M_1(\xi) \, \mathbf{u}_1 + M_2(\xi) \, \mathbf{u}_2 + M_0(\xi) \, \Delta \mathbf{u}, \tag{12.117}$$

where $\Delta\mathbf{u}$ is the hierarchical mid-side displacement vector. The drilling rotation $\omega(\xi)$ is approximated as in eq. (12.108).

We apply these approximations to particular terms of the transverse shear strain of eq. (12.116), and then separate the constant terms, indicated by "0", and the terms depending on ξ, in the following way:

1. The derivative of the current position vector is $\mathbf{x}_{,1} = (\mathbf{y} + \mathbf{u})_{,1} = \mathbf{t}_1 + \mathbf{u}_{,1}$, thus for the assumed shape functions, we obtain

$$\mathbf{u}_{,1} = \frac{2}{L}\mathbf{u}_{,\xi} = \frac{2}{L}\mathbf{u}^0_{,\xi} + \frac{2}{L}\Delta\mathbf{u}(-2\xi), \tag{12.118}$$

where $\mathbf{u}^0_{,\xi} \doteq \frac{1}{2}(\mathbf{u}_2 - \mathbf{u}_1)$ is the constant part of $\mathbf{u}_{,\xi}$ and L is the side length.

2. The forward rotated director is expressed as $\mathbf{a}_2 \doteq \mathbf{Q}_0\mathbf{t}_2 = c\,\mathbf{t}_2 - s\,\mathbf{t}_1$, where the drilling rotation tensor $\mathbf{Q} = c\,(\mathbf{t}_1\otimes\mathbf{t}_1 + \mathbf{t}_2\otimes\mathbf{t}_2) + s\,(\mathbf{t}_2\otimes\mathbf{t}_1 - \mathbf{t}_1\otimes\mathbf{t}_2)$, $s \doteq \sin\omega$ and $c \doteq \cos\omega$. For the assumed shape functions $\omega(\xi) = \omega_0 + \xi\,\omega_{,\xi}$ and we obtain

$$s = \sin(\omega_0 + \xi\,\omega_{,\xi}) = s_0\cos(\xi\,\omega_{,\xi}) + c_0\sin(\xi\,\omega_{,\xi}),$$
$$c = \cos(\omega_0 + \xi\,\omega_{,\xi}) = c_0\cos(\xi\,\omega_{,\xi}) - s_0\sin(\xi\,\omega_{,\xi}),$$

where $s_0 \doteq \sin\omega_0$, $c_0 \doteq \cos\omega_0$, for $\omega_0 \doteq \omega(\xi = 0)$, and the derivative $\omega_{,\xi} = \frac{1}{2}(\omega_2 - \omega_1)$. Hence,

$$\mathbf{a}_2 = c\,\mathbf{t}_2 - s\,\mathbf{t}_1 = \cos(\xi\,\omega_{,\xi})\,\mathbf{a}^0_2 - \sin(\xi\,\omega_{,\xi})\,\mathbf{a}^0_1, \tag{12.119}$$

where $\mathbf{a}^0_2 \doteq c_0\,\mathbf{t}_2 - s_0\,\mathbf{t}_1$ and $\mathbf{a}^0_1 \doteq s_0\,\mathbf{t}_2 + c_0\,\mathbf{t}_1$.

Next, eqs. (12.118) and (12.119) are inserted into eq. (12.116).

Using the conditions (12.115), we obtain the following components of $\Delta\mathbf{u}$:

$$(\Delta\mathbf{u}\cdot\mathbf{a}^0_1) = \frac{L}{8}\,\omega_{,\xi}\left(\mathbf{t}_1 + \frac{2}{L}\mathbf{u}^0_{,\xi}\right)\cdot\mathbf{a}^0_2, \quad (\Delta\mathbf{u}\cdot\mathbf{a}^0_2) = -\frac{L}{4}\,\omega_{,\xi}\left(\mathbf{t}_1 + \frac{2}{L}\mathbf{u}^0_{,\xi}\right)\cdot\mathbf{a}^0_1, \tag{12.120}$$

and the hierarchical mid-side displacement vector can be expressed as

$$\Delta\mathbf{u} = (\Delta\mathbf{u}\cdot\mathbf{a}^0_1)\,\mathbf{a}^0_1 + (\Delta\mathbf{u}\cdot\mathbf{a}^0_2)\,\mathbf{a}^0_2. \tag{12.121}$$

Mid-side displacement vector for small strains. If we assume that strains are small, i.e. (i) $\varepsilon_{12} \approx 0$ and (ii) $\varepsilon_{11} \approx 0$, then we have

$$\mathbf{t}_1 + \frac{2}{L}\mathbf{u}^0_{,\xi} = (\mathbf{y} + \mathbf{u}^L)_{,1} = \frac{2}{L}(\mathbf{y} + \mathbf{u}^L)_{,\xi} = \frac{2}{L}\tfrac{1}{2}(\mathbf{x}_2 - \mathbf{x}_1) \overset{\text{(i)}}{\approx} \frac{2}{L}\tfrac{1}{2}L_c\,\mathbf{a}^0_1 \overset{\text{(ii)}}{\approx} \mathbf{a}^0_1, \tag{12.122}$$

where $\mathbf{u}^L \doteq M_1(\xi)\,\mathbf{u}_1 + M_2(\xi)\,\mathbf{u}_2$. Due to assumption (i), the rotated vector $\mathbf{a}^0_1$ is used instead of the vector which passes through nodes, while by (ii), the current and initial element lengths are equal, i.e. $L_c \approx L$. Because $\mathbf{a}^0_1 \cdot \mathbf{a}^0_1 = 1$ and $\mathbf{a}^0_1 \cdot \mathbf{a}^0_2 = 0$, eq. (12.120) is reduced to

$$(\Delta\mathbf{u} \cdot \mathbf{a}^0_1) \approx 0, \qquad (\Delta\mathbf{u} \cdot \mathbf{a}^0_2) \approx -\frac{L}{4}\,\omega_{,\xi}, \tag{12.123}$$

and the vector of hierarchical displacements of eq. (12.121) is

$$\Delta\mathbf{u} = -\frac{L}{4}\,\omega_{,\xi}\,\mathbf{a}^0_2 = -\frac{L}{8}\,(\omega_2 - \omega_1)\,\mathbf{a}^0_2. \tag{12.124}$$

Assuming, additionally, that the rotations are small, i.e. $\omega_0 \approx 0$, we have $c_0 \approx 1$ and $s_0 \approx 0$ and so $\mathbf{a}^0_2 \doteq c_0\,\mathbf{t}_2 - s_0\,\mathbf{t}_1 \approx \mathbf{t}_2$, i.e. the vector normal to the current boundary $\mathbf{a}^0_2$ is replaced by the vector normal to the initial element boundary $\mathbf{t}^0_2$. Thus, for small rotations, eq. (12.124) yields the classical formula

$$\Delta\mathbf{u} = -\frac{L}{4}\,\omega_{,\xi}\,\mathbf{a}^0_2 = -\frac{L}{8}\,(\omega_2 - \omega_1)\,\mathbf{t}^0_2. \tag{12.125}$$

In numerical calculations we use eq. (12.124) in the form valid for all boundaries,

$$\Delta\mathbf{u}_H \doteq -\frac{L_{JK}}{8}\,(\omega_K - \omega_J)\,\mathbf{n}_{JK}, \qquad H = 5, 6, 7, 8, \tag{12.126}$$

where $J = H - 4$, $K = \operatorname{mod}(H, 4) + 1$, and L_{JK} is the length of the boundary JK. This formula can be directly used in eq. (12.104), which then depends only on corner displacements and on corner drilling rotations.

Remark 1. Note that $\mathbf{n}_{JK}$ is the vector normal to the **current** element boundary, not to the initial one, as in eq. (12.113). Hence, $\mathbf{n}_{JK}$ is a function of nodal displacements, but we freeze this dependence and do not differentiate $\mathbf{n}_{JK}$ w.r.t. nodal displacements. The fact that $\mathbf{n}_{JK}$ is updated is similar to the co-rotational formulation.

Remark 2. The proposed generalization of the classical procedure leads to a new form of Allman shape functions, involving two components of the mid-side displacement. This new form becomes particularly simple for small strains/large rotations. Using this new form, we can exploit the Total Lagrangian description and automatic differentiation. For small rotations, the new form is reduced to the classical one.

Pure bending of Allman quadrilateral. Let us consider the question of whether the Allman shape functions are able to reproduce the analytical solution for the problem of pure bending of a square membrane. We assume that the membrane is 2×2 and the boundaries are parallel to the global basis $\{\mathbf{i}_k\}$ $(k = 1, 2)$, see Fig. 12.8.

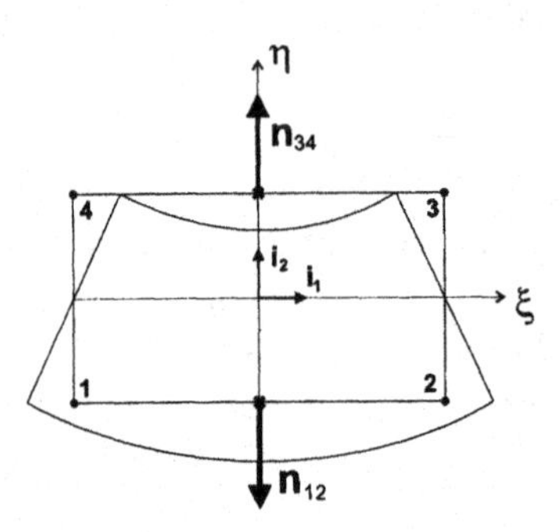

Fig. 12.8 Pure bending of the Allman quadrilateral.

First, we consider the analytical solution for pure bending of a square membrane

$$u(\xi, \eta) = A\,\xi\eta, \qquad v(\xi, \eta) = B\,(1 - \xi^2), \tag{12.127}$$

where A and B are scalar coefficients, see [98] p. 244, Fig. 4.7.2. For the analytical solution, we obtain

- the displacement gradient

$$\nabla \mathbf{u} \doteq \begin{bmatrix} u_{,\xi} & u_{,\eta} \\ v_{,\xi} & v_{,\eta} \end{bmatrix} = \begin{bmatrix} A\eta & A\xi \\ -2B\xi & 0 \end{bmatrix}, \tag{12.128}$$

- the linear strain

$$\boldsymbol{\varepsilon} \doteq \operatorname{sym} \nabla \mathbf{u} = \begin{bmatrix} u_{,\xi} & \frac{1}{2}(v_{,\xi} + u_{,\eta}) \\ \text{sym.} & v_{,\eta} \end{bmatrix} = \begin{bmatrix} A\eta & \frac{1}{2}(A - 2B)\,\xi \\ \text{sym.} & 0 \end{bmatrix}. \tag{12.129}$$

Next, we consider the Allman element. For pure bending, the rotations of corner nodes are as follows:

$$\omega_1 = \omega, \qquad \omega_2 = -\omega, \qquad \omega_3 = -\omega, \qquad \omega_4 = \omega \tag{12.130}$$

and the differences of rotations are

$$\omega_2 - \omega_1 = -2\,\omega, \qquad \omega_3 - \omega_2 = 0, \qquad \omega_4 - \omega_3 = 2\,\omega, \qquad \omega_1 - \omega_4 = 0.$$

From eq. (12.126) we obtain $\Delta \mathbf{u}_6 = \mathbf{0}$ and $\Delta \mathbf{u}_8 = \mathbf{0}$, i.e. the mid-side displacements for the sides which remain straight in pure bending are equal to zero. Thus, we consider only the sides which change the curvature, for which $\mathbf{n}_{12} = -\mathbf{i}_2$, $\mathbf{n}_{34} = \mathbf{i}_2$, and $L_{12} = L_{34} = 2$ and

$$\Delta \mathbf{u}_5 \doteq -\frac{L_{12}}{8}\,(\omega_2 - \omega_1)\,(\mathbf{n})_{12} = -\frac{2}{8}\,(-2\omega)\,(-\mathbf{i}_2) = -\frac{1}{2}\,\omega\,\mathbf{i}_2, \tag{12.131}$$

$$\Delta \mathbf{u}_7 \doteq -\frac{L_{34}}{8}\,(\omega_4 - \omega_3)\,(\mathbf{n})_{34} = -\frac{2}{8}\,(2\omega)\,\mathbf{i}_2 = -\frac{1}{2}\,\omega\,\mathbf{i}_2, \tag{12.132}$$

i.e. their mid-side displacements are equal, in accordance with our intuition. Substituting these expressions into eq. (12.104), we obtain the following form of the hierarchical part of displacements

$$\sum_{H=5}^{8} N_H(\xi,\eta)\;\Delta \mathbf{u}_H = N_5(\xi,\eta)\;\Delta \mathbf{u}_5 + N_7(\xi,\eta)\;\Delta \mathbf{u}_7 = v^A\,\mathbf{i}_2, \tag{12.133}$$

where $v^A(\xi) \doteq -\frac{1}{2}(1-\xi^2)\,\omega = -\frac{1}{2}M_0(\xi)\,\omega$ is the displacement component in the $\mathbf{i}_2$ direction which depends on the bubble function $M_0(\xi)$ and the drilling rotation ω. Hence, eq. (12.104) can be separately written for each component as follows:

$$u(\xi,\eta) = u_0 + \xi\,u_1 + \eta\,u_2 + \xi\eta\,u_3, \tag{12.134}$$

$$v(\xi,\eta) = v_0 + \xi\,v_1 + \eta\,v_2 + \xi\eta\,v_3 \underline{- \frac{1}{2}(1-\xi^2)\,\omega}, \tag{12.135}$$

where the contribution of $v^A(\xi)$ is underlined. For the above approximations, we obtain

- the displacement gradient

$$\nabla \mathbf{u} \doteq \begin{bmatrix} u_{,\xi} & u_{,\eta} \\ v_{,\xi} & v_{,\eta} \end{bmatrix} = \begin{bmatrix} u_1 + \eta\,u_3 & u_2 + \xi\,u_3 \\ v_1 + \eta\,v_3 + \underline{\xi\,\omega} & v_2 + \xi\,v_3 \end{bmatrix}, \tag{12.136}$$

- the linear strain

$$\boldsymbol{\varepsilon} \doteq \mathrm{sym}\nabla\mathbf{u} = \begin{bmatrix} u_{,\xi} & \frac{1}{2}(v_{,\xi}+u_{,\eta}) \\ \mathrm{sym.} & v_{,\eta} \end{bmatrix} = \begin{bmatrix} u_1+\eta\, u_3 & \varepsilon_{12} \\ \mathrm{sym.} & v_2+\xi\, v_3 \end{bmatrix}, \tag{12.137}$$

where $2\varepsilon_{12} = (u_2+v_1)+\xi\, u_3+\eta\, v_3+\underline{\xi\,\omega}$.

The v^A appears only in $v_{,\xi}$ and ε_{12}. Besides, the 11 and 22 components of $\nabla\mathbf{u}$ and $\boldsymbol{\varepsilon}$ are incomplete linear polynomials of ξ and η.

Remark. We can compare eqs. (12.136) and (12.137) for the Allman shape functions with eqs. (12.128) and (12.129) for the analytical solution. Concerning the displacement gradient $(\nabla\mathbf{u})_{12}$, we see that the $\xi\,\omega$ term introduced by v^A is necessary to reproduce the analytical solution. The effect of this term is similar to that of the EADG2 enhancement for the Q4 element,

$$\mathbf{G}^{\xi} = \begin{bmatrix} 0 & q_2\eta \\ q_1\xi & 0 \end{bmatrix}. \tag{12.138}$$

On the other hand, the strain representation is sufficient to reproduce the analytical solution, even without using the Allman shape functions. Nonetheless, the $\xi\omega$ term introduced into ε_{12} by the component v^A positively de-enhances it, similarly to the EADG2 enhancement. The close relation between the Allman shape functions and the EADG2 enhancement is also confirmed by numerical tests.

12.7.2 EADG2x enhancement of Allman quadrilateral

The study of a 2D beam under the in-plane shear load of Sect. 7.2.1, Table 7.1, provides a rational background for using specific enhancing strain modes for the Allman quadrilateral. The strain recovery can be interpreted as a form of strain enhancement in which we add two modes, $\{1,\zeta\}$, to the normal strain E_{33} and where $\{\varepsilon_{33},\kappa_{33}\}$ are multipliers, see eq. (7.85). Only the recovery of κ_{33} and the ζ-mode are important for bending, which is the observation crucial for selecting proper enhancing modes for the Allman element.

The Allman element (standard, without enhancement) identically performs in the above numerical test as the beam without the κ_{33} recovery. Hence, we can enhance the Allman element using the EADG method, see Sect. 11.4.3, and the following two modes,

$$\mathbf{G}^{\xi} = \begin{bmatrix} \xi\, q_1 & 0 \\ 0 & \eta\, q_2 \end{bmatrix}, \tag{12.139}$$

designated as EADG2x. In consequence, the strain components ε_{11} and ε_{22} are enhanced, as shown below.

For simplicity, we consider a square 2×2 element, with the center located at the origin of the Cartesian coordinate system. Then the Cartesian coordinates are equal to the natural coordinates, i.e. $x = \xi$ and $y = \eta$, and the Jacobian matrix is an identity matrix. Then, for the EADG method, eq. (11.71) is reduced to $\tilde{\mathbf{H}} \doteq \mathbf{G}^{\xi}$. For eq. (11.70), the linear strain can be split into two parts

$$\mathbf{E} = \tfrac{1}{2}(\mathbf{F} + \mathbf{F}^T - 2\,\mathbf{I}) = \tfrac{1}{2}[\nabla\mathbf{u} + (\nabla\mathbf{u})^T] + \tfrac{1}{2}(\tilde{\mathbf{H}} + \tilde{\mathbf{H}}^T), \tag{12.140}$$

where the strain enhancement

$$\tfrac{1}{2}(\tilde{\mathbf{H}} + \tilde{\mathbf{H}}^T) = \begin{bmatrix} \xi\, q_1 & 0 \\ 0 & \eta\, q_2 \end{bmatrix}. \tag{12.141}$$

We see that the strain components ε_{11} and ε_{22} are indeed enhanced.

12.7.3 Special techniques for Allman quadrilateral

Even if we use the classical Allman shape functions and do not use the EADG2x enhancement, we still need to implement the two techniques which are described below.

Stabilization of spurious modes. A characteristic feature of Allman quadrilaterals are additional zero eigenvalues.

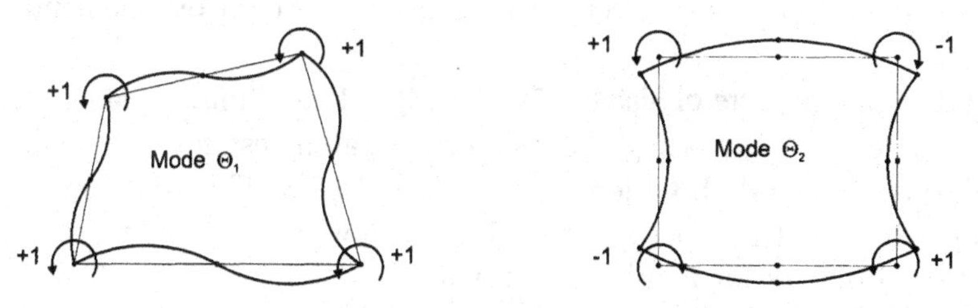

Fig. 12.9 Spurious modes Θ_1 and Θ_2 for Allman's quadrilateral.

1. If the formulation is based on the potential energy functional, then the two spurious modes shown in Fig. 12.9 are obtained. These modes can be eliminated using the penalty method and, e.g., the stabilization functions of [144],

$$P_1 = 10^{-6}\, G\, V\, \Theta_1^2, \qquad \Theta_1 \doteq \frac{1}{4}\sum_{I=1}^{4}(\omega_I - \omega_c), \quad \omega_c \doteq \tfrac{1}{2}(u_{x,y} - u_{y,x})_c, \tag{12.142}$$

$$P_2 = 10^{-3}\, G\, V\, \Theta_2^2, \qquad \Theta_2 \doteq \frac{1}{4}(\omega_1 - \omega_2 + \omega_3 - \omega_4), \tag{12.143}$$

where ω_I are the nodal rotations and V is the element volume. Note that ω_c is identical to that obtained from the linearized drill RC equation at the element's center. It seems, however, that the form of Θ_1 should rather be $\Theta_1 \doteq \left(\frac{1}{4}\sum_{I=1}^{4}\omega_I\right) - \omega_c$.
The tangent matrix yielded by P_2 is as follows:

$$\mathbf{K}^{\text{stab}}_{\omega\omega} = \underbrace{\frac{1}{8}10^{-3}\, G\, V}_{\text{multiplier}} \begin{bmatrix} 1 & -1 & 1 & -1 \\ -1 & 1 & -1 & 1 \\ 1 & -1 & 1 & -1 \\ -1 & 1 & -1 & 1 \end{bmatrix}. \tag{12.144}$$

Its eigenvalues are $\{4, 0, 0, 0\} \times$ multiplier, i.e. only one eigenvalue is non-zero.

2. If the formulation is based on the functional incorporating the drill RC equation, then only one spurious mode is obtained. It is identical to Θ_2 and is eliminated by eq. (12.143). It suffices to enforce the drill RC at one point to obtain the correct rank and the center of element is a natural choice.
Note that eq. (12.143) can be rewritten as $\Theta_2 \doteq \frac{1}{4}\mathbf{h}\cdot\boldsymbol{\omega}_I$, where $\mathbf{h} \doteq [1, -1, 1, -1]$ is the hourglass mode. After implementation of this function, the number of zero eigenvalues is three for our elements.

Adaptation of procedure of [Jetteur, Frey, 1986]. The Allman quadrilaterals have problems with passing the membrane patch test for the boundary conditions **b2** and **b3** described in Sect. 15.2.3. This problem can be circumvented by the procedure of [117]. Below, this procedure is generalized to also work for non-linear strains.

In [117], the part of membrane strains computed from the drilling rotation ω is modified as follows:

$$\tilde{\boldsymbol{\varepsilon}}(\omega) = \boldsymbol{\varepsilon}(\omega) - \frac{1}{A}\int_A \boldsymbol{\varepsilon}(\omega)\, \mathrm{d}A, \tag{12.145}$$

where $\varepsilon(\omega)$ is the strain obtained from the hierarchical displacements

$$\mathbf{u}(\omega_I) \doteq \sum_{H=5}^{8} N_H(\xi,\eta)\ \Delta\mathbf{u}_H \tag{12.146}$$

of eq. (12.104) and A is the element area. We see that, in eq. (12.145), the average value of $\varepsilon(\omega)$ for the element is subtracted from $\varepsilon(\omega)$, with the purpose of minimizing the effect of $\varepsilon(\omega)$ on the constant strains which are checked in the patch test. Consider two cases:

- For linear strains, $\boldsymbol{\varepsilon} \doteq \mathrm{sym}\nabla\mathbf{u}$, and eq. (12.145) can be replaced by

$$\tilde{\mathbf{B}}_I\,\omega_I = \left(\mathbf{B}_I - \frac{1}{A}\int_A \mathbf{B}_I\ \mathrm{d}A\right)\omega_I, \tag{12.147}$$

  where $\mathbf{B}_I \doteq \partial\boldsymbol{\varepsilon}(\omega_I)/\partial\omega_I$, see [235], eq. (5.1)–(5.6).
- For non-linear strains, $\boldsymbol{\varepsilon}(\omega_I)$ is complicated and therefore we replace eq. (12.145) by the formula for the gradient of the displacements depending on ω_I, i.e.

$$\widetilde{\nabla\mathbf{u}}(\omega_I) = \nabla\mathbf{u}(\omega_I) - \frac{1}{A}\int_A \nabla\mathbf{u}(\omega_I)\ \mathrm{d}A, \tag{12.148}$$

  where the integral is evaluated by a 2×2 Gaussian quadrature. This formula enabled our Allman elements to pass the patch tests for the boundary conditions **b2** and **b3**, see Table 15.4.

This procedure slightly changes some of the eigenvalues but does not change the number of zero eigenvalues.

12.7.4 Allman+EADG2x elements

The Allman+EADG2x element is based on the Green strain and is valid for large drilling rotations. It uses the EADG2x enhancement and was developed using the techniques of Sect. 12.7.3. The drill RC was enforced at the element's center.

The extended functionals with drilling rotations are used to formulate two elements:

1. **Element Allman+EADG2x+P** which has the following features:
   a) the penalty method is implemented as specified in Table 12.4,
   b) it uses two additional parameters, which are multipliers of the enhancing modes.

2. **Element Allman+EADG2x+PL** which has the following features:
   a) the Perturbed Lagrange method is implemented as specified in Table 12.5. Only one parameter is used in the representation of the Lagrange multiplier of eq. (12.57), i.e. $T_a(\xi, \eta) \doteq q_0$, so eq. (12.56) becomes
   $$\mathbf{T}_a = \mathbf{J}_{Lc} \begin{bmatrix} 0 & q_0 \\ -q_0 & 0 \end{bmatrix} \mathbf{J}_{Lc}^T. \tag{12.149}$$
   b) it uses three additional parameters: two multipliers of the enhancing modes, and one parameter of the Lagrange multiplier.

These elements have a correct rank and pass the patch test for all types of boundary conditions. As indicated by the numerical results of Table 12.6, these elements have the following features: (i) their rotations converge from above, while displacements converge from below, (ii) they perform quite well for coarse distorted meshes but they are not top performers.

Finally, we note that the Allman-type 2D+drill elements can be used as the membrane part of the four-node shell element with six dofs/node. Typically, they are used in "flat" shell elements, e.g. in [3, 57], due to the lack of Allman shape functions for initially warped elements. For the latter elements, the curvature (warping) correction must be applied, as in [117, 235]; this topic is addressed in Sect. 14.

12.8 Numerical tests

Below are presented the most indicative numerical tests related to the implementation of the drilling rotation; other tests can be found in [43, 57, 56, 266, 102].

12.8.1 Comparison of various elements

All the tested 2D+drill elements have a correct rank and pass the patch test for all types of boundary conditions for the drilling rotation of Table 15.4.

Cook's membrane. The performance of 2D and 2D+drill elements is compared in Cook's membrane test, which is very demanding, see Sect. 15.2.7.

Two meshes are used in computations; a coarse 2×2-element mesh and a fine 32×32-element mesh. The regularizing parameter $\gamma = G$. The vertical displacements and the drilling rotation at point A, see Fig. 15.9, are given in Table 12.6. The results for the penalty (P) method and the Perturbed Lagrange (PL) method are identical, which is indicated as P=PL.

We see that the best coarse mesh performance in the class of the 2D+drill elements is provided by the elements HR7-S+EADG2 and HW18-SS+EADG2. Among the elements using a small number of additional parameters, the EADG4 element is better than the HR5-S element; the converse is true for their 2D counterparts.

Table 12.6 Cook's membrane. Linear test. $\gamma = G$.

Formulation	Element	Mesh 2 × 2		Mesh 32 × 32	
		u_y	ω	u_y	ω
2D	Q4	11.845	-	23.818	-
2D+drill	Q4 (P=PL)	11.173	0.316	23.790	0.876
2D	EADG4	21.050	-	23.940	-
2D+drill	EADG4 (P=PL)	20.940	0.879	23.936	0.891
2D	HR5-S	21.353	-	23.940	-
2D+drill	HR5-S (P=PL)	18.495	0.634	23.911	0.881
2D	HR7-S+EADG2	21.353	-	23.940	-
2D+drill	HR7-S+EADG2 (P=PL)	21.263	0.899	23.936	0.890
2D	HW14-SS	21.353	-	23.940	-
2D+drill	HW14-SS (P=PL)	18.490	0.634	23.911	0.881
2D	HW18-SS	21.353	-	23.940	-
2D+drill	HW18-SS+EADG2 (P=PL)	21.237	0.895	23.936	0.891
2D+drill	Allman+EADG2x (P=PL)	20.253	1.109	23.930	0.899
Ref.		23.81		23.81	

12.8.2 Selection of the value of regularization parameter

Below, we establish the effect of the value of the regularization parameter in order to select the most suitable value for it.

In the tests we use the four-node EADG4 element with the drill RC part, in which we use the Perturbed Lagrange method with the local multipliers, implemented as described in Table 12.5, and eliminated on the element's level.

Straight cantilever beam. This test is described in Sect. 15.2.6. Here the in-plane shear load is considered and four meshes are tested, of either 6×1 or 12×2 elements, and of either rectangular or trapezoidal elements.

The vertical displacements and the drilling rotation at the end of cantilever are shown for $\gamma \in [10^0, 10^{15}]$ in Fig. 12.10 and we note a deterioration of accuracy for trapezoidal elements. For the 12×2 element mesh, the dependence on the regularization parameter γ varies. Three selected values of this parameter, $G, G/10, G/100,$ are marked in this figure by vertical lines. We see that $\gamma = G/100$ yields a slightly better accuracy than $\gamma = G/10$, and clearly better than $\gamma = G$.

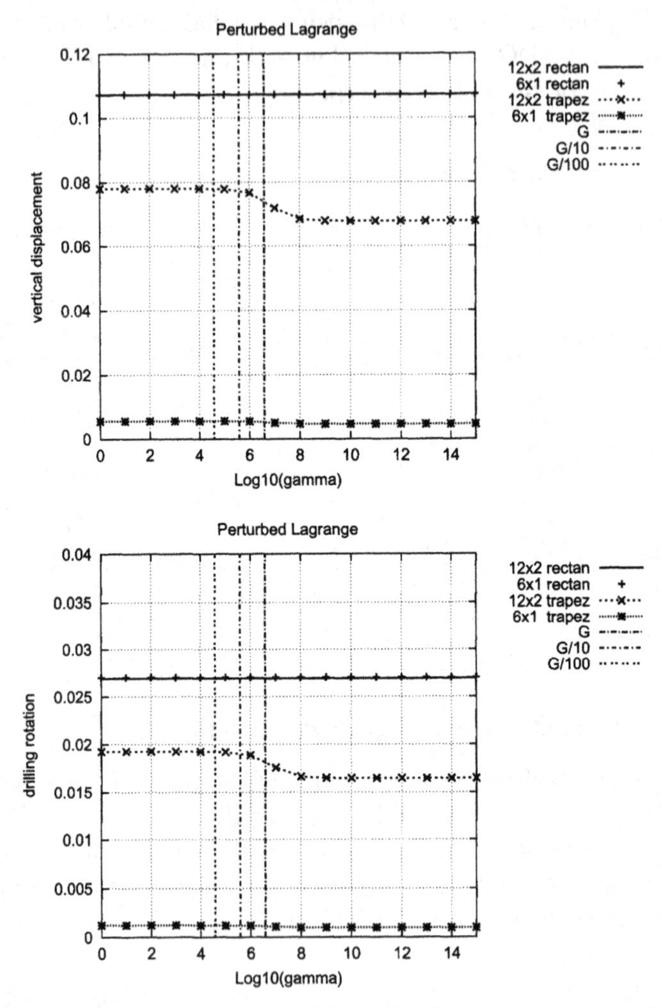

Fig. 12.10 Straight cantilever beam. Effect of γ for various shapes of elements. a) vertical displacement at point A, b) drilling rotation at point A.

Cook's membrane. In this test, elements are skew and tapered and the shear deformation dominates, see Sect. 15.2.7. Three meshes are used in computations: 2×2, 4×4, and the fine 32×32-element mesh which is used for reference.

The vertical displacement and drill rotations are shown for $\gamma \in [10^{-10}, 10^{10}]$ in Fig. 12.11. Three selected values of γ are marked in this figure by vertical lines G, $G/10$ and $G/100$. For the displacement, the conclusion is similar to that of the previous example, i.e. $\gamma = G/100$ yields the best accuracy. For the drilling rotation, the plots are too complicated to be the basis for any conclusion.

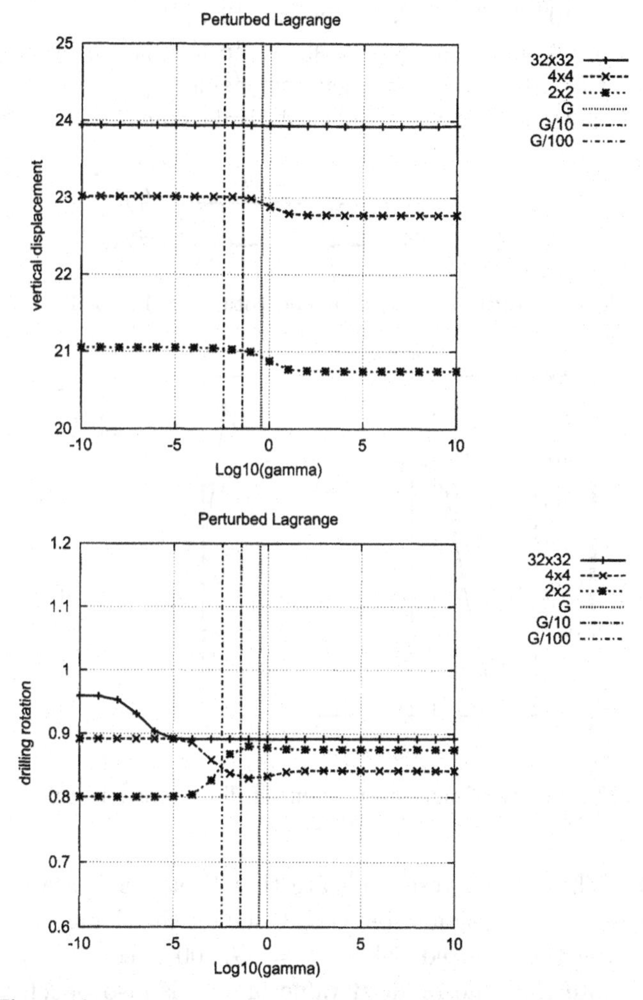

Fig. 12.11 Cook's membrane. Effect of γ for various meshes.
a) Vertical displacement at point A, b) drilling rotation at point A.

Bending of slender cantilever by end drilling rotations. This test checks whether the drilling rotation is correctly linked with displacements and transferred between elements. The drilling rotation $\omega^* = 1.2 \times 10^{-3}$ is prescribed at two tip nodes of a slender cantilever, see Fig. 12.12. The mesh consists of 1×100 elements, and the elements are 1×1 squares. The geometry and data are defined in Sect. 15.3.1.

The vertical displacements are monitored at the tip nodes where the drilling rotations are applied and they are identical for both nodes. The reference value is the Timoshenko beam solution $u_y = 0.06$. The dependence on $\gamma \in [10^0, 10^{15}]$ is shown in Fig. 12.13.

Three selected values of γ are shown in this figure by vertical lines G, $G/10$ and $G/100$. The best accuracy yields $\gamma = G$. An identical conclusion is obtained for the horizontal displacement.

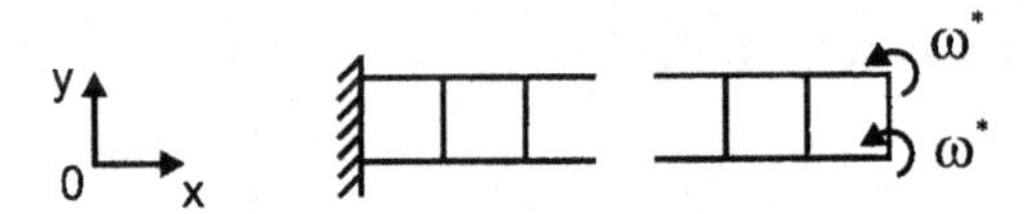

Fig. 12.12 Slender cantilever loaded by end rotations. 100 of 1×1 elements.

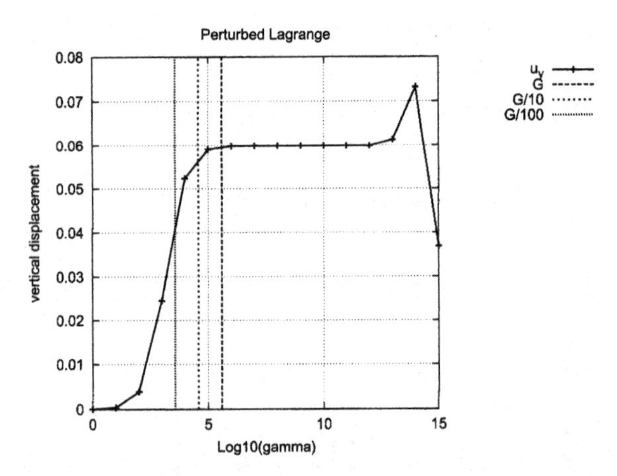

Fig. 12.13 Bending of cantilever by end drilling rotation. Effect of γ.

Conclusion. These three tests indicate that if elements are rectangular, then the value $\gamma = G$ should be used, while for the elements of distorted irregular shape the reduced value $\gamma = G/100$ seems to be optimal. Finally, we note that the reduced value of γ is also beneficial for the warped elements described in Sect. 14.

13

Modification of transverse shear stiffness of shell element

In order to improve the performance of a four-node shell element in bending, the transverse shear energy and stiffness must be treated in a special way. The two main problems are the transverse shear locking (TSL) and the poor performance of very thin elements. These problems also appear for beams, which are simpler and more suitable for analytical studies.

13.1 Treatment of transverse shear stiffness of beams

To identify the problems related to the transverse shear, it suffices to consider the linear kinematics and small rotations of the Timoshenko beam.

Timoshenko beam equations. For the Timoshenko beam, the membrane, transverse shear, and bending strain components are as follows:

$$\varepsilon_{xx} = u_{x,x}, \qquad 2\varepsilon_{zx} = w_{,x} - \theta \doteq \gamma, \qquad \kappa_{xx} = -\theta_{,x}, \tag{13.1}$$

where u is a tangent displacement, w is the normal displacement, and θ is the rotation angle of the middle line of the beam. The strain energy is

$$\mathcal{W} \doteq \int_0^L (\mathcal{W}_\varepsilon + \mathcal{W}_\gamma + \mathcal{W}_\kappa)\, \mathrm{d}x, \tag{13.2}$$

where

$$\mathcal{W}_\varepsilon \doteq \frac{1}{2} EA\, \varepsilon_{xx}^2, \qquad \mathcal{W}_\gamma \doteq k\, 2GA\, \varepsilon_{zx}^2 = \frac{1}{2} k\, GA\, \gamma^2, \qquad \mathcal{W}_\kappa \doteq \frac{1}{2} EI\, \kappa_{xx}^2.$$

Besides, $k = 5/6$ is the shear correction coefficient and L is the beam length. For a rectangular cross-section of the beam of height h and width b, the area is $A = bh$, and the moment of inertia is $I = bh^3/12$.

Transverse shear strain for large rotation beam. Consider a straight (not curved) beam in the 13-plane (XZ-plane). The transverse shear strain of a beam in the ortho-normal basis $\{\mathbf{t}_k\}$ $(k = 1, 3)$ is as follows:

$$\varepsilon_{13} = \tfrac{1}{2}\mathbf{x}_{0,x} \cdot \mathbf{a}_3. \tag{13.3}$$

The form of this strain is identical for the Green strain and the symmetric right stretch strain. The rotation tensor is

$$\mathbf{Q} = c\,(\mathbf{t}_1 \otimes \mathbf{t}_1 + \mathbf{t}_3 \otimes \mathbf{t}_3) + s\,(\mathbf{t}_3 \otimes \mathbf{t}_1 - \mathbf{t}_1 \otimes \mathbf{t}_3), \quad s \doteq \sin\theta, \quad c \doteq \cos\theta, \tag{13.4}$$

where θ is the rotation angle about the axis $\mathbf{t}_2$. Then $\mathbf{a}_3 \doteq \mathbf{Q}\,\mathbf{t}_3 = c\,\mathbf{t}_3 - s\,\mathbf{t}_1$. Besides, $\mathbf{x}_{0,x} = (\mathbf{y}_0 + \mathbf{u}_0)_{,x} = \mathbf{t}_\alpha + \mathbf{u}_{,x}$, where $\mathbf{u} = u\mathbf{t}_1 + w\mathbf{t}_3$. For a straight beam, $\mathbf{t}_{1,x} = \mathbf{t}_{3,x} = \mathbf{0}$, and eq. (13.3) becomes

$$2\varepsilon_{13} = \mathbf{x}_{0,x} \cdot \mathbf{a}_3 = -s - u_{,x}\, s + w_{,x}\, c. \tag{13.5}$$

For small rotations, $\theta \approx 0$, we have $s \approx \theta$ and $c \approx 1$ and neglecting the second order term $u_{,x}\,\theta$, we obtain the linearized transverse shear strain of eq. (13.1).

13.1.1 Reduced integration of transverse shear energy

Transverse shear locking. The transverse shear locking (TSL) is a pathological phenomenon plaguing elements based on the Reissner hypothesis and low-order approximations. It manifests itself in two ways:

1. an artificial over-stiffening of an FE model is observed for coarse meshes. The solution is too small, compared to the analytical solution, see Fig. 13.1. In other words, the solution is "locked".

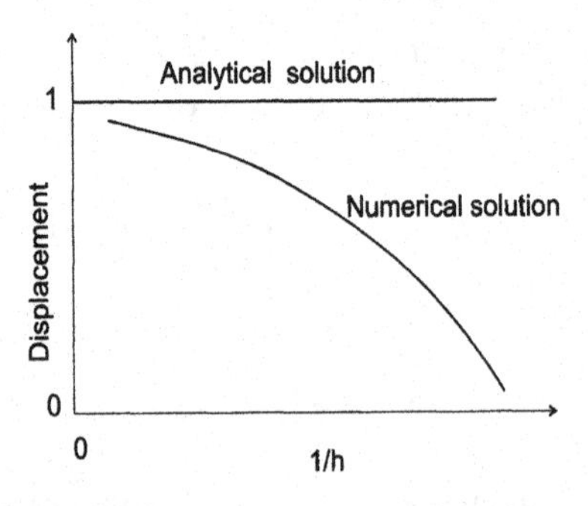

Fig. 13.1 Locking of numerical solution.

2. The rate of the mesh convergence deteriorates and a much denser mesh is necessary to obtain an accurate solution. In the mesh limit, however, the solution is correct.

The TSL is caused by two factors:

1. improper approximation of the transverse shear strain, due to non-matching approximations of particular terms and
2. high values of the (transverse shear stiffness/bending stiffness) ratio for very small thickness h. This ratio is proportional to $12/h^2$, so the thickness is a critical parameter.

We stress that the TSL is not caused by finite computer representations and arithmetic and its presence can be shown in an analytical way, e.g. considering pure bending of a beam element. The first papers on the subject were [226, 227].

Transverse shear locking of two-node beam. Assume that the element's center is located at $x = 0$, see Fig. 13.2. Then $x = (l/2)\,\xi$, where $\xi \in [-1, +1]$, l is the element's length, and we differentiate as follows:

$$(\cdot)_{,x} = (\cdot)_{,\xi} \left(\frac{\mathrm{d}x}{\mathrm{d}\xi}\right)^{-1} = \frac{2}{l}\,(\cdot)_{,\xi}. \tag{13.6}$$

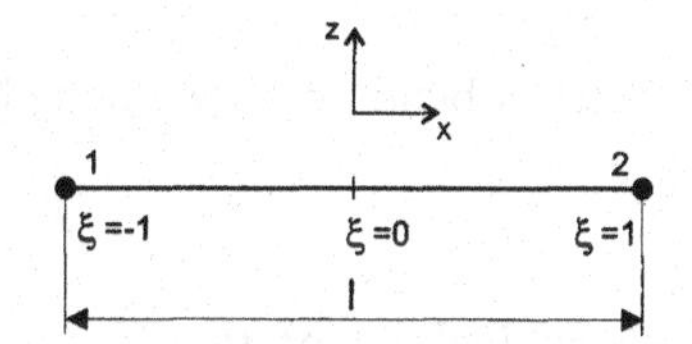

Fig. 13.2 Two-node beam element.

To derive a two-node beam element, we use the approximations

$$w(\xi) = \sum_{I=1}^{2} N_I(\xi)\, w_I, \qquad \theta(\xi) = \sum_{I=1}^{2} N_I(\xi)\, \theta_I, \tag{13.7}$$

where $(\cdot)_I$ designates the nodal values and the linear shape functions are

$$N_1(\xi) \doteq \frac{1}{2}(\xi - 1), \qquad N_2(\xi) \doteq \frac{1}{2}(\xi + 1). \tag{13.8}$$

For these approximations, the components of $\varepsilon_{xz}(\xi)$ of eq. (13.1) are as follows:

$$\theta(\xi) = \tfrac{1}{2}(1-\xi)\,\theta_1 + \tfrac{1}{2}(1+\xi)\,\theta_2, \qquad w_{,x}(\xi) = \frac{1}{L}(w_2 - w_1), \tag{13.9}$$

i.e. $\theta(\xi)$ is a linear function, while $w_{,x}(\xi)$ is a constant function. We see that approximations of these two components do not match up, which has negative consequences.

For pure cylindrical bending of the two-node beam element, see Fig. 13.3, the nodal displacements $w_2 = w_1$ and the nodal rotations $\theta_1 = -\theta$ and $\theta_2 = \theta$. Hence, $\theta(\xi) = \xi\,\theta$ and $w_{,x} = 0$, so we obtain

$$2\varepsilon_{xz}(\xi) = \xi\,\theta, \tag{13.10}$$

which is a linear function of ξ. The analytical value of $\varepsilon_{zx}(\xi)$ for pure bending is zero and is obtained from the above formula only at one point, $\xi = 0$, i.e. at the element's center. This observation is exploited by the reduced integration technique described in the sequel.

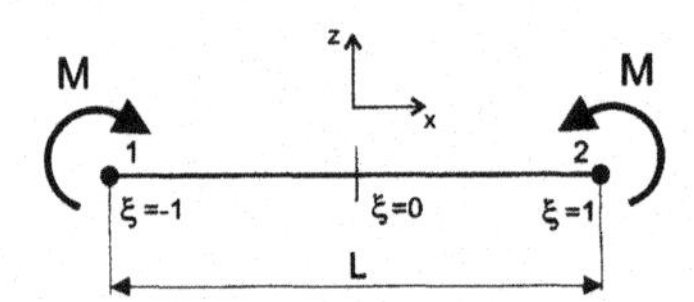

Fig. 13.3 Pure cylindrical bending of two-node beam element.

Remark on transverse shear locking of three-node beam element. Note that the TSL also appears for the three-node beam element based on quadratic shape functions. However, the TSL does not appear for pure bending, but for the transverse loads shown in Fig. 13.4. Two points, $\xi = \pm 1/\sqrt{3}$, at which the approximated $\varepsilon_{zx}(\xi)$ yields analytical values can be found as a solution of a quadratic equation, see [94].

Reduced integration (SRI and URI) of transverse shear energy. To avoid the TSL, we can use the numerical Gauss integration based on the points at which ε_{zx} is correct.

For instance, for a two-node beam, we may use the one-point integration rule using the point $\xi = 0$, while, for the three-node beam, we may

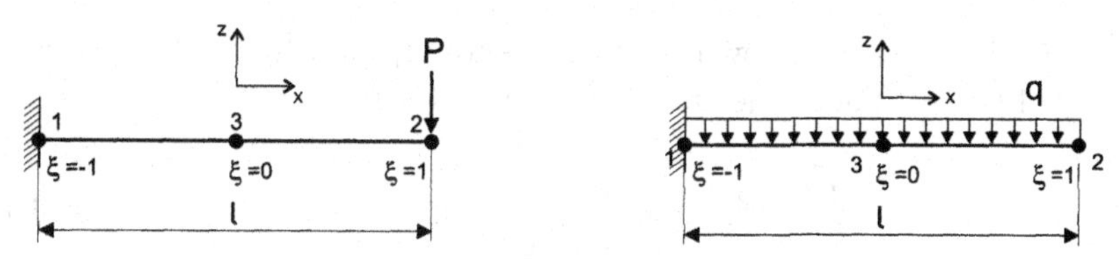

Fig. 13.4 Bending of three-node beam element by force P and distributed load q.

use the two-point integration rule exploiting the points, $\xi = \pm 1/\sqrt{3}$. Fortunately, in both cases, the under-integration does not yield spurious zero eigenvalues of the tangent matrix.

Two forms of the reduced integration are in use. Within the Selective Reduced Integration (SRI) technique, only the transverse shear strain energy is under-integrated. However, for a two-node beam, we can uniformly under-integrate all terms of the strain energy and the element's rank still remains correct. Such a technique is called the Uniform Reduced Integration (URI). If the reduced integration technique yields spurious zero eigenvalues of the tangent matrix, then it must be additionally stabilized.

The SRI technique works very well for beams; the accuracy of the SRI integration and the full integration is compared in [103], Tables I and II, for the example of a cantilever beam loaded by a transverse force.

Why poor approximation of transverse shear strain locks the solution. Consider only the transverse shear and bending strain components in the beam strain energy of eq. (13.2), which can be rewritten as follows:

$$2\mathcal{W}/EI = \int_{-l/2}^{l/2} \left(\kappa_{xx}^2 + \alpha\, \varepsilon_{xz}^2\right) \mathrm{d}x, \qquad \alpha \doteq \frac{k2GA}{EI}. \tag{13.11}$$

For a rectangular cross-section, when $A = bh$ and $I = bh^3/12$,

$$\alpha = \frac{24kG}{Eh^2} = \frac{12k}{(1+\nu)}\frac{1}{h^2}.$$

If the thickness $h \to 0$, then $\alpha \to \infty$, and the component $\alpha\, \varepsilon_{xz}^2 = \alpha\,(\varepsilon_{xz} - 0)^2$ can be interpreted as the penalty term enforcing the condition $\varepsilon_{xz} = 0$. This condition is physically correct for $h \to 0$.

The problem appears when ε_{xz} is not properly approximated within an element because then, not $\varepsilon_{xz} = 0$, but some other condition is enforced. This can be shown for pure bending, for which $2\varepsilon_{xz}(\xi) = \xi\,\theta$, see eq. (13.10). Then, the transverse shear term is

$$\int_{-l/2}^{l/2} \varepsilon_{xz}^2 \, dx = \frac{l}{2}\int_{-1}^{1} \varepsilon_{xz}^2 \, d\xi = \frac{l}{12}\,\theta^2, \tag{13.12}$$

and, eq. (13.11) becomes

$$2\mathcal{W}/EI = \int_{-l/2}^{l/2} \kappa_{xx}^2 \, dx + \alpha\,\frac{l}{12}\,(\theta - 0)^2. \tag{13.13}$$

We see that for $\alpha \rightarrow \infty$, the condition $\theta = 0$ is enforced, which is non-physical and causes an over-stiffened response (locking) of the two-node beam element.

13.1.2 Residual Bending Flexibility (RBF) correction

Introduction. Low-order elements, such as a two-node Timoshenko beam element and a four-node Reissner shell element, seriously lock for the sinusoidal bending shown in Fig. 13.5a, because this form of deformation cannot be properly represented by linear (or bilinear) shape functions. To remedy this problem, we can use the corrected value of the transverse shear stiffness defined as follows:

$$(GA)^* \doteq c_{\mathrm{RBF}}\, GA, \tag{13.14}$$

where c_{RBF} is a scalar coefficient determined by the method of the Residual Bending Flexibility (RBF), which is described below. Note that

1. the RBF correction does not affect the cylindrical (pure) bending shown in Fig. 13.5b, for which the transverse shear strain is zero. This type of bending is improved by the reduced integration of the transverse shear energy or proper sampling of the transverse shear strain.
2. The RBF correction is beneficial for extremely thin elements when the elemental aspect ratio (l/h) is very large. For this case, we also can use the scaling down of [103], discussed in the sequel.

The RBF correction was proposed for beams in [198] and adapted for shells in [138].

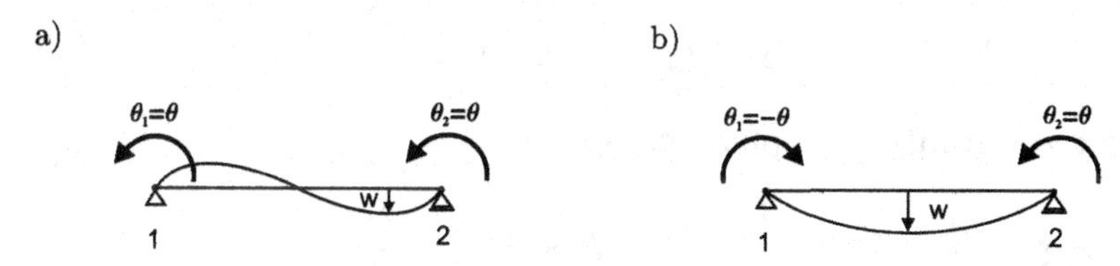

Fig. 13.5 Bending of beam: a) Sinusoidal bending. b) Cylindrical (pure) bending.

RBF correction for beam. For two-node Timoshenko beams, the RBF correction amounts to using, in computations, the corrected value of a transverse shear stiffness

$$(GA)^* = \left(\frac{1}{GA} + \frac{l^2}{12EI} \right)^{-1}, \tag{13.15}$$

where the second term, $l^2/(12EI)$, is designated as the *residual bending flexibility* (RBF) and l is the element's length. For a rectangular cross-section when $A = bh$ and $I = bh^3/12$, we can define the corrected shear modulus

$$G^* \doteq \frac{(GA)^*}{A} = \frac{h^2 EG}{h^2 E + l^2 G} = \left(\frac{1}{G} + \frac{l^2}{h^2 E} \right)^{-1}. \tag{13.16}$$

The RBF term does not vanish since $l \neq 0$, and dominates for $l/h > \sqrt{E/G} = \sqrt{2(1+\nu)}$. If the RBF term strongly dominates, i.e. when $l/h \gg \sqrt{E/G}$, then we can neglect $1/(GA)$ in eq. (13.15), which yields

$$(GA)^* \approx \frac{12EI}{l^2} \quad \text{and} \quad G^* \approx \left(\frac{h}{l} \right)^2 E. \tag{13.17}$$

These formulas are well suited for elements of large (l/h) aspect ratios.

Derivation of the RBF correction for a beam. The derivation below is for a small strain/small rotation beam, but the obtained corrected transverse shear stiffness is subsequently tested also on non-linear problems.

We drive a two-node Discrete Kirchhoff (DK) beam element with the normal displacement approximated by a cubic polynomial

$$w(\xi) = a_0 + a_1 \xi + a_2 \xi^2 + a_3 \xi^3, \qquad \xi \in [-1, +1]. \tag{13.18}$$

Using the boundary conditions: $w(-1) = w_1$, $w(+1) = w_2$, $w_{,\xi}(-1) = (w_{,\xi})_1$, and $w_{,\xi}(+1) = (w_{,\xi})_2$, we can determine the coefficients a_i $(i = 0, 1, 2, 3)$ and rewrite eq. (13.18) as

$$w(\xi) = N_1(\xi)\, w_1 + N_2(\xi)\, w_2 + N_3(\xi)\, (w_{,\xi})_1 + N_4(\xi)\, (w_{,\xi})_2, \tag{13.19}$$

where the Hermitian shape functions are

$$N_1(\xi) \doteq \frac{1}{2} - \frac{3\xi}{4} + \frac{\xi^3}{4}, \qquad N_2(\xi) \doteq \frac{1}{2} + \frac{3\xi}{4} - \frac{\xi^3}{4},$$
$$N_3(\xi) \doteq \frac{1}{4}(1 - \xi - \xi^2 + \xi^3), \quad N_4(\xi) \doteq \frac{1}{4}(-1 - \xi + \xi^2 + \xi^3), \tag{13.20}$$

see Fig. 13.6.

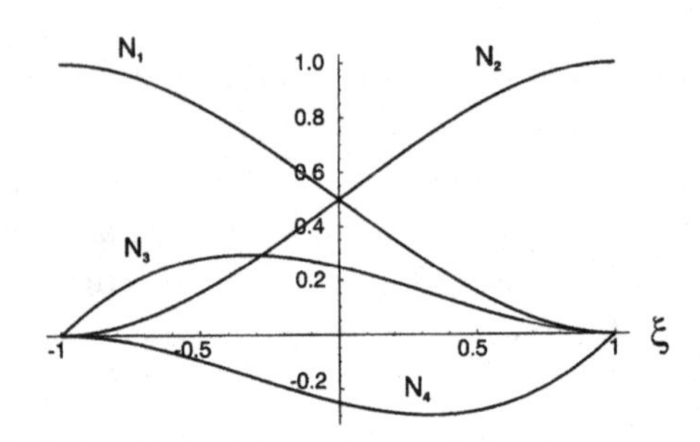

Fig. 13.6 Hermitian shape functions N_1, N_2, N_3, and N_4.

The Kirchhoff constraint, $\varepsilon_{13} = 0$, yields the relation $\theta = w_{,x}$. Applying this constraint to discrete points, namely to the boundary nodes, and using $w_{,x} = (2/l) w_{,\xi}$, we obtain $(w_{,\xi})_1 = (l/2)\, \theta_1$ and $(w_{,\xi})_2 = (l/2)\, \theta_2$. Hence, the normal displacement becomes

$$w(\xi) = N_1(\xi)\, w_1 + N_2(\xi)\, w_2 + \frac{l}{2}\left[N_3(\xi)\, \theta_1 + N_4(\xi)\, \theta_2\right], \tag{13.21}$$

where θ_1 and θ_2 are the nodal rotations. This form of w is used in the sequel, as it conforms with the boundary conditions for the cases shown in Fig. 13.5. Note that we can separate the terms for the cylindrical and sinusoidal bending as follows:

$$\frac{l}{2}\left[N_3(\xi)\, \theta_1 + N_4(\xi)\, \theta_2\right] = \frac{l}{8}[\underbrace{(1 - \xi^2)}_{\text{cylindrical}}\, (\theta_1 - \theta_2) + \underbrace{(-\xi + \xi^3)}_{\text{sinusoidal}}\, (\theta_1 + \theta_2)], \tag{13.22}$$

and if $\theta_1 = \pm\theta_2$, then only one type of bending remains.

Let us now define components of the strain energy of the DK beam element:

1. the bending energy

$$\mathcal{W}_\kappa \doteq \frac{l}{4}\int_{-1}^{+1} EI\,\kappa^2\,\mathrm{d}\xi, \qquad \kappa = -\frac{4}{l^2}w_{,\xi\xi}, \tag{13.23}$$

where the bending strain $\kappa = -\theta_{,x} \approx -w_{,xx}$ for the Kirchhoff constraint.

2. the transverse shear energy

$$\mathcal{W}_\gamma \doteq \frac{l}{4}\int_{-1}^{+1} GA\,\gamma^2\,\mathrm{d}\xi, \qquad \gamma = \frac{EI}{GA}\frac{2}{l}\kappa_{,\xi}, \tag{13.24}$$

where the transverse shear strain $\gamma = -(EI/GA)\,\kappa_{,x}$ is recovered from the equilibrium equation $Q = -M_{,x}$, in which we used $M = EI\,\kappa$ and $Q = GA\,\gamma$.

Using these energies, we can define the ratio of the shear energy to the total energy

$$c \doteq \frac{\mathcal{W}_\gamma}{\mathcal{W}_\kappa + \mathcal{W}_\gamma}. \tag{13.25}$$

For the sinusoidal bending of Fig. 13.5b, we have $w(\xi) = \frac{l}{4}\,\xi(\xi^2-1)\,\theta$ and the above formulas yield

$$\mathcal{W}_\kappa = \frac{6EI}{l}\theta^2, \quad \mathcal{W}_\gamma = \frac{72(EI)^2}{GAl^3}\theta^2, \quad c_{\mathrm{RBF}} \doteq c = \frac{12EI}{12EI + GAl^2}, \tag{13.26}$$

where c_{RBF} does not depend on the rotation θ ! Note that these formulas are for the beam element based on cubic displacements (13.18), but we shall apply the coefficient c_{RBF} to the two-node Timoshenko beam element, which is based on linear displacements and rotation.

Sinusoidal bending of very slender beam. Consider the sinusoidal bending of a very slender beam element, for which $h/l \ll 1$. Then, for the DK beam element based on cubic displacements, the bending energy dominates in eq. (13.26), i.e. $\mathcal{W}_\kappa \gg \mathcal{W}_\gamma$, so

$$\mathcal{W} \doteq \mathcal{W}_\kappa + \mathcal{W}_\gamma \approx \mathcal{W}_\kappa = \frac{6EI}{l}\theta^2, \tag{13.27}$$

where $\mathcal{W}_\gamma$ is neglected. On the other hand, for a two-node Timoshenko element with linear approximations of θ and w, the sinusoidal bending yields

$$\mathcal{W} \doteq \mathcal{W}_\kappa + \mathcal{W}_\gamma = \mathcal{W}_\gamma = \frac{GA\,l}{2}\theta^2, \tag{13.28}$$

i.e. only the transverse shear energy is non-zero. These two energies are equal if $GA = 12EI/l^2$, so we define $(GA)^* \doteq 12EI/l^2$ which has the form of eq. (13.14). For this corrected transverse shear stiffness, the two-node Timoshenko element with linear approximation of w yields almost identical nodal rotations as the DK beam element based on a cubic approximation of w.

13.1.3 Scaling down of transverse shear stiffness

For extremely thin beam and shell elements, we obtain very inaccurate solutions. This is attributed to disparity between the orders of bending and shear terms, which means that, due to the finite computer precision, the bending stiffness is annihilated.

Then, either the RBF correction or the method of scaling down the transverse shear stiffness proposed in [103] can be applied. The most important difference between these two methods is that the scaling down does not pertain to any particular form of deformation, while the RBF method does.

In the method of scaling down the transverse shear stiffness, the annihilation of the bending stiffness is prevented by the following strategy:

1. We find the maximum aspect ratio of an element, $(l/h)_{\max}$, for which the accuracy is still correct. This value is determined by a numerical experiment and $10^4/16$ for beams and $10^5/8$ for plates was found in the cited work.
2. For the aspect ratios which are larger than the maximum aspect ratio, i.e. for $l/h > (l/h)_{\max}$, we scale down the transverse shear stiffness

$$(GA)^* \doteq s\, GA, \qquad s \doteq \left(\frac{h}{l}\right)^2 \left(\frac{l}{h}\right)^2_{\max}. \tag{13.29}$$

The scaling factor s is plotted in Fig. 13.7 and we see that it tends to zero for $l/h \to \infty$. Then the strain energy of a beam is

$$2\mathcal{W} = \int_{-l/2}^{l/2} \left[EI\kappa_{xx}^2 + (GA)^*\,\varepsilon_{xz}^2\right] \,dx. \tag{13.30}$$

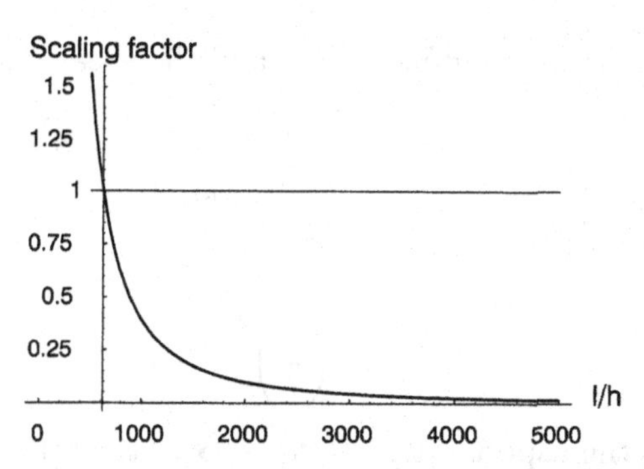

Fig. 13.7 Scaling factor for $(l/h)_{\max} = 10^4/16$.

Alternatively, we can scale down the ratio of the shear stiffness to the bending stiffness α of eq. (13.11),

$$\alpha^* \doteq s\,\alpha, \tag{13.31}$$

so the strain energy of eq. (13.11) becomes

$$2\mathcal{W}/EI = \int_{-l/2}^{l/2} \left(\kappa_{xx}^2 + \alpha^*\,\varepsilon_{xz}^2\right)\,\mathrm{d}x. \tag{13.32}$$

Both above forms of the strain energy are equivalent. Note that the above scaling down is in accord with eq. (13.27) for sinusoidal bending of a very slender beam.

Scaling down parameter s**.** The scaling down parameter s of eq. (13.29) can be obtained by the simple reasoning presented below. The estimation of energy components of a two-node Timoshenko beam element is as follows:

1. the bending energy

$$\mathcal{W}_\kappa \doteq \frac{1}{2}\frac{Eh^3}{12}\int_0^l \theta_{,x}^2\,\mathrm{d}x \approx \frac{1}{2}\frac{Eh^3}{12}\frac{(\theta_2-\theta_1)^2}{l}, \tag{13.33}$$

2. the shear energy

$$\begin{aligned}\mathcal{W}_\gamma &\doteq \frac{1}{2}kGh\int_0^l (w_{,x}-\theta)^2\,\mathrm{d}\xi \\ &\approx \frac{1}{2}kGh\left[\frac{(w_2-w_1)^2}{l} - 2(w_2-w_1)\,\theta_c + \theta_c^2\,l\right],\end{aligned} \tag{13.34}$$

where $\theta_c \doteq \frac{1}{2}(\theta_1+\theta_2)$ is the rotation at the element's center.

The order of the bending stiffness S_κ and the shear stiffness S_γ is as follows:

$$S_\kappa \doteq \frac{\partial \mathcal{W}_\kappa}{\partial \theta_\alpha} \sim \frac{h^3}{l}, \qquad S_\gamma \doteq \frac{\partial \mathcal{W}_\gamma}{\partial \theta_\alpha} \sim hl, \qquad \alpha = 1, 2, \tag{13.35}$$

and their ratio is

$$\frac{S_\kappa}{S_\gamma} = \left(\frac{h}{l}\right)^2. \tag{13.36}$$

We see that, for the aspect ratio $(l/h) \to \infty$, the ratio $(S_\kappa/S_\gamma) \to 0$. Because S_κ is much smaller than S_γ, the bending stiffness is annihilated due to finite computer precision. The effect of disparity between the bending and shear term is alleviated if we scale down as in eq. (13.29).

13.1.4 Numerical tests for beams

Test 1. Transverse shear locking of cantilever. One boundary of a cantilever is fixed while at the other one a vertical force $P = 1$ is applied. The data is as follows: $E = 3 \times 10^6$, $\nu = 0.3$, $L = 100$, $h = 3$, $b = 1$. The two-node Timoshenko beam element is integrated using either one- or two-point Gauss integration of the transverse shear energy.

The mesh convergence for the linear test is shown in Fig. 13.8, where the normalizing value is 4.9416×10^{-2}. We see that the element with the two-point integration converges very slowly while the element using one-point integration converges quickly. The difference is significant, as four elements with one-point integration provide better accuracy than 100 elements with two-point integration.

Test 2. Eigenvalues. Effect of the RBF correction. Two types of two-node Timoshenko beam elements are checked:

1. the element designated as "Linear" is based on linear shape functions. The bending and membrane energy is integrated using either one point at the element's center or two Gauss points, while the shear strain energy is integrated at the element center,
2. the element designated as "Allman" is based on Allman-type shape functions so the displacement vector is approximated as follows:

$$\mathbf{u}_0 = \mathbf{u}_0^L - \frac{l}{8}(1 - \xi^2)(\theta_2 - \theta_1)\,\mathbf{n}, \tag{13.37}$$

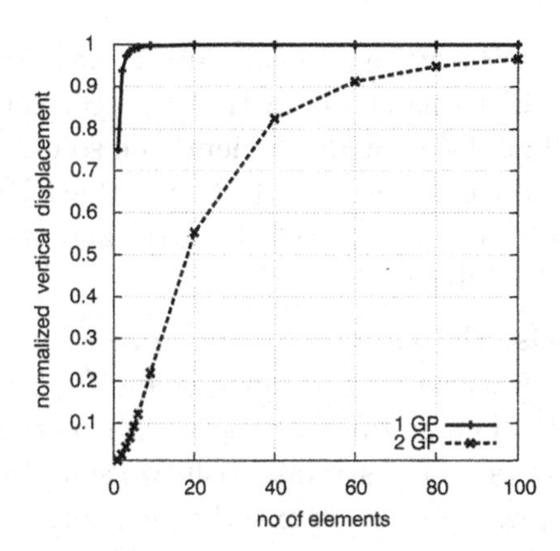

Fig. 13.8 Cantilever loaded by vertical force. Effect of integration rule for transverse shear.

where $\mathbf{u}_0^L$ is approximated by linear shape functions, θ_1, θ_2 are nodal rotations, and $\mathbf{n}$ is a vector normal to the element. (For more details on Allman shape functions, see Sect. 12.7.) The bending and membrane energy is integrated using two Gauss points, while the shear strain energy is integrated by one Gauss point, at the element's center.

We see in Table 13.1 that both elements have identical eigenvalues and the second eigenvalue, which is associated with the transverse shear, is decreased about 33 times by the RBF correction.

Table 13.1 Non-zero eigenvalues of two-node beam elements based on Green strain. $E = 10^6$, $\nu = 0.3$, $h = 0.1$, $l = 1$, $b = 1$.

Shape functions	Non-zero eigenvalues		
no transverse shear stiffness			
Linear	0.2000E+06		0.1667E+03
Allman	0.2000E+06		0.1667E+03
with transverse shear			
Linear	0.2000E+06	0.8013E+05	0.1667E+03
Allman	0.2000E+06	0.8013E+05	0.1667E+03
transverse shear with RBF correction			
Linear	0.2000E+06	0.2424E+04	0.1667E+03
Allman	0.2000E+06	0.2424E+04	0.1667E+03

Test 3. Sinusoidal bending of simply supported beam. The beam is simply supported and loaded by two end moments $M_1 = M_2 = 1$, which gener-

ate the sinusoidal bending of Fig. 13.5a. The results for the Timoshenko beam element ("Linear") with/without the RBF correction are shown in Table 13.2. We see that, for a single element, the solution obtained without the RBF correction is severely locked, but the RBF correction is a perfect remedy. Note that $h/l \in [0.0001, 0.01]$ and the coefficient c_{RBF} assumes values greatly differing from 1!

For comparison, also the solution for three-node beam element is provided. This element is based on parabolic shape functions and uses the two-point Gauss URI. The Assumed Strain (AS) method of [94] for the membrane and transverse shear strains, with two sampling points and the three-point Gauss integration yields exactly the same results.

Table 13.2 Sinusoidal bending by two-node beam element. Effect of RBF correction. $E = 2.11 \times 10^{11}$, $\nu = 0.3$, $h = 0.0002$, $l = 2$, $b = 2$.

	Rotation at node 1			
No of elements	1	2	10	100
no RBF	7.3934E-08	1.7773	2.3460	2.3694
RBF	2.3697	2.3697	2.3697	2.3697
three-node beam	2.3697	2.3697	2.3697	2.3697
h/l of element	0.0001	0.005	0.001	0.01
c_{RBF}	3.119E-08	1.247E-7	3.119E-06	3.119E-04

Test 4. Linear and non-linear cantilever beam. The data is the same as in Test 1. Two finite-rotation two-node Timoshenko beam elements, based on either "Linear" or "Allman" shape functions, and either with or without the RBF correction are used. In the linear test, the load $P = 1$, while in the nonlinear one, the initial load increment $\Delta P = 10$ and the arc-length method is used.

The mesh convergence for the linear test is shown in Fig. 13.9a, where the normalizing value is 4.9416×10^{-2}, and the exact solution is obtained, even for one element. In the non-linear test, see Fig. 13.9b, the reference solution is obtained for 10 elements for which the effect of the RBF correction vanishes. For the mesh of two elements based on "Linear" shape functions and without the RBF correction, the solution is almost exact, while the correction yields a slightly too soft solution. For the same mesh, the solutions for the element based on the "Allman" shape functions are correct only up to a certain load but for the 10-element mesh they are close to the reference solution in the whole range.

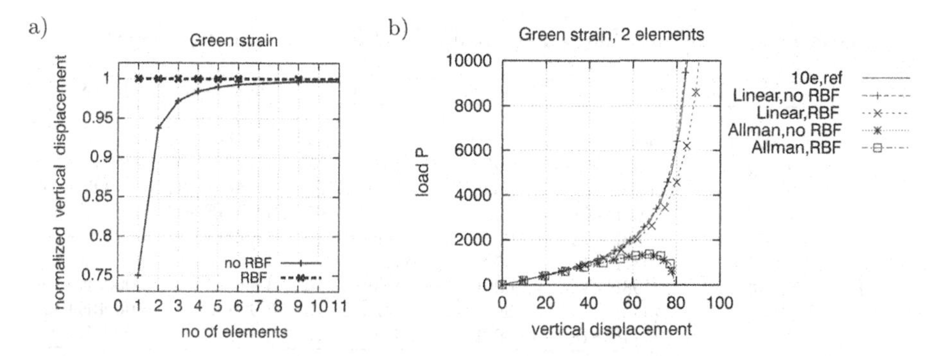

Fig. 13.9 Cantilever loaded by vertical force. Effect of the RBF correction. a) Linear test. Mesh convergence. b) Nonlinear test.

Test 5. Extremely thin cantilever beam. The purpose of this test is to compare the scaling down of the transverse shear stiffness of eq. (13.29) with the RBF correction for the extreme thinness in a linear example involving the transverse shear. The value $(l/h)_{\max} = 10^4/16$ was used for the scaling down.

One boundary of a cantilever is fixed, while at the other one, a vertical force P is applied. The data is as follows: $E = 10^6$, $\nu = 0.3$, $L = 100$, $b = 1$, the load $P = 1$. The two-node Timoshenko beam element based on one-point Gauss integration is used and 100 elements are applied. Hence, the length of a single element is $l = 1$, while the thickness is varied, $h = 10^n$, $n \in [0, -8]$.

The results are given in Table 13.3, where the normal displacement w and the rotation θ at the beam's tip are presented. We see that, for the not modified element, the accuracy is acceptable only for up to $(l/h) = 10^{-4}$, while for the RBF correction and the scaling of eq. (13.29), the accuracy is good even for $(l/h) = 10^{-8}$. We note a 5% error appearing for both these methods for $h = 10^{-5}$.

13.1.5 Curvature correction

Two-node beam elements are typically defined by node positions and are straight. If they are used for curved beams or arches and coarse meshes, then the accuracy can be improved if we account for curvature. The curvature can be defined by specifying either normal vectors at nodes or the height of the arch, which is assumed as circular.

Table 13.3 Thin limit of two-node beam. Modifications of transverse shear.

h	standard		RBF correction		shear scaling, eq. (13.29)	
n	w	θ	w	θ	w	θ
0	4.0002E+00	6.0000E–02	4.0003E+00	6.0000E–02	3.9999E+00	6.0000E–02
–1	3.9999E+03	6.0000E+01	4.0000E+03	6.0000E+01	3.9999E+03	6.0000E+01
–2	3.9999E+06	6.0000E+04	4.0000E+06	6.0000E+04	3.9999E+06	6.0000E+04
–3	4.0003E+09	6.0006E+07	4.0000E+09	6.0000E+07	3.9999E+09	6.0000E+07
–4	4.0239E+12	6.0230E+10	4.0000E+12	6.0000E+10	3.9999E+12	6.0000E+10
–5	2.1911E+14	6.7033E+12	4.2253E+15	6.3521E+13	4.2259E+15	6.3531E+13
–6	1.8068E+16	1.2498E+15	4.0000E+18	6.0000E+16	3.9999E+18	6.0000E+16
–7	2.2518E+17	5.4043E+16	4.0000E+21	6.0000E+19	3.9999E+21	6.0000E+19
–8	–6.9175E+18	-9.2234E+18	4.0000E+24	6.0000E+22	3.9999E+24	6.0000E+22

Consider a 2D circular arch 1-3-2 bent by two opposite horizontal forces P, see Fig. 13.10a. If a straight (not curved) two-node element linking nodes 1 and 2 is used, then the forces P do not cause bending in this element, which is incorrect, comparing to the arch. This can be corrected, e.g., by the method of *rigid links*, which introduces two rigid links 1-A and 2-B, and shifts the straight two-node element to the position defined by points A and B. Then, the forces P cause bending in this element, similarly as in the arch.

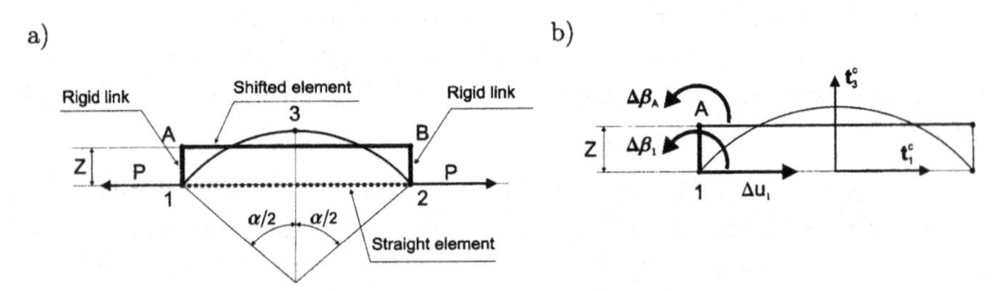

Fig. 13.10 Curvature correction for arch. a) Straight element and rigid links. b) Transformation for rotation.

The displacements and rotation for node A are defined as

$$\Delta u_A = \Delta u_1, \qquad \Delta w_A = \Delta w_1, \qquad \Delta\beta_A = \Delta\beta_1 + Z\,\Delta u_1, \tag{13.38}$$

where only the rotation at node A is corrected and it depends on the rotation at node 1, the horizontal displacement Δu_1 at node 1, and the offset Z, see Fig. 13.10b.

The offset Z corresponds to the length of the rigid links. Its magnitude is arbitrary and must be somehow selected. We have tested two values:

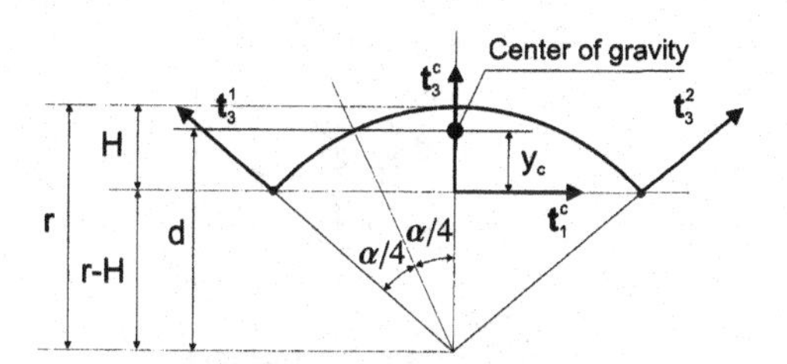

Fig. 13.11 Curvature correction. Center of gravity of arch.

- $Z = H$, where H is the height of arch, see Fig. 13.11, and
- $Z = y_c$, where y_c is the vertical coordinate of the center of gravity of an arch of a constant thickness in the local frame $\{\mathbf{t}_1^c, \mathbf{t}_3^c\}$,

$$y_c = d - (r - H), \qquad d = r\cos(\alpha/4), \tag{13.39}$$

where $\alpha = \alpha_1 + \alpha_2 = \arccos(\mathbf{t}_3^1 \cdot \mathbf{t}_3^c) + \arccos(\mathbf{t}_3^2 \cdot \mathbf{t}_3^c)$. This definition of α involves normal vectors at nodes 1, 2, and at the center, and can also be applied to shapes which are not exactly circular.

Note that the curvature correction must be performed in the local orthonormal basis at the element's center $\{\mathbf{t}_k^c\}$. Besides, the curvature correction slightly impairs a convergence rate of the Newton method, comparing to that for the uncorrected element, but accuracy is improved.

Numerical example. The circular arch is shown in Fig. 13.12. The left boundary is fixed, while at the right one, the horizontal force P is applied and the vertical displacement is constrained to zero. The RBF correction and the curvature correction are tested, using the two-node finite-rotation Timoshenko beam element with one-point integration.

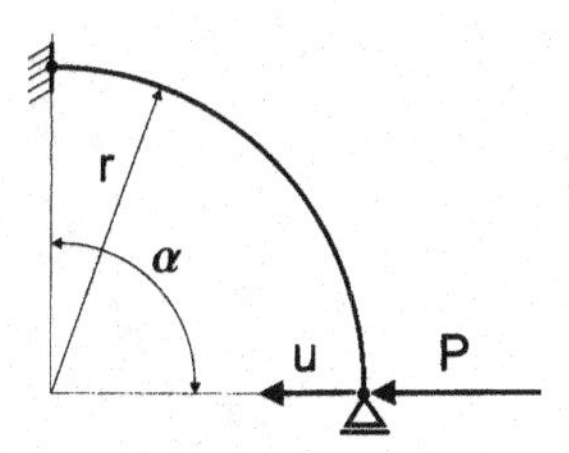

Fig. 13.12 Circular arch. $E = 3 \times 10^6$, $\nu = 0.3$, $r = 300$, $\alpha = 90^o$, $h = 3$.

The mesh convergence in a linear test is shown in Fig. 13.13a, where the normalizing reference value of the horizontal displacement is 3.3422.

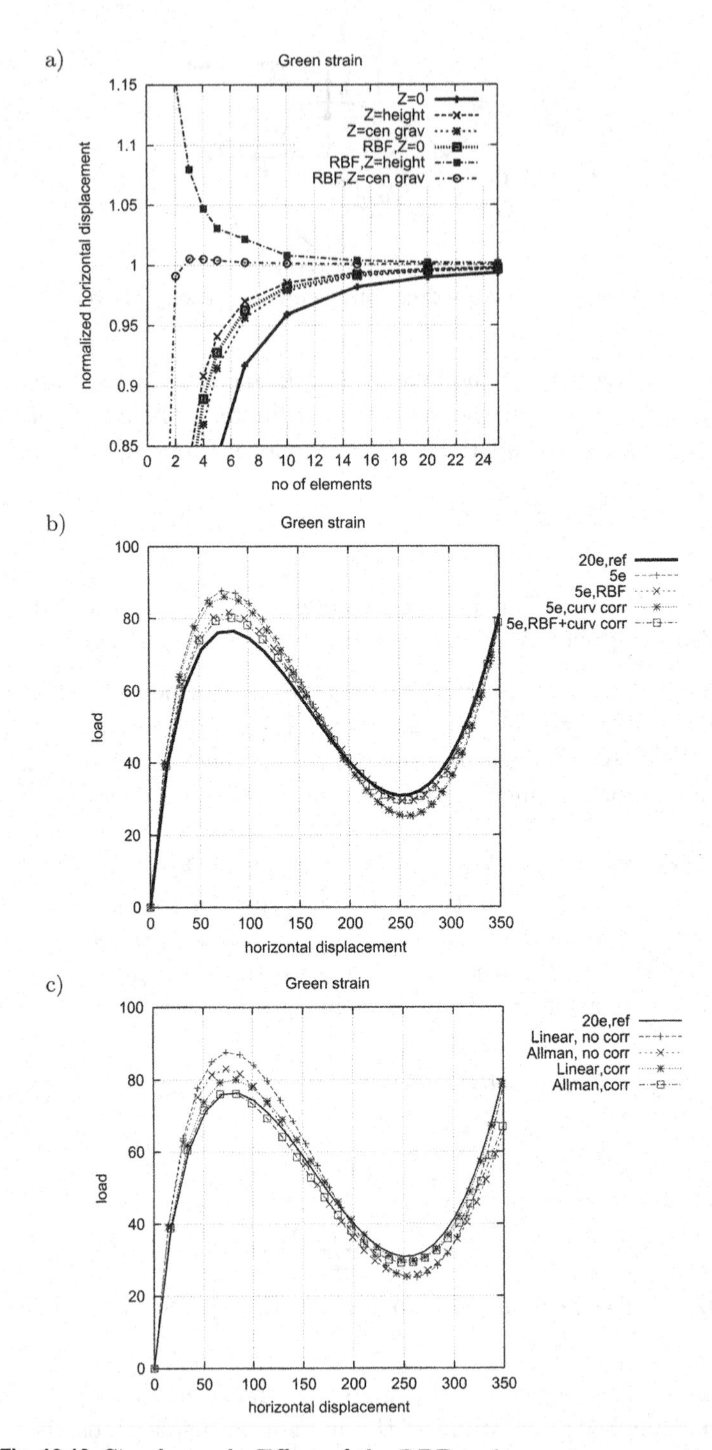

Fig. 13.13 Circular arch. Effect of the RBF and curvature corrections. a) Linear test: mesh convergence. b) Nonlinear test. c) Comparison of two elements "Linear" and "Allman" with/without corrections.

The best accuracy is obtained for both corrections combined together and when the offset $Z = y_c$. For the one-element mesh, the obtained result is not as exact as it was for a straight cantilever, but still the improvement is impressive; one corrected element provides an accuracy comparable with that for 25 uncorrected elements.

The results of a nonlinear test are shown for 5- and 20-element meshes in Fig. 13.13b. The offset $Z = y_c$ is used and its values are as follows: for 20 elements $Z = 0.17346 \approx 0.06h$, while for five elements $Z = 2.7687 \approx 0.92h$, i.e. is of the order of the thickness. The load increment is $\Delta P = 10$.

For five elements, the RBF correction has a stronger effect than the curvature correction but both corrections combined together produce the best result. The RBF correction itself halves the difference between the solutions for 20 and five elements without the RBF. The solution for 20 elements is used for reference, as then the effect of both corrections is negligible.

Finally, we compare the two earlier presented finite-rotation two-node Timoshenko elements, "Linear" and "Allman", the eigenvalues of which are given in Table 13.1. The mesh of five elements is used and solutions are obtained for two cases: (1) no corrections, and (2) with both corrections applied. The reference solution is obtained for a 20-element mesh, and is identical for both elements. Comparing the curves in Fig. 13.13c, we see that the "Allman" element performs better than the "Linear" element, and that the corrections improve the accuracy of both.

13.2 Treatment of transverse shear stiffness of shells

The problems caused by the transverse shear strains, such as the transverse shear locking and a poor performance of very thin elements, also appear for shells, for which we can generalize the techniques developed and tested for beams in Sect. 13.1.

Transverse shear strain for shell. For the Reissner kinematics, the transverse shear strain components in the ortho-normal basis $\{\mathbf{t}_\alpha\}$ are

$$\varepsilon_{\alpha 3} = \varepsilon_{3\alpha} = \tfrac{1}{2}\mathbf{x}_{0,\alpha} \cdot \mathbf{a}_3, \tag{13.40}$$

where the differentiation is performed w.r.t. S^α. The above form of the transverse shear strains is identical for the Green strain and the symmetric right stretch strain.

13.2.1 Selective Reduced Integration

The first remedy which was invented to circumvent the transverse shear locking (TSL) in a four-node plate element was the Selective Reduced Integration (SRI) proposed in [103]. The bending energy was integrated using the 2×2-point Gauss scheme, while the transverse shear energy was integrated by the reduced one-point scheme. Note that the plate element has only three dofs/node, i.e. the normal displacement w, and two tangent rotations θ_α.

However, differently from two-node Timoshenko beams, the under-integration of the transverse shear energy of a plate yields two spurious zero eigenvalues. The associated zero-energy modes are as follows: (i) the hourglass mode $w = \xi\eta$ and $\theta_\alpha = 0$, and (ii) the in-plane twisting mode $\theta_1 = -\eta$, $\theta_2 = \xi$ and $w = 0$, see [103], Fig. 10. According to this paper, the first mode can be removed by 2×2 integration of the $(\partial w/\partial x_\alpha)^2$ term in the transverse shear energy, while the second mode vanishes for a mesh with the rigid body modes removed.

Note that two schemes of integration of the transverse shear strain energy make the SRI complicated and inconvenient for materially nonlinear problems. This provided the motivation for further work, resulting in the ANS technique described in the next section.

For four-node shell elements, the under-integration of the transverse shear energy causes rank deficiency (two spurious zero eigenvalues), similarly as for plate elements, and, for this reason, is not currently used.

13.2.2 Assumed Natural Strain method

Overview. In the Assumed Natural Strain (ANS) method each strain component is treated separately; it is sampled at selected points and approximated over the element domain.

For example, the transverse shear strain $\varepsilon_{13}(\xi, \eta)$ is sampled at two points $(\xi = 0, \eta = \pm 1)$ and approximated by a function which is constant in the ξ-direction and linear in the η-direction. This means that in unidirectional bending in the $\xi\zeta$-plane, the rectangular four-node shell element performs identically as the two-node beam element.

The constant approximation in the ξ-direction corresponds to the reduced one-point integration in the ξ-direction but, when we use the ANS method, we can apply the standard 2×2-point Gauss integration to all terms, including the transverse shear strain energy.

The ANS method was gradually developed in several works, including [138, 104, 139, 18, 19]. The works preceding [18] are well characterized in [141], pp. 401–3. The controversy existed over the optimal position of points in the direction in which the strain is linearly approximated and two values were in use: $\pm 1/\sqrt{3}$ and ± 1. Currently, the latter one is considered as better.

Note that the assumed strain method is also used in nine-node shell elements for which more sophisticated sampling and approximation schemes are used to eliminate the transverse shear and membrane locking, see Sect. 14.4.

Covariant components of transverse shear strains. In eq. (13.40), for the transverse shear strain, the differentiation is performed w.r.t. S^α, but the position vector $\mathbf{x}_0$ is approximated in terms of the natural coordinates $\xi, \eta \in [-1, +1]$. Hence, we have to express the derivatives w.r.t. S^α by the derivatives w.r.t. ξ, η.

Let us form the vector of components of the transverse shear strain of eq. (13.40),

$$\begin{bmatrix} \varepsilon_{13} \\ \varepsilon_{23} \end{bmatrix} = \begin{bmatrix} \frac{1}{2}\mathbf{x}_{0,1} \cdot \mathbf{a}_3 \\ \frac{1}{2}\mathbf{x}_{0,2} \cdot \mathbf{a}_3 \end{bmatrix}, \tag{13.41}$$

where the differentiation is performed w.r.t. S^α and the position vector $\mathbf{x}_0$ is approximated as $\mathbf{x}_0(\xi, \eta) = \sum_{I=1}^{4} N_I(\xi, \eta)\, \mathbf{x}_{0I}$. Then we transform

$$\begin{bmatrix} \frac{1}{2}\mathbf{x}_{0,1} \cdot \mathbf{a}_3 \\ \frac{1}{2}\mathbf{x}_{0,2} \cdot \mathbf{a}_3 \end{bmatrix} = \sum_{I=1}^{4} \begin{bmatrix} \frac{1}{2}N_{I,1}\, \mathbf{x}_{0I} \cdot \mathbf{a}_3 \\ \frac{1}{2}N_{I,2}\, \mathbf{x}_{0I} \cdot \mathbf{a}_3 \end{bmatrix} = \sum_{I=1}^{4} \begin{bmatrix} N_{I,1} \\ N_{I,2} \end{bmatrix} s_I \tag{13.42}$$

where the auxiliary scalar $s_I \doteq \frac{1}{2}\mathbf{x}_{0I} \cdot \mathbf{a}_3$. To calculate the derivatives of shape functions, we can use eq. (2.46),

$$\begin{bmatrix} N_{I,1} \\ N_{I,2} \end{bmatrix} = \mathbf{J}^{-T} \begin{bmatrix} N_{I,\xi} \\ N_{I,\eta} \end{bmatrix}, \tag{13.43}$$

with the Jacobian inverse $\mathbf{J}^{-1}$ defined by eq. (10.56). Then

$$\sum_{I=1}^{4} \begin{bmatrix} N_{I,1} \\ N_{I,2} \end{bmatrix} s_I = \mathbf{J}^{-T} \sum_{I=1}^{4} \begin{bmatrix} N_{I,\xi} \\ N_{I,\eta} \end{bmatrix} s_I = \mathbf{J}^{-T} \begin{bmatrix} \frac{1}{2}\mathbf{x}_{0,\xi} \cdot \mathbf{a}_3 \\ \frac{1}{2}\mathbf{x}_{0,\eta} \cdot \mathbf{a}_3 \end{bmatrix}, \tag{13.44}$$

and the differentiation is performed w.r.t. ξ, η. Hence, eq. (13.41) can be rewritten as

$$\begin{bmatrix} \varepsilon_{13} \\ \varepsilon_{23} \end{bmatrix} = \mathbf{J}^{-T} \begin{bmatrix} \varepsilon^{\xi}_{13} \\ \varepsilon^{\xi}_{23} \end{bmatrix}, \tag{13.45}$$

where

$$\varepsilon^{\xi}_{13} \doteq \tfrac{1}{2}\mathbf{x}_{0,\xi} \cdot \mathbf{a}_3, \qquad \varepsilon^{\xi}_{23} \doteq \tfrac{1}{2}\mathbf{x}_{0,\eta} \cdot \mathbf{a}_3. \tag{13.46}$$

Note that eq. (13.45) transforms the covariant ($\alpha 3$) components of a tensor into Cartesian components, similarly as in eq. (2.27). The covariant components $\varepsilon^{\xi}_{\alpha 3}$ are interpolated within the ANS method in a specific way which is described below.

The ANS method. The Assumed Natural Strain (ANS) method consists of the following steps:

1. The covariant components $\varepsilon^{\xi}_{\alpha 3}$ of eq. (13.46) are evaluated (sampled) at the mid-side points of element edges, at two points for each component, see Fig. 13.14. The sampled values are denoted as $\varepsilon^{M}_{\alpha 3}$, $M = 5, 6, 7, 8$.

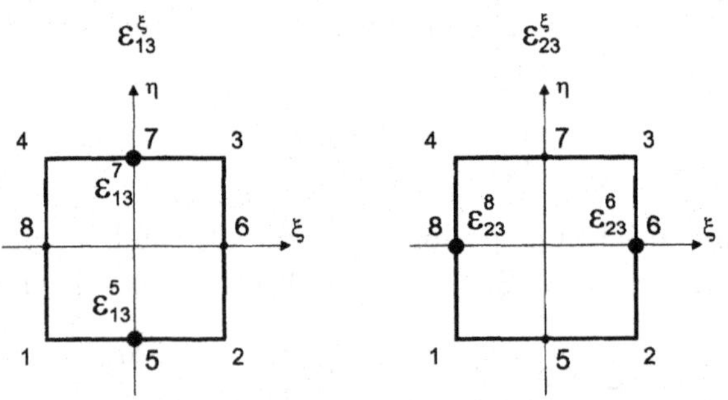

Fig. 13.14 Location of sampling points to evaluate ε^{ξ}_{13} and ε^{ξ}_{23}.

2. The components $\varepsilon^{\xi}_{\alpha 3}$ are approximated over the element domain as follows:

$$\varepsilon^{\xi}_{13}(\xi,\eta) = \tfrac{1}{2}\left[(1-\eta)\,\varepsilon^{5}_{13} + (1+\eta)\,\varepsilon^{7}_{13}\right], \tag{13.47}$$

$$\varepsilon^{\xi}_{23}(\xi,\eta) = \tfrac{1}{2}\left[(1-\xi)\,\varepsilon^{6}_{23} + (1+\xi)\,\varepsilon^{8}_{23}\right], \tag{13.48}$$

where the sampled values $\varepsilon^{M}_{\alpha 3}$ are used. These approximations are constant in the direction in which the derivative is calculated in eq. (13.46) and linear in the other direction, which means that the ANS method is orientation-dependent.

3. At the Gauss integration points, the transverse shear strain $\varepsilon_{\alpha 3}$ is evaluated by eq. (13.45),

$$\begin{bmatrix} \varepsilon_{13} \\ \varepsilon_{23} \end{bmatrix} = \mathbf{J}_{Lc}^{-T} \begin{bmatrix} \varepsilon_{13}^{\xi} \\ \varepsilon_{23}^{\xi} \end{bmatrix}, \tag{13.49}$$

where the Jacobian is local (L) and evaluated at the element center (c). Note that the Jacobian is not approximated by the ANS method ! We have also tested the version with the Jacobian matrix not taken at the center but at the Gauss Points, i.e. using $\mathbf{J}_L$ not $\mathbf{J}_{Lc}$. It also passes the bending patch test and the difference of both solutions in other tests is negligible.

The ANS method effectively removes the transverse shear locking and is used in four-node shell elements as a standard.

Test 1. Unidirectional bending in $\xi\zeta$-plane. Consider a 1×1 square element shown in Fig. 13.15. Nodes 1 and 4 are fixed, while at nodes 2 and 3 we apply: (i) the displacement vector $\mathbf{u} = [0, 0, 0.1]^T$, and (ii) the rotation vector $\boldsymbol{\psi} = [0, 0.01, 0.01]^T$. The rotations are small so the forward-rotated normal vector can be computed as $\mathbf{a}_3 = \mathbf{t}_3 + \boldsymbol{\psi} \times \mathbf{t}_3$.

The transverse shear ε_{23}^{ξ} is equal to zero, while the distribution of ε_{13}^{ξ} obtained for bilinear approximations is shown in Fig. 13.15. In the ANS method, ε_{13}^{ξ} is sampled at points 5 and 7, and approximated by eq. (13.47). Hence, for the unidirectional bending in the $\xi\zeta$-plane, we obtain $\varepsilon_{13}^{\xi}(\xi, \eta)$ which is constant in both directions!

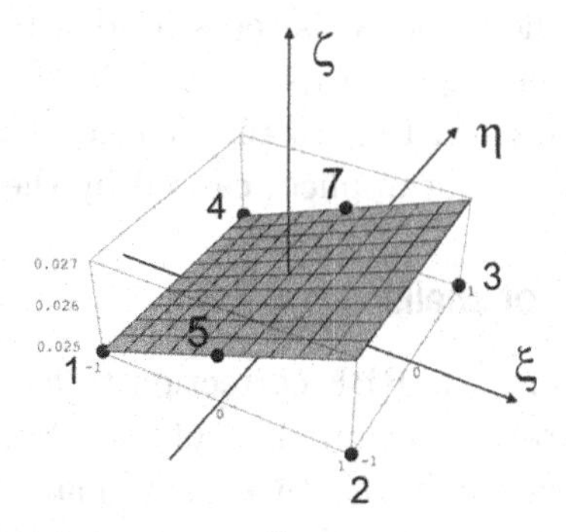

Fig. 13.15 Transverse shear ε_{13}^{ξ} for unidirectional bending.

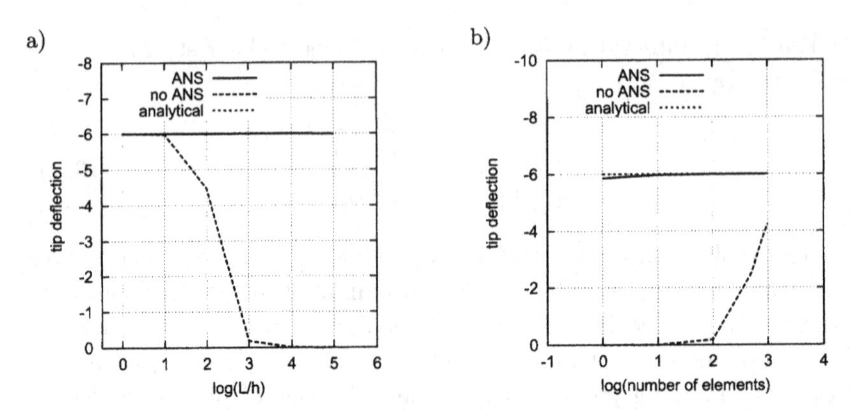

Fig. 13.16 Tip deflection for pure bending. a) Thin limit. b) Mesh convergence.

Test 2. Pure bending of slender cantilever. In this example, we compare results obtained by the same shell element either with or without the ANS procedure.

The data for the cantilever is defined in Sect. 15.3.1 and the cantilever is shown in Fig. 15.13. The mesh of $m \times 1$ four-node shell elements is used.

The bending moment $M = 0.1$ is applied, for which the analytical displacement at the cantilever's tip is $w = ML^2/(2EI) = -6$, where $I = bh^3/12$. Two numerical tests were performed using the four-node shell element:

1. Test of thin limit, in which thickness h was changed, see Fig. 13.16a. 100 elements were used and the moment M was scaled by h^3, to make results independent of thickness. When the ANS procedure is not applied, then the response is too stiff, due to the TSL.
2. Test of mesh convergence with the number of elements m changed. The results are shown in Fig. 13.16b. We see slow convergence when the ANS procedure is not applied, caused by the TSL.

13.2.3 RBF correction for shells

The motivation for using the RBF correction to four-node shell elements is analogous as for two-node beam elements, see Sect. 13.1.2. Note that for the shell element, we can have a two-directional sinusoidal bending, see Fig. 13.17, which can be rendered by bending moments applied to nodes.

The way in which the RBF correction can be applied to four-node shell elements is described in [138], p. 178 and some additional suggestions are

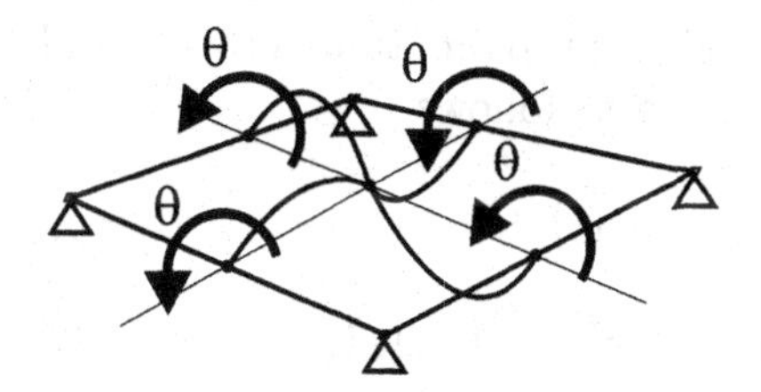

Fig. 13.17 Sinusoidal bending of four-node shell element.

given in [140], p. 104. Usefulness of the RBF correction for higher-order hierarchical p elements is acknowledged in [142], p. 184.

In our implementation of the RBF correction for a four-node bilinear element and an isotropic elastic SVK material, we use the corrected shear modulus G^* of eq. (13.16) separately for each direction, i.e.

$$G_1^* = \left(\frac{1}{G} + \frac{l_1^2}{h^2 E}\right)^{-1}, \qquad G_2^* = \left(\frac{1}{G} + \frac{l_2^2}{h^2 E}\right)^{-1}, \tag{13.50}$$

where l_1 and l_2 are the lengths of vectors connecting opposite mid-side points. The most straightforward implementation is to express the transverse shear strain energy for a single element as follows:

$$\mathcal{W}_\gamma = 2h \int_{-1}^{+1} \int_{-1}^{+1} \left(G_1^*\, \varepsilon_{13}^2 + G_2^*\, \varepsilon_{23}^2\right)\, J\, \mathrm{d}\xi \mathrm{d}\eta, \tag{13.51}$$

where h, G_1^*, and G_2^* are constant over the element. Then the element passes the bending patch test and performs well for bending but, unfortunately, yields erroneous results for twist. This can be observed, e.g., in the linear test of a slender cantilever modelled by one layer of four-node elements, see Sect. 15.3.1. For $h = 0.01$ and twisting by a pair of end forces, the rotation r_x of the tip is too large by about 38%, see the example in the sequel. That is why a more sophisticated approach is needed.

To avoid an excessive twist, it was proposed in [140], p. 104 to apply the full RBF correction to the average values of the (sampled) transverse shear strains, while 4% of the correction to the remaining parts of them. This idea is reconsidered below.

Note that the RBF correction is implemented on top of the ANS method which we use for the transverse shear strain, see Sect. 13.2.2. (Note that the version of the ANS method which we use is different from that used in [138].) The approximations (13.47) and (13.48) of the ANS method can be rewritten as follows

$$\varepsilon_{13}^{\xi}(\xi,\eta) = \tfrac{1}{2}\left(\varepsilon_{13}^{5} + \varepsilon_{13}^{7}\right) + \eta\, \tfrac{1}{2}\left(\varepsilon_{13}^{7} - \varepsilon_{13}^{5}\right) = \varepsilon_{13}^{\mathrm{ave}} + \eta\, \varepsilon_{13}^{\mathrm{d}}, \tag{13.52}$$

$$\varepsilon_{23}^{\xi}(\xi,\eta) = \tfrac{1}{2}\left(\varepsilon_{23}^{6} + \varepsilon_{23}^{8}\right) + \xi\, \tfrac{1}{2}\left(\varepsilon_{23}^{8} - \varepsilon_{23}^{6}\right) = \varepsilon_{23}^{\mathrm{ave}} + \xi\, \varepsilon_{23}^{\mathrm{d}}, \tag{13.53}$$

where, in the final forms, we distinguish the average values "ave" and the differences "d" of the sampled strain components. Then the transverse shear strains of eq. (13.49) are

$$\begin{bmatrix} \varepsilon_{13} \\ \varepsilon_{23} \end{bmatrix} = \mathbf{J}_{Lc}^{-T} \begin{bmatrix} \varepsilon_{13}^{\mathrm{ave}} + \eta\, \varepsilon_{13}^{\mathrm{d}} \\ \varepsilon_{23}^{\mathrm{ave}} + \xi\, \varepsilon_{23}^{\mathrm{d}} \end{bmatrix}, \tag{13.54}$$

where

$$\mathbf{J}_{Lc}^{-1} = \begin{bmatrix} \bar{J}_{11} & \bar{J}_{12} \\ \bar{J}_{21} & \bar{J}_{22} \end{bmatrix} \tag{13.55}$$

is evaluated at the element center i.e. is constant in ξ and η. Hence, the 13-component is a linear function of ξ and η,

$$\begin{aligned} \varepsilon_{13} &= (\bar{J}_{11}\varepsilon_{13}^{\mathrm{ave}} + \bar{J}_{21}\varepsilon_{23}^{\mathrm{ave}}) + \bar{J}_{21}\varepsilon_{33}^{\mathrm{d}}\,\xi + \bar{J}_{11}\varepsilon_{13}^{\mathrm{d}}\,\eta \\ &= \bar{\varepsilon}_{13}^{\mathrm{ave}} + \bar{\varepsilon}_{13}^{\mathrm{d1}}\,\xi + \bar{\varepsilon}_{13}^{\mathrm{d2}}\,\eta, \end{aligned} \tag{13.56}$$

where the parts "ave" and "d" are separated and "d" is additionally split into parts "d1" and "d2" multiplied by ξ and η respectively. The definitions of the $\bar{\varepsilon}_{\alpha 3}$ terms are obvious. As we shall see below, separation of these parts in the strain energy requires additional simplifications.

The Jacobian determinant for a four-node bilinear element is $J(\xi,\eta) = J_0 + J_1\xi + J_2\eta$, see eq. (10.63). Both ε_{13} and J are linear functions of ξ and η and the integration yields

$$\int_{-1}^{+1}\int_{-1}^{+1} \varepsilon_{13}^{2}\, J\, \mathrm{d}\xi \mathrm{d}\eta =$$
$$4(\bar{\varepsilon}_{13}^{\mathrm{ave}})^2 J_0 + \frac{4}{3}\left[(\bar{\varepsilon}_{13}^{\mathrm{d1}})^2 + (\bar{\varepsilon}_{13}^{\mathrm{d2}})^2\right] J_0 + \frac{8}{3}\left[\bar{\varepsilon}_{13}^{\mathrm{ave}}(\bar{\varepsilon}_{13}^{\mathrm{d1}} J_1 + \bar{\varepsilon}_{13}^{\mathrm{d2}} J_2)\right], \tag{13.57}$$

where the "ave" and "d" terms in the last bracket are coupled. Two ways of treating of this coupling can be used.

1. The first way is in the spirit of the suggestion of [140], p. 104. The separation of the "ave" and "d" terms can be achieved by using a simplified form of ε_{13}^2,

$$\varepsilon_{13}^2 \approx (\bar{\varepsilon}_{13}^{\text{ave}})^2 + (\bar{\varepsilon}_{13}^{\text{d1}})^2 \,\xi^2 + (\bar{\varepsilon}_{13}^{\text{d2}})^2 \,\eta^2, \tag{13.58}$$

obtained by omitting the linear terms. The bilinear term is also omitted as it yields zero in integration. For the simplified ε_{13}^2,

$$\int_{-1}^{+1}\int_{-1}^{+1} \varepsilon_{13}^2 \; J \,\mathrm{d}\xi \mathrm{d}\eta \approx \frac{4}{3}\left[3(\bar{\varepsilon}_{13}^{\text{ave}})^2 + (\bar{\varepsilon}_{13}^{\text{d1}})^2 + (\bar{\varepsilon}_{13}^{\text{d2}})^2\right] J_0, \tag{13.59}$$

in which the terms "ave" and "d" of the sampled strains are separated. (Note that the same result of integration is obtained for the full form of ε_{13}^2 and $J(\xi,\eta) \approx J_0$.) Finally, the integrand of the strain energy (13.51) is modified as follows:

$$G_1^* \,\varepsilon_{13}^2 \approx G_1^* \,(\bar{\varepsilon}_{13}^{\text{ave}})^2 + G_{1c}^* \,(\bar{\varepsilon}_{13}^{\text{d1}})^2 \,\xi^2 + G_{1c}^* \,(\bar{\varepsilon}_{13}^{\text{d2}})^2 \,\eta^2, \tag{13.60}$$

where the additionally corrected shear modulus is defined as

$$G_{1c}^* \doteq \left(\frac{1}{G} + a\,\frac{l_1^2}{h^2 E}\right)^{-1}, \qquad a \doteq \frac{c}{c + (1-c)\,(l_1/l_2)^2}, \tag{13.61}$$

where c is a corrective coefficient, designated as ϵ in [138]. Besides, (l_1/l_2) is the element aspect ratio, as the average size of an element in each direction is $l_1 \doteq \frac{1}{2}(-x_1 + x_2 + x_3 - x_4)$ and $l_2 \doteq \frac{1}{2}(-y_1 - y_2 + y_3 + y_4)$, where x_I and y_I are coordinates of nodes $I = 1,2,3,4$ in the local Cartesian basis at the element's center, $\{\mathbf{t}_k^c\}$. Too small values of c can cause problems with the conditioning of the stiffness matrix; the value 0.04 is selected in [138].

Similar expressions can be obtained for ε_{23},

$$G_2^* \,\varepsilon_{23}^2 \approx G_2^* \,(\bar{\varepsilon}_{23}^{\text{ave}})^2 + G_{2c}^* \,(\bar{\varepsilon}_{23}^{\text{d1}})^2 \,\xi^2 + G_{2c}^* \,(\bar{\varepsilon}_{23}^{\text{d2}})^2 \,\eta^2, \tag{13.62}$$

where the additionally corrected shear modulus is defined as

$$G_{2c}^* \doteq \left(\frac{1}{G} + b\,\frac{l_2^2}{h^2 E}\right)^{-1}, \qquad b \doteq \frac{c}{c + (1-c)\,(l_2/l_1)^2}. \tag{13.63}$$

2. In our treatment of the coupling, the full RBF correction is applied to the average values but a fraction of it is applied to the whole remaining

part, not to the differences of the transverse shear strains, as in the previous method. Then we modify the integrand of the strain energy (13.51) as follows:

$$G_1^* \, \varepsilon_{13}^2 \approx G_1^* \, (\bar{\varepsilon}_{13}^{\text{ave}})^2 + G_{1c}^* \left[\varepsilon_{13}^2 - (\bar{\varepsilon}_{13}^{\text{ave}})^2 \right], \tag{13.64}$$

$$G_2^* \, \varepsilon_{23}^2 \approx G_2^* \, (\bar{\varepsilon}_{23}^{\text{ave}})^2 + G_{2c}^* \left[\varepsilon_{23}^2 - (\bar{\varepsilon}_{23}^{\text{ave}})^2 \right], \tag{13.65}$$

where G_{1c}^* and G_{2c}^* are defined in eqs. (13.61) and (13.63).

This formula was implemented for shell elements based on the potential energy; the modifications necessary for the mixed functionals are described below. This method works very well as we can see in the example presented in the sequel.

RBF correction for mixed formulations of shells. For the Hellinger–Reissner functional and the Hu–Washizu functional, the modifications related to the RBF correction are as follows:

Hellinger–Reissner functional. Normally, the stress corresponding to the assumed strain is calculated as $\sigma_{13}^a = 2G \, \varepsilon_{13}^a$, while with the RBF correction, we compute

$$\sigma_{\alpha 3}^a = 2G_\alpha^* \, (\bar{\varepsilon}_{\alpha 3}^a)^{\text{ave}} + 2G_{\alpha c}^* \left[\varepsilon_{\alpha 3}^a - (\bar{\varepsilon}_{\alpha 3}^a)^{\text{ave}} \right], \qquad \alpha = 1, 2, \tag{13.66}$$

where $(\bar{\varepsilon}_{\alpha 3}^a)^{\text{ave}}$ is the average value of the assumed strain.

Hu–Washizu functional. The strain energy corresponding to the assumed strain is calculated as in eq. (13.65), i.e.

$$\mathcal{W}(\varepsilon_{\alpha 3}^a) = G_\alpha^* \left[(\bar{\varepsilon}_{\alpha 3}^a)^{\text{ave}} \right]^2 + G_{\alpha c}^* \left\{ (\varepsilon_{\alpha 3}^a)^2 - \left[(\bar{\varepsilon}_{\alpha 3}^a)^{\text{ave}} \right]^2 \right\}, \quad \alpha = 1, 2, \tag{13.67}$$

while the other parts are not modified.

Linear example: Twisted cantilever. Consider the slender initially flat cantilever of Sect. 15.3.1, see Fig. 15.13, twisted by a pair of vertical transverse forces $P_z = \pm 1$. One layer of four-node shell elements is used and the shell thickness, $h = 0.01$.

The displacements and rotations at the tip node obtained by a linear analysis are presented in Table 13.4. Comparing the results obtained without the RBF correction for the 100×1-element mesh and the 100×9-element mesh, we see that the RBF correction is not needed for the twist.

Because we use the RBF correction to improve the sinusoidal bending, we have to select such a value of c for which the results of twist are unaffected. The value $c = 0$ yields almost exact results, but, unfortunately, then the problem with conditioning of the stiffness matrix appears. This problem disappears for $c = 0.01$; a slightly higher value $c = 0.04$ is suggested in [138].

For reference, we use the solution obtained by the shell element with six dofs/node of FEAP, described in [235]. This is the Discrete Kirchhoff Quadrilateral (DKQ) element, with linear kinematics (small strains and rotations), based on the Allman shape functions and with the bending part of [20].

Table 13.4 Effect of the RBF correction for characteristic values of c. Twist of slender cantilever by forces $P_z = \pm 1$. Shell element EADG4-PL-Warped.

Mesh	RBF correction	Displacement $u_z/10^2$	Rotations $r_x/10^2$	r_y
100 × 1	no	3.8897	7.7788	-3.9001
100 × 9	no	3.8922	7.7828	-3.8987
100 × 1	yes, c=0 (*)	3.8896	7.7787	-3.9001
	yes, c=0.005 (*)	3.8973	7.7933	-3.8995
	yes, c=0.01	3.9050	7.8082	-3.8986
	yes, c=0.04	3.9508	7.8980	-3.8899
	yes, c=0.1	4.0419	8.0784	-3.8633
100 × 1, FEAP		3.8929	7.785	-3.9000

(*) Conditioning problem: D-max/D-min $\approx 10^{10}$

13.2.4 Miscellaneous topics

EAS method for transverse shear strains. Recall that eq. (13.45) specifies a transformation of the covariant transverse shear strains to the local orthonormal basis

$$\begin{bmatrix} \varepsilon_{13} \\ \varepsilon_{23} \end{bmatrix} = \mathbf{J}^{-T} \begin{bmatrix} \varepsilon_{13}^{\xi} \\ \varepsilon_{23}^{\xi} \end{bmatrix}. \tag{13.68}$$

Within the EAS method, we can assume the following representation for the enhanced transverse shear components

$$\begin{bmatrix} \varepsilon_{13}^{\mathrm{enh}} \\ \varepsilon_{23}^{\mathrm{enh}} \end{bmatrix} = \mathbf{J}_c^{-T} \begin{bmatrix} \xi\, q_1 + \xi\eta\, q_2 \\ \eta\, q_3 + \xi\eta\, q_4 \end{bmatrix} \left(\frac{j_c}{j} \right). \tag{13.69}$$

Such a representation was used in [201], eqs. (97) and (98). The role of (j_c/j) is identical to that for the membrane enhancement. The enhancement is added to the compatible part of the transverse shear strains treated by the ANS method.

Discrete Kirchhoff (DK) elements. If the transverse shear is negligible, then we can exploit this fact using the Kirchhoff constraint, $\varepsilon_{\alpha 3} = 0$, at selected discrete points to modify the shape functions. This leads to the so-called Discrete Kirchhoff (DK) family of elements; beams, plates, and shells. This concept is considered in [246, 247].

1. The two-node Discrete Kirchhoff Beam (DKB) is based on a cubic approximation of normal displacement w, see eq. (13.18), and a quadratic approximation of rotation θ. First, the Hermitian-type shape functions are obtained for w which are expressed by values of w and $w_{,\xi}$ at nodes and, next, two DK constraints are applied at these nodes, which yields eq. (13.21). Additionally, one DK constraint is applied at the element center to accommodate the quadratic term of the rotation.
2. The four-node Discrete Kirchhoff Quadrilateral (DKQ) for plates was proposed in [20]. The approach to plates is a natural extension of the concept for the DK beam and was used as the bending part of several shell elements in [117, 113, 235].
3. Several Discrete Kirchhoff Triangle (DKT) plate and shell elements can also be found in the literature.

The DK elements are based on polynomials of relatively high order and perform very well in bending, including sinusoidal bending, and twisting. However, they neglect the transverse shear energy and can be used only for thin shells.

Kirchhoff limit for transverse shear constrained to zero. The formulation based on the Reissner hypothesis can be constrained by enforcing the RC $\text{skew}(\mathbf{Q}^T\mathbf{F}) = \mathbf{0}$ for the $\alpha 3$ components only, which means that the transverse shear strain $\varepsilon_{\alpha 3} = 0$.

As an example, we analyze the cantilever shown in Fig. 15.13; the data is defined in Sect. 15.3.1, but the thickness $h = 10^n$, where $n \in [-3, +3]$. The mesh consists of 100 two-node Timoshenko beam elements. The 13-component of $\text{skew}(\mathbf{Q}^T\mathbf{F}) = \mathbf{0}$ is enforced using the penalty method with $\gamma = 2G\,h \times 10^3$.

The original and constrained Reissner solutions are shown in Table 13.5. The tip displacement w is affected by the constraint and indeed forced to attain the Kirchhoff limit 4×10^{-3n} for $h > 10^{-1}$. Note that w of the original element is bigger than w of the constrained element!

The tip rotation θ is in accordance with the Kirchhoff solution $6 \times 10^{-2-3n}$ for the whole range of h, except for the very thin beam of $h = 10^{-3}$, for which the solution is destroyed by the penalty method.

Table 13.5 Effect of 13-component of $\text{skew}(\mathbf{Q}^T\mathbf{F}) = \mathbf{0}$ on Timoshenko beam.

Thickness n	Solution type	Displacement w	Rotation θ
+3	original	3.1600E–07	6.0000E–11
	constrained	4.1298E–09	6.0000E–11
+2	original	7.1199E–06	6.0000E–08
	constrained	4.0012E–06	6.0000E–08
+1	original	4.0311E–03	6.0000E–05
	constrained	3.9999E–03	6.0000E–05
0	original	4.0002E+00	6.0000E–02
	constrained	3.9999E+00	6.0000E–02
–1	original	3.9999E+03	6.0000E+01
	constrained	3.9999E+03	6.0000E+01
–2	original	3.9999E+06	6.0000E+04
	constrained	4.0009E+06	6.0050E+04
–3	original	4.0003E+09	6.0006E+07
	constrained	5.3823E+08	1.1540E+07
	Kirchhoff limit	4×10^{-3n}	$6 \times 10^{-2-3n}$

14

Warped four-node shell element

In this chapter, we describe various modifications of the four-node shell elements related to the warped element's geometry.

14.1 Definition of warpage

For curved structures, it is often impossible to locate all nodes of a four-node shell element in one plane, and then the initial geometry of the element is warped. The warpage of the four-node element can be defined in several ways, see Fig. 14.1, and can be automatically detected for given positions of nodes by introducing the local elemental basis.

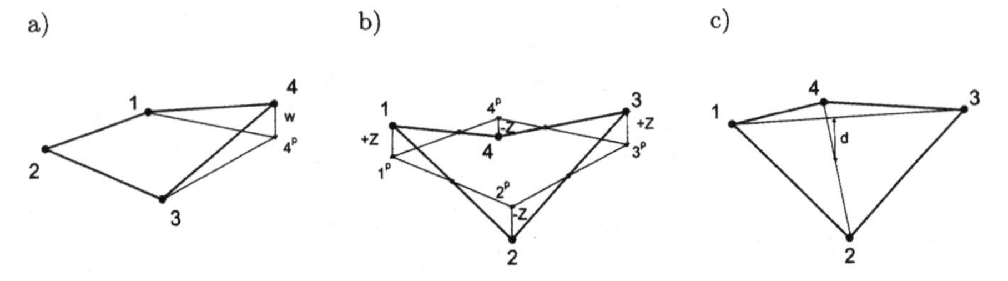

Fig. 14.1 Various warpage parameters of four-node element. a) w is the out-of-planeness of node 4, b) Z is the distance of node from the mean plane, c) d is the minimal distance between diagonals.

Even a moderate warpage can drastically change the solution which, for some load cases, is much more flexible while for others is much stiffer than for a flat element. For instance, for the load cases of Fig. 15.7, the ratio of flat/warped solution is between 1/100 and 7, as shown in Table 14.1.

Both solutions were obtained for the test of Sect. 15.2.5 using a 50 × 50-element mesh. For this reason, warped four-node elements require a special treatment while, for flat elements, we can use the formulations described in earlier sections.

Table 14.1 Ratio of flat/warped displacements at corner 3. $Z = 1$. Mesh 50 × 50 elements.

Load case	1	2	3	4	5	6
Component	u_1	u_1	u_2	u_2	u_3	u_3
Flat/warped ratio	1/10	1/100	1/60	1/10	1	7

The reference surface of a warped four-node shell element is a saddle (h-p) surface, see Fig. 14.2a. To understand how it deforms, we can replace it by a patch of four triangular elements, see Fig. 14.2b. This is also a method to treat warpage (with one additional node introduced) when it is too large to be dealt with by a four-node element.

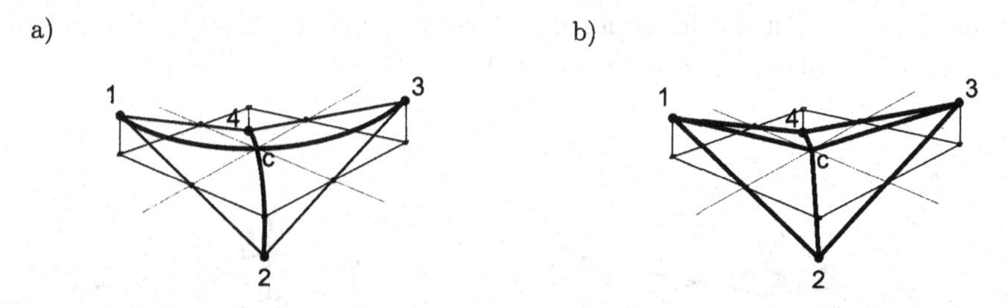

Fig. 14.2 Warped four-node element can be replaced by four triangles.

Relative warpage parameters. The size of warpage can be related to other geometrical parameters of the element:

1. to the thickness of the element

$$\Phi \doteq \frac{Z}{h}, \tag{14.1}$$

2. to the in-plane size of the element, in one of the following ways:
   a)

$$\Psi \doteq \frac{w}{l_{\text{ave}}}, \tag{14.2}$$

where w is the out-of-planeness of node 4 of Fig. 14.1a. Besides, $l_{\text{ave}} \doteq \frac{1}{4}\sum_M l_M$ is the average length of the element's side and $l_M \doteq \|\mathbf{y}_{0A} - \mathbf{y}_{0B}\|$ is the length of the side. M, A and B are defined in eq. (14.7).

b)

$$\Psi \doteq \frac{Z}{\sqrt{A_e}}, \tag{14.3}$$

where Z is the distance of the node from the mean plane, shown in Fig. 14.1b, and A_e is the element's area,

c)

$$\Psi \doteq \frac{d}{\|\mathbf{d}_{13} \times \mathbf{d}_{24}\|}, \tag{14.4}$$

where d is the minimal distance between diagonals $\mathbf{d}_{13}, \mathbf{d}_{24}$ of Fig. 14.1c.

Finally, we note that the scalar product of the corner normal vectors shown in Fig. 10.6 and the normal vector at the element's center can be used to check warping of the element. For instance, in [180], if the angle subtended between these normals exceeds 10 degrees, then a warning is issued and a mesh refinement is recommended. For the element of size 1×1, this limit corresponds to $Z \approx 0.125$.

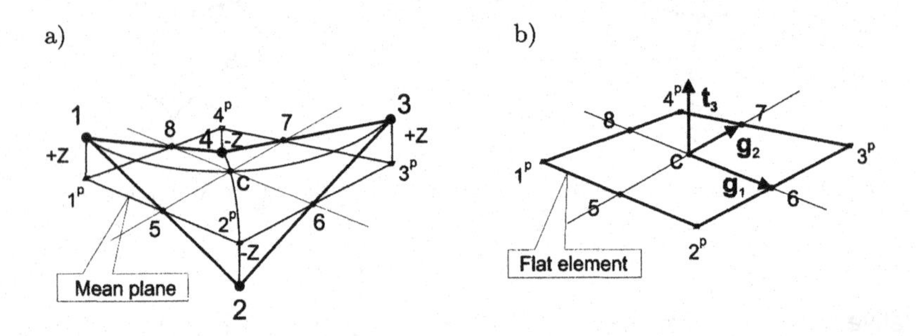

Fig. 14.3 Four-node warped element. a) Warped element and mean plane. b) Substitute flat element.

Mean plane of warped element. Consider the warped shell element in the initial configuration. For this element we can define an auxiliary plane, designated as the mean plane, see Fig. 14.3. The mean plane is defined by the normal vector and a point through which it passes. For this purpose we use the element's center,

$$\mathbf{c} \doteq \frac{1}{4}\left(\mathbf{y}_{01} + \mathbf{y}_{02} + \mathbf{y}_{03} + \mathbf{y}_{04}\right), \tag{14.5}$$

but the mean plane can also pass through two opposite nodes, see [188], Fig. 4. The orientation of this plane can be defined by the orthogonal matrix

$$\mathbf{R}_c = [\mathbf{t}_1^c \,|\, \mathbf{t}_2^c \,|\, \mathbf{t}_3^c], \tag{14.6}$$

where $\mathbf{t}_k^c$ $(k = 1, 2, 3)$ are components of vectors of the element's Cartesian basis located at the element's center. It can be constructed as described in Sect. 10.2, but the tangent natural vectors can also be obtained using the positions of a mid-side points M,

$$\mathbf{y}_{0M} = \tfrac{1}{2}(\mathbf{y}_{0A} + \mathbf{y}_{0B}), \qquad \text{where} \qquad \begin{array}{ccc} \hline M & A & B \\ \hline 5 & 1 & 2 \\ 6 & 2 & 3 \\ 7 & 3 & 4 \\ 8 & 4 & 1 \\ \hline \end{array}\,, \tag{14.7}$$

and $\mathbf{y}_{0A}, \mathbf{y}_{0B}$ are the initial position vectors of the nodes. The tangent natural vectors are defined as

$$\mathbf{g}_1 \doteq \frac{1}{2}(\mathbf{y}_{06} - \mathbf{y}_{08}), \qquad \mathbf{g}_2 \doteq \frac{1}{2}(\mathbf{y}_{07} - \mathbf{y}_{05}), \tag{14.8}$$

and the vector normal to the mean plane is

$$\mathbf{g}_3 \doteq \frac{\bar{\mathbf{g}}_3}{\|\bar{\mathbf{g}}_3\|}, \qquad \text{where} \qquad \bar{\mathbf{g}}_3 \doteq \mathbf{g}_1 \times \mathbf{g}_2. \tag{14.9}$$

Using the natural vectors $\mathbf{g}_k$, we can construct the Cartesian basis located at the element's center $\{\mathbf{t}_k^c\}$ as **Basis 2** of Sect. 10.2.

Calculation of warpage parameter. Having the orthogonal matrix $\mathbf{R}_c$ and the position vector $\mathbf{c}$ of the element's center, we can obtain the local positions of the nodes as follows:

$$\mathbf{y}_{0I}^L \doteq \mathbf{R}_c^T(\mathbf{y}_{0I} - \mathbf{c}), \qquad I = 1, 2, 3, 4. \tag{14.10}$$

The third components of vectors $\mathbf{y}_{0I}^L$ contain the distance d_I of node I from the mean plane. Nodes of the four-node element are equidistant from the mean plane, i.e.

$$[d_1, d_2, d_3, d_4] = Z\,\mathbf{h}, \tag{14.11}$$

where $Z \doteq |\,d_I\,|$ is the warpage parameter, see Fig. 14.3a, and $\mathbf{h} \doteq \{1, -1, 1, -1\}$ is the hourglass vector of eq. (10.6). $Z = 0$ indicates a flat element, while $Z > 0$ a warped element.

To decide whether $Z > 0$ is small or large, we have to use the earlier defined relative warpage parameters Φ and/or Ψ, and some empirical threshold values. For very small values of Φ and Ψ, the warping correction is not necessary, while for very large values, the computations are terminated and the mesh must be corrected. For small and average values, we use one of the two formulations described below.

Two formulations of warped elements. Below are presented two formulations of four-node elements which can be used when nodes are not co-planar.

1. The warped element with certain modifications of formulation is discussed in Sect. 14.2.
2. The substitute flat shell element with the warpage correction is discussed in Sect. 14.3. For the elements which were originally developed as flat, the logic is reversed and the warpage correction is an add-on feature.

Note that the accuracy of the warped elements must be evaluated on a suitable set of benchmarks involving warpage. Very often only the twisted beam example of Sect. 15.3.5 is used for this purpose; in our opinion the one element test of Sect. 15.2.5 with six load cases is more indicative.

14.2 Warped element with modifications

In this section, we consider a four-node warped element, for which warpage is not neglected, but its formulation is modified. Numerical results for the modifications discussed below are given for the warped single element test in Table 15.5 of Sect. 15.2.5.

Green strain. Designate the Green strain in the global reference basis as $\mathbf{E}^G$. Prior to the numerical integration, this strain must be rotated to a local basis. To reduce the over-stiffening, it should be rotated to the local basis at a Gauss point,

$$\mathbf{E}_L \doteq \mathbf{R}^T\, \mathbf{E}_G\, \mathbf{R}, \tag{14.12}$$

using eq.(2.13) and $\mathbf{R} = [\mathbf{t}_1 \,|\, \mathbf{t}_2 \,|\, \mathbf{t}_3]$, where $\mathbf{t}_k$ $(k = 1, 2, 3)$ are components of the vectors of the Cartesian basis at the Gauss point. Note that for a flat element, we can also transform the strain to the local basis at the element's center,

$$\mathbf{E}_{Lc} \doteq \mathbf{R}_c^T \, \mathbf{E}_G \, \mathbf{R}_c, \tag{14.13}$$

using $\mathbf{R}_c \doteq [\mathbf{t}_1^c \,|\, \mathbf{t}_2^c \,|\, \mathbf{t}_3^c]$ for the element's center. Both forms of transformation yield a correct number of zero eigenvalues (6), but the element for the latter transformation is much stiffer.

In-plane shear strain. Consider the deformation gradient at the reference surface, $\mathbf{F}_0 \doteq \mathbf{F}|_{\zeta=0}$ enhanced additively by the matrix $\tilde{\mathbf{H}}$ of the EADG method, see eq. (11.70). Then, the membrane part of the Green strain is

$$\varepsilon_G = \frac{1}{2}\left[(\mathbf{F}_0 + \tilde{\mathbf{H}})^T(\mathbf{F}_0 + \tilde{\mathbf{H}}) - \mathbf{I}\right], \tag{14.14}$$

and is transformed to the local Cartesian basis using $\varepsilon_L \doteq \mathbf{R}^T \varepsilon_G \mathbf{R}$. Denote the 12-component of $\mathbf{F}_0^T\mathbf{F}_0$ as $A \doteq (\mathbf{F}_0^T\mathbf{F}_0)_{12}$. We tested the following four approximations of $A(\xi, \eta)$, where $\xi, \eta \in [-1, 1]$:

1. it was sampled at Gauss Points (standard treatment),
2. it was sampled at the center,

$$A(\xi, \eta) \approx A|_{\xi=0,\eta=0}, \tag{14.15}$$

3. it was calculated as an average of the values sampled at the midsides,

$$A(\xi, \eta) \approx \frac{1}{4}\left(A|_{\xi=0,\eta=-1} + A|_{\xi=1,\eta=0} + A|_{\xi=0,\eta=1} + A|_{\xi=-1,\eta=0}\right), \tag{14.16}$$

4. it was calculated as an average of the values sampled at the corners,

$$A(\xi, \eta) \approx \frac{1}{4}\left(A|_{\xi=-1,\eta=-1} + A|_{\xi=1,\eta=-1} + A|_{\xi=1,\eta=1} + A|_{\xi=-1,\eta=1}\right). \tag{14.17}$$

Note that the elements with these approximations have the correct rank and pass the membrane patch test.

For the last three approximations, the solution significantly improves, which is well seen in the warped element test of Sect. 15.2.5 and Table 15.5, see load cases 3 and 6. The above approximations were also tested in the non-linear twisted beam example of Sect. 15.3.5 and yielded identical results, which is a result of a small warpage.

Drill RC. Designate as $\mathfrak{C}_G$, the $\mathfrak{C} \doteq \text{skew}(\mathbf{Q}^T\mathbf{F})$ in the global reference basis. We can verify that when $\mathfrak{C}_G$ is rotated to the local basis at a Gauss point,

$$\mathfrak{C}_L \doteq \mathbf{R}^T \, \mathfrak{C}_G \, \mathbf{R}, \tag{14.18}$$

then the number of zero eigenvalues is correct (6), even for the drill RC enforced at all four Gauss points. The above transformation is identical as the one for the strain in eq. (14.12). Due to this transformation of $\mathfrak{C}_G$, the response for the load cases 2, 3, and 5 in the one-element test of Sect. 15.2.5 becomes less stiff and can be controlled by the regularization parameter.

Regularization parameter for drill RC. The regularizing parameter for the drill RC can be scaled down to reduce the overstiffening due to warpage. The results for the load cases 2, 3 and 5 in the warped single element test of Sect. 15.2.5 are sensitive to the value of this parameter.

In [203] eq. (28), the regularizing parameter γ is multiplied by a coefficient depending on the element warpage in the following way:

$$c_{warp} \doteq s_1 + (1 - s_1) \exp{(-s_2 \, s_3)}, \tag{14.19}$$

where

$$s_1 \doteq \frac{h^2}{A_e}, \qquad s_3 \doteq \frac{d}{h}.$$

Here, h is the shell thickness, A_e is the element's area, and d is the shortest distance between the diagonals, see Fig. 14.1c. Besides, s_2 is a (positive) scalar to be determined numerically.

The dependence of c_{warp} on s_1 is linear, while on s_2 and s_3 it is exponential. For $d = 0$, we obtain $c_{warp} = 1$, and for $d \to \infty$ we have $c_{\text{warp}} \to s_1$. We note that $c_{\text{warp}} < 1$, when $s_1 < 1$.

Membrane over-stiffening of warped shell element. In the case of a warped (h-p) geometry, even simple loads such as these shown in Fig. 15.7, cause complicated states of deformation. We cannot render pure bending, so it is impossible to detect membrane locking in a similar way as for curved nine-node shell elements, see Sect. 14.4.

For all load cases of the test in Sect. 15.2.5, except load case 3, the accuracy of a single warped element (and $c_{\text{warp}} = 1$) is worse than that of a flat element and it is always too stiff, see Table 15.5.

For the load cases 2, 3, and 5, the membrane response dominates and the change of bending or transverse shear stiffness does not affect these solutions much. For these cases, the solution can be improved by reducing the value of the regularizing parameter for the drill RC ($c_{\text{warp}} < 1$). Besides, by sampling in-plane shear strain at the element's center, we can improve the results for load case 6. This indicates membrane over-stiffening of a warped four-node shell element.

Note, however, that the accuracy for load case 1 cannot be improved in this way, which indicates that other causes of over-stiffening are also present in the element.

Remarks. The warped element has a correct number of zero eigenvalues and passes patch tests. However, in some planar tests, e.g. Cook's membrane of Sect. 15.2.7, it is less accurate than the flat element. Therefore, it should be used only for $Z > 0$.

14.3 Substitute flat element and warpage correction

Below, we discuss the approach in which the warped element is replaced by a flat substitute element. The stiffness matrix and the residual vector are generated for this flat element and, next, the so-called warpage correction is applied to them. Several forms of warpage correction operators were proposed in the literature, but not one is generally accepted.

Note that the warpage corrections used in explicit dynamics are different to these used in statics, see [22, 267].

Substitute flat element. The substitute flat element is defined by projections of nodes of the warped element onto the mean plane. The projected nodes are denoted as $1'$, $2'$, $3'$ and $4'$ in Fig. 14.3. The orientation of the flat element is defined by the orthogonal matrix $\mathbf{R}_c$, while its local position is defined by first two components of vectors $\mathbf{y}_{0I}^L$. The vertical coordinate of the flat element in the local elemental basis is zero.

The stiffness matrix and the residual vector are computed for the flat substitute element and then modified by the so-called warpage operator defined below.

Remark. The curvature correction in the shell element can be also used for another reason; to account for shapes which are not represented by low-order approximations, such as cylindrical or spherical shapes. An example is shown in Fig. 14.8, which shows that using various positions of the element's nodes on a cylindrical surface, we can obtain either a flat element or a warped element but not a cylindrical element.

Warpage operator T. Consider the generalized displacements $\mathbf{q} \doteq \{\mathbf{u}, \boldsymbol{\psi}\}$, where $\mathbf{u}$ are displacements and $\boldsymbol{\psi}$ are rotation vectors. Assume that the relation between the nodal generalized displacements $\mathbf{q}$ for the flat (f) substitute element and the warped (w) element, is provided by some operator $\mathbf{T}$, defined as follows:

$$\mathbf{T}: \quad \mathbf{q}_f = \mathbf{T}\,\mathbf{q}_w \quad \wedge \quad (Z \to 0 \;\Rightarrow\; \mathbf{T} \to \mathbf{I}), \tag{14.20}$$

i.e. it becomes an identity operator when warpage vanishes. To obtain the alternative formula, we can use the Virtual Work equation for nodal forces $\mathbf{f}_w$ of the warped element,

$$\delta\mathbf{q}_w^T\,\mathbf{f}_w = \delta\mathbf{q}_f^T\,\mathbf{T}^{-T}\,\mathbf{f}_w = \delta\mathbf{q}_f^T\,\mathbf{f}_f, \tag{14.21}$$

where the last form is obtained for

$$\mathbf{f}_w = \mathbf{T}^T\,\mathbf{f}_f. \tag{14.22}$$

This relation for nodal forces can also be used to derive the transposed operator $\mathbf{T}^T$.

The equilibrium equation for the flat element can be written as

$$\delta\mathbf{q}_f^T\,(\mathbf{K}_f \Delta\mathbf{q}_f) = -\delta\mathbf{q}_f^T\,\mathbf{r}_f, \tag{14.23}$$

where $\mathbf{K}_f$ is the tangent matrix and $\mathbf{r}_f$ is the residual vector, both for the flat element. Using eq. (14.20), the equilibrium equation can be transformed to

$$\delta\mathbf{q}_w^T\,(\mathbf{K}_w \Delta\mathbf{q}_w) = -\delta\mathbf{q}_w^T\,\mathbf{r}_w, \tag{14.24}$$

where the tangent matrix $\mathbf{K}_w$ and the residual vector $\mathbf{r}_w$ for the warped element are defined as

$$\mathbf{K}_w \doteq \mathbf{T}^T\,\mathbf{K}_f\,\mathbf{T}, \qquad \mathbf{r}_w \doteq \mathbf{T}^T\,\mathbf{r}_f. \tag{14.25}$$

Note that these transformations are performed at the level of the element.

Several forms of the transformation operator are derived in the literature, and the methods used for this purpose either

1. use the equilibrium equations and some simplifying assumptions or
2. use the rigid links concept, similar to that used for beams in Sect. 13.1.5.

These methods are described below.

Correction methods based on equilibrium. Below, we overview the methods based on the equilibrium of nodal forces for membranes and nodal forces and moments for shells. Equation (14.22) is used in these methods, and the transpose operator $\mathbf{T}^T$ is derived.

The first work in which warping was considered is probably [7], in which Chapter II is devoted to twisted shear carrying panels. It was noted that the overall equilibrium equations were not satisfied for such panels, and additional nodal forces were introduced to remedy the situation. The derivation was made in a global reference basis and warping was not clearly defined, but several remarks on the subject were correct from the viewpoint of the next findings.

In [137], the modification for non-planar nodes was performed separately for each edge of the element. The nodal forces of the edge were projected onto the edge and a pair of opposite self-equilibrated forces was used to compensate for the shift of these forces to nodes of a warped element. In [141] p. 439, the transfer of bending moments between warped elements is also considered and the necessity of a corrective vertical moment is indicated. This moment is applied as a couple of horizontal forces normal to the edge. As described in [145] p. 240, the above corrections are abandoned in NASTRAN in favor of a version of the rigid links method.

In [88], the effects of the so-called out-of-planeness of membrane four-node elements was studied and several elements used were tested. Different sets of three nodes were used to define the reference plane of the element (see Fig. 14.1a) resulting in a scatter of results which was quantified for a selected example. Note that this problem vanishes if the mean plane of element is used, see Fig. 14.1b.

In [194], also only membrane forces are considered. The mean plane was defined as in Fig. 14.3, and the element's local basis was used. The corrective vertical forces were introduced to balance the moments generated by the shift of horizontal forces from nodes of a flat element to nodes of a warped element. Two pairs of self-equilibrated vertical forces, $\{W_1, -W_1\}$ and $\{W_2, -W_2\}$, were applied at opposite nodes of the warped element, see Fig. 14.4a.

In [196], nodes 1, 2, and 3 of Fig. 14.1a define the plane of the element, for which the element's basis is constructed. An additional nodal tangent basis is constructed at node 4, and the orthogonal transformation matrix between these two bases is used as $\mathbf{T}$. The set of forces is transformed from one basis to another using $\mathbf{T}$, and the normal force in the nodal basis is assumed to be zero, which is only suitable for membranes. For elements having bending stiffness, it is suggested that the vertical force at node 4 should be introduced, and reacted by the vertical force at node 1.

In [152], the warped shell element is considered. The method of [137] based on equilibrium of edges is scrutinized and the method based on equilibrium of all nodal forces and some form of the virtual work equation is proposed. In our tests, this method yielded the same corrective vertical forces as assumed from the outset in [194]. The corrective vertical moments were used to obtain zero drilling moments at nodes of the warped element and were later balanced by a set of self-equilibrated horizontal forces R applied at all nodes, see Fig. 14.4b. A set of one element tests involving several load cases was proposed, see Sect. 15.2.5, which can be used to evaluate the accuracy of various forms of $\mathbf{T}$.

Finally, in our opinion, all these methods must fail when boundary conditions constraining displacements are applied as, e.g., in the test of Sect. 15.2.5. The reason is that some of the forces are eliminated by boundary conditions so the applied set of forces is not self-equilibrated any longer.

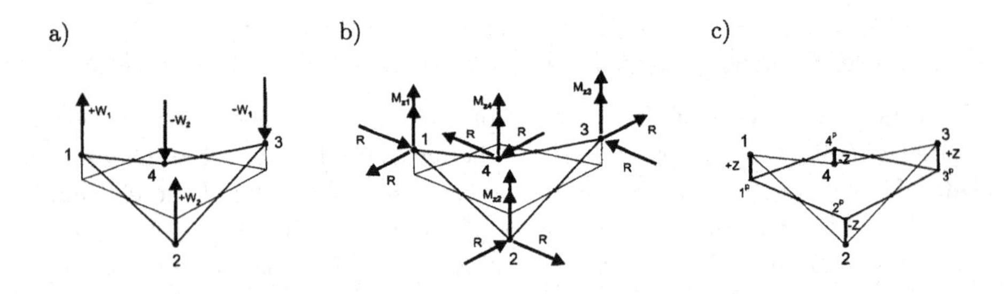

Fig. 14.4 Warpage corrections for four-node element. a) Self-equilibrated vertical forces of [194], b) Vertical moments and balancing self-equilibrated horizontal forces of [152], c) Rigid links.

Correction by the rigid links method. The rigid links method for shells is similar to this described for beams in Sect. 13.1.5. The rigid vertical links are applied at nodes, as shown in Fig. 14.4c.

Consider the warped shell element with six dofs/node in the Cartesian basis at the element's center. The relation between the nodal generalized displacements for flat and warped elements can be written in two forms.

1. Defining the vectors of nodal variables as

$$\mathbf{z}_w = [u_{1w},\, u_{2w},\, u_{3w} \,|\, \psi_{1w},\, \psi_{2w},\, \psi_{3w}]^T,$$
$$\mathbf{z}_f = [u_{1w},\, u_{2w},\, u_{3w} \,|\, \psi_{1w} + Zu_{2w},\, \psi_{2w} - Zu_{1w},\, \psi_{3w}]^T, \quad (14.26)$$

we obtain, in accordance with eq. (14.20),

$$\mathbf{T} \doteq \frac{d\mathbf{z}_f}{d\mathbf{z}_w} = \left[\begin{array}{ccc|ccc} 1 & 0 & 0 & 0 & 0 & 0 \\ 0 & 1 & 0 & 0 & 0 & 0 \\ 0 & 0 & 1 & 0 & 0 & 0 \\ \hline 0 & Z & 0 & 1 & 0 & 0 \\ -Z & 0 & 0 & 0 & 1 & 0 \\ 0 & 0 & 0 & 0 & 0 & 1 \end{array}\right]. \quad (14.27)$$

2. Defining the vectors of nodal variables as

$$\mathbf{z}_f = [u_{1f},\, u_{2f},\, u_{3f} \,|\, \psi_{1f},\, \psi_{2f},\, \psi_{3f}]^T,$$
$$\mathbf{z}_w = [u_{1f} - Z\psi_{2f},\, u_{2f} + Z\psi_{1f},\, u_{3f} \,|\, \psi_{1f},\, \psi_{2f},\, \psi_{3f}]^T, \quad (14.28)$$

we obtain

$$\frac{d\mathbf{z}_w}{d\mathbf{z}_f} = \mathbf{T}^T. \quad (14.29)$$

Note that eq. (14.26) is analogous to eq. (13.38) for the curvature correction for beams.

Equation (14.26) and the operator (14.27) were used, e.g., in [118, 235, 113]. They yield excessive displacements for some load cases in the warped single element test of Sect. 15.2.5 and only in four first cases can the accuracy be improved by using a fraction of Z, see Table 14.2, where the results are normalized by solutions for the 50×50-element mesh.

Table 14.2 Solutions for various lengths of rigid links. Element EADG4+PL.

	Normalized displacements at node 3					
Load case	1	2	3	4	5	6
Component	u_1	u_1	u_2	u_2	u_3	u_3
Z	9.80	0.78	1.23	8.96	1.04	7.71
Z/5	0.40	0.03	0.05	0.36	1.04	7.71

Equation (14.28) was used in [145], but the nodal rotations were replaced by rigid-element rotations, defined as

$$\alpha_1 = w_{,y}, \qquad \alpha_2 = -w_{,x}, \quad (14.30)$$

see [145] eqs. (15) and (16), so the warpage correction involves nodal displacements and their derivatives but not rotations. In this way, a purely membrane response becomes possible, even for a warped element.

14.4 Membrane locking of curved shell elements

As we pointed out in Sect. 14.2, a warped four-node shell element suffers from membrane over-stiffening which, however, is not identical to membrane locking of curved nine-node shell elements. It this section, we discuss the latter phenomenon.

Overview. The membrane locking occurs for curved beams and shells and is manifested as the inability of an element to undergo pure bending. It is caused by disparity of orders of particular terms of membrane strains and its effects are similar as shown in Fig. 13.1. This type of locking is typical for curved three-node beams and nine-node shell elements; the first papers were on curved beams, see [226, 227].

Several remedies are used to circumvent this problem, such as two-level approximations of strains (the so-called Assumed Strain method), and the reduced integration techniques (the SRI or the URI with stabilization), see the survey in [164]. They are all based on the fact that at some points within the element, values of the approximated membrane strains are equal to the analytical ones. To find these points, the pure bending of a beam element shown in Fig. 14.5 can be analyzed.

Membrane locking of 2D beam elements. Typically, an analysis of membrane locking is performed for the simplified form of the membrane strain of the classical shallow beam equations

$$\varepsilon_{xx} = \overline{u}_{,x} + \frac{1}{2}\overline{w}_{,x}^2, \tag{14.31}$$

see [237], p. 384. Note that this is the Green strain component for the assumption that $\overline{u}_{,x}^2$ is small compared to the other terms.

To describe the initial curvature of the beam, we introduce the initial displacements u_0 and w_0, measured from the straight middle axis of the beam. Then the total displacement is expressed as $\overline{u} = u_0 + u$ and $\overline{w} = w_0 + w$, and eq. (14.31) becomes

$$\varepsilon_{xx} = u_{0,x} + u_{,x} + \frac{1}{2}w_{0,x}^2 + w_{0,x}w_{,x} + \frac{1}{2}w_{,x}^2 \approx u_{,x} + w_{0,x}w_{,x}. \tag{14.32}$$

The last form is obtained through the assumption that the curvature does not cause the initial strain, i.e. $\varepsilon_{xx}^0 \doteq u_{0,x} + \frac{1}{2}w_{0,x}^2 = 0$, and that the quadratic term $\frac{1}{2}w_{,x}^2$ can be neglected. This form of membrane strain is used in the analysis of locking of the three-node beam element in [94], pp. 84-87. Here it is modified to enable the analysis of two-node arches based on the Reissner hypothesis.

If we apply the Reissner hypothesis to the curved beam, then the strain ε_{xx} depends on rotation θ of the director. The related strains can be derived in a systematic way, but we take a shortcut and transform eq. (14.32) to the form involving the rotation θ and valid for a circular arch.

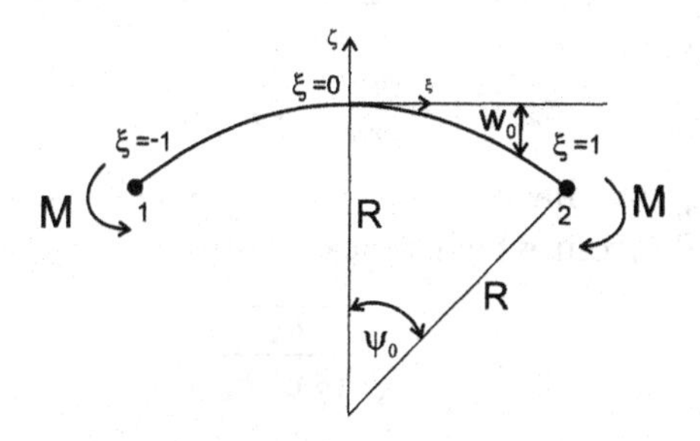

Fig. 14.5 Pure bending of two-node arch element.

Membrane strain of circular arch. Consider a circular arch of radius R and length L as shown in Fig. 14.5. The central angle $\psi_0 = L/(2R)$. Assume that the arch is thin, $h/R \ll 1$, and undergoes small rotations, $\theta \approx 0$. For the shallow circular arch (of small ψ_0), we have

$$w_0 \approx -\frac{x^2}{2R}. \tag{14.33}$$

Using $x = R\sin\psi \approx R\psi = R\psi_0\xi$, we obtain $w_{0,x} = -x/R = -\psi_0\xi$ and we may rewrite eq. (14.32) as

$$\varepsilon_{xx} = u_{,x} + w_{0,x}\, w_{,x} = u_{,x} - \xi\psi_0\,\theta, \tag{14.34}$$

where $\theta \approx w_{,x}$, from the condition that the transverse shear strain of eq. (13.1) is zero for a thin beam.

Elimination of membrane locking of two-node arch element. In the two-node element, u and θ are approximated by a linear polynomial of ξ, while ψ_0 is constant. Hence, the first term in eq. (14.34) is constant, while the second term is a quadratic polynomial of ξ. This disparity in orders of the terms causes membrane locking which we can analyze for pure cylindrical bending.

For the pure bending of the element shown in Fig. 14.5, the nodal displacement and rotation are $u_1 = -u_2$ and $\theta_1 = -\theta_2$. Hence,

$$u(\xi) = \sum_{I=1}^{2} N_I(\xi)\, u_I = u_2\xi, \quad u_{,x} = u_2/x_2, \quad \theta(\xi) = \sum_{I=1}^{2} N_I(\xi)\, \theta_I = \theta_2\xi, \tag{14.35}$$

where we used eq. (13.6) with $2/L \approx -1/x_2$. Using these relations in eq. (14.34), we have

$$\varepsilon_{xx}(\xi) = \frac{u_2}{x_2} - \psi_0\, \theta_2\, \xi^2. \tag{14.36}$$

For pure bending, we have $\varepsilon_{xx} = 0$, which yields the equation $u_2 - x_2\, \psi_0\, \theta_2 \xi^2 = 0$ from which we calculate

$$\xi = \pm\sqrt{\frac{u_2}{x_2\, \psi_0\, \theta_2}}. \tag{14.37}$$

This formula can be expressed in terms of ψ_0 only, noting that

1. the deformation transforms the arc of radius R and angle ψ_0 into the arc of radius R' and angle ψ_0',
2. the length L of the arc is not changed, $L = R\psi_0 \approx R'\psi_0'$,
3. $\psi_0' \approx \psi_0 + \Delta\psi_0$, where the increment $\Delta\psi_0$ is small.

Hence, at node 2 we have

$$u_2 = R' \sin\psi_0' - R\sin\psi_0, \qquad \theta_2 = -\Delta\psi_0, \qquad x_2 = R\sin\psi_0.$$

On use of these relations in eq. (14.37), we obtain

$$\xi(\psi_0) \doteq \lim_{\Delta\psi_0 \to 0} \xi(\psi_0 + \Delta\psi_0) = \pm\sqrt{\frac{1 - \psi_0 \cot\psi_0}{\psi_0^2}}, \tag{14.38}$$

which depends solely on the angle ψ_0, see Fig. 14.6. In the limit case of a straight beam, i.e. when $\psi_0 \to 0$, we obtain

$$\lim_{\psi_0 \to 0} \xi(\psi_0) = \pm\frac{1}{\sqrt{3}}, \tag{14.39}$$

i.e. only at these two points the approximated value of ε_{xx} is correct. Hence, for a circular arch, a location of the points in which the membrane locking is avoided is identical to that for the three-node Bernoulli beam element, see [94].

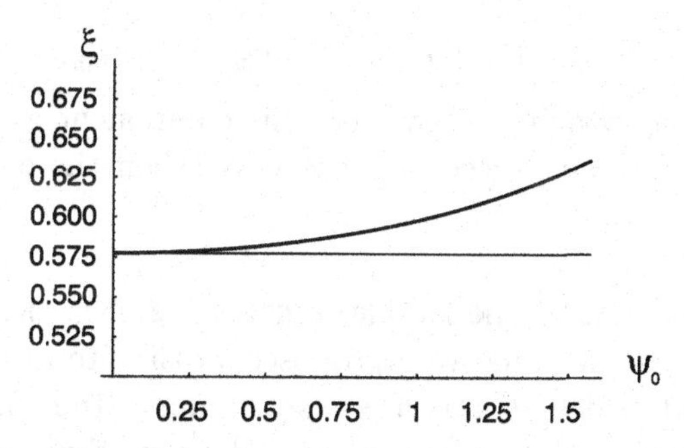

Fig. 14.6 Location of sampling points in two-node arch element. Function $\xi(\psi_0)$.

Membrane strain of curved shallow shell. Consider a curved but shallow shell element. For the simplest set of shell equations of Marguerre [149] or of Donnell-Mushtari-Vlasow in [66], the membrane components of strain tensor are

$$\varepsilon_{xx} = \overline{u}_{,x} + \frac{1}{2}\overline{w}^2_{,x}, \qquad \varepsilon_{yy} = \overline{v}_{,y} + \frac{1}{2}\overline{w}^2_{,y}, \qquad 2\varepsilon_{xy} = \overline{u}_{,x} + \overline{v}_{,y} + \overline{w}_{,x}\overline{w}_{,y}. \tag{14.40}$$

The components ε_{xx} and ε_{yy} are analogous to eq. (14.31) for a curved beam. We can treat the curvature similarly and introduce initial displacements u_0, v_0 and w_0,

$$\overline{u} = u_0 + u, \qquad \overline{v} = v_0 + v, \qquad \overline{w} = w_0 + w, \tag{14.41}$$

measured from the planar middle surface of the shell. Assuming that the strains for the initial displacements are zero, i.e. $\varepsilon^0_{xx} = u_{0,x} + \frac{1}{2}w^2_{0,x} = 0$ and $\varepsilon^0_{yy} = v_{0,x} + \frac{1}{2}w^2_{0,y} = 0$, and neglecting the higher-order terms, $\frac{1}{2}w^2_{,x}$ and $\frac{1}{2}w^2_{,y}$, we obtain

$$\varepsilon_{xx} = u_{,x} + w_{0,x}w_{,x}, \qquad \varepsilon_{yy} = v_{,y} + w_{0,y}w_{,y}. \tag{14.42}$$

Their forms are analogous to eq. (14.32) for the curved beam so they can be a source of membrane locking of a curved nine-node shell element.

For the in-plane shear strain of eq. (14.40), using eq. (14.41), we obtain

$$2\varepsilon_{xy} = u_{0,x} + u_{,x} + v_{0,y} + v_{,y} + (w_{0,x} + w_{,x})(w_{0,y} + w_{,y}), \tag{14.43}$$

where $(w_{0,x} + w_{,x})(w_{0,y} + w_{,y}) = w_{0,x}w_{0,y} + w_{0,x}w_{,y} + w_{,x}w_{0,y} + w_{,x}w_{,y}$. Assuming that the strain for the initial displacements is zero, i.e. $2\varepsilon^0_{xy} = u_{0,x} + u_{0,y} + w_{0,x}w_{0,y} = 0$ and neglecting the higher-order term, $w_{,x}w_{,y}$, we obtain

$$2\varepsilon_{xy} = u_{,x} + v_{,y} + w_{0,x}w_{,y} + w_{,x}w_{0,y}. \tag{14.44}$$

It is difficult to show analytically that this component causes membrane locking. However, if we proceed as if it were, then the element becomes more accurate.

Methods of avoiding membrane locking of nine-node shell elements. Several methods of avoiding locking were proposed in the literature and they all use the points at which strains are exact, found either for a three-node beam or for a planar nine-node element. These methods can be summarized as follows:

1. Uniform reduced integration (URI) combined with stabilization. The URI was proposed in [269], but it yields a rank-deficient stiffness matrix and requires stabilization; various methods of deriving the stabilization matrix were proposed in [25, 167, 26]. Note that the 2×2 Gauss integration applied to the in-plane shear strain ε_{12}, does not yield spurious zero eigenvalues (mechanisms), see [24].
2. Selective reduced integration (SRI), which uses the Gauss points coinciding with the points at which the strains are exact. Its main deficiency is that the membrane and bending strain energy can be decoupled only when the material properties are constant through the thickness or symmetric w.r.t. the mid-surface. This excludes the use of the SRI elements, e.g. to plasticity with several integration points through the thickness. Some SRI elements, according to the literature, exhibit poor mesh convergence in the pinched hemisphere or the pinched cylinder example. But, the 9-SRI element of [164] does not have this deficiency.
3. The assumed strain (AS) method in conjunction with the concept of two-level approximation. This concept consist of sampling the strain components at certain points and extrapolating these values over the element. In [164], the AS method is applied to nine-node shell elements with drilling rotation.
   The AS method was gradually developed for plates and shells in several papers, including [138, 104, 139, 67, 95, 18, 19, 96, 167, 114, 39], and many others. It is also covered in the books [94, 47]. Different variants of the AS method has been developed for:
   a) the transverse shear strains in four-node plate and shell elements, and
   b) the transverse shear strains and the membrane strains in three-node beams and nine-node plate and shell elements. The most often

used locations of the sampling points for nine-node elements are shown in Fig. 14.7.

Generally, the variants of the AS found in the literature differ in the components which are sampled, in the location of sampling points, and in the interpolation functions.

Two-level approximations of the AS method. In the two-level approximations, we have to assume the position of the sampling (tying) points and the form of the interpolation functions. The three sets of the sampling points which are in use are shown in Fig. 14.7 and they are combined with the interpolation functions as presented below. In all the formulas below, $a = \sqrt{1/3}$.

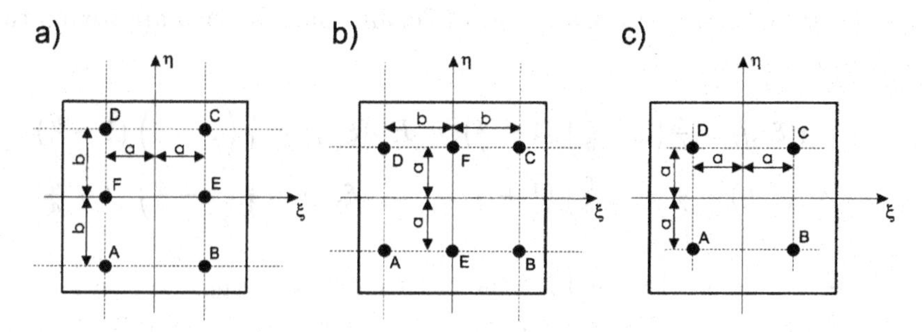

Fig. 14.7 Location of sampling points. Reduced number of points: a) in ξ direction, b) in η direction, c) in both directions.

1. For the strains $\varepsilon_{\alpha\alpha}$ and $\varepsilon_{3\alpha}$ $(\alpha = 1, 2)$, the sampling points are shown in Figs. 14.7a and b, and the set of interpolation functions proposed in [95, 96] is used:

   - for the reduced number of points in the ξ direction, Fig. 14.7a,

$$R_A(\xi,\eta) = \tfrac{1}{4}(1 - \tfrac{\xi}{a})\left[(\tfrac{\eta}{b})^2 - \tfrac{\eta}{b}\right], \quad R_B(\xi,\eta) = \tfrac{1}{4}(1 + \tfrac{\xi}{a})\left[(\tfrac{\eta}{b})^2 - \tfrac{\eta}{b}\right],$$
$$R_C(\xi,\eta) = \tfrac{1}{4}(1 + \tfrac{\xi}{a})\left[(\tfrac{\eta}{b})^2 + \tfrac{\eta}{b}\right], \quad R_D(\xi,\eta) = \tfrac{1}{4}(1 - \tfrac{\xi}{a})\left[(\tfrac{\eta}{b})^2 + \tfrac{\eta}{b}\right],$$
$$R_E(\xi,\eta) = \tfrac{1}{2}(1 + \tfrac{\xi}{a})\left[1 - (\tfrac{\eta}{b})^2\right], \quad R_F(\xi,\eta) = \tfrac{1}{2}(1 - \tfrac{\xi}{a})\left[1 - (\tfrac{\eta}{b})^2\right]. \tag{14.45}$$

   This set is applied to the strains ε_{11} and ε_{31}.

   - for the reduced number of points in the η direction, Fig. 14.7b,

$$
\begin{aligned}
R_A(\xi,\eta) &= \tfrac{1}{4}(1-\tfrac{\eta}{a})\left[(\tfrac{\xi}{b})^2 - \tfrac{\xi}{b}\right], & R_B(\xi,\eta) &= \tfrac{1}{4}(1+\tfrac{\eta}{a})\left[(\tfrac{\xi}{b})^2 - \tfrac{\xi}{b}\right], \\
R_C(\xi,\eta) &= \tfrac{1}{4}(1+\tfrac{\eta}{a})\left[(\tfrac{\xi}{b})^2 + \tfrac{\xi}{b}\right], & R_D(\xi,\eta) &= \tfrac{1}{4}(1-\tfrac{\eta}{a})\left[(\tfrac{\xi}{b})^2 + \tfrac{\xi}{b}\right], \\
R_E(\xi,\eta) &= \tfrac{1}{2}(1+\tfrac{\eta}{a})\left[1 - (\tfrac{\xi}{b})^2\right], & R_F(\xi,\eta) &= \tfrac{1}{2}(1-\tfrac{\eta}{a})\left[1 - (\tfrac{\xi}{b})^2\right].
\end{aligned}
\tag{14.46}
$$

This set is applied to strains ε_{22} and ε_{32}.

In the direction in which the number of points is not reduced, either $b = 1$ or $b = \sqrt{3/5}$ can be used; we prefer the latter value as it has the advantage that the sampling points and the integration points coincide.

2. For the shear strain ε_{12} three approaches are used; we prefer the scheme of [39], in which the reduced number of sampling points is used in both directions, see Fig. 14.7c, and the following approximation functions are used,

$$
\begin{aligned}
R_A(\xi,\eta) &= \tfrac{1}{4}\left(1-\tfrac{\xi}{a}\right)(1-\tfrac{\eta}{a}), & R_B(\xi,\eta) &= \tfrac{1}{4}\left(1+\tfrac{\xi}{a}\right)(1-\tfrac{\eta}{a}), \\
R_C(\xi,\eta) &= \tfrac{1}{4}\left(1+\tfrac{\xi}{a}\right)(1+\tfrac{\eta}{a}), & R_D(\xi,\eta) &= \tfrac{1}{4}\left(1-\tfrac{\xi}{a}\right)(1+\tfrac{\eta}{a}),
\end{aligned}
\tag{14.47}
$$

This scheme is also used for the twisting strain κ_{12}.

For all the above schemes, the strain component ε to which we apply the two-level approximation, is expressed as follows:

$$
\tilde{\varepsilon}(\xi,\eta) = \sum_i R_i(\xi,\eta)\, \varepsilon_i, \tag{14.48}
$$

where $i = A, B, C, D, E, F$ for the schemes with six sampling points and $i = A, B, C, D$ for the scheme with four sampling points, see Fig. 14.7.

The AS method eliminates several types of locking in nine-node elements, including the membrane locking caused by ε_{11} and ε_{22}, the transverse shear locking caused by ε_{31} and ε_{32}, and the over-stiffening in twisting caused by κ_{12}. To remedy the transverse shear locking, the ANS method can also be applied.

The above forms of the two-level approximations were used in the nine-node shell element with drilling rotation of [164]. The element is designated as 9-AS and characterized in Table 14.3. It is integrated by a 3×3-point Gauss scheme.

Table 14.3 Assumed Strain interpolations of nine-node shell element.

Strain components	Approximation scheme
ε_{11}, ε_{13}	eq. (14.45), Fig. 14.7a
ε_{22}, ε_{23}	eq. (14.46), Fig. 14.7b
ε_{12}	eq. (14.47), Fig. 14.7c
κ_{12}	eq. (14.47), Fig. 14.7c

14.5 Remarks on approximation of curved surfaces by four-node elements

When modeling a curved shell by four-node shell elements, we have to consider several issues which are discussed below.

Is the generated element warped of flat? Four nodes of a bilinear element can span two geometries: either a planar element or a hyperbolic-paraboloidal element. This means that, e.g., the cylindrical surface of Fig. 14.8 can be approximated by either flat elements or h-p elements, which certainly affects the results, especially for crude meshes.

Is the curvature similarly approximated? Even when the generated FEs are flat, the approximation of the curvature should also be controlled. Consider the two arcs of different radii, i.e. 1-2-3 and 4-5-6, shown in Fig. 14.9. The finite elements 1-3 and 4-6 are flat, but we can use the same angle α for both of them. Then

$$\frac{h_1}{L_1} = \frac{h_2}{L_2} = \frac{1}{2} \tan \frac{\alpha}{4}, \tag{14.49}$$

i.e. the ratio of the neglected altitude h_i to the element length L_i is identical for both elements. In this sense, the same level of approximation of curvature is provided.

Do we need five or six degrees of freedom per node? The sixth dof at a node is the drilling rotation and it is necessary to connect:

1. shell parts intersecting at large angle. Otherwise, we can have the situation shown in Fig. 14.10a, where the vertical elements remain undeformed for $\omega > 0$, and gaps occur between the horizontal and vertical elements !

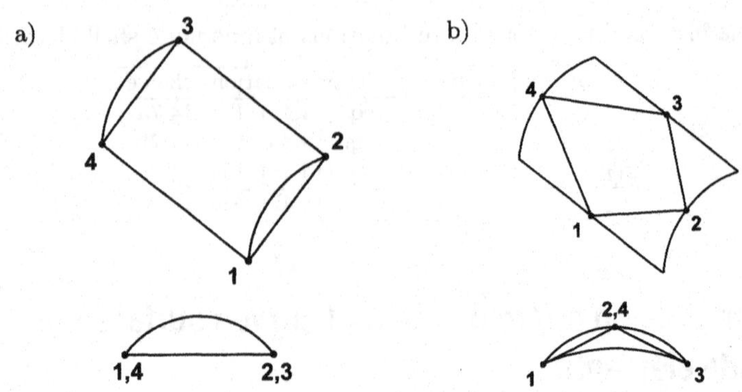

Fig. 14.8 Positions of nodes 1,2,3 and 4 on a cylindrical surface can result in a) a flat element, b) a warped (h-p) element.

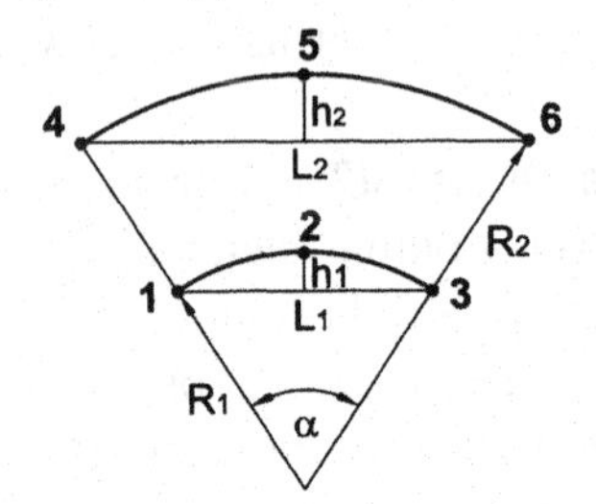

Fig. 14.9 Similar approximation of arcs 1-2-3 and 4-5-6 by flat elements 1-3 and 4-6.

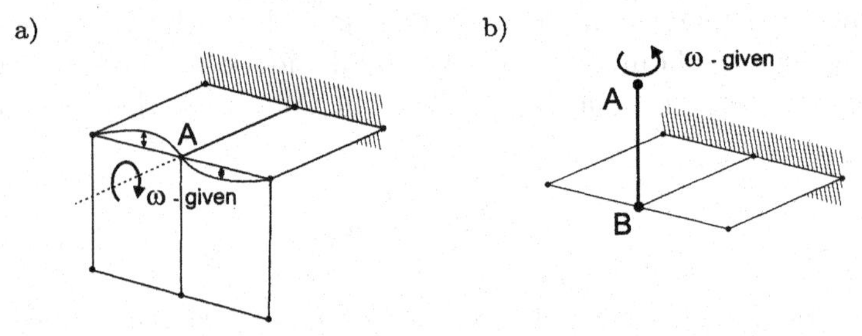

Fig. 14.10 Connection of a) flat shell FEs with five dofs/node, b) beam element A-B and shell FEs with five dofs/node.

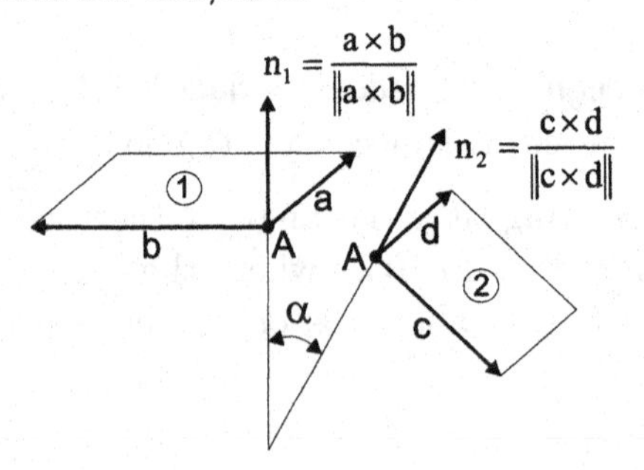

Fig. 14.11 Use six dofs/node if $\alpha = \arccos(\mathbf{n}_1 \cdot \mathbf{n}_2) > 5^o$ at node A.

2. shells with beams, which naturally have six dofs/node. Otherwise, we can have the situation shown in Fig. 14.10b, where the shell remains undeformed for $\omega > 0$!

If the modeled shell is flat or weakly curved, then shell elements with five dofs/node are acceptable. The threshold condition can be formulated in terms of the angle α between the normals of adjacent elements, i.e. between $\mathbf{n}_1$ and $\mathbf{n}_2$ in Fig. 14.11. The shell elements with five dofs/node can be used for up to, say, $\alpha < 5^{\circ}$, while for bigger angles the elements with six dofs/node should be used.

Part V
NUMERICAL EXAMPLES

15 Numerical tests

In this chapter, the numerical tests of our shell FEs are described. The tests ensure correctness of the formulation and verify its quality by comparing the accuracy and robustness with other elements.

Good tests stimulate progress in FE technology, although sometimes are found accidentally as, e.g., the Raasch hook test of Sect. 15.2.10. Popular benchmark problems for shell elements are provided in [143, 233, 51]. The number of tests which are in use for verification of elements is large; only a selection of them can be presented here.

15.1 Characteristics of tested shell elements

Requirements for shell elements. The shell FE should satisfy the following requirements:

1. not contain spurious zero energy eigenvalues,
2. be able to represent the zero strains for rigid body motions and the constant strains, i.e. should satisfy the invariance and convergence requirements [229].
3. be free from locking phenomena,
4. be insensitive to geometrical distortions and fairly accurate for coarse meshes,
5. should not use problem-dependent adjustable factors,
6. enable linking up of various constitutive modules,
7. be versatile, i.e. applicable to thin and thick shells, flat and curved shells, small and large rotations, small and large strains, work for statics and dynamics,
8. be computationally efficient, to enable large-scale computations,

9. have six degrees of freedom per node, to easily connect with beams and shells,
10. be based on a simple formulation, easy to understand, implement, modify, and debug.

Similar lists of features are given e.g. in [91, 19, 267].

To evaluate the performance of elements in linear tests, we can use e.g. the scheme of grading proposed in [143],

Grade	A	B	C	D	E
Error in %	≤ 2	(2,10]	(10,20]	(20,50]	> 50

More precise is to calculate an error for each test, and an average error for all tests. More complicated methods of evaluation are needed for non-linear tests.

Characteristics of our four-node shell elements. Our Reissner-type four-node shell elements have the following features:

1. Six dofs/node, including the drilling rotation. The variables at nodes are the displacements $\Delta \mathbf{u}$ and the canonical rotation vector $\Delta \boldsymbol{\psi}$, both in the reference basis.
2. The element's geometry is specified by positions of nodal points (the normals at nodes are not used). At Gauss points on the mid-surface, the ortho-normal basis $\{\mathbf{t}_i\}$ at equal angles with the natural basis $\{\mathbf{g}_i\}$ is used.
3. Rotations can be finite (unrestricted) and the Green strain is used, i.e. the elements are applicable to finite-deformation problems.
4. The transverse shear strains are modified by the ANS technique and the RBF correction, see Sect. 13.2.
5. Drilling RC are implemented using the Perturbed Lagrange (PL) method, see Sect. 12.3.2,
6. The multipliers of additional modes are eliminated on the element's level and updated by the scheme U2 of Sect. 11.3,
7. Integration scheme: 2×2 Gauss Points in lamina, and analytical or numerical integration over fiber, with either the two-point Gauss or five-point Simpson rule.

The tested shell elements are listed in Table 15.1. These elements were derived using the AD program AceGen [131] and tested within the FE program FEAP [268], for more details see Sect. 10.6. The use of these programs is gratefully acknowledged.

Table 15.1 Tested four-node shell elements.

Element	Characteristics	Ref.
EADG4	Potential Energy + EADG4 enhancement	own
EADG5	Potential Energy + EADG5 enhancement	own

The EADG5 method is implemented as in eqs. (12.76) and (11.71), where

$$\mathbf{G} \doteq \begin{bmatrix} q_1\,\xi & q_3\,\eta + \underline{q_5 \xi \eta} \\ q_2\,\xi & q_4\,\eta \end{bmatrix}. \tag{15.1}$$

The additional bilinear (underlined) term added to the EADG4 enhancement improves accuracy in several examples, especially when the shell is very thin, see Sects. 15.3.5, 15.3.6, 15.3.8, and 15.3.12.

For reference, we use the shell elements listed in Table 15.2. The reference results for nine-node elements were obtained in [163]. The help of Dr P. Panasz in testing is gratefully acknowledged.

Table 15.2 Reference shell elements.

Element	Characteristics	Ref.
Q4	Potential Energy, basic (non-enhanced)	own
S4	four-node	ABAQUS [180]
FEAP, six dofs/node	four-node, Discrete Kirchhoff, linear	FEAP [183]
9-AS	nine-node, Assumed Strain	own [164]
MITC9	nine-node, Assumed Strain	ADINA [182]
S9R5	nine-node, RI, stabilized	ABAQUS [180]

15.2 Elementary and linear tests

15.2.1 Eigenvalues of a single element

Eigenvalues and eigenvectors of the tangent matrix are computed in the following way:

1. A single element is tested because some zero eigenvalues can vanish for a patch of elements. Boundary conditions are not imposed because they can mask the presence of zero eigenvalues.
2. The elements of a square and an irregular shape are tested. For the square element, we use the side length equal to 1. The irregular element

is obtained, e.g., by a shift of node 3 of a square by $\{s, s, w\}$, where s is the in-plane shift and w is the out-of-plane warp, see Fig. 15.1. We use $s = w = 0.5$.

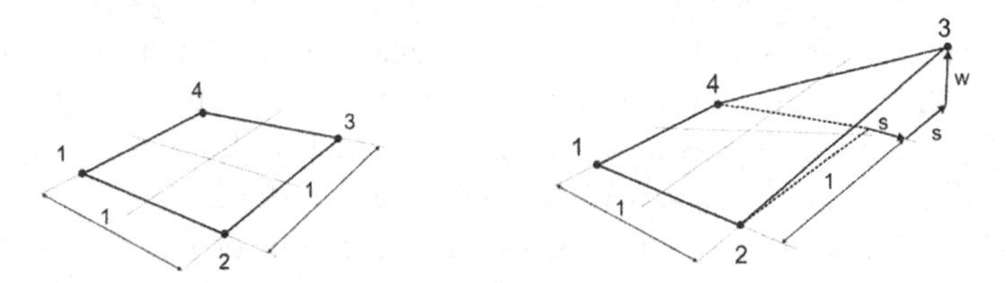

Fig. 15.1 Shapes of shell elements used in the eigenvalue analysis.

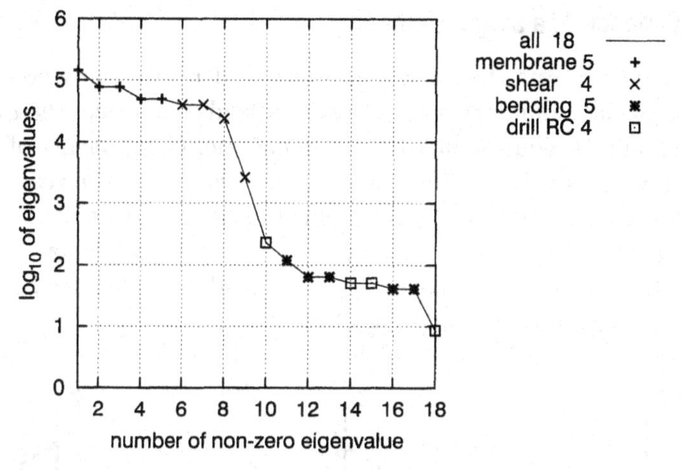

Fig. 15.2 Eigenvalues of a square shell element. $E = 10^6$, $\nu = 0.3$, $h = 0.1$, $\gamma = 0.01G$.

3. The thickness $h = 1$ is not used in the eigenvalue analysis for shell elements, as then $h = h^3$ and eigenvalues for the bending and membrane parts are too close.

Having the eigenvalues of the element, we have to check

1. The number of zero eigenvalues. The shell element of a correct rank has six zero eigenvalues corresponding to the rigid body modes. Additional zero eigenvalues are spurious and should be at least stabilized.
2. The number of positive eigenvalues and negative eigenvalues. The negative eigenvalues appear for mixed formulations in the non-reduced

tangent matrix and indicate a saddle point of a discrete HR or HW functional. The number of them is equal to the number of Lagrange multipliers (stress parameters).

The non-zero eigenvalues of a square shell element Q4 (with no enhancement) are shown in Fig. 15.2, where these associated with the membrane, shear and bending parts and the drill RC functional are indicated. The non-zero eigenvalues are very distinct and the condition number cond $\mathbf{K} \doteq \lambda_{\max}/\lambda_{\min} \approx 16719$ for the given data.

Finally, we note that (1) the strain enhancement or the use a different functional changes the eigenvalues, (2) the number of zero eigenvalues of the elements based on Allman's shape functions depends on the element's shape, see [255].

15.2.2 Invariance of a single element

The invariance means that if we translate and rotate the whole system (i.e. the element, boundary conditions, and loads) in space, then the new solution should be equivalent to the original one. Components of the displacement vectors will be different but the vectors themselves, i.e. their lengths and directions relative to the local basis should be identical.

We can select the positions for which we immediately know the orientation of the element's local basis $\{\mathbf{t}_k\}$ in the global reference basis $\{\mathbf{i}_k\}$, see Fig. 15.3. All our elements pass this test.

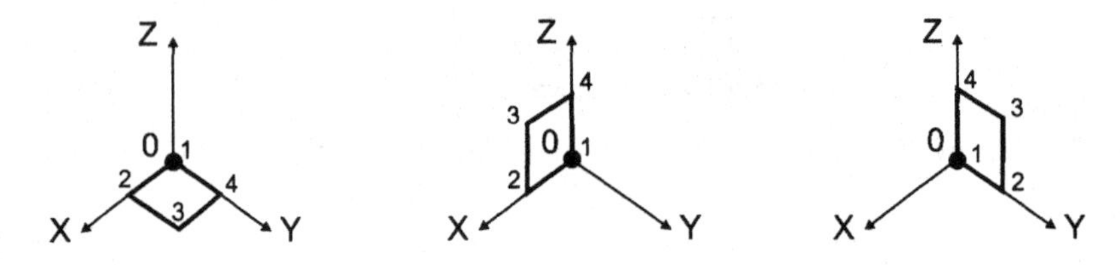

Fig. 15.3 Three positions of the element to test the rotational invariance.

15.2.3 Constant strain patch tests

Even if an individual rectangular element can represent the constant stress and strain states, possibility still exists that a patch (assembly, or group) of distorted elements may not. This property cannot be easily checked analytically, but we can use the so-called "patch test" proposed by B. Irons, see [110, 21, 111]. If an element passes the patch test, then it will behave well in an arbitrary mesh of elements in a constant strain field. Various patches of elements are in use; the five-element patch of [195] is shown in Fig. 15.4.

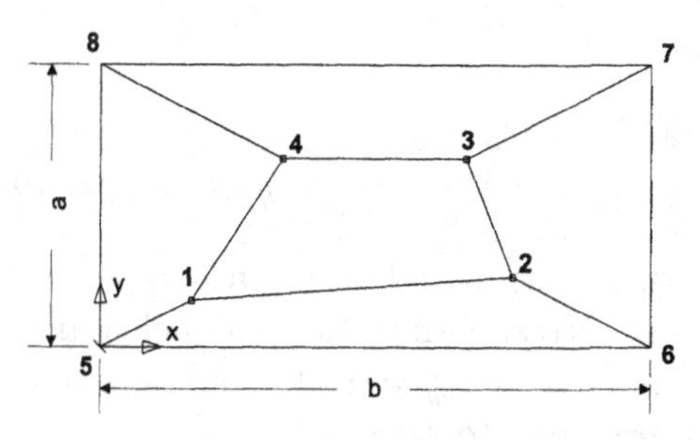

Fig. 15.4 Patch test. $E = 10^6$, $\nu = 0.25$, $h = 0.001$, $a = 0.12$, $b = 0.24$.

Procedure of the patch test. The procedure of the patch test is as follows:

1. Assume algebraic formulae for displacements and rotations, such that they render constant strains in a body (patch of elements).
2. Compute the values of displacements and rotations at all nodes using these algebraic formulae.
3. Use the values of displacements and rotations at the external nodes (5,6,7,8) as the boundary values, and compute a solution at the internal nodes (1,2,3,4) using the tested finite element.
4. For internal nodes (1,2,3,4), compare the computed values with the values calculated by the algebraic formulae; they should be identical.

For shells, the patch test is performed separately for membrane or bending constant strain states.

Algebraic formulae for constant strain tests. We assume the following algebraic formulae for displacements and rotations:

1. Membrane patch test

$$\begin{aligned} &u_x(x,y) = 0.001(x + \tfrac{1}{2}y), \qquad u_y(x,y) = 0.001(\tfrac{1}{2}x + y), \\ &\psi_z(x,y) = 0, \end{aligned} \tag{15.2}$$

while the other displacement and rotation components are equal to zero. The membrane strains corresponding to eq. (15.2) are constant, i.e. $\varepsilon_{xx} = \varepsilon_{yy} = 0.001$ and $\varepsilon_{xy} = 0.0005$, while the bending strains are equal to zero.

2. Bending patch test

$$
\begin{aligned}
&u_z(x,y) = 0.0005(x^2 + xy + y^2), \\
&\psi_x(x,y) = 0.0005(x + 2y), \qquad \psi_y(x,y) = -0.0005(2x + y),
\end{aligned}
\tag{15.3}
$$

while the other displacement and rotation components are set to zero. The bending strains corresponding to these fields are constant over the patch of elements, i.e. $\kappa_{xx} = \kappa_{yy} = 0.001$ and $\kappa_{xy} = -0.0005$, while the membrane strains are equal to zero.

Reference results for the patch tests. The reference displacements and rotations at internal nodes are given in Table 15.3. We use only the elements which pass both patch tests.

Table 15.3 Coordinates of inner nodes and reference results of patch tests.

	Coordinates		Membrane test			Bending test		
Node	x	y	u_x	u_y	ψ_z	w	ψ_x	ψ_y
1	0.04	0.02	5.00E-05	4.0E-05	0	1.400E-06	4.0E-05	-5.00E-05
2	0.18	0.03	1.95E-04	1.2E-04	0	1.935E-05	1.2E-04	-1.95E-04
3	0.16	0.08	2.00E-04	1.6E-04	0	2.240E-05	1.6E-04	-2.00E-04
4	0.08	0.08	1.20E-04	1.2E-04	0	9.600E-06	1.2E-04	-1.20E-04

Remark. The patch test is sometimes treated as equivalent to the consistency condition, which is a necessary condition for mesh convergence. This equivalence has been proven so far only for regular meshes. However, there is no doubt that this test verifies completeness of approximating polynomials and the element's ability to reproduce strains of a specific order. Besides, it is very useful in detecting errors in the element's formulation and/or programming. Various extensions of the constant strain patch test and a historical note are given in [268] p. 250, and [23] p. 461.

Membrane patch test for elements with drilling rotation. Consider the membrane patch test, for which the displacements are defined by eq. (15.2).

For these fields, the linearized drill RC equation yields the drilling rotation equal to zero, i.e.

$$\psi_z(x,y) \doteq \tfrac{1}{2}(u_{x,y} - u_{y,x}) = 0. \tag{15.4}$$

This value was already used in eq. (15.2) and in Table 15.3.

At all boundary nodes of the elements with the drilling rotation, the displacements should be restrained as described earlier, while the drilling rotation can be prescribed in three different ways listed in Table 15.4.

Table 15.4 Three types of boundary conditions for elements with drilling rotation.

Designation	Boundary condition applied to drilling rotations
b1	ψ_z is restrained at all boundary nodes
b2	ψ_z is restrained at one node, e.g. at node No.5, so then $(\psi_z)_5 = 0$
b3	ψ_z is free at all boundary nodes

For all these boundary conditions, we should obtain zero drilling rotations at internal nodes! The condition **b3** is the most demanding.

Finally, we note that to pass the patch test, the strain enhancement must be formulated in a specific way and the Allman elements require the Jetteur–Frey procedure of Sect. 12.7.3.

15.2.4 Distortion test

This test allows us to determine the sensitivity of an element to mesh distortions for in-plane bending. The cantilever is shown in Fig. 15.5a and the end moment is applied as two opposite forces P. The mesh is divided into two parts, and the tilt of their common side is defined by the parameter d. For $d = 0$, the parts are rectangles.

For the mesh of two elements and the data defined in Fig. 15.5a, the vertical displacement u_y and the drilling rotation ψ_z at node 6 are shown in Fig. 15.5b. Both exhibit a similar drop of accuracy.

For the mesh of eight elements shown in Fig. 15.6a, the vertical displacement at node 3 is presented in Fig. 15.6b. The data is taken from [268], where this test is treated as a higher-order patch test. For $d = 1$, the tilt is 45° and the elements are trapezoidal. The reference solutions are obtained using the nine-node elements based on the same nodes.

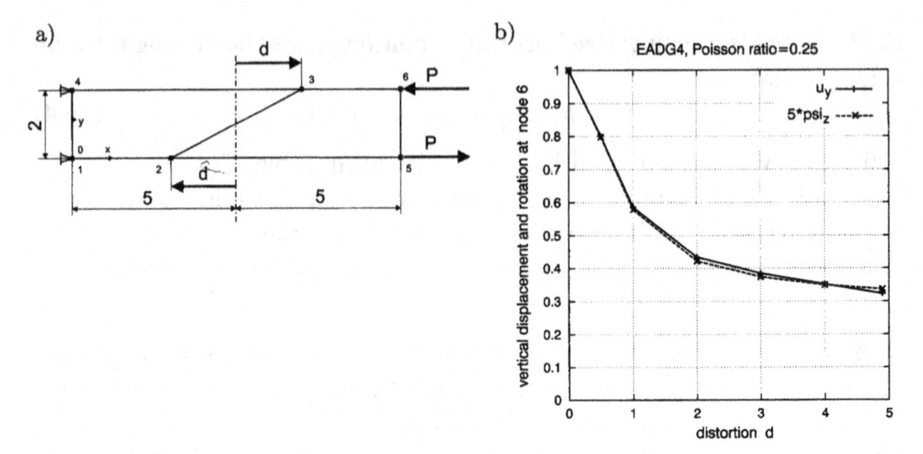

Fig. 15.5 Two-element distortion test. $E = 1500$, $\nu = 0.25$, $h = 1$, $P = 10$. $\gamma = G$. a) Initial geometry and load. b) Vertical displacement and drilling rotation at node 6.

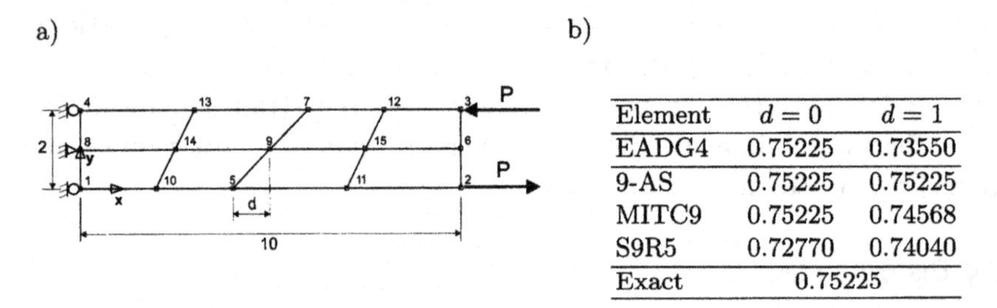

Element	$d = 0$	$d = 1$
EADG4	0.75225	0.73550
9-AS	0.75225	0.75225
MITC9	0.75225	0.74568
S9R5	0.72770	0.74040
Exact	0.75225	

Fig. 15.6 Eight-element distortion test. $E = 100$, $\nu = 0.3$, $h = 10$, $P = 5$, $\gamma = G/1000$. a) Initial geometry and load. b) Linear analysis. Vertical displacement at node 3.

15.2.5 Warped single element

The four-node elements show a serious over-stiffening for a warped initial geometry, see Sect. 14. The purpose of this test is to check the accuracy of a single warped element for selected six load cases, see [152].

The warped element, see Fig. 15.7, is clamped at nodes 1 and 4, and the external forces are applied at nodes 2 and 3. For the given data, the relative warping parameters of Sect. 14 are

$$\Phi \doteq \frac{Z}{h} = 1, \qquad \Psi \doteq \frac{Z}{\sqrt{A_e}} = \frac{1}{100}. \tag{15.5}$$

Six load cases are used which, in a flat element ($Z = 0$), would cause the following deformation: (1) stretch, (2) in-plane bending, (3) in-plane shear, (4) in-plane pinching, (5) out-of-plane shear, and (6) out-of-plane twisting. In the warped element ($Z > 0$), more complex states of deformation occur for these loads.

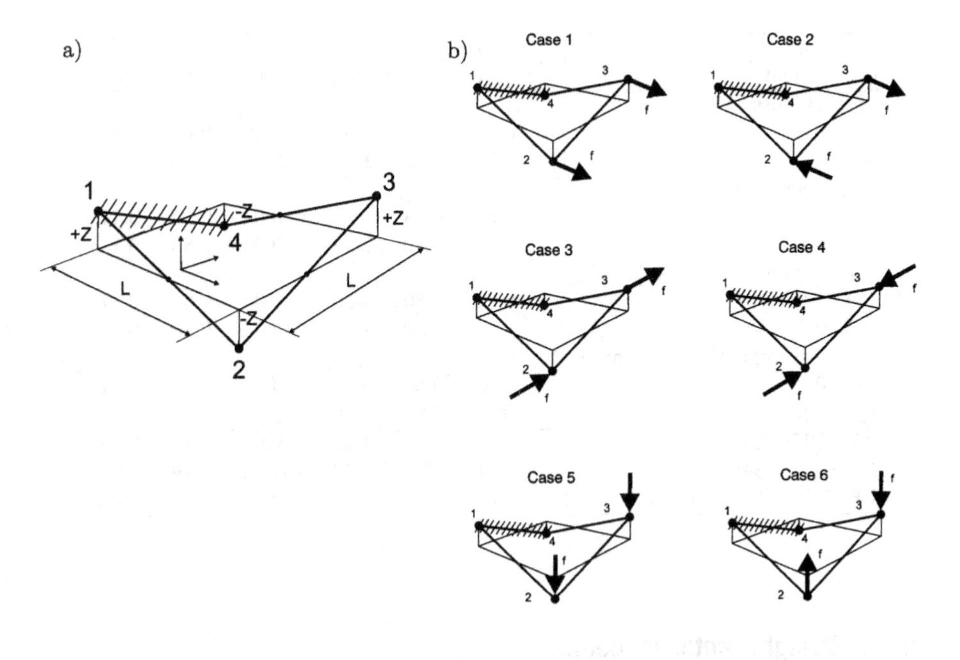

Fig. 15.7 Warped single element. $E = 10^6$, $\nu = 0.3$, $h = 0.1$, $L = 10$, $Z = 0.1$. a) Geometry and boundary conditions, b) Six load cases.

The in-plane membrane shear strain, ε_{12}, can be treated in one of the following four ways: (1) sampled at Gauss Points (standard treatment), (2) sampled at the element's center, (3) calculated as the average of values at the element's midsides, and (4) calculated as the average of values at the element's corners. In all cases, $\mathbf{R}$ is used in transformations instead of $\mathbf{R}_c$, see Sect. 14.2.

The dominant displacement at node 3 is reported in Table 15.5; it is normalized by the 50×50-element mesh solution. Two values of s_2 of c_{warp} are used, see eq. (14.19).

We checked that, for $Z \to 0$, our results converge to the results for the flat element and the effect of c_{warp} vanishes, which is correct. We see that the reference elements show a large scatter of results. This test remains a serious challenge for future research.

Table 15.5 Warped single element test. Normalized dominant displacement at node 3. Shell element EADG4+PL. $\gamma = G$. Warpage $Z = 1$.

Load case	1	2	3	4	5	6
Component	u_1	u_1	u_2	u_2	u_3	u_3
ε_{12} at Gauss Points						
$s_2 = 0.3$	0.02	0.60	1.39	0.02	0.60	0.04
$s_2 = 0.5$	0.02	0.80	1.85	0.02	0.80	0.04
ε_{12} at center						
$s_2 = 0.3$	0.02	0.53	1.97	0.04	0.53	0.11
$s_2 = 0.5$	0.02	0.80	2.98	0.04	0.80	0.11
ε_{12} as average of midside values						
$s_2 = 0.3$	0.02	0.60	1.10	0.04	0.60	0.11
$s_2 = 0.5$	0.02	0.80	1.46	0.04	0.80	0.11
ε_{12} as average of corner values						
$s_2 = 0.3$	0.02	0.66	0.82	0.04	0.66	0.11
$s_2 = 0.5$	0.02	0.80	0.99	0.04	0.79	0.11
ABAQUS, S4	0.01	0.79	1.81	0.02	0.78	0.09
ABAQUS, S4R	0.01	1.43	3.29	-32.64	0.79	5.99
FEAP, shell 6dofs/node	18.56	0.78	1.26	13.54	0.99	8.36

15.2.6 Straight cantilever beam

This test was proposed in [143]. The mesh consists of six elements of three shapes: rectangular, trapezoidal, and parallelogram. The beam is clamped at nodes 1 and 8, and the external loads are applied at nodes 7 and 14, see Fig. 15.8. This test is difficult because the mesh is coarse and distorted and the elements have a large aspect ratio (= 5 for rectangles).

The external loads are applied in four different ways, to render basic deformation modes, such as stretching, in-plane shearing, out-of-plane shearing, and twisting. The total load is always equal to 1. To have the twisting moment $M = 1$, we apply two forces $F_z = \pm 5$.

The solutions at node 14 for different loads and elements are shown in Table 15.6. The reference solutions are taken from [143] Table 3. For the out-of-plane bending by a unit moment, the reference solution is calculated as $\psi_y = M_y L/(EI)$, where $I = bh^3/12$, see [245].

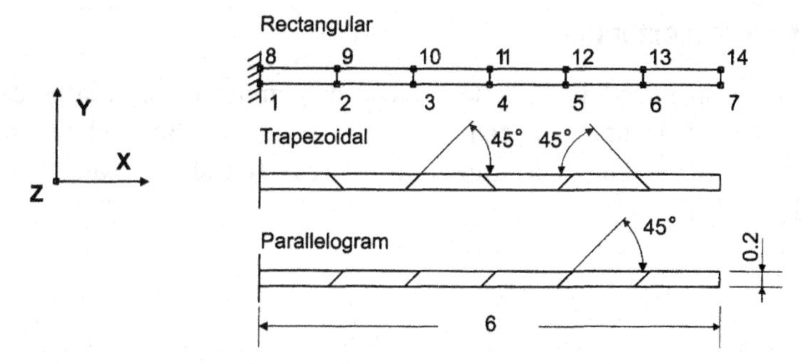

Fig. 15.8 Straight cantilever beam. $E = 10^7,\ \nu = 0.3,\ h = 0.1$.

Table 15.6 Straight cantilever beam. Linear test.

Element	Rectangular	Trapezoidal	Parallelogram
	In-plane shear load $(u_y \times 10)$		
EADG4	1.0737	0.7803	1.0146
S4	0.7704	0.1361	0.1704
9-AS	1.0748	1.0715	1.0749
MITC9	1.0748	1.0646	1.0746
S9R5	0.9344	0.9261	0.9854
Ref.		1.0810	
	Out-of-plane shear load $(u_z \times 10)$		
EADG4	4.2850	4.2886	4.2886
S4	4.2350	4.1860	4.2260
9-AS	4.2958	4.2812	4.2929
MITC9	4.2955	4.2823	4.2930
S9R5	4.3100	4.3110	4.3110
Ref.		4.3210	
	Twisting by pair of forces $(\psi_x \times 100)$		
EADG4	3.0313	3.0345	2.9150
S4	2.5061	2.5479	2.5121
9-AS	3.0307	3.0334	3.0229
MITC9	2.9128	2.9130	2.9033
S9R5	3.0400	3.0400	3.0300
Ref.		3.2080	
	Bending by moment $(\psi_y \times 100)$		
EADG4	3.5919	3.5927	3.5923
S4	3.5835	3.5838	3.5837
9-AS	3.5929	3.5930	3.5344
MITC9	3.5930	3.5928	3.5178
S9R5	3.5995	3.5994	3.6000
Ref.		3.6000	

15.2.7 Cook's membrane

In this test proposed in [54], the shear deformation dominates and the elements are skew and tapered. The membrane is clamped at one end, while at the other end, the uniformly distributed shear load $P = 1$ is applied, see Fig. 15.9.

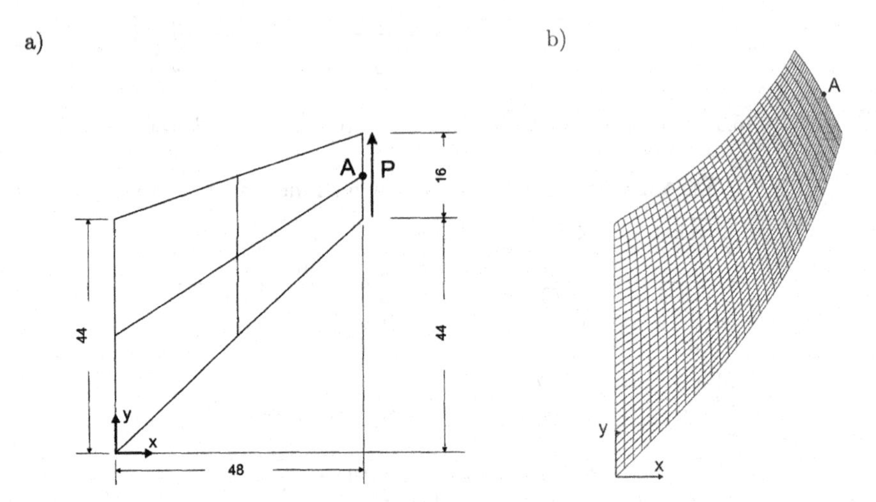

Fig. 15.9 Cook's membrane. $E = 1$, $\nu = 1/3$, $h = 1$. a) Initial geometry and load, b) Deformed configuration (not to scale).

The computed vertical displacement and drilling rotation at node A are presented in Table 15.7.

For four-node elements, the 2×2 and 32×32-element meshes are used, and results for $\gamma = G$ and $\gamma = G/1000$ are presented. We see that the value of γ is important for the coarse mesh. For nine-node elements, the 1×1 and 2×2-element meshes are used. More results for this test is presented in Sect. 12.8, Table 12.6.

Table 15.7 Cook's membrane. Linear analysis

	Mesh 2×2		Mesh 32×32	
Four-node elements	u_y	ω	u_y	ω
$\gamma = G/1000$				
Q4	11.842	0.353	23.818	0.878
EADG4	21.043	0.821	23.940	0.891
$\gamma = G$				
Q4	11.173	0.316	23.790	0.876
EADG4	20.940	0.879	23.936	0.891
S4	20.71	0.715	23.93	0.879
FEAP	20.854	0.655	23.922	0.854
Simo *et al.* [210]	21.124	-	-	-
D-type [109]	20.682	-	-	-
	Mesh 1×1		Mesh 2×2	
Nine-node elements	u_y	ω	u_y	ω
9-AS	21.799	0.807	23.576	0.869
MITC9	22.209	0.704	23.613	0.830
S9R5	26.540	0.852	23.980	0.793
Ref.	23.81		23.81	

15.2.8 Curved beam

The curved beam is clamped at one end and at the other end is loaded by a unit force, which acts either in the in-plane or in out-of-plane direction, see Fig. 15.10a. The elements are trapezoidal. This test was proposed in [143].

a)

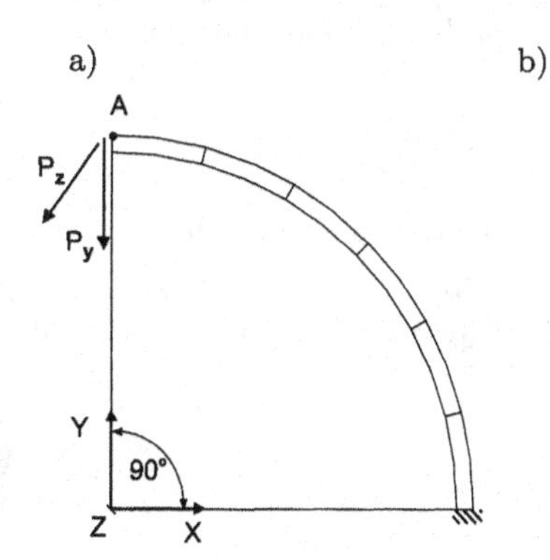

b)

Load	P_y		P_z	
Displacement	$-v \times 100$		$w \times 10$	
Mesh	1×6	1×24	1×6	1×24
EADG4	8.7355	8.8465	4.8776	4.8939
9-AS	8.8236	8.8495	4.8847	4.8944
MITC9	8.9852	9.0041	4.8045	4.8131
S9R5	7.3666	8.8476	4.8940	4.8940
Ref.	8.734		5.022	

Fig. 15.10 Curved beam. $E = 10^7$, $\nu = 0.25$, $h = 0.1$, $R_{\text{int}} = 4.12$, $R_{\text{ext}} = 4.32$. a) Initial geometry and load. b) Results of linear analysis.

The displacement in the direction of force at node A for various meshes and for both types of loads are shown in Fig. 15.10b. One element per beam thickness and a different number of elements in the circumferential direction are used.

15.2.9 Pinched cylinder with end diaphragms

A cylindrical shell is closed at both ends by rigid diaphragms and is pinched by two opposite forces P applied at the middle section. This test involves inextensional bending and complex membrane states of stress. The geometry and data are defined in Fig. 15.11.

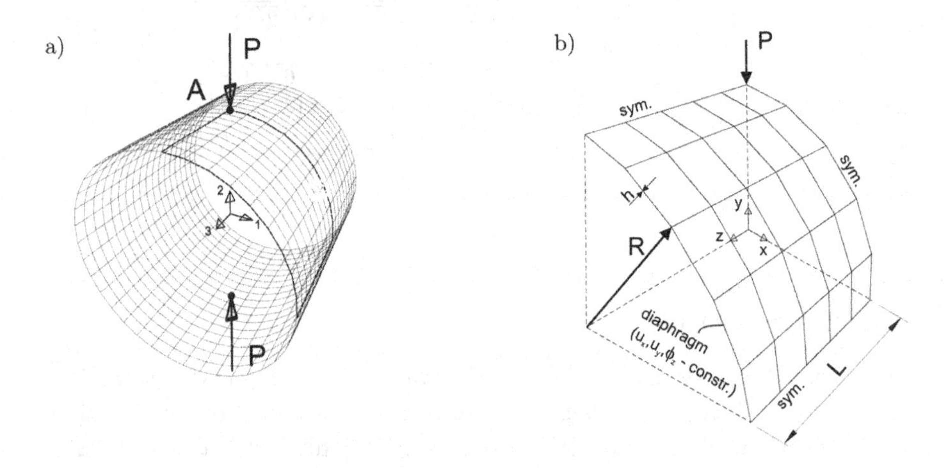

Fig. 15.11 Pinched cylinder with diaphragms. $E = 3 \times 10^6$, $\nu = 0.3$, $h = 3$, $R = 300$, $L = 300$. a) Initial geometry and loads. b) FE mesh for 1/8 of the cylinder.

Table 15.8 Pinched cylinder. Vertical displacement ($\times 10^5$) under the force.

Element/Mesh	4×4	10×10
EADG4	1.3855	1.7548
9-AS	1.4535	1.8194
MITC9	1.3180	1.7979
S9R5	1.3870	1.8040
Ref.	1.8249	

Because of symmetries, only one-eighth of the cylinder is analyzed, see Fig. 15.11b. The 4×4 and 10×10 element meshes are used for four-node elements, and 2×2 and 5×5 element meshes for nine-node elements.

The results of the linear analysis are presented in Table 15.8, where the vertical displacement at point A is presented for four-node and nine-node elements.

15.2.10 Raasch's hook

The hook is a thick curved strip, clamped at one end and loaded by a unit shear load P_z at the other end, see Fig. 15.12.

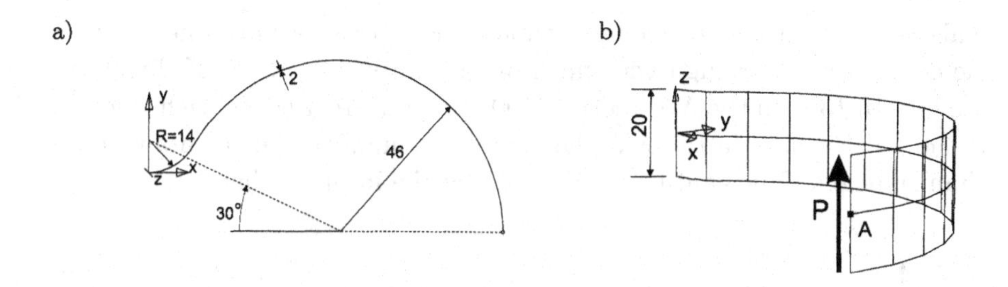

Fig. 15.12 Raasch's hook. $E = 3300$, $\nu = 0.35$, $h = 2$, $P = 1$.

This example was computed by I. Raasch of BMW, and revealed the erroneous behavior of the previous QUAD4 shell element of the MSC/Nastran FE code: (1) displacements were larger than the exact solution, although the element was fully integrated and the convergence should have been from below and (2) the accuracy of the solution deteriorated when the mesh was refined. The source of the problem were the transformation formulas between the nodal and elemental degrees of freedom, see [145].

Table 15.9 Raasch's hook. Vertical displacement at point A.

Element/Mesh	2×14	4×28	8×56
EADG4	4.9478	5.0802	5.1715
9-AS	4.9182	4.9640	5.0135
MITC9	4.8364	4.9163	4.9740
S9R5	4.8350	4.9140	4.9940

The results of the linear analysis are shown in Table 15.9. Note that there exist at least three reference solutions of Raasch's hook problem: (1) in [126], $u_z = 4.9352$, (2) in [92], $u_z = 5.012$, and (3) in [180], $u_z = 5.020$. The letter solution was obtained for the $20 \times 144 \times 2$-element mesh of 3D 20-node elements with reduced integration (C3D20R). The four-node shell elements of [180] (S4 and S4R) yield, for the 20×144-element mesh, the solution which is about 3% above this value, i.e. $u_z = 5.1706$.

15.3 Nonlinear tests

15.3.1 Slender cantilever under in-plane shear

This is a severe test of element capabilities, because only one layer of elements is used through the width of the cantilever, in the Y-direction, see Fig. 15.13a. In the X-direction, 100 elements are used, so each element is the 1×1 square in the XY plane, and the number of elements is more than sufficient. The cantilever is loaded by the in-plane shear force.

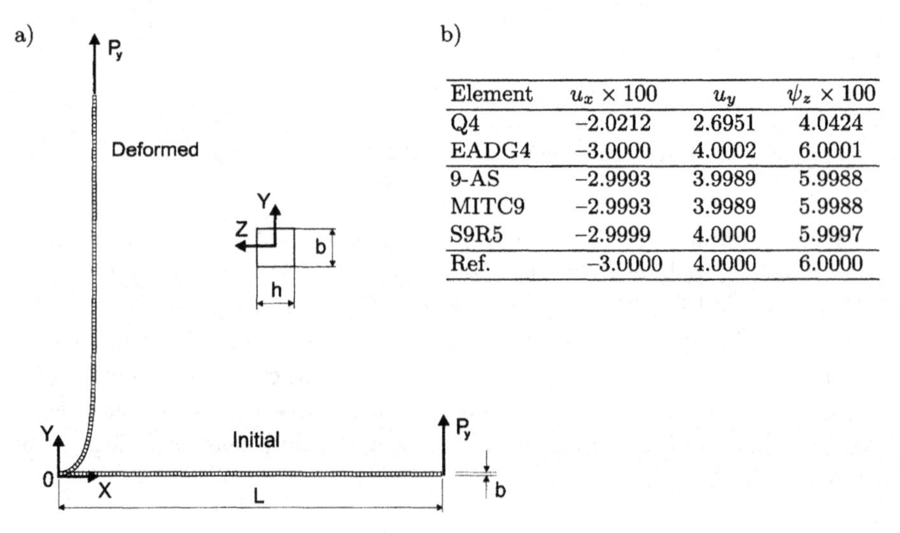

Element	$u_x \times 100$	u_y	$\psi_z \times 100$
Q4	–2.0212	2.6951	4.0424
EADG4	–3.0000	4.0002	6.0001
9-AS	–2.9993	3.9989	5.9988
MITC9	–2.9993	3.9989	5.9988
S9R5	–2.9999	4.0000	5.9997
Ref.	–3.0000	4.0000	6.0000

Fig. 15.13 Slender cantilever. $E = 10^6$, $\nu = 0.3$, $L = 100$, $h = b = 1.0$. a) Initial geometry and load. b) Results of linear analysis.

The results of the linear analysis for the top node of the tip are presented in Fig. 15.13b. For reference, the beam analytical solution is used; the vertical displacement $u_y = PL^3/(3EI)$, and the rotation

$\psi_z = PL^2/(2EI)$, both at the tip, and $I = bh^3/12$. For the standard bilinear Q4 element, the solution is locked, and the errors are about 33%.

In the non-linear test, the rotation of the cantilever's tip is almost 90°, see Fig. 15.13a.

1. To see the effects of the EADG4 enhancement, six steps are made using the arc-length method with the initial $\Delta P_y = 5$. The plots of u_y and $\psi_z = r_z$ at the cantilever's tip are shown in Fig. 15.14a. The finite-rotation Timoshenko beam solution is used for reference. The EADG4 solutions coincide with the beam solutions, while the Q4 curves are shifted. The steps of the arc-length procedure and the final load are much bigger for the EADG4 element than for the Q4 element.

2. To compare the convergence rates of various elements with drilling rotation of Sect. 12, one step of the Newton method is performed for $\Delta P = 40$. The number of iterations N used by the update scheme U2 is given in Fig. 15.14b, where the values of tolerances were $\tau_1 = 10^{-8}$ for the residual norm and $\tau_2 = 10^{-15}$ for the energy norm. The update schemes U1 and U2 are described in Sect. 11.3. We see that the HW element converges about two times faster than the EADG4 element.

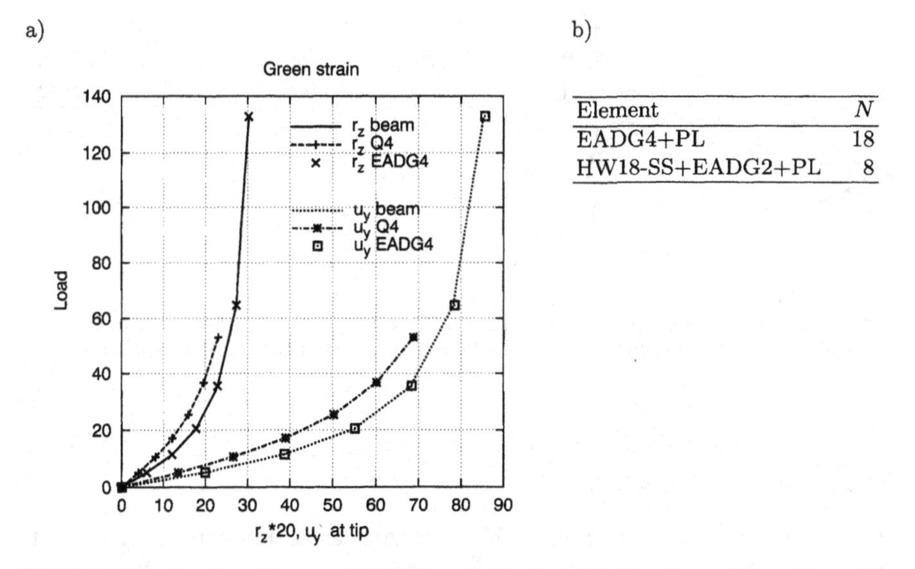

Element	N
EADG4+PL	18
HW18-SS+EADG2+PL	8

Fig. 15.14 Slender cantilever. a) Nonlinear solution. b) Number of iterations for update scheme U2.

15.3.2 Roll-up of a clamped beam

This test can be used to test procedures for finite rotations, as the tip of the beam makes almost five full turns. In dynamics, the examples involving free-body motion better suit this purpose but in statics, this test is indispensable.

The planar straight beam is clamped at one end and loaded by a bending moment at the other end, see Fig. 15.15a. The mesh of 25 four-node shell elements is used and two tests are performed.

1. *Bending into full circle.* The final deformed shape of the beam is shown in Fig. 15.15a and it should be obtained for $M = 2\pi EI/L = 628.319$. For this load, the solution is given in Fig. 15.15b.
2. *Roll-up of a beam into a small circle.* The applied moment is $\Delta M = 30$, and the solution is obtained by the Newton method. The tip of the beam makes almost five full turns (almost 10π radians) and the beam deforms into a small circle, shown to scale in Fig. 15.16a. The tip's displacements and rotation are compared in Fig. 15.15b with the Timoshenko beam solution.

a)

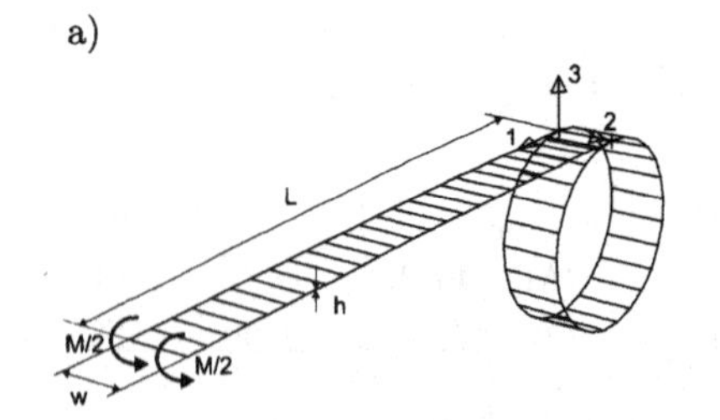

b)

	u_1	u_3	ω_2 [rad]
EADG4	-9.9934	-0.000013	6.2873
exact	-10.0000	0.000000	6.2832

Fig. 15.15 Roll-up of a beam. $E = 12 \times 10^6$, $\nu = 0$, $h = 0.1$, $w = 1$, $L = 10$. a) Initial and final geometry. b) Displacements and rotation of beam's tip.

15.3.3 Torsion of a plate strip

The plate strip shown in Fig. 15.17 undergoes a torsion caused by a twisting moment M. The final twisting rotation is over 180°. This test is computed in [211], but the solution is not reported, only the deformed configuration is shown.

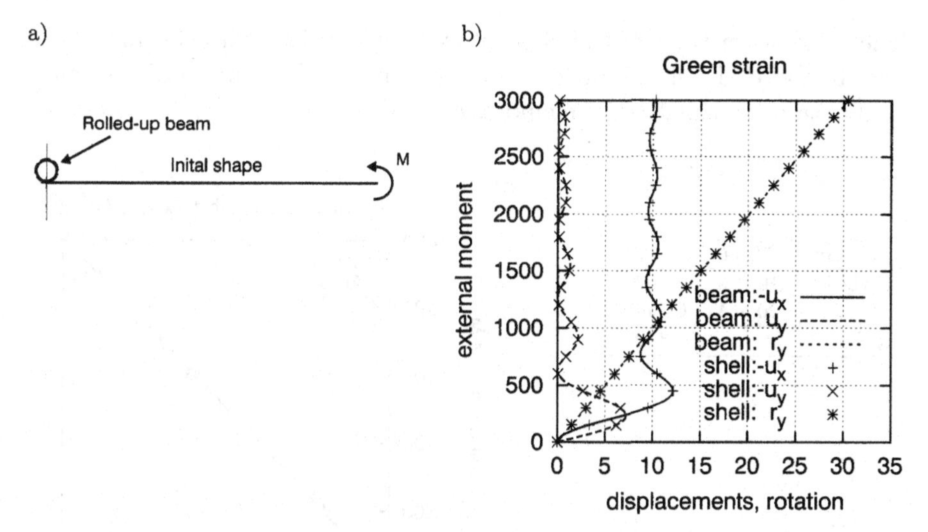

Fig. 15.16 Roll-up of a beam into a small circle. a) Initial and deformed geometry, side view to scale. b) Displacements and rotation of beam's tip.

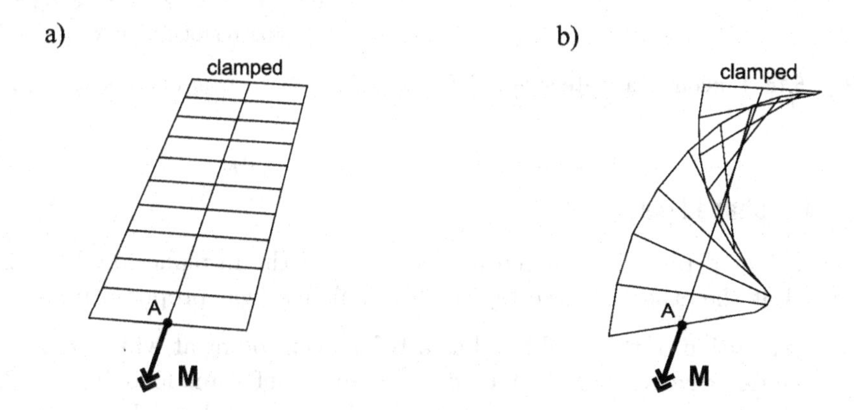

Fig. 15.17 Torsion of a plate strip. Initial and deformed configuration. $E = 1.2 \times 10^7$, $\nu = 0.3$, $L = 1$, width $w = 0.25$, $h = 0.1$.

The linear solution for $M = 1$ is given in Fig. 15.18a. The reference solution for a beam with a rectangular cross-section is as follows:

$$\theta_A = \frac{ML}{KG}, \qquad K = ab^3 \left[\frac{16}{3} - 3.36 \frac{b}{a} \left(1 - \frac{b^4}{12a^4} \right) \right], \tag{15.6}$$

where $a = w/2$, $b = h/2$, see [245].

The non-linear solution, obtained by the Newton method and $\Delta M = 100$, is shown in Fig. 15.18b. The twisting rotation at point A is monitored.

The solution for the EADG4 element exactly coincides with the analytical one. In the EADG4k element, the enhancement of the first-order strains is analogous to this for the membrane strains.

a)

Element	Mesh	θ_A
EADG4	10×2	0.36252
	20×4	0.34055
Ref.		0.31974

b)

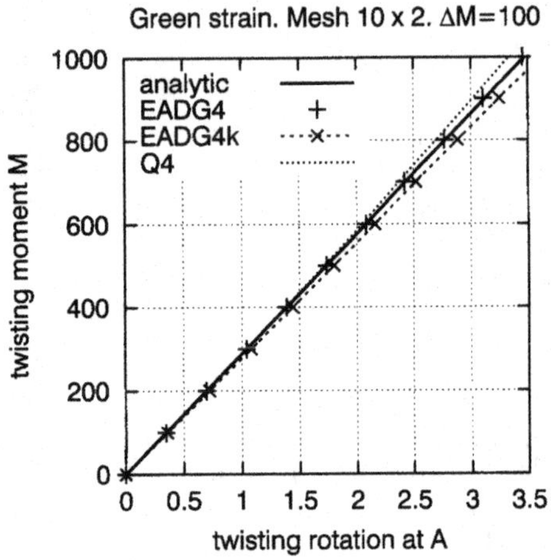

Fig. 15.18 Torsion of a plate strip. a) Linear solutions. b) Non-linear solutions.

15.3.4 L-shaped plate

The L-shaped plate is clamped at one end and the in-plane force P is applied at the other end, see Fig. 15.19. This test was proposed in [6].

The solution of this problem has a bifurcation point at which an out-of-plane deformation occurs. We add a small out-of-plane load $10^{-5} \times P$, and solve the equilibrium problem using the arc-length method, with a small initial $\Delta P = 0.2$, to estimate the bifurcation load. A mesh of 64 elements is used and P is applied at one of three points: A, B and C.

The solution curves, obtained using the EADG4 element, are presented in Fig. 15.20a; the region at the bifurcation points is magnified in Fig. 15.20b. The out-of-plane displacement u_3 is monitored either at the point where the force is applied or at point C. For P applied at point B, i.e. as in [6], from Fig. 15.20b, we can estimate the bifurcation load as $P = 1.137$; the same value was obtained in [211], Table 6.2.2.

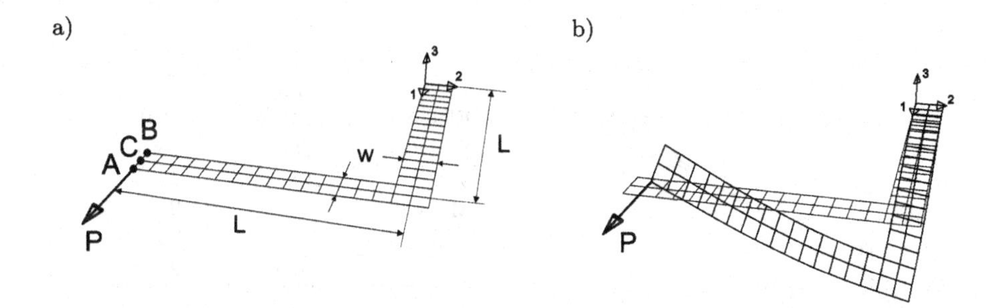

Fig. 15.19 L-shaped plate. $E = 71240$, $\nu = 0.31$, $h = 0.6$, width $w = 30$, $L = 240$. a) Initial geometry. b) Deformed geometry at $P = 1.95$ applied at point A.

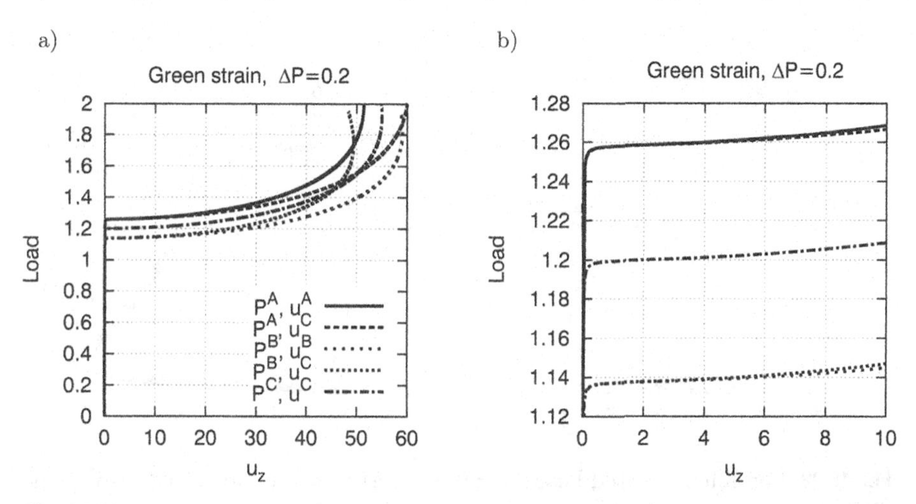

Fig. 15.20 L-shaped plate. a) Out-of-plane displacement. b) Region at bifurcation points.

15.3.5 Twisted beam

The initial geometry of the beam is twisted, see Fig. 15.21a, but the initial strain is equal to zero. The beam is clamped at one end and loaded by a unit force at the other. The force is applied either along the Z axis (in-plane) or along the Y axis (out-of-plane).

This example belongs to the set of tests of [143] and, later, was used in [26] to illustrate the importance of accounting for the variation of the Jacobian through the thickness. Note that flat shell elements have problems with this example.

a)

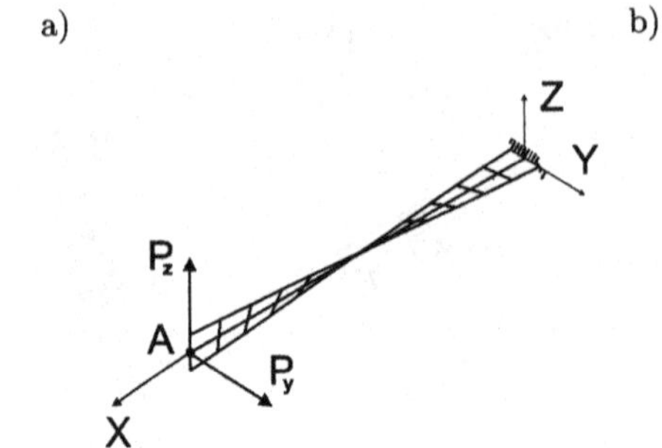

b)

Element	In-plane $u_z \times 10^6$	Out-of-plane $u_y \times 10^6$
EADG4	5.1888	1.2861
9-AS	5.2283	1.2935
MITC9	5.2468	1.2920
S9R5	5.2683	1.2958
Ref.	5.2560	1.2940

Fig. 15.21 Twisted beam. $E = 2.9 \times 10^7$, $\nu = 0.22$, $L = 12$, $w = 1.1$, twist $= 90°$. a) Initial geometry and load. b) Results of linear analysis. Thickness $h = 0.0032$

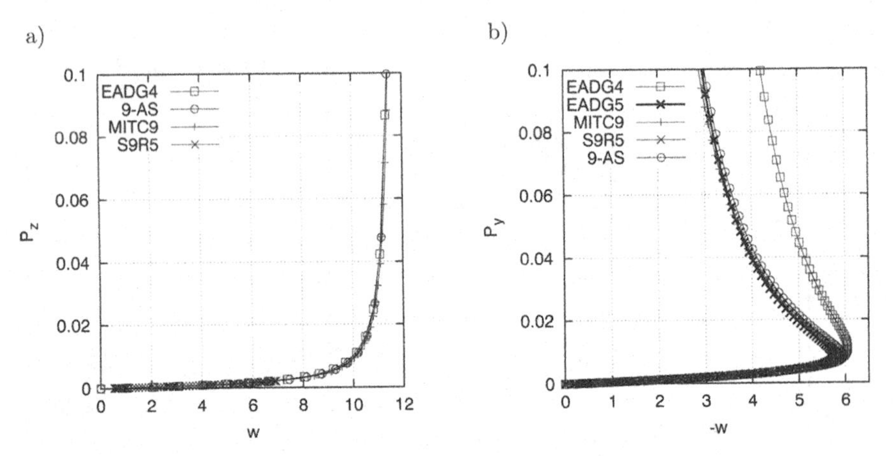

Fig. 15.22 Non-linear twisted beam $h = 0.0032$. a) In-plane force. b) Out-of-plane force.

The 4×24-element mesh was used for four-node elements, and 2×12-element mesh for nine-node elements. The displacement in the direction of force at point A is monitored.

The linear results for the force 1×10^{-6} are presented in Fig. 15.21b. Although the shell is very thin, the results are not corrupted by the membrane locking.

The non-linear load-deflection curves obtained by the Newton method are shown in Fig. 15.22. For the out-of-plane load, the solution from the EADG5 element is very close to the solutions from nine-node elements, but the EADG4 solution differs significantly.

15.3.6 Hinged cylindrical panel

A cylindrical panel of small curvature is loaded by a single force applied centrally, see [28]. The straight edges are hinged and the curved ones are free, see Fig. 15.23a. (This test is also performed for different data, see [187, 211, 200].) Due to the symmetry, only a quarter of the shell is modeled by the 4 × 4-element mesh.

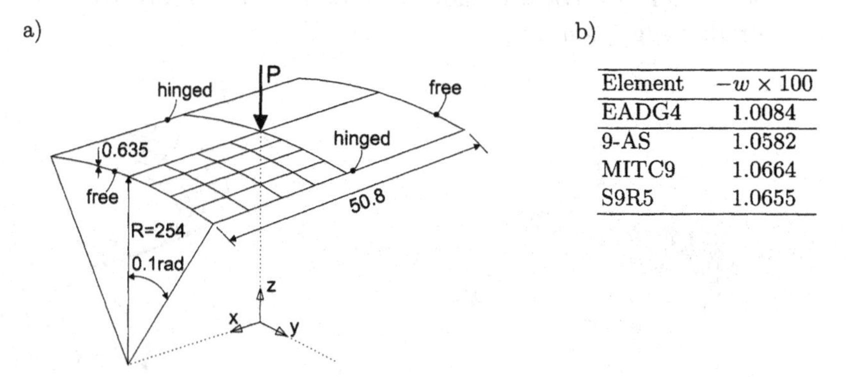

Element	$-w \times 100$
EADG4	1.0084
9-AS	1.0582
MITC9	1.0664
S9R5	1.0655

Fig. 15.23 Hinged cylindrical panel. $E = 310.275$, $\nu = 0.3$. a) Initial geometry and load. b) Linear solution for $P = 0.01$.

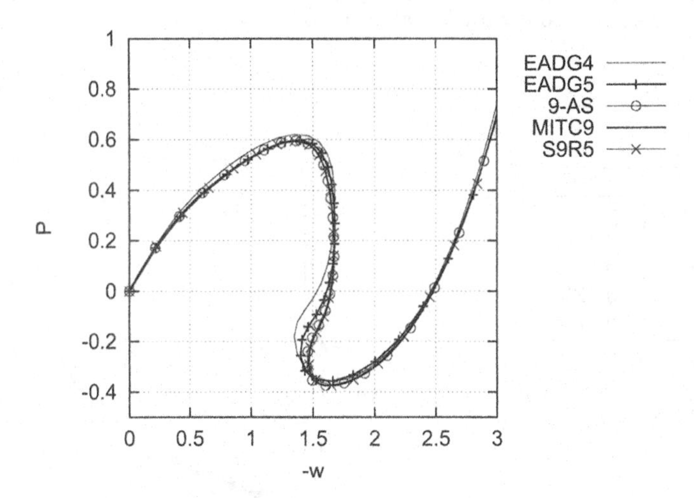

Fig. 15.24 Hinged cylindrical panel. Non-linear solutions.

The linear solution is given in Fig. 15.23b. The nonlinear solution was computed using the arc-length method for the initial $\Delta P = -0.05 \times 4$

and is shown in Fig. 15.24. In both analyses, the vertical displacement w under the force is monitored. We see that the EADG5 element is more accurate than the EADG4 element.

15.3.7 Slit open annular plate

The plate is slit open radially, with one end fully clamped and the other end free, see [14]. It is vertically loaded by forces p, uniformly distributed along the radial edge as in Fig. 15.25a.

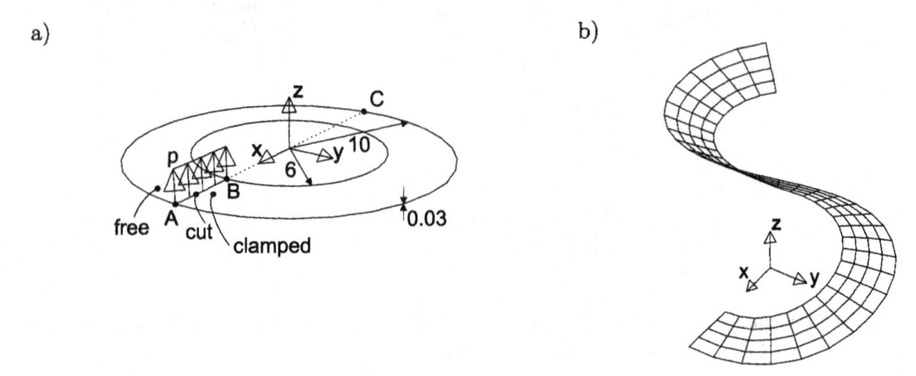

Fig. 15.25 Slit open annular plate. $E = 2.1 \times 10^8$, $\nu = 0$. a) Initial geometry and load. b) Deformed configuration at $p = 7.62$.

a)

Element	w
EADG4	0.1145
S4	0.1144
9-AS	0.1134
MITC9	0.1135
S9R5	0.1147

b)

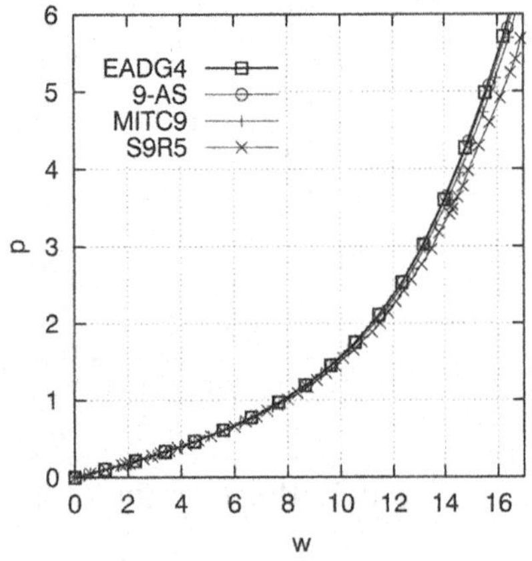

Fig. 15.26 Slit open annular plate. a) Linear solution. b) Non-linear solutions.

A 32 × 4 mesh of four-node elements and 16 × 2 mesh of nine-node elements was used. The linear solution for $p = 0.01$ is given in Fig. 15.26a. The non-linear solution was obtained using the arc-length method with the initial $\Delta p = 0.1$, see Fig. 15.26b. In both analyses, the vertical displacement at point A is monitored.

15.3.8 Pinched hemispherical shell with hole

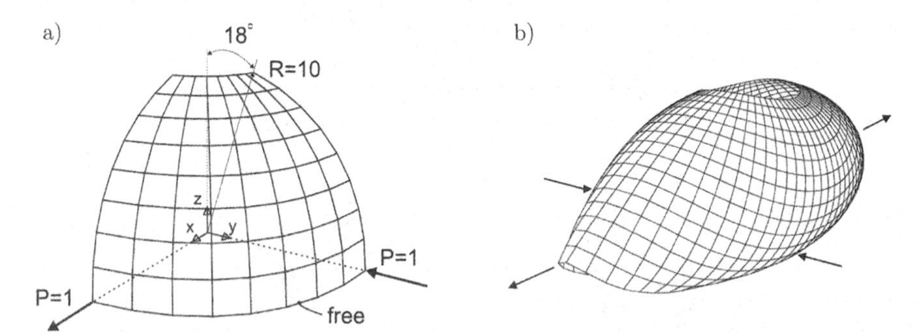

Fig. 15.27 Pinched hemispherical shell. $E = 6.825 \times 10^7$, $\nu = 0.3$, $h = 0.04$. a) Initial geometry and load. b) Deformed configuration at $P = 335$.

A hemispherical shell with an 18° hole is loaded by two pairs of equal but opposite external forces, applied in the plane $z = 0$, along the X and Y axes, see [143]. The shell undergoes strong bending but the deformation is almost in-extensional. Due to a double symmetry, only a quarter of the shell is modeled.

For the mesh shown in Fig. 15.27, the four-node elements are flat and trapezoidal. The 8 × 8 and 16 × 16-element meshes are used for four-node elements, and 4 × 4 and 8 × 8-element meshes for nine-node elements. The membrane locking of nine-node elements can be strong in this example [26].

The results of the linear analysis for four-node and nine-node elements are given in Table 15.10. The displacement at the point where the load is applied and in the direction of the load is reported. The reference value is taken from [143].

In a nonlinear analysis, the Newton method with $\Delta P = 10$ is used. The 16 × 16 mesh of four-node elements is used, and the 8 × 8 mesh of nine-node elements. Due to geometrical non-linearities, the displacements under the inward forces are larger than under the outward forces; the

Table 15.10 Pinched hemispherical shell with a hole. Linear analysis. $h = 0.04$. Displacement $-u_y \times 100$.

Element/Mesh	8×8	16×16
EADG4	9.1360	9.3005
EADG5	9.1299	9.2995
S4	9.2576	9.3018
Taylor [235]	9.4153	9.3501
Simo *et al.* [210]	9.2814	9.2907
D-type [109]	9.3701	9.3487
9-AS	9.3306	9.3473
MITC9	8.1762	8.5687
S9R5	9.3365	9.3513
Ref.	9.4000	

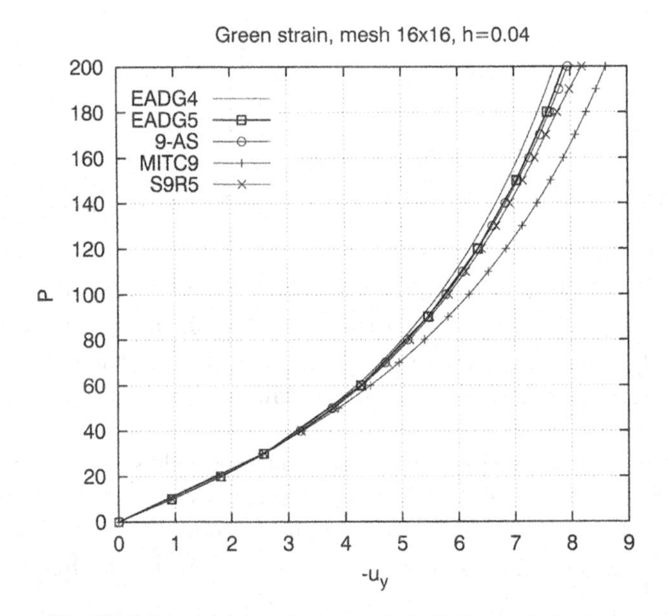

Fig. 15.28 Pinched hemispherical shell. Inward displacement.

former ones are shown in Fig. 15.28. We see that the EADG5 element is more accurate than the EADG4 element and yields the solution almost identical as the 9-AS element.

This difference is more visible for a thinner shell of $h = 0.01$, see Fig. 15.29. The "solid-shell" element based on the Hu–Washizu functional (SS-HW) and the S4 element perform identically. The EADG5 element

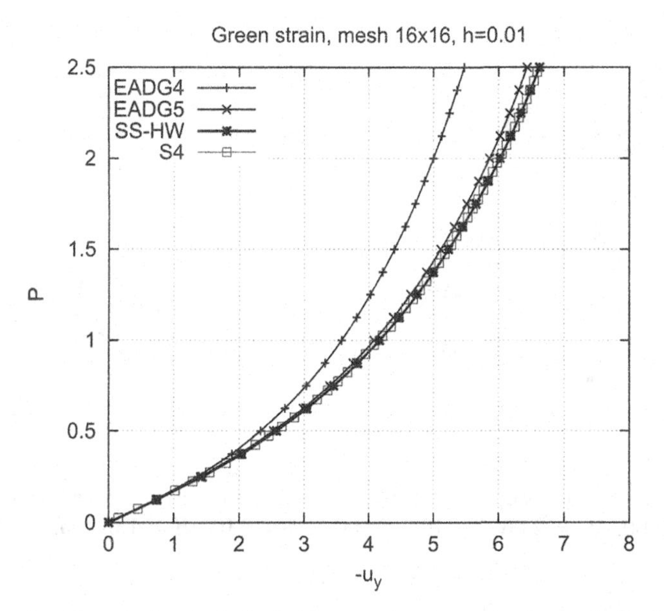

Fig. 15.29 Pinched hemispherical shell. Four-node elements. Inward displacement.

performs as the 9-AS element, while the solution by the EADG4 element is locked.

15.3.9 Pinched clamped cylinder

The cylinder shell is clamped at one end and loaded by two opposite forces P at the other end, see Fig. 15.30. The data of [35] is used; it is slightly different than this of [225].

Due to the symmetry, only a quarter of the cylinder is analyzed and two meshes of 16×16 and 32×32 elements are used. The drilling rotation is constrained at lateral boundaries. The EADG4 element is tested.

The non-zero displacements and rotation at point A obtained by the linear analysis for $P = 1$ are shown in Fig. 15.31a.

The nonlinear solution was obtained using the arc-length method for the initial $\Delta P = 100$. The vertical displacement u_y at point A is shown in Fig. 15.31b. For the displacement equal to the radius, the opposite points A and B come in contact, but we do not account for it.

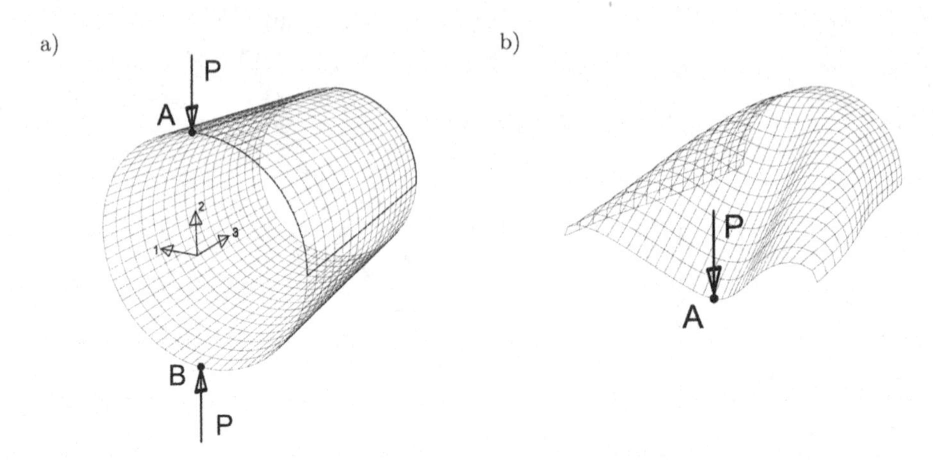

Fig. 15.30 Pinched clamped cylinder. a) Initial geometry. b) Deformed geometry at $P = 1600$. $E = 2.0685 \times 10^7$, $\nu = 0.3$, $h = 0.03$, $R = 1.016$, $L = 3R$.

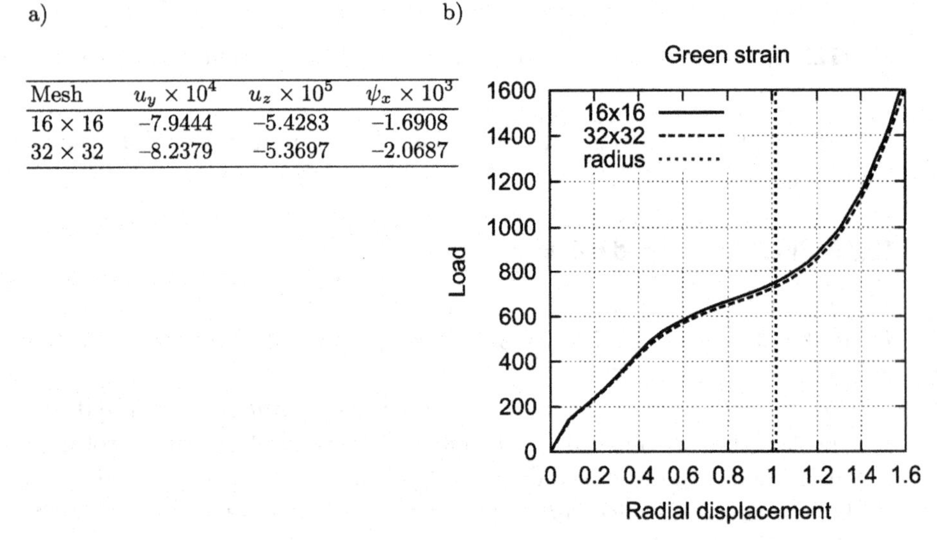

a)

Mesh	$u_y \times 10^4$	$u_z \times 10^5$	$\psi_x \times 10^3$
16×16	-7.9444	-5.4283	-1.6908
32×32	-8.2379	-5.3697	-2.0687

Fig. 15.31 Pinched clamped cylinder. a) Linear solution. b) Non-linear solutions.

15.3.10 Stretched cylinder with free ends

A cylindrical shell is stretched by two opposite forces P applied at the middle section and its boundaries are free, see Fig. 15.32. This is a popular test, see the review of earlier works in [51].

Because of symmetries, only one-eighth of the cylinder is analysed. The EADG4 element and the mesh of 12×8 elements (12 along the circumfer-

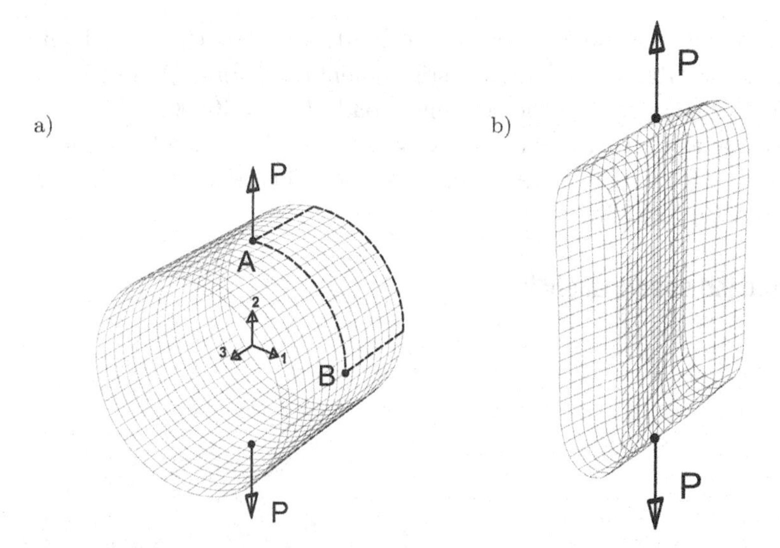

Fig. 15.32 Stretched cylinder. $E = 10.5 \times 10^6$, $\nu = 0.3125$, $h = 0.094$, radius $R = 4.953$, length $L = 10.35$. a) Initial geometry. b) Deformed geometry for load $\lambda = 0.935$.

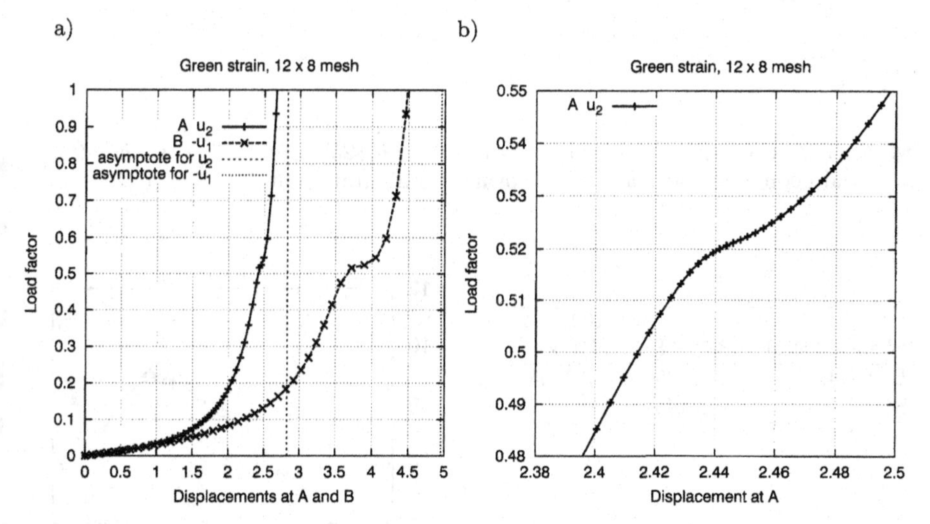

Fig. 15.33 Stretched cylinder. a) Displacement at two points. b) Vicinity of the deflection point.

ence) is used. For $P = 1$, the linear analysis yields $u_y = 1.1339 \times 10^{-3}$ at point A.

The non-linear solution was obtained using the arc-length method with the initial $\Delta P = 100$. The radial displacements at points A and B are shown in Fig. 15.33a, where the reference load $P_{\text{ref}} = 40000$. The maximum radial deflection at point A is $u_2 = (\pi/2 - 1)R = 2.827$, while at point B $u_1 = -R = -4.953$. A vicinity of the deflection point on curve A is shown in Fig. 15.33b.

15.3.11 Pinched spherical shell

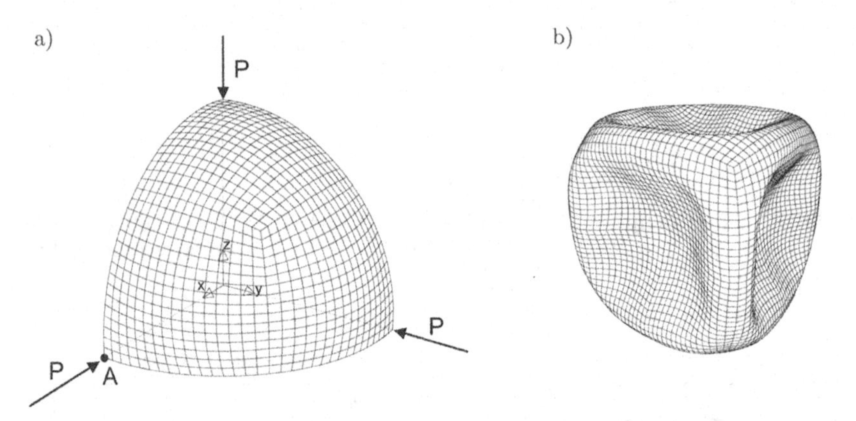

Fig. 15.34 Pinched spherical shell. $R = 10$, $E = 6.825 \times 10^7$, $\nu = 0.3$, $h = 0.2$. a) Initial geometry and load. b) Deformed configuration at $P = 5 \times 10^5$.

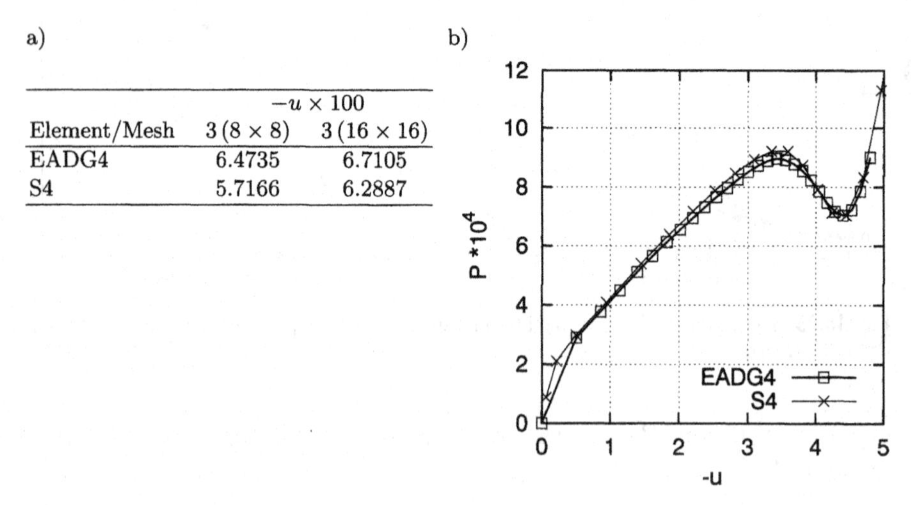

a)

	$-u \times 100$	
Element/Mesh	$3\,(8 \times 8)$	$3\,(16 \times 16)$
EADG4	6.4735	6.7105
S4	5.7166	6.2887

Fig. 15.35 Pinched spherical shell. a) Linear solutions. b) Non-linear solutions.

A full spherical shell is compressed by three pairs of pinching forces P applied along the x, y and z axes. Because of the symmetries, only one-eighth of the sphere is modeled, see Fig. 15.34a.

The mesh is composed of three parts, with either 8×8 or 16×16 elements in each. The radial displacement at point A is monitored.

The linear results obtained for $P = 10^4$ are presented in Fig. 15.35a. The non-linear analysis was performed using the arc-length method with the initial $\Delta P = 10^4$, and the mesh of $3\,(16 \times 16)$ elements was used. The deformed configuration is shown in Fig. 15.34b, while the solution curves for point A are presented in Fig. 15.35b.

15.3.12 Short channel section beam

A short C-beam is fully clamped at one end and loaded by a vertical force P at the other, see Fig. 15.36a. At the clamped end, displacements and rotations are constrained to zero. This test was proposed in [52].

For the four-node elements, a $(2+6+2) \times 36$ mesh is used, where the web is modeled by 36×6 elements and each flange by 36×2 elements. For the nine-node elements, the $(1+3+1) \times 18$ mesh is used.

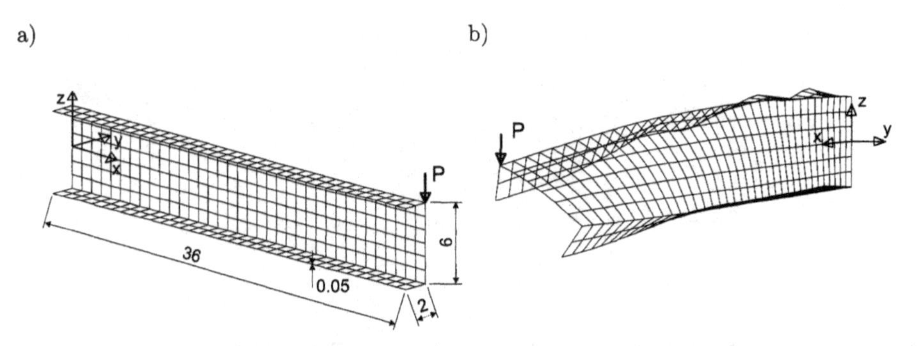

Fig. 15.36 Short channel section beam. $E = 10^7$, $\nu = 0.333$, $h = 0.05$. a) Initial geometry and load. b) Deformed configuration at force $P = 112$.

The linear solutions for $P = 1$ are given in Fig. 15.37a. The reference value was computed in [51], using the $(2+3+2) \times 9$ mesh of the 16-node CAM elements. The non-linear solution is computed using the arc-length method with the initial $\Delta P = 20$ and is shown in Fig. 15.37b. The vertical displacement at the point where the force is applied is monitored. The solution for the EADG5 element is closer to the solutions for nine-node elements than the EADG4 element.

a)

Element	$-w \times 10^3$
EADG4	1.1541
9-AS	1.1556
MITC9	1.2839
S9R5	1.1628
Ref.	1.1544

b)

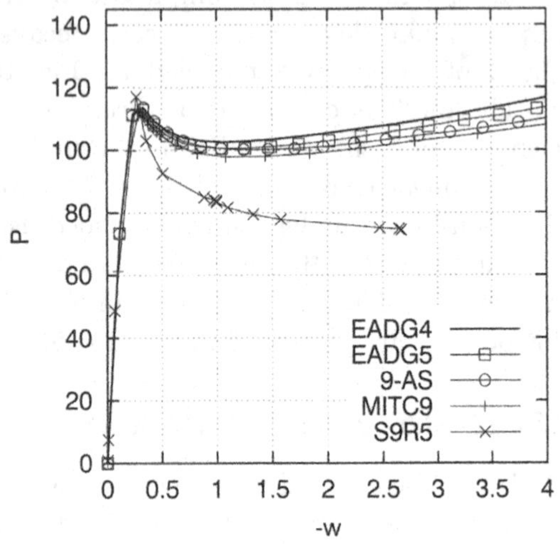

Fig. 15.37 Short channel section beam. a) Linear solutions. b) Non-linear solutions.

15.3.13 Long channel section beam

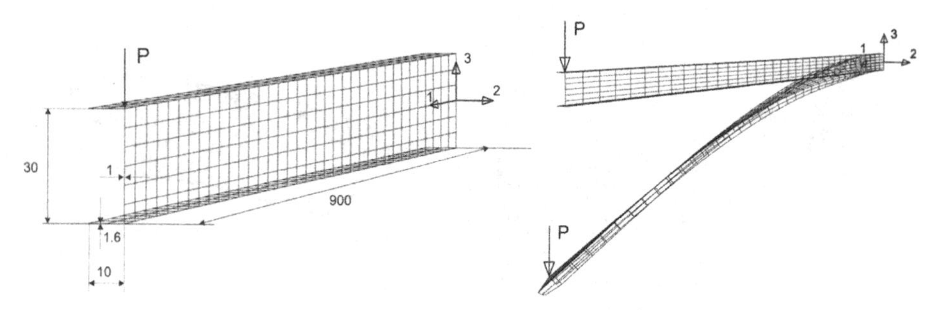

Fig. 15.38 Long channel section beam. $E = 21000$, $\nu = 0.3$.
a) Initial geometry and load. b) Deformed configuration at force $P = 20$.

A long C-beam is fully clamped at one end and loaded by a vertical force P at the other end, see Fig. 15.38. This test was proposed in [243]. The behavior of the long beam is very different from that of the short beam of Sect. 15.3.12, as the global response dominates.

Two meshes are used. The coarse mesh with 360 four-node elements, where each flange is modeled by 36×2 elements and the web by 36×6 elements. The fine mesh is two times denser in each direction, which

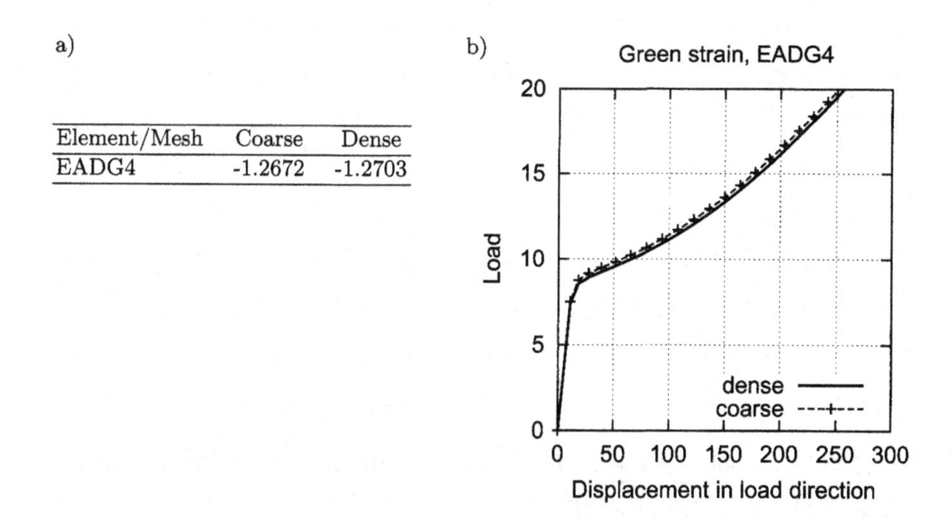
a)

Element/Mesh	Coarse	Dense
EADG4	-1.2672	-1.2703

Fig. 15.39 Long channel section beam. a) Linear solution. b) Non-linear solutions.

yields 1440 elements. At the clamped end, displacements and rotations are constrained.

The linear solution for $P = 1$ is shown in Fig. 15.39a. The nonlinear solution was obtained using the arc-length method with the initial $\Delta P = 5$ and is shown in Fig 15.39b. The vertical displacement at the point where the force is applied is monitored.

15.3.14 Hyperboloidal shell

The hyperboloidal shell is loaded by two pairs of equal but opposite external forces, applied in the symmetry plane $z = 0$, along the X and Y axes. Due to symmetry, only one octant of it is analyzed, see Fig. 15.40. This test was proposed in [15] for laminated shells, but, here, the isotropic SVK material is used.

The octant of the shell is meshed by 16×16 elements. The linear solution for $P = 1$ are shown in Tab.15.41a, and the radial displacement at point B is reported. The nonlinear solution was obtained using the Newton method, with the force increment $\Delta P = 40$ for the EADG4 element and $\Delta P = 12$ for the S4 element. The radial displacements at point A and B are shown in Fig. 15.41b.

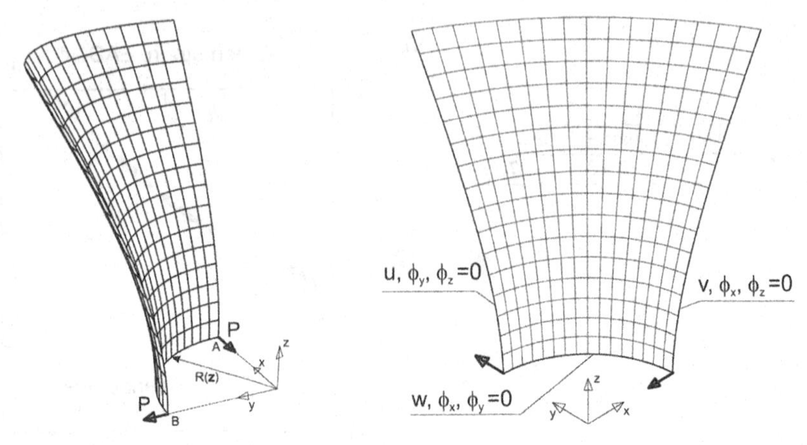

Fig. 15.40 Hyperboloidal shell. $H = 20$, $h = 0.04$, $E = 4.0 \times 10^7$, $\nu = 0.25$, radius $R(z) = 7.5\sqrt{1 + 3\,(z/20)^2}$.

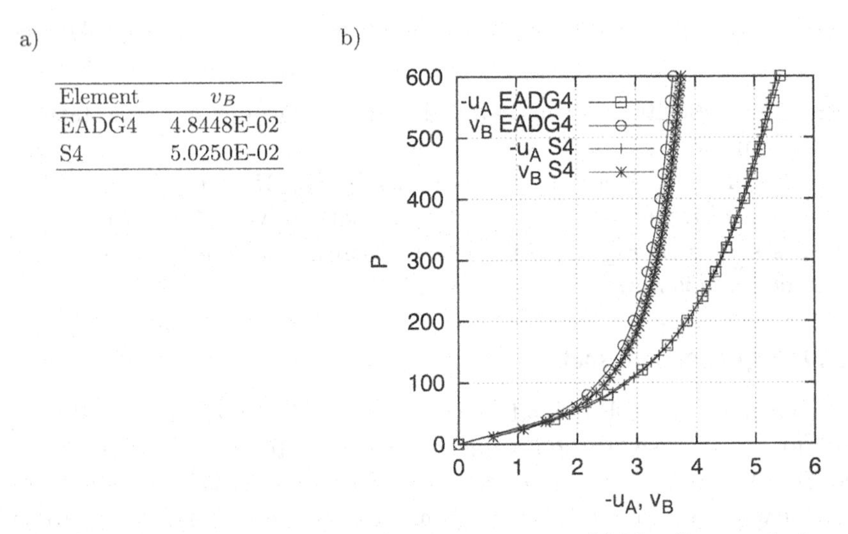

a)

Element	v_B
EADG4	4.8448E-02
S4	5.0250E-02

Fig. 15.41 Hyperboloidal shell. a) Linear solutions. b) Non-linear solutions.

15.3.15 Twisted ring

The ring is twisted by a moment M_x applied at point A and is clamped at the opposite point B, both points on the X-axis, see Fig. 15.42. This test was proposed in [80] and is difficult because finite rotations and twisting are involved, see Fig. 15.43. It is a tough test for the path-following procedure.

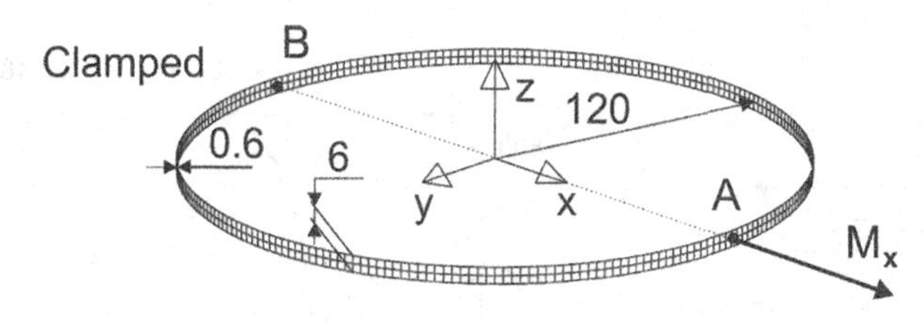

Fig. 15.42 Twisted ring. $E = 2 \times 10^5$, $\nu = 0.3$, $h = 0.6$, width $w = 6$, $R = 120$.

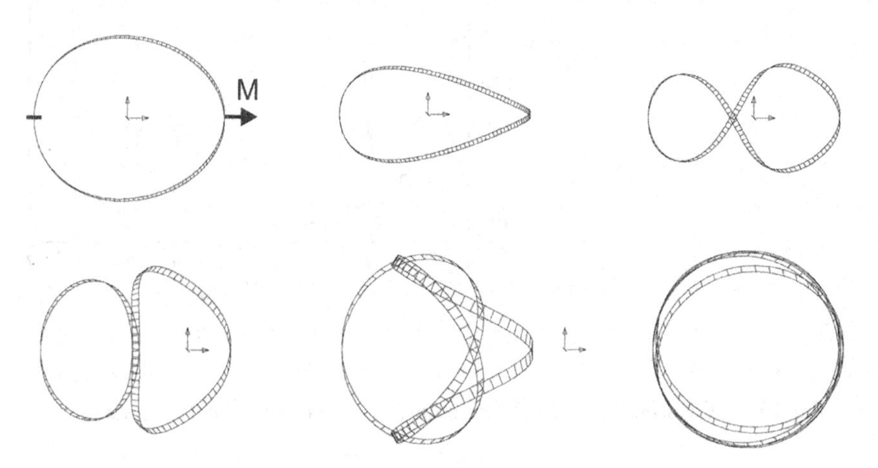

Fig. 15.43 Twisted ring. Subsequent deformed shapes.

According to [80], twisting yields a small ring with a diameter equal to one-third of the original one, and the "final shape is kept without any external force". We experimented with a paper ring and obtained a coil which cannot be flattened.

The mesh of 2×248 four-node elements is used. The twisting moment M_x is applied using a small auxiliary square plate, as in [189]. The plate is normal to the ring and has the size 6×6, and a thickness $10\ h$.

The linear solution for $M_x = 1$ is given in Fig. 15.44a, where the drilling rotation ψ_x at point A is reported.

The non-linear solution was computed using the arc-length method for the initial $M_x^{\text{ref}} = 50$. The drilling rotation r_x and the radial displacement u_x at point A are shown in Fig. 15.44b. Recall that in Sect. 9.3.5, this ring is computed by 3D beam elements and various update schemes for rotations are considered.

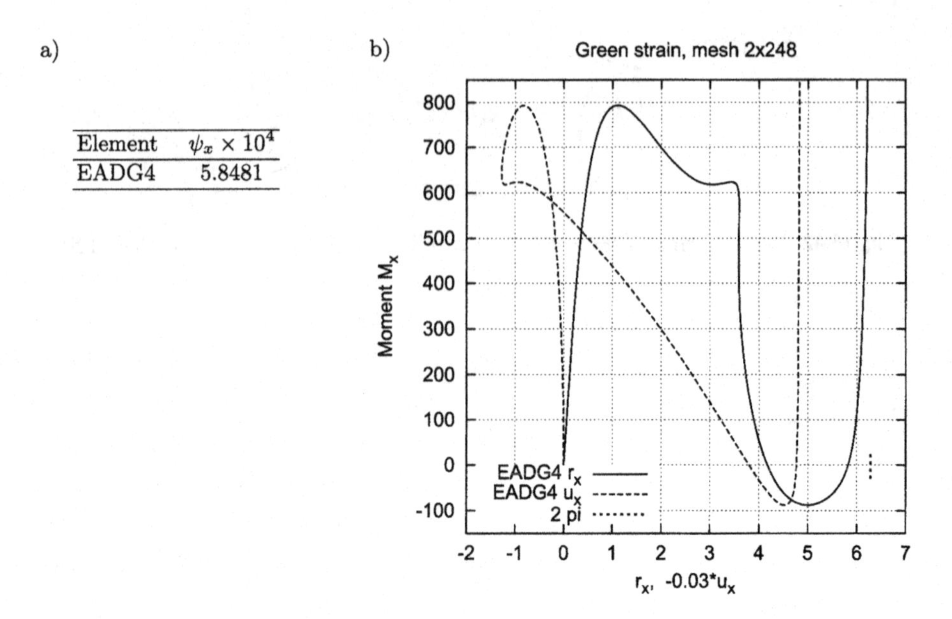

a)

Element	$\psi_x \times 10^4$
EADG4	5.8481

Fig. 15.44 Twisted ring. a) Linear solution. b) Non-linear solutions.

References

1. Allman D.J.: *A compatible triangular element including vertex rotations for plane elasticity analysis.* Computers & Structures, Vol. 19, No. 2, 1–8 (1984)
2. Altman S.L.: *Rotations, Quaternions and Double Groups.* Clarendon Press, Oxford, 1986
3. Aminpour M.A.: *An assumed stress hybrid 4-node shell element with drilling degrees of freedom.* Int. J. Num. Meth. Engng., Vol. 33, 19–38 (1992)
4. Angeles J.: *Rational Kinematics.* Springer-Verlag, 1988
5. Argyris J.: *An excursion into large rotations.* Comput. Methods Appl. Mech. Engng., Vol. 32, 85–155 (1982)
6. Argyris J., Balmer H., Doltsinis J.St., Dunne P.C., Haase M., Kleiber M., Malejannakis G.A., Mlejenek J.P., Muller M., Scharp D.W.: *Finite element method - the natural approach.* Comput. Methods Appl. Mech. Engng., Vol. 17/18, 1–106 (1979)
7. Argyris J.H., Kelsey S.: *Modern fuselage analysis and the elastic aircraft.* Butterworth, London, 1963
8. Argyris J., Poterasu V.F.: *Large rotations revised application of Lie algebra.* Comput. Methods Appl. Mech. Engng., Vol. 103, 11–42 (1993)
9. Atluri S.N., Cazzani A.: *Rotations in computational solid mechanics.* Archives of Computational Methods in Engineering, Vol. 2, No. 1, 49–138 (1995)
10. Atluri S.N., Murakawa H.: *On hybrid finite element models in nonlinear solid mechanics.* In: Bergan P.G. *et al.* (eds.) Finite Elements in Nonlinear Mechanics, Vol. 1, Tapir, Trondheim, 3–40 (1977)
11. Babuška I.: *Error-Bounds for the Finite Element Method,* Numer. Math., Vol. 16, 322–333 (1971)
12. Babuška I.: (edited and typed by L. Vardapetyan. I. Yotow) *On the inf-sup (Babuška–Brezzi) condition,* TICAM Forum No. 5 (October 1996)
13. Badur J., Pietraszkiewicz W.: *On geometrically non-linear theory of elastic shells derived from pseudo-Cosserat continuum with constrained micro-rotations.* In: Pietraszkiewicz W. (ed.) Finite Rotations in Structural Mechanics, 19–32, Springer, Berlin, 1986
14. Basar Y., Ding Y.: *Finite-Rotation Shell Elements for the Analysis of Finite-Rotation Shell Problems.* Int. J. Num. Meth. Engng., Vol. 34, 165–169 (1992)

15. Basar Y., Ding Y., Schultz R.: *Refined shear deformation models for composite laminates with finite rotations.* Int. J. Solids Structures, Vol. 30, 2611–2638 (1993)

16. Bathe K.J.: *Finite element procedures.* Prentice-Hall, Englewood-Cliffs, 1996

17. Bathe K.J.: *The inf-sup condition and its evaluation for mixed finite element methods.* Computers & Structures, Vol. 79, 243–252 (2001)

18. Bathe K-J., Dvorkin E.N.: *A four-node plate bending element based on Mindlin-Reissner plate theory and mixed interpolation.* Int. J. Num. Meth. Engng., Vol. 21, 367–383 (1985)

19. Bathe K-J., Dvorkin E.N.: *A formulation of general shell elements. The use of mixed interpolations of tensorial components.* Int. J. Num. Meth. Engng., Vol. 22, 697-722 (1986)

20. Batoz J.-L., Tahar M.B.: *Evaluation of a new quadrilateral thin plate bending element.* Int. J. Num. Meth. Engng., Vol. 18, 1655–1677 (1982)

21. Bazeley G.P., Cheung Y.K., Irons B.M., Zienkiewicz O.C.: *Triangular elements in bending - conforming and non-conforming solutions.* Proc. 1st Conference on Matrix Methods in Structural Mechanics. Air Force Inst. Tech., Wright-Patterson AF Base, Dayton Ohio, 1966

22. Belytschko T., Leviathan I.: *Physical stabilization of 4-node shell element with one-point quadrature.* Comput. Methods Appl. Mech. Engng., Vol. 113, 321–350 (1994)

23. Belytschko T., Liu W.K., Moran B.: *Nonlinear Finite Elements for Continua and Structures.* Wiley, Chichester, 2000

24. Belytschko T., Liu W.K., Ong J.S.: *Mixed variational principles and stabilization of spurious modes in the 9-node element,* Comput. Methods Appl. Mech. Engng., Vol. 62, 275–292 (1987)

25. Belytschko T., Ong J.S., Liu W.K.: *A consistent control of spurious singular modes in the 9-node Lagrangian element for the Laplace and Mindlin plate equations.* Comput. Methods Appl. Mech. Engng., Vol. 44, 269–295 (1985)

26. Belytschko T., Wong B.L., Stolarski H.: *Assumed strain stabilization procedure for the 9-node Lagrange Shell element.* Int. J. Num. Meth. Engng., Vol. 28, 385–414 (1989)

27. Bergan P.G., Felippa C.A.: *A traingular membrane element with rotational degrees of freedom.* Comput. Methods Appl. Mech. Engng., Vol. 50, 25–60 (1985)

28. Bergan P.G., Horrigmoe G., Krakeland B., Soreide T.H.: *Solution techniques for nonlinear finite element problems.* Int. J. Num. Meth. Engng., Vol. 12, 1677–1696 (1978)

29. Bertsekas D.P.: *Nonlinear programming.* Athena Scentific, 1995

30. Betsch P., Menzel A., Stein E.: *On the parametrization of finite rotations in computational mechanics. A classification of concepts with application to smooth shells.* Comput. Methods Appl. Mech. Engng., Vol. 155, 273–305

(1998)

31. Betsch P., Stein E.: *An assumed strain approach avoiding artificial thickness straining for a non-linear 4-node shell element.* Commun. Numer. Methods Engrg., Vol. 11, 899–909 (1995)

32. Bischoff M., Ramm E.: *Shear deformable shell elements for large strains and rotations.* Int. J. Num. Meth. Engng., Vol. 40, 4427–4449 (1997)

33. deBoer R.: *Vector- und Tensorrechnung für Ingenieure.* Springer, 1982

34. Bowen R.M., Wang C.-C.: *Introduction to vectors and tensors.* Vol. 1 and 2. Plenum Press, 1976

35. Brank B., Peric D., Damjanic F.B.: *On implementation of a non-linear four node shell finite element for thin multi-layered elastic shells.* Comput. Mech., Vol. 16, 341–359 (1995)

36. Brebbia C.A., Connor J.J: *Fundamentals of finite element techniques.* John Wiley and Sons, 1974

37. Brezzi F., Bathe K-J.: *A discourse on the stability conditions for FE formulations,* Comput. Methods Appl. Mech. Engng., Vol. 82, 27–57 (1990)

38. Brezzi F., Fortin M.: *Mixed and hybrid finite element methods.* Springer, New York, 1991

39. Bucalem M.L., Bathe K-J.: *Higher-order MITC general shell elements.* Int. J. Num. Meth. Engng., Vol. 36, 3729–3754 (1993)

40. Buechter N., Ramm E.: *Shell theory versus degeneration - a comparison in large rotation finite element analysis.* Int. J. Num. Meth. Engng., Vol. 34, 39–59 (1992)

41. Buechter N., Ramm E., Roehl D.: *Three-dimensional extension of nonlinear shell formulation based on the EAS concept.* Int. J. Num. Meth. Engng., Vol. 37, 2551–2564 (1994)

42. Bufler H.: *The Biot stresses in nonlinear elasticity and associated generalized variational principles.* Ing.-Arch., Vol. 55, 450–462 (1985)

43. Bufler H.: *On drilling degrees of freedom in nonlinear elasticity and a hyperelastic material description in terms of the stretch tensor. Part I: Theory.* Acta Mechanica, Vol. 113, 21–35 (1995)

44. Cardona A., Geradin M.: *A beam finite element nonlinear theory with finite rotations.* Int. J. Num. Meth. Engng., Vol. 26, 2403–2438 (1988)

45. Cartan E.: *The theory of spinors.* Dover, 1981

46. Chapelle D., Bathe K.J.: *The inf-sup test.* Computers & Structures, Vol. 47, Nos. 4/5, 537–545 (1993)

47. Chapelle D., Bathe K.J.: *The finite element analysis of shells - Fundamentals.* Springer, Berlin, 2003

48. Chernykh K.F.: *Nonlinear theory of isotropically elastic thin shells* (in Russian), Mekh. Tverdogo Tela, Vol. 15, 148–159 (1980)

49. Chernykh K.F.: *The theory of thin shells of elastomers (in Russian),* Advances in Mechanics, Vol. 6, 111–147 (1983)

50. Cheung Y.K., Lo S.H., Leung A.Y.T.: *Finite Element Implementation.* Blackwell Science, Oxford, 1996

51. Chróścielewski J., Makowski J., Pietraszkiewicz W.: *Statics and dynamics of multi-segmented shells. Nonlinear theory and finite element method.* IFTR PAS Publisher, Warsaw, 2004 (in Polish)

52. Chróścielewski J., Makowski J., Stumpf H.: *Genuinely resultant shell finite elements accounting for geometric and material nonlinearity.* Int. J. Num. Meth. Engng., Vol. 35, 63–94 (1992)

53. Chróścielewski J., Witkowski W.: *Four-node semi-EAS element in six-field nonlinear theory of shells.* Int. J. Num. Meth. Engng., Vol. 68, 1137–1179 (2006)

54. Cook R.D.: *Improved two dimensional finite element.* J. Struct. Div. ASCE, Vol. 100, 1851-1863 (1976)

55. Cook R.D.: *On the Allmann triangle and a related quadrilateral element. A plane hybrid element with rotational d.o.f. and adjustable stiffness.* Computers & Structures, Vol. 22, No. 6, 1065–1067 (1986)

56. Cook R.D.: *A plane hybrid element with rotational d.o.f. and adjustable stiffness.* Int. J. Num. Meth. Engng., Vol. 24, 1499–1508 (1987)

57. Cook R.D.: *Four-node 'flat' shell element: drilling degrees of freedom, membrane-bending coupled, warped geometry and behavior.* Computers & Structures, Vol. 50, 549-555 (1994)

58. Cook R.D., Malkus D.S., Plesha M.E.: *Concepts and applications of finite element analysis.* Fourth edition. Wiley, New York, 2001

59. Cosserat E., Cosserat F.: *Theorie des Corps deformables.* Herman, Paris, 1909

60. Crisfield M.A.: *A consistent co-rotational formulation for non-linear three-dimensional beam elements.* Comput. Methods Appl. Mech. Engng., Vol. 81, No. 2, 131–150 (1990)

61. Crisfield M.A.: *Non-linear Finite Element Analysis of Solids and Structures, Vol. 1.* John Wiley, Chichester, 1991.

62. Crisfield M.A.: *Non-linear Finite Element Analysis of Solids and Structures, Vol. 2.* John Wiley, Chichester, 1997.

63. Crisfield M.A., Moita G.F., Jelenic G., Lyons L.P.R.: *Enhanced lower-order element formulations for large strains.* In: Owen D.R.J. and Onate E. (eds.) "Computational Plasticity - Fundamentals and Applications", Pineridge Press, Swansea, 293–320, 1995

64. Crisfield M.A., Moita G.F.: *A unified co-rotational framework for solids, shells and beams.* Int. J. Solids Structures, Vol. 33, No. 20-22, 2969–2992 (1996)

65. Dennis J.E., Schnabel R.B.: *Numerical Methods for Unconstrained Optimization and Nonlinear Equations.* Prentice-Hall, 1983

66. Donnell L.H.: *A new theory for the buckling of thin cylinders under axial compression and bending.* Trans. Am. Soc. Mech. Engrs, Vol. 56, 795 (1934)

67. Dvorkin E.N., Bathe K-J.: *A continuum mechanics based four-node shell element for general nonlinear analysis.* Engng. Comput., Vol. 1, 77–88 (1984)

68. Eberlein R., Wriggers P.: *Finite element concepts for elastoplastic strains and isotropic stress response in shells: theoretical and computational analysis.* Comput. Methods Appl. Mech. Engng., Vol. 171, 243–279 (1999)

69. Felippa C.A.: *Error analysis of penalty function techniques for constraint definition in linear algebraic system.* Int. J. Num. Meth. Engng., Vol. 11, 709–728 (1977)

70. Felippa C.A.: *Iterative procedures for improving penalty function solutions of algebraic systems,* Int. J. Num. Meth. Engng., Vol. 12, 821–836 (1978)

71. Figueiras J.A., Owen P.R.: *Analysis of elasto-plastic and geometrically nonlinear anisotropic plates and shells,* In: Hinton E., Owen D.R.J. (eds.) "Finite Element Software for Plates and Shells" Pineridge Press Ltd., Swansea, 1984

72. Fletcher R.: *Practical methods of optimization. 2nd edition.* J. Wiley and Sons, 1987

73. Fox D.D., Simo J.C.: *A drill rotation formulation for geometrically exact shells.* Comput. Methods Appl. Mech. Engng., Vol. 98, 329–343 (1992)

74. Fraeijs de Veubeke B.: *A new variational principle for finite elastic displacements.* Int. J. Engng. Sci., Vol. 10, 745–763 (1972)

75. Franca L.P.: *An algorithm to compute the square root of a* 3×3 *positive definite matrix.* Computers and Mathematics with Applications, Vol. 18, 459–466 (1989)

76. Geradin M., Cardona A.: *Flexible Multibody Dynamics.* J. Wiley & Sons, 2001

77. Geradin M., Rixen D.: *Parametrization of finite rotations in computational dynamics. A review.* Europ. J. Finite Elem., Special Issue: Ibrahimbegovic A., Geradin M. (eds.), Vol. 4, 497–553 (1995)

78. Goldenveizer A.L.: *Theory of thin elastic shells.* Pergamon Press, 1961

79. Goldstein H.: *Classical Mechanics.* 2nd Edition. Addison Wesley, Reading, Mass. (1980)

80. Goto Y., Watanabe Y., Kasugai T., Obata M.: *Elastic Buckling Phenomenon Applicable to Deployable Rings.* Int. J. Solids Structures, Vol. 29, 893–909 (1992)

81. Green A.E., Adkins J.E.: *Large Elastic Deformations.* 2nd Edition. Oxford Univ. Press, Oxford, 1970

82. Griewank A.: *On Automatic Differentiation, Mathematical Programming: Recent Developments and Applications.* Kluwer Academic Publisher, Dordrecht, 1989.

83. Griewank A.: *Evaluating Derivatives: Principles and Techniques of Algorithmic Differentiation.* SIAM, Philadelphia, 2000.

84. Griewank A., Corliss G.F., ed., *Automatic Differentiation of Algorithms: Theory, Implementation, and Application.* SIAM, Philadelphia, 1991.

85. Gruttmann F., Stein E., Wriggers P.: *Theory and numerics of thin elastic shells with finite rotations.* Ing. Archive, Vol. 59, 54–67 (1989)

86. Gruttmann F., Taylor R.L.: *Theory and finite element formulation of rubber-like membrane shells using principal stretches.* Int. J. Num. Meth. Engng., Vol. 35, 1111–1126 (1992)

87. Gruttmann F., Wagner W., Wriggers P.: *A nonlinear quadrilateral shell element with drilling degrees of freedom.* Archives of Applied Mechanics, Vol. 62, 474–486 (1992)

88. Haftka R.T., Robinson C.J.: *Effect of out-of-planeness of membrane quadrilateral finite elements.* AIAA J., Vol. 11, No. 5, 742–744 (1973)

89. Hallquist J.O., Benson D.J., Goudreau G.L.: *Implementation of a modified Hughes-Liu shell into a fully vectorized explicit finite element code.* Proc. Int. Symp. Finite Element Methods for Nonlinear Problems, Trondheim (1985)

90. Hassenpflug W.C.: *Rotation angles.* Comput. Methods Appl. Mech. Engng., Vol. 105, 111–124 (1993)

91. Hinton E., Abdel Rahman H.H.: *Mindlin plate finite elements.* In: Hinton E., Owen D.R.J. (eds) "Finite Element Software for Plates and Shells", Pineridge Press, Swansea, 157–233, 1984

92. Hoff C.C., Harder R.L., Campbell G., MacNeal R.H., Wilson C.T.: *Analysis of shell structures using MSC/NASTRAN's shell elements with surface normals.* Proc. 1995 MSC World Users' Conf. Universal City, CA, May 8-12, Paper No. 26 (1995)

93. Holzapfel G. A.: *Nonlinear solid mechanics. A continuum approach for engineering.* John Wiley, Chichester, 2000

94. Huang H.-Ch.: *Static and dynamic analyses of plates and shells. Theory, Software and Applications.* Springer, London, 1989

95. Huang H.C., Hinton E.: *A nine node Lagrangian Mindlin plate element with enhanced shear interpolation.* Engng. Comput., Vol. 1, 369–379 (1984)

96. Huang H.C., Hinton E.: *A new nine node degenerated shell element with enhanced membrane and shear interpolation.* Int. J. Num. Meth. Engng., Vol. 22, 73–92 (1986)

97. Hughes T.J.R.: *Numerical implementation of constitutive models: rate independent deviatoric plasticity.* In: Nemat-Nasser S. *et al.* (eds.) "Theoretical Foundations for Large-Scale Computing for Nonlinear Material Behaviour", 29–57. Nijhoff, Dordrecht, 1984

98. Hughes, T.J.R.: *The Finite Element Method. Linear Static and Dynamic Finite Element Analysis.* Prentice-Hall, Englewood Cliffs, 1987

99. Hughes T.J.R., Brezzi F.: *On drilling degrees of freedom.* Comput. Methods Appl. Mech. Engng., Vol. 72, 105–121 (1989)

100. Hughes T.J.R., Liu W.K.: *Nonlinear finite element analysis of shells. Part I. Three-dimensional shells.* Comput. Methods Appl. Mech. Engng., Vol. 26, 331–362 (1981)

101. Hughes T.J.R., Liu W.K.: *Nonlinear finite element analysis of shells. Part II. Two-dimensional shells.* Comput. Methods Appl. Mech. Engng., Vol. 27, 167–182 (1981)

102. Hughes T.J.R., Masud A., Harari I.: *Numerical assessment of some membrane elements with drilling degrees of freedom.* Computers & Structures, Vol. 55, 297–314 (1995)

103. Hughes T.J.R., Taylor R.L., Kanok-Nukulchai W.: *A simple and efficient finite element for plate bending.* Int. J. Num. Meth. Engng., Vol. 11, 1529–1543 (1977)

104. Hughes T.J.R., Tezduyar T.E.: *Finite elements based upon Mindlin plate theory with particular reference to the four-node isoparametric element.* J. Appl. Mech., Vol. 48, 587–596 (1981)

105. Hughes T.J.R., Winget J.: *Finite rotation effects in numerical integration of rate constitutive equations arising in large-deformation analysis.* Int. J. Num. Meth. Engng., Vol. 15, 1862–1867 (1980)

106. Ibrahimbegovic A.: *Stress resultant geometrically nonlinear shell theory with drilling rotations. Part 1. A consistent formulation.* Comput. Methods Appl. Mech. Engng., Vol. 118, 265–284 (1994)

107. Ibrahimbegovic A.: *On the choice of finite rotation parameters.* Comput. Methods Appl. Mech. Engng., Vol. 149, 49–71 (1997)

108. Ibrahimbegovic A., Frey F.: *Stress resultant geometrically nonlinear shell theory with drilling rotations. Part II. Computational aspects.* Comput. Methods Appl. Mech. Engng., Vol. 118, 285–308 (1994)

109. Ibrahimbegovic A., Taylor R.L., Wilson E.: *A robust quadrilateral membrane finite element with drilling degrees of freedom.* Int. J. Num. Meth. Engng., Vol. 30, 445–457 (1990).

110. Irons B.M.: *Numerical integration applied to finite element methods.* Conference on Use of Digital Computers in Structural Engineering, Univ. of Newcastle, 1966

111. Irons B.N., Razzaque A.: *Experience with the patch test.* In: A.R. Aziz (ed.) "Mathematical Foundations of the Finite Element Method with Applications to Partial Differential Equations", 557–587, Academic Press, New York, 1972

112. Iura M., Atluri S.N.: *Formulation of a membrane finite element with drilling degrees of freedom.* Comput. Mech., Vol. 9, 417–428 (1992)

113. Jaamei S., Frey F.R., Jetteur P.: *Nonlinear thin shell finite element with six degrees of freedom per node.* Comput. Methods Appl. Mech. Engng., Vol. 75, 251–266 (1989)

114. Jang J., Pinsky P.M.: *An assumed covariant strain based 9-node shell element.* Int. J. Num. Meth. Engng., Vol. 24, 2389–2411 (1987)

115. Jaunzemis W.: *Continuum mechanics.* MacMillan, London, 1967

116. Jemioło S.: *A study on hyperelastic properties of isotropic materials. Modelling and numerical implementation* (in Polish). Warsaw Technical University, 2002

117. Jetteur P., Frey F.R.: *A four-node Marguerre element for non-linear shell analysis.* Engng. Comput., 276–282 (1986)

118. Jetteur P.: *Improvement of the quadrilateral JET shell element for a particular class of shell problems.* IREM Internal Report 87/1, Ecole Politechnique Federale de Lausanne, February 1987.

119. John F.: *Estimates for the derivatives of the stresses in a thin shell and interior shell equations.* Comm. Pure and Appl. Math., Vol. 18, 235–267 (1965)

120. John F.: *Refined interior equations for thin elastic shells.* Comm. Pure and Appl. Math., Vol. 24, 585–615 (1971)

121. Jones E.: *A generalization of the direct-stiffness method of structural analysis.* AIAA J., Vol. 2, No. 5, 821–826 (1964)

122. Kane T.R., Likins P.W., Levinson D.A.: *Spacecraft Dynamics.* McGraw-Hill, New York, 1983

123. Kanok-Nukulchai W.: *A simple and efficient finite element for general shell analysis.* Int. J. Num. Meth. Engng., Vol. 14, 179–200 (1979)

124. Kirchhoff G.R.: *Mechanik.* Second ed. Berlin, 1877

125. Kleiber M.: *Incremental Finite Element Modeling in Non-Linear Solid Mechanics.* Ellis Horwood, Chichester, 1989

126. Knight N.F. Jr.: *The Raasch Challange for shell elements.* 37th AIAA/ASME/ASCE/AHS/ASC Structures, Structural Dynamics, and Material Conference. Salt Lake City, UT, April 15-17, CP962, 450–460, 1996

127. Koiter W.T.: *A Consistent First Approximation in the General Theory of Thin Elastic Shells.* In: Koiter W.T. (ed.) Proc. of the Symposium in the Theory of Thin Elastic Shells, Delft, August 1959, 12–33. North-Holland, Amsterdam, 1960

128. Koiter W.T.: *Couple-stresses in the theory of elasticity.* Proc. Kon. Ned. Ak. Wet., Ser. B, Vol. 67, No. 1, 17–48 (1964)

129. Koiter W.T.: *On the nonlinear theory of thin elastic shells.* Proc. Kon. Ned. Ak. Wet., Ser. B, Vol. 69, No. 1, 1–54 (1966)

130. Korelc J.: Automatic generation of finite-element code by simultaneous optimization of expressions. Theoretical Computer Science, Vol. 187, 231–248

(1997)

131. Korelc J.: *Multi-language and multi-environment generation of nonlinear finite element codes.* Engineering with Computers, Vol. 18, 312–327 (2002)

132. Korelc J., Wriggers P.: *Improved enhanced strain four-node element with Taylor expansion of the shape functions.* Int. J. Num. Meth. Engng., Vol. 40, No. 3, 407–421 (1997)

133. Laursen T.A.: *Computational contact and impact mechanics.* Springer, Berlin, 2002

134. Lewiński T., Telega J.J.: *Plates, Laminates and Shells. Asymptotic Analysis and Homogenization.* World Scientific Publishing, Series on Advances in Mathematics for Applied Sciences, Vol. 52, 2000.

135. Liu W.K., Ong J.S.-J., Uras R.A.: *Finite element stabilization matrices-a unification approach.* Comput. Methods Appl. Mech. Engng., Vol. 53, 13–46 (1985)

136. Luenberger D.G.: *Linear and Nonlinear Programming.* Second edition. Addison-Wesley, Reading, Mass., 1984

137. MacNeal R.H.: *The NASTRAN theoretical manual.* (1972)

138. MacNeal R.H.: *A simple quadrilateral shell element,* Computers & Structures, Vol. 8, No. 2, 175–183 (1978)

139. MacNeal R.H.: *Derivation of element stiffness matrices by assumed strain distributions.* Nuclear Engineering and Design, Vol. 70, 3–12 (1982)

140. MacNeal R.H.: *The evolution of lower order plate and shell elements in MSC/NASTRAN.* In: Hughes T.J.R., Hinton E. (eds.) "Finite element methods for plate and shell structures", Vol. 1. Element Technology. Pineridge Press, Swansea, UK, 1986

141. MacNeal R.H.: *Finite Elements: Their Design and Performance.* Mechanical Engineering, Vol. 89, Marcel Dekker Inc., New York, 1994

142. MacNeal R.H.: *Perspective on finite elements for shell analysis.* Finite Elements in Analysis and Design, Vol. 30, 175–186 (1998)

143. MacNeal R.H., Harder R.L.: *A proposed standard set of problems to test finite element accuracy.* Finite Elements in Analysis and Design. Vol. 1, 3–20 (1985)

144. MacNeal R.H., Harder R.L.: *A refined four-noded membrane element with rotational degrees of freedom.* Computers & Structures. Vol. 28, No. 1, 75–84 (1988)

145. MacNeal R.H., Wilson C.T., Harder R.L., Hoff C.C.: *The treatment of shell normals in finite element analysis.* Finite Elements in Analysis and Design. Vol. 30, 235–242 (1998)

146. Mäkinen J.: *Critical study of Newmark-scheme on manifold of finite rotations.* Comput. Methods Appl. Mech. Engng., Vol. 191, 817–828 (2001)

147. Makowski J., Stumpf H.: *Finite strains and rotations in shells.* In: Pietraszkiewicz W. (ed.) "Finite Rotations in Structural Mechanics", 175–194, Springer, Berlin, 1986

148. Makowski J., Stumpf H.: *On the "symmetry" of tangent operators in non-linear mechanics.* ZAMM, Vol. 75, No. 3, 189–198 (1995)

149. Marguerre K.: *Zur theorie der gekrummten platte grosser formanderung.* Proc. Int. Congress Appl. Mech., 5–93, 1939

150. *Mathematics at a Glance. A compendium.* VEB Bibliographisches Institut, Leipzig, 1975

151. Morley L.S.D.: *Skew Plates and Structures.* Pergamon Press, Oxford, 1963

152. Naganarayana B.P., Prathap G.: *Force and moment corrections for the warped four-node quadrilateral plane shell element.* Computers & Structures, Vol. 33, 1107–1115 (1989)

153. Naghdi P.M.: *The Theory of Shells and Plates.* In: C. Truesdell (ed.) "Encyclopedia of Physics", Vol. VIa/2 (Flügge S., ed.). Springer, New York, 1972

154. Nagtegaal J.C., Fox D.D.: *Using assumed enhanced strain elements for large compressive deformation.* Int. J. Solids Structures, Vol. 33, 3151–3159 (1996)

155. Nour-Omid B., Wriggers P.: *A note on the optimum choice for penalty parameters.* Commun. Appl. Num. Methods, Vol. 3, 581–585 (1987)

156. Novozhilov V.V.: *The theory of thin shells.* Walters Nordhoff Publ., Groningen, 1959

157. Oden J.T.: *Exterior penalty methods for contact problems in elasticity.* In: Bathe K.J., Stein E., Wunderlich W. (eds.) "Nonlinear Finite Element Analysis in Structural Mechanics", Springer, Berlin, 1980

158. Ogden R.: *Elastic deformations of rubber-like solids,* In: Hopkins H.G., Sewell M.J. (eds.) "Mechanics of Solids", The R. Hill 60th Unniversary Volume. Pergamon Press, Oxford, 499–537, 1981

159. Ogden R.: *Non-Linear Elastic Deformations.* Ellis Horwood, Chichester, 1984

160. Oñate E.: *Structural Analysis with the Finite Element Method. Linear Statics.* Vol. 1: Basis and Solids. Series: Lecture Notes on Numerical Methods in Engineering and Sciences, CIMNE-Springer, 2009

161. Oñate E.: *Structural Analysis with the Finite Element Method. Linear Statics.* Vol. 2: Beams, Plates and Shells. Series: Lecture Notes on Numerical Methods in Engineering and Sciences, CIMNE-Springer, 2010

162. Oñate E., Zienkiewicz O.C.: *A viscous shell formulation for the analysis of thin sheet-metal forming.* Int. J. Mechanical Sciences, Vol. 25 (5), pp. 305-335 (1983)

163. Panasz P.: *Non-linear models of shells with 6 dofs based on two-level approximations* (in Polish). Ph.D. thesis, IPPT PAN, Warsaw, 2008

164. Panasz P., Wisniewski K.: *Nine-node shell elements with 6 dofs/node based on two-level approximations.* Finite Elements in Analysis and Design, Vol. 44, 784–796 (2008)

165. Parisch H.: *An investigation of a finite rotation four node assumed strain element.* Int. J. Num. Meth. Engng., Vol. 31, 127–150 (1991)

166. Parish H.: *A continuum-based shell theory for non-linear applications.* Int. J. Num. Meth. Engng., Vol. 38, 1855–1883 (1995)

167. Park K.C., Stanley G.M.: *A Curved C^0 Shell Element Based on Assumed Natural-Coordinate Strains.* Trans. ASME, Vol. 53, 278–290 (1986)

168. Pian T.H.H.: *Derivation of element stiffness matrices by assumed stress distributions.* AIAA, Vol. 2, 1333–1336 (1964)

169. Pian T.H.H., Chen D.-P.: *Alternative ways for formulation of hybrid stress elements.* Int. J. Num. Meth. Engng., Vol. 18, 1679–1684 (1982)

170. Pian T.H.H., Sumihara K.: *Rational approach for assumed stress finite elements.* Int. J. Num. Meth. Engng., Vol. 20, 1685–1695 (1984)

171. Pietraszkiewicz W.: *Introduction to the Non-Linear Theory of Shells.* Ruhr Universität Bochum, Mitt. Inst. für Mechanik, No. 10, 1977

172. Pietraszkiewicz W.: *Finite rotations and Lagrangean description in the non-linear theory of shells.* Polish Scientific Publisher, Warsaw, 1979

173. Pietraszkiewicz W.: *Lagrangian description and incremental formulation in the nonlinear theory of thin shells.* Int. J. Nonlin. Mech., Vol. 19, 115–140 (1984)

174. Pietraszkiewicz W.: *Geometrically nonlinear theories of thin elastic shells.* Adv. Mech., Vol. 12, 51–130 (1989)

175. Pietraszkiewicz W., Badur J.: *Finite rotations in the description of continuum deformation.* Int. J. Engng. Sci., Vol. 21, No. 9, 1097–1115 (1983)

176. Piltner R.: *An implementation of mixed enhanced finite elements with strains assumed in Cartesian and natural element coordinates using sparse $\bar{B}$-matrices.* Engng. Comput., Vol. 17, No. 8, 933–949 (2000)

177. Piltner R., Taylor R.L.: *A quadrilateral mixed finite element with two enhanced strain modes.* Int. J. Num. Meth. Engng., Vol. 38, 1783–1808 (1995)

178. Piltner R., Taylor R.L.: *A systematic construction of B-bar functions for linear and non-linear mixed-enhanced finite elements for plane elasticity problems.* Int. J. Num. Meth. Engng., Vol. 44, 615–639 (1999)

179. Powell M.J.D.: *A method for nonlinear constraints in minimization problems.* In: R. Fletcher (ed.) "Optimization", Academic Press, New York, 1969

180. Program ABAQUS. Ver.6.6-2.

181. Program AceGen by J. Korelc (http://www.fgg.uni-lj.si/Symech/)

182. Program ADINA. Ver.8.3.1.

183. Program FEAP by R.L. Taylor, Ver.7.4., University of California, Berkeley (http://www.ce.berkeley.edu/rlt)

184. Robinson J.: *The mode-amplitude technique and hierarchical stress elements- a simplified and natural approach.* Int. J. Num. Meth. Engng., Vol. 21, 487–507 (1985)

185. Rall L. B.: *Automatic Differentiation: Techniques and Applications.* Lecture Notes in Computer Science. Vol. 120, Springer, 1981.

186. Ramm E.: *Geometrisch nichtlineare Elastostatik und finite Elemente.* Bericht Nr.76-2, Institute für Baustatik, Universität Stuttgart, 1976

187. Ramm E.: *Strategies for Tracing the Nonlinear Response Near Limit Points.* In: Wunderlich W., Stein E., Bathe K.J. (eds.) Proc. Europe-U.S. Workshop, Bochum 1980, 63–89, Springer, Berlin, 1981

188. Rankin C.C., Nour-Omid B.: *The use of projectors to improve finite element performance.* Computers & Structures, Vol. 30, 257–267 (1988)

189. Rebel G.: *Finite rotation shell theory including drill rotations and its finite element implementation.* Delft University Press, 1998

190. Reissner E.: *The effect of transverse shear deformation on the bending elastic plates.* J. Appl. Mech., Vol. 12, No. 2, Trans. ASME, Vol. 67, June 1945, pp. A-69-77

191. Reissner E.: *Formulation of variational theorems in geometrically nonlinear elasticity.* J. Eng. Mech. Vol. 110, 1377–1390 (1984)

192. Rhiu J.J., Lee S.W.: *A new efficient mixed formulation for thin shell finite element models.* Int. J. Num. Meth. Engng., Vol. 24, 581–604 (1987)

193. Roberts J.E., Thomas J.M.: *Mixed and Hybrid Elements.* In: Ciarlet P.G., Lions J.L. (eds.) "Handbook of Numerical Analysis", Vol. II, Part 2, Elsevier, Amsterdam, 1991.

194. Robinson J.: *A warped quadrilateral strain membrane element.* Comput. Methods Appl. Mech. Engng., Vol. 7, 359–367 (1976)

195. Robinson J., Blackham S.: *An evaluation of lower order membranes as contained in MSC/NASTRAN, ASA and PAFEC FEM Systems.* Robinson and Associates, Dorset, England, 1979

196. Robinson C.J., Blackburn Ch.L.: *Evaluation of a hybrid, anisotropic, multilayered, quadrilateral finite element.* NASA Technical Paper 1236, 1978

197. Rosenberg R.M.: *Analytical Dynamics of Discrete Systems.* Plenum Press, New York, 1977

198. Russell W.T., MacNeal R.H.: *An improved electrical analogy for the analysis of beams in bending.* J. Appl. Mech.(Sept. 1953)

199. Sansour C.: *A theory and finite element formulation of shells at finite deformations involving thickness change: circumventing the use of a rotation tensor.* Archives of Applied Mechanics, Vol. 65, 194–216 (1995)

200. Sansour C., Bednarczyk H.: *The Cosserat surface as a shell model, theory and finite-element formulation.* Comput. Methods Appl. Mech. Engng., Vol. 120, 1–32 (1995)

201. Sansour C., Kollmann F.G.: *Families of 4-node and 9-node finite elements for a finite deformation shell theory. An assessment of hybrid stress, hybrid strain and enhanced strain elements.* Comput. Mech., Vol. 24, 435–447 (2000)

202. Schieck B., Pietraszkiewicz W., Stumpf H.: *Theory and numerical analysis of shells undergoing large elastic strains.* Int. J. Solids Structures, Vol. 29, No. 6, 689–709 (1992)

203. Sze K.Y., Sim Y.S., Soh A.K.: *A hybrid stress quadrilateral shell element with full rotational d.o.f.s.* Int. J. Num. Meth. Engng., Vol. 40, 1785–1800 (1997).

204. Simmonds J.G., Danielson D.A.: *Nonlinear shell theory with finite rotation vector.* Proc. Kon. Ned. Ak. Wet. Series B, Vol. 73, 460–478 (1970)

205. Simmonds J.G., Danielson D.A.: *Nonlinear shell theory with finite rotation and stress function vectors.* J. Appl. Mech., Vol. 39, 1085–1090 (1972)

206. Simo J.C.: *The (symmetric) Hessian for geometrically nonlinear models in solid mechanics: Intrinsic definition and geometric interpretation.* Comput. Methods Appl. Mech. Engng., Vol. 96, 189–200 (1992)

207. Simo J.C.: *On a stress resultant geometrically exact shell model. Part VII: Shell intersections with 5/6-dof finite element formulation.* Comput. Methods Appl. Mech. Engng., Vol. 108, 319–339 (1993)

208. Simo J.C., Armero F.: *Geometrically non-linear enhanced strain mixed methods and the method of incompatible modes.* Int. J. Num. Meth. Engng., Vol. 33, 1413–1449 (1992)

209. Simo J.C., Fox D.D.: *On a stress resultant geometrically exact shell model. Part I: Formulation and optimal parametrization.* Int. J. Num. Meth. Engng., Vol. 72, 267–304 (1989)

210. Simo, J.C., Fox, D.D., Rifai, M.S.: *On a stress resultant geometrically exact shell model. Part II: The Linear Theory; Computational Aspects.* Int. J. Num. Meth. Engng., Vol. 73, 53–92 (1989)

211. Simo J.C., Fox D.D., Rifai M.S.: *On a stress resultant geometrically exact shell model. Part III: Computational aspects of the nonlinear theory.* Int. J. Num. Meth. Engng., Vol. 79, 21–70 (1990)

212. Simo J.C., Pister K.S.: *Remarks on rate constitutive equations for finite deformation problems: computational implications.* Comput. Methods Appl. Mech. Engng., Vol. 46, 201–215 (1984)

213. Simo J.C., Rifai M.S., Fox D.D.: *On a stress resultant geometrically exact shell model. Part IV: Variable thickness shells with through-the-thickness stretching.* Comput. Methods Appl. Mech. Engng., Vol. 81, 91–126 (1990)

214. Simo J.C., Fox D.D., Hughes T.J.R.: *Formulations of finite elasticity with independent rotations.* Int. J. Num. Meth. Engng., Vol. 95, 227–288 (1992)

215. Simo J.C., Hughes T.J.R.: *On the Variational Foundations of Assumed Strain Methods.* J. Appl. Mech., Vol. 53, 51–54, (1986)

216. Simo J.C., Rifai M.S.: *A class of mixed assumed strain methods and the method of incompatible modes.* Int. J. Num. Meth. Engng., Vol. 29, 1595–1638 (1990)

217. Simo J.C., Tarnow N.: *On a stress resultant geometrically exact shell model. Part VI: 5/6 dof treatment.* Int. J. Num. Meth. Engng., Vol. 34, 117–164 (1992)

218. Simo J.C., Taylor R.L.: *Quasi-incompressible finite elasticity in principal stretches. Continuum basis and numerical algorithms.* Comput. Methods Appl. Mech. Engng., Vol. 85, 273–310 (1991)

219. Simo J.C, Vu-Quoc L.: *A three-dimensional finite strain rod model. Part II: Computational aspects.* Comput. Methods Appl. Mech. Engng., Vol. 58, 79–116 (1986)

220. Simo J.C., and Vu-Quoc L.: *On the dynamics of 3-d finite strain rods.* In: "Finite Element Methods for Plate and Shell Structures", Vol. 2. "Formulations and Algorithms". Pineridge Press, Swansea, 1–30 (1986)

221. Simo J.C., Wong K.K.: *Unconditionally stable algorithms for rigid body dynamics that exactly preserve energy and momentum.* Comput. Methods Appl. Mech. Engng., Vol. 31, 19–52 (1991)

222. Simo J.C., Wriggers P., Taylor R.L : *A perturbed Lagrangian formulation for the finite element solution of contact problems.* Comput. Methods Appl. Mech. Engng., Vol. 50, 163–180 (1985)

223. Spilker R.L., Maskeri S.M., Kania E.: *Plane isoparametric hybrid-stress elements: invariance and optimal sampling.* Int. J. Num. Meth. Engng., Vol. 17, No. 10, 1469–96 (1981)

224. Spring K.W.: *Euler parameters and the use of quaternion algebra in the manipulation of finite rotations: a review.* Mechanism and Machine Theory, Vol. 21, No. 5, 365-373 (1986)

225. Stander N., Matzenmiller A., Ramm E.: *An assessment of assumed strain methods in finite rotation shell analysis.* Engng. Comput., Vol. 6, 58–66 (1989)

226. Stolarski H., Belytschko T.: *Membrane locking and reduced integration for curved elements.* J. Appl. Mech. ASME, Vol. 49, 172–176 (1982)

227. Stolarski H., Belytschko T.: *Shear and membrane locking in curved elements.* Comput. Methods Appl. Mech. Engng., Vol. 41, 279–296 (1983)

228. Stolarski H., Belytschko T., Lee S.-H.: *A review of shell finite elements and corotational theories.* Computational Mechanics Advances, Vol. 2, 125–212 (1995)

229. Strang G., Fix G.J.: *An Analysis of the Finite Element Method.* In: G. Forsythe (ed.) "Series in Automatic Computation". Prentice-Hall, Englewood Cliffs, N.J., 1973

230. Struik D.J.: *Lectures on Classical Differential Geometry.* Dover, 1988

231. Stuelpnagel J.: *On the parametrization of three-dimensional rotational group.* SIAM Review, Vol. 6, No. 4, 422–430 (1964)

232. Stumpf H., Makowski J.: *On large strain deformations of shells.* Acta Mechanica, Vol. 65, 153–168 (1986)

233. Sze K.Y., Liu X.H., Lo S.H.: *Popular benchmark problems for geometric nonlinear analysis of shells.* Finite Elements in Analysis and Design, Vol. 40, 1551–1569 (2004)

234. Taylor R.L., Beresford P.J., Wilson E.L.: *A non-conforming element for stress analysis.* Int. J. Num. Meth. Engng., Vol. 10, 1211–1220 (1976)

235. Taylor R.L.: *Finite element analysis of linear shell problems.* In: Whiteman J.R. (ed.) "The Mathematics of Finite Elements and Applications VI. MAFELAP 1987". Academic Press, London, 1988

236. Ting T.C.T.: *Determination of* $\mathbf{C}^{1/2}$, $\mathbf{C}^{-1/2}$ *and more general isotropic tensor functions of* $\mathbf{C}$. J. Elasticity, Vol. 15, 319–323 (1985)

237. Timoshenko S., Woinowsky-Krieger S.: *Theory of Plates and Shells.* McGraw-Hill, New York, 1959

238. Toupin R.A.: *Theories of elasticity with couple-stress.* Arch. Rational Mech. Anal., Vol. 17, 85–112 (1964)

239. Truesdell C., Noll W.: *The Non-Linear Field Theory.* Handbuch der Physik, Vol. III/3, Springer, Berlin, 1965

240. Valid R.: *The nonlinear theory of shells through variational principles.* John Wiley, Chichester, 1995

241. Vu-Quoc L., Tan X.G.: *Optimal solid shells for non-linear analyses of multilayer composites. I. Statics.* Comput. Methods Appl. Mech. Engng., Vol. 192, 975–1016 (2003)

242. Wagner W., Gruttmann F.: *A simple finite rotation formulation for composite shell elements.* Engng. Comput., Vol. 11, 145–176 (1994)

243. Wagner W., Gruttmann F.: *A robust nonlinear mixed hybrid quadrilateral shell element.* Int. J. Num. Meth. Engng., Vol. 64, No. 5, 635–666 (2005)

244. Wan F.Y.M., Weinitschke H.J.: *On shells of revolution with the Love-Kirchhoff hypotheses.* J. Engng. Math., Vol. 22, 285–334 (1988).

245. Warren C.Y.: *Roark's Formulas for Stress and Strain.* 6th Edition. Mc Graw-Hill, New York, 1989

246. Wempner G. A.: *New concepts for finite elements of shells.* Z. Angew. Math. Mech., 48, T174-T176 (1968)

247. Wempner G., Talaslidis D., Hwang C.-M.: *A simple and efficient approximation of shells via finite quadrilateral elements.* J. Appl. Mech. ASME, Vol. 49, No. 1, 115–120 (1982)

248. Wilson E.L., Taylor R.L., Doherty W.P., Ghaboussi J.: *Incompatible displacement models.* In: Fenves S.J., Perrone N., Robinson A.R., Schnobrich W.C. (eds.) "Numerical and Computer Methods in Finite Element Analysis". Academic Press, New York, 43–57 (1973)

249. Wisniewski K.: *A shell theory with independent rotations for relaxed Biot stress and right stretch strain.* Comput. Mech., Vol. 21, No. 2, 101–122 (1998)

250. Wisniewski K., Kowalczyk P., Turska E.: *Analytical DSA for explicit dynamics of elastic-plastic shells.* Comput. Mech., Vol. 39, No. 6, 761–85 (2007)

251. Wisniewski K., Turska E.: *A note on the hyperelastic constitutive equation for rotated Biot stress.* Archives of Mechanics, Vol. 48, No. 5, 947–953 (1996)

252. Wisniewski K., Turska E.: *Kinematics of finite rotation shells with in-plane twist parameter.* Comput. Methods Appl. Mech. Engng., Vol. 190, No. 8-10, 1117–1135 (2000)

253. Wisniewski K., Turska E.: *Warping and in-plane twist parameter in kinematics of finite rotation shells,* Comput. Methods Appl. Mech. Engng., Vol. 190, No. 43–44, 5739–5758 (2001)

254. Wisniewski K., Turska E.: *Second order shell kinematics implied by rotation constraint equation,* J. Elasticity, Vol. 67, 229–246 (2002).

255. Wisniewski K., Turska E.: *Enhanced Allman quadrilateral for finite drilling rotations.* Comput. Methods Appl. Mech. Engng., Vol. 195, No. 44-47, 6086–6109 (2006)

256. Wisniewski K., Turska E.: *Improved four-node Hellinger-Reissner elements based on skew coordinates.* Int. J. Num. Meth. Engng., Vol. 76, 798–836 (2008)

257. Wisniewski K., Turska E.: *Improved four-node Hu-Washizu elements based on skew coordinates.* Computers & Structures, Vol. 87, 407–424 (2009)

258. Wittenburg J.: *Dynamics of Systems of Rigid Bodies.* B.G. Teubner, Stuttgart, 1977

259. Woźniak Cz.: *Nonlinear Theory of Shells.* (in Polish) Polish Scientific Publisher, Warsaw, 1966

260. Woźniak Cz. (ed): *Mechanika Techniczna. Tom VIII. Mechanika sprezystych plyt i powlok.* (in Polish) Wydawnictwo Naukowe PWN, Warsaw, 2001

261. Wriggers P.: *Computational contact mechanics.* Wiley, New York, 2002

262. Wriggers P., Gruttmann F.: *Thin shells with finite rotations formulated in Biot stresses: theory and finite element formulation.* Int. J. Num. Meth. Engng., Vol. 36, 2049–2071 (1993)

263. Wriggers P., Korelc J.: *On enhanced strain methods for small and finite deformations of solids.* Comput. Mech., Vol. 18, No. 6, 413–428 (1996)

264. Wriggers P., Reese S.: *A note on enhanced strain methods for large deformations,* Comput. Methods Appl. Mech. Engng., Vol. 135, 201–209 (1996)

265. Yuan K., Huang Y.-S., Pian T.H.H.: *New strategy for assumed stress for 4-node hybrid stress membrane element.* Int. J. Num. Meth. Engng., Vol. 36, 1747–1763 (1993)

266. Yunus S.M.: *A study of different hybrid elements with and without rotational d.o.f. for plane stress/plane strain problems.* Computers & Structures, Vol. 30, No. 5, 1127–1133 (1988)

267. Zhu Y., Zacharia Th.: *A new one-point quadrature, quadrilateral shell element with drilling degrees of freedom.* Comput. Methods Appl. Mech. Engng., Vol. 136, 165–203 (1996)

268. Zienkiewicz O.C, Taylor R.L.: *The Finite Element Method. Fifth Edition. Vol. 1. The Basis.* Butteworth-Heinemann, Oxford, 2000

269. Zienkiewicz O.C., Taylor R.L., Too J.M.: *Reduced Integration Technique in General Analysis of Plates and Shells'.* Int. J. Num. Meth. Engng., Vol. 3, 275–290 (1971)

270. Zienkiewicz O.C., Wood W.L., Taylor R.L.: *An alternative single-step algorithm for dynamic problems.* Earthquake Engineering and Structural Dynamics, Vol. 8, 31–40 (1980)

Author index

Subject index